Handbook of Research on Nanoelectronic Sensor Modeling and Applications

Mohammad Taghi Ahmadi
Urmia University, Iran

Razali Ismail
Universiti Teknologi Malaysia, Malaysia

Sohail Anwar
Penn State University, USA

A volume in the Advances in Computer and Electrical Engineering (ACEE) Book Series

IGI GLOBAL
DISSEMINATOR OF KNOWLEDGE

www.igi-global.com

Published in the United States of America by
IGI Global
Engineering Science Reference (an imprint of IGI Global)
701 E. Chocolate Avenue
Hershey PA, USA 17033
Tel: 717-533-8845
Fax: 717-533-8661
E-mail: cust@igi-global.com
Web site: http://www.igi-global.com

Library of Congress Cataloging-in-Publication Data

Names: Ahmadi, Mohammad Taghi, editor. | Bin Ismail, Razali, editor. | Anwar, Sohail, editor.
Title: Handbook of research on nanoelectronic sensor modeling and applications / Mohammad Taghi Ahmadi, Razali Ismail, and Sohail Anwar, editors.
Description: Hershey, PA : Engineering Science Reference, [2017] | Series: Advances in computer and electrical engineering | Includes bibliographical references and index.
Identifiers: LCCN 2016023356| ISBN 9781522507369 (hardcover) | ISBN 9781522507376 (eISBN)
Subjects: LCSH: Nanoelectronics. | Quantum electronics. | Carbon nanotubes. | Detectors--Materials.
Classification: LCC TK7874.84 .H37 2017 | DDC 681/.2--dc23 LC record available at https://lccn.loc.gov/2016023356

This book is published in the IGI Global book series Advances in Computer and Electrical Engineering (ACEE) (ISSN: 2327-039X; eISSN: 2327-0403)

British Cataloguing in Publication Data
A Cataloguing in Publication record for this book is available from the British Library.

All work contributed to this book is new, previously-unpublished material. The views expressed in this book are those of the authors, but not necessarily of the publisher.

For electronic access to this publication, please contact: eresources@igi-global.com.

Advances in Computer and Electrical Engineering (ACEE) Book Series

Srikanta Patnaik
SOA University, India

ISSN: 2327-039X
EISSN: 2327-0403

Mission

The fields of computer engineering and electrical engineering encompass a broad range of interdisciplinary topics allowing for expansive research developments across multiple fields. Research in these areas continues to develop and become increasingly important as computer and electrical systems have become an integral part of everyday life.

The **Advances in Computer and Electrical Engineering (ACEE) Book Series** aims to publish research on diverse topics pertaining to computer engineering and electrical engineering. **ACEE** encourages scholarly discourse on the latest applications, tools, and methodologies being implemented in the field for the design and development of computer and electrical systems.

Coverage

- Microprocessor Design
- Algorithms
- Circuit Analysis
- Optical Electronics
- Electrical Power Conversion
- Computer science
- Chip Design
- VLSI Fabrication
- VLSI Design
- Qualitative Methods

IGI Global is currently accepting manuscripts for publication within this series. To submit a proposal for a volume in this series, please contact our Acquisition Editors at Acquisitions@igi-global.com or visit: http://www.igi-global.com/publish/.

The Advances in Computer and Electrical Engineering (ACEE) Book Series (ISSN 2327-039X) is published by IGI Global, 701 E. Chocolate Avenue, Hershey, PA 17033-1240, USA, www.igi-global.com. This series is composed of titles available for purchase individually; each title is edited to be contextually exclusive from any other title within the series. For pricing and ordering information please visit http://www.igi-global.com/book-series/advances-computer-electrical-engineering/73675. Postmaster: Send all address changes to above address.

Titles in this Series

For a list of additional titles in this series, please visit: www.igi-global.com

Field-Programmable Gate Array (FPGA) Technologies for High Performance Instrumentation
Julio Daniel Dondo Gazzano (University of Castilla-La Mancha, Spain) Maria Liz Crespo (International Centre for Theoretical Physics, Italy) Andres Cicuttin (International Centre for Theoretical Physics, Italy) and Fernando Rincon Calle (University of Castilla-La Mancha, Spain)
Engineering Science Reference • copyright 2016 • 306pp • H/C (ISBN: 9781522502999) • US $185.00 (our price)

Design and Modeling of Low Power VLSI Systems
Manoj Sharma (BVC, India) Ruchi Gautam (MyResearch Labs, Gr Noida, India) and Mohammad Ayoub Khan (Sharda University, India)
Engineering Science Reference • copyright 2016 • 386pp • H/C (ISBN: 9781522501909) • US $205.00 (our price)

Reliability in Power Electronics and Electrical Machines Industrial Applications and Performance Models
Shahriyar Kaboli (Sharif University of Technology, Iran) and Hashem Oraee (Sharif University of Technology, Iran)
Engineering Science Reference • copyright 2016 • 481pp • H/C (ISBN: 9781466694293) • US $255.00 (our price)

Handbook of Research on Emerging Technologies for Electrical Power Planning, Analysis, and Optimization
Smita Shandilya (Sagar Institute of Research Technology & Science, India) Shishir Shandilya (Bansal Institute of Research & Technology, India) Tripta Thakur (Maulana Azad National Institute of Technology, India) and Atulya K. Nagar (Liverpool Hope University, UK)
Engineering Science Reference • copyright 2016 • 410pp • H/C (ISBN: 9781466699113) • US $310.00 (our price)

Sustaining Power Resources through Energy Optimization and Engineering
Pandian Vasant (Universiti Teknologi PETRONAS, Malaysia) and Nikolai Voropai (Energy Systems Institute SB RAS, Russia)
Engineering Science Reference • copyright 2016 • 494pp • H/C (ISBN: 9781466697553) • US $215.00 (our price)

Environmental Impacts on Underground Power Distribution
Osama El-Sayed Gouda (Cairo University, Egypt)
Engineering Science Reference • copyright 2016 • 405pp • H/C (ISBN: 9781466665095) • US $225.00 (our price)

Handbook of Research on Advanced Computational Techniques for Simulation-Based Engineering
Pijush Samui (Centre for Disaster Mitigation and Management, VIT University, India)
Engineering Science Reference • copyright 2016 • 616pp • H/C (ISBN: 9781466694798) • US $310.00 (our price)

List of Contributors

Table of Contents

Detailed Table of Contents

One of the most important drawbacks which caused the Silicon based technologies to their technical limitations is the instability of their products at nano-level. On the other side, carbon based materials such as carbon nanotube (CNT) as alternative materials have been involved in scientific efforts. Some of the important advantages of CNTs over silicon components are high mechanical strength, high sensing capability and large surface-to-volume ratio. Many researches have been presented using CNT as a sensing material in various applications to improve the sensor characteristics. In this chapter, the platform of CNT sensors such as transistor-based sensors, chemiresistors, chemicapacitance and resonator sensors are discussed in detail. Using CNT as a sensor platform although has great advantages; it does not have sufficient sensor reliability. Some of these technical challenges of CNT-based sensors including Schottky contact formation and non-selective synthesization have also been pointed out in the chapter.

Graphene is a promising nanomaterial with outstanding physical and electrical properties that offers a wide range of opportunities for advanced applications in nanoelectronics [1-3]. The application of graphene nanoribbon (GNR) in high-speed electronics is being explored extensively because of some excellent properties such as one-atom thickness, mechanical strength, flexibility, high thermal conductivity up to 50 W cm –1 K –1, extremely highcurrent-carrying capacity up to 10 9 A/cm 2, high carrier mobility in excess of 200,000 cm 2 V –1 s –1, high carrier saturation velocity 3 of ~5×10 7cm s –1, and extraordinarily

electronic field, strain and mass sensors as well as field-effect, electrochemical and piezoelectric gas sensors. Graphene is believed to be a fantastic sensor material because of its single atomic layer of graphite with surface.

Industrial activities have polluted the atmosphere very rapidly in these days. There are a various varieties of air pollutants having a strong effect on human health as well as on climate and specially environment, such as nitrogen dioxide (NO2), oxides of carbon (COx) and hydrocarbons. The World Health Organization (WHO) estimates that each year about 4.6million people die directly from causes of air pollution that will be a serious threat to human health. Therefore, there is a growing demand towards highly sensitive, cheap, low consumption, user-friendly devices which could monitor the quality of air indoor and outdoor areas for protecting human health. It is proved to be an efficient and economically feasible alternative for measuring different gas concentrations.

The aim of this project is to study and develop graphene-based DNA sensor model for detection of DNA hybridization application. This includes modeling and simulation of carrieconcentration, conductance, and current-voltage characteristics of graphene-based sensors on the field effect transistor (FET) platform. The main challenge is to validate the developed modelwith the experimental data,sincegraphene is considered as a new emerging material and research is still rapidly taking place with fabrication effort reported so far. In this research, first, numerical model is developed which shows the dependency of current-voltage characteristics on the DNA concentration factor. The iteration method is used for developing the numerical model. The proposed model is simulated utilizing MATLAB software to validate with experimental data of DNA hybridization. The Id-Vg characteristic of the proposed numerical model is depicted for different concentrations of DNA molecules and compared with experimental data for the verification purpose. After determining the accuracy of the models, particle swarm optimization (PSO) technique is used to minimize the error of the numerical model.Then, optimization results are shown. Overally, the accuracy of more than 98% represents an overall error of less than 2\% which is quite acceptable for the optimized numerical model.

In many research fields and industry such as nuclear physics, notably nuclear technology, fusion plasma diagnostics, radiotherapy and radiation protection, it is very substantial that measure fast neutron spectra. For example in nuclear reactor primary generated neutrons have energies around 2 MeV that lie fast neutron category. Also particle accelerators and Am-Be neutron source raise fast neutrons. Therefore a review of silicon based fast neutron detection with proton recoil methods is surveyed. Furthermore Carbon nanoparticles (CNPs) with simple and low cost preparation methods with exceptional electrical properties have been used widely in nanoelectronic applications such as radiation sensors. In this chapter, fast neutron detectors using Carbon based semiconductor, back-to-back Schottky diode type, and polyethylene as convertor are developed and the Am-Be fast neutron source is used in experimental measurements.

Chapter 16
Sensors and Amplifiers: Sensor Output Signal Amplification Systems 423
Amir Fathi, Urmia University, Iran
Sarkis Azizian, K. N. Toosi University of Technology, Iran
Nastaran Sharifan, Tehran University, Iran

Sensors are electrical-mechanical elements which are the interface between environment and electrical systems. The input of sensors is characteristics of the environment for example temperature, pressure and etc. and their output is a small electric voltage or current. Their job is to convert environment characteristics to an electric voltage or current at their outputs. Since the output current or voltage is very small, it must be amplified in order to be suitable for use in electronic systems. In this chapter we completely explain the design procedure and characteristics of sensor amplifiers. The important parameters of sensor amplifiers are input and output resistance, gain, unity gain bandwidth and etc. One of the most important characteristics of amplifiers is the linearity of amplification in a way that it must have uniformity for all amplitude voltages or currents in all frequencies of the bandwidth. For this purpose, first the operational amplifier is completely discussed, then the linearity of feedback operation will be explained.

Chapter 17
Wireless Nanosensor Networks: Prospects and Challenges... 505
Sohail Anwar, Penn State University, USA

Nanotechnology is enabling the development of devices in a scale ranging from one to a few hundred nanometers, which can perform tasks such as sensing, data storing, computing, and actuation. These nano devices will be able to cover larger areas and perform more complex tasks through communication. Wireless nanosensor networks (WNSNs) are collections of nanosensor devices with communication capability. The key components of a WNSN include nano-nodes, nano-router, nano-micro interface, and gateway. WNSNs have numerous potential biomedical, environmental, industrial, and military applications. This chapter provides an overview of the architecture, applications, and issues associated with the development of WNSNs.

Compilation of References 512

About the Contributors 572

Index 576

Preface

Sensor is an important device with capability to detect physical quantities of materials such as gas, liquid and molecules in the surrounding area. Its feedback in the form of electrical parameters can be measured experimentally. Sensor can be defined as a device that responds to environmental changes and converts these physical quantities into various types of signals which can be analyzed or interpreted by the instruments. As a sensor greatly depends on the material characteristics such as properties, performance and sensitivity, choosing the right one should be considered as an essential decision. Graphene has recently become an important material in the disciplinary field of carbon nanoscience and condensed matter physics. It has attracted attention because of unique properties such as high sensitivity, high mobility and biocompatibility. Furthermore, Graphene/Carbon as a novel material with organic nature is intrinsically sensitive to diversity of molecules and atoms.

Handbook of Research on Nanoelectronic Sensor Modeling and Applications is planned to support researchers in nanotechnology related sensor modeling. It is deliberately planned as a research handbook, which illustrates to the researcher the process of sensing phenomenon to formulate physical effect of molecules in the form of electrical response, conventionally investigated by experimental results. The extensive modeling part of the book is focused on the graphene based sensors and extends the technology of electronic devices to nanoscale sensor modeling and simulation. The physics of sensor device in nanoscale area, which reflects quantum mechanical effects with inter-atomic interactions, is explored in this book. The handbook presents valuable research information on nanoscale sensor modeling and simulation. The electrical response of sensing phenomenon in the form of physical models is employed and numerical algorithms are improved to investigate nanoscale sensor devices.

The book is organized into 17 chapters. Chapter 1 introduces the basic ideas related to the carrier statistic on field effect transistor as a sensor platform for understanding the nanoscale sensor working phenomenon. In addition, the application of new material such as carbon nanotube (CNT) in device technology is discussed. In this chapter, the platform of CNT sensors such as transistor-based sensors, chemiresistors, chemicapacitance and resonator sensors are discussed in detail. Using CNT as a sensor platform although has great advantages, it does not have sufficient sensor reliability. Some of the technical challenges related to CNT-based sensors including Schottky contact formation and non-selective synthesization have been outlined in this chapter.

Chapter 2 highlights application of graphene nanoribbon (GNR) in high-speed electronics. Additionally, trilayer graphene nanoribbon (TGN) as one of the most common multilayers of graphene is discussed in this chapter.

Chapter 3 provides details on the compatibility of silicene with existing semiconductor techniques and the need for new materials to continue Moore's law. It is clear that this material can be used as sensor and field effect transistor platforms. Silicene electrical properties such as electrical conductance, carrier concentration, mobility, and magneto-conductance of 2D silicone are modeled which shows that silicene nanoribbons and nanopores can be used in sensor technology. Specially, their use in DNA sequencing and efficient thermoelectric power generation has been described in this chapter.

Chapters 4 and 5 focus on graphene/CNT based gas sensors. It is known that both graphene and CNTs experience changes in their electrical conductance when exposed to different gases, such as, CO2, NO2, and NH3. In these chapters, several gas sensor models are proposed by employing the Field Effect Transistor structure. In the suggested models, different physical properties such as conductance, capacitance, drift velocity, carrier concentration, and the current-voltage (I-V) characteristics of graphene/CNTs have been used to model the sensing mechanism. An Artificial Neural Network model has also been developed for the especial case of a CNT gas sensor exposed by NH3 to provide a platform to check the accuracy of the models.

Chapters 6 through 9 continue the focus on gas sensors. In Chapter 6, a chemiresistive gas sensor based on conducting polymer is considered. Theses gas sensors are commonly based on polyaniline (PANI), polypyrrole (PPy), polythiophene (PTh), and their derivatives as a transducer. Common configuration and response mechanism of these sensors are reported in this section. Some factors that induce selectivity to these sensors are discussed. Different materials (conductor or insulant) can be used as a substrate of polymerization. Type of substrate, selective membranes, surface modification of conducting polymer, and so on can change the response behavior of these sensors. Chapter 7 is divided as follows: In the first part, a description of the main topologies and materials (carbon nanomaterials plus polymers, metals, ceramics or combinations between these groups) used to fabricate gas sensors based on graphene and carbon nanotubes, which are operated by conductance or electrical resistance, is presented. Next, different mathematical models, that can be used to simulate gas sensors based on these materials, are described. In the third part, the impact of the graphene and carbon nanotubes on gas sensors is exemplified with technical advances achieved until now. Finally, an analysis of the role of gas sensors based on carbon nanomaterials in the next decades is provided. In Chapter 8, the focus is on NO2 detection. Additionally, the defect effect on graphene structure and its consequence in sensing is explored. The World Health Organization (WHO) estimates that each year about 4.6 million people die directly from causes of air pollution. This will be a serious threat to human health. Therefore, there is a growing demand towards highly sensitive, cheap, low consumption, user-friendly devices which can monitor the quality of air indoor and outdoor areas for protecting human health. In Chapter 9, the effect of nitrogen dioxide (NO2), oxides of carbon (COx), and hydrocarbons as main air pollutions on carbon based sensors are discussed.

Chapter 10 focuses on the study and development of graphene-based DNA sensor model for detection of DNA. This study includes modeling and simulation of carrier concentration, conductance, and current-voltage characteristics of graphene-based sensors on the field effect transistor (FET) platform. The main challenge is to validate the developed model with the experimental data. In this research, firstly, numerical model is developed which shows the dependency of current-voltage characteristics on the DNA concentration factor. The iteration method is used for developing the numerical model as well.

Chapter 11 focuses on BGNR quantum capacitance variation effect by the prostate specific antigen (PSA) injected electrons into the FET channel as a sensing mechanism. Also, carrier movement in BGNR as modeling parameter is suggested. Additionally, the effect of PSA adsorption and local pH value on injected carriers in the surface of BGNR is modelled. Carrier concentration as a function of control parameters is predicted. Furthermore, changes in charged lipid membrane properties are electrically detected by graphene based electrolyte gated Graphene Field Effect Transistor (GFET). In this chapter, monolayer graphene-based GFET with focus on conductance variation by membrane and thickness is studied.

In Chapter 12, the idea of graphene and CNT based electrical biosensors is discussed. A liquid gated graphene field effect transistor (LG-GFET) based biosensor model is analytically developed for electrical detection of Escherichia coli (E. coli) bacteria. An E. coli absorption effect on the graphene surface is considered in the form of conductance variation. Moreover, the current-voltage characteristic is applied in terms of conductance model to evaluate the performance of the biosensor model. Furthermore, the CNT-FET platform is employed for modeling biosensor in order to detect Glucose. For diagnosing and monitoring the blood glucose level, glucose oxidase (GOx) based enzyme sensors have been immensely used. According to the proposed CNT-FET structure, charge based analytical modeling approach is used. The charge-based carrier velocity model is implemented to study electrical characteristics of CNT-FET. In the presented model, the gate voltage is considered as a function of glucose concentration.

Chapter 13 presents an appropriate structure to ISFET device for the purpose of electrical measurement of different pH buffer solutions. Electrical detection model of each pH value is suggested using conductance modelling of monolayer graphene. In addition, ISFET based on nanostructured SWCNT is studied for the purpose of electrical detection of hydrogen ion concentrations. Electrical detection of hydrogen ion concentrations by modeling the conductance of SWCNT sheets is proposed. pH buffer as a function of gate voltage is assumed and sensing factor is defined.

In Chapter 14, first of all, a method for refractive index investigation of graphene based structures is introduced, and then the effect of carrier density variant in the form of conductance gradient on graphene based SPR sensor is modeled. The molecular properties such as electro-negativity, molecular mass, effective group number, and effective outer shell factor of the molecule are engaged. In addition, each factor effect in the cumulative carrier variation is explored analytically. The refractive index shift equation based on these factors is defined, and related coefficients are proposed. Finally, a semi-empirical model for interpretation of changes in SPR curve is suggested and tested for some organic molecules.

Chapter 15 presents a review of silicon based fast neutron detectors. Specifically, proton recoil methods are surveyed. Furthermore, carbon nanoparticles (CNPs), having simple and low cost preparation methods and exceptional electrical properties, have widely been used in nanoelectronic applications such as radiation sensors. In this chapter, fast neutron detectors using carbon based semiconductor, back-to-back Schottky diode type, and polyethylene as convertor are developed, and the Am-Be fast neutron source is used in experimental measurements.

In Chapter 16, the design procedures and characteristics of sensor amplifiers are explained. The important parameters of sensor amplifiers are input and output resistance, gain, unity gain bandwidth and etc. One of the most important characteristics of amplifiers is the linearity of amplification in a way that it must have uniformity for all amplitude voltages or currents in all frequencies of the bandwidth. For this purpose, firstly, the operational amplifier is completely discussed, and then the linearity of feedback operation is explained.

Chapter 17 describes nanotechnology importance in enabling the development of devices in a scale ranging from one to a few hundred nanometers, which can perform tasks such as sensing, data storing, computing, and actuation. These nano devices will be able to cover larger areas and perform more complex tasks through communication. Wireless nanosensor networks (WNSNs) are collections of nanosensor devices with communication capability. The key components of a WNSN include nano-nodes, nano-router, nano-micro interface, and gateway. WNSNs have numerous potential biomedical, environmental, industrial, and military applications. This chapter provides an overview of the architecture, applications, and issues associated with the development of WNSNs.

Acknowledgment

I would like to express my deep gratefulness to the authors, who have provided magnificent contributions. My especial thanks to my coeditor, Professor Dr. Razali Ismail, who was abundantly helpful and offered invaluable assistance, support and guidance. Also especial thanks to my other coeditor Associate Prof.Dr. Sohail Anwar who initiated the book and provided advice and comments during the preparation of the manuscripts. The following group members are especially thanked for assisting in the editing: S. Meshginghalam, Bahar Meshkinghalam, Hadi Kasani. I would like to gratefully acknowledge the Nanotechnology research center and nanoelectronic research group of Urmia University for providing excellent research environment and support in which to complete a significant portion of the work in this book. More important than any other support, my wife Mariam and my son Mahan provided me with love and understanding. Their constant encouragement and emotional support kept my vigor and life line alive in research.

Mohammad Taghi Ahmadi
Urmia University, Iran

Razali Ismail
Universiti Teknologi Malaysia, Malaysia

Sohail Anwar
Penn State University, Altoona, USA

Chapter 1
CNT as a Sensor Platform

Amir Fathi
Urmia University, Iran

Mina Hassanzadazar
Urmia University, Iran

ABSTRACT

One of the most important drawbacks which caused the Silicon based technologies to their technical limitations is the instability of their products at nano-level. On the other side, carbon based materials such as carbon nanotube (CNT) as alternative materials have been involved in scientific efforts. Some of the important advantages of CNTs over silicon components are high mechanical strength, high sensing capability and large surface-to-volume ratio. Many researches have been presented using CNT as a sensing material in various applications to improve the sensor characteristics. In this chapter, the platform of CNT sensors such as transistor-based sensors, chemiresistors, chemicapacitance and resonator sensors are discussed in detail. Using CNT as a sensor platform although has great advantages; it does not have sufficient sensor reliability. Some of these technical challenges of CNT-based sensors including Schottky contact formation and non-selective synthesization have also been pointed out in the chapter.

INTRODUCTION

Carbon nanotubes (CNTs) can be assumed as a single layer graphene rolled up in the cylindrical form as illustrated in Figure 1 (Dresselhaus, Dresselhaus, & Eklund, 1996). CNTs, especially single-walled carbon nanotubes (SWCNTs), are considered undertaking materials for next generation of electronic applications.

The strength and flexibility of CNTs are the key factors which make them eligible to control other nanoscale compounds. These factors suggest they will have an important role in nanotechnology engineering. During the recent years, they have drawn the attention of IC designers because of their unique electrical properties (Lu & Chen, 2005). Owing to their specific configuration and their special electronic properties (Lu, 2005; Wei, 2001; Treacy, 1996; Zhang, 2004; Hone, 2000), CNTs have been highly considered for the development of new generation of gas and biosensors. Different sensing approaches have been reported to increase the speed and accuracy of sensors using CNTs.

DOI: 10.4018/978-1-5225-0736-9.ch001

Figure 1. Carbon nanotubes structure as a rolled up graphene

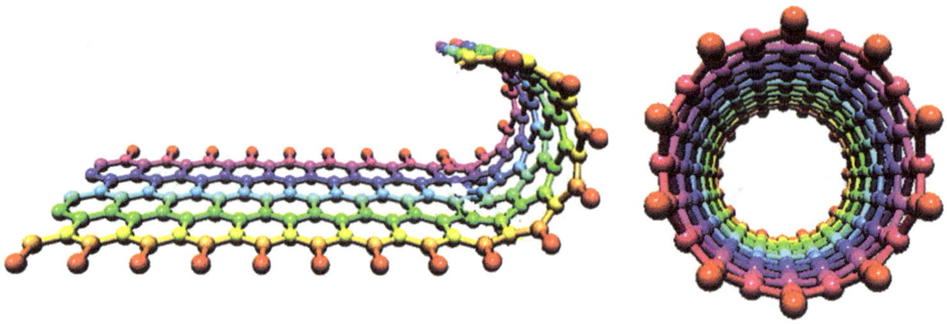

The most currently-used structure among all sensor platforms is transistor-based sensors using CNTs as conducting channels (Martel, 1998; Kong, 2000; Someya, 2003; Collins, 2000). Although CNT-based nanosensors have great advantages but they meet many limitations, such as sometimes low sensitivity, long recovery time and low selectivity (Modi, Koratkar, Lass, Wei & Ajayan, 2003). Several techniques are reported to overcome these limitations to provide fast, stable and analyte-specific sensors. However, the methods still fail to suggest a commercial successful CNT-based sensor design for various applications. In this chapter, the CNT sensors platform, the design hurdles, advantages and some applications of the sensors are demonstrated in detail.

BACKGROUND

A misunderstanding in the origination of CNTs is caused by an editorial written by Monthioux and Kuznetsov (2006). But several unrivaled CNT properties have been found and reported from the standpoint of electrical and elastic modulus, respectively. In the near future, they are expected to play a dominant role in the designation of many nano-material based devices (Yu et al., 2000). A large percentage of research literature ascribes graphitic carbon as the origin of hollow, nanometer-size tubes (Yu et al., 2000). Until 1991, many efforts are done to produce and perceive CNTs under different conditions in order to study their properties. In a research published by Oberlin, Endo, and Koyama (1976) and by means of a vapor-growth technique, hollow carbon fibers with their nanometer-scale diameters are exhibited vividly. Additionally, a single wall of graphene is demonstrated by the authors in a transmission electro microscopic (TEM) image of a nanotube. Later, this image has been attributed as a single-walled nanotube by Endo (Endo & Dresselhaus, 2002).

Nowadays, many primary devices are being fabricated using CNTs, including field-effect transistors (FETs), diodes, single electron transistors, nanoelectrodes, and several others (Postma Henk, 2001; Collins, 2001; Wong, 2002). Their main advantage was 20–30x higher ON current in comparison with Si MOSFETs. This was an important advantage in this field as CNT was displayed to potentially perform better than Si (Javey, Guo, Wang, Lundstrom & Dai, 2003) and it is because of their higher carrier velocity along with ballistic transport (Geetha, 2014; Sharifi, 2015).

CARBON NANOTUBES' PROPERTIES

The electronic structures of CNTs are considerably correlated to chirality (n, m) as defined in Figure 2. Each CNS is distinctively characterized by a roll up vector known as the chiral vector, C_h. This vector is identified by the unit vectors a_1 and a_2 of the graphene lattice shown in Figure 2. Three various types of CNT structures; armchair, zigzag and chiral (helical) CNTs are defined regarding the chiral vector and the angle between C_h and the unit vector of a_1, named as chiral angle. If C_h is along with the *y*-axis then a zigzag nanotube is formed (α=0, m=0) while armchair CNTs have the C_h parallel to the *x*-axis (α=30°, n=m) and the Chiral CNTs have chiral vector with nonequal *n* and *m*(0<α<30) (M.S. Dresselhaus, G. Dresselhaus, Eklund, Saito,1998).

A basic model of SWCNT for the state electronic structure near the Fermi level is that of a graphene sheet with periodic boundary condition. The periodical boundary condition induces the quantised k-vectors that are perpendicular to the axes. Orientation and position of the k-lines near the K-points in a two-dimensional Brillouin zone (BZ) of graphene determines the energy dispersion relation of a CNT (Zólyomi & Kürti, 2004). The wave functions for the CNSs have to satisfy periodic boundary conditions around the circumference of nanotubes, $\vec{k} \cdot \vec{C} = L$. The wave vectors k for one dimensional CNTs are illustrated schematically in the two-dimensional Brillouin zone (BZ) of graphene in Figure 3.

Similarly to graphene, the BZ corners where the conductance and valence bonds meet each other are of the great importance for the electronic properties of CNTs. K_1 and K_2 are the reciprocal lattice vectors along the CNT axis and the circumferential direction of CNT, respectively.K_1 determines the distance between two k-lines and K_2 gives their length. If the k-lines cross a K point in the BZ then a metallic

Figure 2. The chiral vector $\overrightarrow{C_h} = n\hat{a}_1 + m\hat{a}_2$ *is defined in the form of basic vectors:* $a1 = \sqrt{3}a_{cc}\left(\left(\sqrt{3}/2\right)\hat{i} + 0.5\hat{j}\right), a2 = \sqrt{3}a_{cc}\left(\left(\sqrt{3}/2\right)\hat{i} - 0.5\hat{j}\right)$

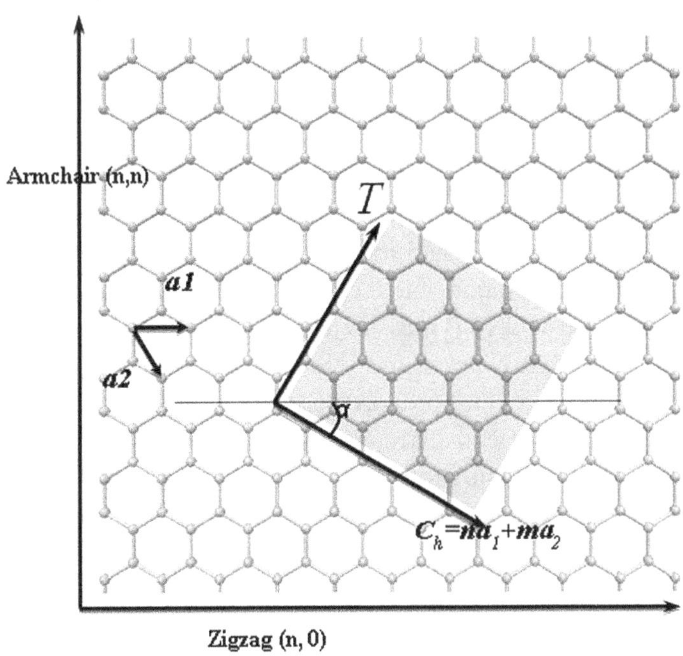

Figure 3. Wave vectors k in the Brillouin zone of a carbon nanotube similar to the graphene sheet for metallic and semiconducting from left to right

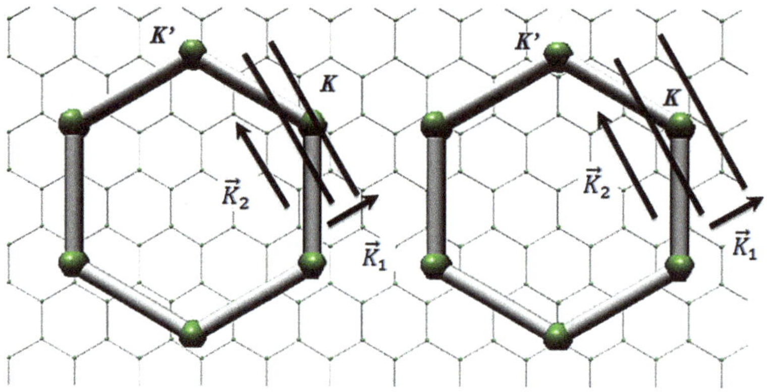

CNT is formed. Whereas in a semiconducting CNT, k-points are positioned at 1/3 or 2/3 of the distance between two adjacent lines as illustrated in Figure 3.

Recent studies have found that SWCNTs can be either metallic or semiconducting based on their inner diameter and chirality. Armchair tubes are metallic, while most of the others are semiconducting except when (n-m)/3 = 2n (i.e. an integer) when they are semimetallic (Treacy,1996; Zhang,2004). The semiconducting characteristics of many CNTs make them suitable candidates for MOS fabrication that has wide application in various types of transistor-based sensors. It has to be noted that semiconducting SWCNTs are considered as p-type semiconductors in which the holes are the major charge carriers.

In addition, CNTs show many superior properties from graphene sheet such as higher tensile strength and elasticity, chemical and thermal stability, potential of chemical functionalization and higher carrier mobility. All these excellent characteristics together with the capability of supporting high current density form the basis for a wide range of CNT applications as nanosensors.

CARBON NANOTUBE-BASED SENSOR PLATFORM

Since CNTs are composed of completely surface atoms, the electrical behavior of CNT-based sensors is considerably changed by the exposure of various molecules. The huge number of interactions between the p-type semiconducting CNT and the sensed analyte affect the electron configuration of CNT. The changes in the density of major charge carriers (holes) lead to the conductance variation of CNT. Therefore, the conductance of CNT is measured before and during the exposure of the analyte and consequently the concentration of the exposed species can be determined upon the conductivity change. This sensing procedure is the basis of most proposed gas and biosensors based on CNTs. However, the method still fails to suggest a high selective sensor without trading other desirable properties such as small size, low power and stable configuration. Many fictionalization methods of CNTs with different materials are also presented in recent years to achieve a perfect CNT-based sensor and improve the adsorption energy and selectivity.

The CNT-FET Sensors Configuration

Among all the existing CNT-based sensors, transistor based sensors are of special interest in which the analyte concentration is directly transduce to an electrical signal. The first CNT field effect transistor have been proposed by Martel and coworkers in 1988 where semiconducting CNTs are employed as conducting channels between source and drain electrodes in FETs (Martel et al,1998). A simple configuration of CNT-FET shown in Figure 4 consists of a p-type semiconducting SWCNT, source (S) and drain (D) electrodes and an additional gate electrode (G) which is separated from the CNT by a thin insulator film. The gate electrode is prepared to provide the ability to control the conductance of CNT through the applied voltage to the gate.

Applying a voltage between source and drain electrodes allows an electric current flow through the p-type CNT channel. Varying the applied voltage to the gate terminal modulates the conductivity of the CNT channel and thus controls the current flow between drain and source.

The applied voltage to the gate terminal imposes an electric field across the gate oxide (usually SiO_2 layer) which serves as an insulator between the gate and CNT channel. The electric field is dependent on both electronic properties of the CNT channel and the interface between the semiconducting CNT and the metal electrodes. The potential barrier between the semiconductor and the metal is referred as Schottky barrier. The thickness and the height of the barrier can be changed by the electric field imposed by the gate voltage which in turn controls the current flow of charge carriers. However, CNT application on transistors provides lower junction voltage in Schottky barrier which means that the ideal diode approximation can be used for it (Fathi, Ahmadi & Ismail, 2016) . Also, the Schottky transistor can be used with normal diodes and transistors along with the connection of metal or silicide layer to a doped

Figure 4. Typical backgate configuration of a CNT-FET with the associated electrodes (source, drain and the backgate electrode) with an individual a network of CNT connecting the source and drain terminals

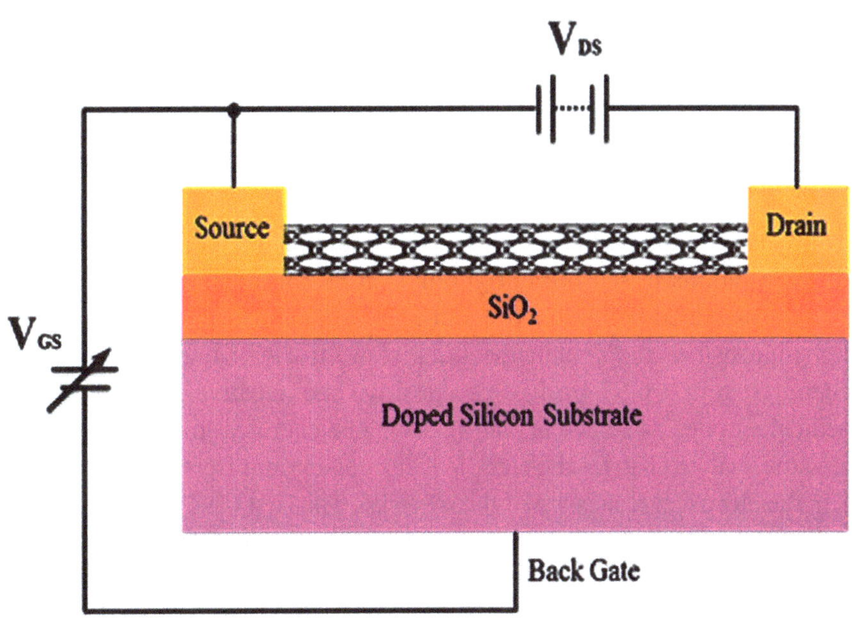

semiconductor layer (Sangtarash et al, 2012). Thus, the thickness of the barrier is the main limiting factor in determining the current flow in CNT-FETs in comparison with the height of barrier. The tunneling process of charge carriers through the barrier is explained in more detail in section 6.1.

Integrating CNTs into Field Effect Transistors

CNT-based devices can be fabricated using different deposition techniques. There are two main approaches for integrating CNT in transistor-based sensors. The first method to bond CNTs across electrodes is to directly grow CNTs on the sensor platform which involves the controlled chemical vapor deposition technique (CVD) (Dai, 2001). CNT growth is performed by Rapid Thermal CVD in two steps:

1. The catalyst metal nanoparticles heating in a furnace while hydrogen gas flows into the chamber for a period of time and the growth step.
2. The growth substrate usually patterns on a heated silicon platform where a thin layer of silicon dioxide is used as the insulator.

The second for CNT integration is to drop-cast a CNT solvent on top of the metal electrodes which may need post-processing in some cases (Sun, Fu, Lin & Hang, 2002). In the method, electrodes are made by sputtering and evaporating metal contacts on the CNT. The drop-casting technique suffers from poor uniformity and it has some limitations in controlling the thickness.

Due to the complicated fabrication process, the products resulting from the above methods have generally low yield. Recently, significant progress has been made in CNT chemistry that can solve the problems of dissolution and dispersion of CNT in different solvents. This establishes alternative methods for achieving desired specifications in a CNT-FET sensor by simply dispensing the dissolved CNT on substrates. Some other techniques such as air brushing, screen printing and inkjet printing have been also used in addition to CNT drop-casting using microsyringes which may sometimes follow by alignment (Valentini, 2003; Li, 2003; Lee, 2006; Kordas, 2006; Ghosh, 2003). These methods offer high reproducible and low cost CNT-based sensor manufacturing in comparison with the CVD-grown CNT-FET sensors and they exhibit higher yields, too.

Other CNT Sensor Configurations

The excellent electrical and structural properties of CNTs have pulled researchers towards more investigations on CNT applications in sensor configuration in recent years. Different types of CNT-based sensors have been demonstrated by several researchers based on resistance or capacitance variation of CNT known as chemiresistors and chemicapacitors, respectively (Kong, 2000; Snow, 2005a; Snow, 2005b). In the chemiresistors, a CNT bridges two positive and negative electrodes or it covers a set of interdigitated metal electrodes as shown in Figure 5(a). The chemical interaction between the sensing material(CNT) and the analyte changes the electrical current amount flowing through the CNT which in leads to changes in the electrical resistance of the material. The first CNT chemiresistor was introduced by Kong et al (2000). The CNT chemiresistor is very similar to the CNT-FET sensors except that in chemiresistors the gate electrode which modulates the resistance of CNT between electrodes does not exist. As a whole, chemiresistors sensors feature lower sensitivity than CNT-FET since the conductance of CNT in a FET can be tuned with the voltages applied to the gate terminal.

Figure 5. (a) Simple schematic of a CNT chemiresistor; (b) simple schematic of a CNT chemicapacitor; (c) front view of a CNT resonator

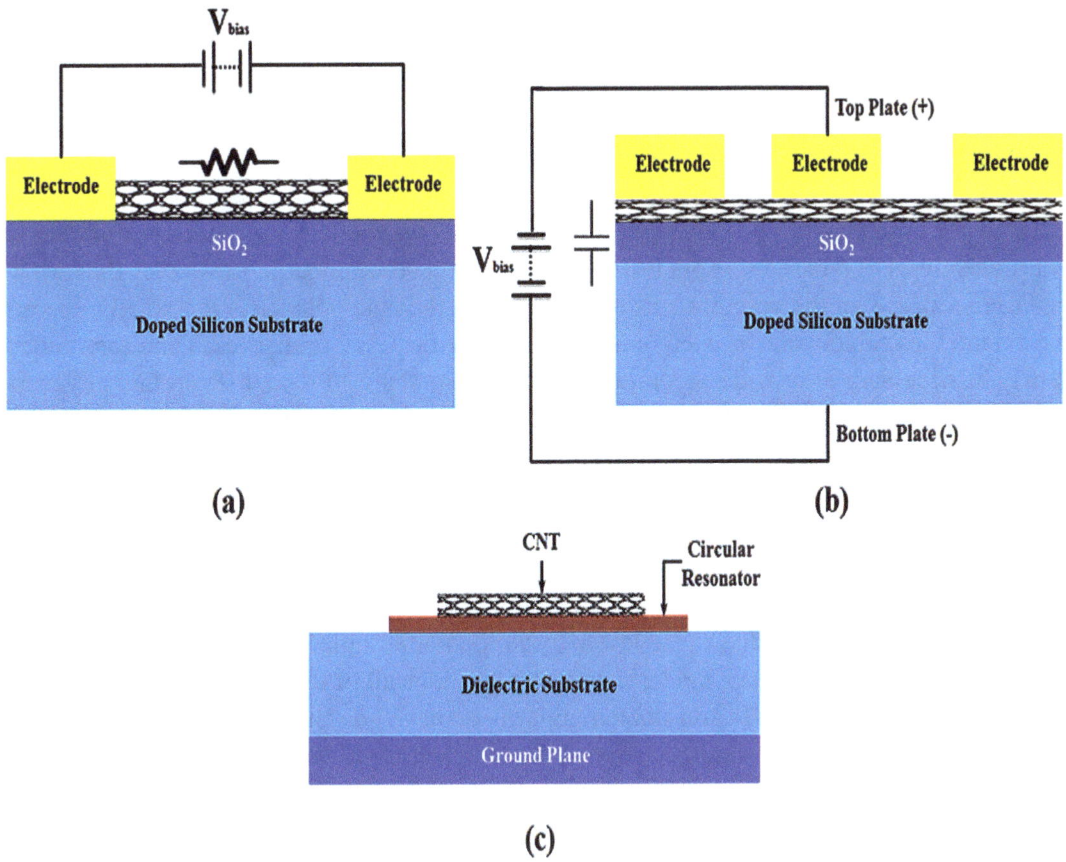

SWCNT-based chemicapacitors proposed by Snow and Perkins (2005) is another technique optimizes the sensitivity and they have low electrical noise. Various types of gas molecules including non-polar gases and vapors can be detected by measuring the capacitance change across the associated oxide (insulator). In the sensor configuration, a SWCNT network integrated on a thick silicon oxide (SiO_2) and the highly doped silicon substrate form a capacitor as illustrated in Figure 5(b).

By applying a voltage between the CNT network and the heavily doped substrate, a large electric field radiates from the CNT which causes the new polarization of adsorbates. The relative change between the old and the new polarization of adsorbates is represented as a change in the capacitance of the sensor. Many studies demonstrate that using the chemicpacitance change improves the sensitivity and the signal-to-noise ratio of the sensor as well as the system reversibility in comparison with chemiresistors. The main outstanding factor of CNT chemicapacitors is that the relative amount of capacitance change is less concentration-dependent than CNT chemiresistors, for most chemical vapors (Kordas, 2006; Ghosh, 2003).

An analytical model for CNT resonant sensors shown in Figure 5(c) has been reported by Chopra and coworkers in which a circular disk electromagnetic resonant circuit has been used that is coated with a single or multi walled CNT by conductive epoxy to design the sensor (Chopra, Pham, Gaillard, Parker

& Rao, 2002). The natural frequency obtained by a CNT based resonant sensor is efficiently high that provides considerable sensitivity for the resonator based sensors. It has been found that the SWCNT sensors are more sensitive than the MWCNT ones that is illustrated as a result of the existence of more adsorption sites for the analyte molecules in SWCNT compared to the MWCNTs.

CNT-BASED SENSOR FUNCTIONALIZATION

One of the major limitations in designing CNT-based sensors is the lack of selectivity for detection of different molecules. CNT-based sensors usually fail to detect some specific molecules that are weakly adsorbed to its surface. In the case, the resulting change in the conductance is too small to be noticed. To some extent, the problem can be overcome by modifying the CNT properties. Therefore, functionalization may be necessary to provide accurate and fast sensors that can recognize certain target species. Functionalization is the addition of some functional particles onto the surface of sensing material by various synthesis methods. The two widely used methods of the surface functionalization of CNTs are covalent and non-covalent functionalization depending on the interaction of attached functional group onto the CNT surface.

The principle of covalent functionalization is based on attaching the functional group onto the sidewall or ends of the CNT typically through esterification and amidation reactions. Since the sidewall of the CNT has low reactivity due to its lower pyrimidization angle, the purification and oxidation process of CNT is done to introduce defect sites along the sidewall of the CNTs. The functional groups are introduced into defective CNTs and stable bonds are achieved on the CNT surface which improves the sensitivity of the sensor. Although covalent functionalization is very stable but it associates with the disruption of the extended sp^2 hybridization which usually decreases the conductance of the CNT (Chopra, 2002; Chen, 1998).

Non-covalent approach employs van der Waals forces and π-π interactions to modify the CNT properties. Non-covalent functionalization is the procedure of supramolcular complexation of CNTs with organic compounds (made by drop-coating or dip-coating methods) without disrupting the basic configuration of CNT. The non-covalent functionalization is of a great concern in many biomedical studies since it enhances the interaction between CNT and biomolecules and enables CNT sensors to perform better in many of the biomedical sensing applications (Star, 2001; Xia, 2003).

CNT SENSOR DESIGN CHALLENGES

Although CNT sensors show potential functionality in many applications, the commercialization of the sensor still remains a challenge. The first limiting problem associates with the difficulty of the pure and contaminated CNT production. There are not many companies that produce appropriate CNT for commercial applications. The next challenge is that the low dimensional geometry of CNT causes structural stability problems.

Another issue concerning about the CNT sensors is the recovery time of the sensors. Due to the fact that the binding energy of the CNT with the target gas analyte is large, the analyte molecules are usually desorbs very slowly from the surface after the exposure. In order to decrease the recovery time of the sensor after the exposure, an external energy source is usually applied to the system. The stabilizing process back to the initial point of the sensor can be done by either local heating of CNT to high temperatures or applying photo-induced desorption like Ultra Violet light illumination. The technique lowers the adsorption energy barrier and causes the fast desorption of the bounded analyte.

As mentioned before, the thickness and the height of the Schottky barrier highly affects the performance of CNT-FET sensors. Schottky barriers at metal and CNT contacts have been studied extensively (Collins, 2000, 2001). In addition to the difficulty of controlling the types, locations, and orientations of SWNTs, a clear realization about the contact between CNTs and macroscopic metal electrodes is needed to be studied. In this section the carrier velocity effect on the CNT based Schottky contact is demonstrated which indicates that complete saturation does not exist in Schottky transistor.

CNT Based Schottky Model

In order to figure out the electrical structure of a CNT, at first its band structure must be obtained. Equation (1) shows the energy band structure of a CNT (Oberlin et al., 1976) which represents a parabolic relationship between E (*eV*) and the momentum k (*nm*):

$$E \approx E_{c0} + \frac{h^2 k_x^2}{2m^*} \tag{1}$$

where k and m* are the wave number and the electron effective mass, respectively. As fundamental parameters, available energy states at a certain energy called Density of state (DOS) has been defined as:

$$DOS = \frac{\Delta n_x}{\Delta E L_x} = \frac{1}{2\pi}\left(E - E_{c0}\right)^{-\frac{1}{2}}\left(\frac{2m^*}{h^2}\right)^{\frac{1}{2}} \tag{2}$$

The carrier concentration in a band is defined by integrating the distribution function over the energy band by employing Fermi probability function in carrier concentration calculation. Schottky-Mott rule is used to foretell the barrier between a metal and semiconductor which is proportional to the difference of the metal-vacuum work function and the semiconductor-vacuum electron affinity. It must be considered that most of the metal-semiconductor interfaces do not pursue this law to the foretoken degree. Therefore, the effect of Fermi level pinning occurs because the nature of these metal-induced gap states and their occupation by electrons tend to pin the center of the band gap to the Fermi level. The current in metal-semiconductor junction is mostly created by majority carriers, based on thermionic emission theory (Sze & Ng, 2007), electrons with energy higher than the height of the barrier will pass the barrier and move towards the barrier. The velocity which is a result of random motion of carriers will affect the current density in the form of:

$$J_s = e\int_1^{N_{\max}} \frac{\int v(E) Dosf(E) dE}{\int Dosf(E) dE} dn \tag{3}$$

which leads to the current density as a function of Fermi integrals.

$$J_s = v_{th} \Im_0(\eta) \ln n \tag{4}$$

But this random motion does not result to the magnitude of zero for a single vector known as the intrinsic velocity. However, velocity can be discussed in degeneracy limits as shown in Figure 6, which specifies Fermi integral with Maxwell approximation as an exponential value for Fermi integrals ($\mathcal{F}_j\left(\eta\right)$) in nondegenerate regime. The average of this intrinsic velocity can be obtained in the form of thermal velocity which indicates temperature dependence motion of carriers as:

$$v_{thermal} = \sqrt{\frac{2k_B T}{\pi m^*}} \tag{5}$$

Nondegenerate limit leads to temperature affected current on Schottky contact as shown in Figure 6:

$$J = en\sqrt{\frac{2k_B T}{\pi m^*}} \tag{6}$$

On the other hand, in degenerate limit carrier velocity does not illustrate temperature dependent behavior although it shows carrier concentration affected movement of the majority carriers.

$$V_f = \frac{\hbar}{4m^*}(n\pi) \tag{7}$$

Figure 6. Inherent velocity in degeneracy limit over the carrier concentration

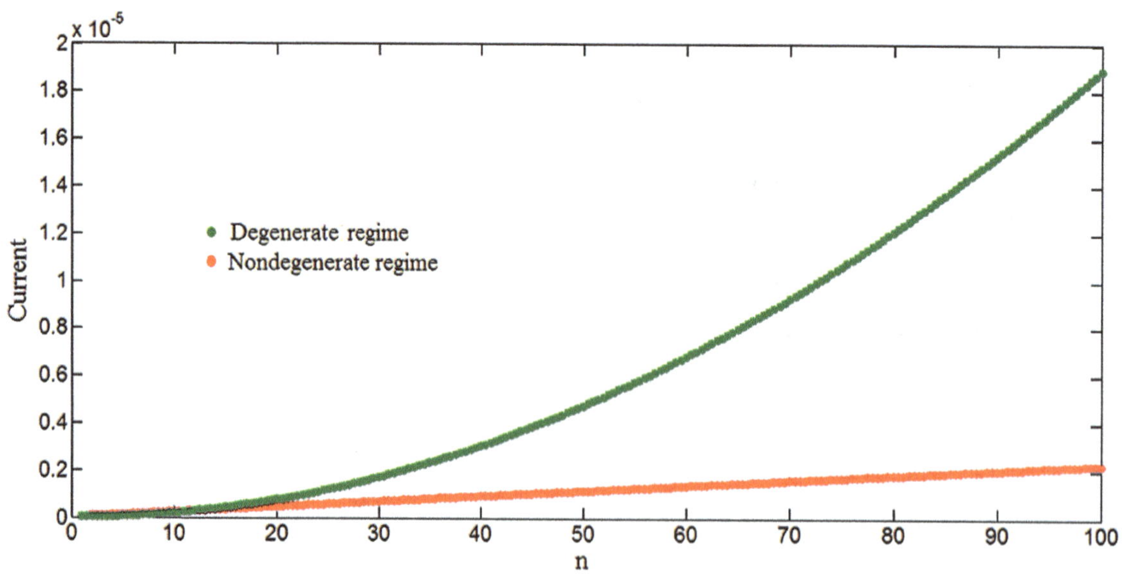

Carrier velocity affected by number of carriers known as Fermi velocity (v_f) guides to a parabolic current density relation in degenerate limit as shown in Figure 6.

$$J = \frac{e\hbar\pi}{8m^*}(n^2) \tag{8}$$

In other words Figure 6 depict that the intrinsic velocity is a linear function of the carrier concentration in nondegenerate regime (with linear changes) and exponentially in degenerate limit.

However, in nondegenerate regime it could be affected by the beginning of a quantum emission (Ahmadi, Ismail, Tan & Arora, 2008). It is required that the velocity is independent of carrier concentration for lower number of carrier as the graph is flat. It seems that most of the previous papers prefer to utilize the thermal velocity for modeling as an independent value of carrier concentration. It is interesting to know that in common the carrier density can be modified in the exponential form as:

$$e^{\frac{J_s}{v_{th}\Im_0(\eta)}} = n \tag{9}$$

In terms of the model, the current of CNTs is increased by increasing the temperature in small applied voltage. It is quite different for high bias voltages which are independent of temperature.

According to the simulation results of Yang et al. (2005) for effect of the temperature variations on I-V curvature, these variations only take effect on triode region.

$$\frac{dJ_s}{d\eta} = v_{th} \ln n \int_0^\infty \frac{-e^{-\eta}}{(1+e^{x-\eta})^2} dx \tag{10}$$

According to the chain rule of differentiation the temperature effect on current-voltage characteristic is explored which indicates that current variation in the saturation region can be seen in Figure 8. The normalized Fermi energy is expressed by $\eta = \left(E_F - E_o\right) / k_B T$ which is a function of temperature that leads to:

$$\frac{dJ_s}{dT} = v_{th} \ln n \int_0^\infty \frac{\eta e^{-\eta}}{T(1+e^{x-\eta})^2} dx \tag{11}$$

For better analysis, the effect of temperature variations on I-V Curvature is investigated and current in CNTST as a function of temperature is reported as shown in Figure 8, in which the green line represents 200K, the yellow line shows 300K and the blue line is used for I-V characteristics of the 400K.

But the simulation results of the work indicate that the temperature variations will affect both of the saturation and triode regions of the transistor. This effect is somehow more sensible in saturation region. Based on the presented model (Equation 9) it seems that the I-V characteristic cannot be affected sharply by number of carriers as shown in Figure 7, conversely degenerate limit indicates rapid variation on I-V characteristics. In other words as illustrated in Figure 6, the current will be increased dramatically by carrier concentration (n). So, contradicts between the results of Figure 6 and Figure 9 can be referred as a size effect in the nanoscale devices, also it need to be explored carefully on the basis of nanoscale device saturation and pinch off effects seems to be questionable in nanoscale size as well.

Based on the proposed model for carrier density the numerical result is investigated and I-V characteristics of a CNT – Schottky transistor is simulated as shown in Figure 7. Since the current doesn't saturate in devices (Ahmadi, Ismail, Tan & Arora, 2009), therefore, the saturation velocity and its constitutive quantities need to be discussed more. By focusing on the curvature in Figure 7, it can be concluded that the presented model for CNT Schottky contact current model indicates the incremental effect of the CNT diameter which can be explained by increasing the current along with the applied voltage.

Based on the presented model of CNTST, an increase in the applied voltage will result to an increment in the current which is independent of thermal velocity for the high number of majority carriers. As discussed completely, the current increases by Fermi velocity in degenerate limit, but the common I-V Curvatures doesn't have considerable changes with the variations in number of carriers and the current doesn't increase sharply by voltage variations over different number of carriers which can be explained in the form of size effect in the nanoscale devices. It is reported that the saturation phenomenon cannot be seen in the nanoscale CNTST comparable to that of conventional FET.

CNT-BASED SENSORS APPLICATION

The use of CNT in sensors provides cost-effective, sensitive and accurate detection devices in a variety of research and applications. A special interest exists in using the CNT-based sensors in biomedical applications owing to their small size and power efficient performance. They can be highly effective as implantable sensors that are used for controlling temperature, blood glucose, heart activity and ac-

Figure 7. The effect of variations in number of carriers on CNTST I-V characteristic; the yellow curve represents n = 10^{10}, while the green curve is for n = 10^{15}. Finally, n = 10^{20} is distinguished by blue dots.

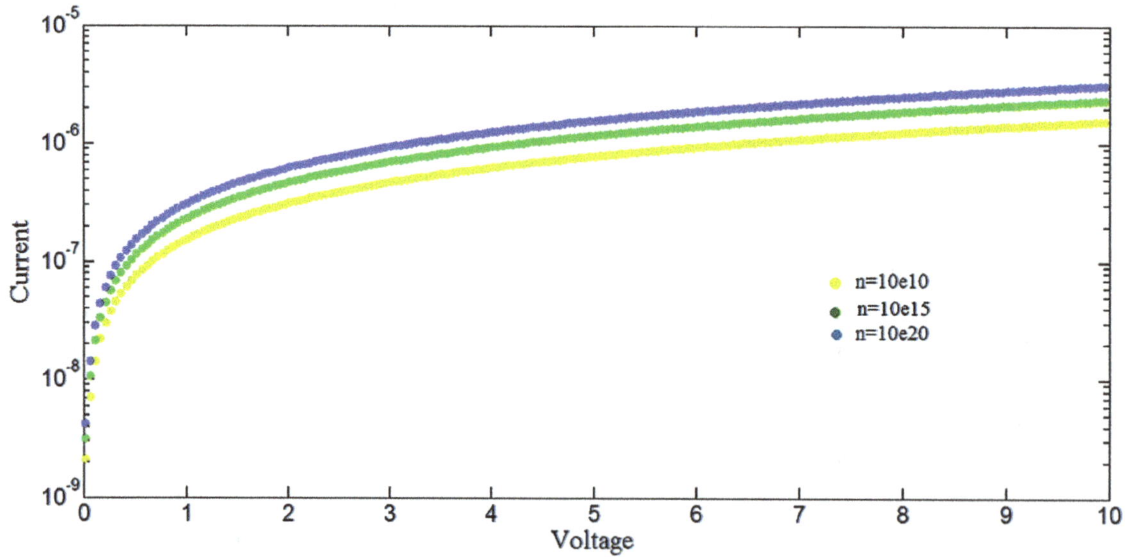

Figure 8. The effect of temperature variations on I-V characteristic clarified by carrier velocity variation

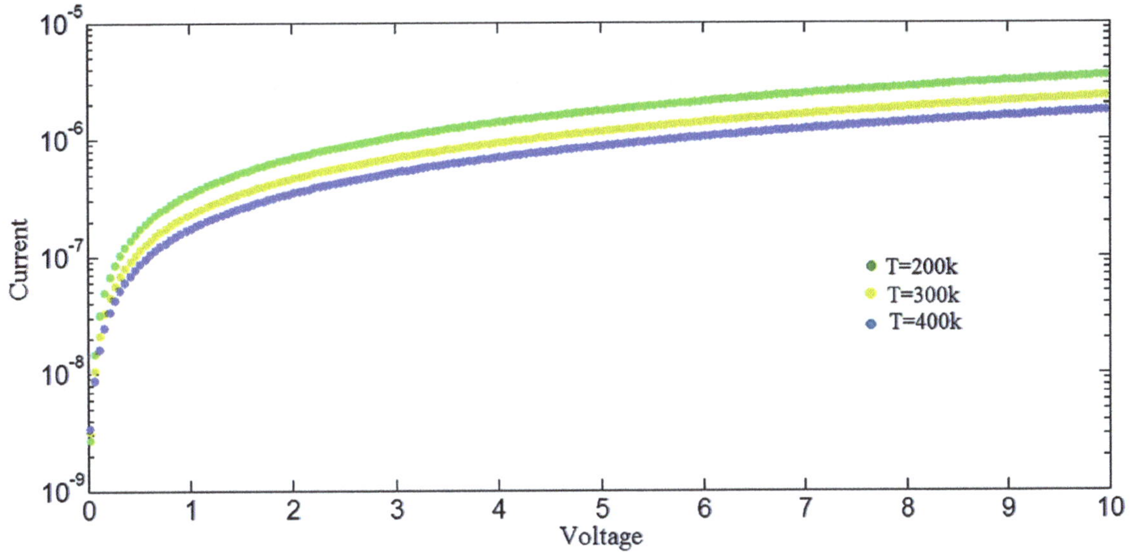

Figure 9. Comparison of proposed model (red dots) with published experimental data (blue line)
Endo & Dresselhaus, 2002.

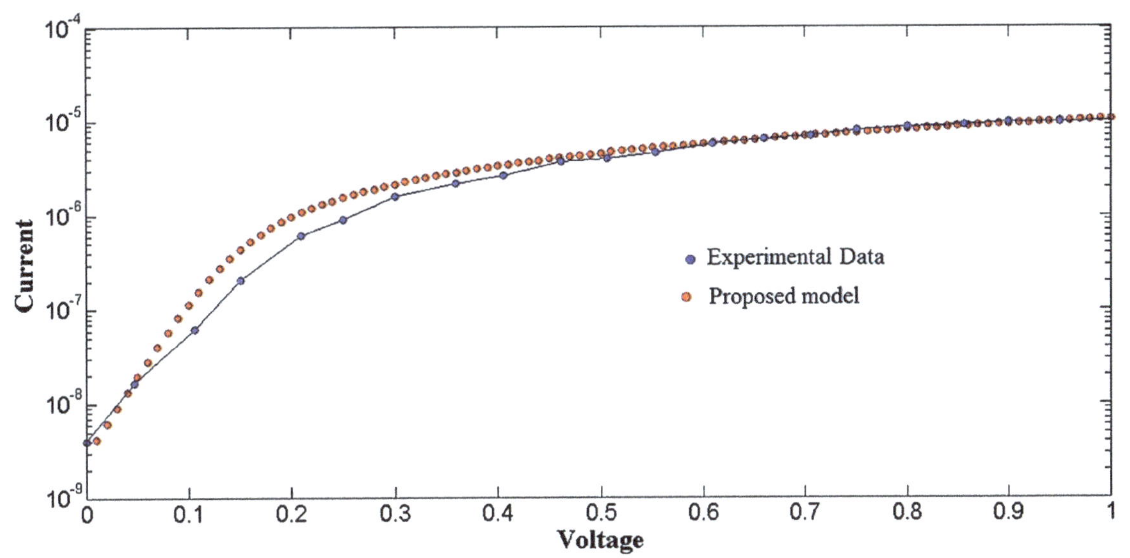

curate gas measurement during surgeries. They also exhibit many advantageous in the performance of immune sensors and DNA biosensors (Wang, 2008; Heller, 2008; Yun, 2007; Sinha, 2006; Liu, 2008; Williams, 2002; Wang, 2004).

The sensors can also be used in many other sensing applications such as metal deposition monitors, chemical reactor monitors, and mass detectors (Wu, 2008; Ahammad, 2009). In these applications, changes in mass varies the conductance, capacitance or natural frequency in FET-based, chemiresistors, chemicapacitors and resonators, respectively. The resulted conductance and capacitance change or the frequency shift is used to detect either presence or concentration of the attached mass. Moreover, different applications of the CNTs for gas detection have been investigated. Several types of gases and agents such as NO_2, O_2, NH_3, N_2, CO_2, CH_4, H_2O, H_2, Ar, Choline, Ethanol, Methanol, Chloromethane and H_2O_2 on both SWCNTs and MWCNTs using different techniques (Shults, 1994; Bolz, 1997; Shandas,2003). The large surface-to-volume ratio and hollow structure of CNTs show highly suitable performance in gas molecules adsorption and storage.

CONCLUSION

This chapter highlighted the specific configuration of the CNT sensors that are highly considered for many applications. Due to the special structural, dynamical and electronic properties of CNTs, they exhibit good sensing performance and also provide very good stability in the sensor. In this chapter special electrical properties of CNTs and the platform of CNT-based sensors have been carefully reviewed. It is found that the use of CNTs as the sensing material in different configurations of sensors improves the selectivity, sensitivity and the sensing dynamic range compared to the conventional silicon based sensors.

However, it is hard to engineer the configuration of the CNT sensor to be highly selective to a specific analyte while remains stable. Some solutions have been reported to moderately overcome the limitations. The selective adsorption of the analyte on the surface of CNT can be accomplished by the functionalization of the CNT. Different functionalization methods with different functionalization materials that make the CNT sensor more selective and sensitive have been also discussed in the chapter. Formation of Schottky barrier in the interface of the metal electrode and the surface of CNT is another main limiting factor that may affect the proper performance of the designed sensors. In order to clarify this effect, an analytical model has been presented to describe this effect more specifically. In general, fabrication of Schottky diode by means of CNT is an open area of research to conquer the limitations of silicon technology. Although broad application categories for CNT sensors have been investigated in many studies, but many of them are still active research field to provide commercially efficient manufacture of CNT-based sensors.

REFERENCES

Ahammad, A. J. S., Lee, J. J., & Rahman, M. A. (2009). Article. *Sensors (Basel, Switzerland)*, *9*(4), 2289–2319.

Ahmadi, M. T., Tan, M. L. P., Ismail, R., & Arora, V. K. (2009). Article. *International. Journal of Nanotechnology*, *6*(7-8), 601–617. doi:10.1504/IJNT.2009.025299

Bolz, A., Lang, V., Merkely, B., & Schaldach, M. (1997). Paper. *Conference Proc. Ann. Int. Conf. IEEE Eng. Med. and Biol*, *5*, 2341.

Chen, J., Hamon, M. A., Hu, H., Chen, Y., Rao, A. M., Eklund, P. C., & Haddon, R. C. (1998). Solution Properties of Single-Walled Carbon Nanotubes. *Science*, *282*(5386), 95–98. doi:10.1126/science.282.5386.95 PMID:9756485

Chopra, S., Pham, A., Gaillard, J., Parker, A., & Rao, A. M. (2002). Carbon-nanotube-based resonant-circuit sensor for ammonia. *Applied Physics Letters*, *80*(24), 4632–4634. doi:10.1063/1.1486481

Collins, P. G., Arnold, M. S., & Avouris, P. (2001). Engineering Carbon Nanotubes and Nanotube Circuits Using Electrical Breakdown. *Science*, *292*(5517), 706–709. doi:10.1126/science.1058782 PMID:11326094

Collins, P. G., Bradley, K., Ishigami, M., & Zettl, A. (2000). Extreme Oxygen Sensitivity of Electronic Properties of Carbon Nanotubes. *Science*, *287*(5459), 1801–1804. doi:10.1126/science.287.5459.1801 PMID:10710305

Dai, H. J. (2001). *Carbon Nanotubes*. Berlin: Springer.

Dresselhaus, M. S., Dresselhaus, G., & Eklund, P. C. (1996). *Science of Fullerenes and Carbon Nanotubes*. San Diego, CA: Academic Press.

Dresselhaus, M. S., Dresselhaus, G., Eklund, P. C., & Saito, R. (1998). Carbon nanotubes. *Physics World*, *11*(1), 33–38. doi:10.1088/2058-7058/11/1/32

Geetha, P., & WahidaBanu, R. S. D. (2014). A compact modelling of double-walled gate wrap around carbon nanotube array field effect transistors. *Journal of Computational Electronics*, *13*(4), 900–916. doi:10.1007/s10825-014-0607-7

Ghosh, S., Sood, A., & Kumar, N. (2003). Carbon Nanotube Flow Sensors. *Science*, *299*(5609), 1042–1044. doi:10.1126/science.1079080 PMID:12532025

Hamon, M. A., Hu, H., Bhowmik, P., Niyogi, S., Zhao, B., Itkis, M. E., & Haddon, R. C. (2001). End-group and defect analysis of soluble single-walled carbon nanotubes. *Chemical Physics Letters*, *347*(1-3), 8–12. doi:10.1016/S0009-2614(01)01035-1

Heller, A., & Feldman, B. (2008). Electrochemical Glucose Sensors and Their Applications in Diabetes Management. *Chemical Reviews*, *108*(7), 2482–2505. doi:10.1021/cr068069y PMID:18465900

Hone, J., Llaguno, M. C., Nemes, N. M., Johnson, A. T., Fischer, J. E., & Walters, D. A. et al. (2000). Electrical and thermal transport properties of magnetically aligned single wall carbon nanotube films. *Applied Physics Letters*, *77*(5), 666–668. doi:10.1063/1.127079

Javey, A., Guo, J., Wang, Q., Lundstrom, M., & Dai, H. (2003). Ballistic carbon nanotube field-effect transistors. *Nature*, *424*(6949), 654–657. doi:10.1038/nature01797 PMID:12904787

Kong, J., Franklin, N. R., Zhou, C., Chapline, M. G., Peng, S., Cho, K., & Dai, H. (2000). Nanotube Molecular Wires as Chemical Sensors. *Science*, *287*(5453), 622–625. doi:10.1126/science.287.5453.622 PMID:10649989

Kong, J., Franklin, N. R., Zhou, C. W., Chapline, M. G., Peng, S., Cho, K. J., & Dai, H. J. (2000). Nanotube Molecular Wires as Chemical Sensors. *Science*, *287*(5453), 622–625. doi:10.1126/science.287.5453.622 PMID:10649989

Kordas, K., Mustonen, T., Toth, G., Jantunen, H., Lajunen, M., & Soldano, C. et al. (2006). Inkjet Printing of Electrically Conductive Patterns of Carbon Nanotubes. *Small*, *2*(8-9), 1021–1025. doi:10.1002/smll.200600061 PMID:17193162

Lee, C. Y., Baik, S., Zhang, J. Q., Masel, R. I., & Strano, M. S. (2006). Charge Transfer from Metallic Single-Walled Carbon Nanotube Sensor Arrays. *The Journal of Physical Chemistry B, 110*(23), 11055–11061. doi:10.1021/jp056425v PMID:16771365

Li, L., Lu, Y., Ye, Q., Cinke, M., Han, J., & Meyyappan, M. (2003). Carbon Nanotube Sensors for Gas and Organic Vapor Detection. *Nano Letters, 3*(7), 929–933. doi:10.1021/nl034220x

Liu, H., Wang, G., Chen, D., Zhang, W., Li, C., & Fang, B. (2008). Fabrication of polythionine/NPAu/MWNTs modified electrode for simultaneous determination of adenine and guanine in DNA. *Sensors and Actuators. B, Chemical, 128*(2), 414–421. doi:10.1016/j.snb.2007.06.028

Lu, X., & Chen, Z. (2005). Curved Pi-Conjugation, Aromaticity, and the Related Chemistry of Small Fullerenes ($<C_{60}$) and Single-Walled Carbon Nanotubes. *Chemical Reviews, 105*(10), 3643–3696. doi:10.1021/cr030093d PMID:16218563

Martel, R., Schmidt, T., Shea, H. R., Hertel, T., & Avouris, P. (1998). Single- and multi-wall carbon nanotube field-effect transistors. *Applied Physics Letters, 73*(17), 2447–2449. doi:10.1063/1.122477

Modi, A., Koratkar, N., Lass, E., Wei, B., & Ajayan, P. (2003). Miniaturized gas ionization sensors using carbon nanotubes. *Nature, 424*(6945), 171–174. doi:10.1038/nature01777 PMID:12853951

Monthioux, M., & Kuznetsov, V. (2006). Who should be given the credit for the discovery of carbon nanotubes? *Carbon, 44*(9), 1621–1623. doi:10.1016/j.carbon.2006.03.019

Oberlin, A., Endo, M., & Koyama, T. (1976). Filamentous growth of carbon through benzene decomposition. *Journal of Crystal Growth, 32*(3), 335–349. doi:10.1016/0022-0248(76)90115-9

Sangtarash, S., Sadeghi, H., Ahmadi, M.T., Ghadiry, M.H., Anwar, S., & Ismail, R. (2012). Article. *Journal of Computational and Theoretical Nanoscience, 9*(10), 1554-1557.

Shandas, R., & Lanning, C. (2003). Article. *Medical & Biological Engineering & Computing, 41*, 416–424. doi:10.1007/BF02348084 PMID:12892364

Sharifi, F., Moaiyeri, M. H., Navi, K., & Bagherzadeh, N. (2015). Quaternary full adder cells based on carbon nanotube FETs. *Journal of Computational Electronics, 14*(3), 762–772. doi:10.1007/s10825-015-0714-0

Shults, M. C., Rhodes, R. K., Updike, S. J., Gilligan, B. J., & Reining, W. N. (1994). A telemetry-instrumentation system for monitoring multiple subcutaneously implanted glucose sensors. *IEEE Transactions on Bio-Medical Engineering, 41*(10), 937–942. doi:10.1109/10.324525 PMID:7959800

Sinha, N., Ma, J., & Yeow, J. T. W. (2006). Carbon Nanotube-Based Sensors. *Journal of Nanoscience and Nanotechnology, 6*(3), 573–590. doi:10.1166/jnn.2006.121 PMID:16573108

Snow, E. S., & Perkins, F. K. (2005). Capacitance and Conductance of Single-Walled Carbon Nanotubes in the Presence of Chemical Vapors. *Nanoletter, 5*(12), 2414–2417. doi:10.1021/nl051669c

Snow, E. S., Perkins, F. K., Houser, E. J., Badescu, S. C., & Reinecke, T. L. (2005). Chemical Detection with a Single-Walled Carbon Nanotube Capacitor. *Science, 307*(5717), 1942–1945. doi:10.1126/science.1109128 PMID:15790850

Someya, T., Small, J., Kim, P., Nuckolls, C., & Yardley, J. T. (2003). Alcohol Vapor Sensors Based on Single-Walled Carbon Nanotube Field Effect Transistors. *Nano Letters*, *3*(7), 877–881. doi:10.1021/nl034061h

Star, A., Stoddart, J. F., Steuerman, D., Diehl, M., Boukai, A., & Wong, E. W. et al. (2001). Preparation and Properties of Polymer-Wrapped Single-Walled Carbon Nanotubes. *Angewandte Chemie*, *113*(9), 1771–1775. doi:10.1002/1521-3757(20010504)113:9<1771::AID-ANGE17710>3.0.CO;2-Y

Sun, Y. P., Fu, K., Lin, Y., & Hang, W. (2002). Functionalized Carbon Nanotubes: Properties and Applications. *Accounts of Chemical Research*, *35*(12), 1096–1104. doi:10.1021/ar010160v PMID:12484798

S. M. Sze & K. K. Ng, (2007). *Physics of semiconductor devices*. Wiley-Blackwell.

Treacy, M. M. J., Ebbesen, T. W., & Gibson, J. M. (1996). Exceptionally high Young's modulus observed for individual carbon nanotubes. *Nature*, *381*(6584), 678–680. doi:10.1038/381678a0

Valentini, L., Cantalini, C., Armentano, I., Kenny, J. M., Lozzi, L., & Santucci, S. (2003). Investigation of the NO[sub 2] sensitivity properties of multiwalled carbon nanotubes prepared by plasma enhanced chemical vapor deposition. *Journal of Vacuum Science & Technology B Microelectronics and Nanometer Structures*, *21*(5), 1996–2000. doi:10.1116/1.1599858

Wang, J. (2008). Electrochemical Glucose Biosensors. *Chemical Reviews*, *108*(2), 814–825. doi:10.1021/cr068123a PMID:18154363

Wang, J., Liu, G., & Jan, M. R. (2004). Ultrasensitive Electrical Biosensing of Proteins and DNA: Carbon-Nanotube Derived Amplification of the Recognition and Transduction Events. *Journal of the American Chemical Society*, *126*(10), 3010–3011. doi:10.1021/ja031723w PMID:15012105

Wei, B. Q., Vajtai, R., & Ajayan, P. M. (2001). Reliability and current carrying capacity of carbon nanotubes. *Applied Physics Letters*, *79*(8), 1172–1174. doi:10.1063/1.1396632

Williams, K. A., Veenhuizen, P. T. M., Torre, B. G., Eritja, R., & Dekker, C. (2002). Nanotechnology: Carbon nanotubes with DNA recognition. *Nature*, *420*(6917), 761. doi:10.1038/420761a PMID:12490938

Wong, H. S. P. (2002). Beyond the conventional transistor. *IBM Journal of Research and Development*, *46*(2.3), 133–168. doi:10.1147/rd.462.0133

Wu, W., Palaniapan, M., & Wong, W. K. (2008). Multiwall carbon nanotube resonator for ultra-sensitive mass detection. *Electronics Letters*, *44*(18), 1060–1061. doi:10.1049/el:20081932

Xia, H., Wang, Q., & Qiu, G. (2003). Polymer-Encapsulated Carbon Nanotubes Prepared through Ultrasonically Initiated In Situ Emulsion Polymerization. *Chemistry of Materials*, *15*(20), 3879–3886. doi:10.1021/cm0341890

Yang, M., Teo, K., Milne, W., & Hasko, D. (2005). Carbon nanotube Schottky diode and directionally dependent field-effect transistor using asymmetrical contacts. *Applied Physics Letters*, *87*(25), 253116. doi:10.1063/1.2149991

Yu, M. F., Lourie, O., Dyer, M. J., Moloni, K., Kelly, T. F., & Ruoff, R. S. (2000). Strength and Breaking Mechanism of Multiwalled Carbon Nanotubes Under Tensile Load. *Science*, *287*(5453), 637–640. doi:10.1126/science.287.5453.637 PMID:10649994

Yun, Y. H., Bange, A., Heineman, W. R., Halsall, H. B., Shanov, V. N., & Dong, Z. et al. (2007). A nanotube array immunosensor for direct electrochemical detection of antigen–antibody binding. *Sensors and Actuators. B, Chemical*, *123*(1), 177–182. doi:10.1016/j.snb.2006.08.014

Zhang, W., Zhu, Z. Y., Wang, F., Wang, T. T., Sun, L. T., & Wang, Z. X. (2004). Chirality dependence of the thermal conductivity of carbon nanotubes. *Nanotechnology*, *15*(8), 936–939. doi:10.1088/0957-4484/15/8/010

Zólyomi, V., & Kürti, J. (2004). First-principles calculations for the electronic band structures of small diameter single-wall carbon nanotubes. *Physical Review B: Condensed Matter and Materials Physics*, *70*(8), 085403. doi:10.1103/PhysRevB.70.085403

Chapter 2
Modeling Trilayer Graphene-Based DET Characteristics for a Nanoscale Sensor

Meisam Rahmani
Universiti Teknologi Malaysia, Malaysia

Ali Hosseingholi Pourasl
Universiti Teknologi Malaysia, Malaysia

Hediyeh Karimi
Swinburne University of Technology, Australia

Komeil Rahmani
Islamic Azad University, Iran

Mohammad Javad Kiani
Islamic Azad University, Iran

Mohammad Taghi Ahmadi
Universiti Teknologi Malaysia, Malaysia & Urmia University, Iran

Razali Ismail
Universiti Teknologi Malaysia, Malaysia

ABSTRACT

Graphene is a promising nanomaterial with outstanding physical and electrical properties that offers a wide range of opportunities for advanced applications in nanoelectronics [1-3].The application of graphene nanoribbon (GNR) in high-speed electronics is being explored extensively because of some excellent properties such as one-atom thickness, mechanical strength, flexibility, high thermal conductivity up to 50 W cm –1 K –1, extremely highcurrent-carrying capacity up to 10 9 A/cm 2, high carrier mobility in excess of 200,000 cm 2 V –1 s –1, high carrier saturation velocity 3 of ~5×10 7cm s –1, and extraordinarily rapid charge-carrier transportwhich is intrinsically ambipolar, meaning that both positive and negative carriers are important [4-9]. Trilayer graphene nanoribbon (TGN) as one of the most common multilayers of graphene is taken into consideration in this study.

INTRODUCTION

Graphene is a promising nanomaterial with outstanding physical and electrical properties that offers a wide range of opportunities for advanced applications in nanoelectronics (Dragoman, 2009; Rahmani et al., 2013; Dong, 2011). The application of graphene nanoribbon (GNR) in high-speed electronics is

DOI: 10.4018/978-1-5225-0736-9.ch002

being explored extensively because of some excellent properties such as one-atom thickness, mechanical strength, flexibility, high thermal conductivity up to 50 W cm $^{-1}$ K $^{-1}$, extremely highcurrent-carrying capacity up to 10 ^{9} A/cm ^{2}, high carrier mobility in excess of 200,000 cm ^{2} V $^{-1}$ s $^{-1}$, high carrier saturation velocity 3 of ~5×10 ^{7}cm s $^{-1}$, and extraordinarily rapid charge-carrier transportwhich is intrinsically ambipolar, meaning that both positive and negative carriers are important (Castro Neto, 2009; Dong, 2010; Xie, 2009; Ahmadi, 2012; Rahmani et al., 2013; Craciun, 2011). Trilayer graphene nanoribbon (TGN) as one of the most common multilayers of graphene is taken into consideration in this study. TGN has attracted enormous attention from material scientists and device engineers because of its exceptional electronic properties (Craciun et al., 2009; Rutter, 2008; Rahmani, 2012). The scaling of field effect transistors (FETs) at nanoscale assures better performance of the device. The phenomenon of downsizing the device dimensions has led to challenges such as short channel effects, leakage current, interconnect difficulties, high power consumption and quantum effects (Rahmani, 2012). Therefore, new materials and device structures are needed as alternatives to overcome these challenges. According to the TGN structure, it can satisfy our main requirement of a channel in nanoscale FETs. Major advances have been attained in understanding device physics and improving device performance for TGNFET (Shengjun, 2010; Rahmani, 2013). In this study, we evaluate the ultimate device performance when the ABA-stacked TGNis applied to the FET channel. Figure 1 shows the cross sectional view of the proposed TGNFET.

Apparently, large-sized ABA-stacked TGN sheet is used as a channel material. The channel region shows the characteristics of the resistor in small values of drain-source voltage as the source and substrate terminals are held at ground potential. The major strength of adopting TGN with ABA stacking as a channel material is due to its strong capability to control the electrostatics and hence expected to reduce the short channel effects that rely on the FET electrostatics (Ouyang, 2009). Moreover, much of the interest surrounding TGN is due to the carrier velocity control at high-electric fields which is essential in GNRFET applications, particularly those based on low-noise and high-frequency operation (Fiori, 2007). In this chapter, scaling behaviors of grapheneFET application on sensor modeling in corporation with analytical device models are analytically studied. In order to utilize TGN in coming nanoscale FET based devices, its carrier velocity in high-field transport and quantum capacitance of TGNFET as important parametersare investigated.This can be done if quantities such as the density of states (DOS) and carrier concentration (n) are obtained. Some existing works on high-field transport discuss the optical phonon influence on the velocity (Scott, 2010; Jena, 2009; Dorgan, 2010; Amin et al., 2010).

Figure 1. (a) ABA-stacked TGN as a one-dimensional material; (b) A schematic of implemented TGN in FET

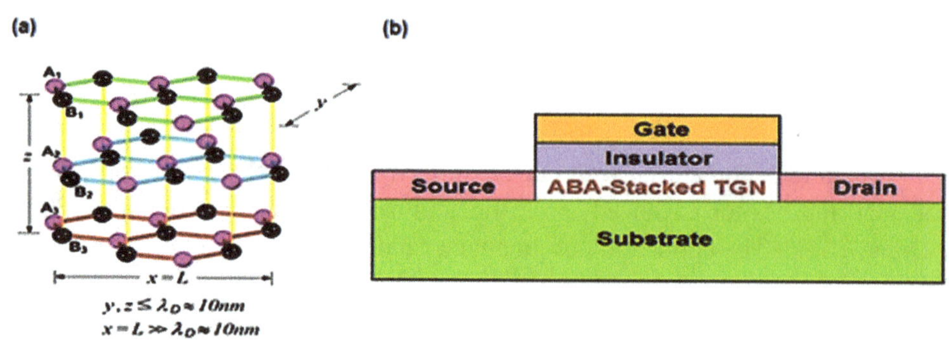

PROPOSED MODELOF CARRIER VELOCITY IN HIGH-FIELD TRANSPORT OF TGNFET

In order to obtain the ABA-stacked TGN band energy, the Tight-binding energy method is adopted. So the band energy throughout the Brillion zone can be calculated as (Guinea, 2006)

$$E_k = \pm(\alpha k - \beta k^3) \tag{1}$$

where k is the wave vector in x direction, $\alpha = \frac{v_f \Delta}{t_\perp \sqrt{2}}$, $\beta = \frac{v_f^3}{t_\perp \sqrt{2\Delta}}$, ν_F is the Fermi velocity, Δ is the applied voltage and $t_\perp$ is the hopping energy. According to the energy relation with wave-vector, the band structure of ABA-stacked TGN is plotted in Figure 2. As illustrated, the low E-k curve exposes overlap between the conduction and valence band structure. By applying a perpendicular external electric field on ABA-stacked TGN surface, the possibility of opening a gap on band structure emerges, which can be controlled by value of applied voltage. In fact, applied electric field effect can be observed in the form of carrier velocity in the channel region of a FET. As revealed in Figure 2, the band overlap increases with increasing the applied voltage. In other words, by increasing the applied voltage, the curvature of the E-K graph increases. It's notable that during this process, not only are the same number of electrons and holes maintained, but also the intrinsic Fermi level will move away from the band gap midpoint.

As a main parameter,the DOS which shows the availability of energy states at each energy levels, is defined as (Rahmani, 2013):

Figure 2. The effect of applied voltage on the curvature of the E-K graph in ABA-stacked TGN (a) V=0.4 V (b) V=0.6 V

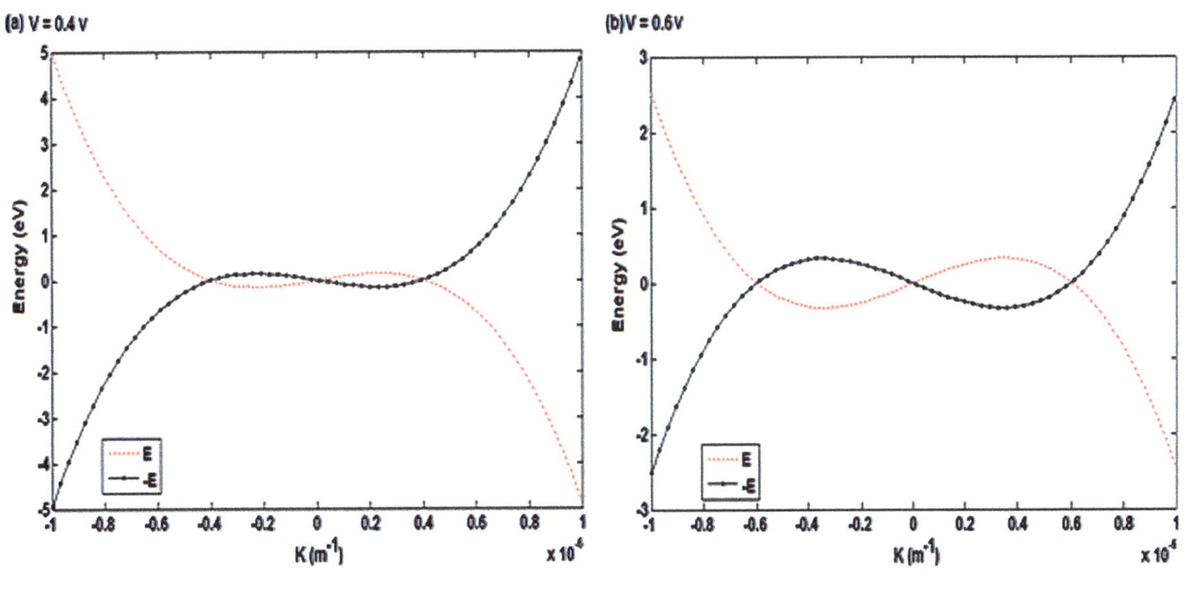

$$DOS(E) = \frac{\Delta n}{l.\Delta E} \tag{2}$$

where Δn is the number of available energy states, l is the length and ΔE is the energy. By employing the wave vector relation in the DOS definition in simplified form (Rahmani, 2012), the proposed model of DOS is obtained as

$$DOS(E) = \frac{1}{A - \frac{B}{\left(DE + \sqrt{F + E^2}\right)^{\frac{2}{3}}} - C\left(DE + \sqrt{F + E^2}\right)^{\frac{2}{3}}} \tag{3}$$

where E is the energy band structure and $A = -6.2832\alpha$, $B = 14.3849\alpha^2\beta$, $C = \frac{2.7444}{\beta}$, $D = -9\beta^2$, $F = \frac{-0.1690\alpha^3}{\beta}$. The carrier transport features for long channel FET model are no longer capable of describing carrier transport perfectly even for sub-100nm FETs. The carrier velocity is a significant parameter that will help predict the carrier mobility and a device performance. Besides that, the control of carrier velocity in high-electric field is very important, particularly when dealing with the high-frequency operation (Betti, 2011). In order to determine the average velocity of carrier in any material, usually the DOS and carrier concentration models are required. To model atheoretical prediction of average velocity for carriers in TGNFET, the expression (4) is used which depends on the density of eigenstates in the channel (Ahmadi, 2008). In fact, the average velocity of the carriers can be calculated as an accumulative velocity of all carriers divided by number of the carriers as

$$\nu = \frac{1}{n}\int_{E_c}^{E_{top}\approx\infty} |\nu| DOS(E) f(E) dE \tag{4}$$

where n is the carrier concentration, $f(E) = 1/\left[1 + \exp\left(E - E_f/k_B T\right)\right]$ is the Fermi-Dirac distribution function, k_B is the Boltzmann constant, E_f is the 1D Fermi energy and T is the temperature. Also, the magnitude of the carrier velocity is defined as $|\nu| = \sqrt{\frac{2(E - E_c)}{m}}$ (m is the carrier effective mass).Consequently, the proposed model of carrier velocity for TGNFET is proportional to the averaged DOS over a few $k_B T$ at Fermi level as

$$\nu = \frac{1}{n_{TGN}} \int_{E_c}^{E_{top}\approx\infty} \frac{\sqrt{\frac{2(E-E_c)}{m}}\, dE}{\left[A - \frac{B}{\left(DE+\sqrt{F+E^2}\right)^{\frac{2}{3}}} - CD\left(E+\sqrt{F+E^2}\right)^{\frac{2}{3}}\right]\left(1+\exp\left(E-E_f/k_BT\right)\right)} \tag{5}$$

We can rewrite equation (5) as

$$\nu = \frac{1}{n_{TGN}}\sqrt{\frac{2}{m}}(k_BT)^{\frac{3}{2}} \int_0^{\eta} \frac{x^{\frac{1}{2}}dx}{\left\{\begin{array}{l} A - \frac{B}{(k_BT)^{\frac{2}{3}}\left\{D(x+E_{CO})+\sqrt{N+(x+E_{CO})^2}\right\}^{\frac{2}{3}}} \\ -C(k_BT)^{\frac{2}{3}}\left\{D(x+E_{CO})+\sqrt{N+(x+E_{CO})^2}\right\}^{\frac{2}{3}} \end{array}\right\}(1+\exp(x-\eta))} \tag{6}$$

where $x = \frac{E-E_c}{k_BT}$, normalized Fermi energy $\eta = \frac{-E_c+E_f}{k_BT}$, $E_{co} = \frac{E_c}{k_BT}$, $N = \frac{F}{(k_BT)^2}$ and the carrier concentration of TGN (n_{TGN}) which has been modeled as (Rahmani, 2012)

$$n_{TGN} = \int_0^{\eta} \frac{(k_BT)dx}{\left\{\begin{array}{l} A - \frac{B}{(k_BT)^{\frac{2}{3}}\left\{D(x+E_{CO})+\sqrt{N+(x+E_{CO})^2}\right\}^{\frac{2}{3}}} \\ -C(k_BT)^{\frac{2}{3}}\left\{D(x+E_{CO})+\sqrt{N+(x+E_{CO})^2}\right\}^{\frac{2}{3}} \end{array}\right\}(1+\exp(x-\eta))} \tag{7}$$

By employing the proposed model described in the preceding section, the velocity of carriers for TGNFET is plotted in Figure 3. The proposed model provides a quantitative description of carrier velocity in terms of normalized Fermi energy and electric field. Apparently, in the low- field limits, the velocity-electric field graph is linear which the ohmic mobility can be obtained. In other words, the result indicates the insensitivity of the ultimate saturation velocity on the low-fieldmobility. The inconsistencyoccurs from the fact that it is unfeasible to measure directly the ultimate saturation velocity that needs aninfinite electric field. This is because there is no device in nanoscale which is able to sustain such ahigh-electric field. Therefore, the ultimate saturation velocity canbe obtained indirectly. In general, the highest measuredvelocity is assigned to be the ultimate saturation velocity whichis lower than the actual saturation velocity (Ahmadi, 2008).

Figure 3. Carrier velocity of TGNFET as a function of electric field (T=300K)

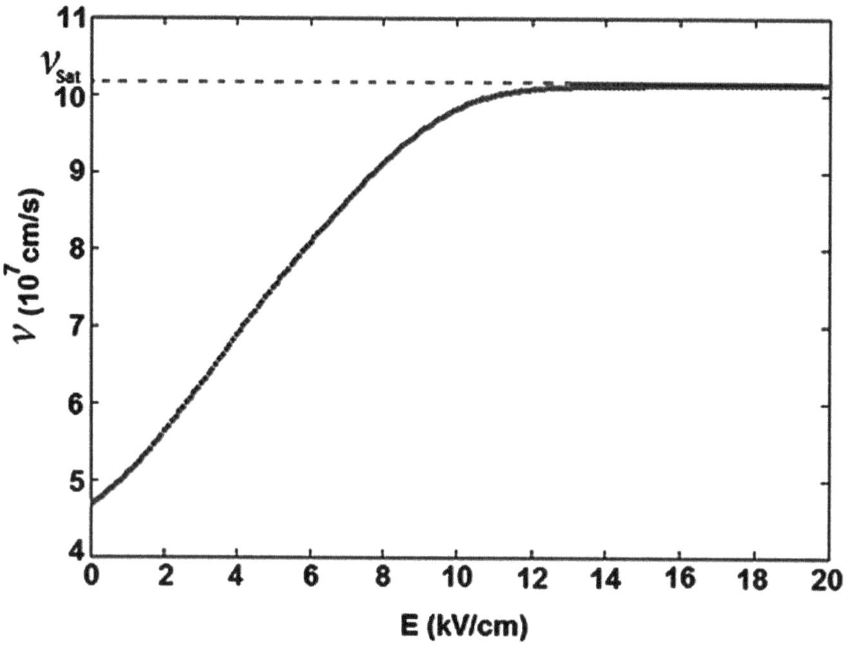

Considering the importance of carrier velocity on the device performance, the proposed model of carrier velocity for TGNFET in the degenerate and nondegenerate regimes is analytically studied. The band gap illustrates a non-degenerate condition as the distance of Fermi level more than $3K_BT$ from either the conduction or valance bands. In this case, due to high difference between $x$and η, 1 can be neglected in comparison with an exponential function.Therefore, the model in the non-degenerate regime is carried out by the following equation as

$$\nu_{ND} = \frac{\displaystyle\int_0^{\eta} \frac{\sqrt{\frac{2}{m}}(k_BT)^{\frac{3}{2}} x^{\frac{1}{2}} \exp(\eta - x)dx}{\left\{A - \frac{B}{(k_BT)^{\frac{2}{3}}\left\{\begin{matrix} D(x+E_{CO}) \\ +\sqrt{N+(x+E_{CO})^2}\end{matrix}\right\}^{\frac{2}{3}}} - C(k_BT)^{\frac{2}{3}}\left\{\begin{matrix} D(x+E_{CO}) \\ +\sqrt{N+(x+E_{CO})^2}\end{matrix}\right\}^{\frac{2}{3}}\right\}}}{\displaystyle\int_0^{\eta} \frac{(k_BT)\exp(\eta - x)dx}{\left\{A - \frac{B}{(k_BT)^{\frac{2}{3}}\left\{\begin{matrix} D(x+E_{CO}) \\ +\sqrt{N+(x+E_{CO})^2}\end{matrix}\right\}^{\frac{2}{3}}} - C(k_BT)^{\frac{2}{3}}\left\{\begin{matrix} D(x+E_{CO}) \\ +\sqrt{N+(x+E_{CO})^2}\end{matrix}\right\}^{\frac{2}{3}}\right\}}} \quad (8)$$

In case of carrier statistics study, the Fermi level in the degenerate approximation is located less than $3K_BT$ away from the conduction and valence bands or situated within a band. In this region, we can neglect the $\exp(x-\eta)$ in comparison with 1 because the value of $x-\eta$ is very small. Consequently, the carrier velocity model in the degenerate regime is proposed as

$$\nu_D = \frac{\displaystyle\int_0^{\eta} \frac{\sqrt{\frac{2}{m}}(k_BT)^{\frac{3}{2}} x^{\frac{1}{2}} dx}{\left[\left\{A - \frac{B}{(k_BT)^{\frac{2}{3}}\left\{\frac{D(x+E_{CO})}{+\sqrt{N+(x+E_{CO})^2}}\right\}^{\frac{2}{3}}} - C(k_BT)^{\frac{2}{3}}\left\{\frac{D(x+E_{CO})}{+\sqrt{N+(x+E_{CO})^2}}\right\}^{\frac{2}{3}}\right\}\right]}}{\displaystyle\int_0^{\eta} \frac{(k_BT)dx}{\left[\left\{A - \frac{B}{(k_BT)^{\frac{2}{3}}\left\{\frac{D(x+E_{CO})}{+\sqrt{N+(x+E_{CO})^2}}\right\}^{\frac{2}{3}}} - C(k_BT)^{\frac{2}{3}}\left\{\frac{D(x+E_{CO})}{+\sqrt{N+(x+E_{CO})^2}}\right\}^{\frac{2}{3}}\right\}\right]}} \tag{9}$$

As shown in Figure 4, the proposed model of carrier velocity for TGNFET is approximated by non-degenerate limit, specifically in low value of electric field. On the other hand, the general model can be approximated by degenerate limit, particularly in highvalueof E.

Accordingly, the simulated result demonstrates that the carrier velocity in the non-degenerate approximationbehaves similar to thermal velocity. On the other hand, carrier velocity in the degenerate approximationis a function of carrier concentration which is strongly unassociated with temperature.In other words, the carrier velocity is the appropriate average of the Fermi velocity for quasi one dimensional (1D) GNR in a degenerate manner. To get a greater insight into the carrier velocity of TGNFET, the carrier velocity versus gate voltage characteristic is analytically studied in this work. To this end, according to the relation between the normalized Fermi energy and applied voltage, the dependence of the carrier velocity on the gate voltage is associated with $\eta \approx \frac{V_G - V_T}{k_BT/e}$ where V_T is the thermal voltage (Datta, 2005). Therefore, the carrier velocity versus gate voltage characteristic of TGNFET is analytically modeled as

Figure 4. Carrier velocity of TGNFET in the non-degenerate and degenerate regimes (T=300K)

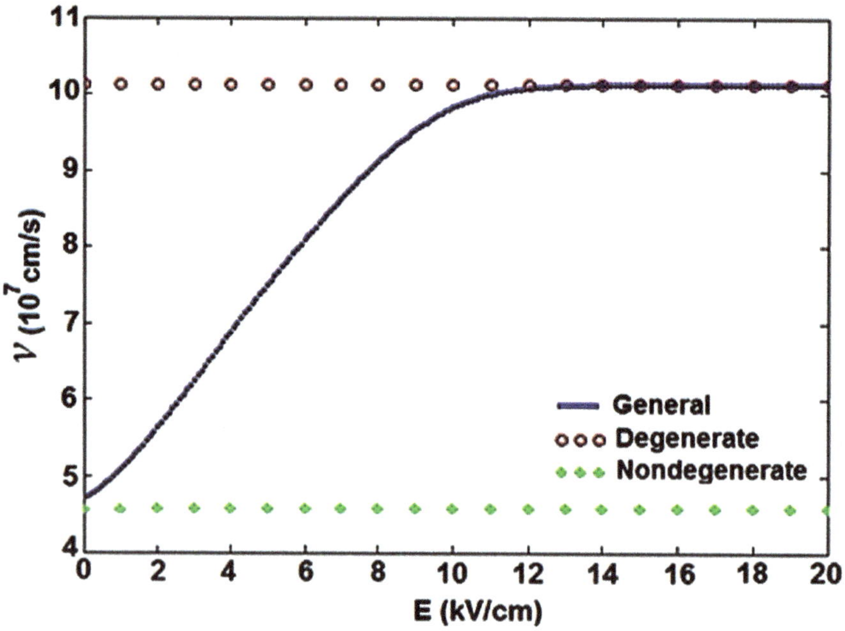

$$\nu = \frac{\displaystyle\int_0^{\eta} \frac{\sqrt{\frac{2}{m}}(k_BT)^{\frac{3}{2}}x^{\frac{1}{2}}dx}{\left\{A - \frac{B}{(k_BT)^{\frac{2}{3}}\left\{\begin{matrix}D(x+E_{CO})\\+\sqrt{N+(x+E_{CO})^2}\end{matrix}\right\}^{\frac{2}{3}}} - C(k_BT)^{\frac{2}{3}}\left\{\begin{matrix}D(x+E_{CO})\\+\sqrt{N+(x+E_{CO})^2}\end{matrix}\right\}^{\frac{2}{3}}\right\}\left(1+\exp\left(x-\frac{V_G-V_T}{k_BT/e}\right)\right)}}{\displaystyle\int_0^{\eta} \frac{(k_BT)dx}{\left\{A - \frac{B}{(k_BT)^{\frac{2}{3}}\left\{\begin{matrix}D(x+E_{CO})\\+\sqrt{N+(x+E_{CO})^2}\end{matrix}\right\}^{\frac{2}{3}}} - C(k_BT)^{\frac{2}{3}}\left\{\begin{matrix}D(x+E_{CO})\\+\sqrt{N+(x+E_{CO})^2}\end{matrix}\right\}^{\frac{2}{3}}\right\}\left(1+\exp\left(x-\frac{V_G-V_T}{k_BT/e}\right)\right)}} \quad (10)$$

Figure 5 shows the characteristic of carrier velocity versus gate voltage for TGNFET which is plotted based on Equation 10. It is of interest to inspect the high carrier velocity versus gate voltage. It is noteworthy that the carrier degeneracy makes the velocity versus gate voltage dependent, and the high values are possible.

Figure 5. Carrier velocity of TGNFET versus gate voltage (T=50K)

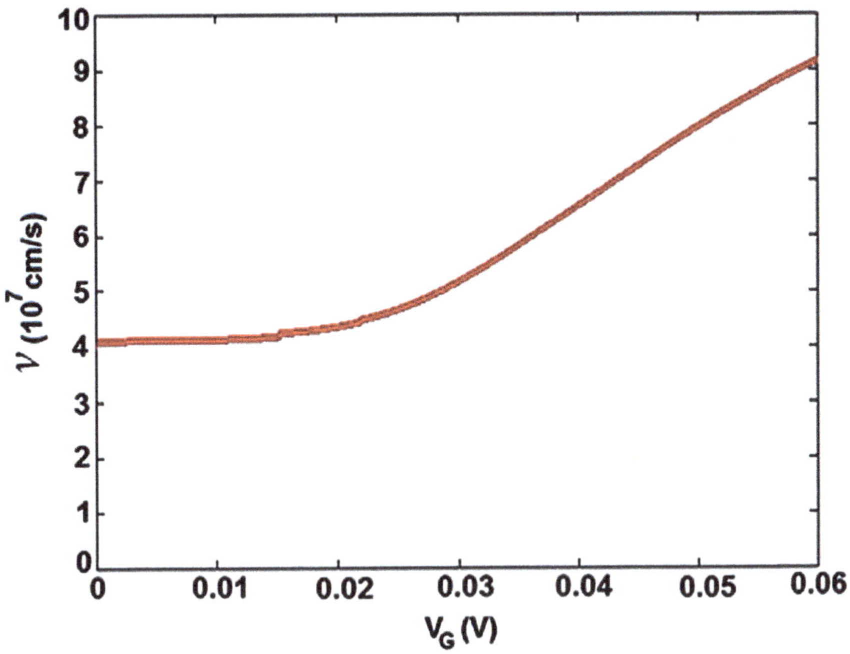

In the device considered in this model, a number of carriers increasingly occupy a higher momentum region. Therefore, as shown in Figure 5, the carrier velocity increases to improve the velocity versus gate voltagecharacteristic. Figure 6 gives a normalized plot of the carriervelocitywhich is a function of carrier concentration. Given an estimate of the inversion layer density at the beginning of the channel, one can use this characteristic to estimate the ballistic on-current (Lundstrom, 2006). Besides that, Figure 6 reveals a rather strong dependence of the carrier velocity on temperature. It can be stated that the carriers follow the degenerate statistics at low temperatures. Therefore, the velocity is limited with the appropriate average of the Fermi velocity which is a function of carrier concentration.As can be seen in Figure 6, the carrier velocityincreases significantly with rising temperature. Moreover, by increasing the carrier concentration, the velocity of carriers decreases. As expected, the velocity is independent of carrier concentration for high values of concentration. In other words, the normalized graphs of velocity versus carrier concentrationare almostunchanged by increasing the concentration.

The carrier transport in GNR is an essential phenomenon that determines the current–voltage characteristic of the device. Furthermore, the temperature has been demonstrated as an effective parameter in carrier velocity of GNR, leading to movement of carriers in a nanoscale FET channel (Winters, 2013; Shishir, 2009). The velocity versus temperaturecharacteristic of TGNFET is shown in Figure 7. The proposed model indicates that carrier velocity in TGNFET is temperature dependent, meaning that by increasing the temperature, the carrier velocity increases. Moreover, the velocity of carriers decreases as the carrier concentration increases.It's notable that as the concentration is increased to embrace the degenerate regime, the carrier velocity tends to be independent of temperature but it depends strongly on carrier concentration.

Figure 6. The effect of temperature on the carrier velocity of TGNFET

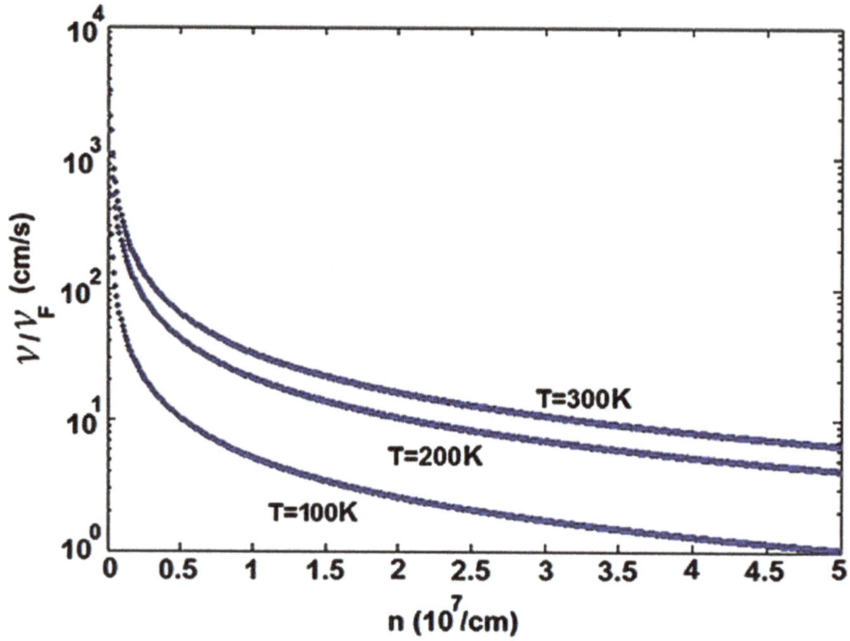

Figure 7. The effect of carrier concentration on the carrier velocity of TGNFET (T=300K)

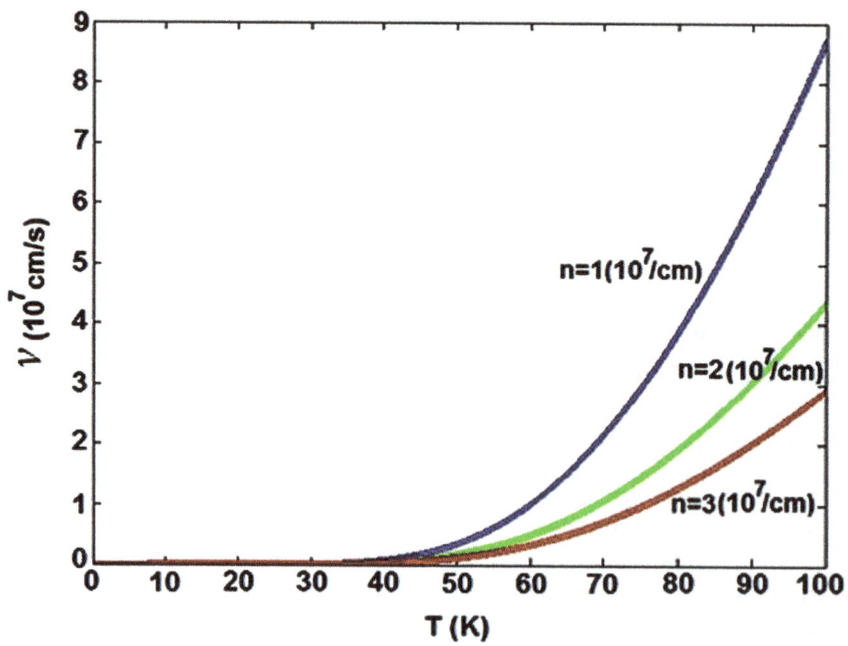

As specified in equation (11), an explicit equation for the electric potential distribution along the device is adopted to drive an analytical model for the contact current (Neamen, 2011). It is noted that in the linear region of the I_d-V_{ds} characteristic, the electric field is approximately constant and the potential drop in the channel is linear,so

$$I_d = nq \cdot \nu = c\nu \cdot V_{ds} \tag{11}$$

where c is the channel capacitance and ν is the velocity of the carriers. The drain current of TGNFET is analytically derived by substitution of the carrier velocity model in equation (11), which is a function of physical and electrical parameters including the carrier effective mass, effective length and width of the channel, velocity of carriers, channel capacitance, applied bias and drain-source voltages. The model provides closed form analytic expression for the current-voltage characteristic of the device. We assumed that the scattering rate is independent of energy, electron density, and spatial distribution. The I_d-V_{ds} characteristic of a TGNFET is plotted as shown in Figure 8. Apparently, the drain current increases significantly with increasing gate-source voltage.It can be concluded thattheproposed TGNFET delivers a high drain current due tothe high carrier velocity.

It is noteworthy that the findings open a new aspect for fabricating high performance FETs with the analytical design, and the structure considered in the proposed model is easier to fabricate in TGN because of the 1D atomistic structure.

Figure 8. I_d-V_{ds} characteristic of TGNFET at different values of V_{gs}

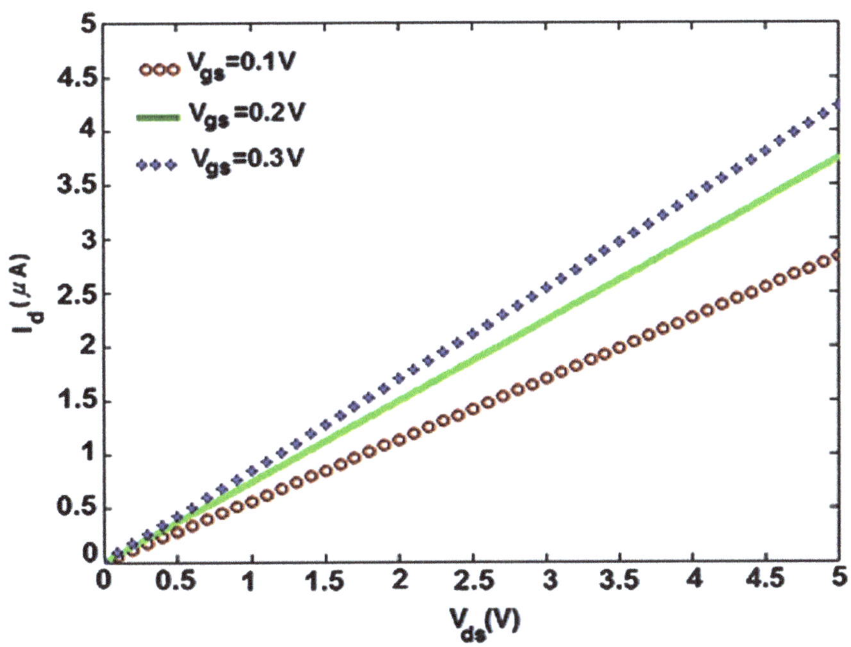

PROPOSED MODEL OF QUANTUM CAPACITANCE FOR TGNFET

The quantum capacitance as one of the most important characteristics of FET formed between the channel and gate. In fact, as the FET is scaled down to nanometer, the oxide layers are expected to be much thinner and have higher values of dielectric constant k, which means that the quantum capacitance will eventually be the dominant source of capacitance. To model a quantum capacitance for TGNFET, the (12) is used which depends on the density of eigenstatesin the channel as (Datta, 2005)

$$C_q = q^2 \int_{-\infty}^{+\infty} dED(E)F_T(E) \tag{12}$$

where q is the electron charge and $F_T(E) = -\frac{df(E)}{dE}$. Consequently, the proposed model of quantum capacitance for TGNFET is proportional to the averaged DOS over a few k_BT at Fermi level as

$$C_q = q^2 \int_{-\infty}^{+\infty} \frac{\exp(x-\eta)dx}{\left\{\begin{bmatrix} A - \dfrac{B}{(k_BT)^{\frac{2}{3}}\left\{D(x+E_{co}) + \sqrt{N+(x+E_{co})^2}\right\}^{\frac{2}{3}}} \\ -C(k_BT)^{\frac{2}{3}}\left\{D(x+E_{co}) + \sqrt{N+(x+E_{co})^2}\right\}^{\frac{2}{3}} \end{bmatrix}(1+\exp(x-\eta))^2)\right.} \tag{13}$$

where $x = \frac{E-E_c}{k_BT}$, normalized Fermi energy $\eta = \frac{-E_c + E_f}{k_BT}$, $E_{co} = \frac{E_c}{k_BT}$ and $N = \frac{F}{(k_BT)^2}$. Equation (13) gives a quantitative description of quantum capacitance in terms of normalized Fermi energy. According to the proposed model (13), ABA-stacked TGNFETquantum capacitance is a function of normalized Fermi energy (η) as shown in Figure 9.

Considering the consequences of quantum capacitance on the device performance, the analytical modeling of quantum capacitance in the degenerate and nondegenerate regimes is also studied. As indicated in Figure10, Fermi level with the distance more than $3K_BT$ from either the conduction or valance band edge within the band gap illustrates a non-degenerate condition . In case of a non-degenerate region, 1 can be neglected in comparison withan exponential function because of high difference between x and η.

Consequently, the quantum capacitance of the device in non-degenerate regime is performed as

$$Cq_{ND} = q^2 \int_{-\infty}^{+\infty} \frac{\exp(\eta-x)dx}{\left\{\begin{bmatrix} A - \dfrac{B}{(k_BT)^{\frac{2}{3}}\left\{D(x+E_{co}) + \sqrt{N+(x+E_{co})^2}\right\}^{\frac{2}{3}}} \\ -C(k_BT)^{\frac{2}{3}}\left\{D(x+E_{co}) + \sqrt{N+(x+E_{co})^2}\right\}^{\frac{2}{3}} \end{bmatrix}\right.} \tag{14}$$

Figure 9. Quantum capacitance of ABA-stacked TGN FET as a function of normalized Fermi energy

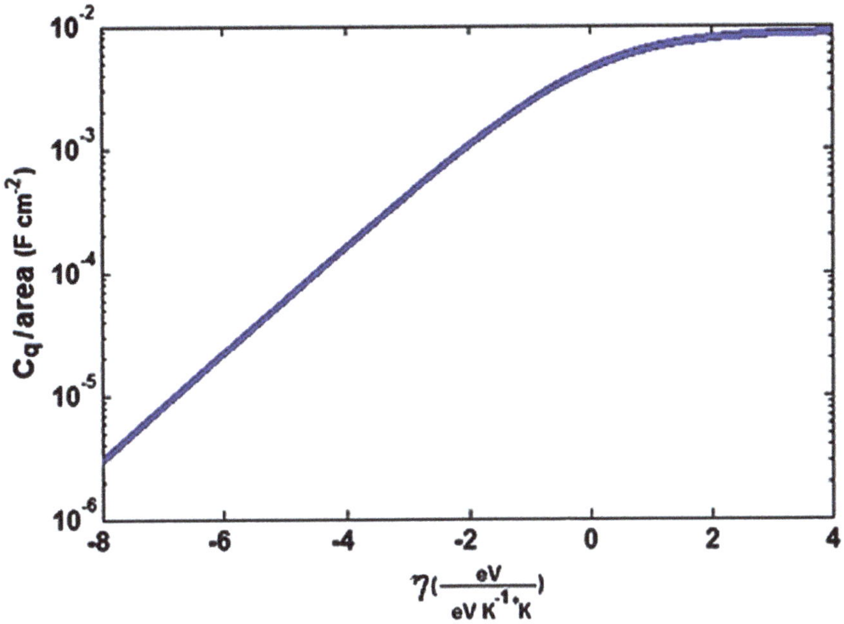

Figure 10. Definition of the degenerate and non-degenerate regions

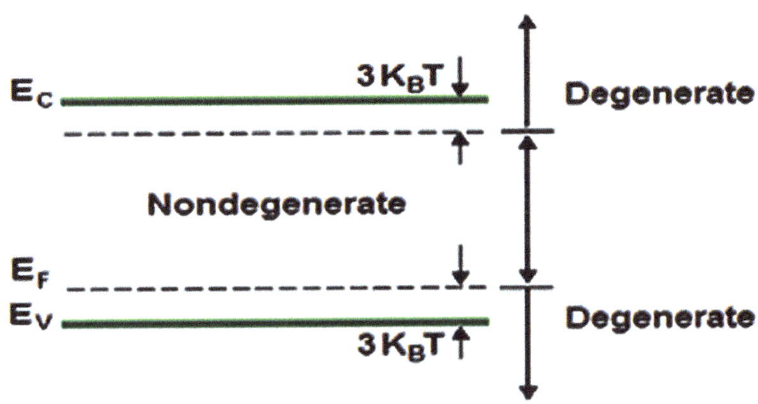

In case the Fermi level is situated less than $3K_BT$ far away from the conduction and valence bands, or located within a band; degenerate approximation will play an important role on the carrier statistics study. In other words, $\exp(x-\eta)$ in comparison with 1 can be neglected because the amount of $x-\eta$ is very small. As shown in Figure 11, ABA-stacked TGNFET quantum capacitance can be approximated by non-degenerate approximation particularly in lower numbers of η $(\eta \leq -1)$. Apparently, the proposedmodel in the degenerate limit is strongly unassociated with normalized Fermi energy. It is notable that quantum capacitance in the degenerate limit has a similar behavior of classical capacitance.

To get a greater insight into the quantum capacitance of ABA-stacked TGNFET, we embark on several important features of the C_q-V_{gs} characteristic. According to the relation between the normalized

Figure 11. Quantum capacitance of ABA-stacked TGN FET innon-degenerate and degenerate regimes

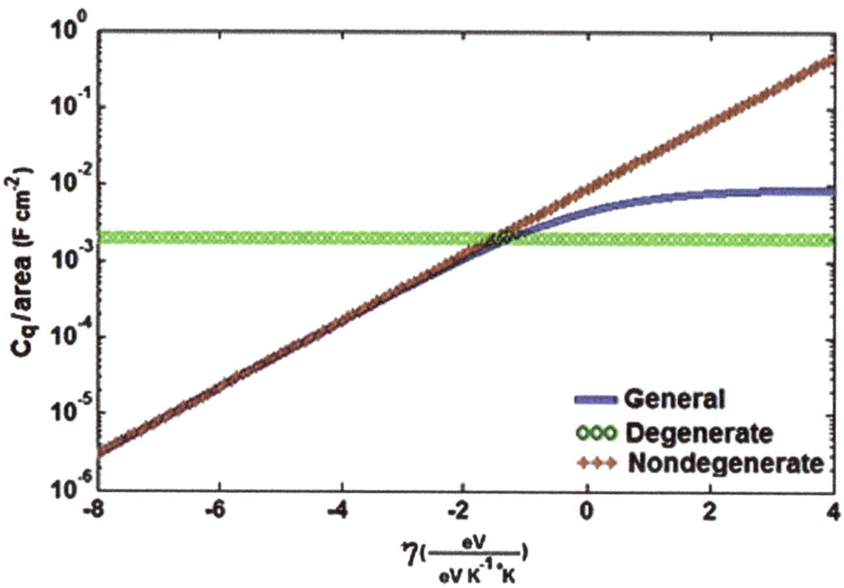

Fermi energy and applied voltage (Datta, 2005), the dependence of the quantum capacitance on the gate-source voltage is associated with $\eta \approx \frac{V_{gs} - V_T}{k_B T / e}$ (V_T is the thermal voltage). Therefore, the quantum capacitance-voltage characteristic of ABA-stacked TGNFET is modeld as

$$C_q = q^2 \int_{-\infty}^{+\infty} \frac{\exp(x - \frac{V_{gs} - V_T}{k_B T / e}) dx}{\left[A - \frac{B}{(k_B T)^{\frac{2}{3}} \left\{ D(x + E_{co}) + \sqrt{N + (x + E_{co})^2} \right\}^{\frac{2}{3}}} - C(k_B T)^{\frac{2}{3}} \left\{ D(x + E_{co}) + \sqrt{N + (x + E_{co})^2} \right\}^{\frac{2}{3}} \right] \left(1 + \exp\left(x - \frac{V_{gs} - V_T}{k_B T / e} \right) \right)^2} \quad (15)$$

Based on (15), the C_q-V_{gs} characteristic of ABA-stacked TGNFET is evaluated as shown in Figure 12. The quantum capacitance has a nonzero minimum value at the Dirac point and a linear increase which is symmetric on both sides of the minimum point with relatively sharp slopes (V-shaped plot).

Temperature dependence of capacitance is stated in parts per million (ppm) per °C. It can be obtained as a linear function but can be obviously non-linear at the extremes of the temperature. It is demonstrated that temperature is an effective factor in capacitance of graphene, leading to movement of carriers in a graphene FET channel (Tan, 2007). Accordingly, the carrier transport in graphene is a vital phenomenon that determined the current–voltage characteristic of the device. Application of the proposed model for the characterization of the quantum capacitance of ABA-stacked TGNFET performance reveals a rather strong dependence of the capacitance on temperature. Figure 13 shows C_q-V_{gs} charactristicof ABA-stacked

Figure 12. Quantum capacitance of ABA-stacked TGN FET versus gate-source voltage

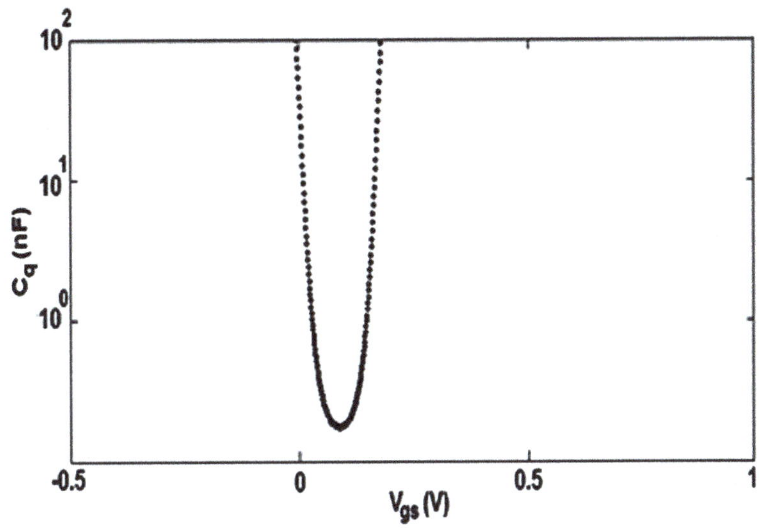

TGNFET indicating the capacitance rises significantly with increasing temperature.The proposed model indicates that ABA-stacked TGNFETcapacitance is temperature dependent near the neutrality point. It can be stated that by decreasing the temperature, the $C_{q\text{'}min}$ value decreases and it will be shifted leftwards which results in the minimum region becoming decreasing round and the slope increases.

In addition, fluctuations in the graph of C_q-V_{gs} could be justified by the manner of the DOS versus the electric field because quantum capacitance is proportional to DOS. But the simulated result (Figure 13)

Figure 13. The effect of temperature on the quantum capacitance of ABA-stacked TGNFET

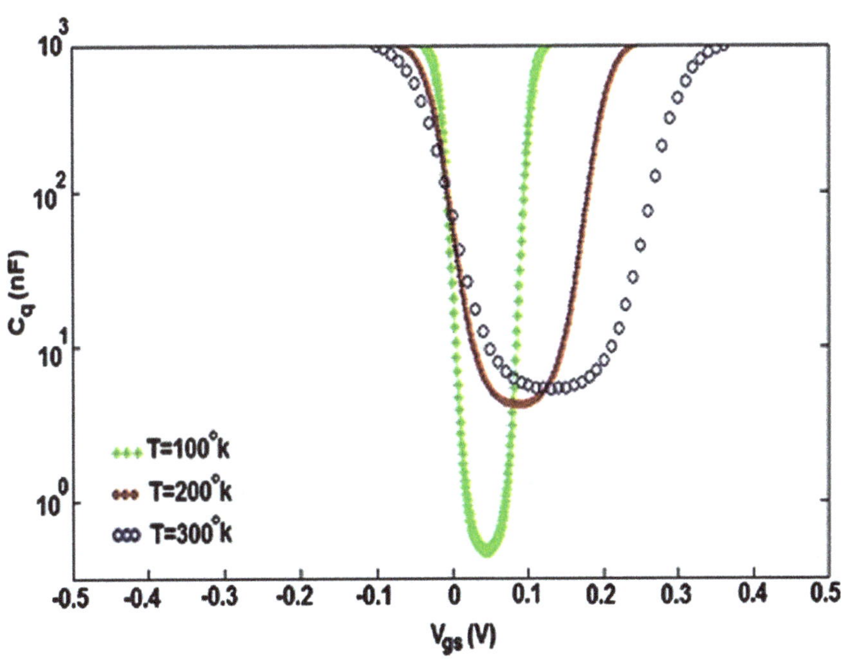

investigates the quantum capacitance dependence of the V_{gs} without fluctuation, this is because the gap doesn't exist in band structure of TGN as it shows a semi-metal behavior (Sedeghi, 2013). To validate the proposed model, comparison between quantum capacitance model of ABA-stacked TGNFET with experimental data is done (Chen, 2008) (Figure 14). Apparently, there is a similar behavior in the C_q-V_{gs} characteristic of single layer graphene and TGN. Comparing C_q on proposed model and published data, it is obvious that quantum capacitance in single layer graphene is higher than multilayer graphene and TGN specifically at the neutrality point which is more prominentin transport of carriers. Besides that, C_q only indicates a pronounced dependence of the V_{gs} for thesingle layer graphene and TGN FETs, while it is independent of the V_{gs} for multilayer of graphene.

As shown in Figure 15, in the structure of nanoscale FET the channel capacitance C can be expressed as aseries combination of a quantum capacitance C_q and an electrostatic capacitance C_E which depends on the dielectric constant as (Datta, 2005)

$$C = \frac{C_E C_q}{C_E + C_q} \tag{16}$$

The electrostatic capacitance per unit area is proportional to the effective dielectric constant εper the distance d between the center of the channeland the gate electrode as (Datta, 2005):

$$C_E = 2C_{in} \tag{17}$$

Figure 14. Comparison between quantum capacitance model with published data of single and multi -layers of graphene
Chen, 2008.

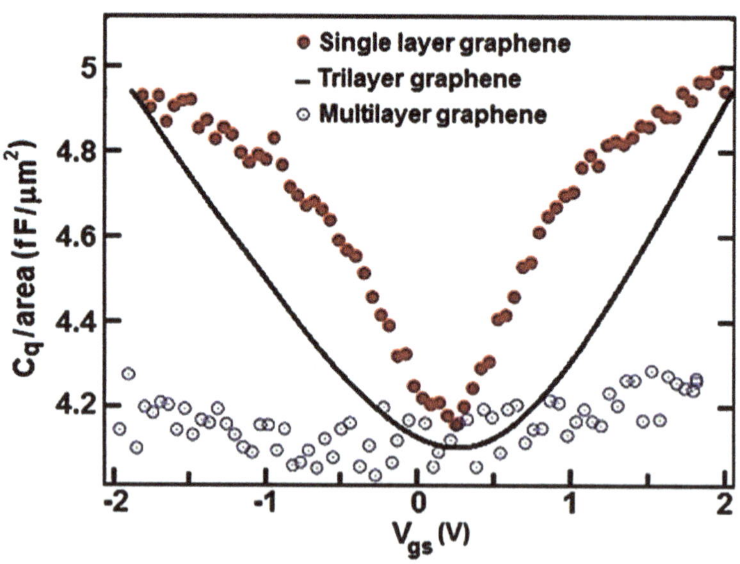

Figure 15. Approximate equivalent circuit representation of the TGNFET capacitor

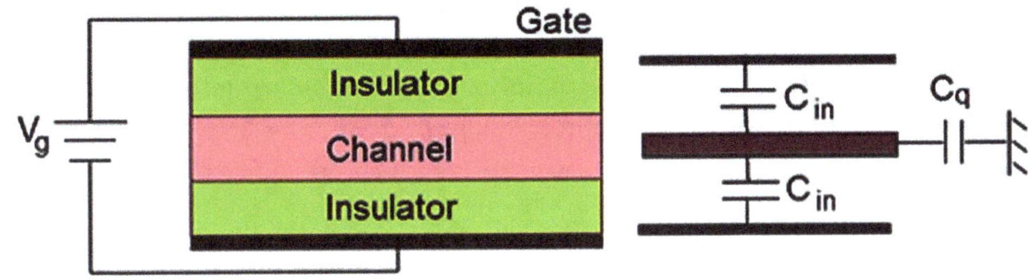

where $C_{in} = \frac{\varepsilon}{d}$. In order to obtain an analytical relation for the contact current, an explicit analytical equation for the electric potential distribution along the device is used as (11) (Lundstrom, 2006). In the linear region of the I_d-V_{ds} characteristic, the potential drop in the channel is linear, and the electric field is approximately constant (Neamen, 2011). By substitution of the quantum capacitance model and electrostatic capacitance in (11), the drain current of ABA-stacked TGNFET is analytically derived as a function of various physical and electrical parameters of the device including the carrier effective mass, effective length and width of the channel, velocity of carriers, applied bias and drain-source voltages. Figure 16(a) illustrates the I_d-V_{ds} characteristic of a double gate TGN FET with 285nmSiO_2 insulator-region and ε= 4, which is plotted based on (8) and (11). It is obvious thatthe device indicates a clear variation of channel current with gate-source voltage dependent. It can be seen that by increasing the V_{gs}, the drain current is increased. As indicated in Figure 16(b), the I_d-V_{gs}characteristic of ABA-stacked TGNFET is plotted which is similar to the C_q-V_{gs} characteristic of the device.

Figure 16. (a) I_d-V_{ds} characteristic of daboule gate TGN FET at different values of V_{gs} (b) I_d-V_{gs} characteristic of the device for V_{ds}= 0.1V (ε=4, d=285 nm)

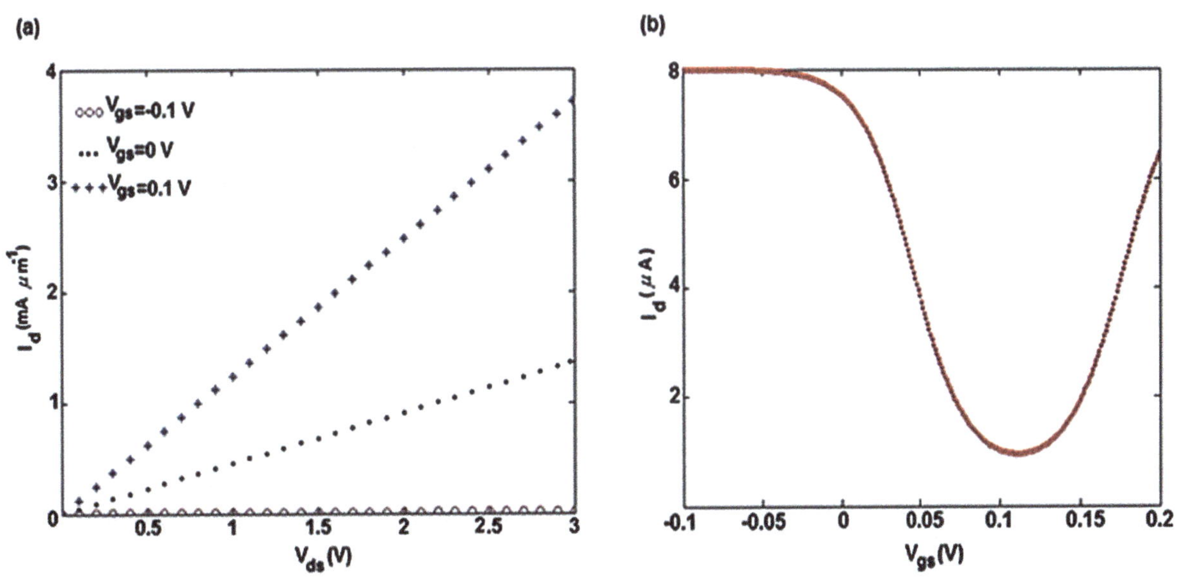

Figure 16 (b) illustrates the band movement of TGN as a function of gate-source voltage. In fact, the resulting VshapedI$_d$-V$_{gs}$ indicates how the gate-source voltage effectsthe transport through the graphene layer.One can clearly demonstrate the ambipolar graphene FET characteristics with a minimum value near to $V_{gs}=0$ that is associated with the Dirac point of graphene. Apparently, high drain currentsare attained for both positive and negative gate-source voltages

CONCLUSION

In this chapter, scaling behaviors of graphene FET application on sensor modeling in corporation with analytical device models are studied. To this end, carrier velocity and quantum capacitance of TGN-FET in the degenerate and non-degenerate regimes are analytically modeled. Furthermore, the effect of temperature and carrier concentration on velocity of carriers is investigated. On the base of simulated results, the plot of quantum capacitance versus gate-source voltage has a symmetric V-shape with a minimum at the Dirac point. In other words, the quantum capacitance increases linearly with gate-source voltage on both sides of the minimum point. The temperature dependence of quantum capacitance is also highlighted, in which minimum capacitance dramatically rises by increasing the temperature. Finally, the carrier velocity and quantum capacitance model are adopted to derive the current-voltage characteristics of TGNFET to evaluate the device performance. The simulated results and the method demonstrated in this work exhibit the potential applications in understanding of TGNFET devices. The comparative study between the proposed models and experimental date indicates that the models can assist in comprehending experiments involving graphene FETs.

REFERENCES

Ahmadi, M. T., Ismail, R., Michael Tan, L. P., & Aroral Vijay, K. (2008). The Ultimate Ballistic Drift Velocity in Carbon Nanotubes. *Journal of Nanomaterials, 769250.* doi:10.1155/2008/769250

Ahmadi, M. T., Rahmani, M., Ghadiry, M. H., & Ismail, R. (2012). Graphene Nanoribbon Homojunction Charactristics. *Sci. Adv. Mater.*, *4*(7), 753–756. doi:10.1166/sam.2012.1367

Amin, N. A., Ahmadi, M. T., Johari, Z., Webb, J. F., Mousavi, M., & Razali, R. (2010). Drift Velocity and Mobility of a Graphene Nanoribbon in a High Magnitude Electric Field.*Fourth Global Conference on Power Control and Optimization*, Sarawak, Malaysia.

Betti, A., Fiori, G., & Iannaccone, G. (2011). Drift velocity peak and negative differential mobility in high field transport in graphene nanoribbons explained by numerical simulations. *Applied Physics Letters*, *99*(24), 242108. doi:10.1063/1.3664091

Castro Neto, A. H., Guinea, F., Peres, N. M. R., Novoselov, K. S., & Geim, A. K. (2009). The electronic properties of graphene. *Reviews of Modern Physics*, *81*(1), 109–162. doi:10.1103/RevModPhys.81.109

Chen, Z., & Appenzeller, J. (2008). Mobility extraction and quantum capacitance impact in high performance graphene field-effect transistor devices. *Electron Devices Meeting, IEEE International, IEDM, 509*, 1-4.

Craciun, M. F., Russo, S., Yamamoto, M., Oostinga, J. B., Morpurgo, A. F., & Tarucha, S. (2009). Trilayer graphene is a semimetal with a gate-tunable band overlap. *Nature Nanotechnology*, *4*(6), 383–388. doi:10.1038/nnano.2009.89 PMID:19498401

Craciun, M. F., Russo, S., Yamamoto, M., & Tarucha, S. (2011). Tuneable electronic properties in graphene. *NanoToday Press*, *6*(1), 42–60. doi:10.1016/j.nantod.2010.12.001

Datta, S. (2005). *Quantum Transport: Atom to Transistor*. New York: Cambridge University Press. doi:10.1017/CBO9781139164313

Dong, L. X., & Chen, Q. (2010). Properties, synthesis, and characterization of graphene. *Frontiers of Materials Science in China*, *4*(1), 45–51. doi:10.1007/s11706-010-0014-3

Dong, X., Huang, W., & Chen, P. (2011). In Situ Synthesis of Reduced Graphene Oxide and Gold Nanoelectronics and Biosensing. Nanoscale Research Letters, 6. doi:10.1007/s11671-010-9806-8

Dorgan, V. E., Bae, M. H., & Pop, E. (2010). Mobility and saturation velocity in graphene on SiO_2. *Applied Physics Letters*, *97*(8), 082112–082113. doi:10.1063/1.3483130

Dragoman, M., & Dragoman, D. (2009). Graphene-based quantum electronics. *Quantum Electronics*, *33*(6), 165–214. doi:10.1016/j.pquantelec.2009.08.001

Fiori, G., Yoon, Y., Hong, S., Jannacconet, G., & Guo, J. (2007). Performance Comparison of Graphene Nanoribbon Schottky Barrier and MOSFETs. *IEEE International Electron Devices Meeting*.

Guinea, F., Castro Neto, A. H., & Peres, N. M. R. (2006). Electronic states and Landau levels in graphene stacks. *Physical Review B: Condensed Matter and Materials Physics*, *73*(24), 245426. doi:10.1103/PhysRevB.73.245426

Jena, D. (2009). A theory for the high-field current-carrying capacity of one-dimensional semiconductors. *Journal of Applied Physics*, *105*(12), 123701–123705. doi:10.1063/1.3147877

Lundstrom, M. S., & Guo, J. (2006). *Nanoscale Transistors: Device Physics, Modeling and Simulation.* Springer Science+Business Media Inc.

Neamen, D. A. (2011). *Semiconductor Physics and Devices*. University of New Mexico.

Ouyang, Y., Dai, H., & Guo, J. (2009). Multilayer Graphene Nanoribbon for 3D Stacking of the Transistor Channel. *IEEE International Electron Devices Meeting*.

Rahmani, M. (2012). *Advanced Nanoelectronics: Trilayer Graphene Nanoribbon Field Effect Transistor Modeling*. Boca Raton, FL: Taylor and Francis.

Rahmani, M., Ahmadi, M. T., Ghadiry, M. H., Anwar, S., & Ismail, R. (2012). Performance of Bilayer Graphene Nanoribbon Schottky Diode in Comparison with Conventional Diodes. *Journal of Computational and Theoretical Nanoscience*, *9*, 1618–1621. doi:10.1166/jctn.2012.2254

Rahmani, M., Ahmadi, M. T., Ismail, R., & Ghadiry, M. H. (2013). Performance of Bilayer Graphene Nanoribbon Schottky Diode in Comparison with Conventional Diodes. *Journal of Computational and Theoretical Nanoscience*, *10*(2), 323–327. doi:10.1166/jctn.2013.2699

Rahmani, M., Ahmadi, M. T., Karimi, F. A., Kiani, M. J., Akbari, E., & Ismail, R. (2013). Analytical Modeling of Monolayer Graphene-based NO_2 Sensor. *Sensor Letters*, *11*(2), 270–275. doi:10.1166/sl.2013.2742

Rahmani, M., Ahmadi, M. T., Karimi, F. A., Saeidmanesh, M., Akbari, E., & Ismail, R. (2013). *Analytical Modeling of Trilayer Graphene Nanoribbon Schottky-Barrier FET for High Speed*. Academic Press.

Rahmani, M., Ismail, R., Ahmadi, M. T., & Ghadiry, M. H. (2013). Quantum Confinement Effect on Trilayer Graphene Nanoribbon Carrier Concentration. *Journal of Experimental Nanoscience*., *794309*. doi:10.1080/17458080

Rutter, G. M., Crain, J., Guisinger, N., First, P. N., & Stroscio, J. A. (2008). Optoelectronic properties of ABC-stacked trilayer graphene. *Journal of Vacuum Science & Technology. A, Vacuum, Surfaces, and Films*, *26*, 938–943. doi:10.1116/1.2944257

Scott, B. W., & Leburton, J. (2010). *High-field carrier velocity and current saturation in graphene field-effect transistors*. 10th IEEE Conference on Nanotechnology (IEEE-NANO), Seoul, South Korea. doi:10.1109/NANO.2010.5697808

Sedeghi, H., Lai, D. T. H., Redoute, J. M., & Zayegh, A. (2013). Classic and Quantum Capacitances in Bernal Bilayer and Trilayer Graphene Field Effect Transistor. *Journal of Nanomaterials*.

Shengjun, Y., Raedt, H. D., & Katsnelson, M. I. (2010). *Electronic Transport in Disordered Bilayer*. Academic Press.

Shishir, R. S., Ferry, D. K., & Goodnick, S. M. (2009). Room-temperature velocity saturation in intrinsic graphene. *Journal of Physics: Conference Series*, *193*, 012118. doi:10.1088/1742-6596/193/1/012118

Tan, Y. W., Zhang, Y., Stormer, H. L., & Kim, P. (2007). Temperature dependent electron transport in graphene. *The European Physical Journal. Special Topics*, *148*(1), 15–18. doi:10.1140/epjst/e2007-00221-9

Winters, M., Hassan, J., Zirath, H., Janze, E., & Rorsman, N. (2013). A temperature dependent measurement of the carrier velocity vs. electric field characteristic for as-grown and H-intercalated epitaxial graphene on SiC. *Journal of Applied Physics*, *113*(19), 193708. doi:10.1063/1.4807162

Xie, Y. E., Chen, Y. P., & Zhong, J. X. (2009). Electron transport of folded graphene nanoribbons. *Journal of Applied Physics*, *106*(10), 103714. doi:10.1063/1.3261757

Chapter 3
Silicene Nanoribbons and Nanopores for Nanoelectronic Devices and Applications

Hatef Sadeghi
Lancaster University, UK

Sara Sangtarash
Lancaster University, UK

ABSTRACT

Given the compatibility of silicene with existing semiconductor techniques, and a need for new materials to continue Moore's low, it is natural to ask if this material can form a platform as field effect transistor. Here we provide analytical models to study the electrical properties of two dimensional silicene such as electrical conductance, carrier concentration, mobility and magneto-conductance. Furthermore, we show that silicene nanoribbons and nanopores can be used as a discriminating sensor for DNA sequencing and for efficient thermoelectric power generation.

INTRODUCTION

After isolation of Graphene, two-dimensional planar sheet of sp^2 bonded carbon atoms in honeycomb lattice, which shows interesting electrical properties with zero band gap in Dirac point, attempts to find one atomic thick structure of conventional materials such as silicon or germanium raised considerable scientific interest (Ni, 2011; Pitcher, 2011; Sadeghi, 2011; Sadeghi, 2014b; Sadeghi, 2013a; Sadeghi, 2012; Sadeghi, 2015a). One atomic thick crystalline form of silicon atoms with sp^2-sp^3 bond arranged in a honeycomb lattice structure called Silicene (Figure 1) is experimentally observed by Nakano, et al for the first time (Nakano, 2006). First principles calculations based on density functional theory (DFT) showed that despite carbon atoms that make two demensional planar sheet, Silicene makes stable bounds in semi two dementional (semi-2D) hexagon form (Cahangirov, 2009). The synthesis of silicene nanoribbons has been demonstrated on silver (111) (Aufray, 2010; Avila, 2013; Chen, 2013a; Chen, 2013b; Chen, 2012; Enriquez, 2012; Feng, 2012; Feng, 2013; Jamgotchian, 2012; Lalmi, 2010; Lin,

DOI: 10.4018/978-1-5225-0736-9.ch003

Figure 1. Silicene molecular structure; the bond length, lattice constant a_0 and buckling d are ~2.3Å, ~3.86Å and ~0.5Å, respectively.

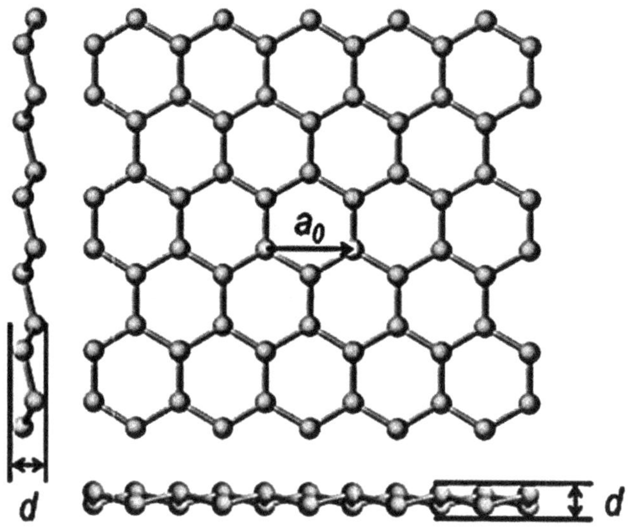

2012; Molle, 2013; Padova, 2013; Vogt, 2012), gold (110) (Tchalal, 2013), iridium (111) (Meng, 2013) and the zirconium diboride (0001) (Fleurence, 2012; Friedlein, 2013) substrates and are predicted to be stable on non-metallic substrates (Kokott, 2014). In contrast with graphene, the buckling of silicene (Majzik, 2013) can open up an energy gap at the Fermi energy E_F of between 300meV (Avila, 2013) and 800meV (Huang, 2013), which can be controlled by an external perpendicular electric field (Drummond, 2012). Silicene is shown to be grown on metallic substrate. Recently Tao et al succeeded to transfer it to the isolating substrate and use original metallic platform used on growing process as the silver electrode to realize room temperature silicene based field effect transistor FET (Tao, 2015).

Given the compatibility of silicene with existing semiconductor techniques, it is highly desirable to utilize it for nanoscale FETs, sensors and thermoelectricity. In this chapter, we try to reveal great potential of this material to continue Moor's law. In this chapter, first we start with the band structure and electronic properties of the 2D silicene and provide analytical model of its carrier behaviour, conductance and magneto conductance (Sadeghi, 2014a). Then we will show how silicene nanoribbons and nanopores could be used as thermoelectric material and DNA sensor with discriminating capability (Sadeghi, 2014c; Sadeghi, 2015b).

Zero gap nature of single layer Graphene does not allow proper cut-off in Graphene FETs. This issue is not pronounced for Silicene since it behaves as a semimetal with variable band gap tuned by external electric field (Guzmán-Verri, 2007; Houssa, 2010; Suzuki, 2010). However, the electronic properties of Silicene nanoribbons and Silicene sheets were found to resemble those of Graphene (Kara, 2012). Similarity of Silicene and Graphene is due to the same group (IV) that both belong. Larger ionic radius in Silicene promotes sp^3 hybridization (Zhao, 2012) implies a small buckling to make Silicene structure more stable whereas sp^2 hybridization in Graphene make it stable in planar form (Drummond, 2012; O'Hare, 2012). This buckling has been reported in the range of 0.36Å to 0.75Å (Drummond, 2012; Huang, 2013; Kamal, 2013; Kim, 2012b; Sadeghi, 2014c; Sadeghi, 2015b) by DFT calculation and 0.75Å in Silicene grown on Ag (Vogt, 2012) experimentally. The gap depends on external electric field

(Drummond, 2012) as well as the amount of the buckling (Huang, 2013) which could be modified by epitaxial strain (Fleurence, 2012). Silicene is stable in a wide doping range (Cheng, 2011; Ding, 2012a; Gao, 2012; Sadeghi, 2015b). Silicene nanoribbons edges do not reveal oxygen reactivity (Padova, 2011; Padova, 2008). Unlike oxygen, hydrogen could strongly affect the electronic structure of the silicene nanoribbons. The behavior of the Silicene could be changed from metallic to magnetic semiconducting and then to nonmagnetic semiconducting (Ding, 2012b; Houssa, 2011; Lew Yan Voon, 2010; Osborn, 2011; Pan, 2012b; Wang, 2012; Zhang, 2011) by hydrogenation. Half-hydrogenation of Silicene breaks the extended π bonding network of Silicene, exhibiting ferromagnetic semiconducting behavior with a band gap of 0.95 eV (Zhang, 2012). Calculations predict direct and indirect gap for zigzag and armchair edge shape Silicene nanoribbon, respectively (Kim, 2012a). Unlike graphite, which consists of weakly held stacks of Graphene layers through dispersion forces, the interlayer coupling in Silicenes found to be very strong (Jose, 2011).

SILICENE ELECTRONIC STRUCTURE

The spectrum of full tight-binding low energy effective Hamiltonian of Silicene in the presence of perpendicular external electric field could be written as Equation (1) (Drummond, 2012; Liu, 2011). The differences between electric potential of top v_{g1} and bottom gates v_{g2} induce electric field which breaks the symmetry between A and B sub lattices and hence opening the gap $\Delta = q\left(v_{g1} - v_{g2}\right)/2$.

$$E = \pm\sqrt{\Delta^2 + \left(\hbar v_F k\right)^2} \tag{1}$$

where $k = \sqrt{k_x^2 + k_y^2}$ is the k vector, $\hbar = 1.054 \times 10^{-34} js$ is the reduced Planck's constant and $v_F = 1.3 \times 10^6 ms^{-1}$ (Chen, 2012; Vogt, 2012) is Fermi velocity in Silicene. Figure 2a shows the band structure of the Silicene in different gate voltages ($v_g = v_{g1} - v_{g2}$). As it is clear band gap increase with gate voltage. It is interesting to note that opening a gap in the presence of electric field is main differences between Graphene and Silicene which is the consequence of the non-planar structure of the Silicene despite Graphene. However, in the absence of gate voltage Silicene behaves like Graphene and its band structure turns to Graphene band structure ($E = \hbar v_F \left|k\right|$). By the aid of DFT calculation using SIESTA method which is explained in the methods section, the band structure of the Silicene in the absence of external electric field could be obtained as Figure 2b which shows zero gap in Dirac point. Since in the absence of external electric field ($\Delta = 0$) the band structure becomes Graphene like gapless linear band structure, in entire of the paper we will discuss the transport of the Silicene in the presence of the external electric field ($\Delta \neq 0$). It is apparent that the Fermi velocity in Graphene and Silicene are different and even in the absence of external electric field they would behave slightly different. However, all analytical formulas for Graphene in the body of literature are applicable for Silicene in the absence of external electric field by taking Silicene Fermi velocity into account.

Figure 2. Silicene band structure (a) low energy effective tight binding Hamiltonian in the absence and presence of perpendicular electrical field, (b) DFT in the absence perpendicular electrical field

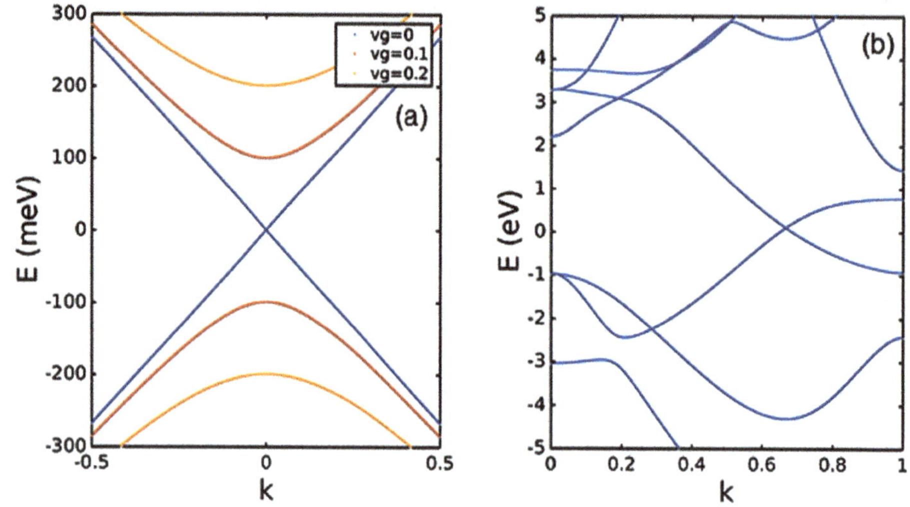

SILICENE FIELD EFFECT TRANSISTOR

1. **Electron Density in Silicene:** Carrier concentration in a band is achieved by integrating the Fermi - Dirac distribution function over energy band (Datta, 2005) as $n = \int D(E) f_F(E) dE$ where *D(E)* and $f_F(E) = \left(1 + e^{(E-E_F)/k_BT}\right)^{-1}$ are available energy states (Density of States) and Fermi-Dirac distribution function, respectively. Density of states could be calculated by summation over *k* space with respect to *E-k* relation as $D(E) = \sum_k \delta(E - E(k))$. By changing the summation into an integral, the density of states of Silicene could be written as:

$$D(E) = \frac{WL}{4\pi^2} \int_{-\pi}^{+\pi} d\theta \int_0^{\infty} kdk\delta(E - E(k)) \tag{2}$$

where the momentum (*k*) can be obtained from *E-k* relation of Silicene (Eq. 1) as $k = \sqrt{\left(E^2 - \Delta^2\right) / \hbar^2 v_F^2}$ which induces $kdk = \left|E\right| dE / \hbar^2 v_F^2$. Substituting *kdk* into Equation (3) and including the spin and the valley degeneracy indicates Silicene Density of States as:

$$D(E) = \frac{2WL\left|E\right|}{\pi \hbar^2 v_F^2} \tag{3}$$

Density of states in cooperation with the Fermi Dirac distribution function indicates the total electron density

$$n = \int_0^{\infty} dE D(E) \Theta\left(\left|E\right| - \Delta\right) f(E)$$

of the Silicene as:

$$n = \int_0^{+\infty} dE \left(\frac{2WL\left|E\right|}{\pi \hbar^2 v_F^2} \right) \Theta\left(\left|E\right| - \Delta\right) \frac{1}{1 + e^{\frac{E - E_F}{k_B T}}} \tag{4}$$

where $\Theta\left(\left|E\right| - \Delta\right) = \begin{cases} 0, & 0 < \left|E\right| < \Delta \\ 1, & otherwise \end{cases}$. By changing the variables as $x = (E - \Delta) / k_B T$ and $\eta = (E_F - \Delta) / k_B T$ Equation (4) could be written as:

$$n = \frac{2WL}{\pi \hbar^2 v_F^2} \left(k_B T \int_0^{+\infty} dx \frac{x k_B T + \Delta}{1 + e^{x - \eta}} \right) \tag{5}$$

The first and second parts of the carrier density model looks like complete Fermi–Dirac integral

$$\Im_j(x) = \Gamma(j+1)^{-1} \int_0^{\infty} \frac{t^j dt}{e^{t-x} + 1}$$

with zero and second order *j*, respectively where Γ is gamma function. Substituting the Fermi Dirac integral into the Equation (5), the total electron density model of the Silicene could be obtained as the Equation (6).

$$n = \frac{2WL}{\pi \hbar^2 v_F^2} \left(\Delta \Gamma(1) k_B T \Im_0(\eta) + \Gamma(2) \left(k_B T\right)^2 \Im_1(\eta) \right) \tag{6}$$

Figure 3 shows the electron density of Silicene in the different gate voltages and temperature for a sample with the width and length of 1µm. The electron density increase with the gate voltage as expected in the conventional semiconductors and the rate of the electron density increment is higher for higher gate voltages. However, increasing the temperature dramatically increases the electron density in the neutrality point despite higher gate voltage where it remains stable.

2. **Conductance of Silicene:** For the Silicene where the electron transport is assumed to be ballistic, the conductance could be calculated using the Landauer's formula in the Ballistic regime (Datta, 2005)

$$G = 2q^2/h \int dE M(E) \left(-df_f(E)\right)$$

Figure 3. Total carrier concentration of the Silicene vs gate voltage in different temperatures based on Eq. (6)

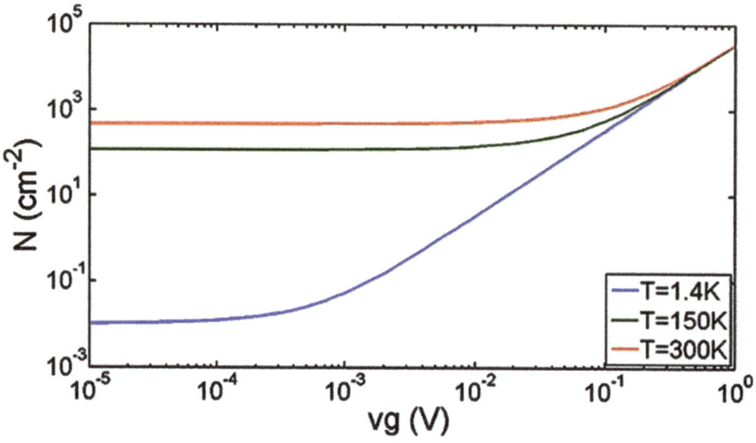

where q is the electron charge, h is the Planck's constant. The number of modes $M(E)$ that are above the cut-off at energy E in the transmission channel could be obtained as:

$$M(E) = \sum_k \delta(E - E(k)) \frac{\pi\hbar}{L} \left| v_x(k) \right| \tag{7}$$

where $v_x(k) = dE / \hbar dk_x$ is the carriers velocity in the x direction which could be calculated for the Silicene as:

$$v_x(k) = \frac{2\hbar v_F^2 k_x}{\sqrt{\Delta^2 + \left(\hbar v_F k\right)^2}} \tag{8}$$

Substituting the carrier's velocity into the Equation (7) and changing the summation to integral by employing $k_x = k\cos(\theta)$ where θ is the angle between x and y directions in 2D channel and including the spin degeneracy, the total number of the modes in the Silicene channel could be written as:

$$M(E) = \frac{WL\pi\hbar}{4\pi^2} \int_{-\pi}^{+\pi} \left|\cos\left(\theta\right)\right| d\theta \int_0^{\infty} kdk\delta(E - E(k)) \left| \frac{2\hbar v_F^2 k_x}{\sqrt{\Delta^2 + \left(\hbar v_F k\right)^2}} \right| \tag{9}$$

Equation (9) could be readily written as the Equation (10) which indicates the analytical description of the number of modes in the Silicene 2D channel.

$$M(E) = \frac{W\left|\sqrt{E^2 - \Delta^2}\right|}{4\hbar v_F} \tag{10}$$

By employing Taylor series, in the presence of external electric field ($\Delta \neq 0$), $\left|\sqrt{E^2 - \Delta^2}\right|$ could be extended as: $\left|\frac{E^2}{2\Delta} - \Delta\right|$. This approximation would be very accurate specifically when $E^2 << 2\Delta$. We know that Fermi-Dirac distribution function always limits the Landauer's formula and therefore transport in the range of few k_BT around the Fermi energy which is *25.8meV* in the room temperature. Meaning that *E* is confined to few k_BT. If we take this explanation into account for $\Delta > 500\,\mu eV$, the differences between $\left|\sqrt{E^2 - \Delta^2}\right|$ and $\left|\frac{E^2}{2\Delta} - \Delta\right|$ would be less than 1 percent in the room temperature (which could be even lower in lower temperatures). Therefore this substitution is perfectly applicable when $\Delta > 500\,\mu eV$ which is a practical value. Hence, using this extension, the number of the modes in combination with the Landauer's formula indicates the conductance of the Silicene as:

$$G = -\frac{2q^2}{h}\frac{W}{4\hbar v_F}\left(\int_0^{+\infty} dE\left|\frac{E^2}{2\Delta} - \Delta\right|\frac{\partial}{\partial E}\left(\frac{1}{1+e^{\frac{E+E_F}{k_BT}}}\right) - \int_0^{+\infty} dE\left|\frac{E^2}{2\Delta} - \Delta\right|\frac{\partial}{\partial E}\left(\frac{1}{1+e^{\frac{E-E_F}{k_BT}}}\right)\right) \tag{11}$$

Employing partial integration methods, changing variables and some simplifications Equation (11) could be written as Equation (12) which indicates analytical conductance model of the Silisece as a 2D channel in Silicene based FET devices.

$$G = \frac{Wq^2}{4\pi\Delta\hbar^2 v_F}\Gamma(2)(k_BT)^2\left(\Im_1(-\eta) - \Im_1(\eta)\right) \tag{12}$$

Figure 4 shows the conductance variation of Silicene versus gate voltage in different temperature between about the absolute zero and room temperature based on Equation (12). The conductance of Silicene is expected to increase with gate voltage. The increment rate is higher around neutrality point and it remains more stable out of the neutrality point (degenerate regime). Minimum conductance increases with temperature, whereas temperature does not affect conductance in degenerate regime.

Employing similar method, for the single layer Graphene with zero gap band structure and linear energy dispersion ($E = \hbar v_{FG}\left|k\right|$), number of mode could be obtained as $M(E) = 2W\left|E\right| / \pi\hbar v_{FG}$ which with respect to Landauer's formula indicates the conductance model of Graphene as:

$$G_G = \frac{2Wq^2}{\pi^2\hbar^2 v_{FG}}\Gamma(1)(k_BT)\left(\Im_1(-\eta) + \Im_1(\eta)\right) \tag{13}$$

Figure 4. The conductance of the Silicene vs gate voltage in different temperatures

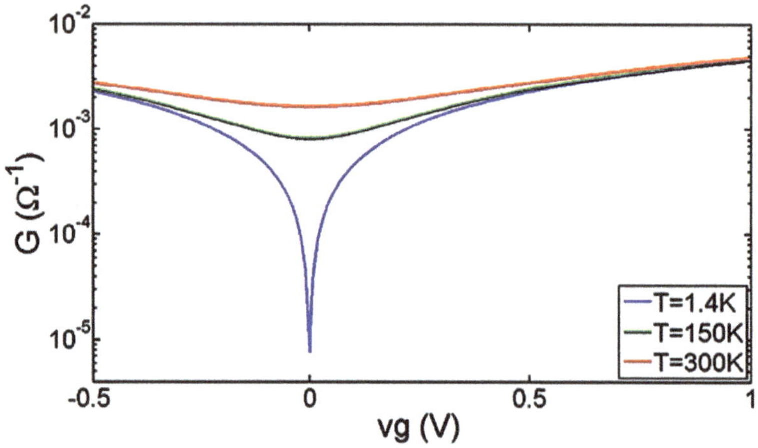

where $v_{FG} = 1 \times 10^6 ms^{-1}$ (Avouris, 2007) for Graphene. Figure 5 shows the conductance of the Graphene and Silicene (with the same size) based on analytical models (Eq 12 and 13) presented in this paper in T=150K and room temperature. It is clear that that the minimum conductance remains almost similar for both Graphene and Silicene which may be explained by the similar band structure in zero gate voltage where the band structure of the Silicene and Graphene are linear. However, out of neutrality point, the conductance of Graphene is about ten times higher than Silicene. Silicene temperature dependence is different than Graphene out of neutrality point as shown in Figure 5. Although conductance of Silicene remains stable with increasing the temperature in degenerate regime, it is reduced for Graphene. However, minimum conductance shows similar trends in both cases as it increases with temperature.

3. **Magneto-Conductance of Silicene:** Magneto-conductance could be calculated by recording the changing of electrical conductance of Silicene in *x* direction when an external magnetic field (*B*)

Figure 5. Comparison of the conductance of the Silicene and single layer Graphene vs gate voltage in T=150K and room temperature with similar size

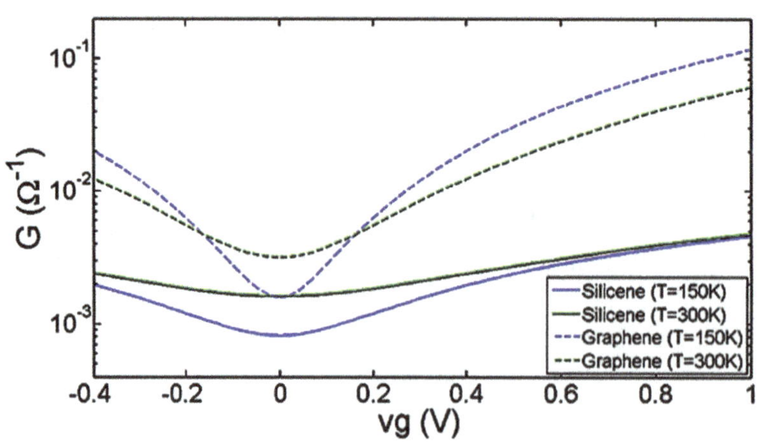

is applied in z direction. Including one types of carriers (electron), Lorentz forces $F_L = q\left(v \times B\right)$ added by electrostatic force $F_{el} = qE$ motions the carriers which could be obtained by $F_L + F_{el} = qv / \mu$ where μ is the mobility. Electronic quality of materials is usually characterized by mobility (μ) of their charge carriers (Morozov, 2008). Charge carrier mobility is the speed (*cm/s*) of the charge carriers in the presence of an applied electric field (*V/cm*) which move in the material in a given direction. Carrier mobility in a material is a function of conductivity and total carrier density as $\mu = \sigma / qn$ where σ is conductivity and n is carrier density. The mobility is the term commonly used along with some form of scattering. Here we deal with the ballistic regime though. Since mobility is depend on the carrier density and conductivity and both of them show gate voltage dependent in Silicene, mobility of the carrier may depend on gate voltage as well. Therefore, it is worth to calculate the upper bound of the carrier mobility in Silicene which would be helpful in calculating the magneto-conductance of the Silicene. In short, mobility in ballistic regime could be translate to upper bound (UB) mobility and taking this explanation into account, employing Equation (6) and Equation (12) UB carrier mobility of Silicene could be obtained as:

$$\mu_{UB} = \frac{qv_F}{8W\Delta} \frac{(k_BT)\left(\Im_1(-\eta) - \Im_1(\eta)\right)}{\left(\Delta\Im_0(\eta) + \left(k_BT\right)\Im_1(\eta)\right)} \tag{14}$$

It is interesting that the UB mobility of the carriers is decreased with gate voltage as shown in Figure 6 plotted based on Equation (14). This dependence is the main reason that we are keen to define the UB carrier motilities since it definitely would provide more accurate model in studying the magneto-conductance rather than assuming it as free parameter. Temperature would affect the UB carrier motilities in very low applied perpendicular external electric filed as shown in Figure 7. However, it is not affected by temperature out of neutrality point $vg > 0.1V$.

Figure 6. The UB carrier mobility of the Silicene vs gate voltage in different temperatures

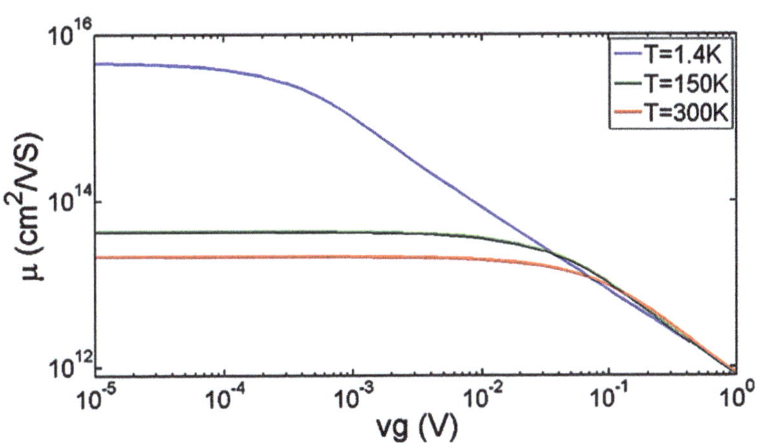

Figure 7. The conductance (a), carrier density (b) and UB carrier mobility (c) of the Silicene vs temperatures in two different gate voltages (0.01, 0.1, 0.2 and 1 V)

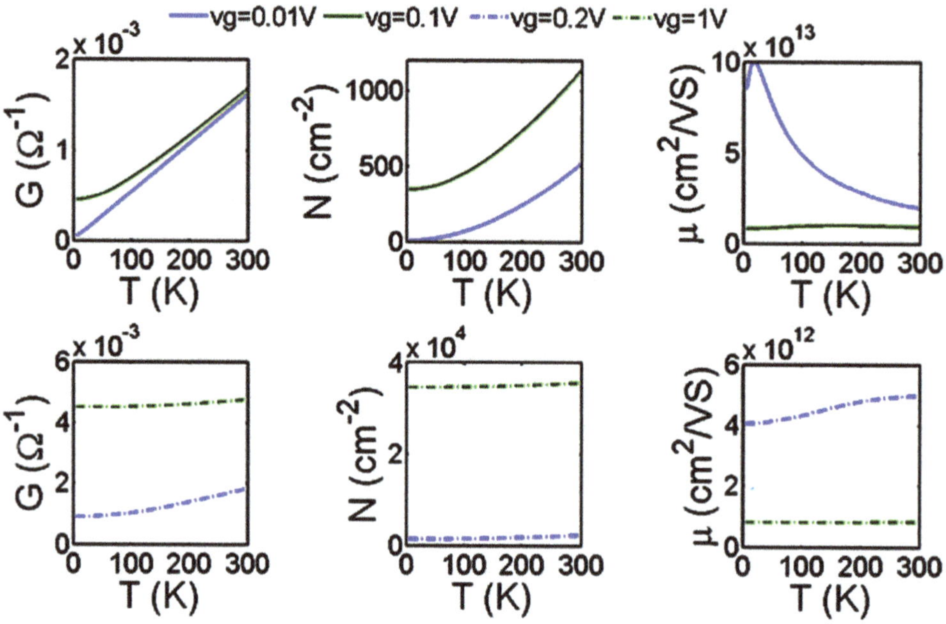

Although conductivity of Silicene is lower than Graphene, calculation shows very high UB carrier mobility (in the order of $10^{13} cm^2V^{-1}S^{-1}$) for Silicene in room temperature which could take place in low carrier concentration. Combination of low carrier concentration and relatively high conductivity in Silicene results high UB carrier mobility. For example, UB carrier mobility as high as ~$0.5\text{x}10^{13}$ ($cm^2V^{-1}S^{-1}$) could be obtained in Silicene with n~10^4 (cm^{-2}) and σ~0.5 (s^{-1}) at $vg = 0.5V$ in room temperature based on Equations (6), (12) and (14). This UB carrier mobility is very promising for future electronic devices.

Temperature dependence of conductance, carrier density and UB mobility of Silicene in different gate voltages $vg = 0.01, 0.1, 0.2, 1V$ has been demonstrated in Figure 7. Total carrier density and conductance are increased by temperature for low gate voltage regime $vg = 0.01, 0.1$ and $0.2V$; despite it is remained stable in higher gate voltages $vg = 1V$.

Although UB carriers mobility are always decreased by gate voltage in any temperature, the temperature that Silicene experience highest UB carrier mobility is varying by gate voltage. For example, highest UB carrier mobility occurs about absolute temperature for $vg = 0.1mV$, whereas it happens in T ≈ 25K for $vg = 10mV$ as shown in Figure 8. Highest UB carrier mobility occurs in higher temperature when gate voltage increased in Silicene. However, increasing of the gate voltage makes the system almost affectless by temperature (in the range of absolute and room temperature) in all cases.

In order to calculate magneto-conductance of Silicene using the definition of magneto-conductance $F_L + F_{el} = qv/\mu$, Equation 15 could be simply written in *x* and *y* directions.

Figure 8. Normalized UB carrier mobility vs temperature in different gate voltages

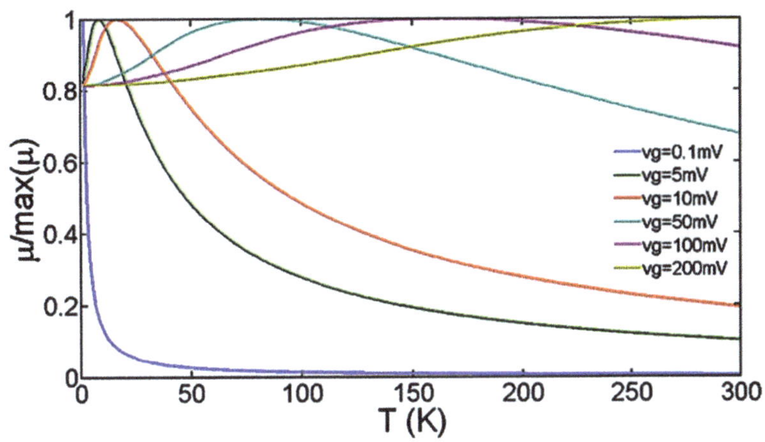

$$\frac{qv_x}{\mu} = q\left(E_x - v_y B\right)$$
$$\frac{qv_y}{\mu} = q\left(E_y + v_x B\right) \tag{15}$$

The matrix form of Equation 15 after algebraic inversion could be written as:

$$\begin{pmatrix} v_x \\ v_y \end{pmatrix} = \frac{-\mu}{1+(\mu B)^2}\begin{pmatrix} 1 & -\mu B \\ \mu B & 1 \end{pmatrix}\begin{pmatrix} E_x \\ E_y \end{pmatrix} \tag{16}$$

The total current is $J = qnv_x$ as there is no current in *y* direction. Therefore electron conductivity $\sigma = J/E$ in the presence of magnetic field (*B*) could be written as:

$$\sigma = \frac{qn\mu}{1+(\mu B)^2}\begin{pmatrix} 1 & \mu B \\ -\mu B & 1 \end{pmatrix} \tag{17}$$

Therefore, longitudinal magneto-conductivity σ_{xx} and transversal magneto-conductivity σ_{xy} could be obtained as:

$$\sigma_{xx} = \frac{qn\mu}{1+(\mu B)^2}, \sigma_{xy} = \frac{qn\mu^2 B}{1+(\mu B)^2} \tag{18}$$

Substituting carrier concentrations and UP mobility of Silicene from Equations (6) and (14), longitudinal magneto-conductance model of Silicene could be obtained as:

$$G_{xx} = \frac{Wq^2}{4\Delta\pi\hbar^2 v_F} \frac{(k_B T)^2 \left(\Im_1(-\eta) - \Im_1(\eta)\right)}{1 + \left(\frac{qv_F}{8W\Delta} \frac{(k_B T)\left(\Im_1(-\eta) - \Im_1(\eta)\right)}{\left(\Delta\Im_0(\eta) + (k_B T)\Im_1(\eta)\right)}\right)^2 B^2} \tag{19}$$

Figure 9 shows the longitudinal magneto-conductance of Silicene versus magnetic field in different gate voltages and room temperature. Longitudinal magneto-conductance is decreased with magnetic field. Like Graphene, increment in gate voltage increases magneto-conductance of Silicene. In addition minimum magneto-conductance of Silicene is less affected by gate voltage rather than higher magnetic fields where conductance more affected by gate voltage.

Total carrier density and upper bound carrier mobility could be obtained for the single layer Graphene with similar methods as

$$n_G = \left(\pi\hbar^2 v_{FG}^2\right)^{-1} \left(k_B T\right)^2 / \left(\Im_1(\eta) - \Im_1(-\eta)\right)$$

and

$$\mu_{UPBG} = qv_{FG}\left(\Im_1(\eta) + \Im_1(-\eta)\right) / \left(\pi W k_B T \left(\Im_1(\eta) - \Im_1(-\eta)\right)\right).$$

So longitudinal magneto-conductance of the single layer Graphene could be obtained as:

$$G_{xxG} = \frac{2Wq^2}{\pi^2\hbar^2 v_F} \frac{(k_B T)\left(\Im_1(\eta) + \Im_1(-\eta)\right)}{1 + \left(\frac{qv_F}{(k_B T)W\pi} \frac{\left(\Im_1(\eta) + \Im_1(-\eta)\right)}{\left(\Im_1(\eta) - \Im_1(-\eta)\right)}\right)^2 B^2} \tag{20}$$

Figure 9. Longitudinal magneto-conductance vs magnetic field in different gate voltages and room temperature

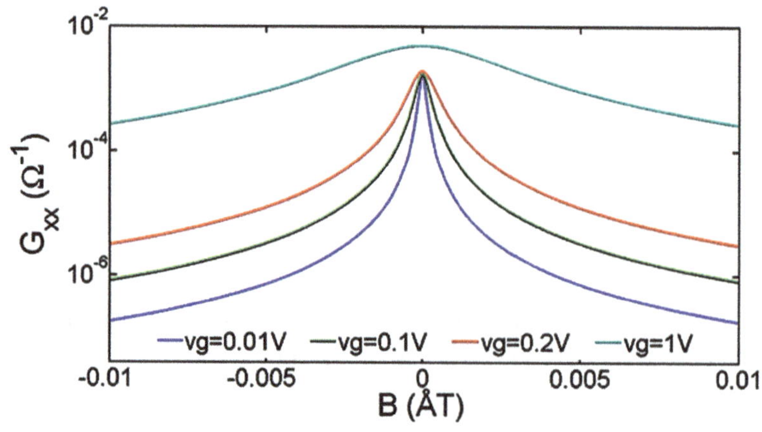

Dashed lines in Figure 10 are illustrated by employing Equation 20. Similar to Silicene, longitudinal magneto-conductance of Graphene is increased with gate voltage and decreased with magnetic field (dashed lines in Figure 10).

In addition, employing Equation 19, longitudinal magneto-conductance of Silicene (solid lines) is compared with Graphene (dashed line). In low gate voltage regime, Silicene and Graphene behave very similar which could be explained by very small gap of Silicene induced in lower gate voltages. However, it is apparent that Graphene shows higher longitudinal magneto-conductance than Silicene in high gate voltage regime and in room temperature. Presented transport models of Silicene can provide a better understanding toward the Silicene channel behavior in field-effect transistor configuration which can be employed to model Silicene based transistors for integrated circuits.

SILICENE NANOPORES FOR BIOLOGICAL SENSING AND THERMOELECTRICITY

1. **DNA Nucleobases Sensing and Sequencing:** In this section, our aim is to demonstrate the potential of the recently-synthesized two-dimensional material silicene as a platform for DNA sequencing and demonstrate that the unique electrical properties of nanoporous silicene allow direct electrical transduction and selective sensing of nucleobases. DNA sequencing (sensing the order of bases in a DNA strand) is an essential step toward personalized medicine for improving human health (Shendure, 2008). Despite recent developments, conventional DNA sequencing methods are still expensive and time consuming (Shendure, 2012). Therefore the challenge of developing accurate, fast and inexpensive, fourth-generation DNA sequencing alternatives has attracted huge scientific interest (McGinn, 2012). All molecular based biosensors rely on a molecular recognition layer and a signal transducer, which converts specific recognition events into optical, mechanical, electro-chemical or electrical signals (Drummond, 2003). Of these, electrical transduction is potentially the fastest and least expensive, because it is compatible with nanoelectronics integration technologies.

Figure 10. Longitudinal magneto-conductance vs magnetic field in different gate voltages and room temperature for Silicene (solid line) and Graphene (dashed line)

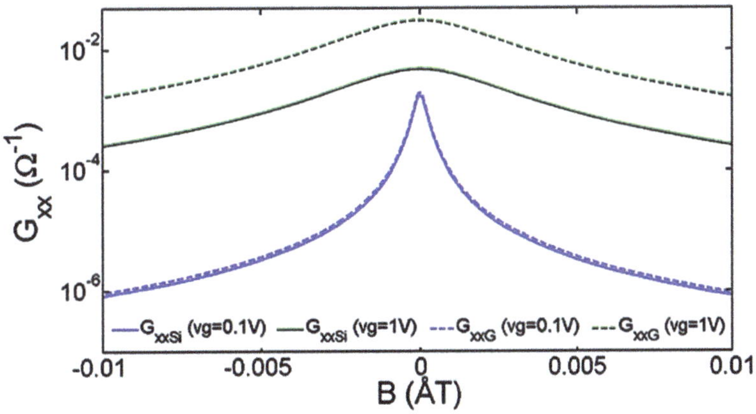

However, attempts to realize such sensors based on silicon platforms, silicon nanowires or graphene (Fritz, 2002; Mohanty, 2008; Xie, 2012) have not yet delivered the required level of selectivity.

Current technology allows the drilling of nanopores with different diameters down to a few angstroms in graphene, Al_2O_3 and TiO_2 based membranes (Kuan, 2012; Liu, 2012; Russo, 2012; Shim, 2012; Spinney, 2012; Taniguchi, 2009; Venkatesan, 2012). Three types of nanopores have been presented in the literature for DNA sensing (Wanunu, 2012). Currently-available solid-state nanopore-based strategies rely on reading the variation of an ionic current through a surrounding fluid due to the translocation of DNA strand through a pore in a solid state membrane (Dekker, 2007; Garaj, 2013; Shim, 2013; Wells, 2012). However, ionic current leakage in the thin membranes, poor signal to noise and difficulties in controlling the speed of translocation through the pore have so far limited the development of this technique (Fyta, 2011; Luan, 2012; Schneider, 2012). In a second approach, biological nanopores (MspA and α-hemolysin) have been employed as recognition sites inside the pore (Manrao, 2012). This method overcomes key technical problems required for real-time, high-resolution nucleotide monophosphate detection (Clarke, 2009), but several outstanding issues need to be addressed, including the sensitivity of biological nanopores to experimental conditions, the difficulty in integrating biological systems into large-scale arrays, the very small (~pA) ionic currents and the mechanical instability of the lipid bilayer that supports the nanopores (Anselmetti, 2012; Cherf, 2012; Venkatesan, 2011; Xie, 2012). As a third approach to nanopore-based sequencing, changes in the electrical conductance of single-layer graphene have been used for DNA sensing (Avdoshenko, 2013; Nelson, 2010; Saha, 2012). In general, direct electrical conductance measurement is more attractive than blockade ionic current measurement, since the response of the former is much faster and the signal to noise is higher. However, monolayer graphene does not show sufficient selectivity (Avdoshenko, 2013; Saha, 2012).

As shown by Sadeghi, et al. (Sadeghi, 2014c; Sadeghi, 2013b), the electrical conductance of silicene nanoribbons containing a nanopore is selectively sensitive to the translocation of DNA bases through the pore. An example of a silicene nanoribbon containing a nanopore is shown in Figure 11. The electrical conductance G of this nanopore-containing silicene ribbon is computed using a first-principles quantum transport method as described in (Sadeghi, 2014c). This involves computing the transmission coefficient T(E) for electrons of energy E passing from a source on the left to a drain on the right through the structure shown in Figure 11.

Figure 11. Schematic of Silicene nanoribbon containing a nanopore for DNA sequencing

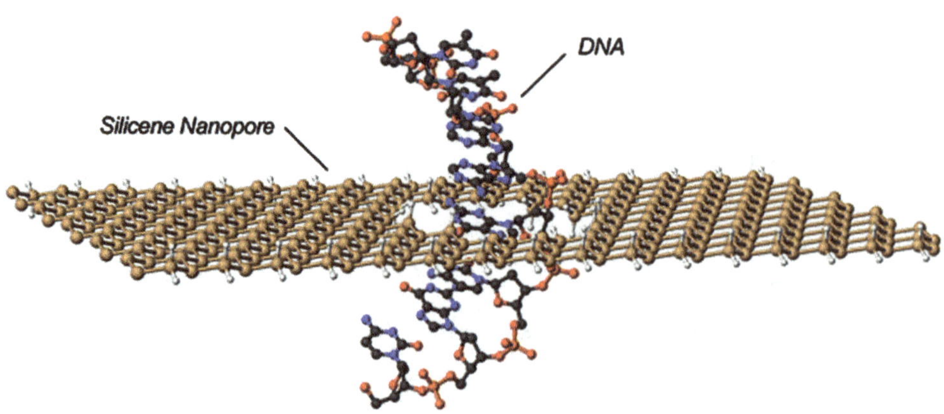

For a perfect nanoribbon (ie in the absence of a nanopore), Figure 12a shows the variation of *T(E)* with energy. In this case, the de Broglie waves of electrons travelling from left to right are not scattered and *T(E)* is an integer equal to the number of open scattering channels available to right-moving electrons. The presence of a sharp feature near the (un-gated) Fermi energy (which we define to be $E_F = 0$) is a consequence of the unique band structure of silicene nanoribbons and is associated with edge states. In the presence of a nanopore, the resulting *T(E)* is shown in Figure 12b. In this case electrons are scattered by the nanopore and *T(E)* is reduced compared to that of the perfect ribbon. Nevertheless the feature near $E = 0$ survives.

In what follows, *T(E)* is computed in the presence of each of the four bases *X=[A, C, G, T]*. Of course the result depends on the orientation of the base within the pore and therefore for each base *X*, also a number (m_{max}) of distinct orientations labeled $m = 1, \ldots m_{max}$ is consider. The resulting transmission coefficients are denoted $T_{X,m}(E)$. To achieve the required selectivity, an appropriate signal-processing method is required. The most appropriate method will depend on the precise experimental setup, but inevitably will involve interrogating $T_{X,m}(E)$ over a range of energies. At low source-drain biases, the electrical conductance is given by the Landauer formula

$$G = G_0 \int_{-\infty}^{+\infty} dE\, T(E)\left(-\partial f / \partial E\right),$$

where $G_0 = 2e^2/h$ is the conductance quantum and

$$f(E) = \left(1 + \exp\left(\left(E - E_F\right) / k_B T\right)\right)^{-1}$$

is the Fermi function. Provided *T(E)* is slowly varying on the scale of 0.025eV (ie on the scale of k_B x room temperature) this equation simplifies to $G = G_0\, T(E_F)$. In a real device, E_F can be varied through the application of an external gate potential and therefore *T(E)* can be sampled over a range of energies. For such a three-terminal device, the $\alpha_{X,m}(E)$ quantity is defined:

Figure 12. (a) Transmission coefficient $T_{bare}(E)$ from left lead to the right lead in the perfect silicene nanoribbon without pore; (b) transmission coefficient $T_{bare}(E)$ from left lead to the right lead in the presence of a pore

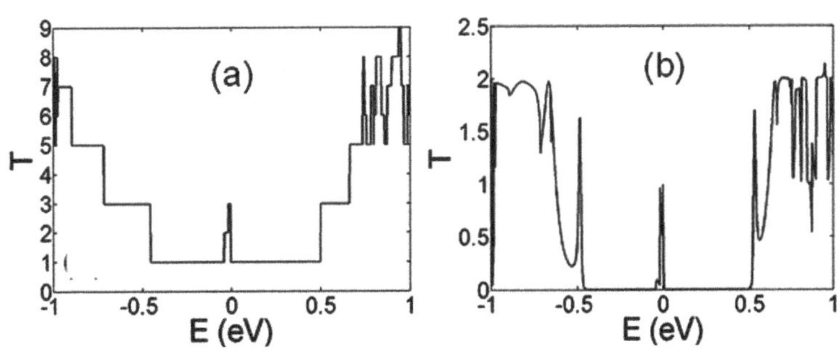

$$\alpha_{X,m}(E_F) = \log_{10}(T_{X,m}(E_F)) - \log_{10}(T_{bare}(E_F)) \tag{21}$$

which is a measure of the differences between $T_{X,m}(E)$ and the transmission of the unoccupied pore in the absence of a base. To differentiate between different bases, we analyze the set of all values of $\alpha_{X,m}(E)$ for energies lying within some convenient range, $E_{min}<E<E_{max}$ and introduce the probability distribution $P_{X,m}(\alpha)$ defined such that $P_{X,m}(\alpha)d\alpha$ is the probability that $\alpha_{X,m}(E)$ lies between α and $\alpha+d\alpha$. Formally this is defined by:

$$P_X(\alpha) = \frac{1}{m_{max}\left(E_{max} - E_{min}\right)} \sum_{n=1}^{m_{max}} \int_{E_{min}}^{E_{max}} dE\, \delta\left(\alpha - \alpha_{X,m}(E)\right) \tag{22}$$

Due to the technological limits in molecular scale electrostatic gating, two terminal devices are more desirable. To show that our proposed device could even selectively sense the nucleobases in two terminal configurations, the following quantity is examined, which can be measured using two-probe geometries:

$$\beta_{X,m}\left(V\right) = \log_{10}\left(I_{X,m}\left(V\right)\right) - \log_{10}\left(I_{bare}\left(V\right)\right) \tag{23}$$

where $I_{X,m}(V)$ is the current through the device at voltage V, in the presence of nucleobase X, with orientation m, defined by:

$$I_{X,m}\left(V\right) = \frac{2e}{h} \int_{E_F - \frac{eV}{2}}^{E_F + \frac{eV}{2}} dE\, T_{X,m}(E) \tag{24}$$

In Equation (23), the quantity $I_{bare}(V)$ is the current through the 'bare' device in the absence of any nucleobase. The probability distribution of the set $\{\beta_{X,m}(V)\}$ for a given base X is then defined by

$$P_X(\beta) = \frac{1}{m_{max}\left(eV_{max} - eV_{min}\right)} \sum_{n=1}^{m_{max}} \int_{eV_{min}}^{eV_{max}} dV\, \delta\left(\beta - \beta_{X,m}(V)\right) \tag{25}$$

The nanopore of Figure 11 has a diameter of 1.7 nm and is created in a zigzag silicene nanoribbon of width 3.2 nm. The edges of the ribbon and the pore are terminated by hydrogen and the structure relaxed to achieve its ground state energy as explained above. Now the transmission coefficient of the nanopore upon translocation of nucleotides inside the pore is considered. For each nucleobase, $T_{X,m}(E)$ for $m_{max}=4$ different orientations is calculated. These are used to obtain $\alpha_{X,m}(E)$ and $\beta_{X,m}(V)$ for $X=A$, G, C, T. For each of the nucleobases, Figure 13a and 13b show plots of the quantity $P_X(\alpha)$ and $P_X(\beta)$, respectively. Well-separated peaks in this figure show selective sensitivity to the nucleotides regardless of their different orientation.

The transmissions coefficients for different orientations are slightly different arise from both pi-pi interactions between the nucleobase and the pore surface states and through electrostatic interactions with the inhomogeneous charge distribution of the nucleobase. When combined together, these lead to

Figure 13. The probability distribution $P_X(\beta)$ of the set $\{\beta_{X,m}\}$ are shown for a given base X where X = A in black, C in red, G in blue and T in green. The maximum voltage employed in the calculation is 0.55 V; meaning that the voltage window is [-0.55 0.55] V.

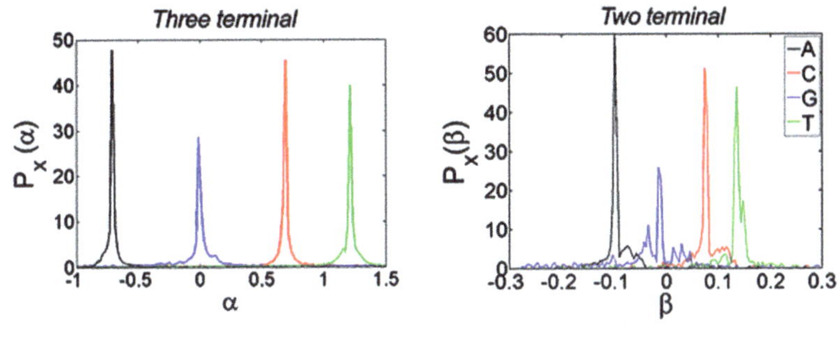

the probability distribution $P_X(\alpha)$ and $P_X(\beta)$ shown in Figure 13. Clearly the transmission coefficients depend on the position and orientation of the nucleobase and therefore the key issue is whether or not this dependency restricts the ability to selectively sense nucleobases within the pore. Figure 13 demonstrates that despite the sensitivity to position and orientation, selective sensing is preserved. Clearly the presence of well-separated peaks demonstrates that through an appropriate signal processing method, the bases can be selectively detected. The heights and positions of the peaks are different for a given base and either of them could be used to select and recognize the base type. This figure demonstrates the excellent potential of silicene nanopores for DNA sequencing.

2. **Silicene Thermal Properties:** In the recent years, the challenge of removing heat from nanoelectronic devices (Balandin, 2011) and thermoelectrically converting waste heat into electricity has attracted huge scientific interest (Nika, 2012), not only due to the questions posed for fundamental science, but also because thermoelectric energy conversion is an essential requirement for the next generation of nanoscale electronic, optoelectronic and photonic devices (Nika, 2012). The efficiency of a thermoelectric device and material is determined by its thermoelectric figure of merit (*ZT*) defined as:

$$ZT = \frac{S^2GT}{\kappa} \tag{26}$$

where *S* is the Seebeck coefficient (thermopower), *G* is the electrical conductance, *T* the temperature and κ the thermal conductance given by $\kappa = \kappa_e + \kappa_p$, where κ_e (κ_p) is the electronic (phononic) contribution to κ (Karamitaheri, 2012). Clearly *ZT* could be enhanced by increasing the power factor (S^2GT) or decreasing the thermal conductance and therefore a high-performance thermoelectric material should possess a large Seebeck coefficient and electrical conductance and simultaneously a low thermal conductance. The search for new materials with enhanced thermal properties continues to intensify, because these factors are correlated and increasing *ZT* to values greater than unity requires a delicate optimization of several material properties.

One promising approach has been to reduce the contribution κ_p of parasitic phonons by nanostructuring materials (Karamitaheri, 2011). Although acoustic phonons are the main heat carriers in bulk crystals, nanostructures can exhibit significantly lower phonon thermal conductances κ_p, due to increased phonon boundary scattering, changes in the phonon density of states and modified phonon dispersion in low-dimensional materials (Nika, 2012). The nanostructuring of materials is also a promising route to increasing the power factor, because it can lead to sharp features in the electronic density of states and the transmission coefficient *T(E)* describing the propagation of electrons of energy *E* through a device. It is well known that the thermopower *S* is controlled by the asymmetry of *T(E)* at the Fermi energy E_F and therefore if E_F is close to such an asymmetry, both *S* and *ZT* will be enhanced. Unfortunately, asymmetries, even if they occur, are not necessarily located near E_F and therefore a method of tuning them is required. In a nanoscale device, one could of course consider introducing a third gate electrode to control such features, but in practice this costs energy and does not solve the problem of designing a new material.

The above considerations suggest that an effective strategy for enhancing thermoelectric properties should:

1. Start from a parent material with low intrinsic thermal conductance,
2. Nanostructure the material to further reduce the thermal conductance,
3. Implement additional nanostructuring to introduce sharp features in *T*(*E*), and
4. Chemically modify the material to move these sharp features towards the Fermi energy.

In this section our aim is to demonstrate that silicene is an ideal material for implementing the four key elements of this strategy, because it is not only CMOS-compatible, but also it possesses a low intrinsic thermal conductance, which can be further reduced by nanostructuring the material to form nanoribbons.

Bulk silicon has a very low *ZT* (≈ 0.01) and therefore as a first step in a strategy for enhancing thermoelectric performance it is natural to focus on silicene. Calculations of thermoelectric properties of armchair and zigzag silicene nanoribbons with and without hydrogen-passivated edges suggest that this material may be attractive for thermoelectric devices (Pan, 2012a; Yang, 2014; Zberecki, 2013; Zhang, 2014), because the thermal conductivity of 2D silicene is predicted to be much smaller than bulk silicon and its counterpart graphene (Xie, 2014), with only ~10% of the total phononic thermal conductivity being due to the out-of-plane acoustic phonons (Bo, 2014; Gu, 2015). Both of these desirable features arise from the presence of small buckling, which breaks the reflection symmetry of the structure (Gu, 2015; Xie, 2014). Further reductions in the phonon contribution arise from additional nanostructuring. For example the thermal conductivity of a silicene nanosheet $\sigma_p^{SiNS} = \left[6 - 30\right] W / mK$. (Bo, 2014; Gu, 2015; Xie, 2014; Zhang, 2014) reduces to $\sigma_p^{ZSiNR} = \left[0.16 - 1.4\right] W / mK$ in zigzag silicene nanoribbons (Pan, 2012a; Yang, 2014; Zberecki, 2013). This could be potentially even lower in porous silicene, since the lattice thermal conductivity is reduced in nanoporous semiconductors (Mingo, 2007) such as nanoporous Bi (Song, 2004), Ge (Lee, 2009), graphene (Chang, 2012; Hu, 2009; Mortazavi, 2014), Bi_2Te_3 (Zhang, 2010) and SiGe (Lee, 2010). Moreover, it is been shown that placing nanopores in bulk silicon greatly reduces the thermal conductivity and enhances *ZT* by the factor of two (Lee, 2008).

Since the low phonon thermal conductance of silicene is well established and therefore elements 1 and 2 of the above strategy are satisfied, in this section we focus steps 3 and 4 and demonstrate that they can be achieved by inserting nanopores into silicene nanoribbons, whose edges are terminated by hydrogen and by introducing adsorbates to tune the position of features in *T*(*E*) relative to E_F. In view of the low value of $\kappa_{p,}$ as shown by Sadeghi et al (Sadeghi, 2015b) the main focus should be on the electron contribution in thermoelectric figure of merit ZT_e (obtained by setting $\kappa_p = 0$ in Equation (26)) and for ribbons containing nanopores with hydrogen-passivated edges, ZT_e could be enhanced to values as high as 160 at room temperature by creating sharp features in the *T(E)* and controlling the Fermi energy or by introducing adsorbates such as Tetrathiafulvalene (TTF) and Tetracyanoquinodimethane (TCNQ) onto the silicene surface. By including the κ_p of silicene nanoribbons, a high value of the full *ZT* of order 3.5 could be achieved.

The electronic contribution to the thermal conductance, the Seebeck and ZT_e are calculated for a silicene nanopore (diameter $\approx 1.3\ nm$) with and without functionalization with different dopant concentration. The optimized lattice constant (a_0) and buckling (*d*) of the silicene nanoribbon shown in Figure 1 are found to be 3.6 Å and 0.53 Å, respectively, similar to that reported elsewhere (Ding, 2009; Pan, 2012a; Sadeghi, 2014c). In silicene nanoribbons a sharp feature in transmission coefficient is obtained around the Fermi energy E_F. Due to its band structure, one open channel around the Fermi energy is expected for silicene. Due to the edge effect the band structure is bending in the corner of the *k*-space. This creates an extra sharp feature. The thermopower is suppressed in silicene nanoribbons where it possess low ZTe due to the background channels (for bulk silicene). Introducing a pore in the silicene blocks the background channel despite the sharp feature around Fermi energy which is due to the edge effect is less affected. This yields a delta-function like transmission coefficient which could show high ZTe (Figure 14c) provided it is asymmetrically located in the optimum position from Fermi energy (Sadeghi, 2015b). For the purpose of improving intrinsic material performance, a method of shifting E_F by doping the silicene nanoribbon is needed. Covalent bonding of dopants to the nanoribbons may adversely affect electronic properties, therefore tuning the Fermi energy need to be considered by adsorbing planar molecules which interact with the surface only through weak pi-pi interactions. The first adsorbed molecule studied here is Tetrathiafulvalene (TTF) which is a strong donor and the second is Tetracyanoquinodimethane (TCNQ), which is a strong acceptor (Anderson, 1973).

In the presence of adsorbates, calculations were performed by relaxing the adsorbates to find their minimum-energy states on the surface of the silicene, which yielded an optimized distance between the silicene and TCNQ of 3.1 Å and 3.54 Å for TTF. For the lowest dopant concentrations, we place a monolayer or sub-monolayer of TTF or TCNQ on the surface of the electrodes, whereas for the highest concentration, two layers above and below the silicene are included. As shown in Figure 14d, with increasing concentration of TTF on silicene the delta function like feature around the Fermi energy moves toward the optimum position yeldis higher ZTe, whereas with increasing concentration of TCNQ the ZTe reduced. This identifies nanoporous silicene as an ideal thermoelectric material with the potential for unprecedented performance.

COMPUTATIONAL METHODS

To find the optimized geometry and ground state Hamiltonian of the structures of interest, we employed the SIESTA (Soler, 2002) implementation of Density Functional Theory (DFT) using the generalized gradient approximation (GGA) of the exchange and correlation functional with the Perdew-Burke-Ernzerhof parameterization (PBE) (Perdew, 1996) a double zeta polarized basis set, a real-space grid defined with a plane wave cut-off energy of 250 Ry and a maximum force tolerance of 40 meV/Ang. The calculation with VDW-DF exchange and correlation functional with BH parameterization was also

Figure 14. Silicene based thermolectic device: (a) sketch of the molecular structure of the Silicene nanopores and device configuration; (b) Tetrathiafulvalene (TTF) and Tetracyanoquinodimethane (TCNQ) molecular structure used as a dopant in this study; (c) electronic thermoelectric figure of merit of silicene nanoribbon with a nanopores; and (d) electronic figure of merit as a function of temperature for a monolayer silicene nanopore (blue), a monolayer silicene nanopore with TCNQ functionalized zigzag silicene nanoribbon electrodes with low concentration of TCNQ (green) and with higher concentration of TCNQ (light green), a monolayer silicene nanopore with TTF functionalized zigzag silicene ribbon electrodes with low concentration of TTF (red) and with higher concentrations of TTF (purple and dark purple).

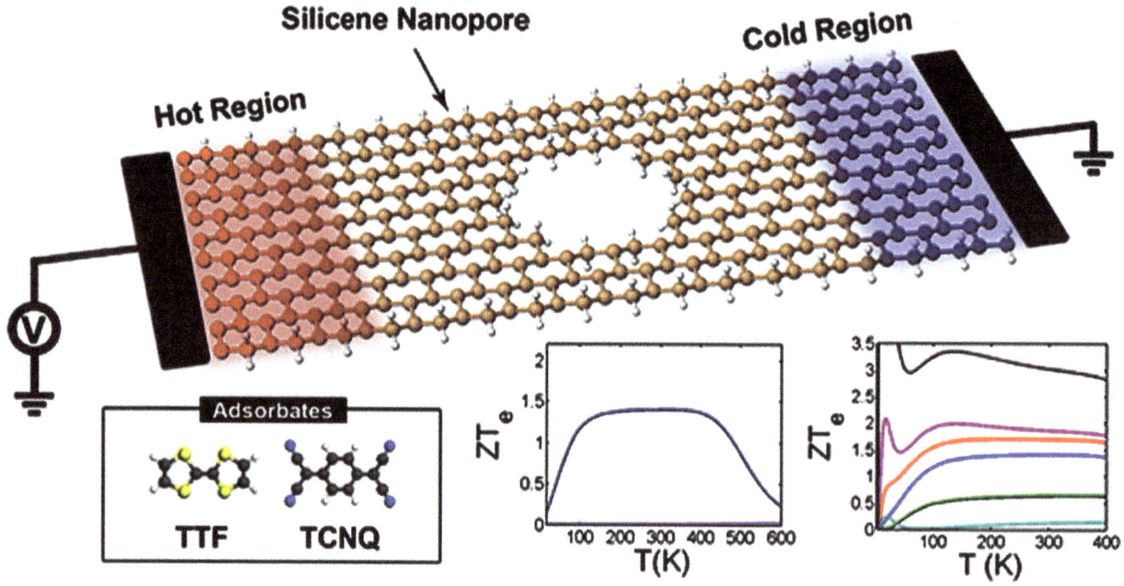

carried out to check the GGA-PBE result in some cases. The BH is the same as DRSLL with some modification (Berland, 2014). From the converged DFT calculation, the underlying mean-field Hamiltonian was combined with our transport code, GOLLUM (Ferrer, 2014) which is an implementation of the non-equilibrium Green's function (NEGF) method. This yields the transmission coefficient $T\left(E\right)$ for electrons of energy E (passing from the source to the drain) via the relation

$$T(E) = Trace\left\{ \Gamma_R(E) G^R(E) \Gamma_L(E) G^{R\dagger}(E) \right\} \tag{27}$$

In this expression,

$$\Gamma_{L,R}\left(E\right) = i\left(\sum\nolimits_{L,R}\left(E\right) - \sum\nolimits_{L,R}^{\dagger}\left(E\right)\right)$$

describes the level broadening due to the coupling between left (L) and right (R) electrodes and the central scattering region, $\sum_{L,R}\left(E\right)$ are the retarded self-energies associated with this coupling and $G^R = \left(ES - H - \sum_L - \sum_R\right)^{-1}$ is the retarded Green's function, where *H* is the Hamiltonian and *S* is overlap matrix (both of them obtained from SIESTA). Thermal properties such as the electrical conductance $G(T)$, the electronic contribution of the thermal conductance $\kappa_e(T)$, the thermo-power $S(T)$ and the Peltier coefficient $\Pi(T)$ of the junction as a function of the temperature are given by (Ferrer, 2014):

$$G(T) = G_0 L_0 \tag{28}$$

$$\kappa_e(T) = \frac{L_0 L_2 - L_1^2}{hTL_0} \tag{29}$$

$$S(T) = -\frac{L_1}{eTL_0} \tag{30}$$

$$\Pi(T) = TS(T) \tag{31}$$

where

$$L_n(T) = \int_{-\infty}^{+\infty} dE \left(E - E_F\right)^n T\left(E\right)\left(-\frac{\partial f\left(E\right)}{\partial E}\right) \tag{32}$$

and $T(E)$ is the transmission coefficient, $f(E)$ is the Fermi-Dirac probability distribution function

$$f\left(E\right) = \left(1 + \exp\left(E - E_F / k_B T\right)\right)^{-1},$$

T is the temperature, E_F is the Fermi energy, $G_0 = 2e^2/h$ is the conductance quantum, e is electron charge and h is the Planck's constant. From these expressions, the electronic contribution to the figure of merit ZT_e is:

$$ZT_e = \frac{L_1^2}{L_0 L_2 - L_1^2} \tag{33}$$

CONCLUSION

Silicene is a single layer of silicon atoms with sp^3 bonds in honeycomb lattice structure. More compatibility of Silicene with current semiconducting technology and some other advantages such as a new and promising alternative for spintronic and nanomagnetic materials have attracted considerable scientific attention in very recent years. In addition, despite single layer Graphene which has zero band gap in

Dirac point, Silicene owns a tunable band gap from zero to semiconducting region, which make it better choice for FET technology. In this chapter we presented the analytical transport model of Silicene, as channel in field effect transistor. The model shows increment in total carrier density and conductance with gate voltage as expected for conventional semiconductors which affected by temperature around neutrality point. Minimum conductance is increased by temperature whereas it remains stable in degenerate regime. High upper bound carrier mobility in the order of 10^{13} in room temperature could be obtained in Silicene which is decreased with gate voltage. Highest UB carrier mobility occurs in higher temperature when gate voltage increased. However, increasing of the gate voltage makes Silicene total carrier density, conductance and UB mobility almost affectless by temperature. Comparing with Graphene conductance, Silicene shows lower conductance in similar size and temperature. Moreover, the longitudinal magneto-conductance of Silicene is expected to decrease with magnetic field and increase with gate voltage based on presented model. In similar condition, Silicene is predicted to show lower longitudinal magneto conductance than Graphene. This model could be employed to model Silicene FET based integrated circuits.

In addition, using first principles calculations combined with quantum scattering theory, we showed that with appropriate signal processing, silicene-based nanopore sensing offers a potential route to selective sensing of DNA nucleobases. Such a sensing platform is a direct electrical sensor and opens the avenues towards fast, cheap and portable DNA sequencing. Moreover we demonstrated a new strategy for improving the thermoelectric performance of silicene-based nanoribbons by investigating the effect of introducing nanopores. We have demonstrated that the thermopower and electronic thermoelectric figure of merit ZT_e can be improved by introducing nanopores and tuning their Fermi energy. As a method of tuning the Fermi energy, we analysed the effect of doping the silicene with either a strong electron donor (TTF) or a strong electron acceptor (TCNQ) and demonstrated that doping by the former tended to decrease the value of ZT_e, because the Fermi energy shift was in an unfavourable direction. On the other hand, we found that doping with TTF increased the room-temperature value to ZT_e to 3.1 and that this value is insensitive to temperature over the range 100K - 400K. This combination of a high temperature-insensitive ZT_e, and the fact that the low phonon thermal conductance renders ZT_e comparable with the full figure of merit ZT, identifies nanoporous silicene as an ideal thermoelectric material with the potential for unprecedented performance.

REFERENCES

Anderson, P., Lee, P., & Saitoh, M. (1973). Remarks on giant conductivity in TTF-TCNQ. *Solid State Communications*, *13*(5), 595–598. doi:10.1016/S0038-1098(73)80020-1

Anselmetti, D. (2012). Nanopores: Tiny holes with great promise. *Nature Nanotechnology*, *7*(2), 81–82. doi:10.1038/nnano.2012.11 PMID:22306893

Aufray, B., Kara, A., Vizzini, S., Oughaddou, H., Leandri, C., Ealet, B., & Le Lay, G. (2010). Graphene-like silicon nanoribbons on Ag(110): A possible formation of silicene. *Applied Physics Letters*, *96*(18), 183102. doi:10.1063/1.3419932

Avdoshenko, S. M., Nozaki, D., Gomes da Rocha, C., González, J. W., Lee, M. H., Gutierrez, R., & Cuniberti, G. (2013). Dynamic and Electronic Transport Properties of DNA Translocation through Graphene Nanopores. *Nano Letters*, *13*(5), 1969–1976. doi:10.1021/nl304735k PMID:23586585

Avila, J., Padova, P. D., Cho, S., Colambo, I., Lorcy, S., & Quaresima, C. et al. (2013). Presence of gapped silicene-derived band in the prototypical (3 × 3) silicene phase on silver (111) surfaces. *Journal of Physics Condensed Matter*, *25*(26), 262001. doi:10.1088/0953-8984/25/26/262001 PMID:23759650

Avouris, P., Chen, Z. H., & Perebeinos, V. (2007). Carbon-based electronics. *Nature Nanotechnology*, *2*(10), 605–615. doi:10.1038/nnano.2007.300 PMID:18654384

Balandin, A. A. (2011). Thermal properties of graphene and nanostructured carbon materials. *Nature Materials*, *10*(8), 569–581. doi:10.1038/nmat3064 PMID:21778997

Berland, K., & Hyldgaard, P. (2014). Exchange functional that tests the robustness of the plasmon description of the van der Waals density functional. *Physical Review B: Condensed Matter and Materials Physics*, *89*(3), 035412. doi:10.1103/PhysRevB.89.035412

Bo, L., Reddy, C. D., Jinwu, J., Hongwei, Z., Julia, A. B., Sergey, V. D., & Kun, Z. (2014). Thermal conductivity of silicene nanosheets and the effect of isotopic doping. *Journal of Physics. D, Applied Physics*, *47*(16), 165301. doi:10.1088/0022-3727/47/16/165301

Cahangirov, S., Topsakal, M., Aktürk, E., Şahin, H., & Ciraci, S. (2009). Two- and One-Dimensional Honeycomb Structures of Silicon and Germanium. *Physical Review Letters*, *102*(23), 236804. doi:10.1103/PhysRevLett.102.236804 PMID:19658958

Chang, P.-H., & Nikolic, B. K. (2012). Edge currents and nanopore arrays in zigzag and chiral graphene nanoribbons as a route toward high-ZT thermoelectrics. *Physical Review B: Condensed Matter and Materials Physics*, *86*(4), 041406. doi:10.1103/PhysRevB.86.041406

Chen, L., Feng, B., & Wu, K. (2013a). Observation of a possible superconducting gap in silicene on Ag(111) surface. *Applied Physics Letters*, *102*(8), 081602. doi:10.1063/1.4793998

Chen, L., Feng, B., & Wu, K. (2013b). Observation of superconductivity in silicene. *Applied Physics Letters*, *102*, 081602. doi:10.1063/1.4793998

Chen, L., Liu, C.-C., Feng, B., He, X., Cheng, P., & Ding, Z. et al. (2012). Evidence for Dirac Fermions in a Honeycomb Lattice Based on Silicon. *Physical Review Letters*, *109*(5), 056804. doi:10.1103/PhysRevLett.109.056804 PMID:23006197

Cheng, Y. C., Zhu, Z. Y., & Schwingenschlögl, U. (2011). Doped silicene: Evidence of a wide stability range.[Europhysics Letters]. *EPL*, *95*(1), 17005. doi:10.1209/0295-5075/95/17005

Cherf, G. M., Lieberman, K. R., Rashid, H., Lam, C. E., Karplus, K., & Akeson, M. (2012). Automated forward and reverse ratcheting of DNA in a nanopore at 5-A precision. *Nature Biotechnology*, *30*(4), 344–348. doi:10.1038/nbt.2147 PMID:22334048

Clarke, J., Wu, H.-C., Jayasinghe, L., Patel, A., Reid, S., & Bayley, H. (2009). Continuous base identification for single-molecule nanopore DNA sequencing. *Nature Nanotechnology*, *4*(4), 265-270. Retrieved from http://www.nature.com/nnano/journal/v4/n4/suppinfo/nnano.2009.12_S1.html

Datta, S. (2005). *Quantum Transport: Atom to Transistor*. New York: Cambridge University Press. doi:10.1017/CBO9781139164313

Dekker, C. (2007). Solid-state nanopores. *Nature Nanotechnology*, *2*(4), 209–215. doi:10.1038/nnano.2007.27 PMID:18654264

Ding, Y., & Ni, J. (2009). Electronic structures of silicon nanoribbons. *Applied Physics Letters, 95*(8), 083115. doi:10.1063/1.3211968

Ding, Y., & Wang, Y. (2012a). Electronic structures of silicene fluoride and hydride. *Applied Physics Letters*, *100*(8), 083102–083102. doi:10.1063/1.3688035

Ding, Y., & Wang, Y. (2012b). Electronic structures of zigzag SiC nanoribbons with asymmetric hydrogen-terminations. *Applied Physics Letters*, *101*(1), 013102–013104. doi:10.1063/1.4732786

Drummond, N. D., Zólyomi, V., & Fal'ko, V. I. (2012). Electrically tunable band gap in silicene. *Physical Review B: Condensed Matter and Materials Physics*, *85*(7), 075423. doi:10.1103/PhysRevB.85.075423

Drummond, T. G., Hill, M. G., & Barton, J. K. (2003). Electrochemical DNA sensors. *Nature Biotechnology*, *21*(10), 1192–1199. doi:10.1038/nbt873 PMID:14520405

Enriquez, H., Vizzini, S., Kara, A., Lalmi, B., & Oughaddou, H. (2012). Silicene structures on silver surfaces. *Journal of Physics Condensed Matter*, *24*(31), 314211. doi:10.1088/0953-8984/24/31/314211 PMID:22820837

Feng, B., Ding, Z., Meng, S., Yao, Y., He, X., & Cheng, P. et al. (2012). Evidence of Silicene in Honeycomb Structures of Silicon on Ag(111). *Nano Letters*, *12*(7), 3507–3511. doi:10.1021/nl301047g PMID:22658061

Feng, B., Li, H., Liu, C.-C., Shao, T., Cheng, P., & Yao, Y. et al. (2013). Observation of Dirac cone warping and chirality effects in silicene. *ACS Nano*, *7*(10), 9049–9054. doi:10.1021/nn403661h PMID:24003914

Ferrer, J., Lambert, C., Garcia-Suarez, V., Zsolt Manrique, D., Visontai, D., & Oroszlani, L. et al. (2014). GOLLUM: A next-generation simulation tool for electron, thermal and spin transport. *New Journal of Physics*, *16*(9), 093029. doi:10.1088/1367-2630/16/9/093029

Fleurence, A., Friedlein, R., Ozaki, T., Kawai, H., Wang, Y., & Yamada-Takamura, Y. (2012). Experimental Evidence for Epitaxial Silicene on Diboride Thin Films. *Physical Review Letters*, *108*(24), 245501. doi:10.1103/PhysRevLett.108.245501 PMID:23004288

Friedlein, R., Fleurence, A., Sadowski, J. T., & Yamada-Takamura, Y. (2013). Tuning of silicene-substrate interactions with potassium adsorption. *Applied Physics Letters*, *102*(22), 221603. doi:10.1063/1.4808214

Fritz, J., Cooper, E. B., Gaudet, S., Sorger, P. K., & Manalis, S. R. (2002). Electronic detection of DNA by its intrinsic molecular charge. *Proceedings of the National Academy of Sciences of the United States of America*, *99*(22), 14142–14146. doi:10.1073/pnas.232276699 PMID:12386345

Fyta, M., Melchionna, S., & Succi, S. (2011). Translocation of Biomolecules Through Solid-State Nanopores: Theory Meets Experiments. *Journal of Polymer Science. Part B, Polymer Physics*, *49*(14), 985–1011. doi:10.1002/polb.22284

Gao, N., Zheng, W. T., & Jiang, Q. (2012). Density functional theory calculations for two-dimensional silicene with halogen functionalization. *Physical Chemistry Chemical Physics*, *14*(1), 257–261. doi:10.1039/C1CP22719J PMID:22083171

Garaj, S., Liu, S., Golovchenko, J. A., & Branton, D. (2013). Molecule-hugging graphene nanopores. *Proceedings of the National Academy of Sciences of the United States of America*, *110*(30), 12192–12196. doi:10.1073/pnas.1220012110 PMID:23836648

Gu, X., & Yang, R. (2015). First-Principles Prediction of Phononic Thermal Conductivity of Silicene: A Comparison with Graphene. *Journal of Applied Physics*, *117*(2), 025102. doi:10.1063/1.4905540

Guzmán-Verri, G. G., & Lew Yan Voon, L. C. (2007). Electronic structure of silicon-based nanostructures. *Physical Review B: Condensed Matter and Materials Physics*, *76*(7), 075131. doi:10.1103/PhysRevB.76.075131

Houssa, M., Pourtois, G., Afanas'Ev, V. V., & Stesmans, A. (2010). Can silicon behave like graphene? A first-principles study. *Applied Physics Letters, 97*(11), 112106 - 112106-112103

Houssa, M., Scalise, E., Sankaran, K., Pourtois, G., Afanas'Ev, V. V., & Stesmans, A. (2011). Electronic properties of hydrogenated silicene and germanene. *Applied Physics Letters*, *98*(22), 223107–223107. doi:10.1063/1.3595682

Hu, J., Ruan, X., & Chen, Y. P. (2009). Thermal Conductivity and Thermal Rectification in Graphene Nanoribbons: A Molecular Dynamics Study. *Nano Letters*, *9*(7), 2730–2735. doi:10.1021/nl901231s PMID:19499898

Huang, S., Kang, W., & Yang, L. (2013). Electronic Structure and Quasiparticle Band Gap of Silicene Structures. *Applied Physics Letters*, *102*(13), 133106. doi:10.1063/1.4801309

Jamgotchian, H., Colignon, Y., Hamzaoui, N., Ealet, B., Hoarau, J., Aufray, B., & Bibérian, J. (2012). Growth of silicene layers on Ag (111): Unexpected effect of the substrate temperature. *Journal of Physics Condensed Matter*, *24*(17), 172001. doi:10.1088/0953-8984/24/17/172001 PMID:22487603

Jose, D., & Datta, A. (2011). Structures and electronic properties of silicene clusters: A promising material for FET and hydrogen storage. *Physical Chemistry Chemical Physics*, *13*(16), 7304–7311. doi:10.1039/c0cp02580a PMID:21359313

Kamal, C., Chakrabarti, A., Banerjee, A., & Deb, S. (2013). Silicene Beyond Mono-layers-Different Stacking Configurations And Their Properties. *Journal of Physics Condensed Matter*, *25*(8), 085508. doi:10.1088/0953-8984/25/8/085508 PMID:23370369

Kara, A., Enriquez, H., Seitsonen, A. P., Lew Yan Voon, L. C., Vizzini, S., Aufray, B., & Oughaddou, H. (2012). A review on silicene — New candidate for electronics. *Surface Science Reports*, *67*(1), 1–18. doi:10.1016/j.surfrep.2011.10.001

Karamitaheri, H., Neophytou, N., Pourfath, M., Faez, R., & Kosina, H. (2012). Engineering enhanced thermoelectric properties in zigzag graphene nanoribbons. *Journal of Applied Physics*, *111*(5), 054501. doi:10.1063/1.3688034

Karamitaheri, H., Pourfath, M., Faez, R., & Kosina, H. (2011). Geometrical effects on the thermoelectric properties of ballistic graphene antidot lattices. *Journal of Applied Physics*, *110*(5), 054506. doi:10.1063/1.3629990

Kim, J., Fischetti, M. V., & Aboud, S. (2012a). *Electronic and transport properties of armchair and zigzag sp3-hybridized silicane nanoribbons.* Paper presented at the Computational Electronics (IWCE), 2012 15th International Workshop on.

Kim, J., Fischetti, M. V., & Aboud, S. (2012b). Structural, electronic, and transport properties of silicane nanoribbons. *Physical Review B: Condensed Matter and Materials Physics*, *86*(20), 205323. doi:10.1103/PhysRevB.86.205323

Kokott, S., Pflugradt, P., Matthes, L., & Bechstedt, F. (2014). Nonmetallic substrates for growth of silicene: An ab initio prediction. *Journal of Physics Condensed Matter*, *26*(18), 185002. doi:10.1088/0953-8984/26/18/185002 PMID:24728001

Kuan, A. T., & Golovchenko, J. A. (2012). Nanometer-thin solid-state nanopores by cold ion beam sculpting. *Applied Physics Letters*, *100*(21), 213104. doi:10.1063/1.4719679 PMID:22711913

Lalmi, B., Oughaddou, H., Enriquez, H., Kara, A., Vizzini, S., Ealet, B., & Aufray, B. (2010). Epitaxial growth of a silicene sheet. *Applied Physics Letters*, *97*(22), 223109. doi:10.1063/1.3524215

Lee, H., Vashaee, D., Wang, D., Dresselhaus, M. S., Ren, Z., & Chen, G. (2010). Effects of nanoscale porosity on thermoelectric properties of SiGe. *Journal of Applied Physics*, *107*(9), 094308. doi:10.1063/1.3388076 PMID:20473369

Lee, J.-H., Galli, G. A., & Grossman, J. C. (2008). Nanoporous Si as an Efficient Thermoelectric Material. *Nano Letters*, *8*(11), 3750–3754. doi:10.1021/nl802045f PMID:18947211

Lee, J.-H., & Grossman, J. C. (2009). Thermoelectric properties of nanoporous Ge. *Applied Physics Letters*, *95*(1), 013106. doi:10.1063/1.3159813

Lew Yan Voon, L. C., Sandberg, E., Aga, R. S., & Farajian, A. A. (2010). Hydrogen compounds of group-IV nanosheets. *Applied Physics Letters*, *97*(16), 163114–163114. doi:10.1063/1.3495786

Lin, C.-L., Arafune, R., Kawahara, K., Tsukahara, N., Minamitani, E., & Kim, Y. et al. (2012). Structure of silicene grown on Ag (111). *Applied Physics Express*, *5*(4), 045802. doi:10.1143/APEX.5.045802

Liu, C. C., Jiang, H., & Yao, Y. (2011). Low-energy effective Hamiltonian involving spin-orbit coupling in silicene and two-dimensional germanium and tin. *Physical Review B: Condensed Matter and Materials Physics*, *84*(19), 195430. doi:10.1103/PhysRevB.84.195430

Liu, S., Zhao, Q., Xu, J., Yan, K., Peng, H., & Yang, F. et al. (2012). Fast and controllable fabrication of suspended graphene nanopore devices. *Nanotechnology*, *23*(8), 085301–085301. doi:10.1088/0957-4484/23/8/085301 PMID:22293107

Luan, B., Stolovitzky, G., & Martyna, G. (2012). Slowing and controlling the translocation of DNA in a solid-state nanopore. *Nanoscale*, *4*(4), 1068–1077. doi:10.1039/C1NR11201E PMID:22081018

Majzik, Z., Tchalala, M. R., Švec, M., Hapala, P., Enriquez, H., & Kara, A. et al. (2013). Combined AFM and STM measurements of a silicene sheet grown on the Ag(111) surface. *Journal of Physics Condensed Matter*, *25*(22), 225301. doi:10.1088/0953-8984/25/22/225301 PMID:23674193

Manrao, E. A., Derrington, I. M., Laszlo, A. H., Langford, K. W., Hopper, M. K., & Gillgren, N. et al. (2012). Reading DNA at single-nucleotide resolution with a mutant MspA nanopore and phi29 DNA polymerase. *Nature Biotechnology*, *30*(4), 349–353. doi:10.1038/nbt.2171 PMID:22446694

McGinn, S., & Gut, I. G. (2012). DNA sequencing – spanning the generations. *New Biotechnology*, (0). doi:10.1016/j.nbt.2012.11.012 PMID:23165096

Meng, L., Wang, Y., Zhang, L., Du, S., Wu, R., & Li, L. et al. (2013). Buckled Silicene Formation on Ir(111). *Nano Letters*, *13*(2), 685–690. doi:10.1021/nl304347w PMID:23330602

Mingo, N., & Broido, D. A. (2007). Thermoelectric power factor of nanoporous semiconductors. *Journal of Applied Physics, 101*(1), 014322. doi:10.1063/1.2405232

Mohanty, N., & Berry, V. (2008). Graphene-Based Single-Bacterium Resolution Biodevice and DNA Transistor: Interfacing Graphene Derivatives with Nanoscale and Microscale Biocomponents. *Nano Letters*, *8*(12), 4469–4476. doi:10.1021/nl802412n PMID:19367973

Molle, A., Grazianetti, C., Chiappe, D., Cinquanta, E., Cianci, E., Tallarida, G., & Fanciulli, M. (2013). Hindering the Oxidation of Silicene with Non-Reactive Encapsulation. *Advanced Functional Materials*, *23*(35), 4340–4344. doi:10.1002/adfm.201300354

Morozov, S. V., Novoselov, K. S., Katsnelson, M. I., Schedin, F., Elias, D. C., Jaszczak, J. A., & Geim, A. K. (2008). Giant intrinsic carrier mobilities in graphene and its bilayer. *Physical Review Letters*, *100*(1), 016602. doi:10.1103/PhysRevLett.100.016602 PMID:18232798

Mortazavi, B., Potschke, M., & Cuniberti, G. (2014). Multiscale modeling of thermal conductivity of polycrystalline graphene sheets. *Nanoscale*, *6*(6), 3344–3352. doi:10.1039/c3nr06388g PMID:24518878

Nakano, H., Mitsuoka, T., Harada, M., Horibuchi, K., Nozaki, H., & Takahashi, N. et al. (2006). Soft synthesis of single-crystal silicon monolayer sheets. *Angewandte Chemie International Edition*, *45*(38), 6303–6306. doi:10.1002/anie.200600321 PMID:16941712

Nelson, T., Zhang, B., & Prezhdo, O. V. (2010). Detection of Nucleic Acids with Graphene Nanopores: Ab Initio Characterization of a Novel Sequencing Device. *Nano Letters*, *10*(9), 3237–3242. doi:10.1021/nl9035934 PMID:20722409

Ni, Z., Liu, Q., Tang, K., Zheng, J., Zhou, J., & Qin, R. et al. (2011). Tunable Bandgap in Silicene and Germanene. *Nano Letters*, *12*(1), 113–118. doi:10.1021/nl203065e PMID:22050667

Nika, D. L., & Balandin, A. A. (2012). Two-dimensional phonon transport in graphene. *Journal of Physics Condensed Matter*, *24*(23), 233203. doi:10.1088/0953-8984/24/23/233203 PMID:22562955

O'Hare, A., Kusmartsev, F. V., & Kugel, K. I. (2012). A Stable "Flat" Form of Two-Dimensional Crystals: Could Graphene, Silicene, Germanene Be Minigap Semiconductors? *Nano Letters*, *12*(2), 1045–1052. doi:10.1021/nl204283q PMID:22236130

Osborn, T. H., Farajian, A. A., Pupysheva, O. V., Aga, R. S., & Voon, L. C. L. Y. (2011). Ab initio simulations of silicene hydrogenation. *Chemical Physics Letters*, *511*(1-3), 101–105. doi:10.1016/j.cplett.2011.06.009

Padova, P., Quaresima, C., Olivieri, B., Perfetti, P., & Le Lay, G. (2011). Strong resistance of silicene nanoribbons towards oxidation. *Journal of Physics. D, Applied Physics*, *44*(31), 312001. doi:10.1088/0022-3727/44/31/312001

Padova, P. D., Leandri, C., Vizzini, S., Quaresima, C., Perfetti, P., & Olivieri, B. et al. (2008). Burning Match Oxidation Process of Silicon Nanowires Screened at the Atomic Scale. *Nano Letters*, *8*(8), 2299–2304. doi:10.1021/nl800994s PMID:18624391

Padova, P. D., Vogt, P., Resta, A., Avila, J., Razado-Colambo, I., & Quaresima, C. et al. (2013). Evidence of Dirac fermions in multilayer silicene. *Applied Physics Letters*, *102*(16), 163106. doi:10.1063/1.4802782

Pan, L., Liu, H. J., Tan, X. J., Lv, H. Y., Shi, J., Tang, X. F., & Zheng, G. (2012a). Thermoelectric properties of armchair and zigzag silicene nanoribbons. *Physical Chemistry Chemical Physics*, *14*(39), 13588–13593. doi:10.1039/c2cp42645e PMID:22965156

Pan, L., Liu, H. J., Wen, Y. W., Tan, X. J., Lv, H. Y., Shi, J., & Tang, X. F. (2012b). Magnetic and electronic properties of silicane with hydrogen vacancies on the surface. *Applied Surface Science*, *258*(24), 10135–10139. doi:10.1016/j.apsusc.2012.06.092

Perdew, J. P., Burke, K., & Ernzerhof, M. (1996). Generalized Gradient Approximation Made Simple. *Physical Review Letters*, *77*(18), 3865–3868. doi:10.1103/PhysRevLett.77.3865 PMID:10062328

Pitcher, G. (2011). Scientists create silicene. *New Electronics*, *44*(7), 8–8.

Russo, C. J., & Golovchenko, J. A. (2012). Atom-by-atom nucleation and growth of graphene nanopores. *Proceedings of the National Academy of Sciences of the United States of America*, *109*(16), 5953–5957. doi:10.1073/pnas.1119827109 PMID:22492975

Sadeghi, H. (2014a). Electrical Transport Model of Silicene as a Channel of Field Effect Transistor. *Journal of Nanoscience and Nanotechnology*, *14*(6), 4178–4184. doi:10.1166/jnn.2014.8914 PMID:24738367

Sadeghi, H., Ahmadi, M. T., Ishak, B., Mousavi, S., & Ismail, R. (2011). Ballistic Conductance Model of Bilayer Graphene Nanoribbon (BGN). *Journal of Computational and Theoretical Nanoscience*, *8*(10), 1993–1998. doi:10.1166/jctn.2011.1915

Sadeghi, H., Ahmadi, M. T., Mousavi, S. M., Ghadiry, M. H., & Ismail, R. (2012). Channel Conductance of ABA Stacking Trilayer Graphene Nanoribbon Field Effect Transistor. *Modern Physics Letters B*, *26*(8), 1250047. doi:10.1142/S0217984912500479

Sadeghi, H., Algaragholy, L., Pope, T., Bailey, S., Visontai, D., & Manrique, D. et al. (2014b). Graphene Sculpturene Nanopores for DNA Nucleobase Sensing. *The Journal of Physical Chemistry B*, *118*(24), 6908–6914. doi:10.1021/jp5034917 PMID:24849015

Sadeghi, H., Bailey, S., & Lambert, C. J. (2014c). Silicene-based DNA nucleobase sensing. *Applied Physics Letters*, *104*(10), 103104. doi:10.1063/1.4868123

Sadeghi, H., Lai, D. T., Redoute, J.-M., & Zayegh, A. (2013a). Classic and quantum capacitances in Bernal bilayer and trilayer graphene field effect transistor. *Journal of Nanomaterials*, *2013*, 1–7. doi:10.1155/2013/127690

Sadeghi, H., Lambert, C. J., Algharagholy, L., Bailey, S. W. D., Ferrer, J., & Garcia-Suarez, V. M. (2013b). *Patent: Nanopore arrangement for DNA sequencing*. WO/2015/092411/A1.

Sadeghi, H., Mol, J. A., Lau, C. S., Briggs, G. A. D., Warner, J., & Lambert, C. J. (2015a). Conductance enlargement in picoscale electroburnt graphene nanojunctions. *Proceedings of the National Academy of Sciences of the United States of America*, *112*(9), 2658–2663. doi:10.1073/pnas.1418632112 PMID:25730863

Sadeghi, H., Sangtarash, S., & Lambert, C. J. (2015b). Enhanced Thermoelectric Efficiency of Porous Silicene Nanoribbons. *Scientific Reports*, *5*, 9514. doi:10.1038/srep09514 PMID:25820162

Saha, K. K., Drndic, M., & Nikolic, B. K. (2012). DNA Base-Specific Modulation of Microampere Transverse Edge Currents through a Metallic Graphene Nanoribbon with a Nanopore. *Nano Letters*, *12*(1), 50–55. doi:10.1021/nl202870y PMID:22141739

Schneider, G. F., & Dekker, C. (2012). DNA sequencing with nanopores. *Nature Biotechnology*, *30*(4), 326–328. doi:10.1038/nbt.2181 PMID:22491281

Shendure, J., & Aiden, E. L. (2012). The expanding scope of DNA sequencing. *Nature Biotechnology*, *30*(11), 1084–1094. doi:10.1038/nbt.2421 PMID:23138308

Shendure, J., & Ji, H. (2008). Next-generation DNA sequencing. *Nature Biotechnology*, *26*(10), 1135–1145. doi:10.1038/nbt1486 PMID:18846087

Shim, J., Humphreys, G. I., Venkatesan, B. M., Munz, J. M., Zou, X., Sathe, C., et al. (2013). Detection and Quantification of Methylation in DNA using Solid-State Nanopores. *Sci. Rep., 3*. Retrieved from http://www.nature.com/srep/2013/130306/srep01389/abs/srep01389.html#supplementary-information

Shim, J., Solovyeva, V., Estrada, D., Banerjee, S., Rivera, J., & Pop, E. (2012). *Graphene Nanopores for Nucleic Acid Analysis*. IEEE.

Soler, J. M., Artacho, E., Gale, J. D., García, A., Junquera, J., Ordejón, P., & Sánchez-Portal, D. (2002). The SIESTA method for ab initio order- N materials simulation. *Journal of Physics Condensed Matter*, *14*(11), 2745–2779. doi:10.1088/0953-8984/14/11/302

Song, D. W., Shen, W.-N., Dunn, B., Moore, C. D., Goorsky, M. S., Radetic, T., . . . Chen, G. (2004). Thermal conductivity of nanoporous bismuth thin films. *Applied Physics Letters, 84*(11), 1883-1885. doi:10.1063/1.1682679

Spinney, P. S., Collins, S. D., Howitt, D. G., & Smith, R. L. (2012). Fabrication and characterization of a solid-state nanopore with self-aligned carbon nanoelectrodes for molecular detection. *Nanotechnology*, *23*(13), 135501–135501. doi:10.1088/0957-4484/23/13/135501 PMID:22421078

Suzuki, T., & Yokomizo, Y. (2010). Energy bands of atomic monolayers of various materials: Possibility of energy gap engineering. *Physica E, Low-Dimensional Systems and Nanostructures*, *42*(10), 2820–2825. doi:10.1016/j.physe.2010.04.031

Taniguchi, M., Tsutsui, M., Yokota, K., & Kawai, T. (2009). Fabrication of the gating nanopore device. *Applied Physics Letters*, *95*(12), 123701, 123701–123703. doi:10.1063/1.3236769

Tao, L., Cinquanta, E., Chiappe, D., Grazianetti, C., Fanciulli, M., Dubey, M., et al. (2015). Silicene field-effect transistors operating at room temperature. *Nat Nano, 10*(3), 227-231. Retrieved from http://www.nature.com/nnano/journal/v10/n3/abs/nnano.2014.325.html#supplementary-information

Tchalal, M. R., Enriquez, H., Mayne, A. J., Kara, A., Roth, S., & Silly, M. G. et al. (2013). Formation of one-dimensional self-assembled silicon nanoribbons on Au (110)-(2x1). *Applied Physics Letters*, *102*(8), 083107. doi:10.1063/1.4793536

Venkatesan, B. M., & Bashir, R. (2011). Nanopore sensors for nucleic acid analysis. *Nature Nanotechnology*, *6*(10), 615–624. doi:10.1038/nnano.2011.129 PMID:21926981

Venkatesan, B. M., Estrada, D., Banerjee, S., Jin, X., Dorgan, V. E., & Bae, M.-H. et al. (2012). Stacked Graphene-Al_2O_3 Nanopore Sensors for Sensitive Detection of DNA and DNA-Protein Complexes. *ACS Nano*, *6*(1), 441–450. doi:10.1021/nn203769e PMID:22165962

Vogt, P., De Padova, P., Quaresima, C., Avila, J., Frantzeskakis, E., & Asensio, M. C. et al. (2012). Silicene: Compelling Experimental Evidence for Graphenelike Two-Dimensional Silicon. *Physical Review Letters*, *108*(15), 155501. doi:10.1103/PhysRevLett.108.155501 PMID:22587265

Wang, X.-Q., Li, H.-D., & Wang, J.-T. (2012). Induced ferromagnetism in one-side semihydrogenated silicene and germanene. *Physical Chemistry Chemical Physics*, *14*(9), 3031–3036. doi:10.1039/c2cp23385a PMID:22286024

Wanunu, M. (2012). Nanopores: A journey towards DNA sequencing. *Physics of Life Reviews*, *9*(2), 125–158. doi:10.1016/j.plrev.2012.05.010 PMID:22658507

Wells, D. B., Belkin, M., Comer, J., & Aksimentiev, A. (2012). Assessing Graphene Nanopores for Sequencing DNA. *Nano Letters*, *12*(8), 4117–4123. doi:10.1021/nl301655d PMID:22780094

Xie, H., Hu, M., & Bao, H. (2014). Thermal conductivity of silicene from first-principles. *Applied Physics Letters, 104*(13), 131906. doi:10.1063/1.4870586

Xie, P., Xiong, Q., Fang, Y., Qing, Q., & Lieber, C. M. (2012). Local electrical potential detection of DNA by nanowire-nanopore sensors. *Nature Nanotechnology*, *7*(2), 119–125. doi:10.1038/nnano.2011.217 PMID:22157724

Yang, K., Cahangirov, S., Cantarero, A., Rubio, A., & D'Agosta, R. (2014). Thermoelectric properties of atomically thin silicene and germanene nanostructures. *Physical Review B: Condensed Matter and Materials Physics*, *89*(12), 125403. doi:10.1103/PhysRevB.89.125403

Zberecki, K., Wierzbicki, M., Barnaś, J., & Swirkowicz, R. (2013). Thermoelectric effects in silicene nanoribbons. *Physical Review B: Condensed Matter and Materials Physics*, *88*(11), 115404. doi:10.1103/PhysRevB.88.115404

Zhang, C.-, & Yan, S.-. (2012). First-Principles Study of Ferromagnetism in Two-Dimensional Silicene with Hydrogenation. *The Journal of Physical Chemistry C*, *116*(6), 4163–4166. doi:10.1021/jp2104177

Zhang, P., Li, X. D., Hu, C. H., Wu, S. Q., & Zhu, Z. Z. (2011). First-principles studies of the hydrogenation effects in silicene sheets. *Physics Letters. [Part A]*, *376*(14), 1230–1233. doi:10.1016/j.physleta.2012.02.030

Zhang, X., Xie, H., Hu, M., Bao, H., Yue, S., Qin, G., & Su, G. (2014). Thermal conductivity of silicene calculated using an optimized Stillinger-Weber potential. *Physical Review B: Condensed Matter and Materials Physics*, *89*(5), 054310. doi:10.1103/PhysRevB.89.054310

Zhang, Y., Xu, G., Han, F., Wang, Z., & Ge, C. (2010). Preparation and Thermoelectric Properties of Nanoporous Bi2Te3-Based Alloys. *Journal of Electronic Materials*, *39*(9), 1741–1745. doi:10.1007/s11664-010-1199-z

Zhao, Z., Tian, F., Dong, X., Li, Q., Wang, Q., & Wang, H. et al. (2012). Tetragonal Allotrope of Group 14 Elements. *Journal of the American Chemical Society*, *134*(30), 12362–12365. doi:10.1021/ja304380p PMID:22803841

Chapter 4
GAS Sensor Modelling and Simulation

Elnaz Akbari
Universiti Teknologi Malaysia, Malaysia

Mohammad Taghi Ahmadi
Universiti Teknologi Malaysia, Malaysia

Zolkafle Buntat
Universiti Teknologi Malaysia, Malaysia

Hediyeh Karimi
Universiti Teknologi Malaysia, Malaysia

Mohsen Khaledian
Universiti Teknologi Malaysia, Malaysia

ABSTRACT

Both graphene and CNTs experience changes in their electrical conductance when exposed to different gases (such as CO2, NO2, and NH3), and they are, therefore, ideal candidates for sensing/measuring applications. In this research, a set of novel gas sensor models employing Field Effect Transistor structure using these materials have been proposed. In the suggested models, different physical properties such as conductance, capacitance, drift velocity, carrier concentration, and the current-voltage (I-V) characteristics of graphene/CNTs have been employed to model the sensing mechanism. An Artificial Neural Network model has also been developed for the special case of a CNT gas sensor exposed to NH3 to provide a platform to check the accuracy of the models. The performance of the models has been compared with published experimental data which shows a satisfactory agreement.

INTRODUCTION

It is a known fact that every physical phenomenon needs a mathematical representation to allow for an in depth study of the general characteristics and behaviour in real conditions. More importantly, since the experimental study of some phenomena are costly, hazardous, time and energy consuming or due to many other restrictive reasons, the need to develop mathematical models with which one can predict the actual behaviour and characteristics of different physical properties, chemical properties, etc. is inevitable. Therefore, to study the electrical properties of graphene and carbon nanotubes when employed in metal oxide semiconductor field effect transistor (MOSFET) based gas sensors, a set of mathematical models have been proposed and developed for different electrical properties. These include conductance,

DOI: 10.4018/978-1-5225-0736-9.ch004

capacitance, and carrier concentration as well as drift velocity. The modelling procedure has used an initial assumption of the direct proportionality of the gate voltage with gas concentration and temperature.

A single model has also been proposed for CNT based gas sensor utilizing an artificial neural network algorithm to provide a benchmark for comparison with the analytical model for the sake of validation and in order to check the plausibility and viability of the model. Both models will later on be compared with the experimental results from previous experiments by other researchers. The steps taken to develop both models have been presented in the following sections after which the results and the comparative study come.

ANALYTICAL CONDUCTANCE MODEL OF GRAPHENE BASED CO_2GAS SENSOR

In this study, a new model for the gas sensor based on the gradient of GNR conductance as a function of gas concentration has been provided and a FET based structure as a modelling platform is suggested. Lastly, the proposed model is evaluated through a comparison study between model and experimental result. The graphene gas sensor is able to operate under ambient conditions and at room temperature. For different concentrations of carbon dioxide (CO_2) adsorbed on graphene surface, changes in the device conductance have been reported(Yoon et al. 2011). By increasing CO_2 concentration gas from 10 to 100 ppm, the conductance of the graphene gas sensor rises. This sensor benefits from advantages such asfast response time, high sensitivity, low power consumption and short recovery time (Yoon et al. 2011). Since the mid-20th century, CO_2 and other greenhouse gases emissions have increased daily due to the increasing industrial activities. This phenomenon is known as global warming which is caused by the increase of the average temperature of near-surface air as well as oceans on earth. By monitoring and subsequently decreasing CO_2 emissions released by fuel combustion in vehicle engines and burning coal in power plants, the greenhouse gas concentration is expected to decrease. This can illustrate the need for developing low-cost, highly sensitive, adjustable sensors that can be employed to monitor and control the CO_2 concentration in industrial exhaust gases.

With the aim of enhancing the sensing performance, numerous attempts have been made to develop and improve gas sensors. In this study the modelling of features of a CO_2 gas sensor based on graphene application has been carried out. The sensing characteristics of the gas sensor have also been evaluated during its operation on different temperature and concentration values. Efforts were made to determine whether graphene is a viable candidate to be used in a wide range of sensory applications(Bradley 2012, Wei et al. 2012).

The underlying operational principle for MOSFET (Figure 1) is based on the flow of electrons between source and drain electrodes which is controlled by the gate voltage. The efficiency of electron transport is the important quantity which has been defined as conductance (Hwang et al. 2013, Wouter Olthuis 2007).

According to the Landauer formula, graphene conductance, G, is proportional to the carrier transmission probability from one electrode (source) to another (drain) (Datta 1997)equated as:

$$G = \frac{2q^2}{h} T \qquad (1)$$

Figure 1. Proposed structure of gas sensor based graphene

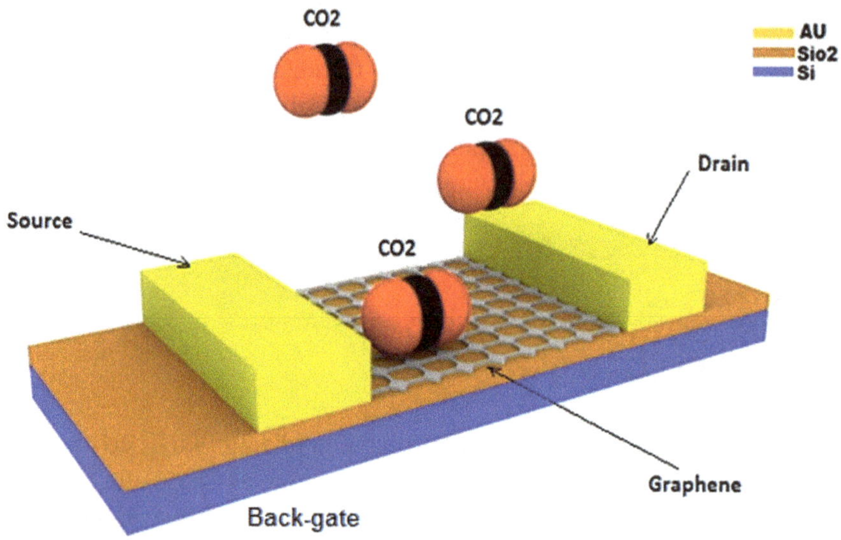

where h is the Planck constant, q is the electron charge and T is the total transmission probability. Taylor expansion has been used to investigate a parabolic relationship involving energy and wave vector(Berger et al. 2006).

$$E(k) \approx \frac{E_g}{2} + \frac{E_g}{4\beta^2} k_x^2 \tag{2}$$

where β is the quantized wave vector given by(Brey and Fertig 2006). The wave vector in parabolic part of the band energy can be extracted as(Ahmadi et al. 2012):

$$k = \sqrt{\frac{4E}{3a_{c-c}t} - 2\beta^2} \tag{3}$$

in which $a_{c-c} = 1.42\,\overset{o}{A}$ represents carbon-carbon(C-C) bond length, and $t = 2.7(ev)$ is the nearest neighbour C-C tight binding overlap energy. The conductance on large channel following the Ohmic scaling law based on Landauer formula can then be obtained as (Ahmadi et al. 2012):

$$G = \frac{6q^2}{hL}\left(a_{c-c} t \pi k_B T\right)^{\frac{1}{2}} \left[\int_0^{+\infty} \frac{x^{-1/2}}{\left(1+e^{x-\eta}\right)} dx + \int_0^{+\infty} \frac{x^{-1/2}}{\left(1+e^{x+\eta}\right)} dx \right] \tag{4}$$

where $x = \frac{E - E_g}{k_B T}$ and the normalized Fermi energy is $\eta = \frac{E_F - E_g}{K_B T}$ in which E_g the band gap energy, K_B Boltzmann constant, *T* is room temperature and E_F is Fermi energy. The performance of CO_2 gas sensor based graphene nanostructure is demonstrated by the conductance characteristic. It has been shown that the conductance changes when the graphene gas sensor is exposed to CO_2 (Yoon et al. 2011). A model has been proposed based on the reported experimental data and the relationship between conductance, gas concentration and temperature.

$$G_{wg} = G_{wogT} + G_{wgF} \tag{5}$$

When the sensor is exposed to gas in different temperatures, two parameters G_{wogT} and G_{wgf} can be defined for its conductance. The first parameter is assumed to represent conductivity changes which depend on *T* parameter. The second parameter G_{wgF} is based on different values of gas concentration with constant temperature. The performance of CO_2 graphene based gas sensor can be demonstrated by the conductance characteristic. It is shown that the conductance of the graphene gas sensor experiences changes when exposed to CO_2 gas at 22°C, 40°C and 60°C, and the ratio of conductance changes to various concentrations of CO_2.

As V_g results in altering conductance of channel, one of the parameters that has a strong influence on gas sensor conductance is gas concentration, thus we can write:

$$V_g \propto F \rightarrow V_g = \alpha_2 F \tag{6}$$

Finally, the conductance model of gas sensor is modified as:

$$G = \alpha_1 T + \frac{6q^2}{hL}\left(a_{c-c} t \pi k_B T\right)^{\frac{1}{2}} \left[\mathcal{I}_{-\frac{1}{2}}\left(\left(\alpha_2 F - V_T\right)q\right) / K_B T + \mathcal{I}_{-\frac{1}{2}}\left(\left(-\alpha_2 F - V_T\right)q\right) / K_B T\right] \tag{7}$$

As presented in Figures 2, 3, and 4 by increasing the *F* parameter, further reaction between the sensor surface and gas molecules will occur, holding the fact that the gas concentration is influenced by the increased number of carriers.

A benchmark of the proposed model coupled with experimental model is shown in Figures 2, 3 and 4 which implies that at the higher temperatures, conductivity dramatically increases when concentration is amplified. Based on the analytical model, α_1 is suggested as the temperature controlling parameter. The analytical model based on the extracted data shows that the rate of changes in conductivity depending on temperature gives better results by:

$$\alpha_1 = aLn\left(T\right) - b \tag{8}$$

where, from the extracted data, parameters a and b are calculated as a=3.4777 and b=1.7725. Also α_2 is introduced as a controlled parameter of gas concentration and shows the rate of changes in conductivity depends on gas concentration which is given by:

Figure 2. Conductance changes vs. gas concentration at T=22°C
Yoon et al. 2011.

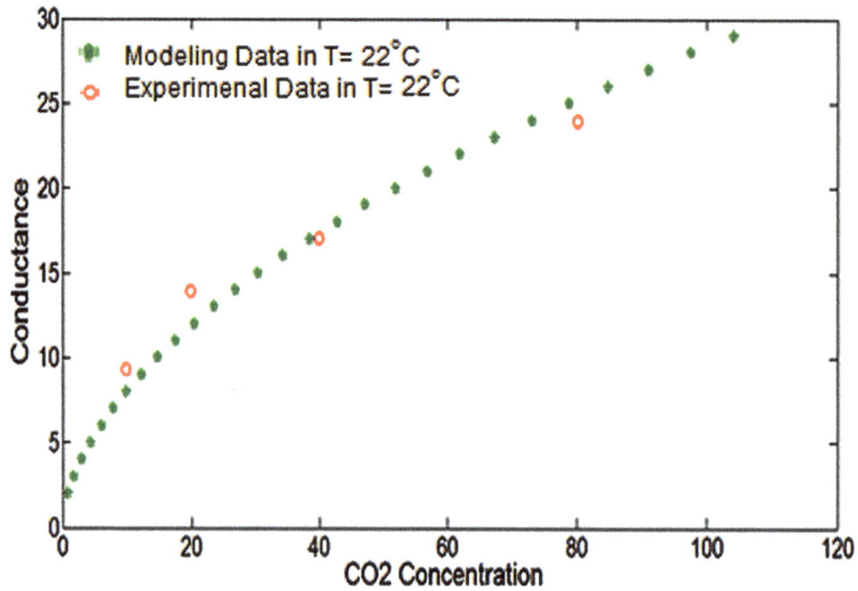

Figure 3. Conductivity changes vs. gas concentration at T=40°C
Yoon et al. 2011.

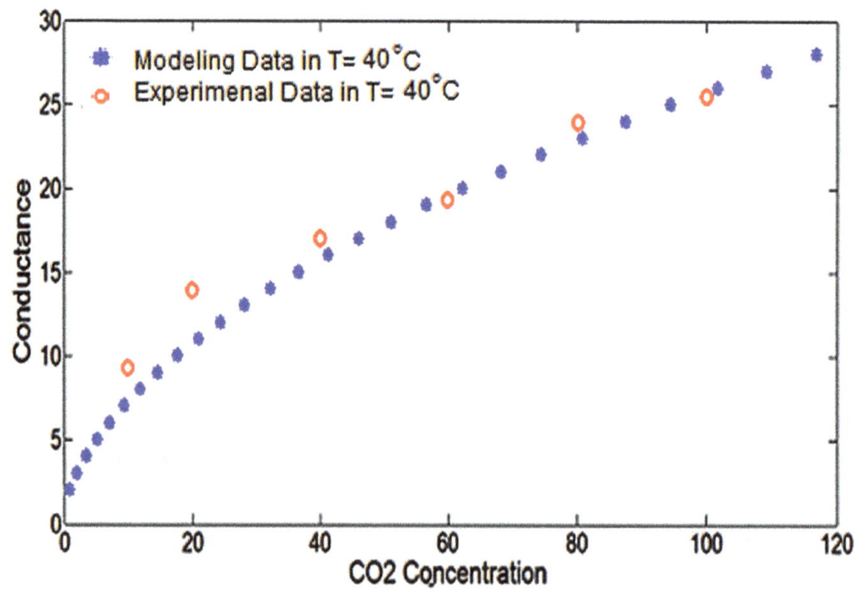

Figure 4. Conductivity changes vs. gas concentration in T=60°C
Yoon et al. 2011.

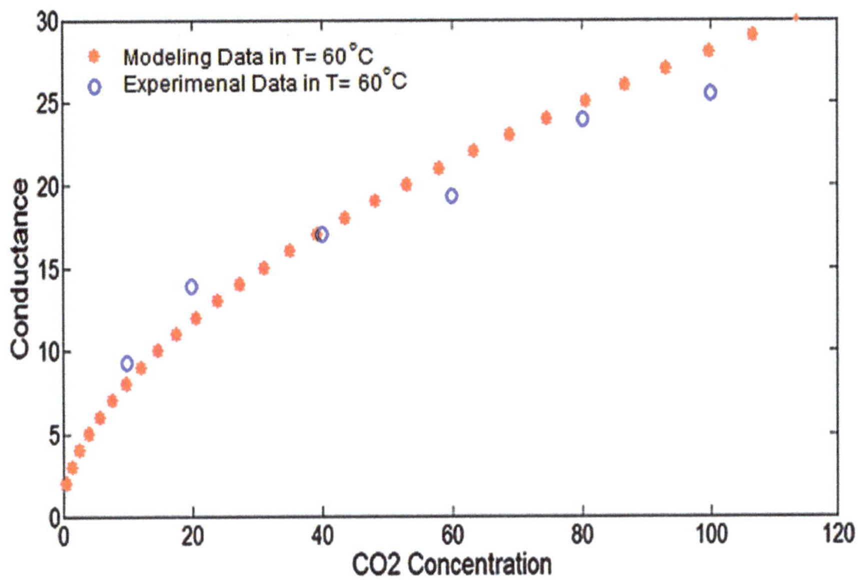

$$\alpha_2 = cLn\left(F\right) + d \tag{9}$$

where in the same manner, the constant parameters have been calculated as: c=4.1211 and d=1.6931

As shown in Figure 5, the performance of gas sensor based on graphene nanostructure is modelled by conductance characteristics. Conductance modelling for different temperatures for gas sensor based graphene is evident in this figure.

Analytical Capacitance Model of Graphene Based Co_2 Gas Sensor

In this section, an analytical model of CO_2 gas sensor based on graphene conductance variation is presented. Capacitance gradient created between the channel and gate is an important property of FET devices considered in this work. Gas concentration variations and their effect on capacitance as a modelling platform are taken into account.

The quantum capacitance characteristic of a monolayer graphene in nanoscale is applied to assess the CO_2 gas detection capability in graphene based gas sensors. In gas sensor modelling, the most important property of FET devices, i.e. their capacitance created between the channel and the gate is employed.

A typical structure of FET based CO_2 sensor is shown in Figure 6, using metallic source and drain contacts and graphene nanoribbon as channel. It has been approved by experimental methods that conductance of graphene is a function carrier density and mobility. In the other words, changes in electron density and/or charge carriers by adsorption of gas molecules or ions in graphene change the conductance (Schwierz 2010a, Sadeghi et al. 2012).

We begin the modelling by carrier concentration given as:

Figure 5. Comparison of conductance modelling for different temperatures
Yoon et al. 2011.

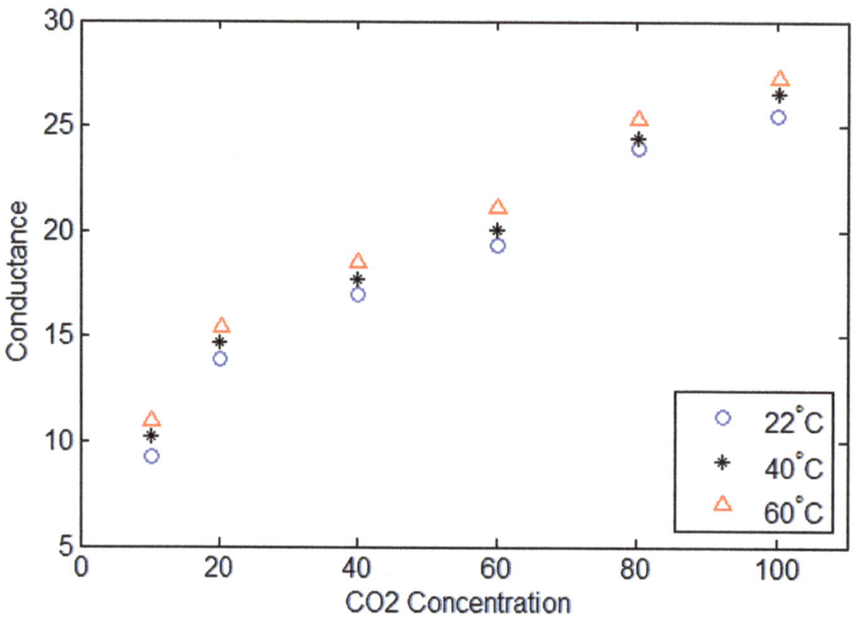

Figure 6. The proposed structure of gas sensor based graphene

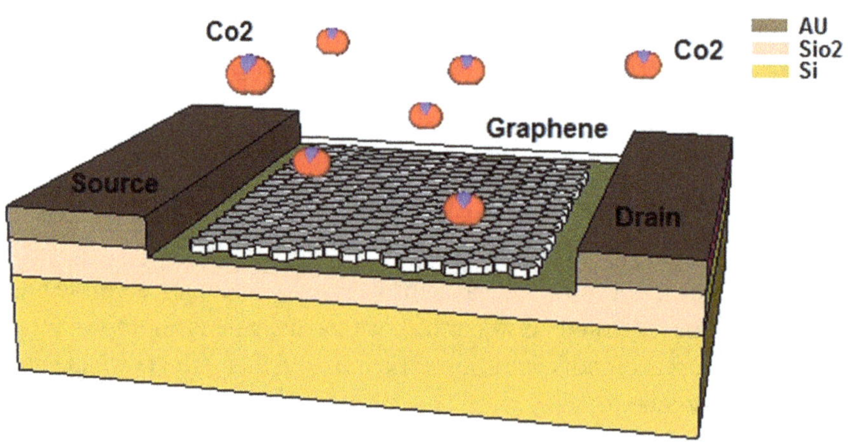

$$n = \int D(E) f(E) d(E) \quad (10)$$

where density of state (DOS) is written as:

$$D(E) = \frac{1}{3\pi t a_{c-c}} \frac{E}{\sqrt{E^2 - \left(\frac{E_g}{2}\right)^2}} \quad (11)$$

E_g the band gap energy which can be written for monolayer graphene as:

$$E_g = 2\pi\sqrt{3}\left(\frac{p}{N+1} - \frac{2}{3}\right) \tag{12}$$

in which *t = 2.7 (eV)* is the nearest neighbor *C-C* narrow binding overlap energy and *p* is an integer number, and the Fermi probability function *f (E)* is defined as:

$$f\left(E\right) = \frac{1}{\exp\left(\frac{E - E_F}{k_B T}\right) + 1} \tag{13}$$

Then the gas concentration is written as:

$$n = \int \frac{1}{3\pi t a_{c-c}} \frac{E}{\sqrt{E^2 - \left(\frac{E_g}{2}\right)^2}} \cdot \frac{1}{\exp\left(\frac{E - E_F}{k_B T}\right) + 1} \cdot d\left(E\right) \tag{14}$$

where *Eg* is band energy and *E* is energy. Now the quantum capacitance can be calculated as:

$$C_q = \frac{\delta Q}{\delta V} \tag{15}$$

where

$$\delta Q = e \cdot \delta n_{Total} \tag{16}$$

In which *Q* is the charge measured in coulombs, *e* is electron charge and $n_{Total} = n + n'$ where, *n* is GNR's intrinsic carrier concentration and *n'* is the carrier concentration when the CO_2 gas is injected. $\partial V = \frac{\partial E}{e}$ is the voltage applied to the device and C_q is given as (Shylau, Kłos and Zozoulenko 2009):

$$C_q = \frac{e \partial n_{Total}}{\frac{\partial E}{e}} \tag{17}$$

which becomes by a simple mathematical rearrangement:

$$C_q = e^2 \frac{\partial n_{Total}}{\partial E} \tag{18}$$

Using the definition of n_{total}, the following equation can be written for quantum capacitance:

$$C_q = e^2 \frac{\partial n}{\partial E} + e^2 \frac{\partial n'}{\partial E} \tag{19}$$

Carrier concentration (*n'*) is a function of gas concentration and temperature, for which the equation is given as:

$$n' = \alpha F_g + \beta T \tag{20}$$

where *F* and *T* are the carrier concentration and temperature parameters, respectively, and α and β are constant values. As a result, the quantum capacitance can be written as:

$$C_q = e^2 \cdot \frac{1}{3\pi t a_{c-c}} \frac{E}{\sqrt{E^2 - \left(\frac{E_g}{2}\right)^2}} \cdot \frac{1}{\exp\left(\frac{E - E_F}{k_B T}\right) + 1} + e^2 \alpha \frac{\partial F}{\partial t} \frac{\partial t}{\partial E} + e^2 \beta \frac{\partial T}{\partial t} \frac{\partial t}{\partial E} \tag{21}$$

where ∂F is the gas concentration variation and ∂T is a different temperature value; therefore, the conductance characteristic is given as:

$$G = \left(e^2 \cdot \frac{1}{3\pi t a_{c-c}} \frac{E}{\sqrt{E^2 - \left(\frac{E_g}{2}\right)^2}} \cdot \frac{1}{\exp\left(\frac{E - E_F}{k_B T}\right) + 1} + e^2 \alpha \frac{\partial F}{\partial t} \frac{\partial t}{\partial E} + e^2 \beta \frac{\partial T}{\partial t} \frac{\partial t}{\partial E} \right) \cdot V_{ds} \cdot \mu_e \tag{22}$$

In Equation 22, conductance is calculated from quantum capacitance, drain-source voltage (V_{ds}), and electron mobility. The operation of CO_2 gas sensor based on graphene can be characterized by its conductivity. As it was previously presented, when the gas sensor based on graphene is exposed to CO_2 at various temperatures, such as 22°C, 40°C and 60°C, the rate of conductivity will change due to he consequent different gas concentrations. As shown in Figures 7, 8 and 9, by increasing gas concentration parameter (*F*), more reactions among the gas molecules and the surface of sensor will happen.

It has been shown in Figures 7, 8, 9 that the conduction is enhanced via increasing temperature and gas concentration. The comparison between the proposed model and experimental pattern has been demonstrated in both of which increasing temperature and gas concentration causes conductivity enhancement. According to the analytical model, α_1 is proposed as the controlling parameter of gas concentration. The analytic model based on data extracted can be written as follows:

Figure 7. Conductivity changes vs. gas concentration in T=22°C
Yoon et al. 2011.

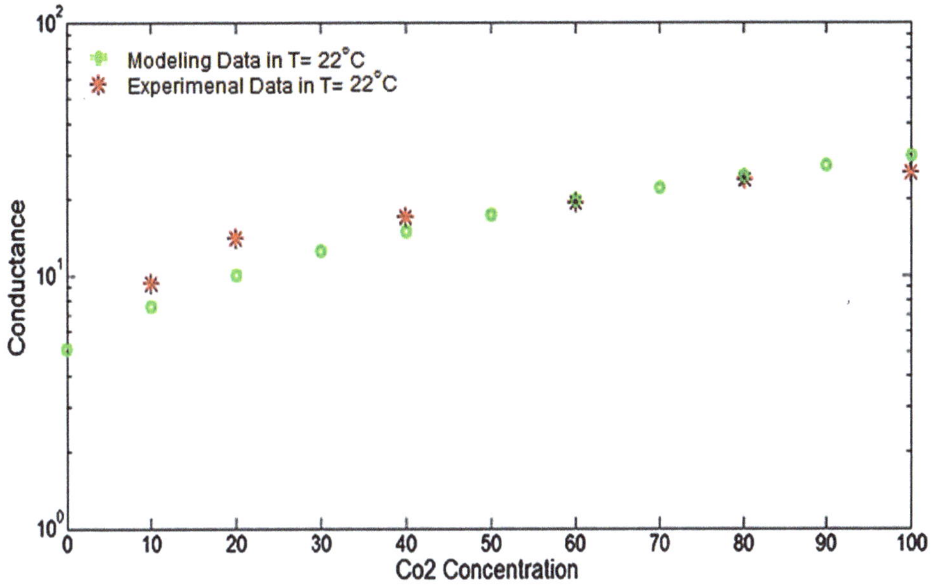

Figure 8. Conductivity changes vs. gas concentration in T=40°C
Yoon et al. 2011.

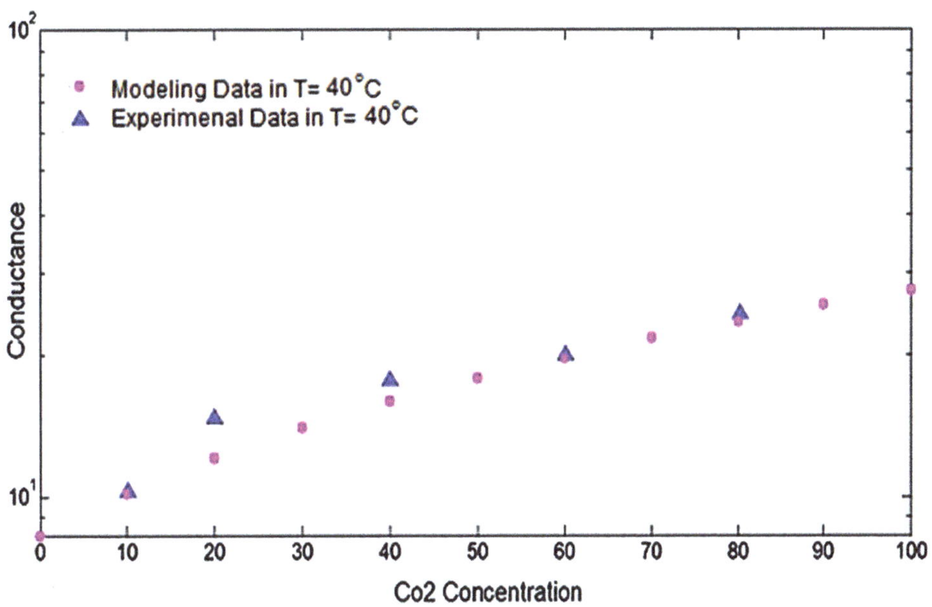

Figure 9. Conductivity changes vs. gas concentration in T=60°C
Yoon et al. 2011.

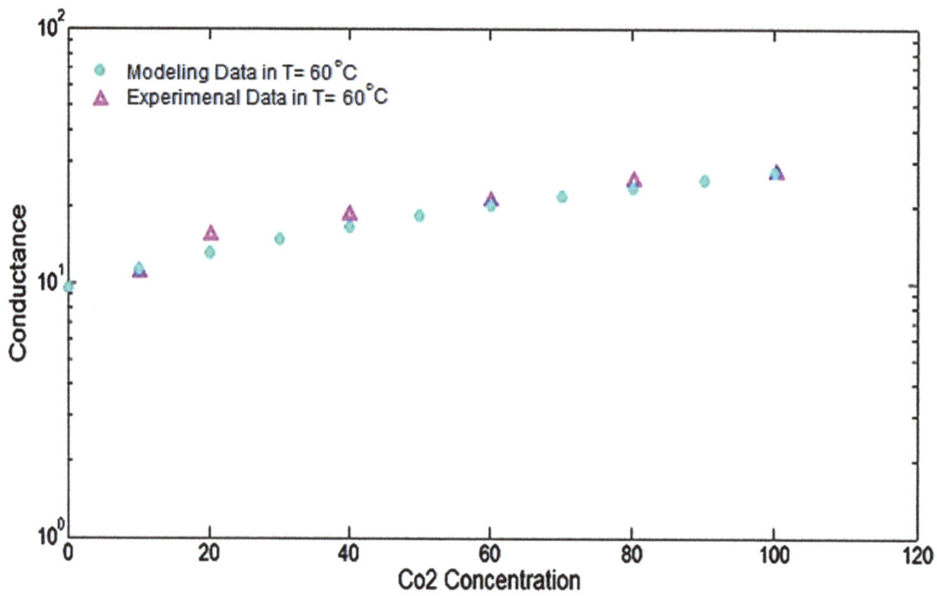

$$\alpha = 28.377Ln\left(F\right) + 19.954 \tag{23}$$

According to the analytical model, the electrical conductivity will improve by increasing the amount of gas concentration. According to the extracted data parameters and are calculated as a=28.377 and b=19.954.

Furthermore, temperature is another controlling parameter that is shown by α_2 and has influence on the rate of conductivity. As demonstrated in the following equation, c and d are two constant parameters obtained from the extracted data:

$$\beta = -0.243Ln\left(T\right) + 0.9184 \tag{24}$$

In the same method as before, c and d are calculated as: c =-0.243 and d=0.9184

Analytical Modelling of Bilayer Graphene Based on No_2 Gas Sensor Carrier Velocity

Bilayer graphene (BLG) is the stack of two graphene layers, and due to its special features such as electrical, physical and optical properties, it is known as an appropriate material to be used in nanotechnology especially in the sensor area (Das, Chakraborty and Sood 2011, Lam, Lee and Liang 2009).

Twisted configuration and Bernal stacking are two common structures of bilayer graphene. In the former, the layers are rotating toward each other while in the latter, half of the carbon atoms in one layer are standing over the other half (Mousavi et al. 2011, Mohammad Javad Kiani, 2012). The electrical and optical features of bilayer graphene can be influenced by orientation and stacking order. The AA-stacked configuration is metallic, but the AB-stacked, which is in our focus, behaves like a semiconductor material (Hatami et al. 2011, Yu et al. 2010). In AB structure of bilayer graphene with hexagonal carbon lattice, the atoms located on top layer of BLG are A_1 and A_2 whereas atoms on the bottom layer of BLG are labelled B_1 and B_2 as shown in Figure 10.

One of the most interesting properties in BLG is its band structure. Through theoretical studies it has been assumed that, by applying a perpendicular electric field, the band gap can be induced by reducing the asymmetry of two graphene layers in the BLG (Mousavi et al. 2011). The controllable bandgap is one of the most excellent properties of BLG that makes it a promising material in nanotechnology. In Figure 11(a), the band structure of the unbiased BLG with no external perpendicular electric fields is shown. It demonstrates that the BLG is a zero gap semiconductor with four parabolic bands, where two inner bands contact each other near the Dirac points at zero energy and the two outer bands are separated by the inter-layer hopping energy, ± t (Amin et al. 2011, Ahmadi et al. 2010b).

In the biased BLG, as it is shown in Figure 11(b), by applying a perpendicular electric field a bandgap is opened which is given by (Nilsson et al. 2008):

$$E_g = \frac{Vt_\perp}{\sqrt{V^2 + t_\perp^2}} \tag{25}$$

where $V=V_1-V_2$, is the potential energy difference between first and second layers and V_1 and V_2 are the potential energy of first and second layers, respectively. The parameter V can be controlled externally; therefore, by adjusting V, the band gap can be tuned (Ahmadi et al. 2012).

Figure 10. The schematic of bilayer graphene (AB stacked configuration)

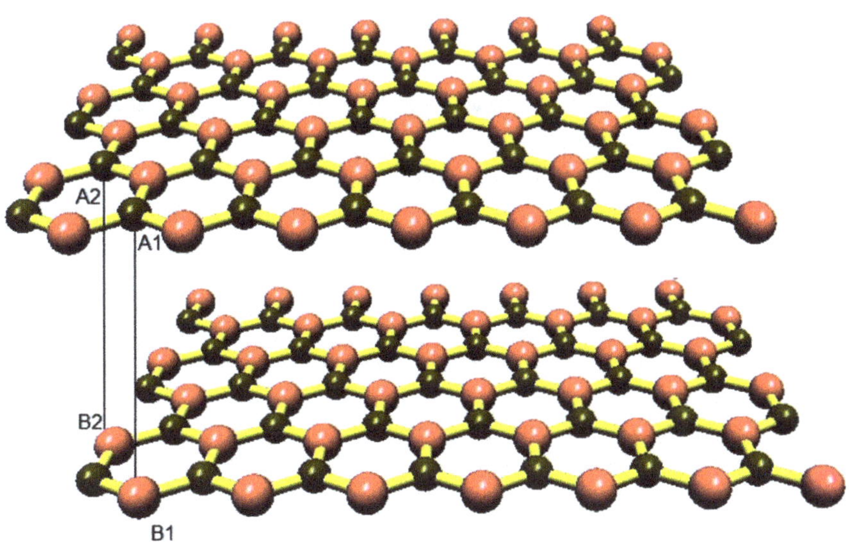

Figure 11. Band structure of BLG near the Dirac points for, a) V = 0 (unbiased BLG), b) V≠ 0 (biased BLG)
Ahmadi et al. 2012.

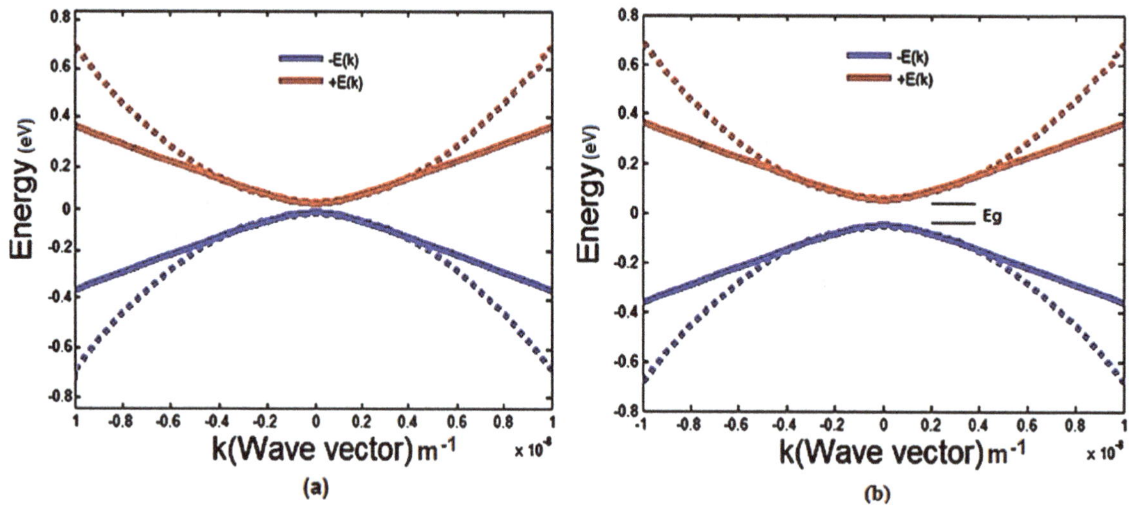

The Proposed Model

The sensitivity of graphene to the miniature applied voltage can be used in sensor technology. In this model, the bilayer graphene as a substrate of gas sensor has been used. As can be seen in the Figure 12, it looks similar to the conventional field effect transistors (FET) which include a source metal, a drain metal, a silicon back gate and a gate insulator (Wouter Olthuis 2007). A graphene channel connects the source and drain electrodes, and a dielectric barrier layer (SiO_2) separates the gate from the channel, SiO_2 is used under the graphene as a dielectric layer while silicon acts as a back gate. When gas molecules attach to the surface or edges of CNT, carrier concentration will change. Due to this variability, the drain source current is a measurable parameter. This platform is employed in our model as a FET based sensor structure (schwierz 2010b).

The threshold voltage V_{TH} of a MOSFET is usually defined as the gate voltage where an inversion layer forms at the interface between the insulating layer (oxide) and the substrate (body) of the transistor. MOSFET is a device used for amplifying or switching electronic signals. When the gate-source voltage is smaller than the threshold voltage $V_{GS}<V_{TH}$, (there will be no conduction between the source and the drain and the switch will be off. In contrast, when $V_{GS}>V_{TH}$, the gate will attract electrons, and the switch will be on.

The tight-binding method has been used to calculate the biased energy of BLGs, which indicates energy dispersion of BLG as (Martina Cheli 2009, Rahmani et al. 2013):

$$E(K) = \frac{V_1 + V_2}{2} \pm \sqrt{\in_k^2 + \frac{V^2}{4} + \frac{t_\perp^2}{2} \pm \frac{1}{2}\sqrt{4\left(V^2 + t_\perp^2\right)\in_k^2 + t_\perp^4}} \tag{26}$$

Figure 12. Proposed structure of gas sensor based on bilayer graphene

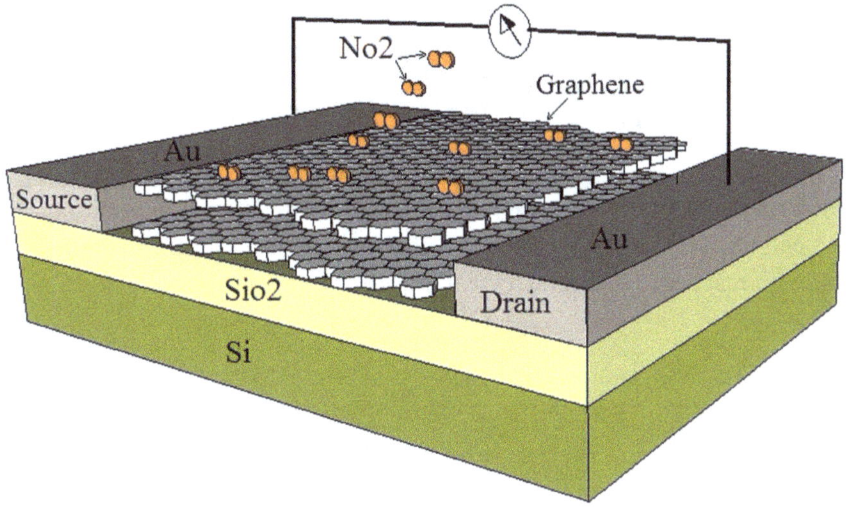

where ε_k is electron's dispersion in monolayer graphene and $t_\perp = 0.32ev$ is the inter-layer hopping energy. The wave vector in which the smallest gap is observed is given by:

$$k_g = \frac{V}{2\nu_F \hbar}\sqrt{\frac{V^2 + 2t_\perp^2}{V^2 + t_\perp^2}} \tag{27}$$

where the $v_f \approx 1 \times 106 m \cdot s^{-1}$ is the Fermi velocity and ħ is the reduced Planck's constant. Near k_g the energy dispersion can be written as (J. Nilsson 2006).

$$E(k) = \frac{E_g}{2} + \frac{\hbar^2}{2m^*}\left(|k| - k_g\right)^2 + \frac{V_{1+}V_2}{2} \tag{28}$$

where $k = k_x \hat{l} + k_y \hat{J}$ and m* is the effective mass in BLG. The density of State for BLG which indicates the number of states for each interval of energy at each energy level that can be occupied by electrons can be written as:

$$DOS = \frac{" n}{A" E} = \left(\frac{\hbar^2}{m^*}\frac{\left(k - k_g\right)}{k}2\pi\right)^{-1} = \frac{km^*}{2\pi\hbar^2\left(k - k_g\right)} \tag{29}$$

The velocity of electrons is directly proportional to the value of DOS at any instance. Average velocity of carriers in first sub band in bilayer graphene can be obtained by accumulative velocity of all carriers divided by number of carriers. The carrier drift velocity has been reported to be formulated in the form (Ahmadi et al. 2008):

$$\nu = \int \frac{|\nu| DOS(E) F(E) dE}{n} \tag{30}$$

From the kinetic energy principle $|\nu| = \sqrt{\frac{2E}{m}}$, $DOS(E)$, is density of states, and $F(E) = \frac{1}{1+e^{\frac{E-E_f}{K_B T}}}$ is the Fermi-Dirac distribution function which gives the probability of occupation of a state at any energy level and n is carrier concentration (Ahmadi and Ismail 2010). Hence, the integral of the numerator in Equation 30 with respect to E can be rewritten as:

$$\int |\nu| DOS(E) F(E) dE = \frac{1}{\pi\hbar^2} \int \frac{Ek}{(k-k_g)} \frac{1}{1+e\hat{}\left(\frac{E-E_f}{k_B T}\right)} dE \tag{31}$$

where $k = \pm \frac{2m^*(E-E_c)^{1/2}}{\hbar^2} + k_g$, giving

$$\int |\nu| DOS(E) F(E) dE = \frac{1}{\pi\hbar^2} \left(\int \frac{E}{1+e^{\frac{E-E_f}{K_B T}}} dE + \frac{k_g \hbar}{\sqrt{2m^*}} \int \frac{(E-E_c)^{-\frac{1}{2}}}{1+e^{\frac{E-E_f}{K_B T}}} dE \right) \tag{32}$$

It has been attempted to write Equation 32 in a form which can be solved using Fermi integrals. The obtained equation is as follows:

$$\int |\nu| DOS(E) F(E) dE = \frac{(K_B T)^2}{\pi\hbar^2} \int \frac{x}{1+e^{(x-\eta)}} dx + \frac{K_B T E_c}{\pi\hbar^2} \int \frac{dx}{1+e^{(x-\eta)}} + \frac{K_B T k_g \hbar}{\pi\hbar^2 \sqrt{2m} \sqrt{K_B T}} \int \frac{x^{-1/2}}{1+e^{(x-\eta)}} dx \tag{33}$$

In which the order of in the numerator of the integrands are 1, 0, and -1/2; thus making the integrals equivalent to Fermi integrals of order 1, 0, and -1/2, respectively. This can finally give Equation 34 as (Kim and Lundstrom 2008):

$$\nu = \left[\frac{(K_B T)^2}{\pi\hbar^2} f(1) + \frac{K_B T E_c}{\pi\hbar^2} f(0) + \frac{K_B T k_g}{\pi\hbar\sqrt{2mK_B T}} f\left(-\frac{1}{2}\right) \right] \Big/ n \tag{34}$$

In which f(1), f(0) and f(-1/2) are Fermi integrals of order 1, 0 and -1/2, respectively. For details on the derivation and mathematical representation of Fermi integrals, the reader is referred to (Kim and Lundstrom 2008). Velocity is evaluated in Figure 13 based on Equation 34 (Saeidmanesh et al. 2013).

As can be seen in Figure 14, when the sensor is exposed to gas, according to the chemical reaction between graphene and gas molecules, graphene experiences a change in the velocity of its carriers which can in turn cause alterations in the current and channel voltage. In other terms, the electron exchange between the gas and the surface of the graphene creates new carriers which change the velocity of electrons (Xia et al. 2009).

It can be concluded that the velocity of electrons in the presence of gas is equal to that of the no gas state plus the velocity change when graphene is exposed to gas.

$$\nu_{withgas} = \nu_{withoutgas} + \nu_{gasinjected} \tag{35}$$

From this, the current-voltage relation can be derived as follows:

$$I = nqvA \tag{36}$$

where n is charge carrier density, q is electrical charge, V is drift velocity of the charge carriers and A is the area which the charges are moving. In Figure 15, the current voltage characteristic of intrinsic graphene exposed to ambient air only, as well as the I-V characteristic of BLG based NO_2 sensor for graphene under 100ppm and 500ppm NO_2 concentration is plotted. By current voltage characteristic of

Figure 13. The velocity of BLG
Ko et al. 2010.

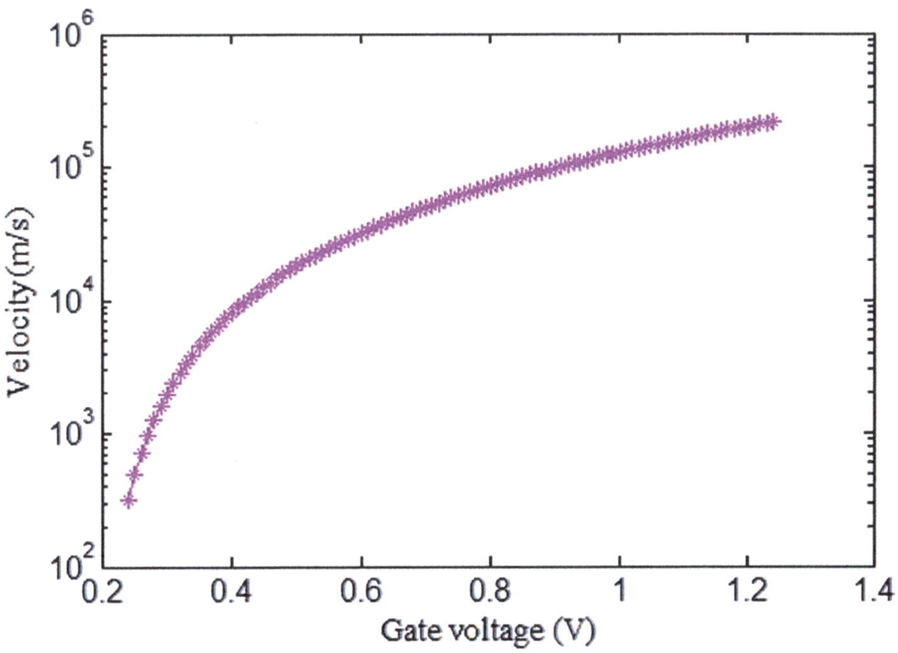

Figure 14. Schematic of NO_2 adsorption processes by surface area of graphene

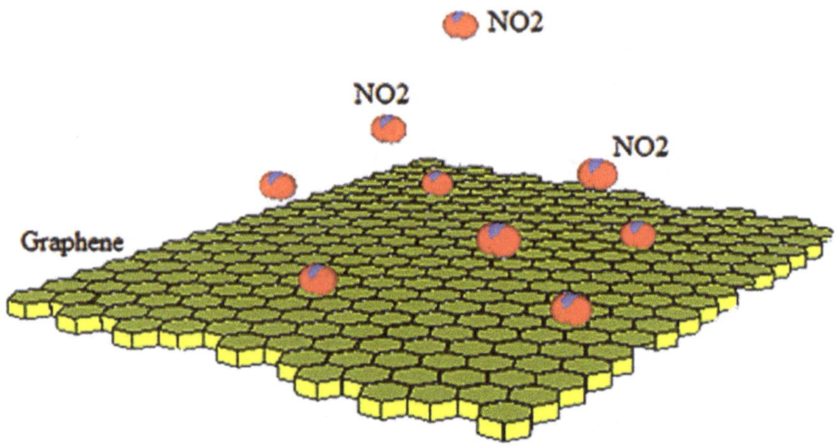

presented model, it is demonstrated that current of gas sensor raise with increasing the NO_2 concentration. Therefore, it is notable that based on the supposed model sensor, I-V characteristic is controlled by NO_2 concentration.

When the sensor is exposed to the gas the density of state can be divided into two parts; one is the density of state without gas $DOS_{WOG,}$ and the second parameter is the density of state with gas proportional to αF, which depends on different values of NO_2 gas concentration.

Figure 15. The comparison of current–voltage characteristics for 100ppm and 500ppm and intrinsic graphene based on proposed model
Ko et al. 2010.

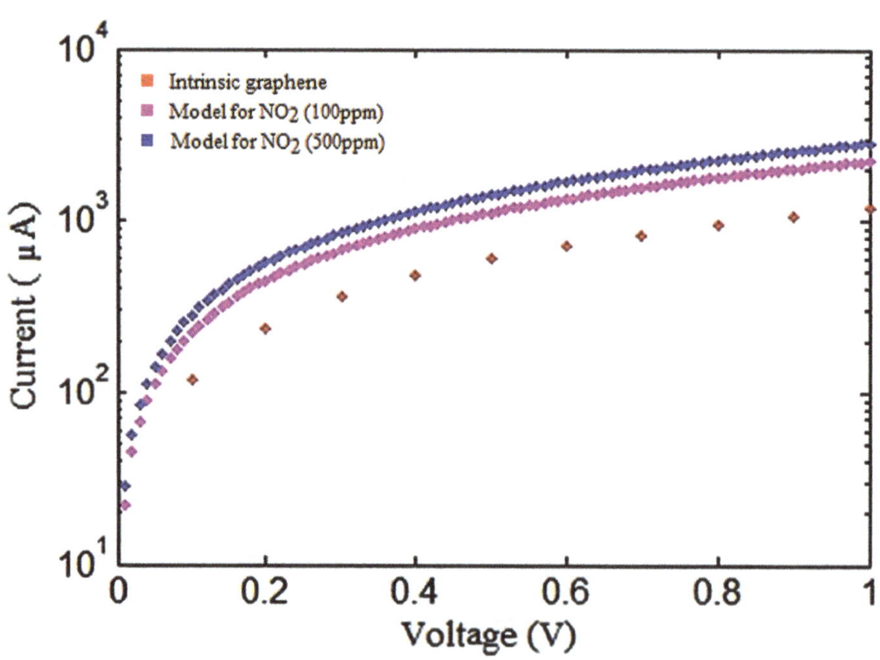

$$DOS = DOS_{WOG} + DOS_{gasinjected} \tag{37}$$

where

$$DOS_{gasinjected} \approx \alpha F \tag{38}$$

In this study, α is the control parameter, i.e. by heuristically changing the values of α, the author has attempted to set the results from the proposed model as close to the experimental results as possible. According to the relation between control parameter and density of state we can write:

$$\nu_{total} = \left[\begin{array}{l} \frac{(K_B T)^2}{\pi \hbar^2} f(1) + \frac{K_B T E_c}{\pi \hbar^2} f(0) + \frac{K_B T k_g}{\pi \hbar \sqrt{2 m K_B T}} f\left(-\frac{1}{2}\right) \\ + \frac{2\alpha F}{m^*} \left(K_B T\right)^2 f(1) + \frac{2\alpha F}{m^*} K_B T E_c f(0) \end{array} \right] \Big/ n \tag{39}$$

where n is the carrier concentration equal to $n=10^{16}$ for two dimensional bilayer graphene (M. Saeid-manesh 2013(In press)).

The velocity of the graphene-based FET devices is influenced by the number of carriers changing in the channel. FET-based graphene with high sensitivity was applied to detect the NO_2 gas, based on velocity variations. As depicted in Figures 16 and 17 the velocity of channel will change due to adsorption of the NO_2 to the surface of FET channel. The performance of graphene based gas sensor under exposure to 100 ppm and 500 ppm of NO_2 gas is evaluated and the analytical results of the proposed model for gas sensor with appropriate parameters are compared with the experimental data extracted from (Ko et al. 2010) which shows a good agreement. It is evident in the figure that the points calculated and obtained from our model satisfactorily coincide with the measurement data.

In the suggested model, different carrier concentrations are demonstrated in the form of αparameter, which is presented in Table 1.

According to the analytical model, αis proposed as the controlling parameter of gas concentration. The analytic model based on data extracted can be written as follows:

$$\alpha = a e^{bx} \tag{40}$$

Referring to the analytical model, the velocity will be enhanced as the amount of gas concentration increases. According to the extracted data, parameters a and b are calculated as a=0.370 and b=-1.57.

Table 1. Different F values with α parameter

Carrier Concentration (F)	Parameter (α)
100 ppm	0.077
500 ppm	0.016

Figure 16. The current–voltage characteristics for 100ppm
Ko et al. 2010.

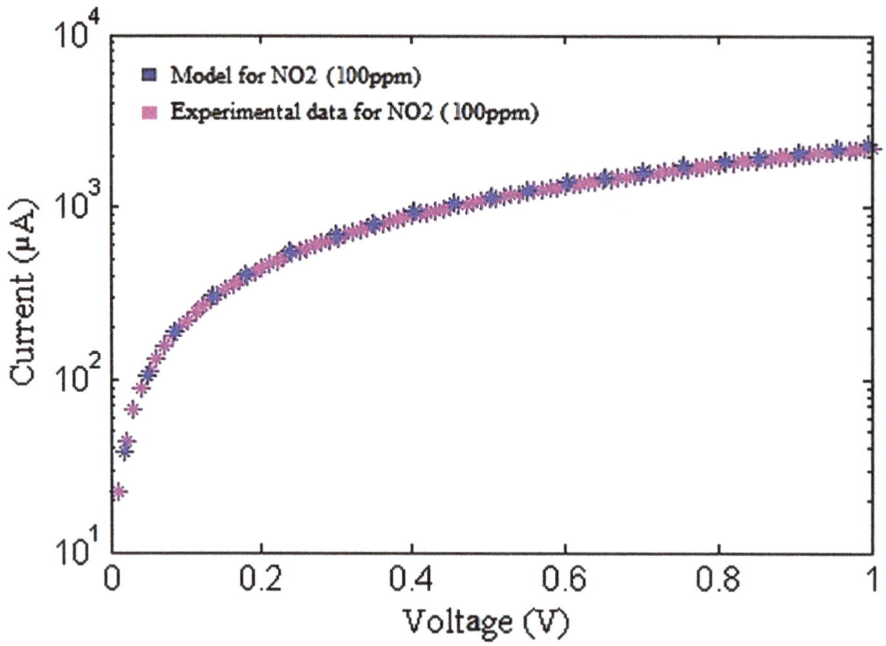

Figure 17. The current–voltage characteristics for 500ppm
Ko et al. 2010.

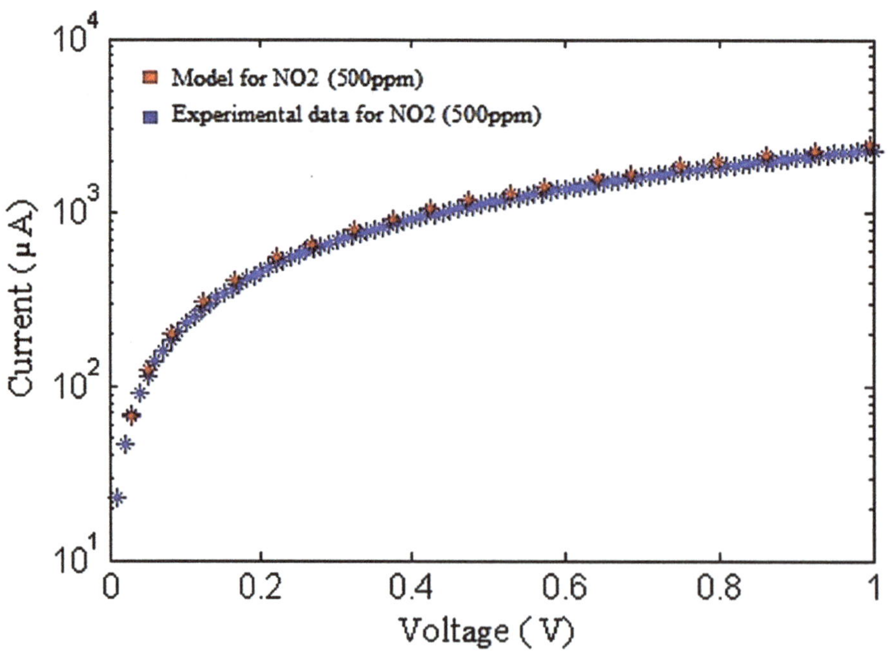

It is evident in Figure 18 that the I–V characteristic curve can be controlled by the carrier concentration factor. The current rises as a result of increase in the carrier concentration.

ARTIFICIAL NEURAL NETWORK SIMULATION AND ANALYTICAL MODELLING OF CURRENT-VOLTAGE CHARACTERISTICS IN CARBON NANOTUBES BASED NH_3 GAS SENSOR

In this study, an artificial neural network (ANN) algorithm has also been incorporated to obtain another model for the I-V characteristic in which the experimental data extracted from a recent work by N. Peng *et al.* (2009) has been used as the training data set. The comparative study of the results from ANN as well as the analytical models with the experimental data in hand show a satisfactory agreement that validates the proposed models.

In order to carry out a comparative study, it has been attempted to implement an artificial intelligence method to develop another model. Artificial neural network as one of the most accurate and powerful intelligent schemes has been chosen as the tool in this step. The results obtained from the constructed ANN are then compared to those from the analytical model as well as the experimental data to check which approach provides better levels of accuracy (Svozil, Kvasnicka and Pospichal 1997, Zurada 1992).

ANN is an intelligent algorithm which has been developed based on an analogy to biological nervous system. Various types and structures of ANNs have been employed in scientific analytical studies among which, the most common and comprehensible is one consisting of interconnected group of neurons which are mathematical operator units called Perceptrons (Abolgasim 2008, Patterson 1998).

Figure 18. I-V characteristic of different values of NO_2 carrier concentration
Ko et al. 2010.

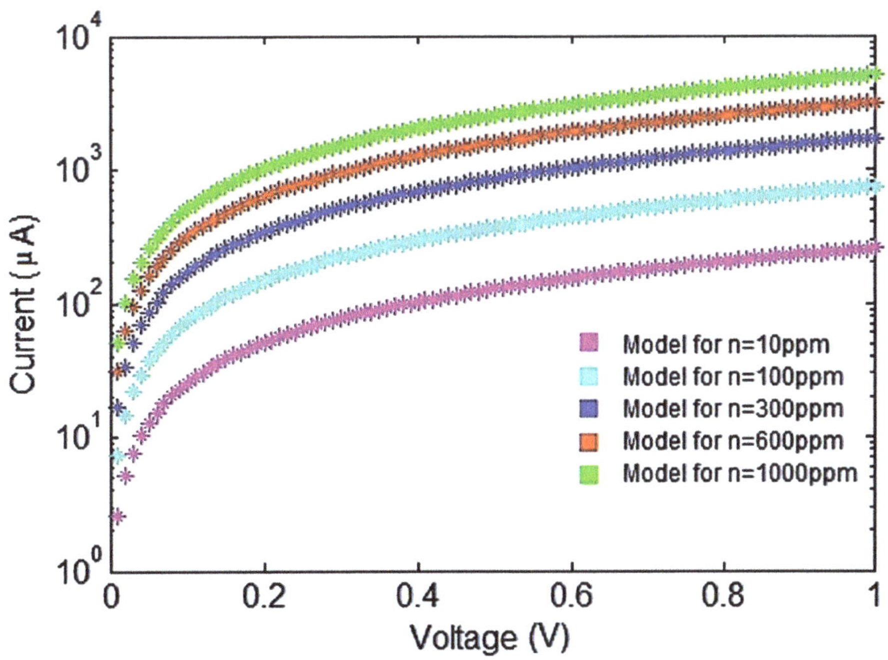

A schematic of the typical structure of an ANN comprising perceptrons is provided in Figure 19. Each input value to a perceptron is multiplied by a weight and added to a bias value. The general mechanism of ANN includes a process in which a set of input data are introduced to the network. Through a series of mathematical calculations based on predefined "activation functions" for each neuron or "node", an output value is given by the ANN. In order for the ANN to learn to compute the optimum output value, a training data set is introduced to the network before the actual inputs are given to the artificial neural network (Zhang 2000, Uhrig 1995).

Based on a predefined learning algorithm, the ANN updates the weights and bias corresponding to each node using the error between the calculated results from the actual inputs and the desired output (Grossi and Buscema 2007, Yao 1999). In this study, Feed-forward structure for the neural network with back propagation learning algorithm has been implemented. The experimental data has been used as the training data set has been employed for validation and testing. The results show satisfactory agreement between the results from the proposed model with the experimental counterparts.

However, the first step to take in this regard is to develop the corresponding analytical model which defines the I-V characteristics of the CNT based gas sensor.

Analytical Model

It has been attempted to model the CNT band structure beginning with modelling the single layer graphene band structure. Employing the Taylor series expansion near the Fermi points, the energy dispersion relation can be derived as follows (Ahmadi et al. 2010c, Ahmadi et al. 2010b).

Figure 19. Simple artificial neural networks

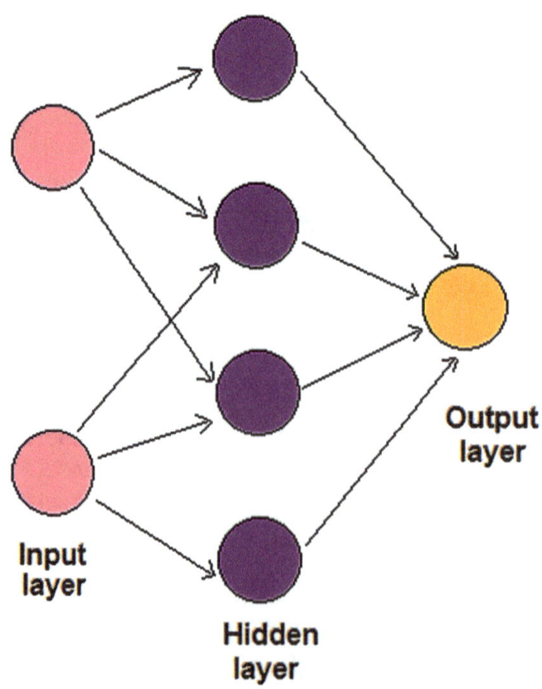

$$E\left(k\right) = \pm \frac{t3a_{c-c}}{2}\sqrt{\left(\frac{2}{3d}\right)^{2} + k_{x}^{2}} \tag{41}$$

where $a_{c-c} = 1.42\overset{o}{A}$ represents the length of carbon-carbon(C-C) bond, is the CNT diameter, t=2.7eV denotes the nearest neighbour C-C tight binding overlap energy, and the ±symbol has been included to account for conductance and valence bands. One can simply write for the first band gap energy $E_{G} = 2a_{c-c}t/d = 0.8$ (eV.nm/d(nm). Also, since the band structure is parabolic near the k=0 points, it can be written as:

$$E\left(k\right) \approx \frac{E_{G}}{2} + \frac{\int^{2}k_{x}^{2}}{2m^{*}} \tag{42}$$

where $\hbar$ is the reduced Plank constant, m* is the effective mass of the CNT depending upon the tube diameter, k_x represents longitudinal wave vector component (Ahmadi et al. 2010d, Anantram and Leonard 2006). The number of conduction channels in the energy E is defined as:

$$M\left(E\right) = 2\frac{\Delta E}{\Delta kL} = \frac{3a_{c-c}t}{L}\left(\frac{4E}{3a_{c-c}t} - \frac{8}{9d}\right)^{\frac{1}{2}} \tag{43}$$

where L denotes the channel length. Two major factors contribute to the conductance effect on large channels, enabling it to follow the Ohmic scaling law based on Landauer formula. The first factor independent of length is the interface resistance. The second one results from the fact that the relation between the conductance and the width is nonlinear and is dependent upon the number of modes in the conductor. However, these modes are the quantized parameters in the Landauer formula in which both factors are interrelated as demonstrated by Equation (44) (Ahmadi et al. 2010a):

$$G = \frac{2q^{2}}{h}\int_{-\infty}^{+\infty} dEM\left(E\right)T\left(E\right)\left(-\frac{df}{dE}\right) \tag{44}$$

where h represents the Plank constant, q denotes the electron charge and T is the transmission probability of an injected electron through the channel approximated as T(E)= 1 in ballistic channels (Datta 2002). Owing to the fact that the express df/dE is noticeable only near the Fermi energy, the conductance can be obtained by considering the Fermi–Dirac distribution functionas (Peres, Neto and Guinea 2006).

$$G = \frac{2q^{2}}{h}\frac{3a_{c-c}t}{L}\left(\frac{4}{3a_{c-c}t}\right)^{\frac{1}{2}}\int_{-\infty}^{+\infty}\left(E - \frac{2a_{c-c}t}{3d^{2}}\right)^{\frac{1}{2}} d\left(-\frac{1}{1+e^{(E-E_{F})/K_{B}T}}\right) \tag{45}$$

Changing the integral boundaries as follows, Equation 45 can be rewritten as:

$$G=\frac{4q^2}{hL}\left(3a_{c-c}t\pi k_BT\right)^{\frac{1}{2}}\left[\int_0^{+\infty}\frac{x^{-1/2}}{\left(1+e^{x-\eta}\right)}dx+\int_0^{+\infty}\frac{x^{-1/2}}{\left(1+e^{x+\eta}\right)}dx\right] \tag{46}$$

where $x=(E-E_g)/k_BT$ and the normalized Fermi energy is given by $\eta=(E_F-E_g)/K_BT$. This equation can be numerically solved by incorporating the partial integration method (Dingle and Dingle 1973, Zaharah et al. 2010, Kažukauskas et al. 2008). The general model for the conductance of carbon nanotube-based gas sensor can be derived similar to that of silicon based model proposed by (Gunlycke, Areshkin and White 2007).

$$G=\frac{4q^2}{hL}\left(3a_{c-c}t\pi k_BT\right)^{\frac{1}{2}}\left[\mathcal{I}_{-\frac{1}{2}}\left(\eta\right)+\mathcal{I}_{-\frac{1}{2}}\left(-\eta\right)\right] \tag{47}$$

The conductance characteristic demonstrates the performance of NH_3 gas sensor based on CNT nanostructure. It has been revealed that when the CNT gas sensor is exposed to $NH_{3,}$ the conductance changes (Yoon et al. 2011). A model has been proposed based on the reported experimental data and the relationship between conductance, gas concentration and temperatures follows (Xia et al. 2009):

$$G_{wg}=G_{wog}+G_{wgT}+G_{wgF} \tag{48}$$

When the sensor is exposed to the gases in different temperatures, we can define three parameters for conductance, namely G_{wog}, G_{wgT} and G_{wgF}. The first parameter, G_{WOG} is the conductance without gas; GwgT is assumed as the conductivity changes depending on T parameter, and the last parameter, $G_{wgF,}$ is based on different values of gas concentration with constant temperature. It is shown that when CNT gas sensor is exposed to NH_3, the conductance levels changes with respect to temperature and varying concentrations. As E_g results in varying conductance of channel, the parameters that have a strong influence on gas sensor conductance are the gas concentration as well as gas temperature. It has been shown that E_g depends on temperature and gas concentration; therefore, one can write:

$$\begin{Bmatrix}E_g\propto F\\E_g\propto T\end{Bmatrix}\Rightarrow E_g=T+F \tag{49}$$

$$G_{wog}=\frac{4q^2}{hL}\left(3a_{c-c}t\pi k_BT\right)^{\frac{1}{2}}\left[\mathcal{I}_{\frac{-1}{2}}\left(\left(E_g-E\right)_T*q\right)/k_BT+\mathcal{I}_{\frac{-1}{2}}\left(-\left(E_g-E_T\right)*q\right)/k_BT\right] \tag{50}$$

Finally, Equations 49 and 50 are employed to obtain the Conductance model of gas sensor as Equation 51:

$$G_{wg} = \frac{4q^2}{hL}\left(3a_{c-c}t\pi k_B T\right)^{\frac{1}{2}} \left[\begin{array}{l} \mathcal{I}_{\frac{-1}{2}}\left(\left(T + F - E_T\right) * q\right) / k_B T \\ + \mathcal{I}_{\frac{-1}{2}}\left(-\left(T + F - E_T\right) * q\right) / k_B T \end{array} \right] \quad (51)$$

Based on the current-voltage characteristic of graphene based FET devices, the performance of the gas sensor can be evaluated by Equation 52. Assuming that the source and substrate terminals are kept in ground potential, and applying a small voltage between source and drain (V_{DC}), the channel region experiences a flow of electrons. Moreover, the relationship between current and conductance can be replaced by Fermi-Dirac integral shape of general conductance model of SWCNT as:

$$I = \left[\begin{array}{l} \frac{4q^2}{hL}\left(3a_{c-c}t\pi k_B T\right)^{\frac{1}{2}} \left(\mathcal{I}_{\frac{-1}{2}}\left(\left(T + F - E_T\right) * q\right) / k_B T \right. \\ \left. + \mathcal{I}_{\frac{-1}{2}}\left(-\left(T + F - E_T\right) * q\right) / k_B T \right) \end{array} \right] * \left(V_{gs} - V_t\right) \quad (52)$$

where V_{gs} is the voltage between the gate and the source and V_t denotes the threshold voltage. I-V characteristic of the proposed model in comparison with experimental results is depicted in Figures (a_1) to (g_1). An increase in the current can be associated to the charge transfer between CNT and NH_3 molecules when the NH_3 molecules are the donors. This phenomenon is also known as chemical doping by gas molecules. It clearly gives an illustration of the fact that there is a good agreement between the proposed model and the extracted data (Peng et al. 2009). In the suggested model, different temperature and concentration values are demonstrated in the form of α and β parameters, respectively, to create an agreement with reported data, which is tabulated in Table 2.

According to the analytical model, α is suggested as the temperature control parameter and is obtained by iteration method. The analytical model based on the extracted data in our study shows that the rate of changes in conductivity depending on temperature gives better results by:

$$\alpha = aLn\left(T\right) - b \quad (53)$$

Table 2. Different temperature and concentration values with α and β parameters

T(°C)	F (ppm)	α	β
25	500	-4	0.03
50	500	-2	0.03
100	500	-1	0.03
150	500	-0.8	0.03
200	100	-0.5	0.01
200	200	-0.5	0.02
200	500	-0.5	0.03

Parameters a and b are extracted as a=0.012 and b=0.046. Also, β defined as a controlled parameter of gas concentration which calculated by iterative method and shows the rate of change in conductivity depends on gas concentration given by:

$$\beta = cLn\left(F\right) - d \tag{54}$$

where the constants are calculated in the same manner as c=1.622 and d=8.814.

ANN Based Model

The proposed artificial neural network has been developed incorporating a network comprising three layers: one input, one output and one hidden layer. The hidden layer consisted of three nodes and feed-forward structure has been employed for the ANN. MATLAB software was utilized for programming and the experimental data were used as the training data set as well as testing data. The built-in neural network tool in MATLAB randomly selects part of the input data for training and the rest is employed for testing. The values for the weights and bias are also randomly chosen by the software and updated in each epoch using the back-propagation learning algorithm. For each set of input data, the corresponding plots of the I-V points as well as the regression graph were plotted. The results are provided in Figures 20 and 21.

The diagrams depicting the I-V characteristic of CNT corresponding to different gas temperatures at 500 ppm concentration are illustrated in Figures 20a to d. The values associated with the analytical model, as well as ANN are compared with those extracted from experimental study.

As a consequence of the chemical interaction between the NH3 molecules and the resultant adsorption on the CNT surface which causes electrical charge to be transferred between them and hence changes the carrier concentration, the channel conductivity varies during the process. As observed from Figures 20a to d corresponding to temperatures of 25, 50, 100, and 150 degrees, respectively, the conductance as a measure of the I-V characteristic has increased at higher temperatures. It is also evident from Figures 20a to d that the proposed ANN model gives to hand better and more accurate estimates of the actual CNT performance in the presence of gas than those provided by the analytical model. This is verified by the fact that the regression values during the calculations of I-V points with ANN are remarkably close to 1. Figures 21a to c depict the I-V characteristics of CNT at 200 degrees in gas concentrations equal to 100, 200, and 500 ppm, respectively.

Physical and chemical phenomena similar to the previous experiments occur in these cases. The illustrations reveal the fact that when the gas concentration is higher, the CNT conductivity increases. Also in these cases, satisfactory agreement between the ANN results as well as the outstanding value of regression almost equal to 1 prove the ANN model to be superior to the analytical counterpart. This has been shown in Figures 21a to c.

Figure 20a. Comparative study of proposed analytical and ANN models with experimental data under 500ppm, at: 25°C, and corresponding ANN regression graph
Peng et al. 2009.

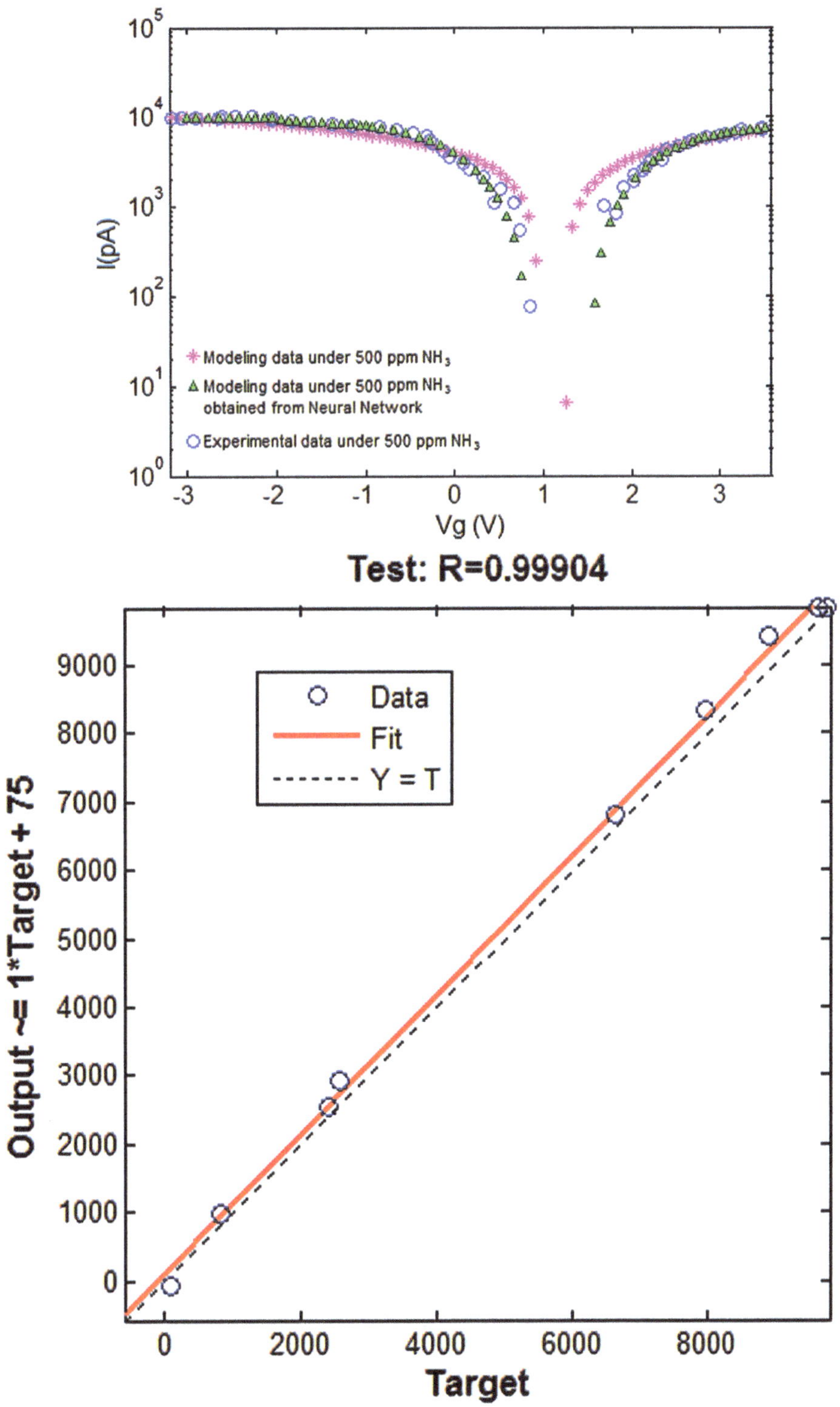

Figure 20b. Comparative study of proposed analytical and ANN models with experimental data under 500ppm, at: 50°C, and corresponding ANN regression graph
Peng et al. 2009.

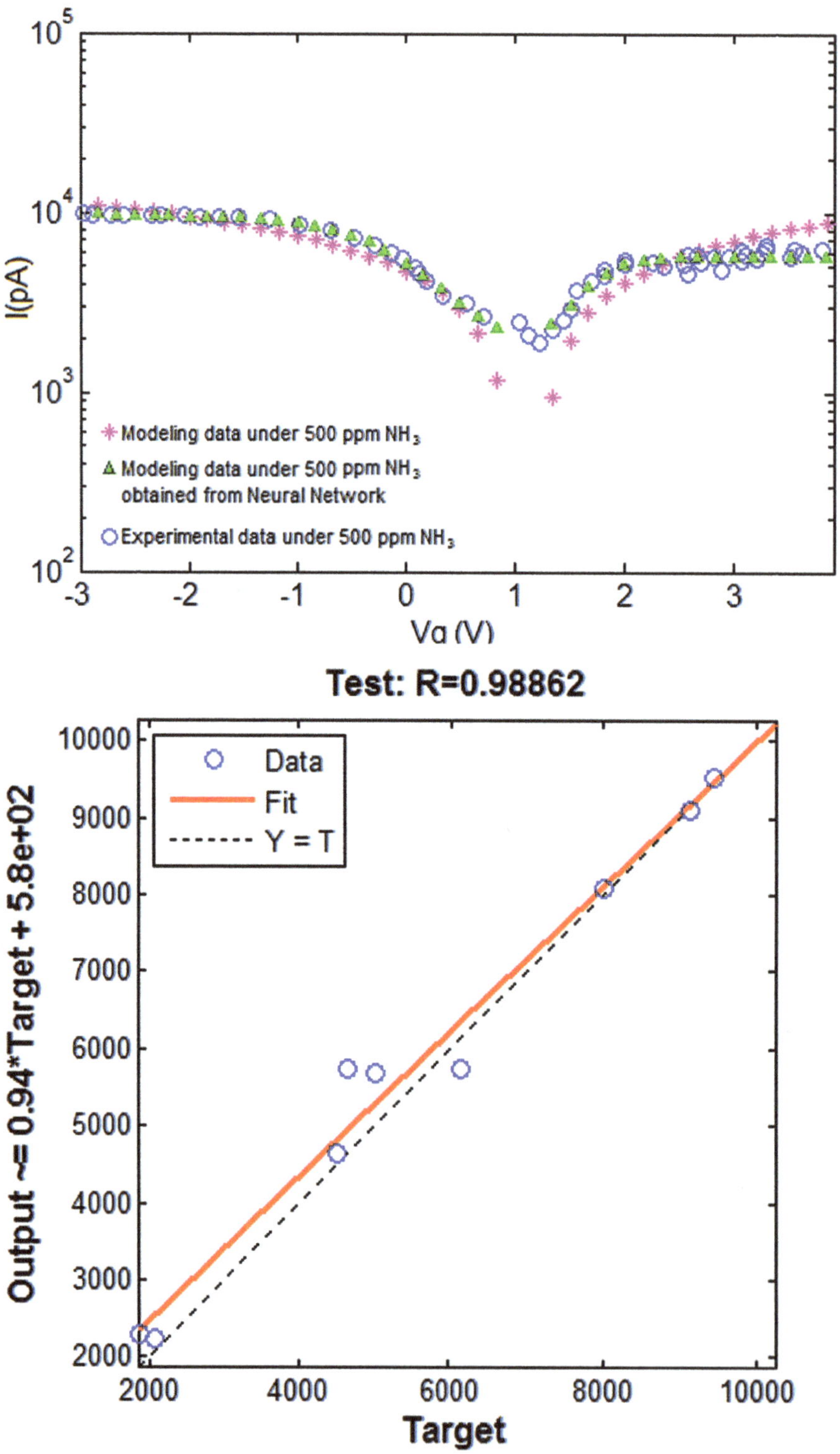

Figure 20c. Comparative study of proposed analytical and ANN models with experimental data under 500ppm, at: 100°C, and corresponding ANN regression graph
Peng et al. 2009.

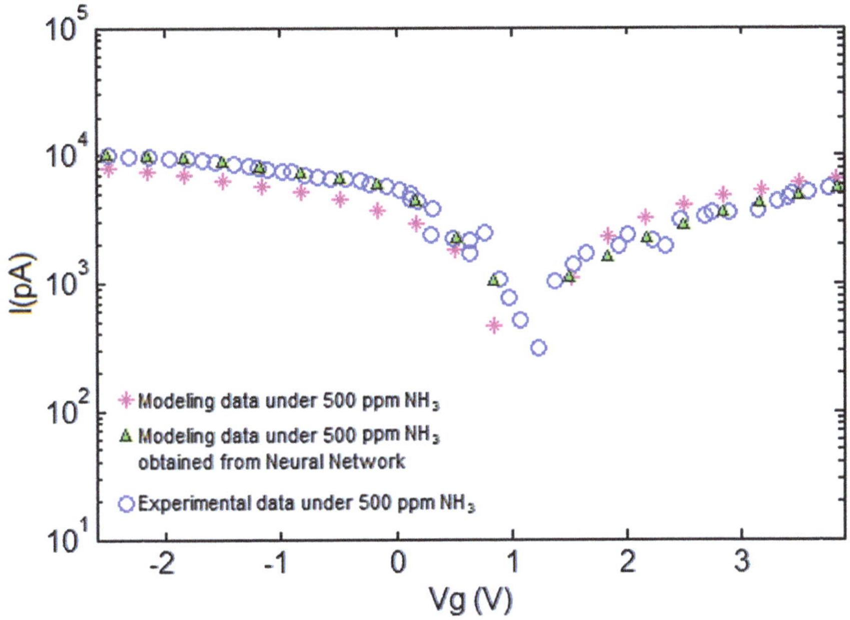

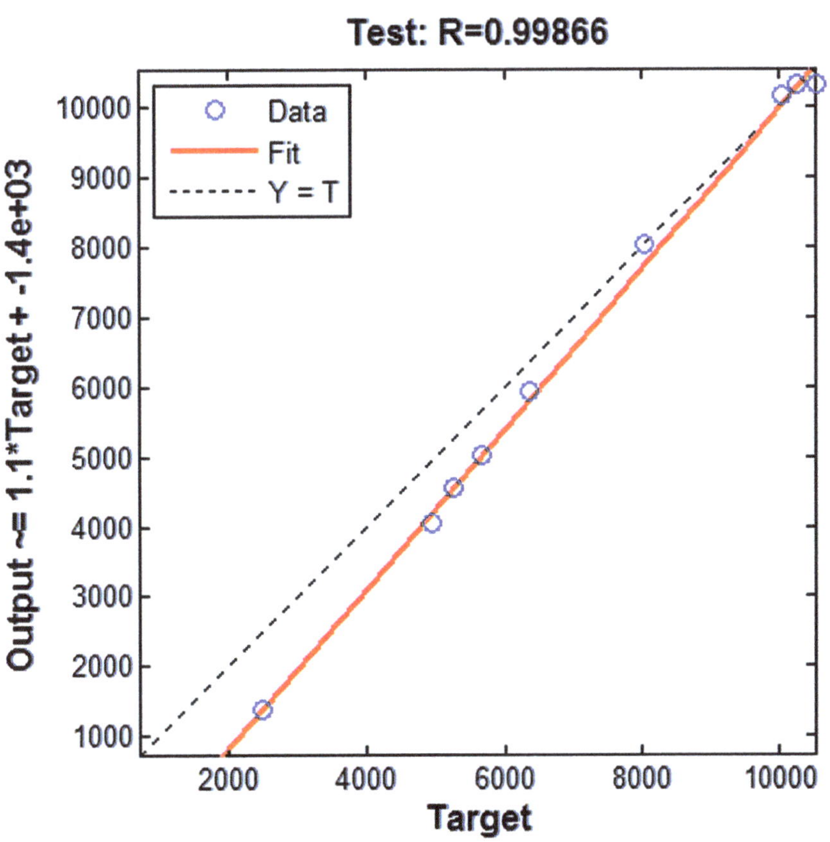

Figure 20d. Comparative study of proposed analytical and ANN models with experimental data under 500ppm, at: 150°C, and corresponding ANN regression graph
Peng et al. 2009.

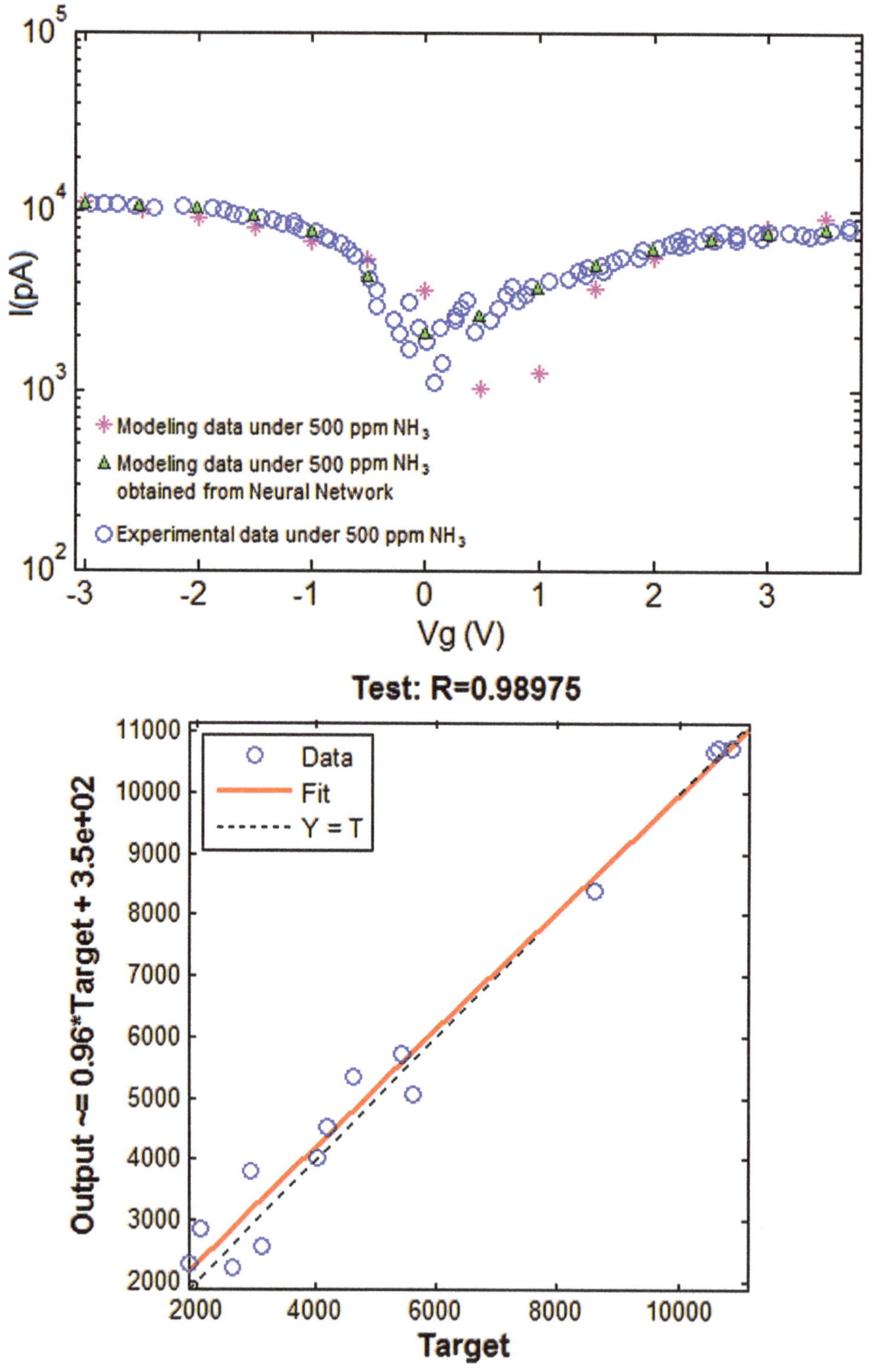

Figure 21a. Comparative study of proposed analytical and ANN models with experimental data at T=200°C under 100 ppm, and the corresponding ANN regression graph
Peng et al. 2009.

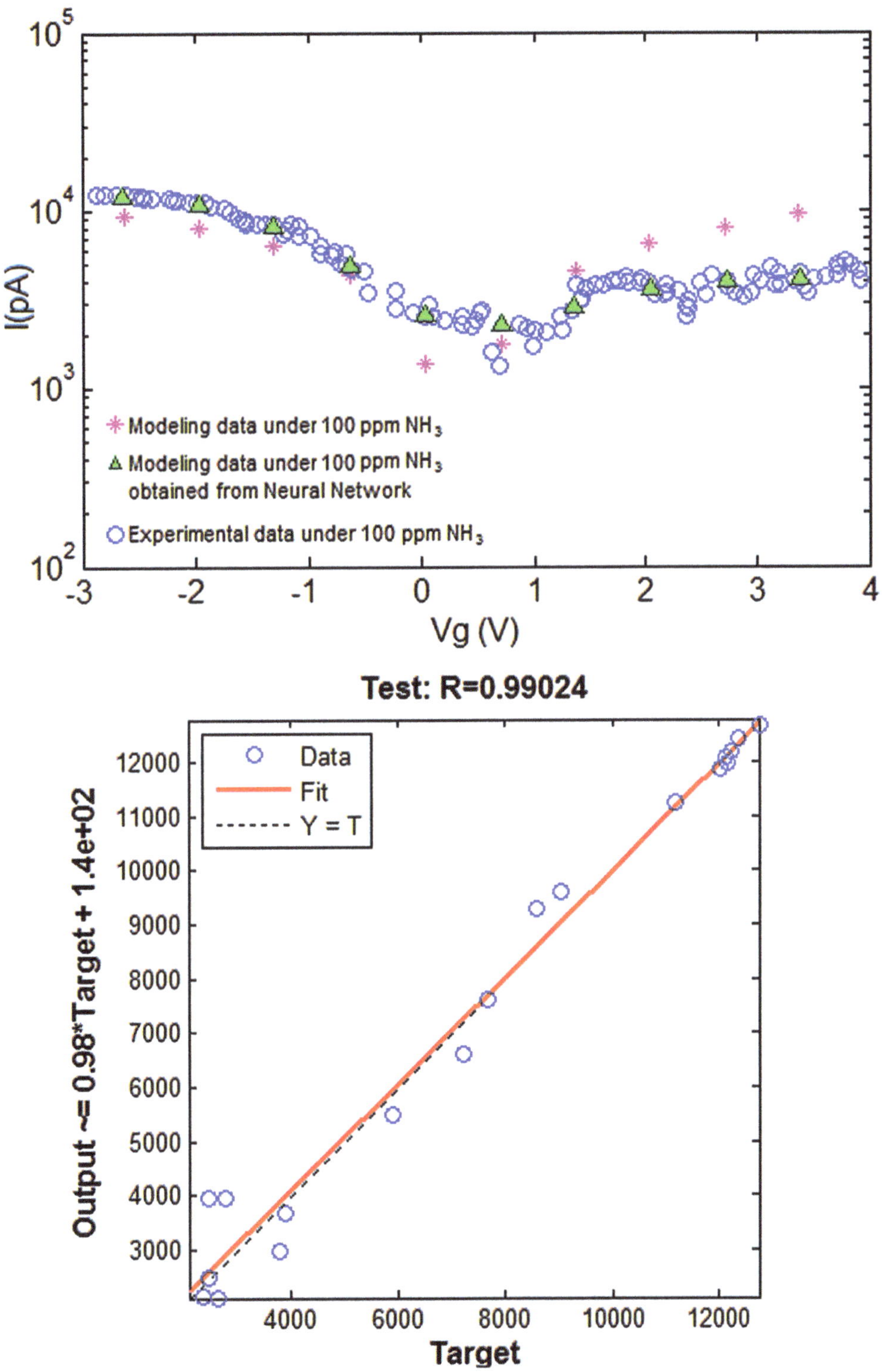

Figure 21b. Comparative study of proposed analytical and ANN models with experimental data at T=200°C under 200ppm, and the corresponding ANN regression graph
Peng et al. 2009.

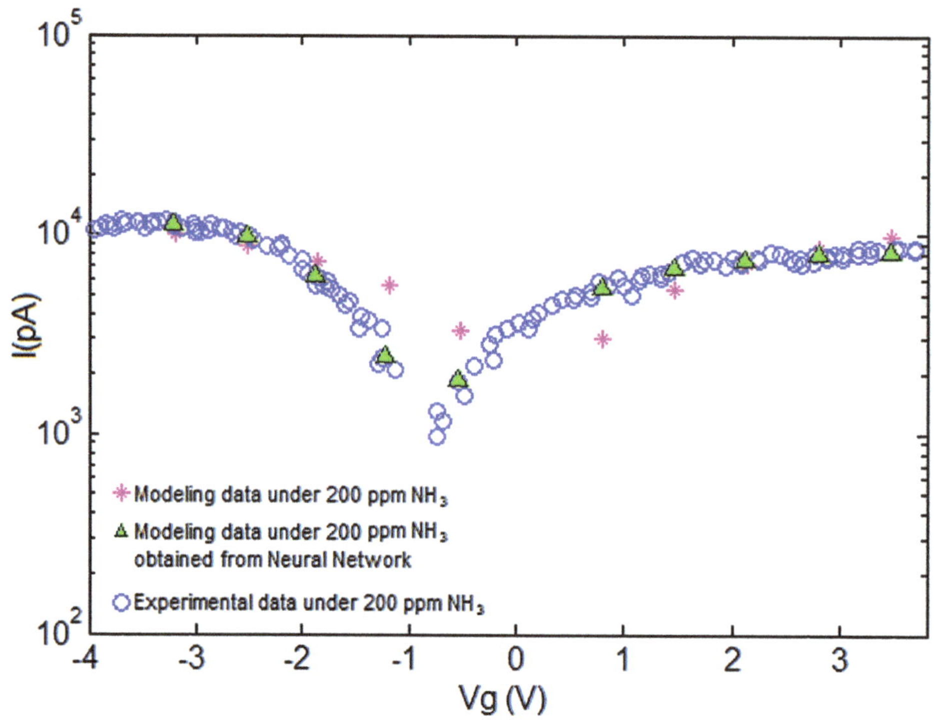

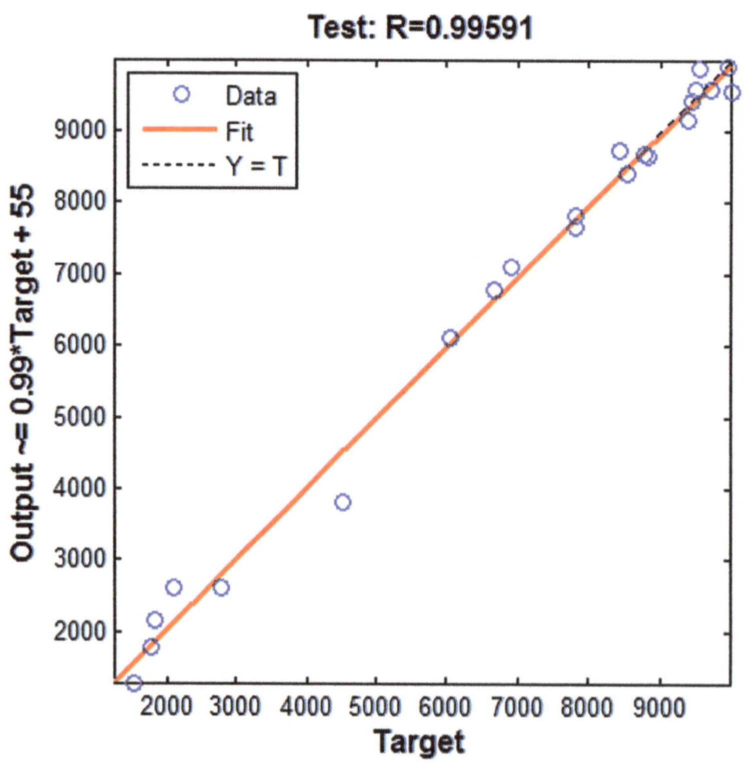

Figure 21c. Comparative study of proposed analytical and ANN models with experimental data at T=200°C under 500ppm, and the corresponding ANN regression graph
Peng et al. 2009.

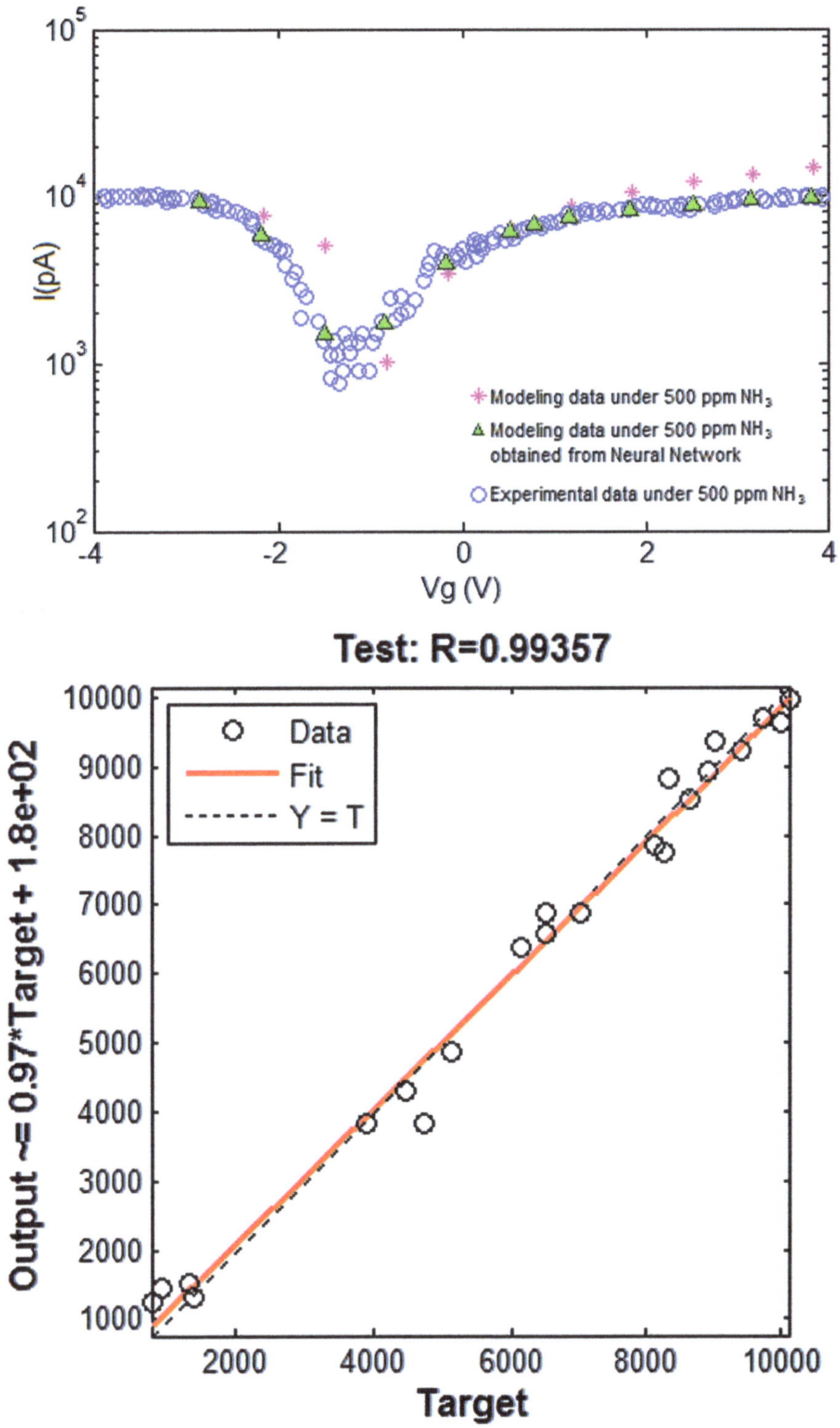

AN ANALYTICAL APPROACH TO EVALUATE THE PERFORMANCE OF GRAPHENE AND CARBON NANOTUBES FOR NH_3 GAS SENSOR APPLICATIONS

CNTs were first discovered by Sumio Iijima in 1991 (Iijima 1991)and have ever since been extensively studied. Single-walled carbon nanotube (SWCNT) is formed by rolling up a honeycomb lattice of single atomic carbon sheet, i.e. graphene along a specific axis (Karamdel et al. 2008), known as chiral direction. The diameter of a typical CNT is around a few nanometers and its length can be over a micrometer; making it distinctly one-dimensional (1D) in its conductance with propagating quantum waves in the quasi free direction along the length of tube. Standing quantum wave are formed in the periphery of the tube forming the cylinder. A CNT is known to have a very high electrical and thermal conductivity as well as Young's modulus giving it the mechanical strength. The applications of CNTs are broad due to their compact structure; finding applications in transistors, sensors, solar cells, fuel cells, etc. (Yoneda).

Andre Geim and Konstantin Novoselov (Geim and Novoselov 2007) discuss several applications of graphene, as one of carbon's allotropes, which can be described as a single atomic layer of graphite. In this material, a two-dimensional honeycomb structure of sp^2-bonded carbon atoms is tightly packed in a lattice structure (Mousavi 2011). Due to its zero band gap energy, graphene has high electron mobility at room temperature. Electron transfer in graphene is 100 times faster than that in silicon. A zero band gap with massless Dirac fermions makes graphene theoretically lossless, making it a perfect two-dimensional (2D) semiconductor (Yang et al. 2007, Son, Cohen and Louie 2006, Amiri, Afroozeh and Bahadoran 2011). Due to the abovementioned outstanding characteristics, graphene and CNT are being used as possible candidates for high performance gas sensors. When integrated in the sensor circuit as illustrated in Figure 22, the detection signals are obtained through the changes in the I-V characteristics of graphene/CNT when exposed to gas. Operational amplifiers amplify these signals that can be changed to digital format for digital signal processing.

As presented in Figure 23(a, b), the structure of the proposed gas sensors that use CNT/graphene as the conducting channel looks quite similar to the conventional metal-oxide semiconductor field effect transistor (MOSFET) which is comprised of source and drain electrodes with the gate insulator inducing the channel of carriers and a silicon back gate (Heller et al. 2008, Sivasathya and Thiruvadigal 2013) to augment the carrier density or adjust the threshold voltage. A CNT/graphene channel connects the source and the drain electrodes, and the gate is separated from the channel by a dielectric barrier layer. In most studies, silicon is used as the back gate while SiO_2 is employed to act as a dielectric layer (Uchida, Saitoh and Kobayashi 2008, Sivasathya and Thiruvadigal 2013). When gas molecules are in contact with the surface of CNT/graphene, carrier concentration will change due to the variability of the current in the

Figure 22. Schematic of gas sensor

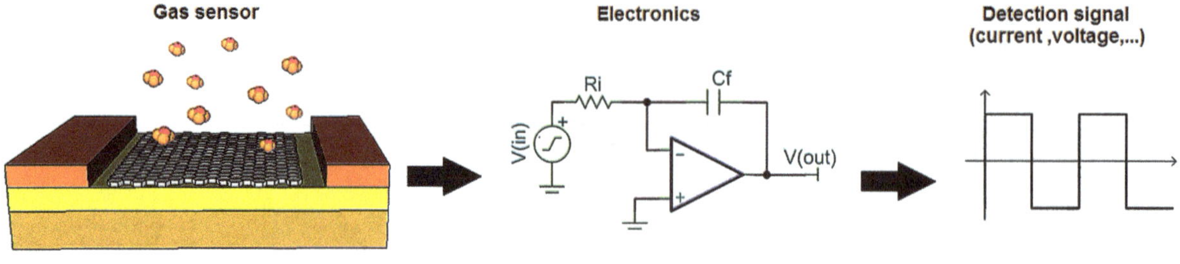

Figure 23. (a)FET-based structure for gas sensor with a CNT channel; (b) graphene channel

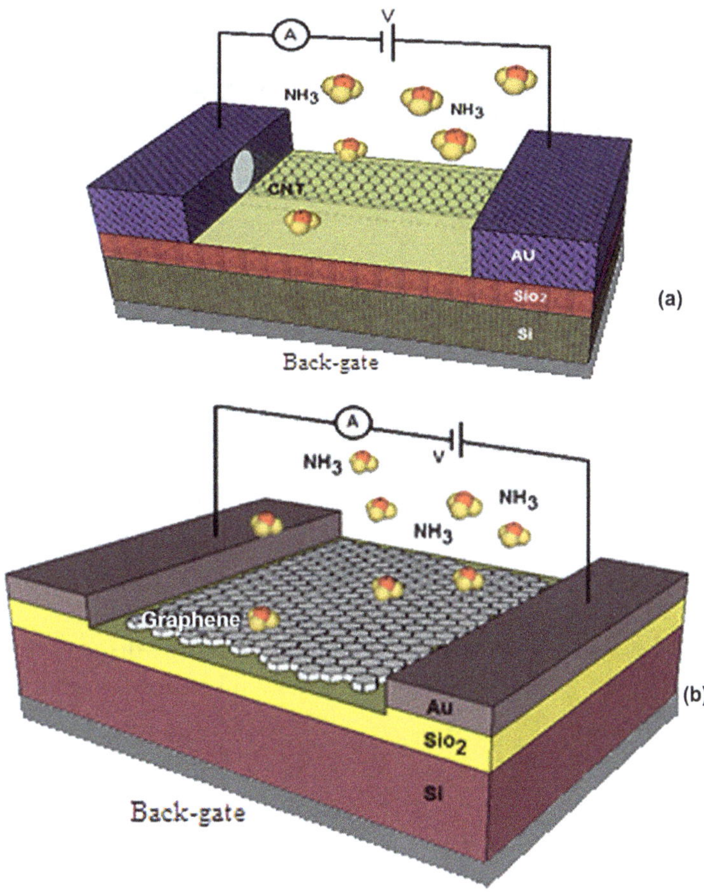

drain and the source which is a measurable parameter (Ding et al. 2009, Zhang et al. 2013, I. S. Amiri et al. 2014, Wang et al. 2012, Amiri and Ali 2013, Amiri et al. 2012).

The best gas sensor has high sensitivity which is capable of sensing even one atom or molecule of gas (Ding et al. 2006, Kiani et al. 2013). Numerous recent theoretical studies on gas molecular adsorption on CNT/graphene have been reported on NO_2, H_2O, NH_3, CO, and NO molecules that are physically adsorbed on the pristine CNT/graphene (Cho et al. 2008, Lich Quang et al. 2013).

Figure 24 illustrates a schematic representation of CNTs when electron donor NH_3 gas molecules are in the atmosphere around the sensor. In such conditions, NH_3 molecules are adsorbed on the surface of the CNT channel while donating electrons to it. This process causes quite significant change in the electrical properties of the CNT. These strong adsorption effects stem from the inherent properties of gas molecules and the bonding characteristics between these molecules and the CNT (Arakelyan et al. 2013, Xie et al. 2012). It is always important to obtain p-type and/or n-type semiconducting CNT to incorporate them in a complementary logic. p-n junction is a result of this complementarity. n-type and p-type nanoscale field effect transistors can be formed for implementation by applying positive or negative gate voltage and can be useful from the application perspectives (McEuen, Fuhrer and Park 2002).

Figure 24. Schematic of NH_3 sensing mechanism employing gas adsorption phenomenon

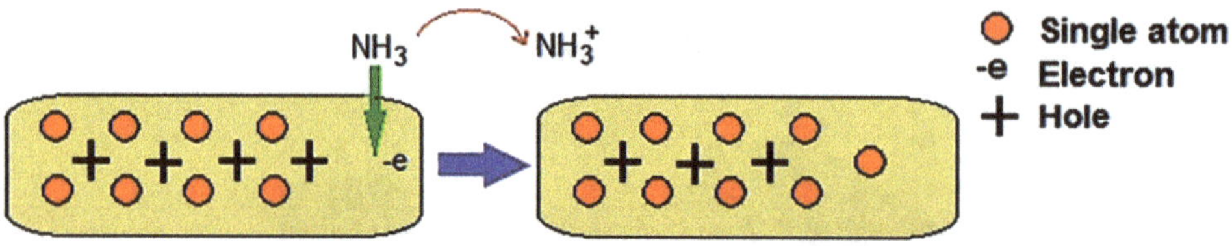

Gas molecules can modulate the electronic structure of graphene in diverse ways. The adsorption of CO_2 and O_2 converts the system to p-type semiconductor while the adsorption of NH_3 leads to n-type behaviour. Similar to CNTs, these rich adsorption effects are caused by the intrinsic property of the gas molecules and the bonding characteristics between gas molecules and graphene (Huang et al. 2008a). The resulting p-type and n-type semiconducting behaviour might be detected in experiment by applying and modulating gate voltage. Among all gas molecules considered, obviously NH_3 molecule adsorption can greatly enhance the conductance (Lich Quang et al. 2013, Huang et al. 2008b).

Proposed Model for CNT

We attempt to model the CNT conductance by considering the energy dispersion relation, and deriving the final model using the Taylor series expansion near the Fermi points, as follows (Ahmadi et al. 2008, Dresselhaus et al. 2000):

$$E(k) = \pm \frac{t3a_{c-c}}{2}\sqrt{\left(\frac{2}{3d}\right)^2 + k_x^2} \tag{55}$$

where the (±) signs have been included to account for the valence and conductance bands. a_{c-c}=0.142nm represents carbon-carbon bond length, *d* denotes CNT diameter and t=2.7(eV) is the nearest neighbour C-C tight binding overlap energy. For the first band gap energy, we can simply write $E_G = \frac{2a_{c-c}t}{d} = \frac{0.8eV.nm}{d(nm)}$. In addition, since the band structure is parabolic near the *k=0* points, we can write for energy:

$$E(k) \approx \frac{E_G}{2} + \frac{\hbar^2 k_x^2}{2m^*} \tag{56}$$

where *ħ* is the reduced Plank constant, k_x represents longitudinal wave vector component along the length of the tube and m* denotes the CNT's effective mass depending on the tube diameter (Ahmadi et al. 2010d, Anantram and Leonard 2006). The number of conduction channels can be written as:

$$M(E) = 2\frac{\Delta E}{\Delta k L} = \frac{3a_{c-c}t}{L}\left(\frac{4E}{3a_{c-c}t} - \frac{8}{9d^2}\right)^{1/2} \tag{57}$$

where L denotes the channel length. Two major factors contribute to the conductance effect on large channels, enabling it to follow the Ohmic scaling law based on Landauer formula. The first factor, which is independent of length, is the interface resistance. The second one results from the fact that the relation between the conductance and the width is nonlinear and is dependent upon the number of modes in the conductor. However, these modes are the quantized parameters in the Landauer formula in which both factors are interrelated as demonstrated below (Ahmadi et al. 2010a):

$$G = \frac{2q^2}{h}\int_{-\infty}^{+\infty} M(E)T(E)\left(-\frac{df}{dE}\right)dE \tag{58}$$

where h represents the Planck's constant, q denotes the electron charge and T is the transmission probability of an electron injected through the channel approximated as T(E)=1 in ballistic channels (Datta 2002). $f(E)$ is the Fermi-Dirac distribution with df/dE exploding to a delta function near the Fermi energy for degenerate statistics. The conductance can be obtained as follows (Peres et al. 2006):

$$G = \frac{2q^2}{h}\frac{3a_{c-c}t}{L}\left(\frac{4}{3a_{c-c}t}\right)^{1/2}\int_{-\infty}^{+\infty}\left(E - \frac{2a_{c-c}t}{3d^2}\right)^{\frac{1}{2}} d\left(\frac{1}{1+e^{(E-E_F)/K_BT}}\right) \tag{59}$$

Equation 59 can be re-written as:

$$G = \frac{4q^2}{hL}\left(3a_{c-c}t\pi k_BT\right)^{\frac{1}{2}}\left[\int_0^{+\infty}\frac{x^{-1/2}}{1+e^{x-\eta}}dx + \int_0^{+\infty}\frac{x^{-1/2}}{1+e^{x+\eta}}dx\right] \tag{60}$$

where $x = (E - E_g) / k_BT$ and the normalized Fermi energy is given by $\eta = (E_F - E_g) / K_BT$. This equation can be numerically solved by incorporating the partial integration method (Dingle and Dingle 1973, Zaharah et al. 2010, Kažukauskas et al. 2008). The general model for the conductance of carbon nanotube-based gas sensor can be derived similar to that of silicon-based model proposed by Gunlycke (Gunlycke et al. 2007).

$$G = \frac{4q^2}{hL}\left(3a_{c-c}t\pi k_BT\right)^{\frac{1}{2}}\left[\mathcal{I}_{-\frac{1}{2}}(\eta) + \mathcal{I}_{-\frac{1}{2}}(-\eta)\right] \tag{61}$$

is the Fermi-Dirac integral of order j. The conductance characteristic demonstrates the performance of NH_3 gas sensor based on CNT nanostructure. It has been revealed that when the CNT gas sensor is exposed to NH_3, the conductance changes (Yoon et al. 2011). We have proposed a model based on the reported experimental data and the relationship between conductance, gas concentration and temperatures as follows (Xia et al. 2009):

$$G_{wg} = G_{wog} + G_{wgT} + G_{wgF} \tag{62}$$

When the sensor is exposed to the gases in different temperatures, we can define three components for conductance, namely G_{wog}, G_{wgT} and G_{wgF}. The first component G_{wog}, is the conductance without the presence of gas; G_{wgT} is assumed as the conductivity changes in the presence of gas depending on temperature and the last component, G_{wgF}, is based on different values of gas concentration at constant temperature (Jilin Xia 2009). The conductance changes with temperature and various concentrations when CNT gas sensor is exposed to NH_3.E_g is dependent on temperature and gas concentration. Therefore, we can write:

$$\left\{\begin{matrix} E_g \propto F \\ E_g \propto T \end{matrix}\right\} \Rightarrow E_g = T + F \tag{63}$$

Writing $\eta = (E_F - E_g) / k_B T$ explicitly, we obtain

$$G_{wog} = \frac{4q^2}{hL}\left(3a_{c-c} t\pi k_B T\right)^{\frac{1}{2}}\left[\mathcal{I}_{\frac{-1}{2}}\left(\frac{E_F - E_g}{k_B T}\right) + \mathcal{I}_{\frac{-1}{2}}\left(\frac{E_g - E_F}{k_B T}\right)\right] \tag{64}$$

Equations (63) and (64) are combined to obtain the conductance of gas sensor as:

$$G_{wg} = \frac{4q^2}{hL}\left(3a_{c-c} t\pi k_B T\right)^{\frac{1}{2}}\left[\mathcal{I}_{\frac{-1}{2}}\left(\frac{E_F - T - F}{k_B T}\right) + \mathcal{I}_{\frac{-1}{2}}\left(\frac{T + F - E_F}{k_B T}\right)\right] \tag{65}$$

The Fermi-Dirac integral plays a significant role in the modeling of semiconductor's behavior. So, the following expansion of Fermi-Dirac integral is taken into consideration:

$$\mathcal{I}_j\left(\eta_F\right) = 2\eta_F^{j+1}\sum_{n=0}^{\infty}\frac{t_{2n}}{\Gamma\left(j+2-2n\right)\eta_F^{2n}} + \cos\left(\pi j\right)\sum_{n=1}^{\infty}\frac{\left(-1\right)^{n-1} e^{-n\eta_F}}{n^{j+1}} \tag{66}$$

where

$$t_0 = \frac{1}{2}, t_n = \sum_{\mu=1}^{\infty}\left(-1\right)^{\mu-1} \Big/ \mu^n = \left(1 - 2^{1-n}\right)\varsigma\left(n\right),$$

and $\varsigma\left(n\right)$ is the Riemann Zeta function. In the degenerate limit $\left(\eta_F \gg 0\right)$ which is the operation regime for the nanoscale devices, the expressions for the Fermi-Dirac integral can be obtained from Equation 66 as $F_j\left(\eta_F\right) \to \eta_F^{j+1} / \Gamma\left(j+2\right)$. Accordingly, the Fermi-Dirac integral of order -1/2 can be simplified as (Kim and Lundstrom 2008):

$$\mathcal{I}_{-\frac{1}{2}}\left(\eta_F\right) \to \frac{2\eta_F^{1/2}}{\sqrt{\pi}} \tag{67}$$

Based on the current-voltage characteristic of graphene-based FET devices, the gas sensor performance can be evaluated through Equation 68. Assuming that the source and substrate terminals are kept in ground potential, by applying a small voltage between source and drain (V_{DS}), the channel region experiences a flow of electrons. Moreover, the relationship between current and conductance can be replaced by Fermi-Dirac integral of general conductance model of SWCNT as:

$$I=\left[\frac{4q^2}{hL}\left(3a_{c-c}t\pi k_BT\right)^{\frac{1}{2}}\left[\mathcal{I}_{\frac{-1}{2}}\left(\frac{E_F-T-F}{k_BT}\right)+\mathcal{I}_{\frac{-1}{2}}\left(\frac{T+F-E_F}{k_BT}\right)\right]\right]\left(V_{gs}-V_t\right) \tag{68}$$

where V_{gs} is the gate-source voltage and V_t is the threshold voltage.

Proposed Model for Graphene

The underlying operational principle in MOSFET is based on the electron flow between the source and drain electrodes which can be controlled by the gate voltage. According to Landauer formula, there is a direct proportionality between conductance *G* and the transmission probability T of carriers from one electrode to another demonstrated by (Ahmadi et al. 2010b):

$$G=\frac{2q^2}{h}T \tag{69}$$

Taylor expansion is used to investigate a parabolic relationship involving energy and wave vector (Berger et al. 2006).

$$E(k)\approx\frac{E_g}{2}+\frac{E_g}{4\beta^2}k_x^2 \tag{70}$$

where β is the quantized wave vector given by (Brey and Fertig 2006). The wave vector in the parabolic part of the band energy can be extracted as:

$$k=\sqrt{\frac{4E}{3a_{c-c}t}-2\beta^2} \tag{71}$$

The conductance on large channel following the Ohmic scaling law based on Landauer formula can then be obtained as:

$$G=\frac{6q^2}{hL}\left(a_{c-c}t\pi k_BT\right)^{\frac{1}{2}}\left[\int_0^{+\infty}\frac{x^{-1/2}}{1+e^{x-\eta}}dx+\int_0^{+\infty}\frac{x^{-1/2}}{1+e^{x+\eta}}dx\right] \tag{72}$$

where $x = \frac{E - E_g}{k_B T}$ and the normalized Fermi energy is $\eta = \frac{E_F - E_g}{k_B T}$.

The performance of NH_3 gas sensor based on graphene nanostructure is demonstrated by its conductance characteristic. It has been shown that the conductance changes when the graphene gas sensor is exposed to NH_3 (Peng et al. 2009).

The corresponding formula equating the I-V characteristic of the graphene channel can then be written as:

$$I = \frac{6q^2}{hL}\left(a_{c-c} t\pi k_B T\right)^{\frac{1}{2}}\left[\mathcal{I}_{\frac{-1}{2}}\left(\frac{E_F - T - F}{k_B T}\right) + \mathcal{I}_{\frac{-1}{2}}\left(\frac{T + F - E_F}{k_B T}\right)\right]\left(V_{gs} - V_t\right) \tag{73}$$

Results and Discussion

Figure 25 illustrates the assessments of the gas sensor performance based on CNT and graphene nano structures considering their current-voltage characteristics when they are exposed to the NH_3(Peng et al. 2009). Also shown is the experimental data (Peng et al. 2009). The agreement is good except near the minimum where Dirac point is shifted to positive gate voltage.

Charge transfer is involved within the sensing mechanism of graphene and CNT-based gas sensors. This occurs during the interaction of gas molecules with the graphene and CNT surfaces. The conductivities of both channel media are modified through this interaction. The phenomenon is likely to occur as a result of the interaction of NH_3 molecules with the carbon on the surface of graphene/CNT. Thus, electrons move from NH_3 molecules to these materials. Figures 25 (a, b, c, d) illustrate the I-V characteristics of the graphene/CNT gas sensors corresponding to temperatures of 25, 50, 100, and 150°C, respectively. At the first three temperatures, for the minimum values of current, there have not been significant changes in the gate voltage. This can imply that at temperatures below, the gas is reluctant to be adsorbed on graphene/CNT. However, the currents corresponding to each gate voltage value have risen at higher temperatures in all these four cases.

Also, as shown in Figure 25(d), at 150°Cthe gate voltage becomes more negative. NH_3 is an electron donating agent and it leaves electrons on the channel. This causes the graphene/CNT Fermi level to move toward their conduction band edges, making the threshold voltage V_{th} more negative. Thus, it can be said that this shift toward negative gate voltage is caused by the adsorption of NH_3on the graphene/CNT channel at this temperature.

Figure 26 (a, b, c) illustrates the I-V characteristics of the proposed models for graphene and CNT in comparison with results for a CNT based experiment. An increase in the current can be associated with the charge transfer between NH_3molecules and graphene/CNT where the NH_3 molecules operate as the donor. This phenomenon is also known as chemical doping by gas molecules. The sensitivity can be observed from the response of graphene/CNT-based gas sensors under 100ppm, 200ppm and 500ppm NH_3. A decreasing trend in the gate voltage similar to that for 150°C can be seen at 200°C. It can be concluded that at temperatures above 150°C, the NH_3 adsorption and the consequent electron donating behaviour increases, which causes further shift of the gate voltage toward negative values. The figure gives a clear illustration of the fact that there is a good agreement between the proposed models

Figure 25. I-V characteristics of Graphene and CNT after exposure to NH_3 under F=500 ppm at (a) T=25°C, (b) T=50°C, (c) T=100°C, (d) T=150°C

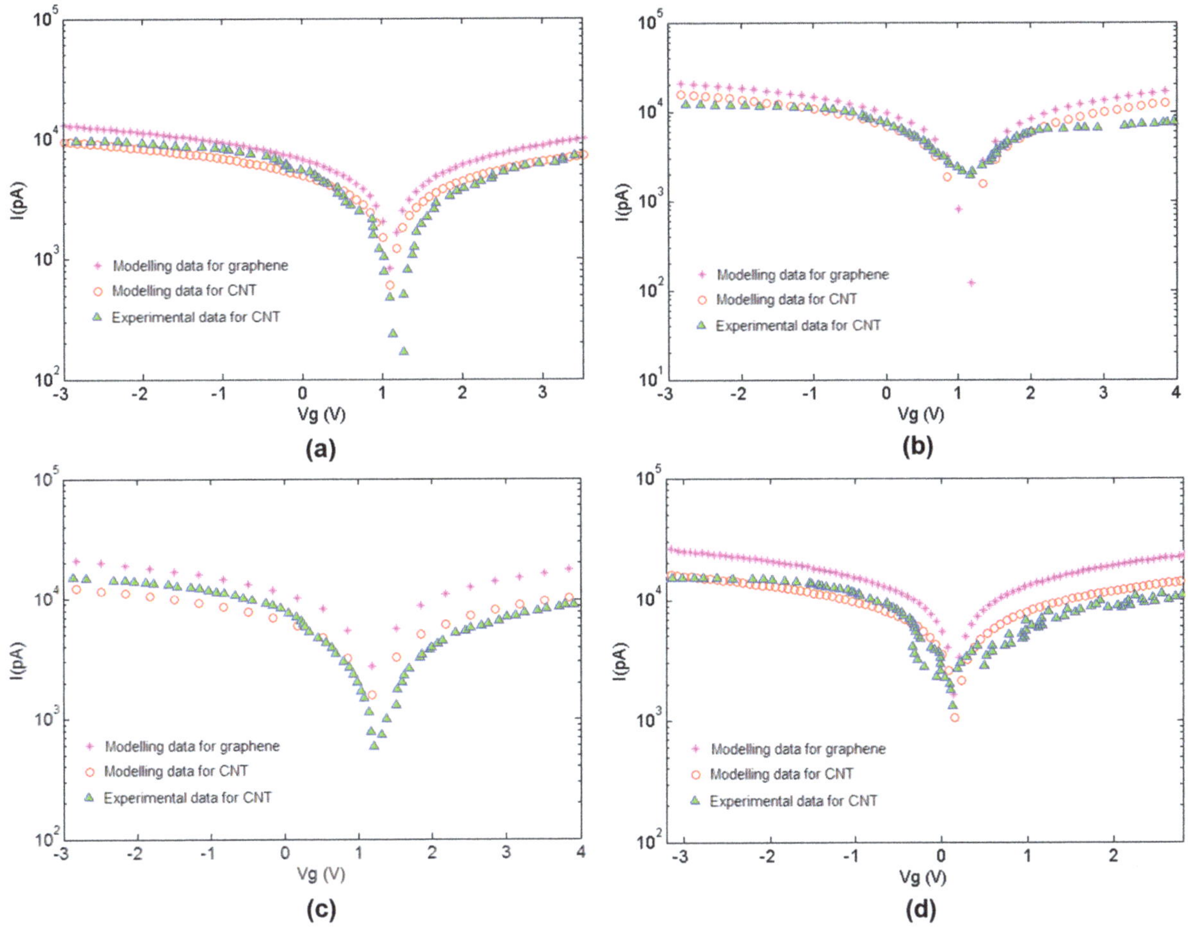

and extracted data (Peng et al. 2009). In the suggested models, different values of temperature and gas concentration are demonstrated in the terms of δ and λ parameters, respectively, as presented in Table 3.

Referring to the analytical models,has been introduced as the temperature control parameter obtained by iteration method. The analytical models in our study show that the rate of changes in conductivity depending on temperature can be expressed by the following equation:

$$\delta = aLn\left(T\right) - b \tag{74}$$

Here, parameters and are calculated as and . The parameter has been defined as a control parameter of gas concentration also calculated by iterative method and shows that the rate of changes in conductivity depends on gas concentration for which the equation can be written as:

$$\lambda = cLn\left(F\right) - d \tag{75}$$

Figure 26. I-V characteristics after exposure to NH_3 for graphene and CNT at T=200 °C and under (a) F=100ppm, (b) F=200ppm, (c) F=500ppm
Peng et al., 2009.

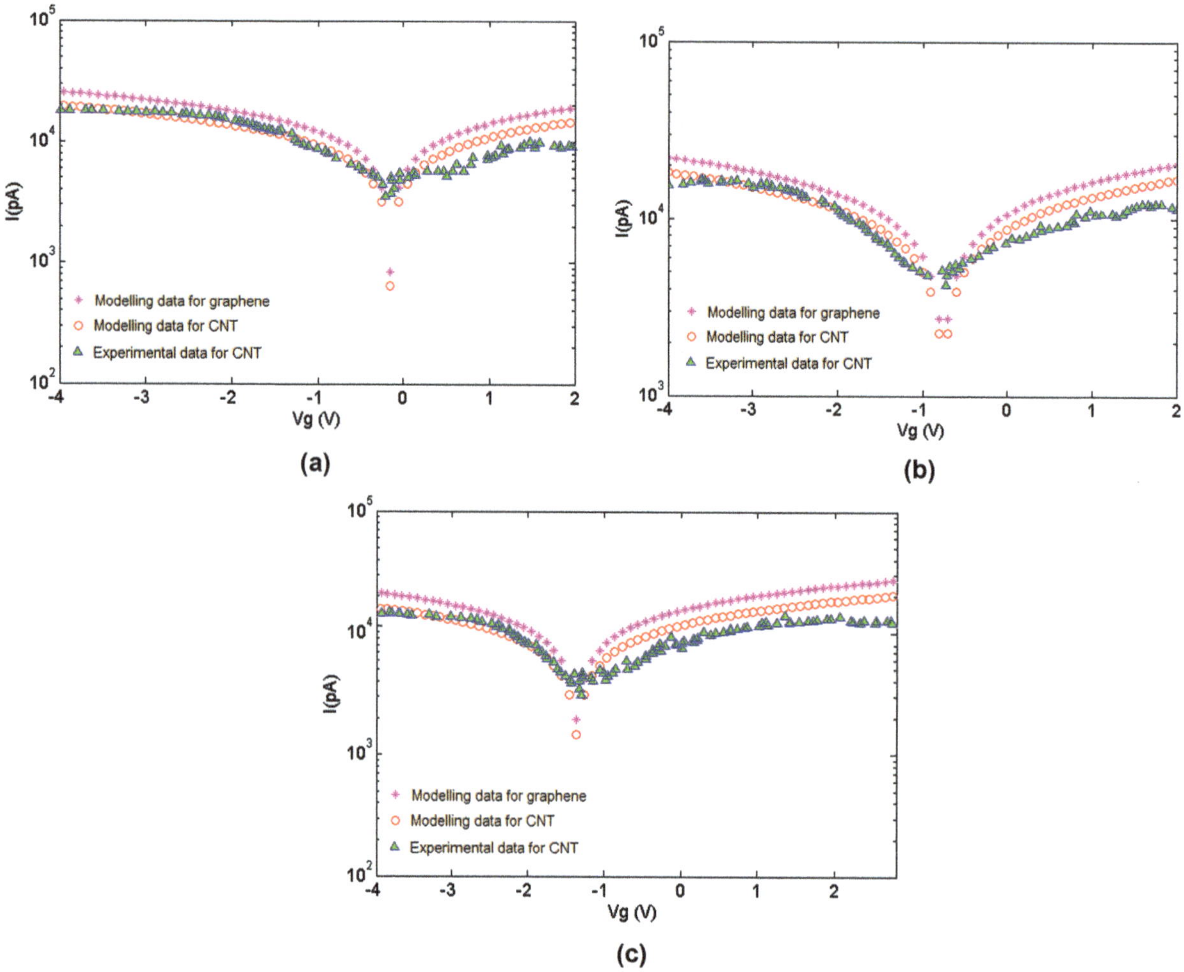

Table 3. Different temperature and concentration values with respective and parameters

T(°C)	F(ppm)	δ	λ
25	500	-3.65	0.027
50	500	-2.35	0.027
100	500	-1.45	0.027
150	500	-0.95	0.027
200	100	-0.7	0.005
200	200	-0.7	0.012
200	500	-0.7	0.027

where the constants are calculated in the same manner as c=1.4129 and d= 8.0494. Outstanding properties such as high sensitivity as well as remarkable carrier transport features make both graphene and CNTs promising candidates for use in nanosensors. It has been observed that these materials experience a measurable change in conductance levels when exposed to NH_3. This interesting feature has been suggested to be employed in gas detection systems.

The modelling and simulation results obtained in this chapter have proven the viability of the proposed models. However, for the sake of better understanding of the electrical properties of carbonaceous materials, a set of experiments have been done in an attempt to produce carbonaceous materials using arc discharge decomposition of methane gas which will be presented in appendix.

The following chapter will present a conclusion of the project along with a list of some possible future works.

CONCLUSION

In the field effect transistor based structure, making an initial assumption that the gate voltage is directly proportional to the gas concentration, different models have been proposed and developed for the gas sensors. It has been demonstrated that both graphene and CNTs experience measureable fluctuations in conductance, capacitance, carrier concentration and drift velocity levels when exposed to gases such as CO_2, NO_2 and NH_3. In the analytical procedures, various electro physical properties and phenomena have been considered and various control parameters (α,β,δ, λ) are incorporated and calculated by iteration method. The models have been validated and compared with experimental data previously published in the literature. The comparative studies show satisfactory agreement and hence the models have been proven valid. In addition to the analytical model, an artificial neural network model was also developed for the single case of I-V characteristics of a CNT based gas sensor with FET structure. The ANN model employs feed forward back propagation algorithm and uses the experimental data as the training data set. Both models are able to produce good results with satisfactory agreement with the extracted experimental data. The ANN model, however, has proved to be able to produce more accurate results than those by the analytical counterpart.

Numerous great progresses have been made in the study of carbon nanomaterials over the recent decades. Yet, it is still a young field with high potentials of activity in countless research areas. This study has focused mainly on the use of graphene and CNTs in gas sensing mechanisms. However, the research findings may pave the way for following works:

1. Derivation and development of CNT/graphene-based gas sensor models using carrier mobility.
2. Development of models for multilayer graphene based gas sensors.
3. Investigation of the impact of channel length on the sensing parameters and sensor accuracy

REFERENCES

Abolgasim, A. A. (2008). *Classification of capital expenditures and revenue expenditures using neural network model*. Universiti Utara Malaysia.

Ahmadi, M. T., & Ismail, R. 2010. Graphene Nanoribbon Fermi energy model in parabolic band structure. In *Intelligent Systems, Modelling and Simulation (ISMS), 2010 International Conference on*, (pp. 401-405). IEEE. doi:10.1109/ISMS.2010.78

Ahmadi, M. T., Ismail, R., Tan, M. L. P., & Arora, V. K. (2008). The ultimate ballistic drift velocity in carbon nanotubes. *Journal of Nanomaterials*, *2008*, 1–8. doi:10.1155/2008/769250

Ahmadi, M. T., Johari, Z., Amin, N. A., Fallahpour, A. H., & Ismail, R. (2010a). Graphene nanoribbon conductance model in parabolic band structure. *Journal of Nanomaterials*, *2010*, 12. doi:10.1155/2010/753738

Ahmadi, M. T., Johari, Z., Amin, N. A., Fallahpour, A. H., & Ismail, R. (2010b). Graphene nanoribbon conductance model in parabolic band structure. *Journal of Nanomaterials*, *2010*, 1–4. doi:10.1155/2010/753738

Ahmadi, M. T., Johari, Z., Amin, N. A., & Ismail, R. (2012). Band energy effect on carrier velocity limit in graphene nanoribbon. *Journal of Experimental Nanoscience*, *7*(1), 62–73. doi:10.1080/17458080.2010.501826

Ahmadi, M. T., Johari, Z., Amin, N. A., Mousavi, S. M., & Ismail, R. (2010c). Carbon nanotube conductance model in parabolic band structure. In *Semiconductor Electronics (ICSE), 2010 IEEE International Conference on*, (pp. 256-259). IEEE. doi:10.1109/SMELEC.2010.5549582

Ahmadi, M. T., Johari, Z., Amin, N. A., Mousavi, S. M., & Ismail, R. (2010d). Carbon nanotube conductance model in parabolic band structure. IEEE.

Amin, N. A., Ahmadi, M. T., Johari, Z., Mousavi, S. M., & Ismail, R. (2011). Effective mobility model of graphene nanoribbon in parabolic band energy. *Modern Physics Letters B*, *25*(10), 739–745. doi:10.1142/S0217984911026073

Amiri, I. S., Afroozeh, A., & Bahadoran, M. (2011). Simulation and analysis of multisoliton generation using a panda ring resonator system. *Chinese Physics Letters*, *28*, 104205. doi:10.1088/0256-307X/28/10/104205

Amiri, I. S., Ahsan, R., Shahidinejad, A., Ali, J., & Yupapin, P. P. (2012). Characterisation of bifurcation and chaos in silicon microring resonator. *IET Communications*, *6*(16), 2671–2675. doi:10.1049/iet-com.2011.0773

Amiri, I. S., & Ali, J. (2013). Optical buffer application used for tissue surgery using direct interaction of nano optical tweezers with nano cells. *Quantum Matter*, *2*(6), 484–488. doi:10.1166/qm.2013.1086

Amiri, I. S., Nikoukar, A., Shahidinejad, A., Anwar, T., & Ali, J. (2014). Quantum Transmission of optical tweezers via fiber optic using half-panda system. *Life Science Journal*, *10*, 391–400.

Anantram, M., & Leonard, F. (2006). Physics of carbon nanotube electronic devices. *Reports on Progress in Physics*, *69*(3), 507–561. doi:10.1088/0034-4885/69/3/R01

Arakelyan, V. M., Aleksanyan, M. S., Hovhannisyan, R. V., Shahnazaryan, G. E., Aroutiounian, V. M., Hernadi, K., & Forro, L. et al. (2013). Gas sensors made of multiwall carbon nanotubes modified by tin dioxide. *Journal of Contemporary Physics-Armenian Academy of Sciences*, *48*(4), 176–183. doi:10.3103/S1068337213040063

Berger, C., Song, Z., Li, X., Wu, X., Brown, N., Naud, C., & Marchenkov, A. N. et al. (2006). Electronic confinement and coherence in patterned epitaxial graphene. *Science*, *312*(5777), 1191–1196. doi:10.1126/science.1125925 PMID:16614173

Bradley, D. (2012). Graphene gas sensor CARBON. *Materials Today*, *15*(6), 233–233. doi:10.1016/S1369-7021(12)70105-0

Brey, L., & Fertig, H. (2006). Electronic states of graphene nanoribbons studied with the Dirac equation. *Physical Review B: Condensed Matter and Materials Physics*, *73*(23), 235411. doi:10.1103/PhysRevB.73.235411

Cho, T. S., Lee, K. J., Kong, J., & Chandrakasan, A. R. (2008) The design of a low power carbon nanotube chemical sensor system. *2008 45th ACM/IEEE Design Automation Conference*.

Das, A., Chakraborty, B., & Sood, A. K. (2011). Probing single and bilayer graphene field effect transistors by raman spectroscopy. *Modern Physics Letters B*, *25*(08), 511–535. doi:10.1142/S0217984911025973

Datta, S. (1997). *Electronic transport in mesoscopic systems*. Cambridge university press.

Datta, S. (2002). *Electronic Transport in Mesoscopic Systems*. Cambridge, UK: Cambridge University Press.

Ding, L., Wang, S., Zhang, Z. Y., Zeng, Q. S., Wang, Z. X., Pei, T., & Peng, L. M. et al. (2009). Y-Contacted high-performance n-type single-walled carbon nanotube field-effect transistors: Scaling and comparison with sc-contacted devices. *Nano Letters*, *9*(12), 4209–4214. doi:10.1021/nl9024243 PMID:19995085

Ding, W. D., Hayashi, R., Suehiro, J., Zhou, G. B., Imasaka, K., & Hara, M. (2006). Calibration methods of carbon nanotube gas sensor for partial discharge detection in SF6. *IEEE Transactions on Dielectrics and Electrical Insulation*, *13*(2), 353–361. doi:10.1109/TDEI.2006.1624280

Dingle, R. B., & Dingle, R. (1973). *Asymptotic expansions: Their derivation and interpretation*. Academic Press London.

Dresselhaus, M. S., Dresselhaus, G., Eklund, P., & Rao, A. (2000). *Carbon nanotubes*. Springer. doi:10.1007/978-94-011-4038-6_9

Geim, A. K., & Novoselov, K. S. (2007). The rise of graphene. *Nature Materials*, *6*(3), 183–191. doi:10.1038/nmat1849 PMID:17330084

Grossi, E., & Buscema, M. (2007). Introduction to artificial neural networks. *European Journal of Gastroenterology & Hepatology*, *19*(12), 1046–1054. doi:10.1097/MEG.0b013e3282f198a0 PMID:17998827

Gunlycke, D., Areshkin, D., & White, C. (2007). Semiconducting graphene nanostrips with edge disorder. *Applied Physics Letters*, *90*(14), 142104. doi:10.1063/1.2718515

Hatami, H., Abedpour, N., Qaiumzadeh, A., & Asgari, R. (2011). Conductance of bilayer graphene in the presence of a magnetic field: Effect of disorder. *Physical Review B: Condensed Matter and Materials Physics*, 83.

Heller, I., Janssens, A. M., Mannik, J., Minot, E. D., Lemay, S. G., & Dekker, C. (2008). Identifying the mechanism of biosensing with carbon nanotube transistors. *Nano Letters*, *8*(2), 591–595. doi:10.1021/nl072996i PMID:18162002

Huang, B., Li, Z., Liu, Z., Zhou, G., Hao, S., Wu, J., & Duan, W. et al. (2008a). Adsorption of gas molecules on graphene nanoribbons and its implication for nanoscale molecule sensor. *The Journal of Physical Chemistry C, 112*(35), 13442–13446. doi:10.1021/jp8021024

Huang, B., Li, Z., Liu, Z., Zhou, G., Hao, S., Wu, J., & Duan, W. et al. (2008b). Adsorption of gas molecules on graphene nanoribbons and its implication for nanoscale molecule sensor. *The Journal of Physical Chemistry C, 112*(35), 13442–13446. doi:10.1021/jp8021024

Hwang, W. S., Zhao, P., Tahy, K., Nyakiti, L. O., Wheeler, V. D., Myers-Ward, R., . . . Haensch, W. (2013). *Graphene nanoribbon field-effect transistors on wafer-scale epitaxial graphene on sic substrates.* arXiv preprint arXiv:1310.6823

Iijima, S. (1991). Helical microtubules of graphite carbon. *Nature, 354*(6348), 56–58. doi:10.1038/354056a0

Jilin Xia, F. C. (2009). Measurement of the quantum capacitance of graphene. *Nature Nanotechnology.* doi:10.1038/NNANO.2009.177

Karamdel, J., Ahmadi, M. T., Majlis, B. Y., Dee, C. F., & Ismail, R. (2008). Formulation and Simulation for Electrical Properties of a (5,3) Single Wall Carbon Nanotube. *ICSE: 2008 IEEE International Conference on Semiconductor Electronics Proceedings.*

Kažukauskas, V., V. Kalendra, C. Bumby, B. Ludbrook & A. Kaiser (2008) Electrical conductivity of carbon nanotubes and polystyrene composites. *Physica Status Solidi (c), 5*, 3172-3174.

Kiani, M. J., Ahmadi, M., Akbari, E., Karimi, H., & Che Harun, F. (2013). Graphene Nanoribbon Based Gas Sensor. *Key Engineering Materials, 553*, 7–11. doi:10.4028/www.scientific.net/KEM.553.7

Kim, R., & Lundstrom, M. (2008). Notes on Fermi-Dirac integrals (2nd ed.). Academic Press.

Ko, G., Kim, H. Y., Ahn, J., Park, Y. M., Lee, K. Y., & Kim, J. (2010). Graphene-based nitrogen dioxide gas sensors. *Current Applied Physics, 10*(4), 1002–1004. doi:10.1016/j.cap.2009.12.024

Lam, K.-T., Lee, C., & Liang, G. (2009). Bilayer graphene nanoribbon nanoelectromechanical system device: A computational study. *Applied Physics Letters*, 95.

Lich Quang, N., Pho Quoc, P., Huyen Ngoc, D., Chien Duc, N., & Lam Huu, N. (2013). Enhancement of NH3 gas sensitivity at room temperature by carbon nanotube-based sensor coated with co nanoparticles. *Sensors (Basel, Switzerland), 13*(2), 1754–1762. doi:10.3390/s130201754 PMID:23364198

Martina Cheli, G. F. a. G. I. (2009). *A semi-analytical model of bilayer-graphene field effect transistor.* Academic Press.

McEuen, P. L., Fuhrer, M. S., & Park, H. (2002). Single-walled carbon nanotube electronics. *Nanotechnology. IEEE Transactions on, 1*, 78–85.

Mohammad Javad Kiani, M. T. A. (2012). Analytical modeling of bilayer graphene based biosensor. *Journal of Computational and Theoretical Nanoscience (CTN).*

Mousavi, H. (2011). Graphene as gas sensors. *Communications in Theoretical Physics, 56*(2), 373–376. doi:10.1088/0253-6102/56/2/30

Mousavi, S. M., Ahmadi, M. T., Webb, J. F., Sadeghi, H., Nilghaz, A., Amin, A., & Ismail, R. et al. (2011). Bilayer graphene nanoribbon carrier statistics in the degenerate regime. In *Proceedings of the Fourth Global Conference on Power Control and Optimization*. doi:10.1063/1.3592463

Nilsson, J. (2006) Electronic properties of graphene multilayers. *Physical Review Letters*. doi:10.1103/PhysRevLett.97.266801

Nilsson, J., Neto, A. C., Guinea, F., & Peres, N. (2008). Electronic properties of bilayer and multilayer graphene. *Physical Review B: Condensed Matter and Materials Physics*, *78*(4), 045405. doi:10.1103/PhysRevB.78.045405

Patterson, D. W. (1998). *Artificial neural networks: theory and applications*. Prentice Hall PTR.

Peng, N., Zhang, Q., Chow, C. L., Tan, O. K., & Marzari, N. (2009). Sensing mechanisms for carbon nanotube based NH3 gas detection. *Nano Letters*, *9*(4), 1626–1630. doi:10.1021/nl803930w PMID:19281216

Peres, N., Neto, A. H. C., & Guinea, F. (2006). Conductance quantization in mesoscopic graphene. *Physical Review B: Condensed Matter and Materials Physics*, *73*(19), 195411. doi:10.1103/PhysRevB.73.195411

Rahmani, M., Ismail, R., Ahmadi, M. T., Kiani, M. J., Saeidmanesh, M., Karimi, F., & Rahmani, K. et al. (2013). The effect of bilayer graphene nanoribbon geometry on Schottky-Barrier diode performance. *Journal of Nanomaterials*.

Sadeghi, H., Ahmadi, M. T., Mousavi, S. M., & Ismail, R. (2012). Channel conductance of aba stacking trilayer graphene nanoribbon field-effect transistor. *Modern Physics Letters B*, 26.

Saeidmanesh, M. (2013). Perpendicular electric field effect on bilayer graphene carrier statistic. *JCTN*.

Saeidmanesh, M., Ahmadi, M., Ghadiry, M., Akbari, E., Kiani, M., & Ismail, R. (2013). Perpendicular electric field effect on bilayer graphene carrier statistic. *Journal of Computational and Theoretical Nanoscience*, *10*(9), 1975–1978. doi:10.1166/jctn.2013.3158

Schwierz, F. (2010a). Graphene transistors. *Nature Nanotechnology*, *5*(7), 487–496. doi:10.1038/nnano.2010.89 PMID:20512128

Schwierz, F. (2010b). graphene transistors. *nature*. *Nanotechnology*. doi:10.1038/NNANO.2010.89

Shylau, A., Kłos, J., & Zozoulenko, I. (2009). Capacitance of graphene nanoribbons. *Physical Review B: Condensed Matter and Materials Physics*, *80*(20), 205402. doi:10.1103/PhysRevB.80.205402

Sivasathya, S., & Thiruvadigal, D. J. (2013). Ab Initio study of carbon nanotube transistor-based gas sensor for NO2 detection. *Asian Journal of Chemistry*, *25*, S411–S413.

Son, Y.-W., Cohen, M. L., & Louie, S. G. (2006). Energy gaps in graphene nanoribbons. *Physical Review Letters*, *97*(21), 216803. doi:10.1103/PhysRevLett.97.216803 PMID:17155765

Svozil, D., Kvasnicka, V., & Pospichal, J. (1997). Introduction to multi-layer feed-forward neural networks. *Chemometrics and Intelligent Laboratory Systems*, *39*(1), 43–62. doi:10.1016/S0169-7439(97)00061-0

Uchida, K., Saitoh, M., & Kobayashi, S. (2008). Carrier transport and stress engineering in advanced nanoscale transistors from (100) and (110) transistors to carbon nanotube fets and beyond. *IEEE International Electron Devices Meeting 2008, Technical Digest*.

Uhrig, R. E. 1995. Introduction to artificial neural networks. In *Industrial Electronics, Control, and Instrumentation, 1995.,Proceedings of the 1995 IEEE IECON 21st International Conference on*. IEEE.

Wang, Y., Huang, B.-C., Zhang, M., & Woo, J. C. S. (2012). Optimizing the fabrication process for high performance graphene field effect transistors. *Microelectronics and Reliability*, *52*(8), 1602–1605. doi:10.1016/j.microrel.2011.09.036

Wei, X.-L., Chen, Y.-P., Liu, W.-L., & Zhong, J.-X. (2012). Enhanced gas sensor based on nitrogen-vacancy graphene nanoribbons. *Physics Letters. [Part A]*, *376*(4), 559–562. doi:10.1016/j.physleta.2011.10.055

Wouter Olthuis, E. F. (2007). *Sensing with FETs - once, now and future*. UT Publications.

Xia, J. L., Chen, F., Li, J. H., & Tao, N. J. (2009). Measurement of the quantum capacitance of graphene. *Nature Nanotechnology*, *4*(8), 505–509. doi:10.1038/nnano.2009.177 PMID:19662012

Xie, H., Sheng, C., Chen, X., Wang, X., Li, Z., & Zhou, J. (2012). Multi-wall carbon nanotube gas sensors modified with amino-group to detect low concentration of formaldehyde. *Sensors and Actuators. B, Chemical*, *168*, 34–38. doi:10.1016/j.snb.2011.12.112

Yang, L., Park, C.-H., Son, Y.-W., Cohen, M. L., & Louie, S. G. (2007). Quasiparticle energies and band gaps in graphene nanoribbons. *Physical Review Letters*, *99*(18), 186801. doi:10.1103/PhysRevLett.99.186801 PMID:17995426

Yao, X. (1999). Evolving artificial neural networks. *Proceedings of the IEEE*, *87*(9), 1423–1447. doi:10.1109/5.784219

Yoneda, T. (n.d.). *Carbon nano tube (CNT) group temperature-change detection element for hydrogen gas sensor, has CNT group arranged perpendicular with respect to collector electrode surfaces and electrically connected between collector electrodes*. Nippon Valqua Ind Ltd.

Yoon, H. J., Jun, D. H., Yang, J. H., Zhou, Z., Yang, S. S., & Cheng, M. M.-C. (2011). Carbon dioxide gas sensor using a graphene sheet. *Sensors and Actuators. B, Chemical*, *157*(1), 310–313. doi:10.1016/j.snb.2011.03.035

Yu, T., Lee, Briggs, Nagabhirava, & Yu. (2010). *Reliability study of bilayer graphene - material for future transistor and interconnect*. IEEE.

Zaharah, J., Mohammad Taghi, A., Desmond Chang Yih, C., Aziziah, A. N., & Razali, I. (2010). Modelling of graphene nanoribbon fermi energy. *Journal of Nanomaterials*.

Zhang, J., Zhang, R., Wang, X., Feng, W., Hu, P., O'Neill, W., & Wang, Z. (2013). Fabrication of highly oriented reduced graphene oxide microbelts array for massive production of sensitive ammonia gas sensors. *Journal of Micromechanics and Microengineering*, *23*(9), 095031. doi:10.1088/0960-1317/23/9/095031

Zhang, X.-S. (2000). Introduction to artificial neural network. In Neural Networks in Optimization, (pp. 83-93). Springer. doi:10.1007/978-1-4757-3167-5_5

Zurada, J. M. (1992). *Introduction to artificial neural systems*. Academic Press.

Chapter 5
Graphene-Based Gas Sensor Theoretical Framework

Elnaz Akbari
Universiti Teknologi Malaysia, Malaysia

Hediyeh Karimi
Universiti Teknologi Malaysia, Malaysia

Aria Enzevaee
Universiti Teknologi Malaysia, Malaysia

Mohammad Taghi Ahmadi
Universiti Teknologi Malaysia, Malaysia

Zolkafle Buntat
Universiti Teknologi Malaysia, Malaysia

ABSTRACT

Both graphene and CNTs experience changes in their electrical conductance when exposed to different gases (such as CO2, NO2, and NH3), and they are, therefore, ideal candidates for sensing/measuring applications. In this research, a set of novel gas sensor models employing Field Effect Transistor structure using these materials have been proposed. In the suggested models, different physical properties such as conductance, capacitance, drift velocity, carrier concentration, and the current-voltage (I-V) characteristics of graphene/CNTs have been employed to model the sensing mechanism. An Artificial Neural Network model has also been developed for the special case of a CNT gas sensor exposed to NH3 to provide a platform to check the accuracy of the models. The performance of the models has been compared with published experimental data which shows a satisfactory agreement.

CARBON ALLOTROPES

Carbon is tetravalent chemical and non-metallic element which has a high level of stability. Carbon has atomic number six with four electrons in the valence band that tend to form an octet state through covalent bonding. This form of atomic bond is very stable. Carbon plays a crucial role in people's lives. Because of its stable feature, carbon has occurred in different natural isotopic structures. In addition, carbon comes in different allotropes which are all necessary in different types of applications especially in nanotechnology as is going to be emphasized in this research (Kinoshita, 1988; Diederich & Rubin, 1992). An illustration of carbon allotropes, i.e. graphite, graphene, fullerene and CNT has been provided

DOI: 10.4018/978-1-5225-0736-9.ch005

in Figure 1a to d (Hirsch, Eigler, & Heinemann, 2013). Each allotrope is obtained from a specific carbon structure possessing various electrical properties associated with its structural features (Singh et al., 2011).

One of the aforementioned carbon allotropes is fullerene with 0 dimension (0D) structure as illustrated in Figure 2. An important use of fullerene is various types of sensors (Giannopoulos, 2014; Elhaes & Ibrahim, 2013). Electronic sensors based on fullerene are sensitive to any adsorbed molecule due mainly to the fact that each of the atoms in fullerene is a surface atom. Electrons can move freely on the surface of fullerene (Yang, Zhang, & Xu, 2013; Wu et al., 2013b; Korotcenkov, 2013).

Another carbon allotrope which possesses 1 dimension (1D) structure, namely carbon nanotube, has attracted the attention of all researchers in the sensor development technology over the past few decades (Cole & Zook, 2006; Sayago et al., 2013). In 1991, Sumio Iijima, a Japanese researcher, discovered the multi-walled carbon nanotube (abbreviated as MWCNT) and two years later, he found single-walled carbon nanotubes (abbreviated as SWCNT) (Saito, 1998; Iijima, 1991). Since the flow of electrons through carbon nanotubes depends heavily upon the side wall function, the sensitivity can be manipulated by controlling the defect locations (Bellucci, 2005; Sevik et al., 2011). Research in the area of CNTs has been greatly influenced by the discovery of MWCNTs and SWCNTs (Gulotty et al., 2013). Figures 3 and 4 depict schematics of these two CNT structures, respectively (Baughman, Zakhidov, & de Heer, 2002).

Figure 1. Carbon allotropes

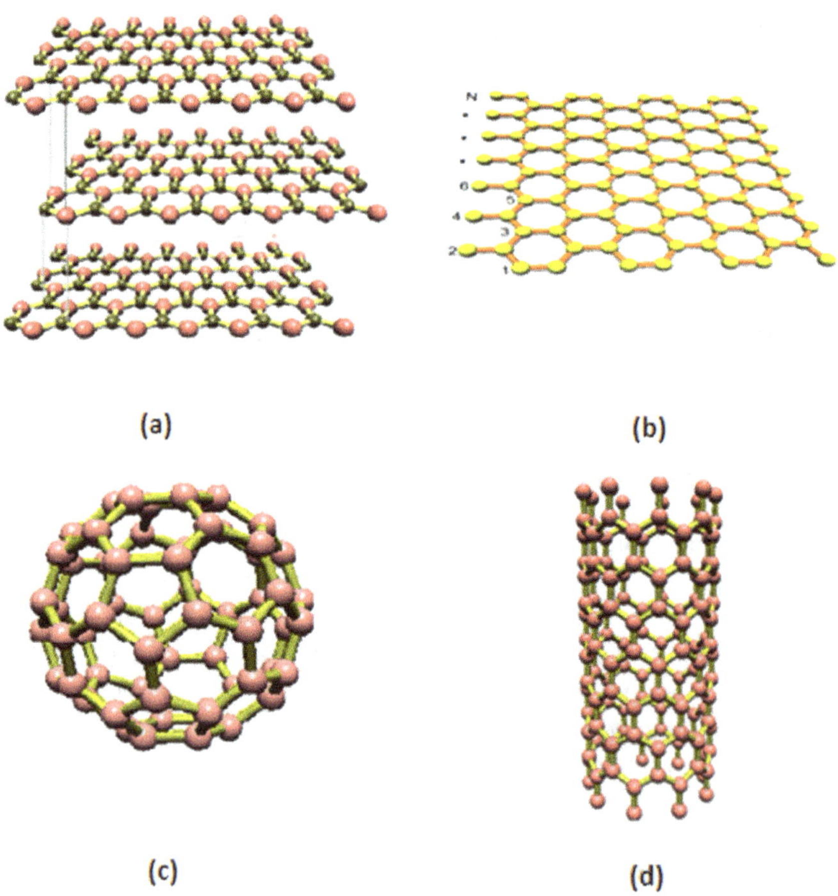

Figure 2. Fullerene structure

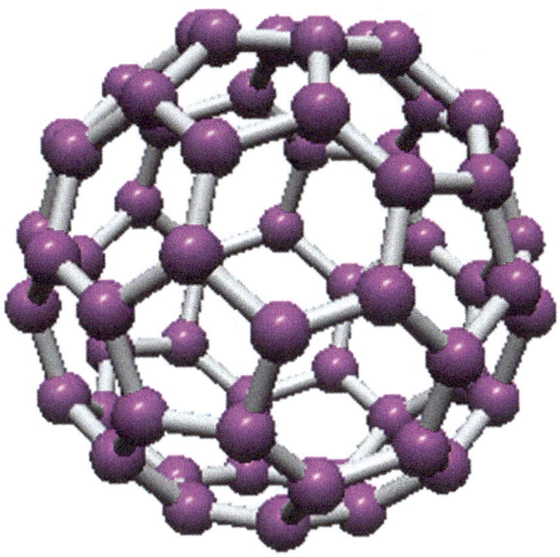

Figure 3. Schematic of a single-walled carbon nanotube

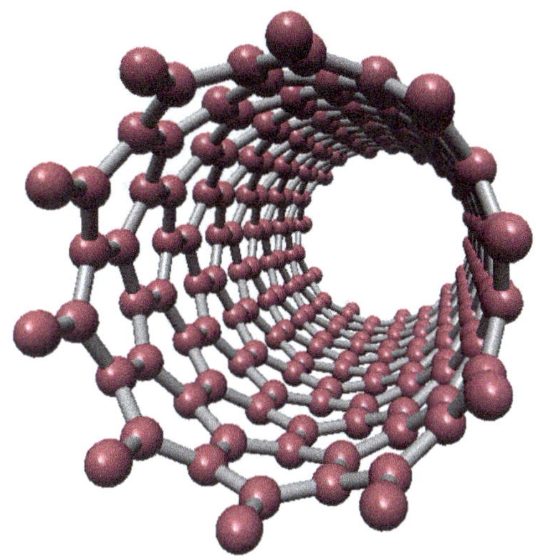

Figure 4. Schematic of multi-walled carbon nanotubes

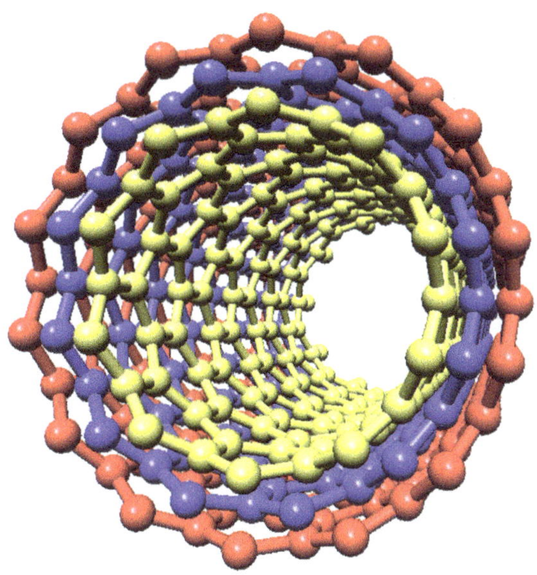

However, using CNT in nanoscale devices is not suitable due to various technical and physical limitations. This problem was solved after a two dimensional carbon allotrope, named graphene, was discovered (Geim & Novoselov, 2007; Geim, 2009).This fantastic scientific finding was considered as a huge breakthrough in the nanosensor area (Rao & Sood, 2013).

From the structure perspective, graphene is comprised of a single graphite sheet with sp^2 hybridized carbon atoms. A two dimensional graphene structure with honeycomb configuration is illustrated in Figure 5(Zhu et al., 2010). Because it is made up of low electrical noise materials, graphene-based sensor is accurate measurement device with rapid response and high sensitivity (Neto et al., 2009). Graphene has

Figure 5. Schematic of graphene structure

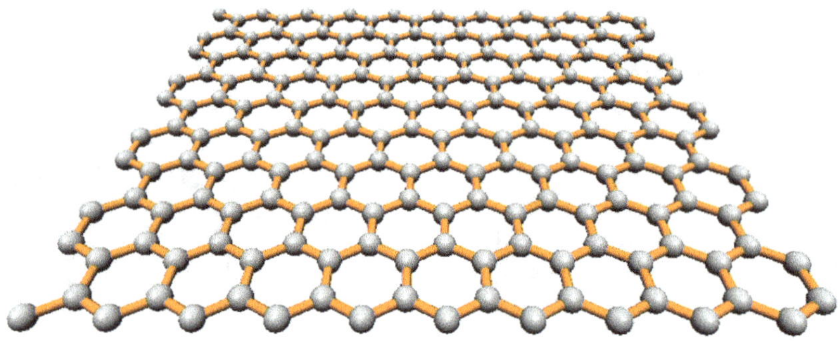

become an ideal material for use in nano-electronic applications due to its high stability and excellent carrier charge mobility (Novoselov et al., 2012; Akbari et al., 2013b).

Graphene nanoribbons (GNRs) are narrow graphene strips produced by standard lithographic techniques (Chen et al., 2007). Figure 6 shows a clear structure of GNRs. General physical properties in GNRs are almost similar to those of carbon nanotubes except for the fact that GNR has long reactive edges which cause it to sense the molecules adsorbed on its surface more accurately and easier (Jiao et al., 2009; Han et al., 2007). The sensor effectiveness depends greatly on the material characteristics such as mechanical/electrical properties, performance and sensitivity, therefore choosing the right materials and components is an important decision (Yu et al., 2011; Kiani et al., 2013b; Grassi et al., 2008).

A brief introduction on each of the mentioned carbon allotropes is presented in the following sections.

GRAPHENE

Carbon-based material has been explored extensively to accommodate the advancing technology nowadays. The discovery of a single atomic sheet of graphite layer or graphene by Andre Geim in 2004 has attracted so many interests in the communities of researchers and technology due to its superb electronic properties (Geim & Novoselov, 2007; Geim, 2009). A single layer graphene is sp^2-bonded carbon atoms

Figure 6. Schematic of unzipped CNT also known as GNR

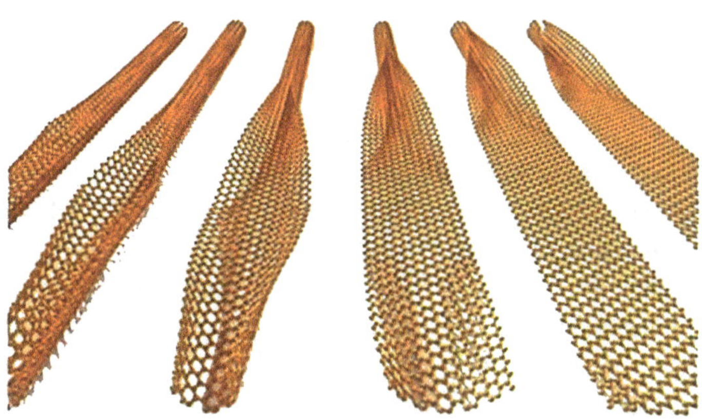

arranged in the form of hexagonal lattice (honeycomb lattice) in a two-dimensional (2D) structure (Rao & Sood, 2013)comprising a thin layer of single carbon atoms. The length of the bond between each two carbon atoms has been measured to be approximately 0.142nm (Adam et al., 2007a). Figure 7 shows the atomic arrangement of graphene.

Furthermore, carbon has various forms such as 0D as known as buckyballs, 1D referring to carbon nanotubes and 3D which is graphite as depicted in Figure 8. Because of its unique physical, electrical and optical properties, graphene can serve as an ideal material for use in nano-electronic technology allowing for a numerous areas of application. This motivates researchers to further study the feasibility and plausibility of using graphene in electrical applications (Acik & Chabal, 2011).

Graphene Band Structure

Due to its gapless structure, the two dimensional graphene cannot serve as a suitable candidate for use in transistors. The valence and conduction bands intersect at a single point in Brillouin zone. The six corners in this zone are named the Dirac points. Since there are no energy bandgaps in graphene, it possesses metallic characteristics and is known to be a zero-gap semi-metal. Figure 9 depicts the graphene band structure (Han et al., 2007).

It is required to produce a band gap in graphene to get remarkable electrical characteristics in specific applications. Through the application of lateral confinements, one can induce the energy band gap (Amin et al., 2011; Reich et al., 2002).

Characteristics of Graphene

Comparing with most conventional materials, graphene exhibits unique electronic properties. Graphene has a very high charge carrier mobility which can exceed 200, 000 cm^2/Vs at room temperature (Kim et al., 2009; Farmer et al., 2009). According to experimental transport measurement results, due to the extraordinary feature of high electron mobility, graphene has become an extremely attractive material for use in future nano-electronic devices particularly in sensor applications (Anderson & Nowak, 2009).

Figure 7. Arrangement of carbon atoms in graphene and C-C bond length
Adam et al., 2007b.

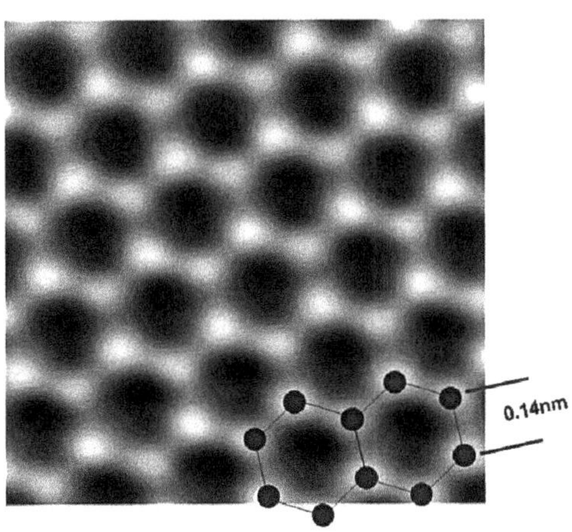

Figure 8. Carbon based materials in 0D fullerene, 1D carbon nanotube and 3D graphite
Geim & Novoselov, 2007.

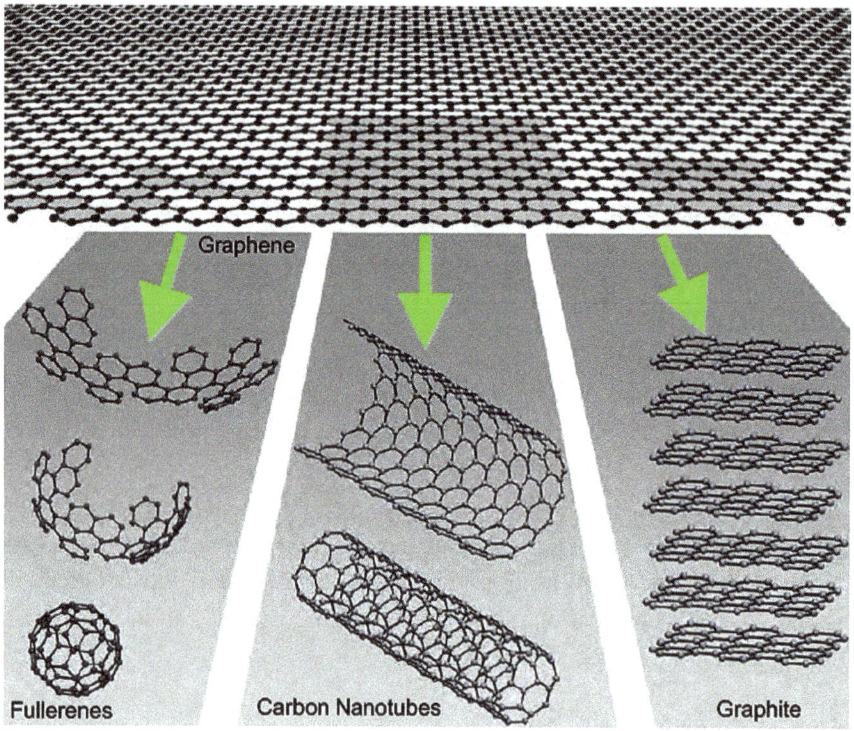

Figure 9. Band structure of grapheme
Liu, Liu & Zhu, 2011.

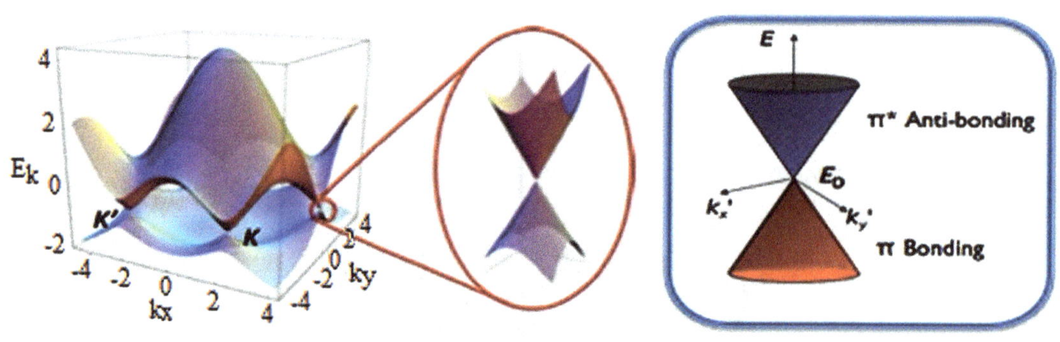

Moreover, graphene is considered to have the lowest resistivity among other carbon allotropes. Graphene conductivity can grow up to 5300Wm^{-1} K^{-1} (Murali, 2012; Balandin, Kotchetkov, & Ghosh, 2009). Since there is a reciprocal relation between resistivity and conductivity, the lower the former, the higher the latter. Graphene is more sensitive compared to carbon nanotubes (CNTs) due mainly to its large surface area (2630m^2/g) (Yang & Murali, 2010; Xu et al., 2012).

GRAPHENE NANORIBBONS (GNR)

As previously mentioned, graphene is a two dimensional carbon based material with no energy band gap (Huang et al., 2011). This leads it to be known as zero-gap semi-metal. This makes graphene a quasi-metallic material which is therefore not good for use in transistors. Yet, we can obtain the band gap energy through lateral confinement (Han et al., 2007). This is possible by forming graphene into narrow graphene strips as known as GN Rusing standard lithography technique (Kan, Li, & Yang). Figure 10 illustrates how two dimensional graphene sheet can be formed into GNR (Gu et al., 2011; Brey & Fertig, 2006).

Recently, experiments on the graphene nanoribbon FET have demonstrated that it is difficult to change and control graphene channel conductivity under ambient conditions (Fiori & Iannaccone, 2007). The major challenge in this regard is how to fabricate extremely small nanowires with adequate energy band gap at room temperature. It has been shown, based on theoretical works, that graphene nanoribbon width is inversely proportional to its energy band gap (Son, Cohen, & Louie, 2006). In order to obtain a minimum band gap at room temperature, the maximum width is allowed to be 10nm (Hwang et al., 2013). The width and the edge effects are crucial factors in determining the material performance (Huang et al., 2008b; Wang & Dai, 2010).

Geometries of Graphene Nanoribbon

According to quantum requirements, a band gap is needed for graphene to become semiconducting materials. One possible solution to generate a band gap in graphene is to form it into graphene nanoribbons. To obtain these, one may need to cut graphene sheets into narrow strips which possess certain electrical properties based on the cutting direction and angle (Wang et al., 2008). One can also consider

Figure 10. Transformation of graphene to GNR

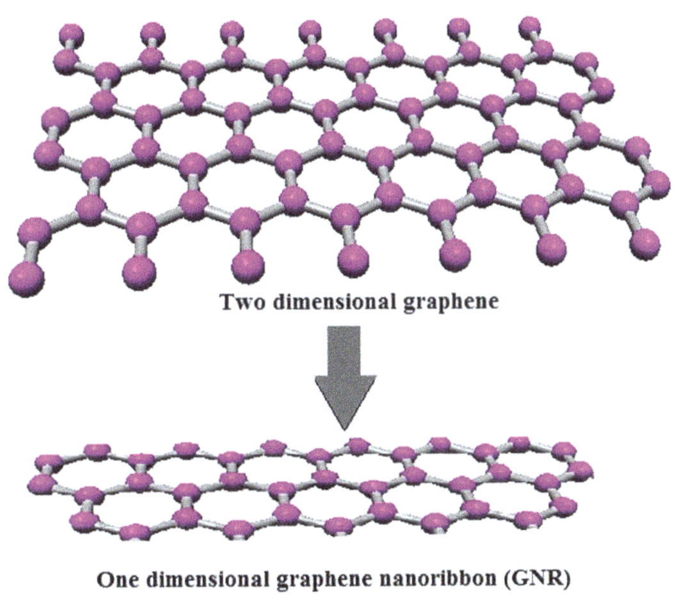

Figure 11. (a) Zigzag, (b) armchair GNR atomic structure
Wakabayashi et al., 1999.

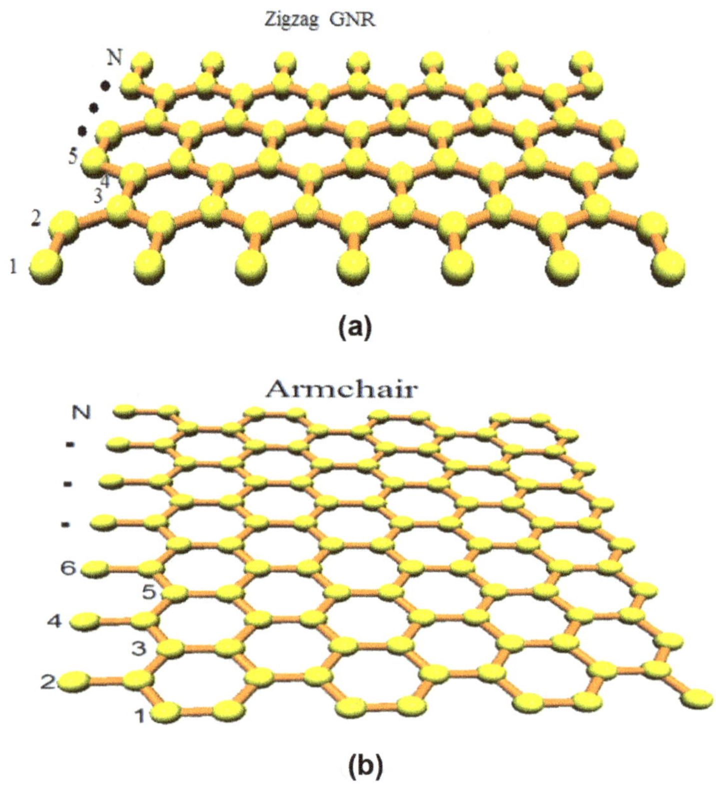

GNRs as unzipped single-walled carbon nanotubes. They can be formed in either zigzag or armchair configurations as shown in Figure 11a and b (Dalosto & Tinte, 2011; Owens, 2008).

Depending on their structure width, the zigzag GNRs are predicted to always exhibit metallic properties while armchair ones may show either semiconducting or metallic properties depending on the width of nanoribbon (Lu et al., 2009; Yazyev, 2013). The GNR electrical behaviour is influenced by the number of dimmer lines, N, in the way that when N=3p and N=3p+1, armchair graphene nanoribbons behave as semiconductors, and when N=3p+2, they exhibit metallic characteristics where p is any integer(Moradian, Mohammadi, & Ghobadi, 2008; Harada, Sato, & Yokoyama, 2013).Different types of edge geometries have been shown to impact the electronic transport in GNRs with a specific width causing them to exhibit like semiconducting behaviour or metallic properties (Kan *et al.*).

GNRs have become suitable materials for integrated circuit interconnect to replace copper owing to their excellent thermal and electrical conductivities as well as low noise effects (Chen et al., 2007; Xu et al., 2010). The GNR high quality has attracted many researchers and technologists to further study, understand and use it in the related fields. Among the largest of such areas is the construction of graphene based nanosensor. GNR-based sensor has recently become the potential device providing numerous benefits in semiconductor field (Jung, 2011).

CARBON NANOTUBE

The carbon nanotube is another carbon allotrope that has been widely explored since early 1990s. It is a cylindrical tubular carbon structure that is confined in diameter as the diameter must be between 0.7~10.0 nm (Haddon, 2002; Xu et al., 1999). This is less than the de Broglie's wavelength that made it a 1D structure (Saito et al.,1998). Research also indicated that the CNT exhibited high mobility as well as ballistic transport (Wang & Lundstrom, 2003). In addition, CNT can be well-defined as metallic, semiconductor or semi-metallic depending on its chirality and diameter (Anantram & Leonard, 2006). CNTs possess remarkably unique geometrical structures giving them outstanding characteristics which will be discussed in detail in the following section.

Structure of Carbon Nanotubes

A carbon nanotube (CNT) is a hollow cylinder made from rolling hexagonal lattice carbon (graphite) sheet. By rolling the graphite sheets in different directions, three types of carbon nanotubes can be created:

- Armchair,
- Zigzag, and
- Chiral (Dresselhaus, Dresselhaus, & Saito, 1995) (see Figure 12).

The chirality determines whether the carbon nanotubes are metallic or semi-conducting. Armchair tubes are metallic whereas zigzag and chiral tubes are semi-conducting (Chen et al., 1998). Carbon nanotubes have two forms; single-walled and multi-walled (Martel et al., 1998). A set of several single-walled carbon nanotubes in coaxial configuration create multi-walled carbon nanotubes. The SWCNT diameter is typically between 0.6 and 2.0nm (Jin et al., 2013). For MWCNTs, the inner diameter is typically larger than 2nm and outer diameter is smaller than 100nm (Ye et al., 2005). The length of nanotubes

Figure 12. Illustration showing the armchair, chiral, or zigzag configurations in nanotubes

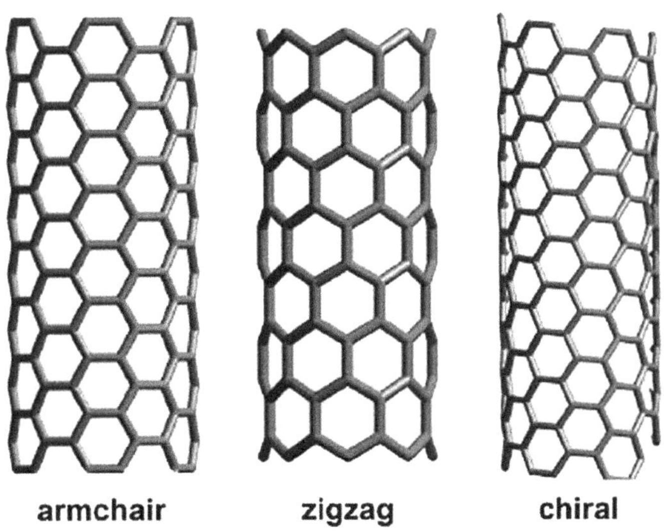

is between 10μm and100μm which causes nanotubes to have one-dimensional structure with a very high aspect ratio (Ganesh , Laurent, Flahaut & Peigney, 2010).

CNTs have excellent mechanical, electrical and thermal properties because of their bonding structure of carbon atoms. A carbon atom in a CNT possesses six electrons two of which are located in the first orbital and the other four fill the sp^2 orbital (Dresselhaus et al., 2000, Baughman et al., 2002).

The large surface area of a CNT makes it an excellent material for molecular adsorption. A number of gas molecules such as O_2, CO_2, NH_3 and NO_2 are reported to be possibly adsorbed by CNTs (Zhao et al., 2002; Sonawane et al.). CNT, as a one-dimensional conductor, allows for many applications in electronic devices such as sensors (Goldoni et al., 2003; Li et al., 2003; Cole & Zook, 2005), actuators (Baughman et al., 1999; Baughman, 2000; Hughes & Spinks, 2005)and transistors (Javey et al., 2003; Postma et al., 2001; Franklin et al., 2013). Sensors in this regard are indeed the most important and thus require more careful consideration. For this reason, the following section and the corresponding sub-sections have been dedicated to the basics of the working principles of a sensor.

SENSORS

Sensor is a term used for devices that can measure specific physical quantities and convert them into a readable electrical signal for an observer or an instrument (Pallas-Areny, Webster, & Areny, 2001). Figure 13 shows the basic sensing process model.

Measurement is generally the process of comparing an input physical quantity to a device (which is able to convert it into an electrical signal) with a predefined standard quantity (Kreibig, Gartz, & Hilger, 1997). In a sensor, the measurement accuracy and effectiveness are very important factors in instrument selection for different types of studies. Therefore, it is necessary for researchers to fully understand the sensor performance. Sensor efficiency is determined by its physical characteristics and has been reported in the works of many researchers (Patra et al., 2008).

Sensors come in many different types based on the intended material that they are expected to detect, as well as mechanisms of detection. They can be classified as electromagnetic sensors, mechanical sensors, thermal sensors, etc. The trend in sensor manufacturing and production is heading toward devices with higher sensitivity, better selectivity and faster response time; also those which are easier to fabricate, more portable, remotely capable and more cost-effective are more desirable (Fraden, 2010). In addition to the main sensing function, the sensor is expected to keep track of various ambient factors such as temperature and humidity, time, location and event history. Rapid improvements in nanoscience and engineering, as well as faster and more advanced compact integrated electronics, are helping

Figure 13. Sensing process model

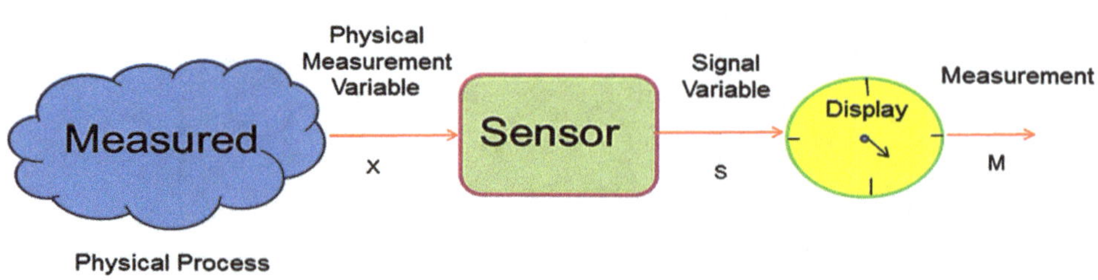

these requirements be met (Frank, 2013). In this regard, many new materials like graphene and carbon nanotubes are now becoming available for the design fabrication of nanosensors (Patra et al., 2008; Wang & Arash, 2014).

Nanoscale Sensor

As previously mentioned about Moore's Law, transistor sizes have been rapidly scaled down every two years (Schaller, 1997). Thus, latest technology has also been driven toward smaller dimensions, i.e. nanoscale (Islam, 2008). As the name shows, a nanosensor is a sensor built in nanoscale (Riu, Maroto, & Rius, 2006). The main job of the nanosensor is to first receive data on the atomic scale, convert it to a readable electrical signal, and then transfer it through a simple way for the sake of experimental analysis. Not only do nanosensors enhance the sensing performance by integrating more functions into a single small package, but they also reduce the device power consumption (Panigrahi, Mahapatra, & Samantaray). Furthermore, they have competitive sensitivity and response time compared to the conventional macro scaled sensors (Assadi & Farshi, 2011). Gas sensors, in this regard, possess great importance and have therefore been focused on in the following section.

Significance of Gas Sensor Application

One motivation for the detection of toxic gases is the need for environmental monitoring of gas pollutants such as NO_x, CO and CO_2. For example, nitrogen dioxide (NO_2), a member of the family of nitrogen oxide (NO_x) gases, is an important marker of air quality standards for the NO_x gases. NO_2 forms naturally in the atmosphere by lightning and is also generated from water, plants or sand. However, only 1% of the atmospheric NO_2 is formed naturally and the rest is primarily generated from fuel combustion processes as a by-product. Upon reaction with water, NO_2 can form HNO_3which is the main component in acid rain and poses a significant threat to the environment. Due to its increasing population and industrial development, Los Angeles has suffered from significant air pollution consisting of NO_x,CO, O_3 and other oxidizers since the 1940s leading to the first air pollution control act in 1955(Hays & Hays, 1989). Gases are critical reactants for numerous manufacturing processes, so an accurate and low-cost method to determine gas composition is crucial (Li & Liu, 2009).

NO_x is harmful to human respiration system at 3ppm concentration. Studies have shown that NO_2enhances the effects of allergens (Kozlov et al., 2013)as well as bronchial reactivity. NO_2 is a deep-lung irritant which can cause pulmonary edema(exposure to 50ppm NO_2) and is fatal at high concentrations (>100ppm inhaled). Recent epidemiological studies suggest a strong correlation between the development of chronic lung disease and respiratory symptoms with long-term exposure to NO_2(Koehler et al., 2013). NO_2 concentrations near vehicles are two to three times higher than that in the atmosphere of a city. People who spend long times near vehicles are at high risk for exposure to NO_2, putting them at risk for respiratory problems, especially asthma. Therefore, an accurate and sensitive toxic gas sensor for NO_2 is needed (Mustafa & Tierney, 1978; Anderson et al., 2012).

GRAPHENE FET BASED GAS SENSOR

Due to its two-dimensional nature, graphene has a maximum surface to volume ratio and is highly suitable for sensors based on surface interactions, such as gas sensors (Massera et al., 2012; Bradley, 2012). In addition, due to its low density of defect states and metallic conductivity, graphene has small 1/f noise and low Johnson noise (Schedin et al., 2007; Balandin, 2013). Its low noise and high mobility together lead to high signal-to-noise ratio (SNR) which is critical for the detection limit in the sensor. Empirical findings show that electron mobility of graphene sheet which is weakly temperature reliant can be as high as to15,000 cm^2. $V^{-1}.s^{-1}$ (Yin et al., 2009)with carrier density of 10^{12} cm^{-2} (Rana et al., 2009)at room temperature. Specifically, the extraordinary mobility of carriers in graphene as much as 200,000 $cm^2.V^{-1}s^{-1}$ (Fuhrer, Lau, & MacDonald, 2010)has been used to explain its high sensitivity, which both provide extremely low noise sensing at room temperature. Due to its single atomic layer of graphite with surface, graphene is adjustable through the direct interaction between each atom in the structure and the sensing environment (Hill, Zhu, & Pan, 2011). Therefore, the interaction between the surface dopants and absorbents can be maximized (Ao et al., 2008).

Graphene based nanostructure provides a special physical uniqueness which can be used in many types of applications including some classes of sensors like optical, magnetic, electronic field, strain and mass sensors as well as field effect, electrochemical and piezoelectric gas sensors (Yin et al., 2013; Zhou et al., 2009). It has been evident to date that graphene based sensors can be achieved with best sensor performance out of graphene (Jeong et al., 2010; Chung et al., 2012).

The prototype FET structures have indicated excellent performance for transistors, interconnects, electromechanical switches, infrared emitters and biosensors as shown in Figure 14 (Schwierz, 2010a). Due to its excellent electronic properties, graphene has become a promising new material for nanotechnology particularly in the sensor area (Liang et al., 2007; Olthuis et al., 2007).

As one can see in the Figure 15, it looks similar to the famous metal-oxide semiconductor field effect transistor (MOSFET) which is comprised of a drain metal, a source metal, a silicon back gate and a gate insulator. A graphene channel connects the source and drain electrodes, and a dielectric barrier layer (SiO_2) separates the gate from the channel. In most studies, SiO_2 layer is used beneath the graphene to act as a dielectric layer while silicon serves as the back gate (Sven Kochmann, 2012). When gas molecules

Figure 14. Device schematic
Romero et al., 2009; Schwierz, 2010b.

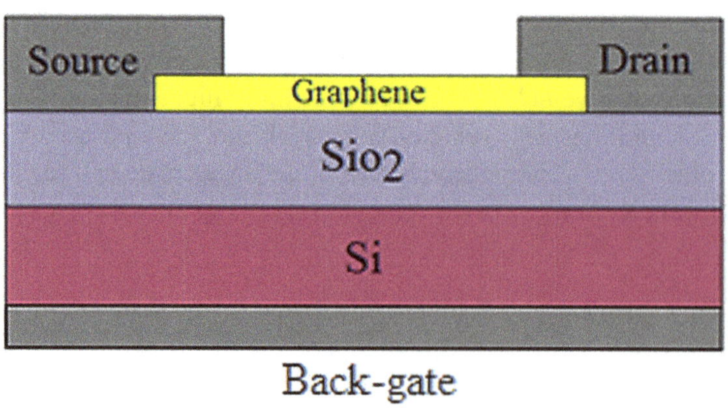

Figure 15. Sensor structure
Singh et al., 2011.

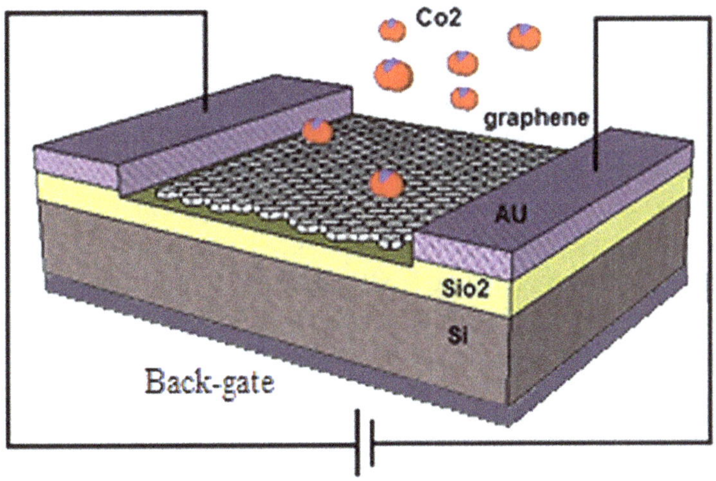

attach to the surface or edge of graphene, carrier concentration will change due to the variability of the drain source current which is a measurable parameter (Zhou et al., 2011).

Theoretical studies have reported that the gas induces various alterations of different magnitudes on resistivity depending on the gas type and whether it is an electron donor (CO, NH_3 and C_2H_6O) or an electron acceptor (NO_2, H_2O) (Hu et al., 2012, Leenaerts, Partoens, & Peeters, 2008, Zhang et al., 2009). Previous experimental results have demonstrated that the graphene charge carrier concentration is increased with increasing gas molecules adsorption (Carrillo, Rangel, & Magaña, 2009). From the chemical perspective, gas adsorption can cause the number of holes to increase if the gas is an acceptor or can make the number of electrons to surge provided that the gas is a donor (Ratinac et al., 2010). According to the relation between the conductivity and charge carrier density or carrier mobility, the adsorbed molecules change the conductivity of the graphene (Hwang, Adam, & Sarma, 2007).

Schedin et al. demonstrated the first graphene based gas sensor detectingNO_2, H_2O, CO and NH_3. The graphene was produced using mechanical exfoliation and was fabricated on a SiO_2 substrate with a back gate. Due to the observed conductance changes, their results suggest that NO_2 as well as H_2O are acceptors, and CO and $NH_3$are donors in graphene. Upon exposure to gas molecules, the carrier density in graphene changed while the mobility was independent of molecular adsorption. (Schedin et al., 2007).

A carbon dioxide gas sensor with a graphene sheet as the sensing medium fabricated by mechanical cleavage was developed by H.J. Yoon et al (2011). It was shown that the graphene sensor can operate at room temperature and in ambient conditions. Conductance changes of the device were measured for various CO_2 gas concentrations adsorbed on graphene surface. These measurements showed that when the concentration of CO_2 gas is increased from 10 to 100ppm, graphene conductance in the gas sensor increases linearly. High sensitivity, low power consumption, fast response time (sensor response in less than 10s), as well as short recovery time were among the advantages of this sensor.

Nitrogen dioxide gas sensors with graphene based structure were proposed, developed and studied by Ko et al. (2010). They showed that NO_2 molecules could be selectively adsorbed/desorbed by graphene at room temperature. The chemical doping with NO_2 molecules altered the conductivity of the graphene layers. By monitoring the current–voltage characteristics at various NO_2 gas concentrations, this phenomenon was quantified. They found that adsorption was faster than desorption, and they attributed this to the reaction occurring on the graphene layer surface. The sensitivity in a 100ppm ambient NO_2 was 9%. Graphene based gas sensors showed good selectivity as well as reversibility, fast response and high sensitivity.

Arash and Wang in 2012 investigated the potential usability of single-layered sheets of graphene as nano-sensors in the detection of noble gases through wave propagation using molecular dynamics simulations. The simulation results in their work indicated that the nanosensor was able to distinguish noble gas atoms with a recognizable sensitivity (Arash & Wang, 2012).

Rahmani et al. (2013) have studied the current-voltage characteristic in a Schottky-barrier FET as a function of its physical properties including GNR length, thickness of gate insulator, effective mass, and electrical parameters including the applied bias voltage and Schottky barrier height. In their study, they considered and analytically modelled the scaling behaviour of a Schottky-barrier FET employing graphene in a trilayer configuration.

Karimi et al. (Abadi et al., 2013)have presented field effect transistor based on nanostructured graphene for the purpose of electrical detection of deoxyribonucleic acid (DNA) hybridization. By modelling the conductance of graphene sheets, they proposed a method of electrical detection of DNA hybridization. Assuming DNA concentration as a function of gate voltage, they defined the sensing factor.

Kiani et al. (2013a) have shown modelling of bilayer graphene nanoribbon (BGNR) sensor. Based on their presented model, they considered BGNR quantum capacitance variation effect by the prostate specific antigen (PSA) injected electrons in to the FET channel as a sensing mechanism.

Akbari et al. (2013a) have demonstrated an analytical model for graphene based CO_2 sensor. They provide new idea of gas sensor model based on the gradient of GNR conductance as a function of gas concentration. They incorporated a field effect transistor based structure for the modelling platform.

In the thesis in hand, graphene as well as single-walled CNT structures have been proposed as the channel medium in the FET-based sensor.

Graphene Sensing Mechanism

There are various kinds of harmful gases to the organic life which are difficult to observe and sense. Therefore, a sensor or detection system is a necessary component to detect these gases in order to prevent them from jeopardizing human, animal and plant life. Among all gas sensors, the best sensor would be defined as the one that can detect even one atom or molecule of the chemical or gas(Rivera et al., 2010; Lee et al. 2011, Pearce et al., 2011; Ho et al., 2013; Wu et al.; 2013a; He et al., 2012). Latest studies on molecular adsorption of gases on graphene surface show that NH_3, H_2O, NO_2, NO, and CO molecules are adsorbed on pristine graphene. A schematic of a graphene gas sensor proposed by Ko et al. is illustrated in Figure 16. It has been evident that NH_3 and CO molecules act as donors while H_2O and NO_2 act as acceptors, which are consistent with previous experiments (Huang et al., 2008a; Leenaerts et al., 2008).

Gases such as CO_2, CO, NO, NO_2, and O_2, have electron withdrawing capability, while NH_3 is an electron-donating functional molecule (Peng & Li, 2013). This is consistent with our conclusion that there is a much stronger interaction between gas molecules and graphene edges than its surface. Significantly, gas molecules can modulate the electronic structure of graphene in diverse ways: the adsorption of CO and NO will induce deep defect states in the band gap; the adsorption of NH_3 leads to n-type behaviour while the adsorption of CO_2 and O_2 transfers the system to p-type semiconductor (Zhang et al., 2009). These rich adsorption effects are originated from the intrinsic property of gas molecules and the bonding

Figure 16. Graphene sensor
Ko et al., 2010.

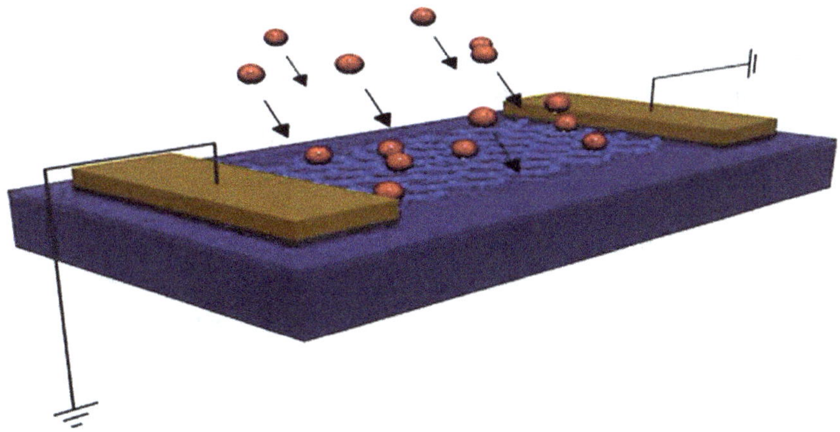

characteristics between the gas molecules and the graphene. The resulting p-type and n-type semiconducting behaviour might be detected in experiment by applying gate voltage and could be useful from the viewpoint of applications, since it is always important to obtain n-type and/or p-type semiconducting graphene for use in nanoscale electronic devices, such as the p-n junction as well as n-type and p-type nanoscale field effect transistors (Huang et al., 2008b).

Zhang et al. proposed to increase graphene sensitivity by introducing dopants and defects and studied these using first principle simulations. In their simulations the adsorption energy was computed for small gas molecules on graphene. All gas molecules exhibited much stronger adsorption on the defective or doped graphene than on the pristine graphene (Zhang et al., 2013).

Huang et al. investigated the impact of edge states on a GNR-based gas sensor. They confirmed that CO, NO_2, O_2 and CO_2 have electron-withdrawing properties and NH_3serves as electron-donating molecule with 10-atom width GNR. Different gas molecules will induce different effect (Huang et al., 2011).

CNT FET BASED GAS SENSOR

One of the areas where nanotechnology has shown potential for immediate commercialization is the use of nanotechnology enabled state-of-the-art sensors for environmental monitoring, medical diagnostics, food quality surveillance etc. Among the nanostructures, carbon nanotubes have been the most attractive material of nanotechnology community since their discovery (Jimenez-Cadena, Riu, & Rius, 2007; Mahmood, Burhanudin, & Hamid, 2013).

The aforementioned properties along with additional features such as small size and high electrical sensitivity make CNTs ideal candidates for use as nanosensors (Riu et al., 2006). Experimental and theoretical studies reveal that these nanometer sized CNTs exhibit unique electrical characteristics making them metallic or semiconducting based on their radial dimensions and chiralities (Panzer et al., 2008). More specifically, SWCNTs have the proven ability to sense different small molecules such as NO_2, NH_3, HCl, Cl_2, etc (Zhao et al., 2002). One suitable configuration for using carbon nanotubes in measuring devices is one in which the CNT is placed as an electrical wire between two electrodes. By applying a specific gate bias voltage, CNT conductance is used as the measured variable (Allen, Kichambare, & Star, 2007; Someya et al., 2003; Kim et al., 2012). Mechanical deformations (Suehiro et al., 2005) and/or chemical doping can significantly affect conductance as the CNT electrical properties are heavily dependent upon the atomic structure. Such variation in electrical properties can be easily detected by measuring the electric current. This makes CNTs valuable minute sensors able to detect changes in their environment as they have already been employed in highly sensitive electronic molecule detection (Chen et al., 2003; Star et al., 2003).

It is a well-known fact that the characteristics of carbon nanotubes are strongly dependent on their physical properties such as diameter and chirality (Pesetski et al., 2006; Baughman et al., 1999). For instance, carbon nanotubes can be either single-walled or multi-walled with varying inherent band gaps. Based on the chirality of their structure, single-walled nanotubes are either semiconductors or metallic conductors. As can be seen in Figure 17, semiconducting SWCNTs can be used in the fabrication of FET devices able to operate at room temperature and in ambient conditions (Wang et al., 2007; Ding et al., 2009).

Several phenomenal thermal, mechanical and electronic properties of CNTs have been utilized in the design of gas sensors. The sensors which have been developed so far use different preparing methods of CNTs in sensors, as well as different mechanisms of interaction between analytes and the CNT. In gas sensor manufacturing, the most commonly used features are the alterations in CNTs electrical properties resulting from their interaction with analytes. This phenomenon has been used for detection of hydrogen (Santangelo et al., 2013), ammonia (Han et al., 2013; Teerapanich et al., 2013), and nitrogen dioxide (Leghrib et al., 2011; Zhang et al., 2008; Cantalini et al., 2003).

Figure 17. Carbon nanotube based FET structure for gas detection

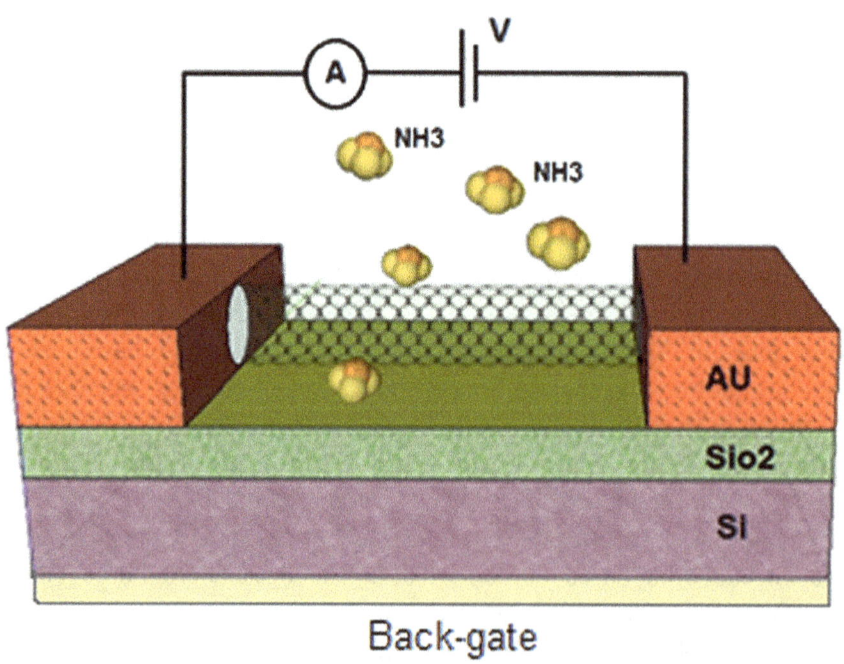

ARTIFICIAL NEURAL NETWORK (ANN)

Artificial neural networks is an intelligent algorithm which has been developed based on an analogy to biological nervous system (Yegnanarayana, 2004; Ünal et al., 2013). Various types and structures of artificial neural networks have been employed in scientific analytical studies among which, the most common and comprehensible is one consisting of interconnected group of neurons which are mathematical operator units called Perceptrons (Abolgasim, 2008; Patterson, 1998; Aldrich & Auret, 2013).

A schematic of the typical structure of an ANN comprising perceptrons is provided in Figure 18. Each input value to a perceptron is multiplied by a weight and added to a bias value. The general mechanism of ANN includes a process in which a set of input data are introduced to the network. Through a series of mathematical calculations based on predefined "activation functions" for each neuron or "node", an output value is given by the ANN. In order for the ANN to learn to compute the optimum output value, a training data set is introduced to the network before the actual inputs are given to the artificial neural network (Zhang, 2000; Uhrig, 1995; Kupny, 2011; Hamker, 2009).

Based on a predefined learning algorithm, the ANN updates the weights and bias corresponding to each node using the error between the calculated results from the actual inputs and the desired output (Grossi & Buscema, 2007; Yao, 1999). In our study, feed-forward structure for the neural network with back propagation learning algorithm has been implemented. The experimental data has been used as the training data set has been employed for validation and testing. The results show satisfactory agreement between the results from the proposed model with the experimental counterparts.

Figure 18. Simple artificial neural networks

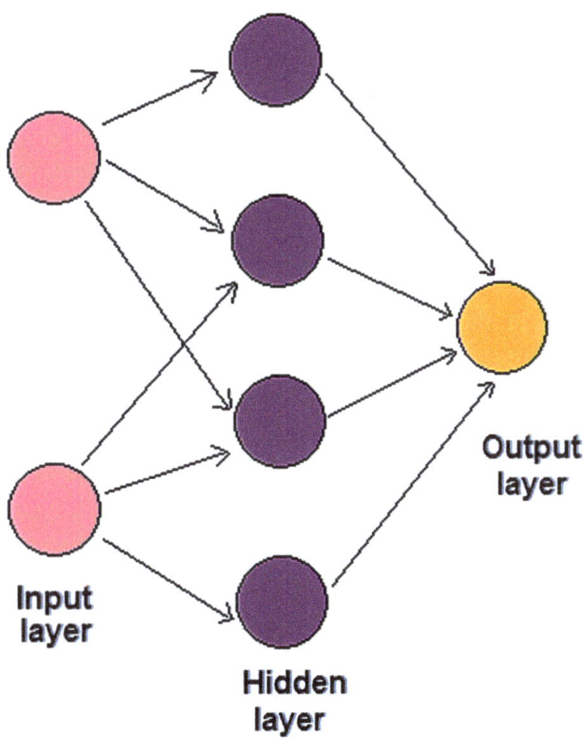

Mohsen Hayati et al. have applied artificial neural network in the simulation and modelling of CNT metal–oxide-semiconductor field-effect transistors. Their simulation was based on an artificial neural network model that reduces the time required for computation while attempts to keep the physics-based model accuracy same way as non-equilibrium Green's function (NEGF) formalism (Hayati, Rezaei, & Seifi, 2010).

Amir Amani et al. (2011)have described ANN principles employed for dealing with certain nanomaterials properties. They have evaluated available publications on nanotechnology and nanoscience which have used the advantages of ANNs. They have provided in their study, a brief description on the significance of nanotechnology. Common ANN applications have been summarized in their paper and it has been attempted to identify different areas in nanotechnology and nanoscience where successful application of artificial neural networks can be envisaged.

SerhanYamacli et al. (2010)aimed to develop voltage-dependent resistance models of metallic CNTs for computer aided design(CAD) tools. Using first principles simulations, they first obtained the resistance of CNT interconnects and they developed a model for voltage dependence of the resistance using ANN. Using self-consistent non-equilibrium Green's function formalism together with density functional theory the authors attempt to calculate the voltage-dependent resistance of metallic CNTs.

ARC DISCHARGE METHOD

In conventional plasma processing techniques, the use of vacuum is a great burden for industrial scale applications. In new scenarios, Atmospheric Pressure Plasmas (APP) has gained a great interest for material processing as it is a cost-effective and energy-saving technique (Machala et al., 2004). Arc discharge plasmas fall in the category of non-thermal plasma which has a property to generate unique reactive species that are independent of reaction temperature. These species are used to initiate chemical reactions which have lower temperatures than normal thermo-chemical reactions (Tao et al., 2011). These attributes of non-thermal plasmas have an attraction for surface modification, diamond-like carbon processing (DLC), deposition of carbon nanotube(CNT), fullerene and grapheme production (Subrahmanyam et al., 2009). High current electric arc discharge synthesis is the first recognized technique after Iijima (1991) (Journet & Bernier, 1998)who produced the first MWCNTs by this method (Hutchison et al., 2001). DC Arc discharge technique is yet believed to be the cost-effective method to produce carbon allotropes/CNTs in bulk quantities (Ando & Zhao, 2006). Some approaches of arc discharge plasmas like gliding arc process (Sagas et al., 2011), corona arc process (Zhao et al., 2006) and dielectric barrier discharge (DBD) corona process (Vincent-Randonnier et al., 2007) are also used abundantly for methane processing. Arc discharge plasmas are also used for chemical conversion of hydrocarbon gases like conversion of methane to methanol and acetylene (Kim et al., 2003, Mishra et al., 2004). Among hydrocarbon gases, methane is a cheap precursor which has been used in many studies such as direct conversion into synthetic fuels like methanol (Beloqui Redondo, Troussard, & van Bokhoven, 2012; Nozaki, 2011; Hijikata, Ogawa & Miyakawa, 1999), acetylene (Zhu et al., 2002) and for production of hydrogen (Li, 2008). These arc discharge techniques are used with some solid catalysts to provide an environment for homogeneous gas phase reaction that help to reduce the capital and operating costs of the high temperature (Tu & Whitehead, 2012, Chavadej et al., 2005; Indarto et al., 2007).

Optical emission spectroscopy (OES) appears as the most proper tool for in-situ diagnostics of plasma parameter especially the electron temperature and density (Abdel-Baky, El-Ghazaly, & Mansour, 2000;

Schmiedt, Kaňka, Hrachová , Gordillo-Vázquez, Camero, & Gomez-Aleixandre, 2006). The literature on OES diagnostics of arc discharge methane plasmas is rare and relatively scattered (Gordillo-Vázquez et al., 2006; Beulens et al., 1994; Jun-Feng et al., 2009). Catherine *et al.* (Barholm-Hansen, Bentzon, & Hansen, 1994; Krcma et al., 2010; Kraus et al., 2002)have compared the OES data of low-pressure methane plasmas generated by low-frequency (2550 kHz) (Nozaki et al., 2001)and radio frequency (13.56 MHz) (Massereau-Guilbaud, Géraud-Grenier, & Plain, 2009) power sources. OES diagnostics of atmospheric pressure micro-plasma jet using methane mixture (Ar/CH4) with noble gases has been studied by Yanguas-Gil et al. (Sternschulte, Schreck, & Stritzker, 2002; Patacsil, Malapit, & Ramos, 2006; Bousrih et al., 1995). So far, literature about OES diagnosis of high voltage arc discharge pure methane plasma especially generated at atmospheric pressure is scarce.

In the experimental study conducted in this work, atmospheric pressure high voltage arc discharge plasma is produced during flow of pure methane 99.99% into tubular reactor. Methane plasma is characterized by OES in the wavelength range of 350-750 nm to deduce qualitative information of the emitted species. Details on the technical procedure as well as the experimental tests and the corresponding results will be presented in the appendix.

CONCLUSION

In the analytical procedures, various electro physical properties and phenomena have been considered and various control parameters (α, β, δ, λ) are incorporated and calculated by iteration method. The models have been validated and compared with experimental data previously published in the literature. The comparative studies show satisfactory agreement and hence the models have been proven valid.

In addition to the analytical model, an artificial neural network model was also developed for the single case of I-V characteristics of a CNT based gas sensor with FET structure. The ANN model employs feed forward back propagation algorithm and uses the experimental data as the training data set. Both models are able to produce good results with satisfactory agreement with the extracted experimental data. The ANN model, however, has proved to be able to produce more accurate results than those by the analytical counterpart.

For the sake of providing a more profound comprehensive insight, a set of experimental studies has been conducted. These empirical studies incorporate high voltage arc discharge methane decomposition to produce carbonaceous materials between two graphite electrodes. Different electrode configurations as well as different gaps between them were employed in the arc discharge carbon deposition processes. The materials obtained in these experiments have been inspected using high resolution optical microscopy as well as scan electron microscopy (SEM) to study the material structure in micro scale.

In addition, the current-voltage characteristics of the C-strands obtained were tested and evaluated in order to assess their potential ability to replace the channel in gas sensors. Applying a DC voltage to the C-strands in successive time periods after the growth, the corresponding currents passing through the conductive channel has been measured and recorded. Two main observations were made during the experimental study. Firstly, the electrode configuration and the gaps between them have direct impact on the fineness and the conductivity of the carbonaceous material. Secondly, the conductivity of the C-strand changes in the presence of gas, i.e. when the channel is exposed to gas, due to the adsorption of gas atoms on the surface of the C-strand, its conductivity increases. Therefore, as a general conclusion, this study proposes that both graphene and CNTs can be used in gas detection sensors presenting high sensitivity and empowering the sensor manufacturers to produce accurate sensing mechanisms in extremely smaller sizes.

REFERENCES

Abadi, H. K. F., Yusof, R., Eshrati, S. M., Naghib, S., Rahmani, M., Ghadiri, M., & Ahmadi, M. et al. (2013). Current–voltage modeling of graphene-based DNA sensor. *Neural Computing & Applications*, 1–5.

Abdel-Baky, A., El-Ghazaly, M., & Mansour, M. (2000). Spectroscopic and electric probe measurements of plasma parameters in the negative glow region of helium discharge plasma. *Journal of Quantitative Spectroscopy & Radiative Transfer*, *67*(3), 217–225. doi:10.1016/S0022-4073(99)00209-5

Abolgasim, A. A. (2008). *Classification of capital expenditures and revenue expenditures using neural network model*. Universiti Utara Malaysia.

Acik, M., & Chabal, Y. J. (2011). Nature of graphene edges: A review. *Japanese Journal of Applied Physics*, 50.

Adam, S., Hwang, E., Galitski, V., & Sarma, S. D. (2007a). A self-consistent theory for graphene transport. *Proceedings of the National Academy of Sciences of the United States of America*, *104*(47), 18392–18397. doi:10.1073/pnas.0704772104 PMID:18003926

Adam, S., Hwang, E. H., Galitski, V. M., & Das Sarma, S. (2007b). A self-consistent theory for graphene transport. *Proceedings of the National Academy of Sciences of the United States of America*, *104*(47), 18392–18397. doi:10.1073/pnas.0704772104 PMID:18003926

Akbari, E., Ahmadi, M., Kiani, M., Feizabadi, H. K., Rahmani, M., & Khalid, M. (2013a). Monolayer graphene based co2 gas sensor analytical model. *Journal of Computational and Theoretical Nanoscience*, *10*(6), 1301–1304. doi:10.1166/jctn.2013.2846

Akbari, E., Ahmadi, M., Yusof, R., Ghadiry, M., & Saeidmanesh, M. (2013b). Gas concentration effect on channel capacitance in graphene based sensors. *Journal of Computational and Theoretical Nanoscience*, *10*(10), 2449–2452. doi:10.1166/jctn.2013.3229

Aldrich, C., & Auret, L. (2013). Artificial neural networks. In Unsupervised Process Monitoring and Fault Diagnosis with Machine Learning Methods, (pp. 71-115). Springer. doi:10.1007/978-1-4471-5185-2_3

Allen, B. L., Kichambare, P. D., & Star, A. (2007). Carbon nanotube field-effect-transistor-based biosensors. *Advanced Materials*, *19*(11), 1439–1451. doi:10.1002/adma.200602043

Amani, A., & Mohammadyani, D. (2011). *Artificial neural networks: applications in nanotechnology. In Artificial neural networks—application*. Rijeka: INTECH.

Amin, N. A., Ahmadi, M. T., Johari, Z., Mousavi, S. M., & Ismail, R. (2011). Effective mobility model of graphene nanoribbon in parabolic band energy. *Modern Physics Letters B*, *25*(10), 739–745. doi:10.1142/S0217984911026073

Anantram, M., & Leonard, F. (2006). Physics of carbon nanotube electronic devices. *Reports on Progress in Physics*, *69*(3), 507–561. doi:10.1088/0034-4885/69/3/R01

Anderson, B. A., & Nowak, E. J. (2009). *Graphene-based field effect transistor has pair of graphene layers located on pair of sidewalls of silicon carbide fin. Anderson B a.* Nowak E J.

Anderson, H., Atkinson, Bremner, Carrington, & Peacock. (2012). *Quantitative systematic review of short term associations between ambient air pollution (particulate matter, ozone, nitrogen dioxide, sulphur dioxide and carbon monoxide), and mortality and morbidity*. Report to Department of Health revised following first review.

Ando, Y., & Zhao, X. (2006). Synthesis of carbon nanotubes by arc-discharge method. *New Diamond and Frontier Carbon Technology*, *16*, 123–138.

Ao, Z., Yang, J., Li, S., & Jiang, Q. (2008). Enhancement of CO detection in Al doped graphene. *Chemical Physics Letters*, *461*(4-6), 276–279. doi:10.1016/j.cplett.2008.07.039

Arash, B., & Wang, Q. (2012). Detection of gas atoms with graphene sheets. *Computational Materials Science*, *60*, 245–249. doi:10.1016/j.commatsci.2012.03.053

Assadi, A., & Farshi, B. (2011). Stability analysis of graphene based laminated composite sheets under non-uniform inplane loading by nonlocal elasticity. *Applied Mathematical Modelling*, *35*(9), 4541–4549. doi:10.1016/j.apm.2011.03.020

Balandin, A. A. (2013). *Review of the low-frequency 1/f noise in graphene devices.* arXiv preprint arXiv:1307.4797

Balandin, A. A., Kotchetkov, D., & Ghosh, S. (2009). *Lateral graphene heat spreaders for electronic and optoelectronic devices and circuits*. Google Patents.

Barholm-Hansen, C., Bentzon, M., & Hansen, J. B. (1994). Optical emission spectroscopy during growth of diamond-like carbon from a methane plasma. *Diamond and Related Materials*, *3*(4-6), 564–568. doi:10.1016/0925-9635(94)90225-9

Baughman, R. H. 2000. Carbon nanotube actuators. In *APS Meeting Abstracts*, 10001. doi:10.1126/science.284.5418.1340

Baughman, R. H., Cui, C., Zakhidov, A. A., Iqbal, Z., Barisci, J. N., Spinks, G. M., & Rinzler, A. G. et al. (1999). Carbon nanotube actuators. *Science*, *284*(5418), 1340–1344. doi:10.1126/science.284.5418.1340 PMID:10334985

Baughman, R. H., Zakhidov, A. A., & de Heer, W. A. (2002). Carbon nanotubes--the route toward applications. *Science*, *297*(5582), 787–792. doi:10.1126/science.1060928 PMID:12161643

Bellucci, S. (2005). Carbon nanotubes: Physics and applications. *Physica Status Solidi (c)*, *2*, 34-47.

Beloqui Redondo, A., Troussard, E., & van Bokhoven, J. (2012). Non-oxidative methane conversion assisted by corona discharge. *Fuel Processing Technology*, *104*, 265–270. doi:10.1016/j.fuproc.2012.05.021

Beulens, J., Gastineau, C., Guerrassimov, N., Koulidiati, J., & Schram, D. (1994). Atomic and molecular emission spectroscopy on an expanding argon/methane plasma. *Plasma Chemistry and Plasma Processing, 14*(1), 15–42. doi:10.1007/BF01448735

Bousrih, S., Ershov-Pavlov, E., Megy, S., & Baronnet, J.-M. (1995). Hydrogen/argon plasma jet with methane addition. *Plasma Chemistry and Plasma Processing, 15*(2), 333–351. doi:10.1007/BF01459702

Bradley, D. (2012). Graphene gas sensor carbon. *Materials Today, 15*(6), 233–233. doi:10.1016/S1369-7021(12)70105-0

Brey, L., & Fertig, H. (2006). Electronic states of graphene nanoribbons studied with the Dirac equation. *Physical Review B: Condensed Matter and Materials Physics, 73*(23), 235411. doi:10.1103/PhysRevB.73.235411

Cantalini, C., Valentini, L., Lozzi, L., Armentano, I., Kenny, J., & Santucci, S. (2003). NO< sub> 2</sub> gas sensitivity of carbon nanotubes obtained by plasma enhanced chemical vapor deposition. *Sensors and Actuators. B, Chemical, 93*(1-3), 333–337. doi:10.1016/S0925-4005(03)00224-7

Carrillo, I., Rangel, E., & Magaña, L. (2009). Adsorption of carbon dioxide and methane on graphene with a high titanium coverage. *Carbon, 47*(11), 2758–2760. doi:10.1016/j.carbon.2009.06.022

Chavadej, S., Supat, K., Lobban, L. L., & Mallinson, R. G. (2005). Partial oxidation of methane and carbon dioxide reforming with methane in corona discharge with/without Pt/KL catalyst. *Journal of Chemical Engineering of Japan, 38*(3), 163–170. doi:10.1252/jcej.38.163

Chen, J., Hamon, M. A., Hu, H., Chen, Y., Rao, A. M., Eklund, P. C., & Haddon, R. C. (1998). Solution properties of single-walled carbon nanotubes. *Science, 282*(5386), 95–98. doi:10.1126/science.282.5386.95 PMID:9756485

Chen, R. J., Bangsaruntip, S., Drouvalakis, K. A., Wong Shi Kam, N., Shim, M., Li, Y., & Dai, H. et al. (2003). Noncovalent functionalization of carbon nanotubes for highly specific electronic biosensors. *Proceedings of the National Academy of Sciences of the United States of America, 100*(9), 4984–4989. doi:10.1073/pnas.0837064100 PMID:12697899

Chen, Z., Lin, Y.-M., Rooks, M. J., & Avouris, P. (2007). Graphene nano-ribbon electronics. *Physica E, Low-Dimensional Systems and Nanostructures, 40*(2), 228–232. doi:10.1016/j.physe.2007.06.020

Chung, M. G., Kim, D. H., Lee, H. M., Kim, T., Choi, J. H., Seo, D. K., & Kim, Y. H. et al. (2012). Highly sensitive NO2 gas sensor based on ozone treated graphene. *Sensors and Actuators. B, Chemical, 166*, 172–176. doi:10.1016/j.snb.2012.02.036

Cole, B. E., & Zook, D. J. (2005). *Carbon nanotube sensor.* Google Patents.

Cole, B. E., & Zook, D. J. (2006). *Carbon nanotube sensor.* Google Patents.

Dalosto, S. D., & Tinte, S. (2011). Fluctuation effects of the electric field induced by water on a graphene dot band gap. *The Journal of Physical Chemistry C, 115*(11), 4381–4386. doi:10.1021/jp109297p

Diederich, F., & Rubin, Y. (1992). Synthetic approaches toward molecular and polymeric carbon allotropes. *Angewandte Chemie International Edition in English, 31*(9), 1101–1123. doi:10.1002/anie.199211013

Ding, L., Wang, S., Zhang, Z. Y., Zeng, Q. S., Wang, Z. X., Pei, T., & Peng, L. M. et al. (2009). Y-contacted high-performance n-type single-walled carbon nanotube field-effect transistors: Scaling and comparison with sc-contacted devices. *Nano Letters*, *9*(12), 4209–4214. doi:10.1021/nl9024243 PMID:19995085

Dresselhaus, M., Dresselhaus, G., & Saito, R. (1995). Physics of carbon nanotubes. *Carbon*, *33*(7), 883–891. doi:10.1016/0008-6223(95)00017-8

Dresselhaus, M. S., Dresselhaus, G., Eklund, P., & Rao, A. (2000). *Carbon nanotubes*. Springer. doi:10.1007/978-94-011-4038-6_9

Elhaes, H., & Ibrahim, M. (2013). Fullerene as sensor for Halides: Modeling approach. *Journal of Computational and Theoretical Nanoscience*, *10*(9), 2026–2028. doi:10.1166/jctn.2013.3164

Farmer, D. B., Chiu, H.-Y., Lin, Y.-M., Jenkins, K. A., Xia, F., & Avouris, P. (2009). Utilization of a buffered dielectric to achieve high field-effect carrier mobility in graphene transistors. *Nano Letters*, *9*(12), 4474–4478. doi:10.1021/nl902788u PMID:19883119

Fiori, G., & Iannaccone, G. (2007). Simulation of graphene nanoribbon field-effect transistors. *Electron Device Letters, IEEE*, *28*(8), 760–762. doi:10.1109/LED.2007.901680

Fraden, J. (2010). *Handbook of modern sensors*. Springer. doi:10.1007/978-1-4419-6466-3

Frank, R. (2013). *Understanding smart sensors*. Artech House.

Franklin, A. D., Koswatta, S. O., Farmer, D. B., Smith, J. T., Gignac, L., Breslin, C. M., & Haensch, W. et al. (2013). Carbon nanotube complementary wrap-gate transistors. *Nano Letters*, *13*(6), 2490–2495. doi:10.1021/nl400544q PMID:23638708

Fuhrer, M. S., Lau, C. N., & MacDonald, A. H. (2010). Graphene: Materially better carbon. *MRS Bulletin*, *35*(04), 289–295. doi:10.1557/mrs2010.551

Ganesh, E. (n.d.). *Single walled and multi walled carbon nanotube structure, synthesis and applications*. Academic Press.

Geim, A. K. (2009). Graphene: Status and prospects. *Science*, *324*(5934), 1530–1534. doi:10.1126/science.1158877 PMID:19541989

Geim, A. K., & Novoselov, K. S. (2007). The rise of graphene. *Nature Materials*, *6*(3), 183–191. doi:10.1038/nmat1849 PMID:17330084

Giannopoulos, G. I. (2014). Fullerenes as mass sensors: A numerical investigation. *Physica E, Low-Dimensional Systems and Nanostructures*, *56*, 36–42. doi:10.1016/j.physe.2013.07.024

Goldoni, A., Larciprete, R., Petaccia, L., & Lizzit, S. (2003). Single-wall carbon nanotube interaction with gases: Sample contaminants and environmental monitoring. *Journal of the American Chemical Society*, *125*(37), 11329–11333. doi:10.1021/ja034898e PMID:16220955

Gordillo-Vázquez, F. J., Camero, M., & Gomez-Aleixandre, C. (2006). Spectroscopic measurements of the electron temperature in low pressure radiofrequency Ar/H2/C2H2 and Ar/H2/CH4 plasmas used for the synthesis of nanocarbon structures. *Plasma Sources Science & Technology*, *15*(1), 42–51. doi:10.1088/0963-0252/15/1/007

Grassi, R., Gnudi, A., Gnani, E., Reggiani, S., & Baccarani, G. (2008). Graphene nanoribbons FETs for high-performance logic applications: Perspectives and challenges. *2008 9th International Conference on Solid-State and Integrated-Circuit Technology.*

Grossi, E., & Buscema, M. (2007). Introduction to artificial neural networks. *European Journal of Gastroenterology & Hepatology*, *19*(12), 1046–1054. doi:10.1097/MEG.0b013e3282f198a0 PMID:17998827

Gu, F., Zhang, J.-H., Yang, L.-J., & Gu, B. (2011). Molecular dynamics simulation of resonance properties of strain graphene nanoribbons. *Acta Physica Sinica*, 60.

Gulotty, R., Castellino, M., Jagdale, P., Tagliaferro, A., & Balandin, A. A. (2013). Effects of functionalization on thermal properties of single-wall and multi-wall carbon nanotube–polymer nanocomposites. *ACS Nano*, *7*(6), 5114–5121. doi:10.1021/nn400726g PMID:23672711

Haddon, R. C. (2002). Carbon nanotubes. *Accounts of Chemical Research*, *35*(12), 997–997. doi:10.1021/ar020259h PMID:12484786

Hamker, F. (2009). *Artificial neural networks*. Academic Press.

Han, J.-W., Kim, B., Li, J., & Meyyappan, M. (2013). A carbon nanotube based ammonia sensor on cotton textile. *Applied Physics Letters*, *102*(19), 193104, 193104–4. doi:10.1063/1.4805025

Han, M. Y., Özyilmaz, B., Zhang, Y., & Kim, P. (2007). Energy band-gap engineering of graphene nanoribbons. *Physical Review Letters*, *98*(20), 206805. doi:10.1103/PhysRevLett.98.206805 PMID:17677729

Harada, N., Sato, S., & Yokoyama, N. (2013). Theoretical Investigation of graphene nanoribbon field-effect transistors designed for digital applications. *Japanese Journal of Applied Physics*, *52*(9R), 094301. doi:10.7567/JJAP.52.094301

Hayati, M., Rezaei, A., & Seifi, M. (2010). CNT-MOSFET modeling based on artificial neural network: Application to simulation of nanoscale circuits. *Solid-State Electronics*, *54*(1), 52–57. doi:10.1016/j.sse.2009.09.027

Hays, S. P., & Hays, B. D. (1989). *Beauty, health, and permanence: Environmental politics in the United States, 1955-1985*. Cambridge University Press.

He, Q., Wu, S., Yin, Z., & Zhang, H. (2012). Graphene-based electronic sensors. *Chemical Science*, *3*(6), 1764–1772. doi:10.1039/c2sc20205k

Hijikata, K., K. Ogawa & N. Miyakawa (1999). Methanol conversion from methane and water vapor by electric discharge (effect of electric discharge process on methane conversion). *Heat Transfer—Asian Research, 28*, 404-417.

Hill, C. M., Zhu, Y., & Pan, S. (2011). Fluorescence and electroluminescence quenching evidence of interfacial charge transfer in poly (3-hexylthiophene): Graphene oxide bulk heterojunction photovoltaic devices. *ACS Nano*, *5*(2), 942–951. doi:10.1021/nn1022457 PMID:21294531

Hirsch, A., Eigler, R., & Heinemann, F. W. (2013). Selective oxidative conversion of Triaryldihydro [C59N] fullerenes: A model case for carbon allotrope oxygenation. *Chemical Communications*.

Ho, K.-I., Liao, J.-H., Huang, C.-H., Su, C.-Y., & Lai, C.-S. (2013). Electrical probing of multi-ions solution by using graphene-based sensor. In *Nanoelectronics Conference (INEC), 2013 IEEE 5th International*, (pp. 37-39). IEEE. doi:10.1109/INEC.2013.6465946

Hu, N., Wang, Y., Chai, J., Gao, R., Yang, Z., Kong, E. S.-W., & Zhang, Y. (2012). Gas sensor based on p-phenylenediamine reduced graphene oxide. *Sensors and Actuators. B, Chemical*, *163*(1), 107–114. doi:10.1016/j.snb.2012.01.016

Huang, B., Li, Z., Liu, Z., Zhou, G., Hao, S., Wu, J., & Duan, W. et al. (2008a). Adsorption of gas molecules on graphene nanoribbons and its implication for nanoscale molecule sensor. *The Journal of Physical Chemistry C*, *112*(35), 13442–13446. doi:10.1021/jp8021024

Huang, X., Yin, Z., Wu, S., Qi, X., He, Q., Zhang, Q., & Zhang, H. et al. (2011). Graphene-based materials: Synthesis, characterization, properties, and applications. *Small*, *7*(14), 1876–1902. doi:10.1002/smll.201002009 PMID:21630440

Hughes, M., & Spinks, G. M. (2005). Multiwalled carbon nanotube actuators. *Advanced Materials*, *17*(4), 443–446. doi:10.1002/adma.200401076

Hutchison, J., Kiselev, N., Krinichnaya, E., Krestinin, A., Loutfy, R., Morawsky, A., & Terekhov, S. et al. (2001). Double-walled carbon nanotubes fabricated by a hydrogen arc discharge method. *Carbon*, *39*(5), 761–770. doi:10.1016/S0008-6223(00)00187-1

Hwang, E., Adam, S., & Sarma, S. D. (2007). Carrier transport in two-dimensional graphene layers. *Physical Review Letters*, *98*(18), 186806. doi:10.1103/PhysRevLett.98.186806 PMID:17501596

Hwang, W. S., Zhao, P., Tahy, K., Nyakiti, L. O., Wheeler, V. D., Myers-Ward, R., . . . Haensch, W. (2013). *Graphene nanoribbon field-effect transistors on wafer-scale epitaxial graphene on SiC substrates.* arXiv preprint arXiv:1310.6823

Iijima, S. (1991). Helical microtubules of graphitic carbon. *Nature*, *354*(6348), 56–58. doi:10.1038/354056a0

Indarto, A., Choi, J., Lee, H., & Song, H. K. (2007). A brief catalyst study on direct methane conversion using a dielectric barrier discharge. *Journal-Chinese Chemical Society Taipei*, *54*(4), 823–828. doi:10.1002/jccs.200700120

Islam, M. S. (2008). The all pervading nanosensors. *International Journal of Nanotechnology*, *5*(4/5), 393–401. doi:10.1504/IJNT.2008.017445

Javey, A., Guo, J., Wang, Q., Lundstrom, M., & Dai, H. (2003). Ballistic carbon nanotube field-effect transistors. *Nature*, *424*(6949), 654–657. doi:10.1038/nature01797 PMID:12904787

Jeong, H. Y., Lee, D.-S., Choi, H. K., Lee, D. H., Kim, J.-E., Lee, J. Y., & Choi, S.-Y. et al. (2010). Flexible room-temperature NO gas sensors based on carbon nanotubes/reduced graphene hybrid films. *Applied Physics Letters*, *96*(21), 213105. doi:10.1063/1.3432446

Jiao, L., Zhang, L., Wang, X., Diankov, G., & Dai, H. (2009). Narrow graphene nanoribbons from carbon nanotubes. *Nature*, *458*(7240), 877–880. doi:10.1038/nature07919 PMID:19370031

Jimenez-Cadena, G., Riu, J., & Rius, F. X. (2007). Gas sensors based on nanostructured materials. *Analyst (London)*, *132*(11), 1083–1099. doi:10.1039/b704562j PMID:17955141

Jin, S. H., Dunham, S. N., Song, J., Xie, X., Kim, J.-, Lu, C., & Felts, J. et al. (2013). Using nanoscale thermocapillary flows to create arrays of purely semiconducting single-walled carbon nanotubes. *Nature Nanotechnology*, *8*(5), 347–355. doi:10.1038/nnano.2013.56 PMID:23624697

Journet, C., & Bernier, P. (1998). Production of carbon nanotubes. *Applied Physics. A, Materials Science & Processing*, *67*(1), 1–9. doi:10.1007/s003390050731

Jun-Feng, Z., Xin-Chao, B., Qiang, C., Fu-Ping, L., & Zhong-Wei, L. (2009). Diagnosis of methane plasma generated in an atmospheric pressure dbd micro-jet by optical emission spectroscopy. *Chinese Physics Letters*, *26*(3), 035203. doi:10.1088/0256-307X/26/3/035203

Jung, J. (2011). Nonlocal exchange effects in zigzag-edge magnetism of neutral graphene nanoribbons. *Physical Review B: Condensed Matter and Materials Physics*, 83.

Kiani, M. J., Ahmadi, M., Akbari, E., Karimi, H., & Che Harun, F. (2013b). Graphene nanoribbon based gas sensor. *Key Engineering Materials*, *553*, 7–11. doi:10.4028/www.scientific.net/KEM.553.7

Kim, B., Lee, J., Namgung, S., Kim, J., Park, J. Y., Lee, M.-S., & Hong, S. (2012). DNA sensors based on CNT-FET with floating electrodes. *Sensors and Actuators. B, Chemical*, *169*, 182–187. doi:10.1016/j.snb.2012.04.063

Kim, S.-S., Lee, H., Na, B.-K., & Song, H. K. (2003). Reaction pathways of methane conversion in dielectric-barrier discharge. *Korean Journal of Chemical Engineering*, *20*(5), 869–872. doi:10.1007/BF02697290

Kinoshita, K. (1988). *Carbon: Electrochemical and physicochemical properties*. Academic Press.

Ko, G., Kim, H. Y., Ahn, J., Park, Y. M., Lee, K. Y., & Kim, J. (2010). Graphene-based nitrogen dioxide gas sensors. *Current Applied Physics*, *10*(4), 1002–1004. doi:10.1016/j.cap.2009.12.024

Koehler, C., Thielen, S., Ginzkey, C., Hackenberg, S., Scherzed, A., Burghartz, M., & Kleinsasser, N. H. et al. (2013). Nitrogen dioxide is genotoxic in urban concentrations. *Inhalation Toxicology*, 1–7. PMID:23701639

Korotcenkov, G. (2013). *Handbook of gas sensor materials*. Springer. doi:10.1007/978-1-4614-7165-3

Kozlov, V., Starik, A., Titova, N., & Vedishchev, I. Y. (2013). On mechanisms of formation of environmentally harmful compounds in homogeneous combustors. *Combustion, Explosion, and Shock Waves*, *49*(5), 520–535. doi:10.1134/S0010508213050031

Kraus, M., Egli, W., Haffner, K., Eliasson, B., Kogelschatz, U., & Wokaun, A. (2002). Investigation of mechanistic aspects of the catalytic CO2 reforming of methane in a dielectric-barrier discharge using optical emission spectroscopy and kinetic modeling. *Physical Chemistry Chemical Physics*, *4*(4), 668–675. doi:10.1039/b108040g

Krcma, F., Klohnova, K., Polachova, L., & Horvath, G. (2010). Optical emission spectroscopy of abnormal glow discharge in nitrogen-methane mixtures at atmospheric pressure. *Publications de l'Observatoire Astronomique de Beograd*, *89*, 371–374.

Kreibig, U., Gartz, M., & Hilger, A. (1997). Mie resonances: Sensors for physical and chemical cluster interface properties. *Berichte der Bunsengesellschaft fur Physikalische Chemie*, *101*(11), 1593–1604. doi:10.1002/bbpc.19971011107

Kupny, S. (2011). *Artificial neural networks*. Academic Press.

Laurent, C., Flahaut, E., & Peigney, A. (2010). The weight and density of carbon nanotubes versus the number of walls and diameter. *Carbon*, *48*(10), 2994–2996. doi:10.1016/j.carbon.2010.04.010

Lee, H. K., Lee, J., Choi, N. J., Moon, S. E., Lee, H., & Yang, W. S. (2011). Efficient reducing method of graphene oxide for gas sensor applications. In G. Kaltsas & C. Tsamis (Eds.), *Eurosensors Xxv*. doi:10.1016/j.proeng.2011.12.219

Leenaerts, O., Partoens, B., & Peeters, F. (2008). Adsorption of H_ {2} O, NH_ {3}, CO, NO_ {2}, and NO on graphene: A first-principles study. *Physical Review B: Condensed Matter and Materials Physics*, *77*(12), 125416. doi:10.1103/PhysRevB.77.125416

Leghrib, R., Felten, A., Pireaux, J. J., & Llobet, E. (2011). Gas sensors based on doped-CNT/SnO2 composites for NO2 detection at room temperature. *Thin Solid Films*, *520*(3), 966–970. doi:10.1016/j.tsf.2011.04.186

Li, J., Lu, Y., Ye, Q., Cinke, M., Han, J., & Meyyappan, M. (2003). Carbon nanotube sensors for gas and organic vapor detection. *Nano Letters*, *3*(7), 929–933. doi:10.1021/nl034220x

Li, M., & Liu, Y. (2009). Underground coal mine monitoring with wireless sensor networks. *ACM Transactions on Sensor Networks*, *5*(2), 10. doi:10.1145/1498915.1498916

Li, X.-S. L., Shi, C., Xu, Y., Wang, Y.-N., & Zhu, A.-M. (2008). Stable kilohertz spark discharges for high-efficiency conversion of methane to hydrogen and acetylene. *Journal Of Physics -London- D Applied Physics, 41*(17).

Liang, G., Neophytou, N., Nikonov, D. E., & Lundstrom, M. S. (2007). Performance projections for ballistic graphene nanoribbon field-effect transistors. *Electron Devices. IEEE Transactions on*, *54*(4), 677–682. doi:10.1109/TED.2007.891872

Liu, H., Liu, Y., & Zhu, D. (2011). Chemical doping of graphene. *Journal of Materials Chemistry*, *21*(10), 3335–3345. doi:10.1039/C0JM02922J

Lu, Y. H., Wu, R. Q., Shen, L., Yang, M., Sha, Z. D., Cai, Y. Q., & Feng, Y. P. et al. (2009). Effects of edge passivation by hydrogen on electronic structure of armchair graphene nanoribbon and band gap engineering. *Applied Physics Letters*, 94.

Machala, Z., Marode, E., Laux, C. O., & Kruger, C. H. (2004). DC glow discharges in atmospheric pressure air. *Journal of Advanced Oxidation Technologies*, *7*, 133–137.

Mahmood, S., Burhanudin, Z. A., & Hamid, N. H. (2013). Field emission model of CNT based ionization gas sensor. In M. R. Mahmood, T. Soga, M. H. Mamat, Z. Khusaimi, & A. M. Nor (Eds.), *Nanosynthesis and Nanodevice* (pp. 135–143). doi:10.4028/www.scientific.net/AMR.667.135

Martel, R., Schmidt, T., Shea, H., Hertel, T., & Avouris, P. (1998). Single-and multi-wall carbon nanotube field-effect transistors. *Applied Physics Letters*, *73*(17), 2447. doi:10.1063/1.122477

Massera, E., La Ferrara, V., Miglietta, M., Polichetti, T., Nasti, I., & Di Francia, G. (2012). Gas sensors based on graphene:Comparison of two different fabrication approaches. *Chimica Oggi-Chemistry Today*, *30*, 29–31.

Massereau-Guilbaud, V., Géraud-Grenier, I., & Plain, A. (2009). Determination of the electron temperature by optical emission spectroscopy in a 13.56 MHz dusty methane plasma: Influence of the power. *Journal of Applied Physics*, *106*(11), 113305, 113305–113307. doi:10.1063/1.3267292

Mishra, L. N., Shibata, K., Ito, H., Yugami, N., & Nishida, Y. (2004). Conversion of methane to hydrogen via pulsed corona discharge. *Journal of Natural Gas Chemistry*, *13*, 82–86.

Moradian, R., Mohammadi, Y., & Ghobadi, N. (2008). Investigation of gas sensing properties of armchair graphene nanoribbons. *Journal of Physics Condensed Matter*, 20.

Murali, R. (2012). *Graphene nanoelectronics: From materials to circuits*. Springer. doi:10.1007/978-1-4614-0548-1

Mustafa, M. G., & Tierney, D. F. (1978). Biochemical and metabolic changes in the lung with oxygen, ozone, and nitrogen dioxide toxicity. *The American Review of Respiratory Disease*, *118*, 1061–1090. PMID:369411

Neto, A. C., Guinea, F., Peres, N., Novoselov, K. S., & Geim, A. K. (2009). The electronic properties of graphene. *Reviews of Modern Physics*, *81*(1), 109–162. doi:10.1103/RevModPhys.81.109

Novoselov, K., Fal, V., Colombo, L., Gellert, P., Schwab, M., & Kim, K. (2012). A roadmap for graphene. *Nature*, *490*, 192–200. doi:10.1038/nature11458 PMID:23060189

Nozaki, T. (2011). *Electron Driven Microchemistry for Green Technology: Direct Conversion of Methane to Methanol*. Academic Press.

Nozaki, T., Unno, Y., Miyazaki, Y., & Okazaki, K. (2001). Optical diagnostics for determining gas temperature of reactive microdischarges in a methane-fed dielectric barrier discharge. *Journal of Physics. D, Applied Physics*, *34*(16), 2504–2511. doi:10.1088/0022-3727/34/16/316

Olthuis, W., Faber, Krommenhoek, & van den Berg. (2007). *Sensing with FETs-once, now and future*. Academic Press.

Owens, F. J. (2008). Electronic and magnetic properties of armchair and zigzag graphene nanoribbons. *The Journal of Chemical Physics*, *128*(19), 194701. doi:10.1063/1.2905215 PMID:18500880

Pallas-Areny, R., J. G. Webster & R. Areny. (2001). *Sensors and signal conditioning*. Wiley.

Panzer, M. A., Zhang, Mann, Hu, Pop, Dai, & Goodson. (2008). Thermal properties of metal-coated vertically aligned single-wall nanotube arrays. *Journal of Heat Transfer-Transactions of the Asme, 130*.

Patacsil, C., Malapit, & Ramos. (2006). Optical emission spectroscopy of low temperature CVD diamond. *J Plasma Fusion Res Ser*, *7*, 145–149.

Patra, M. K., Manzoor, K., Manoth, M., Negi, S. C., Vadera, S. R., & Kumar, N. (2008). Nanotechnology applications for chemical and biological sensors. *Defence Science Journal*, *58*(5), 636–649. doi:10.14429/dsj.58.1686

Patterson, D. W. (1998). *Artificial neural networks: Theory and applications*. Prentice Hall PTR.

Pearce, R., Iakimov, T., Andersson, M., Hultman, L., Spetz, A. L., & Yakimova, R. (2011). Epitaxially grown graphene based gas sensors for ultra sensitive NO2 detection. *Sensors and Actuators. B, Chemical*, *155*(2), 451–455. doi:10.1016/j.snb.2010.12.046

Peng, Y., & Li. (2013). Ammonia adsorption on graphene and graphene oxide: A first-principles study. *Frontiers of Environmental Science & Engineering*, 1-9.

Pesetski, A. A., Baumgardner, J. E., Folk, E., Przybysz, J. X., Adam, J. D., & Zhang, H. (2006). Carbon nanotube field-effect transistor operation at microwave frequencies. *Applied Physics Letters*, 88.

Postma, H. W. C., Teepen, T., Yao, Z., Grifoni, M., & Dekker, C. (2001). Carbon nanotube single-electron transistors at room temperature. *Science*, *293*(5527), 76–79. doi:10.1126/science.1061797 PMID:11441175

Rahmani, M., Ahmadi, M. T., Abadi, H. K. F., Saeidmanesh, M., Akbari, E., & Ismail, R. (2013). Analytical modeling of trilayer graphene nanoribbon Schottky-barrier FET for high-speed switching applications. *Nanoscale Research Letters*, *8*(1), 1–13. doi:10.1186/1556-276X-8-55 PMID:23363692

Rana, F., George, P. A., Strait, J. H., Dawlaty, J., Shivaraman, S., Chandrashekhar, M., & Spencer, M. G. (2009). Carrier recombination and generation rates for intravalley and intervalley phonon scattering in graphene. *Physical Review B: Condensed Matter and Materials Physics*, *79*(11), 115447. doi:10.1103/PhysRevB.79.115447

Rao, C. N. R., & Sood. (2013). *Graphene: Synthesis, properties, and phenomena*. Wiley.

Ratinac, K. R., Yang, W., Ringer, S. P., & Braet, F. (2010). Toward ubiquitous environmental gas sensors capitalizing on the promise of graphene. *Environmental Science & Technology*, *44*(4), 1167–1176. doi:10.1021/es902659d PMID:20099803

Reich, S., Maultzsch, J., Thomsen, C., & Ordejon, P. (2002). Tight-binding description of graphene. *Physical Review B: Condensed Matter and Materials Physics*, *66*(3), 035412. doi:10.1103/PhysRevB.66.035412

Riu, J., Maroto, A., & Rius, F. X. (2006). Nanosensors in environmental analysis. *Talanta*, *69*(2), 288–301. doi:10.1016/j.talanta.2005.09.045 PMID:18970568

Rivera, I. F., Joshi, & Wang. (2010). Graphene-based ultra-sensitive gas sensors. In *2010 IEEE Sensors*, (pp. 1534-1537).

Romero, H. E., Joshi, P., Gupta, A. K., Gutierrez, H. R., Cole, M. W., Tadigadapa, S. A., & Eklund, P. C. (2009). Adsorption of ammonia on graphene. *Nanotechnology*, 20. PMID:19468162

Sagas, J. C., Neto, A. H., Maciel, H., & Lacava, P. (2011). Basic characteristics of gliding-arc discharges in air and natural gas. *Plasma Science. IEEE Transactions on*, *39*, 775–780.

Saito, R. (1998). *Physical properties of carbon nanotubes*. Academic Press.

Santangelo, S., Faggio, G., Messina, G., Fazio, E., Neri, F., & Neri, G. (2013). On the hydrogen sensing mechanism of Pt/TiO2/CNTs based devices. *Sensors and Actuators. B, Chemical, 178*, 473–484. doi:10.1016/j.snb.2013.01.005

Sayago, I., Fernandez, M., Fontecha, J., Horrillo, M., Terrado, E., Seral-Ascaso, A., & Munoz, E. (2013). Carbon nanotube-based SAW sensors. In *Electron Devices (CDE), 2013 Spanish Conference on*, (pp. 127-130). IEEE. doi:10.1109/CDE.2013.6481359

Schaller, R. R. (1997). Moore's law: Past, present and future. Spectrum, IEEE, 34, 52-59.

Schedin, F., Geim, A., Morozov, S., Hill, E., Blake, P., Katsnelson, M., & Novoselov, K. (2007). Detection of individual gas molecules adsorbed on graphene. *Nature Materials*, *6*(9), 652–655. doi:10.1038/nmat1967 PMID:17660825

Schmiedt, L., Kaňka, & Hrachová. (n.d.). *Optical emission spectroscopy–Passive method for plasma diagnostics in dc glow oxygen discharge*. Academic Press.

Schwierz, F. (2010a). Graphene transistors. *Nature Nanotechnology*, *5*(7), 487–496. doi:10.1038/nnano.2010.89 PMID:20512128

Schwierz, F. (2010b). graphene transistors. *Nature Nanotechnology*. doi:10.1038/NNANO.2010.89

Sevik, C., Sevinçli, H., Cuniberti, G., & Cagın, T. (2011). Phonon engineering in carbon nanotubes by controlling defect concentration. *Nano Letters*, *11*(11), 4971–4977. doi:10.1021/nl2029333 PMID:21967464

Singh, V., Joung, D., Zhai, L., Das, S., Khondaker, S. I., & Seal, S. (2011). Graphene based materials: Past, present and future. *Progress in Materials Science*, *56*(8), 1178–1271. doi:10.1016/j.pmatsci.2011.03.003

Someya, T., Small, J., Kim, P., Nuckolls, C., & Yardley, J. T. (2003). Alcohol vapor sensors based on single-walled carbon nanotube field effect transistors. *Nano Letters*, *3*(7), 877–881. doi:10.1021/nl034061h

Son, Y.-W., Cohen, M. L., & Louie, S. G. (2006). Energy gaps in graphene nanoribbons. *Physical Review Letters*, *97*(21), 216803. doi:10.1103/PhysRevLett.97.216803 PMID:17155765

Sonawane, M. R., Habale, Nagare, & Gharde. (n.d.). *Interaction of O2, CO2, NO2 and SO2 on Si-doped Carbon Nanotube*. Academic Press.

Star, A., Han, T. R., Gabriel, J. C. P., Bradley, K., & Grüner, G. (2003). Interaction of aromatic compounds with carbon nanotubes: Correlation to the Hammett parameter of the substituent and measured carbon nanotube FET response. *Nano Letters*, *3*(10), 1421–1423. doi:10.1021/nl0346833

Sternschulte, H., Schreck, M., & Stritzker, B. (2002). In situ characterisation of CVD diamond growth under H< sub> 2</sub> S addition by optical emission spectroscopy, mass spectroscopy and laser reflection interferometry. *Diamond and Related Materials*, *11*(3-6), 296–300. doi:10.1016/S0925-9635(01)00675-6

Subrahmanyam, K., Panchakarla, L., Govindaraj, A., & Rao, C. (2009). Simple method of preparing graphene flakes by an arc-discharge method. *The Journal of Physical Chemistry C*, *113*(11), 4257–4259. doi:10.1021/jp900791y

Suehiro, J., Zhou, G. B., Imakiire, H., Ding, W. D., & Hara, M. (2005). Controlled fabrication of carbon nanotube NO2 gas sensor using dielectrophoretic impedance measurement. *Sensors and Actuators. B, Chemical*, *108*(1-2), 398–403. doi:10.1016/j.snb.2004.09.048

Sven Kochmann, T. H., & Wolfbeis. (2012). Graphenes in chemical sensors and biosensors. *TrAC Trends in Analytical Chemistry, 39.*

Tao, X., Bai, M., Li, X., Long, H., Shang, S., Yin, Y., & Dai, X. (2011). CH< sub> 4</sub>–CO< sub> 2</sub> reforming by plasma–challenges and opportunities. *Progress in Energy and Combustion Science*, *37*(2), 113–124. doi:10.1016/j.pecs.2010.05.001

Teerapanich, P., Myint, M. T. Z., Joseph, C. M., Hornyak, G. L., & Dutta, J. (2013). Development and improvement of carbon nanotube-based ammonia gas sensors using ink-jet printed interdigitated electrodes. *IEEE Transactions on Nanotechnology*, *12*(2), 255–262. doi:10.1109/TNANO.2013.2242203

Tu, X., & Whitehead, J. (2012). Plasma-catalytic dry reforming of methane in an atmospheric dielectric barrier discharge: Understanding the synergistic effect at low temperature. *Applied Catalysis B: Environmental*, *125*, 439–448. doi:10.1016/j.apcatb.2012.06.006

Uhrig, R. E. (1995). Introduction to artificial neural networks. In *Industrial Electronics, Control, and Instrumentation, 1995.,Proceedings of the 1995 IEEE IECON 21st International Conference on*, (pp. 33-37). IEEE.

Ünal, M., Ak, A., Topuz, V., & Erdal, H. (2013). Artificial neural networks. In Optimization of PID Controllers Using Ant Colony and Genetic Algorithms, (pp. 5-17). Springer.

Vincent-Randonnier, A., Larigaldie, S., Magre, P., & Sabel'nikov, V. (2007). Plasma assisted combustion: Effect of a coaxial DBD on a methane diffusion flame. *Plasma Sources Science & Technology*, *16*(1), 149–160. doi:10.1088/0963-0252/16/1/020

Wakabayashi, K., Fujita, M., Ajiki, H., & Sigrist, M. (1999). Electronic and magnetic properties of nanographite ribbons. *Physical Review B: Condensed Matter and Materials Physics*, *59*(12), 8271–8282. doi:10.1103/PhysRevB.59.8271

Wang, D., Yu, Z., McKernan, S., & Burke, P. J. (2007). Ultrahigh frequency carbon nanotube transistor based on a single nanotube. *IEEE Transactions on Nanotechnology*, *6*(4), 400–403. doi:10.1109/TNANO.2007.901179

Wang, J., & Lundstrom, M. (2003). Ballistic transport in high electron mobility transistors. *Electron Devices. IEEE Transactions on*, *50*(7), 1604–1609. doi:10.1109/TED.2003.814980

Wang, Q., & Arash, B. (2014). A review on applications of carbon nanotubes and graphenes as nanoresonator sensors. *Computational Materials Science*, *82*, 350–360. doi:10.1016/j.commatsci.2013.10.010

Wang, X., & Dai, H. (2010). Etching and narrowing of graphene from the edges. *Nature Chemistry*, *2*(8), 661–665. doi:10.1038/nchem.719 PMID:20651729

Wang, X., Ouyang, Y., Li, X., Wang, H., Guo, J., & Dai, H. (2008). Room-temperature all-semiconducting sub-10-nm graphene nanoribbon field-effect transistors. *Physical Review Letters*, *100*(20), 206803. doi:10.1103/PhysRevLett.100.206803 PMID:18518566

Wu, S., He, Q., Tan, C., Wang, Y., & Zhang, H. (2013a). Graphene-based electrochemical sensors. *Small*, *9*(8), 1160–1172. doi:10.1002/smll.201202896 PMID:23494883

Wu, T. C., Chen, M. K., Lee, Y. W., Kuo, M. Y., & Wu, Y. T. (2013b). Bowl-shaped fragments of C70 or higher fullerenes: Synthesis, structural analysis, and inversion dynamics. *Angewandte Chemie International Edition*, *52*(4), 1289–1293. doi:10.1002/anie.201208200 PMID:23225309

Xu, C., Wei, B., Ma, R., Liang, J., Ma, X., & Wu, D. (1999). Fabrication of aluminum–carbon nanotube composites and their electrical properties. *Carbon*, *37*(5), 855–858. doi:10.1016/S0008-6223(98)00285-1

Xu, G., Bai, J., Torres, C., Song, E., Tang, J., Zhou, Y., & Wang, K. et al. (2010). Nanowire-mask based fabrication of high mobility and low noise graphene nanoribbon short-channel field-effect transistors. In *Device Research Conference (DRC)*. IEEE. doi:10.1109/DRC.2010.5551935

Xu, M., Gao, Y., Yang, X., & Chen, H. (2012). Unique synthesis of graphene-based materials for clean energy and biological sensing applications. *Chinese Science Bulletin*, *57*(23), 3000–3009. doi:10.1007/s11434-012-5128-9

Yamacli, S., & Avci, M. (2010). Neural network modeling of voltage-dependent resistance of metallic carbon nanotube interconnects: An< i> ab initio</i> study. *Expert Systems with Applications*, *37*(12), 8014–8018. doi:10.1016/j.eswa.2010.05.089

Yang, M.-Q., Zhang, N., & Xu, Y.-J. (2013). Synthesis of Fullerene–, Carbon Nanotube–, and Graphene–TiO2 Nanocomposite Photocatalysts for Selective Oxidation: A Comparative Study. *ACS Applied Materials & Interfaces*, *5*(3), 1156–1164. doi:10.1021/am3029798 PMID:23324025

Yang, Y. X., & Murali, R. (2010). Impact of size effect on graphene nanoribbon transport. *IEEE Electron Device Letters*, *31*(3), 237–239. doi:10.1109/LED.2009.2039915

Yao, X. (1999). Evolving artificial neural networks. *Proceedings of the IEEE*, *87*(9), 1423–1447. doi:10.1109/5.784219

Yazyev, O. V. (2013). A guide to the design of electronic properties of graphene nanoribbons. *Accounts of Chemical Research*, *46*(10), 2319–2328. doi:10.1021/ar3001487 PMID:23282074

Ye, J.-S., Liu, X., Cui, H. F., Zhang, W.-D., Sheu, F.-S., & Lim, T. M. (2005). Electrochemical oxidation of multi-walled carbon nanotubes and its application to electrochemical double layer capacitors. *Electrochemistry Communications*, *7*(3), 249–255. doi:10.1016/j.elecom.2005.01.008

Yegnanarayana, B. (2004). *Artificial neural networks*. PHI Learning Pvt. Ltd.

Yin, G., Liang, Y., Jiang, F., Chen, H., Wang, P., Mizuseki, H., & Kawazoe, Y. (2009). Polarization-induced switching effect in graphene nanoribbon edge-defect junction. *The Journal of Chemical Physics*, *131*(23), 234706. doi:10.1063/1.3273312 PMID:20025341

Yin, P., Kim, T.-H., Choi, J.-W., & Lee, K. (2013). Prospects for graphene-nanoparticle-based hybrid sensors. *Physical Chemistry Chemical Physics*, *15*(31), 12785. doi:10.1039/c3cp51901e PMID:23828095

Yoon, H. J., Jun, D. H., Yang, J. H., Zhou, Z., Yang, S. S., & Cheng, M. M.-C. (2011). Carbon dioxide gas sensor using a graphene sheet. *Sensors and Actuators. B, Chemical*, *157*(1), 310–313. doi:10.1016/j.snb.2011.03.035

Yu, S.-S., Wang, Zheng, & Jiang. (2011). Mechanical and electron-transport properties of graphene nanoribbons under tensile strain: A first-principles study. *Physica Status Solidi A-Applications and Materials Science, 208*, 2328-2331.

Zhang, T., Mubeen, S., Myung, N. V., & Deshusses, M. A. (2008). Recent progress in carbon nanotube-based gas sensors. *Nanotechnology*, *19*(33), 332001. doi:10.1088/0957-4484/19/33/332001 PMID:21730614

Zhang, X.-S. (2000). Introduction to artificial neural network. In Neural Networks in Optimization, (pp. 83-93). Springer. doi:10.1007/978-1-4757-3167-5_5

Zhang, Y.-H., Chen, Y.-B., Zhou, K.-G., Liu, C.-H., Zeng, J., Zhang, H.-L., & Peng, Y. (2009). Improving gas sensing properties of graphene by introducing dopants and defects: A first-principles study. *Nanotechnology*, *20*(18), 185504. doi:10.1088/0957-4484/20/18/185504 PMID:19420616

Zhang, Y.-H., Han, L.-F., Xiao, Y.-H., Jia, D.-Z., Guo, Z.-H., & Li, F. (2013). Understanding dopant and defect effect on H< sub> 2</sub> S sensing performances of graphene: A first-principles study. *Computational Materials Science*, *69*, 222–228. doi:10.1016/j.commatsci.2012.11.048

Zhao, G.-B., John, S., Zhang, J.-J., Wang, L., Muknahallipatna, S., Hamann, J. C., & Plumb, O. A. et al. (2006). Methane conversion in pulsed corona discharge reactors. *Chemical Engineering Journal*, *125*(2), 67–79. doi:10.1016/j.cej.2006.08.008

Zhao, J., Buldum, A., Han, J., & Lu, J. P. (2002). Gas molecule adsorption in carbon nanotubes and nanotube bundles. *Nanotechnology*, *13*(2), 195–200. doi:10.1088/0957-4484/13/2/312

Zhou, M., Lu, Y.-H., Cai, Y.-Q., Zhang, C., & Feng, Y.-P. (2011). Adsorption of gas molecules on transition metal embedded graphene: A search for high-performance graphene-based catalysts and gas sensors. *Nanotechnology*, 22. PMID:21869463

Zhou, Y., Bao, Q., Tang, L. A. L., Zhong, Y., & Loh, K. P. (2009). Hydrothermal dehydration for the "green" reduction of exfoliated graphene oxide to graphene and demonstration of tunable optical limiting properties. *Chemistry of Materials*, *21*(13), 2950–2956. doi:10.1021/cm9006603

Zhu, A., Zhang, X., Li, X., & Gong, W. (2002). "Beyond-thermal-equilibrium" conversion of methane to acetylene and hydrogen under pulsed corona discharge. *Science in China Series B: Chemistry*, *45*(4), 426–434. doi:10.1360/02yb9055

Zhu, Y., Murali, S., Cai, W., Li, X., Suk, J. W., Potts, J. R., & Ruoff, R. S. (2010). Graphene and graphene oxide: Synthesis, properties, and applications. *Advanced Materials*, *22*(35), 3906–3924. doi:10.1002/adma.201001068 PMID:20706983

Chapter 6
Chemiresistive Gas Sensors Based on Conducting Polymers

Sajad Pirsa
Urmia University, Iran

ABSTRACT

Chemiresistive gas sensor based on conducting polymer is a type of sensors that presents gas sensors with excellent characters; low-cost fabrication, fast detection, simultaneous determination (array gas sensor), portable devices and so. Theses gas sensors are commonly based on polyaniline (PANI), polypyrrole (PPy), polythiophene (PTh) and their derivatives as a transducer. Common configuration and response mechanism of these sensors are reported in this section. Some factors that induce selectivity to these sensors are discussed. Different materials (conductor or insulant) can be used as a substrate of polymerization. Type of substrate, selective membranes, surface modification of conducting polymer and so can change response behavior of these sensors.

INTRODUCTION

Recently, there is very attention to the application of conducting polymers as chemical sensors, biochemical sensors, gas detection devices and other types of gas sensors. Many researchers studied about conducting polymers as a transducer in sensor devices (Arshak, Moore, Lyons, & Clifford, 2004). Conducting polymers (CPs) have some excellent characters like; easy, fast and uniform polymerization, electrical conductance, stability and so. These polymers as a transducer or absorbent in different sensors interact with gas molecules and are used for detection and determination of poisonous and pollutant gasses, so gas sensor based on conducting polymers is used to control air pollution (Freund & Lewis, 1995). Polypyrrole (PPy), polyaniline (PAANI) and polythiophene (PTh) are common conducting polymers that generally possess an extended p-electron conjugation system along with a polymer backbone (Gardner & Bartlett, 1995). These polymers and their derivatives like N -methyl pyrrole, N-Phenyl pyrrole and so as protective materials are used to protect oxidizable metals due to their high electrical conductivity and stability. They are coated on the different substrates like; metals, semimetals, plastics, textiles and so (Mahmoudian, Alias, Basirum, & Ebadi, 2011; Tüken, Tansuğ, Yazıcı, & Erbil, 2007). There are different methods to

DOI: 10.4018/978-1-5225-0736-9.ch006

synthesize of these polymers but, chemical synthesis of conducting polymers provided a condition to coat them on the insulant surfaces like textiles and plastics (Redondo, Sa´nchez, Garcı´a, Raso, Tortajada, & Gonza´lez-Tejera, 2001). These polymers have excellent electrical, chemical, thermal and mechanical properties that these properties depend on different parameters, on the other hand, different parameters can change these properties. Many humidities and PH sensors based on conducting polymers are reported, because these polymers resistance is changed in different PH and humidity condition (Skotheim, 1986). Conducting polymers in the other forms were also prepared successfully like; Composites, copolymers and double layers that have a conductive matrix. Copolymers or composites of conducting polymer have different properties than conducting polymer, so chemical, physical, mechanical properties and stability of these polymers are changeable and controllable (Selapinar, Toppare, Akbulut, Yalcin, & Suzer, 1995; Song, Wang, & Yang, 2011). Currently, the focus is on the development of selective sensors for various organic solvent vapors/gaseous molecules. For this purpose, synthesize of CPs in different composites has been developed quickly, because different composite can induce the CPs to grow in certain manners and hence result in CPs with ordered morphology and porosity which will show superior properties, and using of different composites cause to design some selective gas sensor (Matsuguchi, Sugiyama, & Sakai, 2002; Ratcliffe, 1990, pp. 257–262). Surface modification of gas sensors is a method to obtain selective gas sensor based on CPs with excellent chemical characteristics. For this purpose, the nanocomposite of CPs by semi metals and surface modification of CPs by semi-metals have been developed, because semi-metals can induce the selective interaction of CPs and gas samples (Amaya, Saio, & Hirao, 2007; Li, Wang, Cao, Yuan, & Yang, 2008; Do & Chang, 2001). Their poor selectivity and strong interference with humidity are the major disadvantages of conducting polymer based Chemiresistive gas sensors. An array of conducting polymer-based gas sensor is a suitable method to overcome poor selectivity of chemoresistive gas sensors (Ulrich, Nataliya, & Vladimir, 2008).

CONDUCTING POLYMERS

Conducting polymers against conventional polymers possess some metal characters like; magnetic, electrical and optical properties but, have some conventional polymers character like process ability, mechanical properties and so, also called "Synthetic metals or Organic metals" (Bartlett & Ling-Chung 1989). In the most chemical compounds among normal covalent bonds, valence electrons are tightly held and shared between the atoms that these electrons don't act as a charge carrier (Bhadra & Khastgir, 2008). However, compared to σ-electron the π-electrons are relatively free in double and triple c-c bonds. The interaction of neighboring π-orbitals results in delocalizing of π-electrons in a conjugated double bond system. Thus, π-electrons in the conjugated system can mobile freely over all molecules in the result of π-electron delocalization (Tzamalis, Zaidi, Homes, & Monkman, 2002; Anderson, Mattes, Reiss, & Kaner, 1991; Aphesteguy & Jacobo, 2004).

- The conducting polymer is a type of polymer with a backbone of π–conjugated electrons.
- Upon oxidation and doping, the polymer develops delocalized electrons and conducts electricity.

Polypyrrole (PPy), polyaniline (PANI), polythiophene (PTh) and their composites, copolymers, and derivatives, as conducting polymers have been used in the gas sensor device as the transducer in new recent years (Miasik, Hopper, & Tofield, 1986).

Some characters of these polymers that cause to using of them as a transducer in the Chemiresistive gas sensor are listed following:

1. These polymers are polymerized by chemical or electrochemical methods very fast and simple.
2. Environmentally is not poisonous.
3. They can be polymerized on the surface of metals (Like platinum, Cu and so…) and insulant substrates like textiles, plastics and so.
4. Copolymerization and structural derivations as two methods are used to modify conducting polymer molecular chain structure.

Several conducting polymers used as a transducer in the Chemiresistive gas sensor are represented in Figure 1.

Figure 1. Several conducting polymers

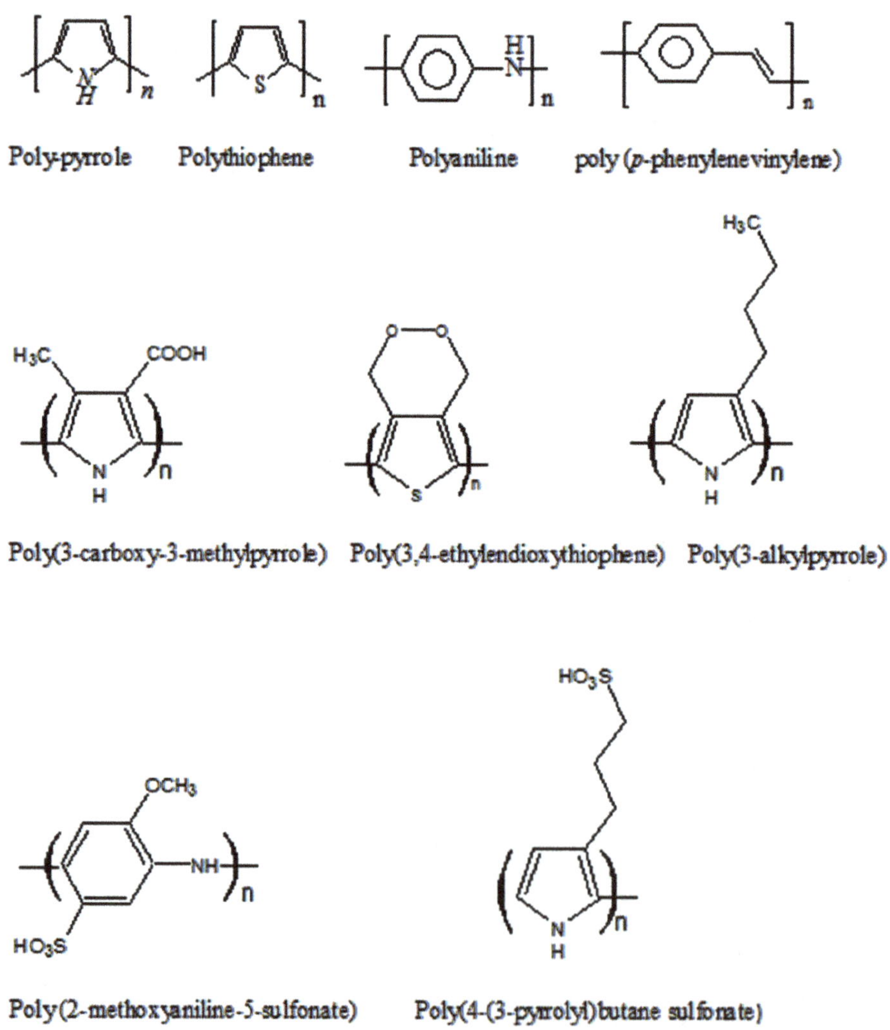

Conductivity of Conducting Polymers

Conductivity, the opposite of resistivity is defined as a measurement of the number of ions per unit volume and their average velocity in the direction of a unit applied field.

Calculation:

$$\sigma = \frac{\pi d R}{2} \tag{1}$$

$$R = \frac{V}{I} \tag{2}$$

R = Resistance (ohm.), V= Voltage (volt), I = Current (ampere),

d = thickness of polymer, $\pi = 31414$, $\sigma = \frac{S}{cm}$

Conducting polymers in pure condition have the conductivity low< 10^{-5} S cm^{-1}. Doping process is a manner to enhance the conductivity of conducting polymers.

Doping

Doping in conducting polymers occurs by electron transferring to conjugated polymer chain with an electron donor agent that doping leads to the wide variety of interesting and important phenomena. Both electrochemical and chemical processes are used to doping conducting polymers. Chemical doping is a straightforward process: Internal reduction/oxidation of conducting polymers can be done by acid-base protonation (Michalska & Lewenstam, 2000). Figure 2 shows chemical doping of polyaniline.

Mechanism of Doping

The conductivity of an organic polymer depends on the state of oxidation or reduction. The polymer may lose (oxidation) or gain (reduction) electrons, resulting in a change in the electronic structure that allows it to conduct electricity. Conjugation in polymer plays an important role for conductivity. Complete doping yields high-quality materials. However, Inhomogeneous doping causes intermediate doping level (MacDiarmid & Epstein, 1995; Michalska & Lewenstam, 2000). Conducting polymers are not changed by the doping because doping is reversible.

Conducting polymers are fundamental interest in the doping that causes to initially attractive of these polymers. However, these polymers naturally are in the non-doped semiconducting state that as a plastic electronic device can be used in different fields like chemiresistive gas sensor (Pirsa & Alizadeh, 2011).

Synthesis of Conducting Polymers

Both chemical and electrochemical manner corresponding to monomers are used to synthesize conducting polymers. Electrochemical polymerization only is done in solution phase and several electrochemical

Figure 2. Chemical doping of polyaniline

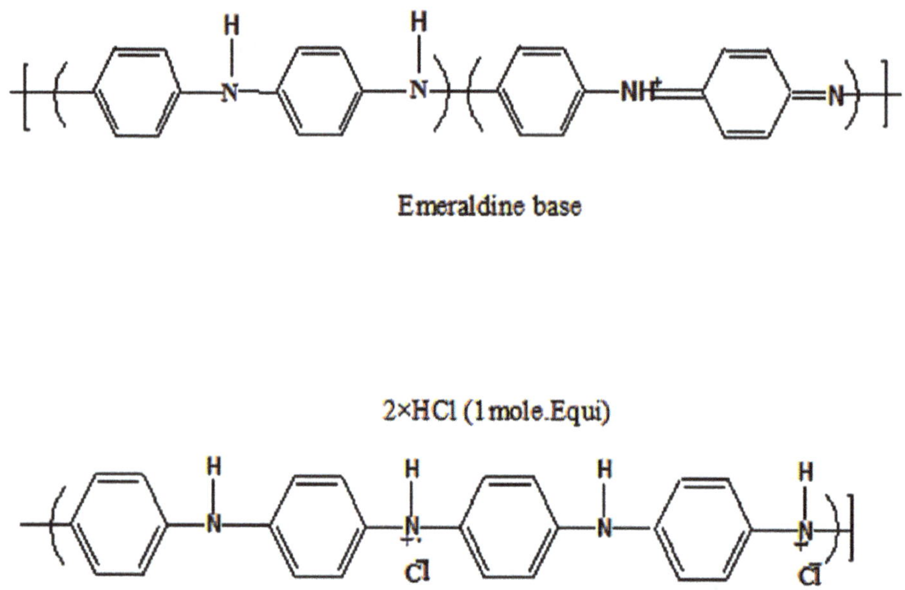

techniques like a cyclic voltameter (CV), galvanostatic, potentiostat and other potentiodynamic methods are used, but chemical polymerization can be done by vapor phase and solution phase polymerization.

CHEMICAL POLYMERIZATION

1. **Chemical Polymerization in Solution Phase:** Chemical solution polymerization is done in the mixing of monomer and oxidant in solution in the presence of substrate (insulant or conductive substrate), and in the case of PANI, to produce a linear structured polymer product, proton acid medium is necessary (MacDiarmid & Epstein, 1995; Michalska & Lewenstam, 2000; Pirsa & Alizadeh, 2011).
2. **Chemical Vapor Phase Polymerization (CVP):** In chemical vapor phase polymerization, firstly oxidant agent coated on the surface of the substrate and then is exposed to monomer vapor. Polymerization is done so fast and uniformly (MacDiarmid & Epstein, 1995; Michalska & Lewenstam, 2000; Pirsa & Alizadeh, 2011).

Oxidants Type

Ferrumchloride ($FeCl_3$), hydrogen peroxide (H_2O_2), potassium dichromate ($K_2Cr_2O_7$), cerium sulfate, ammonium persulfate and so on are some oxidants that have been used to chemical synthesize of CPs. The organic and aqueous medium can be used for synthesizing.

Electrochemical Polymerization

In the electrochemical synthesis techniques, a system with three electrodes is needed. In the electrochemical polymerization, conducting polymers are synthesized on the working electrode. Except working electrode, a counter electrode and a reference electrode are needed in electrochemical systems. Different working electrodes can be used in systems like; indium tin oxide (ITO) glass, stainless steel and platinum (commonly used) (Hernandez-Perez, Morales, Batina, & Salmon, 2001; Patil, Mahajan, More, & Patil, 1999; Janata, 2002. PP. 109–120; Akundy, Rajagopalan, & Iroh, 2002; Peng, Zhang, Soeller, & Travas-Sejdic, 2009; Ramanavicius, Ramanaviciene, & Malinauskas, 2006).

- **Polypyrrole:** Polypyrrole as an important conducting polymer has some excellent characters such as good stability in different environments, thermal property, chemical, mechanical and physical properties and so. During the synthesis of PPy, its chemical and physical properties can be changed by changing of anion dopants species.

Many organic and inorganic materials are combined with PPy and PPy composites are formed. Gas-sensing properties, chemical stability, and strength of PPy are changed by composition formation (Gangopadhyay & De, 2001). Some polypyrrole composites with inorganic compounds like; WO_3, Ag, Fe_3O_4, SnO2, etc. and with organic compounds like; carbon nanofibers, Poly vinyl alcohol, polyester, etc. were reported (Ram, Yavuz, & Aldissi, 2005).

PPy can be doped /undoped by electron donor/electron acceptor agents. An oxidation agent (chemical or electrochemical agents) can remove some electrons from PPy backbones cause to enhance positive charges on backbones result in cation radicals (polaron) acting as the charge carriers (Akundy, & Iroh, 2001). PPy doping process is reversible and can be returned to the undoped state by chemical or electrochemical reductions. PPy oxidation process demonstrated in Figure 3.

- **Polyaniline:** Synthesis of polyaniline can be done by the chemical or electrochemical method. Oxidation of monomer (aniline) to polymer salt in acidic medium is done. Some factors influencing the Polymerization (Li, Jiang, Wu, Chen, & Li, 2000). Polymerization should be done less than pH 3, to prevent deprotonation of polyaniline. Monomer: Oxidant ratio, temperature, washing sequence and reaction time are some parameters that affect chemical polymerization of polyaniline and cell geometry, supporting electrolyte, monomer: dopant anion ratio affects electrochemical polymerization (Chiang & MacDiarmid, 1986). Generally, oxidation (electron accepting) or reduction (electron donor) changes the electrical conductivity of the PANI. The doping process of PANI is different from PPy. Polyaniline base form has a general structure called Emeraldine base (blue). In the emeraldine base, Oxidized and reduced units are repeated alternating (Figure 4).

PANI structure is formed from two types of units, including, 1- Quiniod and 2- Benzenoid. Reduction/oxidation of these unites transforms them to each other. PANI only when the ratio of Benzenoid: Quiniod=3:1 is conductor. Figure 5 illustrates doping process of aniline by protonic acid (Hua & Gao-quan, 2007).

Figure 3. Oxidation doping of PPy
Hua & Gaoquan, 2007.

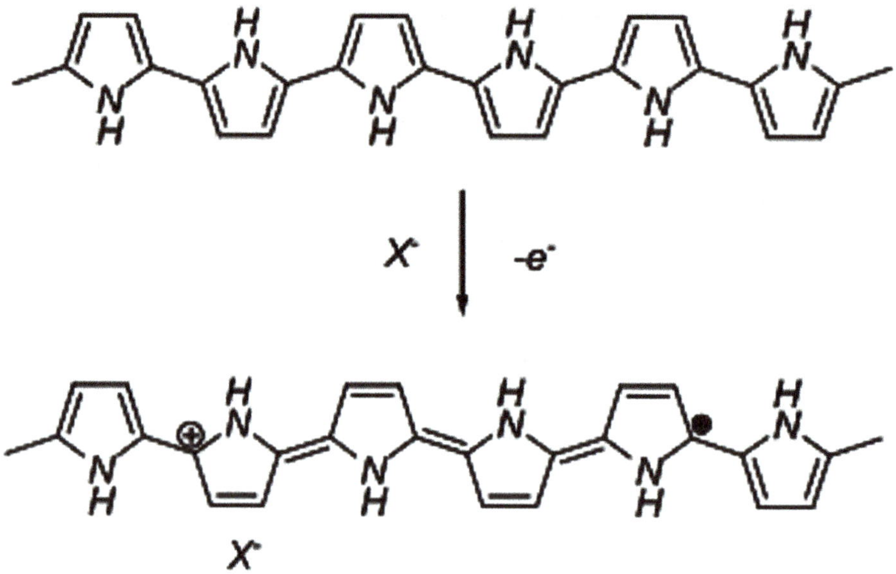

Figure 4. Polyaniline base forms

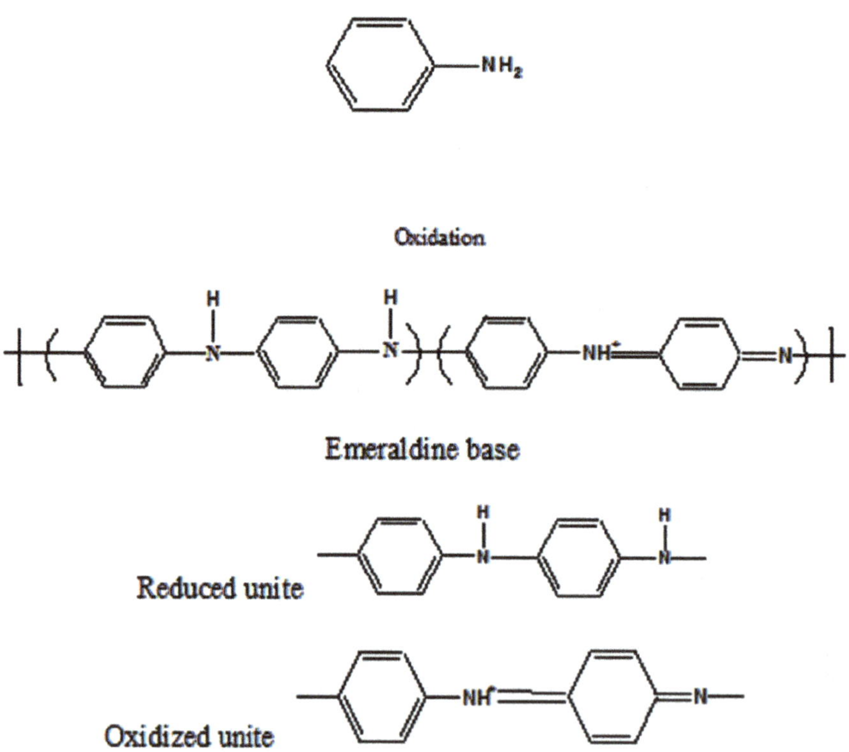

Figure 5. Protonic acid doping process of PANI
Hua & Gaoquan, 2007.

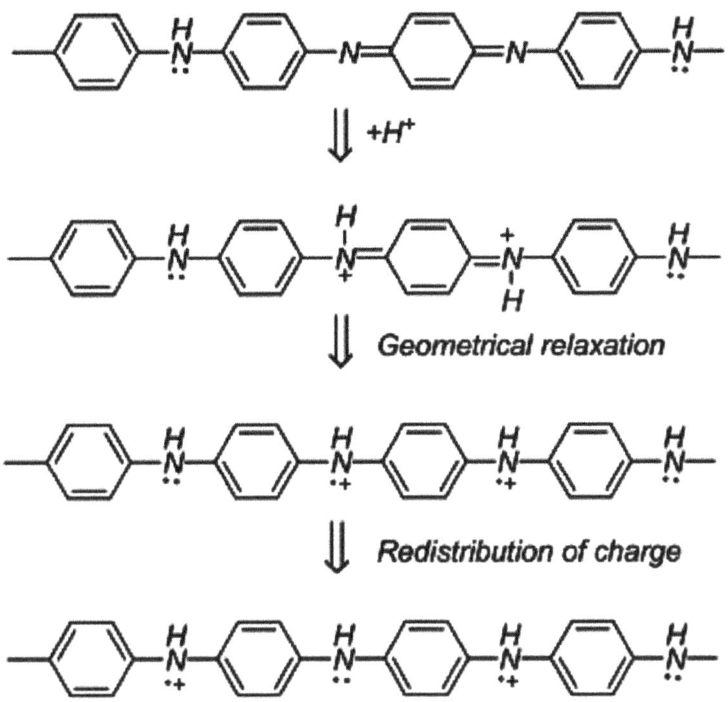

- **Polythiophene:** Polymerization of thiophene has been done by three methods:
    - Chemical oxidation polymerization,
    - Electrochemical polymerization, and
    - Metal-catalyzed coupling reactions.

Electrochemical synthesis of polythiophene is done by a three electrodes that polymerization is done by applying of a potential in a solution containing working electrode, thiophene and an electrolyte (Mhamed, 2005. PP. 474–479). Electrochemical synthesis of PTh cause to produce polymers with varying degrees of regioregularity and undesirable alpha-beta linkages, so electrochemical synthesis is suitable when isolation and purification of polymers is not important. The chemical polymerization method by ferric chloride in chloroform can provide unsubstituted thiophene.

There are little literature reports about synthesize, morphology, sensing ability, conductivity and other properties of polythiophene, unlike polyaniline and polypyrrole. Chemical synthesize of polythiophene in aqueous medium is done by phase transfer catalyst (PTC) (Lakshmi, Anju Dhillon, Avasthi, Azher, & Siddiqui, 2010). Phase transfer catalyst is used as a template for polymerization of PTh and can control size and morphology of polymer. Bromothiophenes are used to synthesize regioregular PTh by catalytic cross-coupling reactions that cause to synthesize polymers with varying degrees of regioregularity (Richard, Rajasekhar, & Subramania, 2009). Figure 6 depicts polythiophene doping.

Figure 6. Polythiophene doping

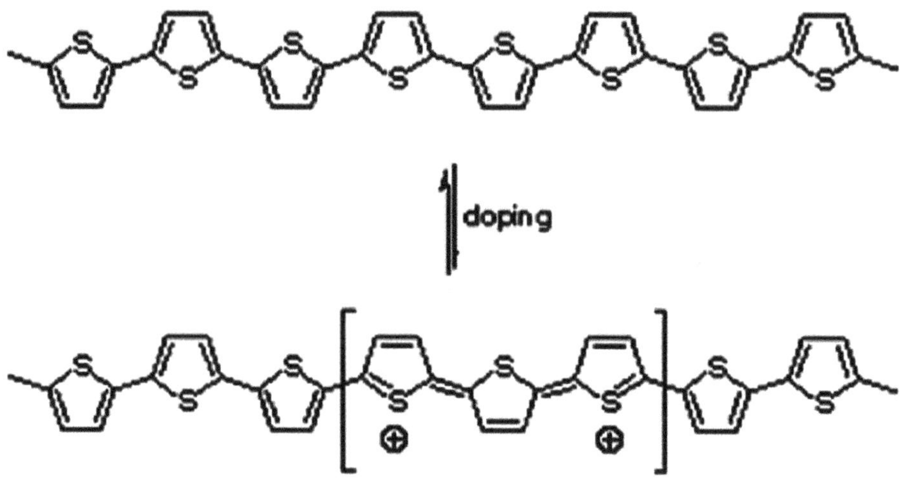

Chemical and Electrochemical Synthesis Advantage and Disadvantages

Chemical synthesis of conducting polymers has two important advantages compare with electrochemical synthesis: various monomers and suitable catalysts can be used for polymerization (Ulrich, Nataliya, & Vladimir, 2008). Some advantages and disadvantages of polymerization methods summarized following:

Chemical Polymerization-Chemical (Solution and Vapor) Powders, Coatings, Colloids

- Advantages:
    - Suitable for bulk synthesis,
    - Cheap – no expensive hardware,
    - CEP doped with reduced form of oxidant:
        - Co-dopant may be required,
        - Dedoping (by reduction or ion exchange).
    - Disadvantages:
        - No control over the oxidizing power of oxidant,
        - Poor control over molecular weight, Temperature, Reactant addition rate.

Electrochemical Polymerization

- Advantages:
    - Full control of oxidation potential,
    - Novel potential/current programming,
    - Molecular weight controlable,
    - Conductivity control able.
- Disadvantages:
    - High cost,

- Heterogeneous reaction,
- Requires suitable electrode substrate,
- Extremely sensitive to electrolyte composition,
- Requires process control.

Some important advantages of electrochemical synthesis of conducting polymers are including; thickness of polymer layer are controllable, the imprinted molecules polymer can be synthesized, morphology and porosity of conducting polymers can be changed by using of different electrochemical techniques (Potyrailo & Mirsky, 2008).

CHEMICAL GAS SENSORS

Sensor is a device that can detect the physical stimulus, such as heat, light, stretch, pressure, bending, motion, and so or chemical stimulus, such as gas interaction, humidity, biological agents and so. A chemical gas sensor is an instrument that can detect and determine type and concentration of gas samples in different mediums. Physical properties of a chemical gas sensor like electrical conductance can be changed by exposing to gaseous chemical compounds (Jon, 2005). There is a relationship between properties changing and gas concentration. A Chemiresistive gas sensor is a sensor with conducting polymer based transducer (Pirsa& Alizadeh, 2010). The general structure of a chemical gas sensor is shown in Figure 7.

Gas Sensor Characteristics

All gas sensors have some characters as following:

Figure 7. General structure of a chemical gas sensor

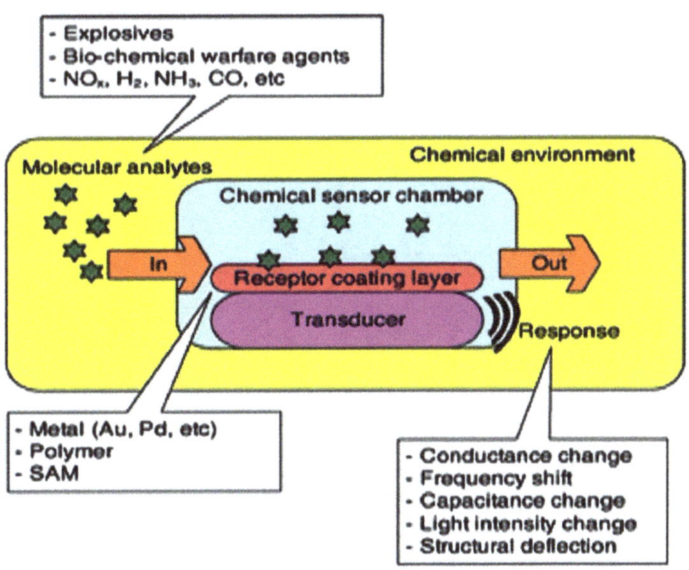

- **Sensitivity:**
    - **Calibration Sensitivity:** Calibration Sensitivity is the ratio of the change in sensor output in response to the change in sensor input in the presence of target gas. By definition if a sensor output y is related to the input x by the function y = f(x), then the sensitivity $S(x_a)$ at point x_a is

$$S(x_a) = \frac{dy}{dx} \int x = x_a \tag{3}$$

    - **Normalized Relative Response:** For a Chemiresistive gas sensor sensitivity is the same of normalized relative response (NRR)

$$S(\%) = NRR = \frac{R_{gas} - R_0}{R_0} \times 100 \tag{4}$$

where R_0 denotes resistance of sensor without gas interaction (initial resistance) and R_{gas} denotes resistance of sensor exposed to gas sample (real-time resistance)

- **Response Time:** When a gas is being exposed to the sensor, response time is generally defined as a time that signal of gas sensor reaches to 90% of maximum signal (figure 8).
- **Recovery Time:** The time that signal of a gas sensor fall to 10% of baseline after removal of the gas is defined as sensor recovery time (Figure 8).
- **Selectivity:** Selectivity is the ability of a gas sensor to detect a target gas without being affected by the presence of other interference gasses. Most gas sensors are sensitive to a family of gasses and it is difficult to produce a sensor specific to only one gas. Moreover, temperature and humidity may also affect sensor performance. A common practice for manufactures of gas sensors is to provide data indicating the changes of different gasses on the sensor output in the presence of common

Figure 8. Response time and recovery time of a gas sensor

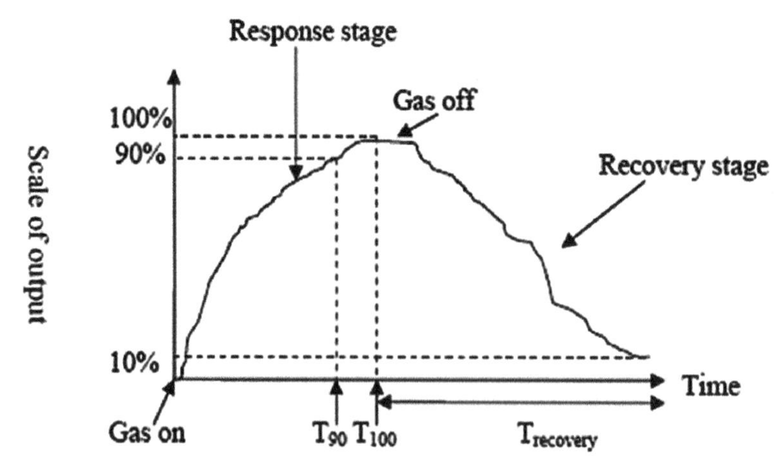

interference gasses. The common problem of the gas sensor based on conducting polymer is temperature and humidity that strongly affect conducting polymer resistance.

- **Repeatability:** Repeatability is defined as the ability of a sensor to repeat the measurements of gas concentrations when the same measure and is applied to it consecutively under the same conditions
- **Linearity:** The closeness between the calibration curve and a specified straight line is defined as linearity (Figure 9). Generally, the initial response of a gas sensor is nearly linear and it tends to saturate as the gas concentration increases. Gas concentration is expressed as volume percent (%) or ppm (by volume). These are unitless expressions as they simply express the ratio of gasses in relation to the balance or carrier gas, such as air or nitrogen (N2).

Classification of Gas Sensors Based on Sensing Principle

Gas sensors based on sensing principle are classified into different types. Table 1 shows different types of gas sensors, but in this section gas sensor based on conducting polymers and response type based on resistance (or conductance) change (Chemiresistive sensors) is studied.

Conducting Polymer-Based Chemiresistive Gas Sensor

Sensing Principles

Sensing principle of all chemical sensors is transforming of analyte concentration to detectable signals. The typical output signals are resistance change, acoustic variables, voltage change and so. Chemiresistive conducting polymer-based gas sensor transform concentrations of gasses to signal (resistance change). The correlation between resistance change and gas concentration is a parameter that helps to determination the concentration of unknown gas. After exposing the gas, the conducting polymer layer as active sensing material (transducer) of the Chemiresistive sensor interacted with the analyte, which causes the

Figure 9. Linearity of a sensor

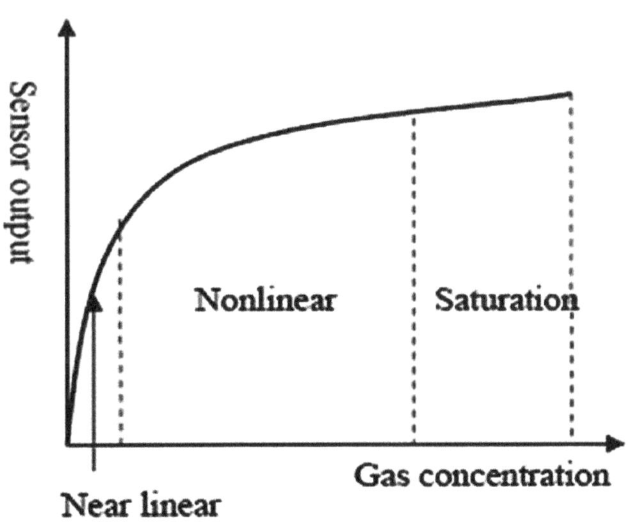

Table 1. Classification of gas sensors based on sensing principle

Type of Sensor	Principle of Operation
Optical	
Fluorescence	Evanescent field excitation or fluorescence enhancement
Reflective	Changes in thickness of polymer sensing layer when exposed to VOC
Absorption	Absorption of VOCs by porphyrins
Thermal	
Pellistor	Change in temperature of catalytic material
Gravimetric	
Quartz crystal microbalances (QCM)	Change in frequency of oscillating crystal in proportion to mass
Surface acoustic wave	Change in frequency of oscillating crystal in proportion to mass
Flexural acoustic wave	Change in frequency of oscillating crystal in proportion to mass
Electrochemical: Chemiresistive	
Metal oxide semiconductors	Change in resistance
Conducting Polymers	Change in resistance
Electrochemical: Potentiometric	
Metal oxide field effect transistor	Change in voltage measurements
Amperometric	Change in current measurements

resistance changes of the conducting polymer. The interactions between conducting polymers and gas samples are multiform, according to different analytes (electron donor, electron acceptor or non both of them) (Hua & Gaoquan, 2007; Ulrich, Nataliya, & Vladimir, 2008). The following sections describe the interaction between conducting polymers and gas samples.

Conducting Polymer and Gas Samples Interaction

There is chemical and physical interaction between gas samples and conducting polymers. Electroactive samples like NH_3 interact chemically and non electroactive samples like benzene interact physically with conducting polymer layer in Chemiresistive gas sensors. Gas samples interaction with a sensor is different according to transducer type of gas sensors. In a conducting polymer-based gas sensor interaction of gas samples and polymer is based on physical and chemical interactions that are described following (Hua & Gaoquan, 2007):

Chemical Reactions between Analytes and Conducting Polymers

Conducting polymers resistance strongly depends on their doping levels. As described above, the doping level of these polymers can be changed by electron transferring from analytes to them and inverse. The amount of doping level change depends on sample type and concentration, so this provides a simple technique to detect the analytes type and concentration. For example, according to our research (Nguyen & Potje-Kamloth, 1999; Bhat, Gadre, & Bambole, 2003; An, Jeong, Hwang, & Lee, 2004) about a PPy-Ag gas sensor, PPy-Ag is a p-type nanocomposite that is doped /undoped by redox reactions; therefore, its doping level can be changed by transferring electrons from or to the solvent vapors. The resistance of PPy-Ag changes by electron transferring. This process occurred when the PPy-Ag film exposed in amines, H_2S and other gasses that have redox activity. Some analytes like I_2 that have electron accepting ability can decrease the electron density on the surface of PPy-Ag aromatic rings that cause to enhance the electrical conductivity of the PPy-Ag (as a p-type semiconductor), and when an electro-donor solvent vapors like an amine interacts with PPy-Ag an opposite process will occur (Pirsa& Alizadeh, 2010). Our results showed that all analyts expose to the sensor, increasing the resistance of the sensor but H_2O decrease the resistance. In the exposing of solvent vapor to the sensor, the active sensing material of the sensor (PPy-Ag layer) interacted with the vapor, which causes the physical property changes of the sensing material. The resistance and physical properties of PPy-Ag like morphology and porosity strongly depend on the type of dopants type and the doping levels. Dimethyl Hydrazine (DMH) as an electron donor compound can transfer electrons to PANI (figure 10). PANI conductivity level as a p-type conducting polymer can be decreased by interaction with DMH. However, washing of DMH from PANI surface can return basic conductivity of PANI. PANI doping/dedoping by DMH was confirmed in our new research about DMH determination by PANI gas sensor (Pirsa, Afshar Asl, Khani, & Allahverdi Pur, 2013).

The electrons on –NH– of DMH are transferred to polyaniline and base form of PANI is formed. The electron donor and electron acceptor compounds chemically can dop/dedop all conducting polymers like PPy, PTh and PANI films. H_2O as a proton donor agent can enhance protons (H^+) on the p-type conducting polymer ring or can remove electrons from the aromatic rings of conducting polymers (Figure 11). When some electrons leave a p-type conducting polymer chain like polypyrrole doping level is enhanced, so the resistance of these polymers enhances too (Jain, Chakane, Samui, Krishnamurthy, & Bhoraskar, 2003).

Figure 10. Dimethyl Hydrazine transfers electrons to PANI

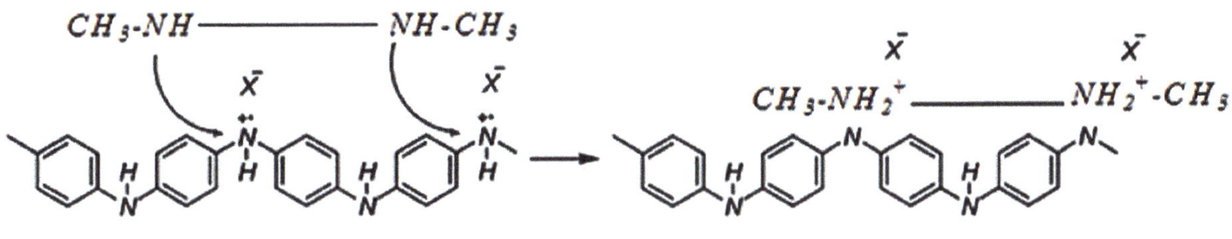

Figure 11. H_2O enhance protons (H^+) on the p-type conducting polymers ring

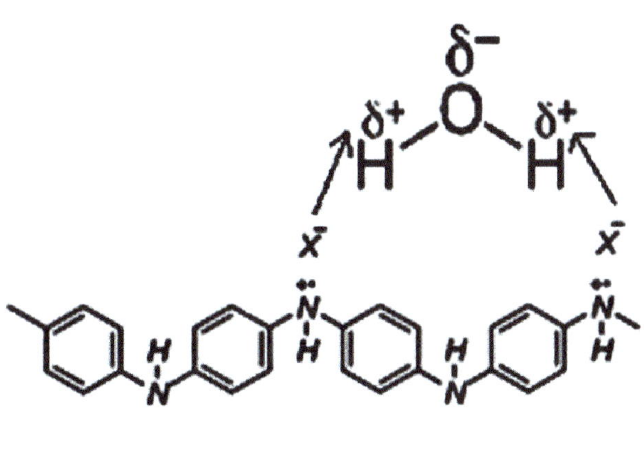

It is important to attend that doping/dedoping of conducting polymers by interaction with electron acceptor/donor compounds help us to determine simultaneously of them. For example, it is possible to the determination of water content of volatile organic compound or NH_3 percent of water medium is possible.

Absorption of Gas Sample and Swelling of Conducting Polymers

Benzene, chloroform, and some other important organic solvents are not chemically active at room temperature and under mild conditions. These chemical solvents don't have any important chemical reaction with conducting polymers, but these solvents can cause to swelling of conducting polymers by physical interaction and surface absorption (Gardner, Bartlett, & Pratt, 1995; Hua & Gaoquan, 2007). Physical absorption and swelling do not change the oxidation levels of conducting polymers but can change sensing properties of conducting polymers. According to this suggestion absorption of gas samples on the surface of conducting polymers depend on chemical properties of the gas, conducting polymer porosity, morphology, temperature, carrier gas flow rate and some other factors. Firstly absorption happens then absorbed gas samples can diffuse into the polymer bulk and swell it, so molecular weight and diffusion constant are important parameters in absorption step. Our new research about PPy-Ag sensor results confirmed the suggestion of chemical and physical reaction between gas samples and conducting polymers. We believe that in our study all solvents including benzene, toluene, methanol, ethanol and so don't have a chemical reaction with PPy that cause increasing of PPy-Ag resistance by absorbing and swelling the PPy-Ag, but H_2O has chemical interaction with PPy-Ag, interaction of H^+ in water by PPy-Ag causes to increase the doping level of PPy-Ag, so electrical conductance of the PPy-Ag is enhanced (Gardner, Bartlett, & Pratt, 1995).

The Configurations and Sensing Principles of Chemiresistive Sensors

Chemiresistive sensors are the most common type of gas sensors. Chemiresistive gas sensors are fabricated have some advantages like; low cost, good reproducibility, high sensitivity, operate simply, detect fast, safe to the environment and accurate determination of trace amounts of gas samples. In a Chemiresistive gas sensor electric resistance is sensitive to the chemical environment, used substrate, temperature, humidity, pressure and so (Janata & Josowicz, 2003; Hua & Gaoquan, 2007).

Figure 12 shows a common scheme of a conducting polymer-based gas sensor that consists of two electrodes and a layer of conducting polymer. The substrate of Chemiresistive can be a fiber (conductor or insulant), a plate, interdigitated electrode or other suitable shapes. Figure 13 shows the scheme of a Chemiresistive gas sensor device with interdigitated and fiber sensor. Interaction of conducting polymer surface and gas samples is the important step in gas detection. Conducting polymer resistance increased or decreased by interaction with all of the samples. The resistance change as the output of the sensor transducer is measured. In this type of gas sensors a simple multimeter and microcomputer are needed to collect and analysis data, a constant voltage applied to the sensor, and resistance change as a signal measured. As the mentioned above a fiber, plate or interdigitated substrate can be used as a transducer in chemiresistive sensors, but to improve the performance and sensitivity of sensors, interdigitated substrate is widely used (Liu, AguilarHernandez, PotjeKamloth, & Liess, 1997; Hua & Gaoquan, 2007).

Figure 12. Configuration of chemiresistive gas sensor

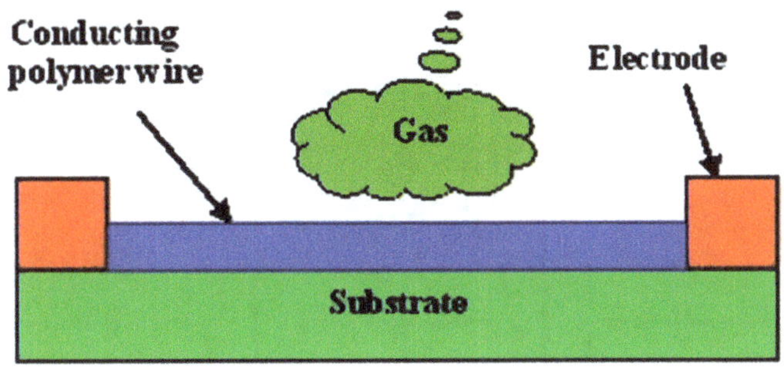

Figure 13. Scheme of a chemiresistive gas sensor device with interdigitated and fiber sensor-dark patterns are conducting polymers in the two types sensor
Pirsa, Alizadeh, Zandi, Almasi, & Heidari, 2014; Alizadeh, Ataei, & Pirsa, 2015.

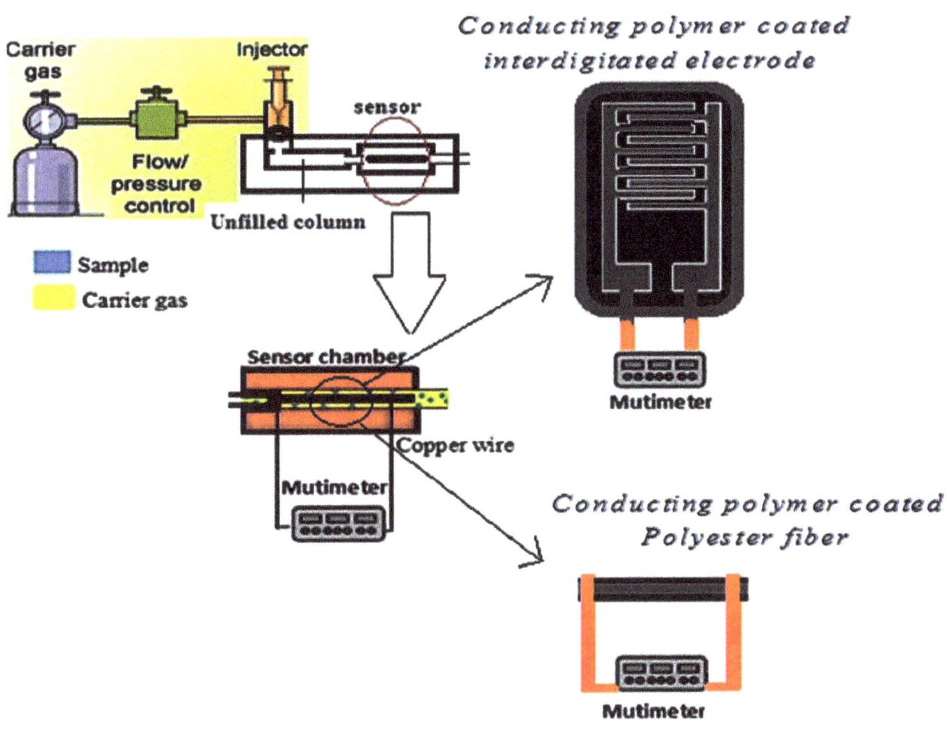

The configuration of the chemiresistive gas sensor is presented with an equivalent circuit diagram in Figure 14 (Chen, Josowicz, & Janatax, 2004; Janata, 2003. PP. 864-869). The overall resistance of the sensor will be changed by changing in any parts of the sensor. Three components have an important role in conducting polymer conductivity:

$$\frac{1}{\sigma} = \frac{1}{\sigma_c} + \frac{1}{\sigma_h} + \frac{1}{\sigma_i} \tag{5}$$

Figure 14. Equivalent circuit diagram of the device shown in Figure 12

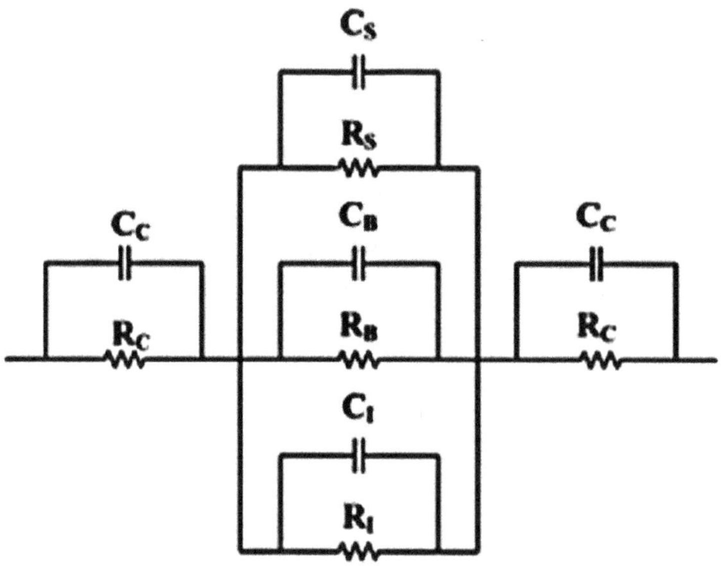

σ = Overall conductivity,
σ_c = Intermolecular conductivity,
σ_h = Intramolecular conductivity,
σ_i = Ionic conductivity.

Reduction/oxidation and acidic/basic interaction of conducting polymers with gas samples can change doping levels of them. Doping level change can alter σ_c as an intermolecular conductivity. Intra-chain distance of polymer chains can be changed by changing crystallinity, swelling the polymer, forming H-bonds and dipolar-dipolar interactions that intramolecular conductivity (σ_h) affected by Intrachain distance of polymer. The interaction between the ions and conducting polymers can effect on the mobility of counter ions that alter ionic conductivity (σ_i) (Chen, Josowicz, & Janatax, 2004; Janata, 2003. PP. 864-869; Alizadeh, Ataei, & Pirsa, 2015; Hua & Gaoquan, 2007).

Selectivity in Gas Sensors

Our previous studies and other research studies demonstrated that PANI and PPy film synthesized by chemical phase polymerization exhibit good sensitivity to VOCs. The sensors based on these polymers are more sensitive to polar compounds than apolar compounds that cause to limit using of these sensors to detect and analysis of apolar compounds. However, all gas sensors based on conducting polymer have some problems that next research should be based on these problems. These problems are

1. **Poor Selectivity:** These sensors don't have significant capability to detect selectively one gas in the presence of other gasses.
2. **High Humidity Interference:** Relatively all gas sensors based on conducting polymers are so sensitive to water (H2O) molecules. This is so good when water determination or humidity deter-

mination is our purpose, but in the case of other gas sample detection in the environment, water interference (humidity interference) decrease sensor sensitivity to detection.

3. **Temperature Dependence of Conducting Polymer:** Conducting polymer resistance change by temperature changing. It will be useful when conducting polymers used as a temperature sensor, but when it is used as a gas sensor, the temperature should be fixed. However the normalizing response of gas sensor based on conducting polymer can help us.

However, there are some methods to increase selectivity of the gas sensor based on conducting polymer and decrease humidity interference in the detection and determination of other gas samples.

Some methods are suggested to induce selectivity in gas sensor and decrease water interference following:

Anion Dopant Effect

According to what discussed above conducting polymer has poor selectivity for detection of gas samples. Anion dopant change in conducting polymer backbone is a simple method to change polymer character. Different anion dopants with polar or apolar character can induce polarity to the surface of conducting polymers. According to this fact that polar surface attracts polar compounds and apolar surface attracts apolar compounds, diffusion of anion dopants with different polarity in the conducting polymer backbone can be an appropriate way to improve selectivity. According to our new research (Pirsa, Alizadeh, 2012), five sulfonate anions, including $HSO3^-$, para-toluene sulfonate (PTS), dodecyl benzene sulfonate (DBS), dodecyl sulfonate (DS) and 5-sulfo salicylate (SS) were used to dope polypyrrole film. Sulfonated doped polypyrrole samples (PPy-S) have been prepared by polymerization of pyrrole on surfaces of the polyester fibers in the presence of an oxidizing agent. The effects of the dopant type on the conductivity and response patterns of the PPy-S as a gas sensor for the different gasses were reported. The responsivity of the PPy-S gas sensors for various volatile organic compounds (VOCs) was also reported. The PPy-HSO3 exhibits a lower detection limit to the dimethyl sulfoxide (DMSO) and could also be successfully applied as a highly selective sensor for detection of DMSO. Although the sensors respond to all tested gas samples, but response pattern of sensors to gas samples was changed. PPy-S sensors response towards polar aprotic compounds has an important difference to nonpolar compounds. The most intense responses are observed for polar aprotic compounds. Figure 15 shows the NRRs of five PPy-S sensors to different volatile organic compounds that prepared with various sulfonate anion dopant. As can be seen, the responses pattern and gas-sensing abilities of the sensor were dependent on anion dopant type. The rapid interaction of DMSO and sulfate is reported in the literature (Pirsa, Alizadeh, 2012; Zhu, Nicovich, & Wine, 2003; Kishore & Asmus, 1989), which may cause a selective response of PP-S sensors to DMSO. In contrast, benzene, chloroform and acetone molecules without electron donating groups cannot interact with polymer matrix efficiently like polar ones due to their non-polar nature, they are likely to act as barriers among polymer chains. The detection limit of PPy-S sensors to DMSO decreases in the following order: PPy-DS, PPy-DBS, PPy-PTS > PPy-SS > PPy-HSO3. The PPy-S sensors with multi-charge anion (PPy-HSO3 and PPy-SS) show lower detection limits to DMSO. This may be explained by the fact that in these cases, most interaction occurs with DMSO. Effects of different sulfonate anion dopants on solubility, electrical and thermal properties of polypyrrole have been reported in the literature (Pirsa & Alizadeh, 2012; Zhu, Nicovich, & Wine, 2003; Kishore & Asmus, 1989; Shen & Wan, 1998; Ahin, Aydın, Arslan Udum, Pekmez, & Yıldız, 2004).

Figure 15. The response behavior of PPy-S sensors different gases
Pirsa & Alizadeh, 2012.

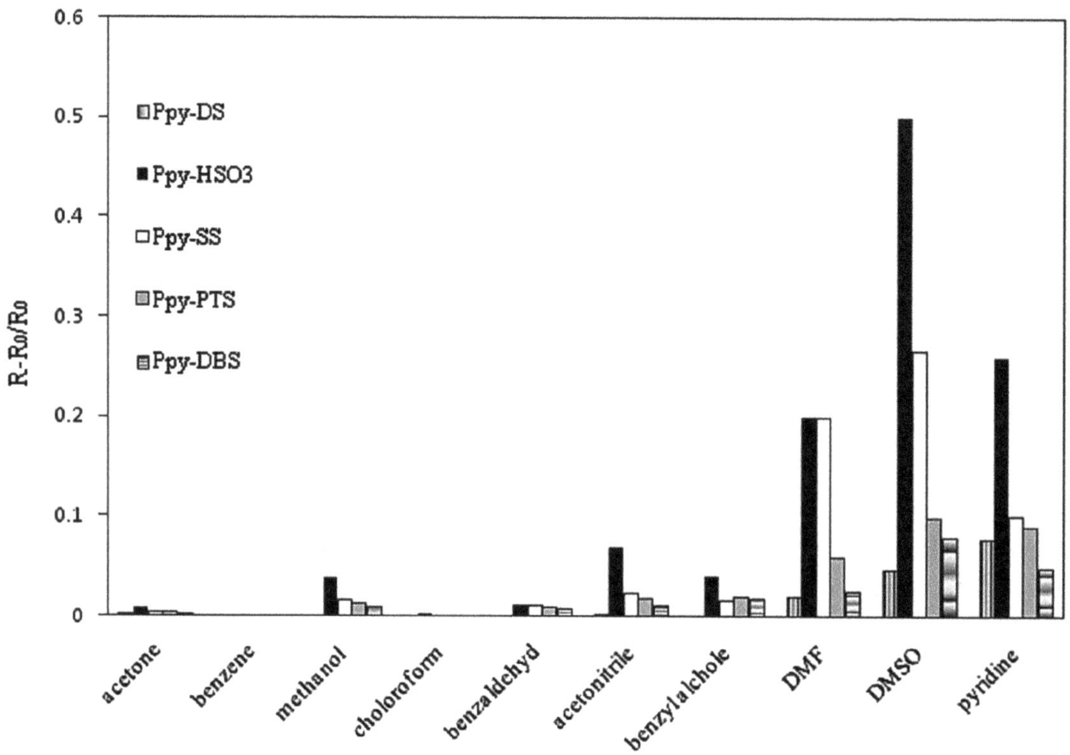

Surface Modification Effect

Surface modification of gas sensors is another method to obtain selective gas sensor based on conducting polymers with excellent chemical characteristics.

For this purpose our new two types of research are reported:

1. **Surface Modification of PPy by CuO:** A selective hydrogen peroxide gas sensor based on nano-sized polypyrrole fiber modified by CuO nanoparticles (PPy-m-CuO) was reported in our new work (Pirsa, Zandi, Almasi, & Hasanlu, 2015). Electrically conductive polymers have been prepared by 1-polymerization of pyrrole on the surface of polyester fiber, 2-modification of the surface of synthesized polymer in the stage 1 by CuO nanoparticles and 3-Polypyrrole-CuO nanocomposite (PPy-CuO) by solution synthesize. The sensing behavior of PPy, PPy-m-CuO and PPy-CuO fibers in the presence of different volatile organic compounds (VOCs) such as NH_3, H_2O, H_2O_2, Methanol and so were reported. Effects of CuO nanoparticles on the sensor behavior of PPy-m-CuO and PPy-CuO fibers were reported (Pirsa, Zandi, Almasi, & Hasanlu, 2015). The PPy-m-CuO exhibits a good selectivity to the H_2O_2. Figure 16 shows the NRRs of PPy, PPy modified by CuO (PPy-m-CuO) and PPy/CuO nanocomposite (PPy-CuO) sensors to different volatile organic compounds. Results show that responses pattern and selectivity of the sensor were changed by surface modification of PPy by CuO and formation of PPy-CuO nanocomposite. Results show that selectivity of sensor can be changed by using of CuO nanoparticles in PPy-CuO composite sensor. PPy-m-CuO sensor

has the best response (selectivity) to H_2O_2. It is obvious that using of CuO particles as a modifier of PPy has more effect than CuO particles as a PPy composite to increase PPy sensor response to H_2O_2 (Pirsa, Zandi, Almasi, & Hasanlu, 2015).

2. **Surface Modification of PPy by Crown Ethers:** Previously it is reported that crown ethers interact with alkyl amine by hydrogen bonding. Gas-phase basicity of some homologous series of alkyl amines and macrocyclic size effects were studied by the research group of Dr. Alizadeh (Alizadeh, Shahdousti, Nabavi, & Tabrizchi, 2011). According to this research 15C5 bind selectivity to normal alkyl amines relative to iso-alkyl amines (Alizadeh, Shahdousti, Nabavi, & Tabrizchi, 2011; Alizadeh, Pirsa, Mani-Varnosfaderani, & Alizadeh, 2015). In recent research, some crown ethers including 15C5, B15C5, 18C6, and DCy18C6 were used to modify polypyrrole film. Crown ether modified polypyrrole (PPy-C) gas sensor have been prepared by physical precipitation of crown ethers on surfaces of the polypyrrole fiber (Alizadeh, Pirsa, Mani-Varnosfaderani, & Alizadeh, 2015). The effect of the type of crown ethers on the response patterns of the PPy-C sensors was reported. Figure 17 shows the NRRs of four PPy-C sensors to amines. As can be seen, the modified sensors exhibited different response patterns to different amines. This observation implies that the modification of the PPy fiber surface with crown ethers induce a remarkable change in the response pattern of the sensors (Alizadeh, Pirsa, Mani-Varnosfaderani, & Alizadeh, 2015).

Backbone Change of Conducting Polymers

However, conducting polymers like polypyrrole, polyaniline, and polythiophene-based sensor selectivity is so poor, but these conducting polymer derivatives can detect selectively some samples.

Another way to change sensor properties of conducting polymers is changing of branch groups on the main chain of these polymers. For example, in pyrrole if –H is replaced by $–CH_3$ or –Phenyl group,

Figure 16. The response behavior of PPy, PPy-m-CuO and PPy-CuO sensors different gas samples
Pirsa, Zandi, Almasi, & Hasanlu, 2015.

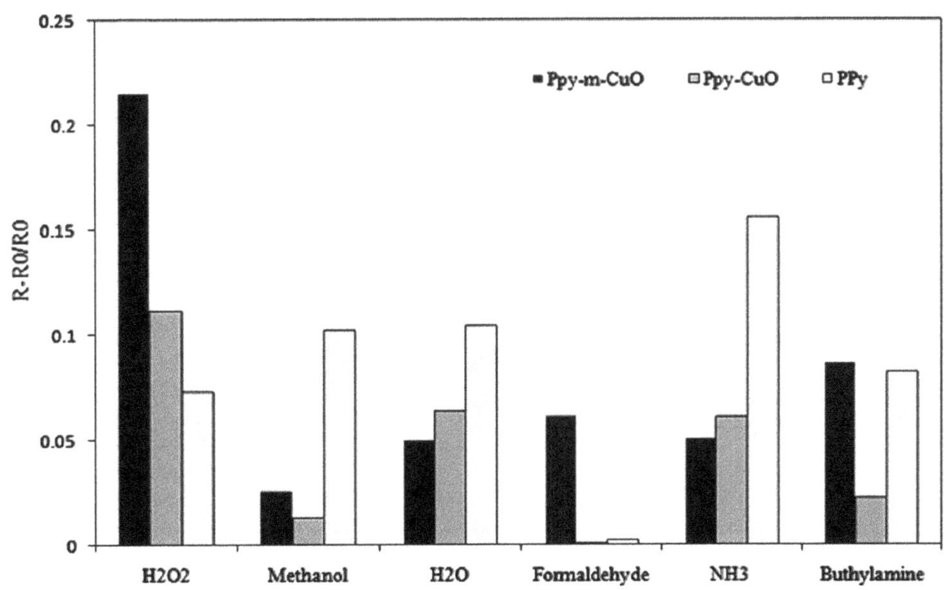

Figure 17. The response behavior of PPy sensors modified by different crown ethers to different alkylamines
Alizadeh, Pirsa, Mani-Varnosfaderani, & Alizadeh, 2015.

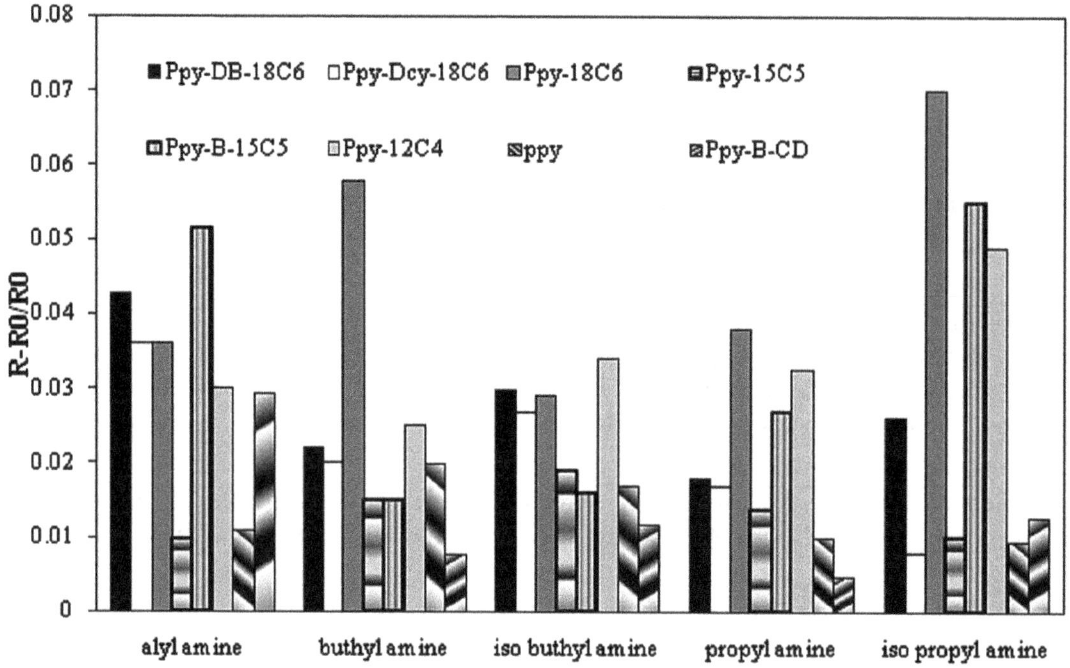

polarity and hydrophilic character of the polymer will be changed. These derivatives on the main chain of conducting polymers have advantages when compared to conducting polymers, due to branch groups which can create some effects like; hydrophobic, magnetic, morphological and so. In the case of Poly N-PhenylPyrrole, the role of phenyl group on the conductivity, stability, thermal and sensor properties of polypyrrole backbone was studied previously (Mangeney, Lacroix, Chane, & Aeiyach, 2000). For example in our new researches (Pirsa, Alizadeh, Zandi, Almasi, & Heidari, 2014; Alizadeh, Ataei, & Pirsa, 2015) to study the influence of $-CH_3$ an –Ph groups on the pyrrole chain in the interaction of the polymer and gas analytes and the role of these group on the morphology and electrical resistance, poly(N-phenyl pyrrole) (P-NPhPy), Poly(N-methyl pyrrole) (P-NMPy) and poly(N-methyl pyrrole-co-pyrrole) (P-NMPy-co-Py) coatings were chemically-deposited on polyester fibers. P-NPhPy and P-NMPy-co-Py sensors were fabricated from these fibers. The Effects of the functional group (–H, $-CH_3$ and –Ph) on pyrrole chain and dopant type (Cl and DBS) in response behavior of sensors were reported. Figure 18 shows the NRRs of three P-NMPy-co-Py and P-NPhPy sensors to different volatile organic compounds that prepared with DBS and Cl anion dopants in solutions. As can be seen, the responses pattern and gas-sensing abilities of the sensor were dependent on the anion dopant type and polymer backbone (Pirsa, Alizadeh, Zandi, Almasi, & Heidari, 2014; Alizadeh, Ataei, & Pirsa, 2015). However, according to our results some groups including $-CH_3$ and –Ph in polypyrrole backbone and anion dopant type can change morphology, porosity and sensor property (selectivity) of polypyrrole based gas sensor.

Figure 18. The response behavior of P-NMPy-co-Py and P-NPhPy sensors different gases
Pirsa, Alizadeh, Zandi, Almasi, & Heidari, 2014; Alizadeh, Ataei, & Pirsa, 2015.

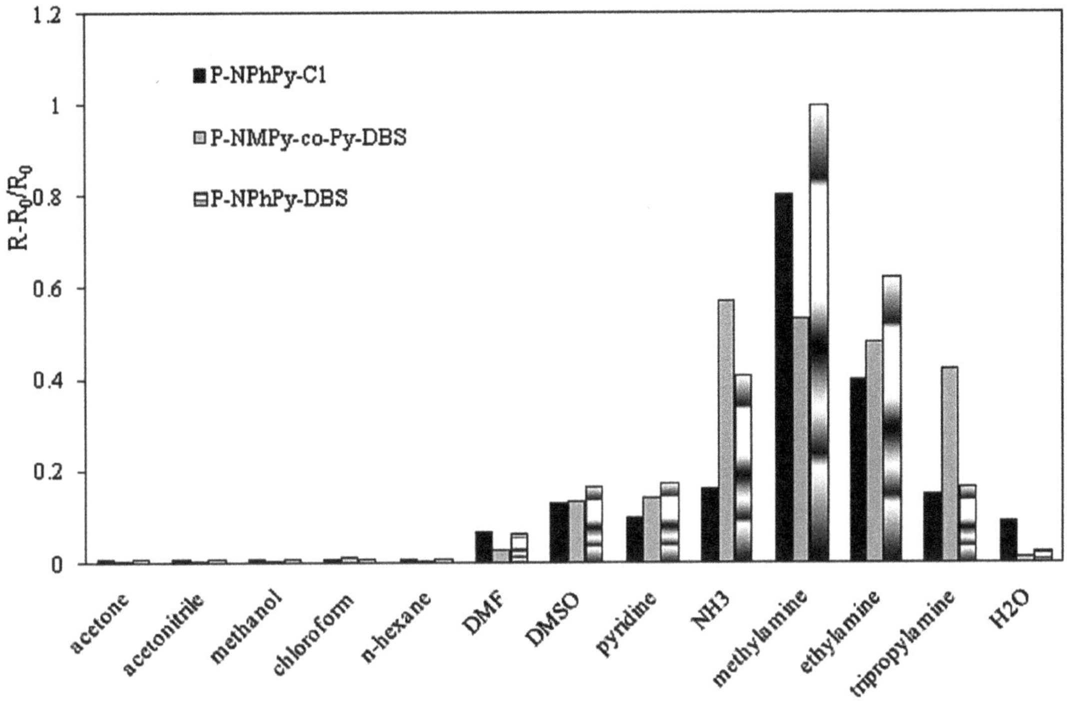

Membrane and Substrate Effect

The different substrate could be used to synthesize of conducting polymers in the gas sensor. Hydrophobe substrate can decrease water interference and can induce selectivity to the sensor. Furthermore, in the same sensor substrate and same conducting polymer using of the selective membrane can induce selectivity in conducting polymer-based gas sensor.

To improve selectivity and decrease water interference of PPy sensor some selective gas sensors based on polypyrrole (PPy) fiber surrounded with polytetrafluoroethylene (Teflon), and polypropylene membranes (PPy-m) were reported in our new researches (Pirsa, Heidari, & Lotfi, 2016). Two membranes, including polytetrafluoroethylene and polypropylene membrane, used to surround polypyrrole fibers. Polypyrrole fiber (PPy-f) samples have been prepared by polymerization of pyrrole on the surfaces of the polyester and polytetrafluoroethylene fibers. The effects of the used membrane type and substrate type of fiber on the response patterns of the PPy-m as a gas sensor to the different gasses was reported (Pirsa, Heidari, & Lotfi, 2016).

Figure 19 shows the NRRs of five PPy and PPy-m sensors that prepared with Teflon and polyester fibers in solution, and vapor phase polymerization of different volatile organic compounds. As can be seen, the response pattern and gas-sensing abilities were dependent on the substrate type, polymerization type and type of membrane. Different polymerization phases, hydrophobic substrates, and membranes have a significant effect on response pattern, water interference, and selectivity.

Figure 19. The response behavior of PPy and PPy-m sensors to different gas samples
Pirsa, Heidari, & Lotfi, 2016.

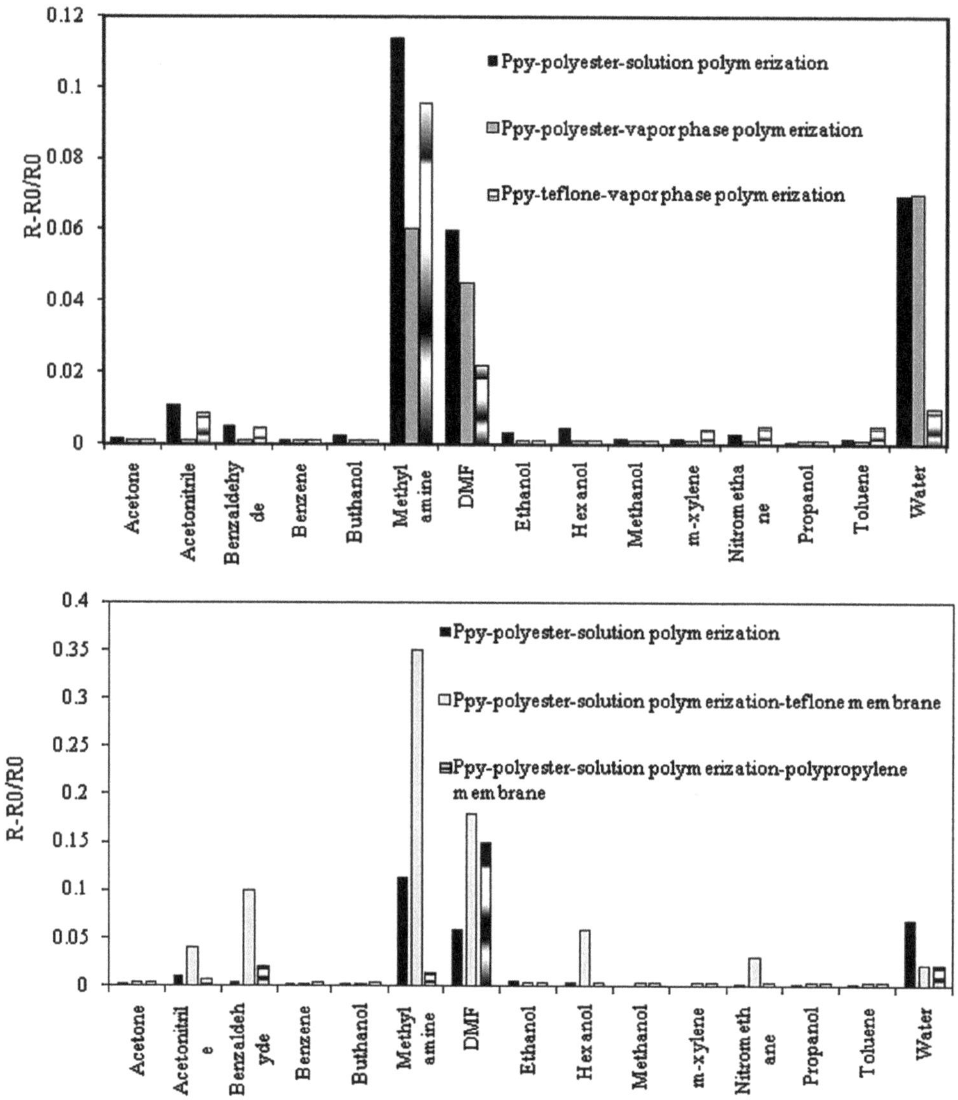

Simultaneous Determination of Some Gases by Conducting Polymers Based Sensors

Determination of Water Content of Organic Solvents by PPy-Ag Sensor

Exposing of different solvent vapors to PPy-Ag gas sensor showed that all of the solvents increasing the resistance of PPy-Ag, but H_2O decrease PPy-Ag resistance, so we can use the PPy-Ag sensor for fast determination of water content of solvents in several seconds.

Figure 20 shows acetone solvent by the different content of water respectively. Each peak has decreased section that is in the result of water and sensor interaction and increase section that is in the result of solvent and sensor interaction. Results show that increasing of the water content of solvents cause for more decreasing of the decreased resistance section while an increased resistance section is relatively fixed. Results show that there is linear respect between sensor resistance decreasing (decreased section) and the water concentration of solvents, so the water content of unknown solvent can be easily determined (Pirsa, 2016).

Simultaneous Determination of Water and Solvent Concentration in a Vapor Phase

The PPy-Ag sensor can recognize simultaneously a solvent and water content of it according to this fact that all solvents cause increasing of PPy-Ag resistance and water cause to decrease PPy-Ag sensor resistance. The PPy-Ag gas sensor was tested with different concentration of mixed solvent and water (Pirsa, 2016).

Figure 21 shows the typical response behavior of the PPy-Ag sensor to the mixed solution of ethanol-water in different concentration. Results show that exposing of mixed gas samples to the sensor that cause decreasing or increasing of resistance when the analyte is eluted by the carrier gas (N_2) sensor resistance return to the base line and in subsequent cycle tests it is reversible. Results show that in the case of ethanol-water mixed, increasing mixed concentration cause to more decreasing of decreased resistance section and more increasing of increased resistance section, but in the case of chloroform-water mixed, increasing mixed concentration don't cause to decrease of decreased resistance section while cause to more increasing of increased resistance section, that is refered to this fact that in chloroform-water mix, water cannot be dissolved in chloroform, so injected solution to system relatively don't has water. It can be seen that electrical resistance decreased upon exposure to water vapor and increased upon exposure to solvent vapor, and recovered when flushed with N_2 flow (Pirsa, 2016).

Figure 20. Resistance response and reproducibility of the PPy-Ag sensor upon exposure to Acetone with different water content
Pirsa, 2016.

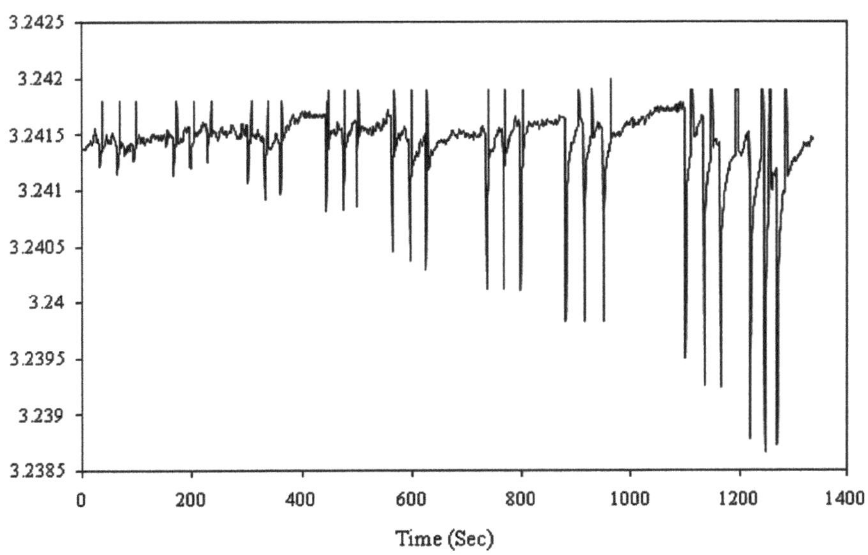

Figure 21. Resistance response and reproducibility of the PPy-Ag sensor upon exposure to mixed solution of Ethanol and water with different concentration
Pirsa, 2016.

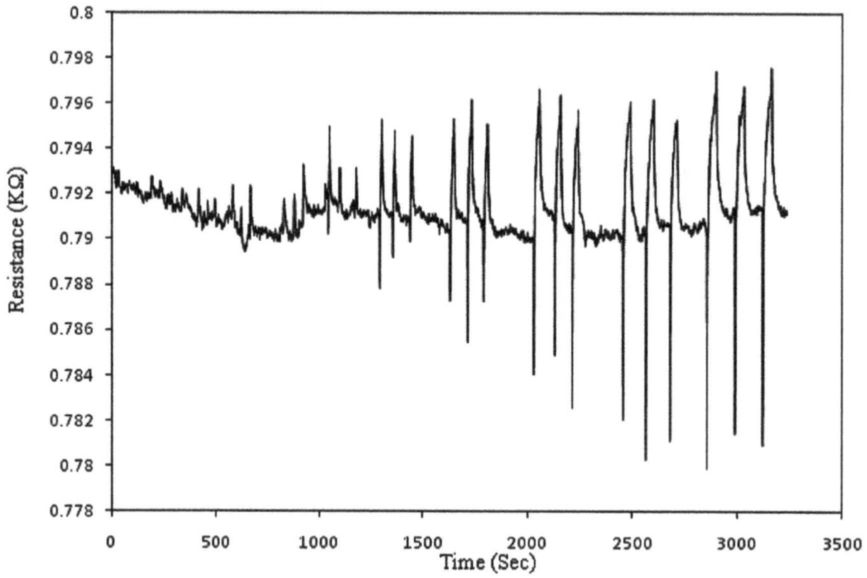

Array Gas Sensor Based on Conducting Polymer

New researches focused on the systems that can detect and analysis of mixed gasses. These systems are non-destructive and have some important advantages like; low cost, operate fast, have portable devices, operate at the atmospheric condition and so. These devices that can determine the concentration of each gas in the mixed gas sample are called array sensor and termed electronic noses (Ulrich, Nataliya, & Vladimir, 2008; Alizadeh, Pirsa, Mani-Varnosfaderani, & Alizadeh, 2015). It is necessary to design suitable software for analysis of multi signals of mixed gasses in array sensors. In the array gas sensor based on conducting polymers, the sensing principle is based on the measurement of the resistance change of the conducting polymers. The response of conducting polymeric based array sensor to different gas samples has been reported by J.A. Morales and his coworkers (Ulrich, Nataliya, & Vladimir, 2008; Alizadeh, Pirsa, Mani-Varnosfaderani, & Alizadeh, 2015). Gas samples can introduce to array sensors by static or dynamic methods. Pattern recognition techniques are common statistical analysis methods to discrimination and classification of signals (data) that are recorded by array gas sensor in the presence of mixed gas samples. Moisture and other gas interferences in the detection by the chemoresistive gas sensor based on conducting polymers affect reproducibility of the sensor and detect ability, discrimination, and determination of gas samples, so array gas sensor by using pattern recognition can dissolve these problems. Our new study about changing of PPy sensor response pattern by modifying of the PPy surface with some crown ethers including 15C5, B15C5, 18C6 and DCy18C6 was reported as mentioned above. An array of sensors has been constructed using four modified polypyrrole sensors (Figure 22). The arrays were used for simultaneous determination of four types of amine including butylamine, isobutylamine, propylamine and isopropyl amine. The principal component analysis (PCA) has been used to test the orthogonality of the responses of the sensors. The results of the PCA technique

Figure 22. Scheme of open-tubular array PPy gas sensors system
Alizadeh, Pirsa, Mani-Varnosfaderani, & Alizadeh, 2015.

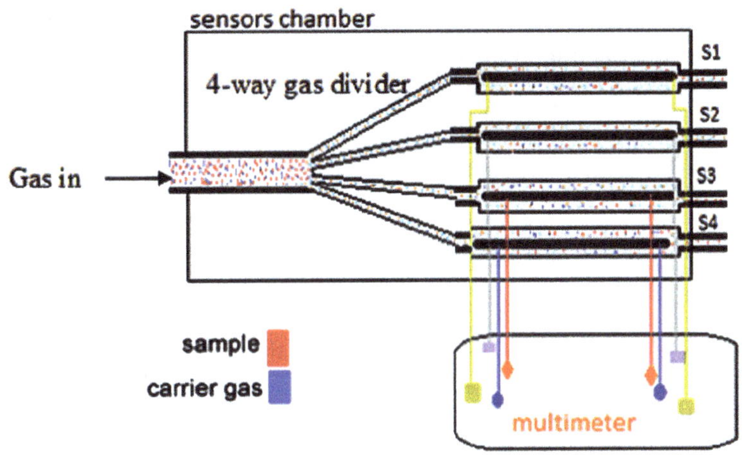

revealed that the developed sensors are orthogonally sensitive to different types of amines. Sixteen mixtures of amines were prepared and the responses of the array sensor were recorded for each mixture. A perceptron artificial neural network with 4-2-4 architecture has been used for correlating the responses of sensors to the concentration of amines in the gas phase. The developed multivariate model was used for simultaneous determination of amines (Alizadeh, Pirsa, Mani-Varnosfaderani, & Alizadeh, 2015).

CONCLUSION

Conducting polymers (CP) by their conductivity and chemical features are so important compounds in gas detection devices and Chemiresistive gas sensors. Several methods are reported to polymerization of these polymers, chemical and electrochemical synthesize. Chemical synthesize can be done in solution phase or in the vapor phase that in two methods and oxidant used for initial polymerization of the polymer. Electrochemical polymerization only is done in solution phase and oxidation agent provided by electrochemical methods. Gas samples and conducting polymer interaction that causes to the resistance change of sensor can be done by chemical or physical interaction. Doping/dedoping of conducting polymer result in a resistance change of them, doping/dedoping mechanism of these polymers are commonly based on electron donating or electron accepting by gas samples. Chemical change of conducting polymer by chemically active materials (like NH_3) and swelling of them by inactive gas samples (like benzene) is common theories about them. CP as a transducer of Chemiresistive sensors can act as only transducer, membrane, linker and a receptor layer or even an electronic circuit for data proceeding. CPs as a transducer of Chemiresistive sensors have the capability to interact with almost many of gas samples and could not discriminate gas samples so selectivity of gas sensors based on conducting polymers is poor. There are several methods to improve selectivity by surface modification, added membrane layer, used substrates, morphology change and so. Response patterns of conducting polymer based Chemiresistive sensor can be changed by several methods. Several gas sensors with a different response pattern can help us to use it as an array gas sensor to detect and analysis of mixed gas samples. According to the discussion above, further works in conducting polymer-based gas sensor mainly include following several aspects. First, it is not clear yet what is a real interaction between conducting polymer and gas samples. Many researchers studied these mechanisms, but all of their reports are presumptions, so it is necessary to investigate further about interactions between the analyte and conducting polymer. Second, the resistance baseline of these polymers is not constant. These sensors initial resistance (resistance baseline) is strongly depending on humidity, temperature and atmospheric condition that cause to poor repeat and reversibility of these sensors (in the several days, weeks or month).

REFERENCES

Ahin, Y. S., Aydın, A., Arslan Udum, Y., Pekmez, K., & Yıldız, A. (2004). Electrochemical synthesis of sulfonated polypyrrole in FSO3H/acetonitrile solution. *Journal of Applied Polymer Science*, *93*, 526–533. doi:10.1002/app.20473

Akundy, G. S., & Iroh, J. (2001). Polypyrrole coatings on aluminum synthesis and characterization. *Polymer*, *42*(24), 9665–9669. doi:10.1016/S0032-3861(01)00529-8

Akundy, G. S., Rajagopalan, R., & Iroh, J. O. (2002). Electrochemical deposition of polyaniline-polypyrrole composite coatings on aluminum. *Journal of Applied Polymer Science*, *83*(9), 1970–1977. doi:10.1002/app.10123

Alizadeh, N., Ataei, A., & Pirsa, S. (2015). Nanostructured conducting polypyrrole film prepared by chemical vapor deposition on the interdigital electrodes at room temperature under atmospheric condition and its application as gas sensor. *Journal of the Indian Chemical Society*, *12*, 1585–1594.

Alizadeh, N., Pirsa, S., Mani-Varnosfaderani, A., & Alizadeh, M. S. (2015). Design and fabrication of open-tubular array gas sensors based on conducting polypyrrole modified with crown ethers for simultaneous determination of alkylamines. *IEEE Sensors Journal*, *15*(7), 4130–4136. doi:10.1109/JSEN.2015.2411515

Alizadeh, N., Shahdousti, P., Nabavi, S., & Tabrizchi, M. (2011). Atmospheric pressure gas-phase ammonium/alkyl ammonium exchange studies of some crown ethers complexes using ion mobility spectrometry: A thermodynamic investigation and collision cross section measurements. *International Journal of Mass Spectrometry*, *308*(1), 18–25. doi:10.1016/j.ijms.2011.07.001

Amaya, T., Saio, D., & Hirao, T. (2007). Template synthesis of polyaniline/Pd nanoparticles and its catalytic application. *Tetrahedron Letters*, *48*(15), 2729–2732. doi:10.1016/j.tetlet.2007.02.059

An, K. H., Jeong, S. Y., Hwang, H. R., & Lee, Y. H. (2004). Enhanced sensitivity of a gas sensor incorporating single-walled carbon nanotube-polypyrrole nanocomposites. *Advanced Materials*, *16*(12), 1005–1009. doi:10.1002/adma.200306176

Anderson, M. R., Mattes, B. R., Reiss, H., & Kaner, R. B. (1991). Conjugated polymer films for gas separations. *Science*, *252*(5011), 1412–1415. doi:10.1126/science.252.5011.1412 PMID:17772913

Aphesteguy, J. C., & Jacobo, S. E. (2004). Composite of polyaniline containing iron oxides. *Physica B, Condensed Matter*, *354*(1-4), 224–227. doi:10.1016/j.physb.2004.09.053

Arshak, K., Moore, G. M., Lyons, J., & Clifford, S. (2004). A review of gas sensors employed in electronic nose applications. *Sensor Review*, *24*(2), 181–198. doi:10.1108/02602280410525977

Bartlett, P. N., & Ling-Chung, S. K. (1989). Conducting polymer gas sensors part III: Results for four different polymers and different vapors. *Sensors and Actuators*, *20*(3), 287–292. doi:10.1016/0250-6874(89)80127-1

Bhadra, S., & Khastgir, D. (2008). Extrinsic and intrinsic structural change during heat treatment of polyaniline. *Polymer Degradation Stability, 93,* 1094–9.

Bhat, N. V., Gadre, A. P., & Bambole, V. A. (2003). Investigation of electropolymerized polypyrrole composite film: Characterization and application to gas sensors. *Journal of Applied Polymer Science*, *88*(1), 22–29. doi:10.1002/app.11641

Chen, H., Josowicz, M., & Janatax, J. (2004). Chemical effects in organic electronics. *Chemistry of Materials*, *16*(23), 4728–4735. doi:10.1021/cm040101m

Chiang, J., & MacDiarmid, A. (1986). Polyaniline: Protonic acid doping of the emeraldine form to the metallic regime. *Synthetic Metals*, *13*(1-3), 193–205. doi:10.1016/0379-6779(86)90070-6

Do, J. S., & Chang, W. B. (2001). Amperometric nitrogen dioxide gas sensor: Preparation of PAn/Au/SPE and sensing behavior. *Sensors and Actuators. B, Chemical*, *72*(2), 101–107. doi:10.1016/S0925-4005(00)00607-9

Freund, M. S., & Lewis, N. S. (1995). A chemically diverse conducting polymer based electronic nose. *Proceedings of the National Academy of Sciences of the United States of America*, *92*(7), 2652–2656. doi:10.1073/pnas.92.7.2652 PMID:11607521

Gangopadhyay, R., & De, A. (2001). Conducting polymer composites: Novel materials for gas sensing. *Sensor and Actuator B*, *77*(1-2), 326–329. doi:10.1016/S0925-4005(01)00719-5

Gardner, J. W., & Bartlett, P. N. (1995). Application of conducting polymer technology in microsystems. *Sensors and Actuators. A, Physical*, *51*(1), 57–66. doi:10.1016/0924-4247(95)85004-X

Gardner, J. W., Bartlett, P. N., & Pratt, K. F. E. (1995). Modeling of gas-sensitive conducting polymer devices. *IEE Proceedings. Circuits, Devices and Systems*, *142*(5), 321–333. doi:10.1049/ip-cds:19952170

Hernandez-Perez, T., Morales, M., Batina, N., & Salmon, M. (2001). Effect of the electrosynthesis method on the surface morphology of the polypyrrole film an atomic force microscopy study. *Journal of the Electrochemical Society*, *148*(5), 369–375. doi:10.1149/1.1365143

Hua, B., & Gaoquan, S. (2007). Gas sensors based on conducting polymers. *Sensors (Basel, Switzerland)*, *7*(3), 267–307. doi:10.3390/s7030267

Jain, S., Chakane, S., Samui, A. B., Krishnamurthy, V. N., & Bhoraskar, S. V. (2003). Humidity sensing with weak acid-doped polyaniline and its composites. *Sensors and Actuators. B, Chemical*, *96*(1-2), 124–129. doi:10.1016/S0925-4005(03)00511-2

Janata, J. (2002). Electrochemical sensors and their impedances. *Analytical Chemistry*, *32*, 109–120.

Janata, J. (2003). Electrochemical microsensors. *Proceedings of the IEEE*, *91*(6), 864–869. doi:10.1109/JPROC.2003.813576

Janata, J., & Josowicz, M. (2003). Conducting polymers in electronic chemical sensors. *Nature Materials*, *2*(1), 19–24. doi:10.1038/nmat768 PMID:12652667

Jon, W. (2005). *Sensor Technology Handbook*. Oxford, UK: Elsevier Inc.

Kishore, K., & Asmus, K. D. (1989). Radical cations from one-electron oxidation of aliphatic sulfoxides in aqueous solution – A radiation chemical study. *Journal of the Chemical Society*, *2*, 2079–2084.

Lakshmi, G. B. V. S., & Anju Dhillon, D. K. (1672–1673). Avasthi, c., Azher, M., & Siddiqui, M. (2010). Synthesis and characterization of thin films of poly(3-methyl thiophene) by rf-plasma polymerization. *Materials Letters*, 64.

Li, D., Jiang, Y., Wu, Z., Chen, X., & Li, Y. (2000). Self-assembly of polyaniline ultrathin films based on doping-induced deposition effect and applications for chemical sensors. *Sensors and Actuators. B, Chemical*, *66*(1-3), 125–127. doi:10.1016/S0925-4005(00)00315-4

Li, Y., Wang, H. C., Cao, X. H., Yuan, M. Y., & Yang, M. J. (2008). A composite of polyelectrolyte grafted multi-walled carbon nanotubes and in situ polymerized polyaniline for the detection of low concentration triethylamine vapor. *Nanotechnology*, *19*(1), 015503. doi:10.1088/0957-4484/19/01/015503 PMID:21730534

Liu, D. M., Aguilar-Hernandez, J., Potje-Kamloth, K., & Liess, H. D. (1997). A new carbon monoxide sensor using a polypyrrole film grown on an interdigital-capacitor substrate. *Sensors and Actuators. B, Chemical*, *41*(1-3), 203–206. doi:10.1016/S0925-4005(97)80296-1

MacDiarmid, A. G., & Epstein, A. J. (1995). Secondary doping in polyaniline. *Synthetic Metals*, *69*(1-3), 85–92. doi:10.1016/0379-6779(94)02374-8

Mahmoudian, M. R., Alias, Y., Basirum, W. J., & Ebadi, M. (2011). Poly (N-methyl pyrrole) and its copolymer with o-toluidine electrodeposited on steel in mixture of DBSA and oxalic acid electrolytes. *Current Applied Physics*, *11*(3), 368–375. doi:10.1016/j.cap.2010.08.006

Mangeney, C., Lacroix, J., Chane, K., & Aeiyach, S. (2000). Macroscopic modulation of the π-electron density of pendant groups grafted on conductive polymers. *Synthetic Metals*, *108*(3), 237–243. doi:10.1016/S0379-6779(99)00260-X

Matsuguchi, M., Io, J., Sugiyama, G., & Sakai, Y. (2002). Effect of NH3 gas on the electrical conductivity of polyaniline blend films. *Synthetic Metals*, *128*(1), 15–19. doi:10.1016/S0379-6779(01)00504-5

Mhamed, C. (2005). Synthesis and characterization of poly(thiophene sulfides) prepared via chemically initiated oxidative polymerization methods. *Synthetic Metals*, *155*(3), 474–479. doi:10.1016/j.synthmet.2005.06.023

Miasik, J., Hopper, A., & Tofield, B. (1986). Conducting polymer gas sensors. *Journal of the Chemical Society*, *82*, 1117.

Michalska, A., & Lewenstam, A. (2000). Potentiometric selectivity of p-doped polymer films. *Analytica Chimica Acta*, *406*(2), 159–169. doi:10.1016/S0003-2670(99)00774-6

Nguyen, V. C., & Potje-Kamloth, K. (1999). Electrical and chemical sensing properties of doped polypyrrole/gold Schottky barrier diodes. *Thin Solid Films*, *338*(1-2), 142–148. doi:10.1016/S0040-6090(98)01060-8

Patil, S., Mahajan, J. R., More, M. A., & Patil, P. P. (1999). Electrochemical synthesis of poly(o-methoxyaniline) thin films: Effect of post treatment. *Materials Chemistry and Physics*, *58*(1), 31–36. doi:10.1016/S0254-0584(98)00245-4

Peng, H., Zhang, L. J., Soeller, C., & Travas-Sejdic, J. (2009). Conducting polymers for electrochemical DNA sensing. *Biomaterials*, *30*(11), 2132–2148. doi:10.1016/j.biomaterials.2008.12.065 PMID:19147223

Pirsa, S. (in press). Fast Determination of water content of some organic solvents by smart sensor based on PPy-Ag nanocomposite. *Nanoscience & Nanotechnology-Asia.*

Pirsa, S., Afshar Asl, A., Khani, A., & Allahverdi Pur, A. (2013). Fabrication of 1, 1-dimethylhydrazine gas sensor based on nano structure conducting polyaniline. *Journal of Sciences. Islamic Republic of Iran*, *24*, 209–215.

Pirsa, S., & Alizadeh, N. (2010). Design and fabrication of gas sensor based on nanostructure conductive polypyrrole for determination of volatile organic solvents. *Sensors and Actuators. B, Chemical*, *147*(2), 461–466. doi:10.1016/j.snb.2010.03.026

Pirsa, S., & Alizadeh, N. (2011). Nanoporous conducting polypyrrole gas sensor coupled to a gas chromatograph for determination of aromatic hydrocarbons using dispersive liquid–liquid microextraction method. *IEEE Sensors Journal*, *11*(12), 3400–3405. doi:10.1109/JSEN.2011.2159970

Pirsa, S., & Alizadeh, N. (2012). A selective DMSO gas sensor based on nanostructured conducting polypyrrole doped with sulfonate anion. *Sensors and Actuators. B, Chemical*, *168*, 303–309. doi:10.1016/j.snb.2012.04.027

Pirsa, S., Alizadeh, N., Zandi, M., Almasi, H., & Heidari, H. (2014). Chemically synthesize nanostructure polypyrrole derivatives and fabrication of gas sensor based on synthesized polymers. *Sensor Letters*, *12*(12), 1–8. doi:10.1166/sl.2014.3386

Pirsa, S., Heidari, H., & Lotfi, J. (2016). Design selective gas sensors based on nano sized polypyrrole/ polytetrafluoroethylene and polypropylene membranes. *IEEE Sensors Journal*, *16*(9), 2922–2928. doi:10.1109/JSEN.2016.2527712

Pirsa, S., Zandi, M., Almasi, H., & Hasanlu, S. (2015). Selective hydrogen peroxide gas sensor based on nanosized polypyrrole modified by CuO nanoparticles. *Sensor Letters*, *13*(7), 1–6. doi:10.1166/sl.2015.3506

Potyrailo, R. A., & Mirsky, V. M. (2008). Combinatorial and high-throughput development of sensing materials. *Chemical Reviews*, *108*(2), 770–813. doi:10.1021/cr068127f PMID:18211102

Ram, M. K., Yavuz, O., & Aldissi, M. (2005). NO_2 gas sensing based on ordered ultrathin films of conducting polymer and its nanocomposite. *Synthetic Metals*, *151*(1), 77–84. doi:10.1016/j.synthmet.2005.03.021

Ramanavicius, A., Ramanaviciene, A., & Malinauskas, A. (2006). Electrochemical sensors based on conducting polymer-polypyrrole. *Electrochimica Acta*, *51*(27), 6025–6037. doi:10.1016/j.electacta.2005.11.052

Ratcliffe, N. M. (1990). Polypyrrole-based sensor for hydrazine and ammonia. *Analytica Chimica Acta*, *239*, 257–262. doi:10.1016/S0003-2670(00)83859-3

Redondo, M. I., Sanchez de la Blanca, E., Garcı'a, M. V., Raso, M. A., Tortajada, J., & Gonzalez-Tejera, M. J. (2001). FTIR study of chemically synthesized poly (N-methylpyrrole). *Synthetic Metals*, *122*(2), 431–435. doi:10.1016/S0379-6779(00)00563-4

Richard, S., Rajasekhar, M., & Subramania, A. (2009). Synthesis of polythiophene nanoparticles by surfactant assisted dilute polymerization method for high performance redox supercapacitors. *International Journal of Electrochemical Science*, *4*, 1289–1301.

Selapinar, F., Toppare, L., Akbulut, U., Yalcin, T., & Suzer, S. (1995). A conducting composite of polypyrrole as a gas sensor. *Synthetic Metals*, *68*(2), 109–116. doi:10.1016/0379-6779(94)02299-E

Shen, Y., & Wan, M. (1998). In situ doping polymerization of pyrrole with sulfonic acid as a dopant. *Synthetic Metals*, *96*(2), 127–132. doi:10.1016/S0379-6779(98)00076-9

Skotheim, T. A. (1986). *Handbook of Conducting Polymers*. New York: Marcel Dekker.

Song, P., Wang, Q., & Yang, Z. (2011). Ammonia gas sensor based on PPy/ZnSnO3 nanocomposites. *Materials Letters*, *65*(3), 430–432. doi:10.1016/j.matlet.2010.10.087

Tüken, T., Tansuğ, G., Yazıcı, B., & Erbil, M. (2007). Poly(N-methyl pyrrole) and its copolymer with pyrrole for mild steel protection. *Surface and Coatings Technology*, *202*(1), 146–154. doi:10.1016/j.surfcoat.2007.05.022

Tzamalis, G., Zaidi, N. A., Homes, C. C., & Monkman, A. P. (2002). Doping-dependent studies of the Anderson–Mott localization in polyaniline at the metal–insulator boundary. *Physical Review B: Condensed Matter and Materials Physics*, *66*(8), 1–7. doi:10.1103/PhysRevB.66.085202

Ulrich, L., Nataliya, V., & Vladimir, M. (2008). Conducting polymers in chemical sensors and arrays. *Analytical Chimica Acta, 6,* 1–26.

Zhu, L., Nicovich, J. M., & Wine, P. H. (2003). Temperature-dependent kinetics studies of aqueous phase reactions of SO_4^- radicals with dimethylsulfoxide, dimethyl-sulfone, and methanesulfonate. *Journal of Photochemistry and Photobiology A Chemistry*, *157*(2-3), 311–319. doi:10.1016/S1010-6030(03)00064-9

Chapter 7
Modeling, Design, and Applications of the Gas Sensors Based on Graphene and Carbon Nanotubes

Rafael Vargas-Bernal
Instituto Tecnológico Superior de Irapuato, Mexico

ABSTRACT

Gas sensing continues attracting research communities due to its potential applications in the sectors military, industrial and commercial. A special emphasis is placed on the use of carbon nanomaterials such as carbon nanotubes and graphene, as sensing materials. The chapter will be divided as follows: In the first part, a description of the main topologies and materials (carbon nanomaterials plus polymers, metals, ceramics or combinations between these groups) used to fabricate gas sensors based on graphene and carbon nanotubes that are operated by conductance or resistance electrical, is realized. Next, different mathematical models that can be used to simulate gas sensors based on these materials are presented. In the third part, the impact of the graphene and carbon nanotubes on gas sensors is exemplified with technical advances achieved until now. Finally, it is provided a prospective analysis on the role of the gas sensors based on carbon nanomaterials in the next decades.

INTRODUCTION

Gas sensing continues attracting research communities due to its potential applications such as detecting air pollutants, determining the toxic gas leakage in facilities, monitoring and quantifying in specific gas generation during chemical reactions of interest, etc. In nowadays, gas sensors are fabricated using a multitude of materials in accordance with the global research effort reported in the scientific literature in last decades. Today's modern world requires the sensing at ultra-low concentration of gases, which have led to the necessity of developing ultrasensitive and ultrafast electronic sensors. Different types of gas sensors can be developed using distinct principles such: optical, magnetic, thermal, acoustic,

DOI: 10.4018/978-1-5225-0736-9.ch007

and electrical. Electrical gas sensors transduce the interaction between the gas molecules and sensing materials as a change in the electrical conductance or electrical resistance. The main parameters to be controlled by designers of gas sensors consist into five different technical aspects: selectivity, sensitivity, stability, recovery time, and response time. In addition, it is desirable that concentrations by below of parts of million can be detected, to achieve the better gas sensor. Among all possible materials to fabricate gas sensors, carbon nanomaterials such as carbon nanotubes and graphene represent the better option to develop the sensing that allow fulfills the aforementioned requirements. Carbon nanomaterials with semiconducting behavior are sensitive to the surrounding environment thanks to their capacity of absorption of gas molecules, which modifies the current flowing through of the sensing material as a function of the applied voltage between a pair of electrodes. Graphene and carbon nanotubes present exceptional properties: electrical, chemical, electrochemical and optical; these qualities make them ideal candidates for use in gas sensors. Moreover, nanostructured materials offer a huge number of expectations to enhance and modulate the properties of the gas sensing, by providing properties such as defined structure, high chemical stability, high surface area and a good thermal conductivity. The combination of polymers, metals or ceramic materials with carbon nanomaterials for the preparation of gas sensors has led to novel areas of research, due to their excellent sensitivity and selectivity to specific gases, thanks to large surface-area-to-volume ratio found in nanomaterials such as carbon nanotubes and graphene. In the case of polymers, these materials can be assembled either by polymer-wrapped carbon materials, placing layer-by-layer of polymer and carbon materials, as composite materials based on polymer, and carbon nanomaterials, as well as molecularly imprinted polymers (MIPs) with carbon nanomaterials. The use of polymers not only improves the dispersion of the carbon nanomaterials in them, but also enhances redox behavior and biocompatibility, and provides additional properties such as photoelectric or swelling capacity. In the case of ceramic materials, they use chemical linkages such as functional groups between carbon nanomaterials and materials such as clay, nitrides, and metal oxides, to achieve a composite material with good properties to be used in gas sensors. Carbon nanomaterials and metal oxide nanoparticles are combined to increase the selectivity and reduce the response and recovery times of the gas sensors. Various transition metals (Pt and Au) can be embedded in carbon nanomaterials for creating functional structures to be used as sensing material in gas sensors. The synergistic combination of metal nanoparticles and carbon nanomaterials modulates the electron properties of carbon nanomaterials, leading to enhancement of selectivity and sensitivity in gas sensors. The use of mathematical models, for predicting the behavior of the gas sensors based on carbon nanotubes and graphene before that they are fabricated in the industry at large quantities, allow realizing comparisons between theoretical results with those obtained from the experimental works of other researchers around the world. In recent years, some researchers have developed mathematical approaches based on algorithms of artificial neural network (ANN), support vector regression (SVR) as well as analytical modeling to simulate the behavior of gas sensors based on carbon nanomaterials. The main purpose of this chapter consists in discussing the more recent advances on the modeling, design and applications of the gas sensors based on carbon nanomaterials, since these materials will facilitate the outstanding development of the next-generation of gas sensors in concentrations below the level of parts per million (ppm), that is, parts per billion (ppb) or even parts per trillion (ppt), in the future decades. The chapter will be divided as follows: In the first part, a description of the main topologies and materials (carbon nanomaterials plus polymers, metals, ceramics or combinations between these groups) used to fabricate gas sensors based on graphene and carbon nanotubes that work by means of electrical conductance or electrical resistance is realized. Next, different mathematical models that can be used to simulate gas sensors based on these materials are presented separately. In the third part, the impact of the graphene and carbon nanotubes on gas sensors is exemplified with technical advances achieved until now. Finally, a prospective on the role of the gas sensors based on graphene and carbon nanotubes in the next decades is provided.

Topologies and Materials Used for Gas Sensors Based on Carbon Nanomaterials

The search for materials, such as carbon nanomaterials that can realize gas sensing by adsorption at room temperature, is of great interest. Carbon nanotubes and graphene possess electronic properties, which are very sensitive to the chemical environment around of them. In particular, when these materials are exposed to gases, its presence produces in them significant changes in their electrical resistance, thermoelectric power, and local density of states. In addition, resistive sensors have a low recovery time and a faster response time. The more surprising of all is that these electronic parameters can be tuned while electrical conductance is maintained below of the achieved level by a carbon nanomaterial with metallic electrical behavior in whose case the gas sensing, it is not more possible. In addition, structural qualities such as size, large surface area, and hollow geometry, facilitate the gas adsorption.

The sensing mechanism in gas sensors based on carbon nanotubes and/or graphene consists in realizing charge transfer between adsorbed gas molecules into surface of the sensing material (carbon nanotube or graphene) or by means of gas-induced changes at the interface between the carbon nanomaterials and their metal contacts (Boyd, 2014). In particular, Schottky barriers are produced between the carbon nanomaterials and their metal electrodes found at the contacts, and thus, a modulation of voltage produces a change in the sensitivity of the gas sensor. Changes in local potential barriers due to the attachment of functional groups in the surface of carbon nanomaterials operate as the main mechanism for single molecule detection. In addition, the electrical conduction of a gas sensing material can be increased by means of charge hopping from one carbon nanotube to the gas molecules and onto a second carbon nanotube in the case of CNT-CNT junctions or graphene sheet-graphene sheet junctions that have been exfoliated (Boyd, 2014).

Carbon nanotubes have a distortion of the electron clouds around their outside surface presented as a rich π-electron conjugation, which favor their electrochemical activity making them sensitive to charge transfer and chemical doping effects by different molecules (Zhang, 2008). Graphene has a high electrical conductivity due to the *p*-orbital de-localized electrons found in the π-bonds formed among neighboring atoms of carbon (Smith, 2015). Thus, these electrons have high mobility and are very sensitive to modifications of their environment as is the presence of different single gas molecules.

Gases such as O_2, NO_2, Br_2, I_2, NH_3, and other gases have been detected using carbon nanotubes as sensing materials. Graphene has been used to detect gases such as O_2, SO_2, CO_2, N_2O, NO, NH_3, and NO_2 (Pu, 2014). Graphene is more sensitive to NO_2, while NO is better adsorbed in rGO (reduced graphene oxide). Moreover, NH_3 is moderately attracted to both graphene and rGO. Resistive gas sensors used in the configuration of field-effect transistor are very attractive, due to their miniaturized size, high sensitivity, and portability (Pu, 2014). Gas molecules tend to be absorbed on sites in the sensing material either crystalline (graphene or carbon nanotubes) or amorphous (rGO, reduced graphene oxide or α-CNTs, amorphous carbon nanotubes) with higher binding energies and higher charge transfer, since these sites favor the sensitivity of the gas sensor. Gas molecules interact with sensing materials, through van der Waals attractions, and with this adsorption, the electronic structure will be altered, and its electrical resistance will change due to the charge transfer realized to the sensing material. This transfer of functionalizing chemicals electrically modifies to the sensing material either into an *n*-type (reducing atmospheres) or *p*-type material (oxidizing atmospheres). The stronger the interaction, more gas molecules will be adsorbed. Volatile organic gases such as tetrahydrofuran (THF) and dichloromethane (DCM) can be easily detected using bi-dimensional (2D) graphene oxide (GO) nanosheets with high sensitivity (Some, 2013). Physical and/or chemical modification of carbon nanomaterials offers a great promise in the gas sensor performance enhancement (Mao, 2014).

Two different topologies can be distinguished for fabricating resistive gas sensors based on graphene and carbon nanotubes:

1. Chemoresistive (chemical resistive) or chemoconductive (chemical conductive) devices, and
2. Field-effect transistors (Zhang, 2008; Vargas-Bernal, 2012).

The first type consists in simple two-terminal device that detects gases and produce electrical conductance/resistance changes as its output (Cooper, 2014). They are the easiest to build, test, and calibrate. The second type consists in three-terminal devices that act as a chemical sensor. It has a configuration similar to a MOSFET transistor (with terminals called gate, source, and drain), where gate terminal changes its electrical charge when it detects molecules of gas (Mao, 2014). The use of carbon nanomaterials have led to devices with very high sensitivity, ultra-low detection limit and promising selectivity (Basu, 2012). Carbon nanotubes and graphene allow a wide range of selectivity thanks to their molecular platforms, which can be chemically controlled by functional chemical groups and whose performance is highly tunable to be exploited in gas sensing (Rigoni, 2014).

A chemiresistor is a two-terminal device whose sensing material changes its electrical resistance in response to changes in the chemical environment around of it. Sensing materials such as metal oxide semiconductors, conductive polymers, composites, as well as, nanomaterials as graphene, carbon nanotubes and nanoparticles, have been used to fabricate chemiresistors. In particular, chemical resistive sensors based on graphene are relatively new and they present excellent sensitivity and selectivity to detect vapour-phase molecules (Joshi, 2010; Schedin, 2007; Cooper, 2014). In the case of chemical resistive sensors based on carbon nanotubes, they were introduced by first time in year 2000 (Kong, 2000). These devices are sensors with qualities such as simple, small, sensitive, with the capacity of detecting gases in air, soil, or water. In addition, these have no moving parts, and are electrically powered with a direct-current (DC) voltage source to determine the difference of electrical resistance of the sensing material, when it is subjected to the air or to a gas of interest. Moreover, this topology of gas sensors is the most largely produced and commercialized. Graphene has better response when it has morphology of films or ribbons in chemiresistors (Joshi, 2010).

Four different topologies of field-effect transistors based on carbon nanotubes or graphene have been developed and/or proposed:

1. Back-gated FETs,
2. Top-gated FETs,
3. Wrap-around gate FET, and
4. Suspended FET (Vargas-Bernal, 2012).

A study more detailed about of these topologies can be found in Vargas-Bernal, 2012. A gas sensor based on graphene using the basic topology of a back-gated field-effect transistor is shown in Figure 1.

Figure 1. Basic topology based on a field-effect transistor for a gas sensor based on graphene

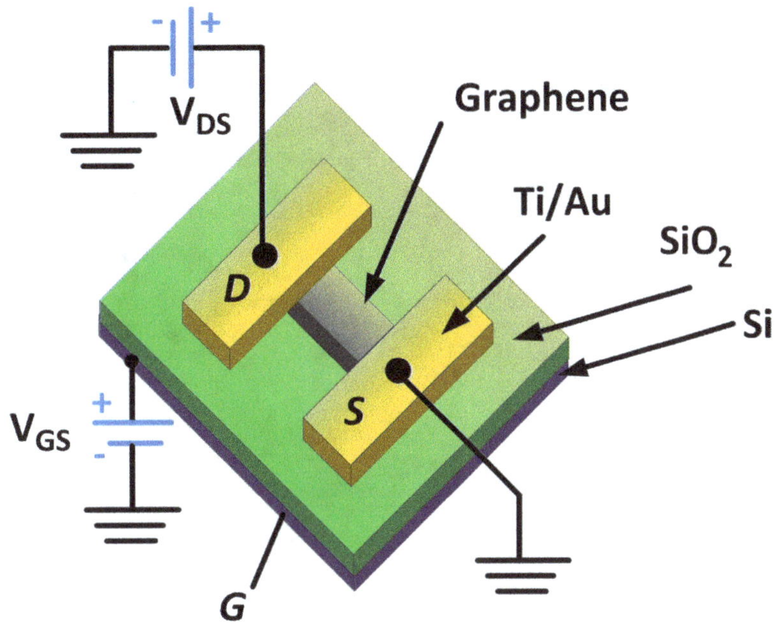

Drain and source terminals are illustrated by Ti/Au, and the gate is formed by the capacitor composed of Si, a gate oxide SiO_2, and the graphene which represents the channel that interconnects the terminals of drain and source, in the transistor.

A gas sensor detects the presence of gases in a volumetric space using the surface area of a material. The surface area-to-volume ratio of carbon nanotubes and graphene is very high, and therefore, it is the major contribution of the carbon nanomaterials to favor the parameters involved with the performance of a gas sensor. The basic distribution of gas molecules on graphene and carbon nanotubes in a gas sensor is depicted in Figure 2 and 3, respectively.

Novel materials that are being used to develop new gas sensors are 1D metal oxide nanostructures, carbon nanotubes, fullerenes, graphene, semiconductor quantum dots, and metal nanoparticles (Korotcenkov, 2013; Korotcenkov, 2014). A complete report about the main physical properties of graphene and

Figure 2. Basic distribution of gas molecules on graphene, in a gas sensor

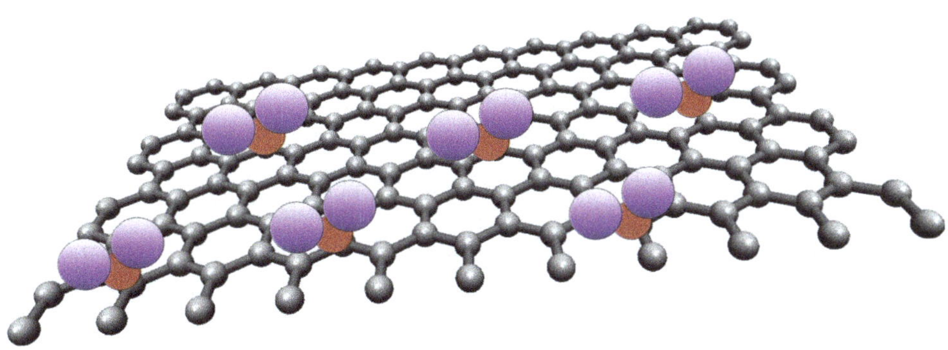

Figure 3. Basic distribution of gas molecules on carbon nanotube, in a gas sensor

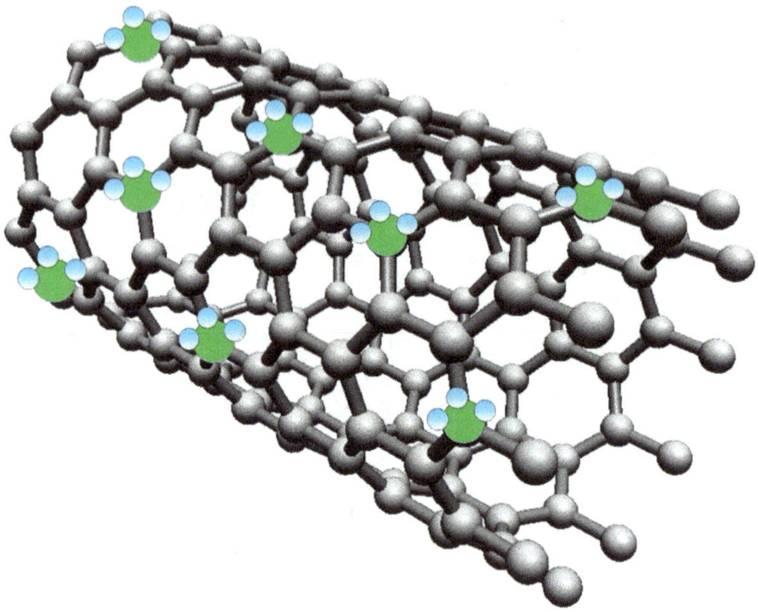

carbon nanotubes associated with the development of gas sensors have been developed (Wu, 2012). In this study, some particular qualities can be exploited in the development of gas sensors, and which favor or not the choice of graphene or carbon nanotubes as sensing material. Graphene has a higher experimental, specific surface area (1500 m^2/g versus 1300 m^2/g), thermal conductivity (4840-5300 W/mK versus 3500 W/mK), and bandgap (semimetallic and/ zero-gap semiconducting versus metallic and semiconducting) than carbon nanotubes. Nanotubes are categorized as single-walled nanotubes (SWNTs) and multi-walled nanotubes (MWNTs) (Vargas-Bernal, 2012). A SWNT is a wrapped one-atom-thick layer of graphite called graphene into a seamless cylinder The way the graphene sheet is wrapped is represented by a pair of indices (n, m) called chiral indices. If $m = 0$, the nanotubes are called zigzag nanotubes, and if $n = m$, the nanotubes are called armchair nanotubes. Otherwise, they are called chiral. SWNTs have one shell or wall and whose diameter ranging from 0.4 to 4 nm, while MWNTs contain several concentric shells and their diameter ranging from several nanometers to tens of nanometers. The electrical properties of the SWNTs can be either metallic or semiconducting materials depending on their chirality, that is, the direction in which they get rolled up. However, MWNTs are always metallic materials. In addition, volume electrical conductivity is fixed to 2000 S/cm for graphene and it is structure-dependent for the carbon nanotube (single wall nanotubes (SWNT) versus multi wall nanotubes (MWNT), and metallic, moderate semiconducting, or semiconducting). In particular, graphene functionalized by oxygen generates species such as O_2-, OH- y OOH- (Chen, 2013).

Gases can be adsorbed by carbon nanomaterials such as onions, cages, nanofibers, nanographites, nanoflakes, foams, nanocomposites, graphenes, hybrid carbon nanomaterials, and more (Terranova, 2012). Carbon nanomaterials can use a wide variety of materials to chemically functionalize and to increase chemical detection of gases such as polymers (Badhulika, 2014), metals (Cho, 2014), metal oxides (Liu, 2015), etc. to modify the surface, and it allows chemical adsorption of either reducing atmospheres and/ or oxidizing atmospheres. These materials provide a high surface area, which increases their sensitivity and decreases their response time to analyte gases.

Performance parameters such as selectivity, lower operating temperature, manufacturability, and lower power consumption can be easily achieved using polymers. Conducting polymers and their derivatives are used in chemical sensors thanks to properties such as electrochemical activity and environmental stability over a wide pH range (Badhulika, 2014). Materials such as poly(3,4-ethylenedioxythiophene) doped with poly(styrene sulfonic acid) (PEDOT:PSS) are used as functional groups using morphologies such as nanowires or thin films in gas sensing. These polymers have allowed the sensing of methanol (CH_3OH), ethanol (CH_3CH_2OH), and methyl ketone (MEK) (CH_3COCH_3) at room temperature. Carbon nanotubes-polyethylene oxide composite films are sensitive and present repeatability to organic vapors such as toluene to room temperature with environmental and industrial applications (Zhou, 2014). Graphene/PANI (polyaniline) nanocomposites synthesized using chemical oxidative polymerization has been used to sense NH_3 for a wide range of concentrations from 1 to 6400 ppm (Wu, 2013). Conducting polymers exploit the high sensitivity, fast response, low cost and operation at room temperature to be used in gas sensors. Volatile compounds such as alcohols, amines, aromatic and ketones at room temperature can be sensed by functionalized carbon nanotubes with layers of metalloporphyrins (MPP) of zinc and manganese tetraphenylporphyrin chloride in a chemiresistor (Penza, 2010), which present an increment in their sensitivity and selectivity to different gases. Gas sensing devices based on p-phenylenediamine (PPD) reduced graphene oxide have been fabricated to detect dimethyl methylphosponate (DMMP) with better repeatability that their traditional counterparts (Hu, 2012).

Similar to many gas sensors based in other materials different to carbon nanomaterials, these sensors also respond to temperature and humidity (Ong, 2002). The effect of humidity and temperature can be eliminated by independently measuring humidity and temperature and performing a humidity-temperature calibration. Carbon nanotubes have a strong attraction to the humidity, and therefore a response time of the gas sensors is presented in high humidity environments. Several authors have studied the influence of humidity on graphene-based devices (Yang, 2012; Yao, 2012; Bi, 2013; Some, 2013). The mechanism of sensing of water on graphene consists in the electrostatic interaction between the water (containing a dipole) and the graphene (Smith, 2015). Water changes the electrical conductivity of the graphene by acting as a dopant for sensing material. Gas sensors operate better under atmospheric conditions where there is negligible cross-sensitivity from competing gases. The humidity sensing mechanism can be explained as the set of interactions of the polar H_2O molecules. Atmospheres such as water can increase or decrease the carrier scattering depending of the dopant used in sensing nanomaterial and the substrate impurities found in the carbon nanomaterial (Yang, 2012). Graphene oxide-silicon bilayers have been used as humidity sensors in a wide detection range of 10-98% RH and these exhibit high humidity sensitivity, good repeatability, small humidity hysteresis and clear and fast response-recovery (Yao, 2012). Functional chemical groups can be divided into two groups: hydrophilic and hydrophobic. When a hyphophilic group is used in the surface of a sensing nanomaterial, then a gas sensor for detect strong acidic and basic environments, water or humidity is fabricated (Some, 2013). In the other hand, hydrophobic groups are used to block water in the detection of gases and these chemical moieties favor the detection of volatile organic gases. The adsorption of water on graphene has been reported that it implies long response times (3-5 min) (Bi, 2013). The use of graphene oxide improves the performance over the entire RH range to detect humidity thanks to the oxygen presented and the hydrophilic groups generated. As the humidity increases, more water molecules are adsorbed on the surface of the sensing material. Likewise, a decrease in humidity will cause water molecules to be desorbed from the surface (Smith, 2015). Gas sensors based on single-walled carbon nanotubes (SWCNs) functionalized with indium tin oxide nanoparticles have been implemented with the aim of tailoring the sensor selectivity with respect to the relevant interfering effects of humidity (Rigoni, 2014). The presence of humidity dramatically alters the working conditions of the gas sensors, biasing the expected value of gas concentration to certain value; therefore, it is necessary design sensing materials capable of realizing hydrophobic actions under water presence. Pristine SWNTs show an increase in their electrical resistance with the relative humidity, while in ITO-SWNTs blends a decrease in the sensor electrical resistance is observed. This variation in electrical resistance is directly related with the discrimination of detecting the presence of water or another gas in the atmosphere. Water displays a reducing character and it works with doping character on carbon nanomaterials. Sensing materials that present a decreasing electrical resistivity are classified as n-type materials, while p-type materials have increasing electrical resistivity.

Volatile organic compounds such as benzene, toluene, methanol, ethanol, and acetone can be detected using functionalized carbon nanomaterials (Hafaiedh, 2013). Pristine carbon nanomaterials are not responsive to benzene adsorption since it has molecule weakly interaction with the sensing material (π-π stacking in which no significant electrical charge is exchanged). Fortunately, the functionalization of oxygen plasma-treated carbon nanomaterials by means of noble metals such as rhodium, ruthenium, nickel, gold, palladium, etc. can be very suitable to increase their sensitivity to volatile organic compounds including benzene (Leghrib, 2010; Zanolli, 2011). Benzene preferentially will bind to the structure carbon nanomaterial-noble metal thanks to the delocalized π-electron ring, i.e., with the benzene ring parallel to a noble metal surface (Zanolli, 2011). Therefore, noble metals with strong interaction with carbon are more suitable for benzene detection, since the sensing mechanism relies on the binding of the gas molecule to the nanocluster of noble metal through van der Waals forces. Metal nanoparticles donate or accept a significant amount of electrical charge upon adsorption of a target molecule in these hybrid sensing materials. The plasma treatment enables the cleaning, activation, functionalization and metal decoration of carbon nanomaterials in a single step, which offers flexibility for tuning the interfacial physicochemical properties of the resulting hybrid materials (Leghrib, 2010). Hybrid materials such as annealing Metalloporphyrins (FeTPP and CoTPP)-functionalized aligned SWNTs can improve the π-delocalization leading to better gas sensing characteristics to detect benzene, toluene and xylene (Rushi, 2014).

Modeling of Gas Sensors Based on Carbon Nanotubes and Graphene

A mathematical model is a description of a system or device using mathematical concepts and language. Mathematical models are used in the natural sciences, social sciences, and engineering disciplines. These may help to explain a system or device and to study the effects of different components, parameters, or variables, and to make predictions about of its behavior.

Some studies have been realized to model the performance of gas sensors as it is discussed now. A thermal model for determining the behavior of the metallic resistance heater used in gas sensors has been developed (Pike, 1997). Electrical equivalent models of the semiconducting gas sensors using Pspice have been developed to predict electrical response, with the change of gas concentration or operating temperature (Llobet, 2001). An ab initio study of the electron transport on gas sensors based on carbon nanotubes suggest that the physisorption of gases will decrease or will increase their electrical conductance, depending of the energy implied in the adsorption of the gas into sensing materials (Sadrzadeh, 2008). Models for analyzing and optimizing the kinetics of Langmuir adsorption-desorption kinetics of the gas sensors based on carbon nanotubes and graphene have been proposed (Liang, 2013). An optimal operating temperature is dominant at high temperatures; the electrical potential applied to the material improves the selectivity to modify the gate voltage required in a field-effect transistor configuration, and it reduces the density of defects in the sensing materials. The effects of the temperature, film thickness and response time, on sensitivity of the gas sensors has been modeled (Selvaraj, 2014). A mathematical model for gas sensing using thin films and based on diffusion equations where are involved the reaction processes and analytical expressions to chemical concentrations, using actual and equivalent models, have been deduced.

In this section, some of the parameters related with gas sensors based on carbon nanomaterials are modeled using mathematical expressions. The electrical conductance of the graphene can be mathematically expressed as (Peres, 2006):

$$G = \frac{2\left(2n+1\right)q^2}{h} \tag{1}$$

where n is the number of electrically active channels, q is the electrical charge of the electron, and h is the Planck constant (Akbari, 2014a; Akbari, 2014b; Akbari, 2014c; Akbari, 2014d; Akbari, 2014e; Vargas-Bernal, 2014; Vargas-Bernal, 2015). The behavior of G in function of n is illustrated in Figure 4. At larger values of the number of electrically active channels in carbon nanomaterials, higher is the electrical conductance of the sensing material. Thus, it can be concluded that carbon nanomaterials require an adequate electrical conductivity in the gas sensor when gas to detect is present.

Figure 4. Behavior of the electrical conductance as a function of the number of electrical active channels in the graphene

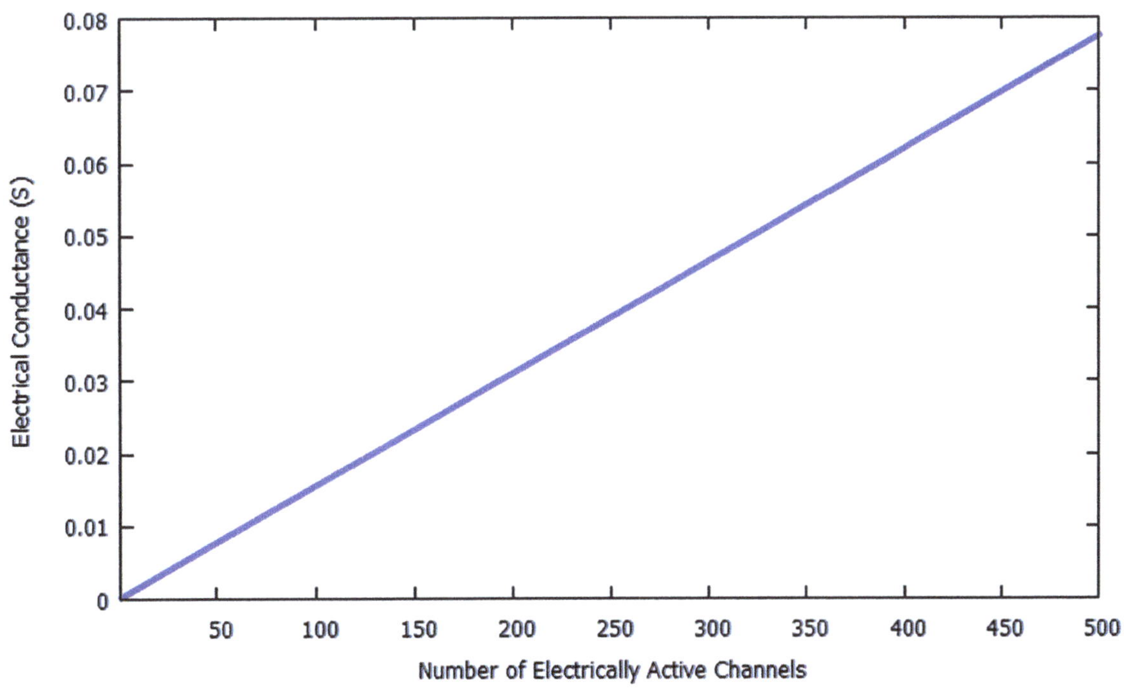

The band gap energy of a carbon nanotube can be mathematically expressed as (Johari, 2010; Akbari, 2014a; Akbari, 2014c):

$$E_G = \frac{2ta_{C-C}}{d} \tag{2}$$

where t is defined as the carbon-carbon (C-C) nearest neighbor tight binding overlap energy and it is equal to 2.7 eV, a_{C-C} is carbon-carbon (C-C) bond length, and d denotes the diameter of the carbon nanotube (Akbari, 2014a; Akbari, 2014b; Akbari, 2014c; Akbari, 2014d; Akbari, 2014e). The behavior of E_G in function of the diameter d is shown in Figure 5. At larger values of the diameter of the carbon nanotube, lower is the bandgap energy of the sensing material. Thus, it can be concluded that carbon nanotubes can detect gases better at median diameters, and therefore, it is highly desired guarantying the electrical conductivity in the gas sensor when gas to detect is present.

Some parameters related with the behavior of the gas sensors are strategic variables with the aim of knowing the influence of the carbon nanotubes and graphene. The rate of change in electrical conductivity (δ) can be mathematically expressed as (Akbari, 2014a; Akbari, 2014b; Akbari, 2014c; Akbari, 2014d; Akbari, 2014e):

$$\delta = aLn\left(T\right) - b \tag{3}$$

Figure 5. Behavior of the band gap as a function of the diameter of carbon nanotubes

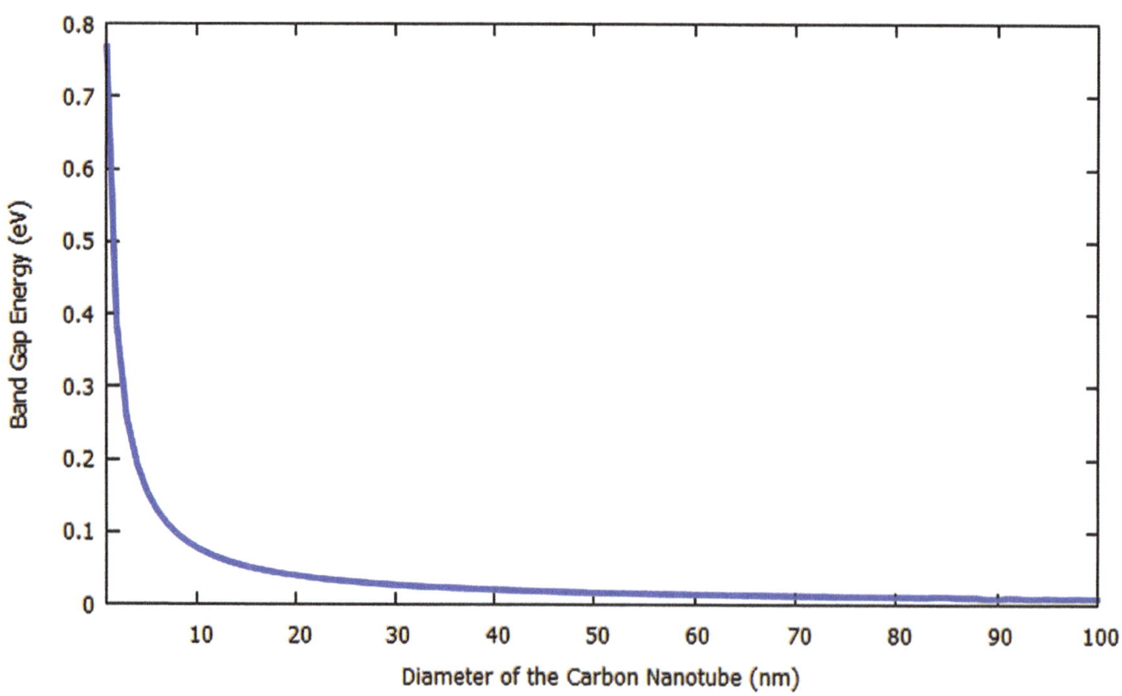

where a and b are experimentally determined constants and T is the operating temperature of the gas sensor. The behavior of δ in function of T is depicted in Figure 6. At higher operating temperature, lower is the rate of change in electrical conductivity in the gas sensing material. Thus, it can be concluded that carbon nanomaterials can detect gases better at low temperatures, and therefore, it is highly desired to reduce electrical power required in the gas sensors. The control parameter of the gas concentration is mathematically expressed as:

$$\lambda = cLn\left(F\right) - d \tag{4}$$

where c and d are experimentally determined constants and F is the concentration of the gas in parts per million. The behavior of λ in function of F is shown in Figure 7. At higher gas concentration, lower is the control parameter of the gas concentration in the gas sensing material. Thus, it can be concluded that carbon nanomaterials can detect better gases at low concentrations, and therefore, parts-per-billion levels can be achieved easily.

Some other researchers have found that the gas sensing is mainly due to the number of contacts among carbon nanotubes, and the correlation between the electrical conductance and the gas concentrations (Boyd, 2014; Dube, 2015). The sensitivity of the gas sensors based on carbon nanomaterials increases with the value of the mass density of the gas to be detected by the gas sensor (Arash, 2015), as it was shown through of the elastic shell model based on molecular dynamics (MD) simulations.

Figure 6. Rate of change in electrical conductivity versus operating temperature of gas sensors based on carbon nanomaterials

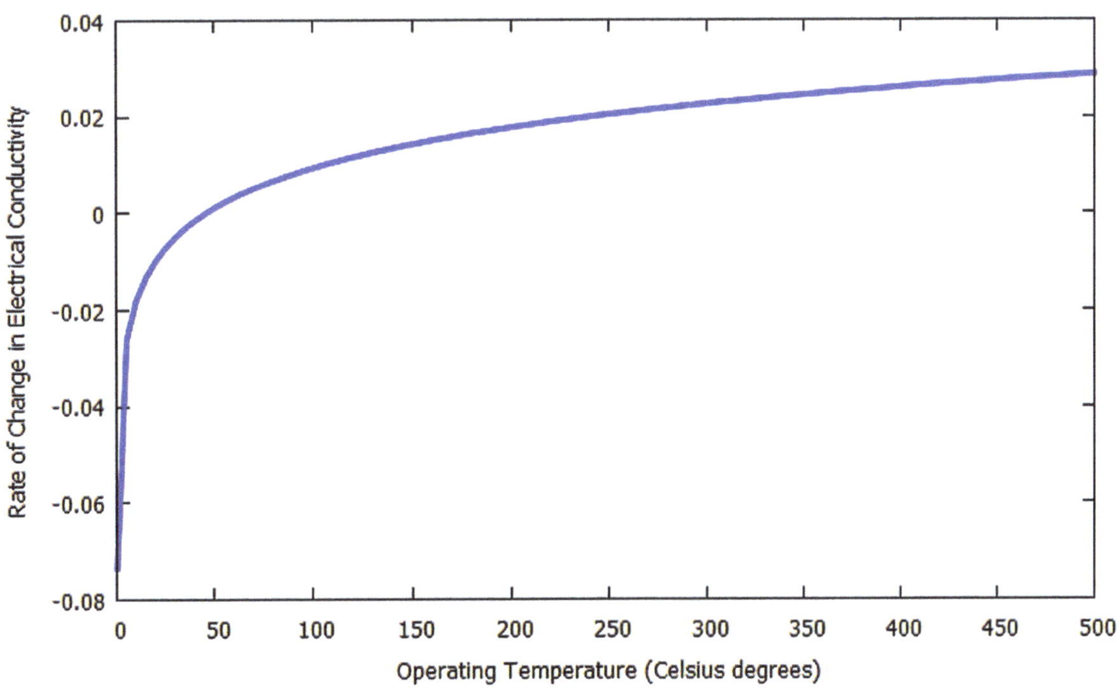

Figure 7. Control parameter of gas concentration as a function of the gas concentration for gas sensors based on carbon nanomaterials

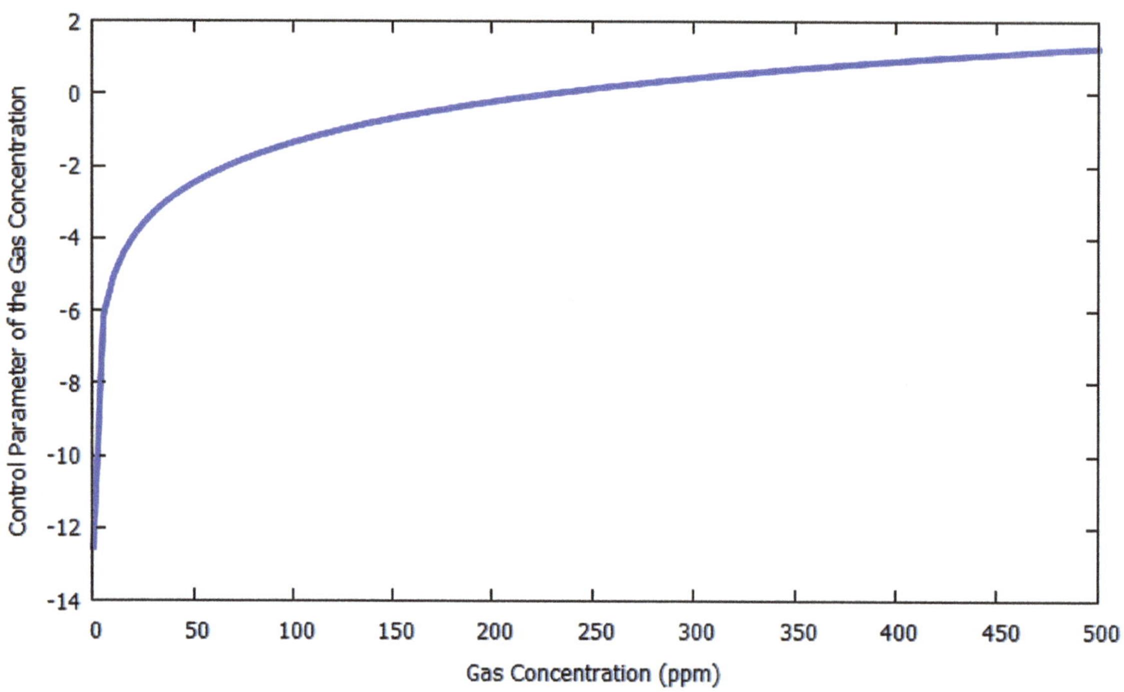

Impact of the Graphene and Carbon Nanotubes on Gas Sensing

Carbon nanotubes are used either as individual nanotubes, bundles of nanotubes, or composites (a combination of insulating, chemically sensitive polymer and carbon nanotubes (electrical nanofiller)) as the sensing material in chemiresistors (chemical resistors) or chemiFETs (chemical field-effect transistors) (Zhang, 2008). The enormous advantage of carbon nanomaterials is that they can be integrated directly onto flexible plastic substrates to be applied in selective chemical detection at the ppb-level (part-per-billion) (carbon nanotubes) or at the ppt-level (parts-per-trilion) (graphene) using low-power, and them can be exploited in commercial disposable devices.

Pristine carbon nanomaterials have low sensitivity to analytes due to low adsorption energy or low affinity, lack of selectivity, irreversibility or long recovery times (Zhang, 2008). The sensitivity and selectivity of carbon nanomaterials can be engineered by employing different techniques both to create defects and graft functional groups in their surface by means of controlled ways (Llobet, 2013). The sensing performance related with the sensitivity, selectivity and response time of a gas sensor can be improved through the rational functionalization of carbon nanomaterials such as graphene and carbon nanotubes by covalent and non-covalent methods through conducting polymers, metal oxide semiconductors, and metal nanoparticles. Covalently functionalized CNTs are based on esterification or amidation of carboxylic acid groups that were introduced as defects to the CNTs during acid treatment used for their purification. Non-covalent functionalization exploits the supramolecular complexation using adsorptive and wrapping forces such as van der Waals and π-stacking interactions for avoiding the destruction of the physical properties of the CNTs. Functionalized CNTs offer a higher sensitivity and a better selectivity compared to pristine CNTs. Alternative routes to improve the selectivity in ChemFETs have been proposed such as diversification of the source/drain metal electrodes, and exploitation of the desorption time of the different gases (Bondavalli, 2009).

Graphene possesses a two-dimensional highly crystalline structure with a thickness of a single-atom. Pristine graphene is capable of detecting gas molecules at extremely low concentrations with detection limits as low as 158 parts-per-quadrillion (ppq) of NO, to gas molecules at room temperature under in-situ UV light illumination using a simple device with two terminals (Chen, 2012). This material offers a high surface-to-volume ratio, low electrical-noise, and, exceptional electrical and thermal transport properties. In addition, an improvement until 300% in sensitivity to gases, it was found for graphene over carbon nanotubes, under similar conditions. The extreme sensitivity of the graphene is due to ease of activation of the adsorption centers, the presence of high binding energy, and thus, the response at very low concentration levels is inherent. Graphene is used in different derivatives:

- Reduced graphene oxide (rGO),
- Pristine graphene, or
- In layers.

Graphene derivatives offer a high surface area, chemically reactive sites, and the tunability of chemical and electronic properties (Varghese, 2015b) over metal oxides, carbon nanotubes and conducting polymers.

A gas sensor using nanocomposites based on graphene and ZnO using morphologies of thick films have been fabricated (Anand, 2014). This sensor decreased the optimum operation temperature and increased the sensitivity thanks to the increment of electrical conductivity generated by the interaction between the *p*-type graphene and *n*-type zinc oxide. Metal oxide semiconductors have low cost, high sensitivity and simple fabrication techniques. Among disadvantages associated with these materials are high operating temperature, wide bandgap, and high electrical resistance (KΩ-MΩ). The mixing of graphene into metal oxide semiconductors reduces the operating temperature, recovery time, and concentration level to be detected by the gas sensors.

Metals as aluminum can be used to dope to the graphene and improve sensitivity to the CO due to the large quantity of acceptor states generated with the introduction of metallic nanoparticles (Ao, 2008). The physicochemical modification with metals of the graphene can be exploited to modify its semi-metallic behavior for producing a bandgap, and therefore, electronic properties can be tunable by the ribbon width and length of the sheet (Shao, 2014). For example, chromium can be used to change the adsorption energy, charge transfer, and density of states (DOS) of the pristine graphene, to increase efficiency of adsorption to gases such as SO_2.

The density of states (DOS) per unit of volume in a material describes the number of states that are available for a particular morphology of material and it is used to determine the carrier concentrations and energy distribution of carriers available. In the case of carbon nanotubes, they have one-dimensional morphology and a density of states discrete and continuum. In addition, graphene is two-dimensional and a density of states continuum by ranges of energy but it presents several asymptotes (Davies, 2005).

Novel electronic properties can be produced thanks to the charge-transfer between electron donor molecules and the electron acceptor molecules with carbon nanomaterials such as carbon nanotubes and graphene (Rao, 2010). The molecular charge transfer produces the transformation of semiconducting nanotubes to metallic nanotubes, and vice versa, and a semi-metallic behavior to a semiconducting behavior in the case of the graphene. Both changes in electrical properties in these materials allow the optimization of gas sensing properties for those sensors based on carbon nanotubes and graphene.

Gas molecules act as charge donors at being weakly physisorbed on the surface of a sensing material (Umadevi, 2014). The gas adsorption on these sites is determined by the binding energy, or adsorption energy, of the gas molecule and availability of sites where gas can be captured (Wang, 2009). Some of these sites are not available for certain gases due to the dimension of the gas molecule and the diameter of the site of the adsorption. In the case of carbon nanotubes, interior pores are accessible only when the single wall carbon nanotubes (SWCNTs) are uncapped or if they have defects on the tube walls. In the case of carbon nanotubes, interior pores are accessible only when the single wall carbon nanotubes (SWCNTs) are uncapped or have defects on the tube walls. When two or more layers or sheets of graphene are piled up, it is feasible that the functionalization of them can open the possibility of increasing the gas adsorption, through of the availability of sites generated and a higher binding energy. The traditional functionalization of carbon nanotubes and graphene can be realized using covalent and non-covalent methods by means of materials such as polymers and metals (Zhang, 2008). The single wall carbon nanotubes (SWCNT) have a better sensitivity compared with multiple wall carbon nanotubes (MWCNT) in excellent agreement with the experimental measurements. Therefore, single wall carbon nanotubes (SWCNTs) have higher number of adsorption sites and more effective coverages than multiple wall carbon nanotubes (MWCNTs) (Picaud, 2009). The value of the density of junctions carbon nanotube to carbon nanotube, establishes the origin of the main response mechanism of the gas sensor; either it comes of the number of nanotubes crossing between themselves or the number of interconnections among carbon nanotubes and electrodes (Boyd, 2014).

The graphene is decorated with metal nanoparticles to increase the sensitivity, selectivity, limit of detection, or a combination of these properties (Gutés, 2012). Platinum (Pt) forms smaller nanoparticles with a lower density, while gold (Au) and palladium (Pd) have similar reactivities, with higher nanoparticle densities and large diameters, when these materials are compared with the platinum. Some studies on the synergistic effect of metals in gas sensors based on graphene have been realized (Cho, 2014). Palladium nanoparticles (Pd NP) increase the number of electrons and holes contained in the graphene, leading to better sensitivity into NH_3. In change, aluminum nanoparticles (Al NP) decrease the number of electrons and holes contained in the graphene, and therefore, a better sensitivity into NO_2 is obtained. Chemiresistors based on functionalized multiple-wall carbon nanotubes (MWCNTs) with metallic nanoclusters of Fe, Co, Au, Pt and Pd, with high sensitivity to NO_2, H_2S, NH_3, and CO (Penza, 2008). Moreover of sensitivity, a fast response, reversibility, good repeatability and range detection limit by under of parts per million (ppm) can be achieved using these materials (Penza, 2007a; Penza, 2007b). Langmuir-Blodgett films of composite materials based on single-walled carbon nanotubes decorated with palladium have been used as sensing material to detect hydrogen (H_2) in nitrogen (N_2) atmosphere at room temperature showing a reversible detection and very fast response time (Lee, 2013a). Single-wall carbon nanotubes (SWCNTs) can be functionalized with Au nanoparticles and deposited on transparent and flexible plastic substrates, such as PET have been used to detect ammonia at levels of 255 ppb at room temperature (Lee, 2013b). Nanocomposites based on graphene and palladium nanoparticles, were deposited layer-by-layer on gold electrodes with the aim of increasing sensitivity to hydrogen (H_2) and nitrogen dioxide (NO_2), and was observed that it is inversely proportional to the number of bilayers of graphene (Lange, 2011). Hydrogen sensors based on Pt nanoparticles and, graphene or carbon nanotubes were manufactured with the aim of studying their sensitivity (Kaniyoor, 2009). They found that sensors based on graphene are more sensitive than those based on carbon nanotubes to detect the hydrogen. Graphene either monolayer, bilayer, or tri-layer can be functionalized with gold and platinum nanoparticles to sense organic vapors such as acetic acid, ethanol, and acetone with a high sensitivity at room temperature (Gautam, 2012). Platinum nanoparticles sputtered on the surface of the multiwall carbon nanotubes can be used for detecting very low concentrations of hydrogen gas at room temperature (Dhall, 2015). A single layer graphene decorated with palladium nanoparticles for detecting hydrogen (H_2) offers mechanical flexibility, durability and high response (Chung, 2012). Epitaxial graphene covered with platinum has been used to sense hydrogen gas given that platinum acts as dopant and increases the conductance of the graphene (Chu, 2011). In addition, these materials offer a robust and repeatable response to the hydrogen. When carbon nanomaterials are chemically doped, they increase sensitivity, minimizing unwanted effects, and these allow tuning the selectivity (Llobet, 2013). Particularly, graphene has lower detection limits that large-diameter carbon nanotubes due to lower noise levels. Smaller, faster and more sensitive gas sensors can be achieved using graphene and its oxides through of the topology using chemiresistors (Basu, 2012).

Hybrid materials based on graphene, carbon nanotubes and metal oxides are being developed to sensing NO_2 at room temperature (Liu, 2015). Aqueous dispersions of mixtures based on rGO (reduced graphene oxide), carbon nanotubes, and SnO_2 are deposited as dip-coating onto ceramic substrates, previously printed on Au electrodes to obtain chemiresistors. This material presents high response, fast response, high recovery rate, high selectivity, and a good stability to NO_2 with respect to versions of gas sensors based on pure rGO and that of the sensors based on rGO and SnO_2. Different depletions layers are formed in the hybrid materials, which increases the possibility of that different oxygen adsorbates O^{2-}, O^- or O_2^- can be used as adsorption centers of gases on surface of carbon nanomaterials (Vargas-Bernal, 2014). In the search of flexible gas sensors, hybrid architectures based on ZnO nanorods and free-standing graphene sheets can be used to detect gases such as ethanol with high sensitivity (Yi, 2011). Hybrid materials based on doped carbon nanotubes and SnO_2 can be used to detect volatile organic compounds such as methanol and ethanol with high selectivity and sensitivity (Wongchoosuk, 2010). Composite nanofibers based on SnO_2 and multiwall carbon nanotubes are more sensitive to carbon monoxide (CO) at room temperature that its counterparts based only on SnO_2 (Yang, 2007). An effective monitoring of the air quality implies the detection of gases in the range by under of parts per billion, which can be achieved by means of nanomaterials. Carbon nanotubes functionalized with indium tin oxide were deposited on plastic substrates to fabricate a chemiresistor. These were used to detect water vapor and ammonia with a shorter recovery time and higher selectivity towards acetone and ethanol (Rigoni, 2014). Composite materials based on Co_3O_4, polyethyleneimine and carbon nanotubes have been used to sense carbon monoxide (CO), and ammonia (NH_3) at different gas concentrations (Lin, 2015). These novel materials present high responses, lower detection limit, and short response time with respect to traditional materials used in gas sensors. Graphene-$ZnFe_2O_4$ composite has been used to design gas sensors for detecting acetone with the aim of reducing its operating temperature, and exhibit good selectivity and reproducibility to acetone vapor (Liu, 2013). A composite material based on SnO_2 nanoflakes and graphene layers was synthesized to design gas sensors for NH_3 sensing with high response magnitude, fast response, good reversibility and repeatability (Lin, 2012). Composites based on MWNTs and tin oxide nanoclusters, were used for sensing NO_2 and CO at room temperature with good selectivity, and they had a better performance that their traditional counterparts in the ranges of ppb and ppm, respectively (Leghrib, 2010). Carbon-nanotube sheet decorated with cobalt oxide (CO_3O_4) has been used to detect H_2 at room temperature to form chemiresistors with high response, reliable reversibility, fast response, short recovery times, and stable repeatability (Jung, 2014). SnO_2 doped with metal oxides such as PtO_2,

PdO, La_2O_3, CuO, and Fe_2O_3 and thin films of multiwalled carbon nanotubes (MWCNTs) have been prepared to detect liquid petroleum gases (LPG) and ethanol, to higher sensitivity and their selectivity is associated with the quantity of a particular dopant (Hieu, 2010). ZnO nanorods were deposited on graphene films to develop gas sensors of H_2S in oxygen at room temperature with a high sensitivity (Cuong, 2010). SnO_2 nanoparticles highly dispersed on the CNTs surface at room temperature have been used to detect ethanol, methanol and H_2S (Mendoza, 2014). These materials are more sensitive than their counterparts based only on CNTs or SnO_2.

Hybrid materials based on carbon nanomaterials such as carbon nanotubes and graphene as well as in noble metals, metal oxides, or conducting polymers offer very interesting materials to be used in gas sensors, due to their higher sensitivity, selectivity, lower response times, and, reduced recovery times; with respect to the conventional materials found in commercial use (Meng, 2015). Competitive materials at carbon nanomaterials are transition metal dichalcogenides (TMDs): hexagonal boron nitride (h-BN), sulfides (molybdenum disulfide (MoS_2), tungsten disulfide (WS_2), etc.), selenides (molybdenum diselenide ($MoSe_2$), tungsten diselenide ($MoSe_2$), etc.), and tellurides (molybdenum ditelluride ($MoTe_2$), tungsten ditelluride (WTe_2), etc.); graphyne, borophene, germanene, silicene, stanine, or phosphorene (Varghese, 2015a).

A high degree of performance can be achieved when carbon nanomaterials are used as sensing material, in their versions purified, modified and functionalized, in gas sensors using the topology of field-effect transistors. This topology provides unprecedented chemical sensitivity to different types of gases. Field-effect transistors based on carbon nanotubes were developed (Peng, 2008; Bondavalli, 2009), and they operate based on the modulation of the Schottky barrier height at the contacts with extreme sensitivity, response times and recovery times very reduced, good efficiency at room temperature, low power consumption, and CMOS compatibility. Until now, it continues being a priority completely understanding the chemical sensing mechanisms of the graphene in field-effect transistors (Kumar, 2013). Field-effect transistors based on graphene were fabricated (Rumyantsev, 2012; Zhang, 2012; Kumar, 2013). Defect sites in graphene and carbon nanotubes form low-energy sorption sites for analyte molecules. However, other researchers have found that there is a significantly influence of the defect sites on Poole-Frenkel conduction regime, which implies that electrons jump through these defects, leading to an improved sensitivity and a reversibility of these sensors (desorption of the previously gases adsorbed) (Salehi-Khojin, 2010; Kumar, 2013). This regime can be achieved until a critical electrical potential is reached, which allows overcoming the effect of hopping of the electron between defect sites. It suggests that a partial overlap of the HOMO and LUMO levels, of the graphene is presented when gas sensors are operating, and a very high sensitivity for very low concentration levels of gases is achieved.

Future Trends in Gas Sensors Based on Carbon Nanotubes and Graphene

Carbon nanotubes and graphene represent emerging, sensing materials, which will be used in gas sensors developed in XXI century. Both materials offer high sensitivity, stability, selectivity, short response and recovery times, low power consumption, lower operating temperature, reduced sizes, low cost, and extended lifetimes for a wide range of environments and applications. The use of functionalization methods and surface-modification technologies will continue being used with the aim of increasing the chemical detection with respect to the pristine-type carbon nanomaterials. The use of hybrid materials will be exhaustive in the searching of materials with highly tunable sensing properties where carbon nanomaterials can be widely used in combination with metals, ceramics and polymers. Additionally, self-assembly techniques will be key issue to develop arrays of gas sensors based on carbon nanomaterials with different levels of sensing. These materials will open the doors for fabricating electronic noses to lowest cost, where different gases can be detected at the same time in a unique device, which has been very complex of achieving previously, even with different sensing materials in the same system. Bottom-up approach will be used, as the main technique to build hybrid materials and self-assembly materials, with properties of gas sensing at concentration levels of parts-per-quadrillion (ppq). Moreover, the possibility of designing these materials atom-by-atom leads to smart arrays capable of offering multifunctionality and integrated properties at the nanoscale level.

In the coming years, nanomaterials such as carbon nanotubes and graphene will continue to favor qualities such as:

1. High adsorptive capacity, due to large surface-area-to-volume ratio,
2. Tunable electrical properties under the exposure to gases,
3. Modulation of the performance parameters, through of the composition and size of the sensing material, and
4. The manufacturability as chemiresistor or field-effect transistor, that can be incorporated in the design of integrated circuits, including a complete instrumentation system.

CONCLUSION

In the search for solutions to reduce the size of the electronic devices; decrease in electrical power used in its supply power; reducing the operating temperature; and optimize their performance in sensitivity, selectivity, response time and recovery time, carbon nanomaterials, such as carbon nanotubes and graphene, offer very interesting alternative materials to be used as sensing materials and these are one of the best options to develop gas sensors in the coming decades.

Through the simulation of mathematical models of gas sensors, it is possible to deduce some conclusions about the behavior of carbon nanomaterials, as sensing materials. At larger values of the number of electrically active channels in carbon nanomaterials, higher is the electrical conductance of the sensing material. Thus, it can be concluded that carbon nanomaterials require an adequate electrical conductivity in the gas sensor, when gas to detect is present. At larger values of the diameter of the carbon nanotube, lower is the bandgap energy of the sensing material. Thus, it can be concluded that carbon nanotubes can detect gases better at median diameters, and therefore, it is highly desired guarantying the electrical conductivity in the gas sensor when gas to detect is present. At higher operating temperature, lower is the rate of charge in electrical conductivity in the gas sensing material. Thus, it can be concluded that carbon nanomaterials can detect gases better at low temperatures, and therefore, it is highly desired to reduce electrical power required in the gas sensors. At higher gas concentration, lower is the control parameter of the gas concentration in the gas sensing material. Thus, it can be concluded that carbon nanomaterials can detect better gases at low concentrations, and therefore, parts-per-billion levels can be achieved easily.

REFERENCES

Akbari, E., Arora, V. K., Enzevaee, A., Ahmadi, M. T., Saeidmanesh, M., Khaledian, M., & Yosof, R. et al. (2014a). An analytical approach to evaluate the performance of graphene and carbon nanotubes for NH_3 gas sensor applications. *Beilstein Journal of Nanotechnology*, *5*, 726–734. doi:10.3762/bjnano.5.85 PMID:24991510

Akbari, E., Buntat, Z., Ahmad, M. H., Enzevaee, A., Yousof, R., Iqbal, S. M. Z., & Karimi, H. et al. (2014b). Analytical calculation of sensing parameters on carbon nanotube based gas sensors. *Sensors (Basel, Switzerland)*, *14*(3), 5502–5515. doi:10.3390/s140305502 PMID:24658617

Akbari, E., Buntat, Z., Ensevaee, A., Ebrahimi, M., Yazdavar, A. H., & Yusof, R. (2014c). Analytical modeling and simulation of I-V characteristics in carbon nanotube gas sensors using ANN and SVR methods. *Chemometrics and Intelligent Laboratory Systems*, *137*, 173–180. doi:10.1016/j.chemolab.2014.07.001

Akbari, E., Yousof, R., Ahmadi, M. T., Kiani, M. J., Rahmani, M., Feiz Abadi, H. K., & Saeidmanesh, M. (2014e). The effect of concentration on gas sensor model based on graphene nanoribbon. *Neural Computing & Applications*, *24*(1), 143–146. doi:10.1007/s00521-013-1463-2

Akbari, E., Yusof, R., Ahmadi, M. T., Enzevaee, A., Kiani, M. J., Karimi, H., & Rahmani, M. (2014d). Bilayer graphene application on NO_2 sensor modelling. *Journal of Nanomaterials*, *534105*. doi:10.1155/2014/534105

Anand, K., Singh, O., Singh, M. P., Kaur, J., & Singh, R. C. (2014). Hydrogen sensor based on graphene/ZnO nanocomposite. *Sensors and Actuators. B, Chemical*, *195*, 409–415. doi:10.1016/j.snb.2014.01.029

Ao, Z. M., Yang, J., Li, S., & Jiang, Q. (2008). Enhancement of CO detection in Al doped graphene. *Chemical Physics Letters*, *461*(4-6), 276–279. doi:10.1016/j.cplett.2008.07.039

Arash, B., & Wang, Q. (2013). Detection of gas atoms with carbon nanotubes. *Scientific Reports*, *3*, 1782. doi:10.1038/srep01782

Badhulika, S., Myung, N. V., & Mulchandani, A. (2014). Conducting polymer coated single-walled carbon nanotube gas sensors for the detection of volatile organic compounds. *Talanta*, *123*, 109–114. doi:10.1016/j.talanta.2014.02.005 PMID:24725871

Basu, S., & Bhattachayya, P. (2012). Recent developments on graphene and graphene oxide based solid state gas sensors. *Sensors and Actuators. B, Chemical*, *173*, 1–21. doi:10.1016/j.snb.2012.07.092

Bi, H., Yin, K., Xie, X., Ji, J., Wan, S., Sun, L., & Dresselhaus, M. S. et al. (2013). Ultrahigh humidity sensitivity of graphene oxide. *Scientific Reports*, *3*, 2714. doi:10.1038/srep02714 PMID:24048093

Bondavalli, P., Legagneux, P., & Pribat, D. (2009). Carbon nanotubes based transistors as gas sensors: State of the art and critical review. *Sensors and Actuators. B, Chemical*, *140*(1), 304–318. doi:10.1016/j.snb.2009.04.025

Boyd, A., Dube, I., Fedorov, G., Paranjape, M., & Barbara, P. (2014). Gas sensing mechanism of carbon nanotubes: From single tubes to high-density networks. *Carbon*, *69*, 417–423. doi:10.1016/j.carbon.2013.12.044

Chen, G., Paronyan, T. M., & Harutyunyan, A. R. (2012). Sub-ppt gas detection with pristine graphene. *Applied Physics Letters*, *101*(5), 053119. doi:10.1063/1.4742327

Chen, N., Li, X., Wang, X., Yu, J., Wang, J., Tang, Z., & Akbar, S. A. (2013). Enhanced Room Temperature Sensing of Co_3O_4-intercalated reduced graphene oxide based gas sensors. *Sensors and Actuators. B, Chemical*, *188*, 902–908. doi:10.1016/j.snb.2013.08.004

Cho, B., Yoon, J., Hahm, M. G., Kim, D.-H., Kim, A. R., Kahng, Y. H., & Ko, H. C. et al. (2014). Graphene-based gas sensor: Metal decoration effect and application to a flexible device. *Journal of Materials Chemistry C*, *2*(27), 5280–5285. doi:10.1039/c4tc00510d

Chu, B. H., Lo, C. F., Nicolosi, J., Chang, C. Y., Chen, V., Strupinski, W., & Ren, F. et al. (2011). Hydrogen detection using platinum coated graphene grown on SiC. *Sensors and Actuators. B, Chemical*, *157*(2), 500–503. doi:10.1016/j.snb.2011.05.007

Chung, M. G., Kim, D.-H., Seo, D. K., Kim, T., Im, H. U., Lee, H. M., & Kim, Y. H. et al. (2012). Flexible hydrogen sensors using graphene with palladium nanoparticle decoration. *Sensors and Actuators. B, Chemical*, *169*, 387–392. doi:10.1016/j.snb.2012.05.031

Cooper, J. S., Myers, M., Chow, E., Hubble, L. J., Cairney, J. M., Pejcic, B., & Raguse, B. et al. (2014). Performance of graphene, carbon nanotube, and gold nanoparticle chemiresistor sensors for the detection of petroleum hydrocarbons in water. *Journal of Nanoparticle Research*, *16*(1), 2173. doi:10.1007/s11051-013-2173-5

Cuong, T. V., Pham, V. H., Chung, J. S., Shin, E. W., Yoo, D. H., Hahn, S. H., & Kohl, P. A. et al. (2010). Solution-Processed ZnO-chemically converted graphene gas sensor. *Materials Letters*, *64*(22), 2479–2482. doi:10.1016/j.matlet.2010.08.027

Davies, J. H. (2005). *Low-dimensional semiconductors: An introduction*. New York, NY, USA: Cambridge University Press.

Dhall, S., & Jaggi, N. (2015). Highly dispersed platinum sputtered multiwall carbon nanotubes based hydrogen gas sensor at room temperature. *Sensors and Actuators. A, Physical*, *224*, 50–56. doi:10.1016/j.sna.2015.01.010

Dube, I., Jiménez, D., Fedorov, G., Boyd, A., Gayduchenko, I., Paranjape, M., & Barbara, P. (2015). Understanding the electrical response and sensing mechanism of carbon-nanotube-based gas sensors. *Carbon*, *87*, 330–337. doi:10.1016/j.carbon.2015.01.060

Gautam, M., & Jayatissa, A. H. (2012). Detection of organic vapors by graphene films functionalized with metallic nanoparticles. *Journal of Applied Physics*, *112*(11), 114326. doi:10.1063/1.4768724

Gutés, A., Hsia, B., Sussman, A., Mickelson, W., Zettl, A., Carraro, C., & Moboudian, R. (2012). Graphene decoration with metal nanoparticles: Towards easy integration for sensing applications. *Nanoscale*, *4*(2), 438–440. doi:10.1039/C1NR11537E PMID:22147241

Hafaiedh, I., Elleuch, W., Clement, P., Llobet, E., & Abdelghani, A. (2013). Multi-walled carbon nanotubes for volatile organic compound detection. *Sensors and Actuators. B, Chemical*, *18*, 344–350. doi:10.1016/j.snb.2013.03.020

Hieu, N. V., Duc, N. A. P., Trung, T., Tuan, M. A., & Chien, N. D. (2010). Gas-sensing properties of tin oxide doped with metal oxides and carbon nanotubes: A competitive sensor for ethanol and liquid petroleum gas. *Sensors and Actuators. B, Chemical, 144*(2), 450–456. doi:10.1016/j.snb.2009.03.043

Hu, N., Wang, Y., Chai, J., Gao, R., Yang, Z., Kong, E. S.-W., & Zhang, Y. (2012). Gas sensor based on p-phenylenediamine reduced graphene oxide. *Sensors and Actuators. B, Chemical, 163*(1), 107–114. doi:10.1016/j.snb.2012.01.016

Johari, Z., Ahmadi, M. T., Chek, D. C. Y., Amin, N. A., & Ismail, R. (2010). Modelling of graphene nanoribbon fermi energy. *Journal of Nanomaterials, 909347*. doi:10.1155/2010/909347

Joshi, R. K., Gomez, H., Alvi, F., & Kumar, A. (2010). Graphene films and ribbons for sensing of O_2, and 100 ppm of CO and NO_2 in practical conditions. *The Journal of Physical Chemistry C, 114*(14), 6610–6613. doi:10.1021/jp100343d

Jung, D., Ham, M., & Lee, G. S. (2014). Room-temperature gas sensor using carbon nanotube with cobalt oxides. *Sensors and Actuators. B, Chemical, 204*, 596–601. doi:10.1016/j.snb.2014.08.020

Kaniyoor, A., Jafri, R. I., Arockiadoss, T., & Ramaprabhu, S. (2009). Nanostructured Pt decorated graphene and multi walled carbon nanotube based room temperature hydrogen gas sensor. *Nanoscale, 1*(3), 382–386. doi:10.1039/b9nr00015a PMID:20648277

Kong, J., Franklin, N. R., Zhou, C., Chapline, M. G., Peng, S., Cho, K., & Dai, H. (2000). Nanotube molecular wires as chemical sensors. *Science, 287*(5453), 622–625. doi:10.1126/science.287.5453.622 PMID:10649989

Korotcenkov, G. (2013). Handbook of gas sensor materials: Properties, advantages and shortcomings for applications volume 1 conventional approaches. Springer. doi:10.1007/978-1-4614-7165-3

Korotcenkov, G. (2014). Handbook of gas sensor materials: Properties, advantages and shortcomings for applications volume 2 new trends and techniques. Springer. doi:10.1007/978-1-4614-7388-6

Kumar, B., Min, K., Bashirzadeh, M., Farimani, A. B., Bae, M.-H., Estrada, D., & Salehi-Khojin, A. et al. (2013). the role of external defects in chemical sensing of graphene field-effect transistors. *Nano Letters, 13*(5), 1962–1968. doi:10.1021/nl304734g PMID:23586702

Lange, U., Hirsch, T., Mirsky, V. M., & Wolfbeis, O. S. (2011). Hydrogen sensor based on a graphene-palladium nanocomposite. *Electrochimica Acta, 56*(10), 3707–3712. doi:10.1016/j.electacta.2010.10.078

Lee, J.-H., Kang, W.-S., Najeeb, C. K., Choi, B.-S., Choi, S.-W., Lee, S. S., & Kim, J.-H. (2013a). A hydrogen gas sensor using single-walled carbon nanotube Langmuir-Blodgett films decorated with palladium nanoparticles. *Sensors and Actuators. B, Chemical, 188*, 169–175. doi:10.1016/j.snb.2013.06.066

Lee, K., Scardaci, V., Kim, H.-Y., Hallam, T., Nolan, H., Bolf, B. E., & Duesberg, G. S. et al. (2013b). Highly sensitive, transparent, and flexible gas sensors based on gold nanoparticle decorated carbon nanotubes. *Sensors and Actuators. B, Chemical, 188*, 571–575. doi:10.1016/j.snb.2013.07.048

Leghrib, R., Felten, A., Demoisson, F., Reniers, F., Pireaux, J.-J., & Llobet, E. (2010). Room-temperature, selective detection of benzene at trace levels using plasma-treated metal-decorated multiwalled carbon nanotubes. *Carbon, 48*(12), 3477–3484. doi:10.1016/j.carbon.2010.05.045

Liang, S.-Z., Chen, G., Harutyunyan, A. R., Cole, M. W., & Sofo, J. O. (2013). Analysis and optimization of carbon nanotubes and graphene sensors based on adsorption-desorption kinetics. *Applied Physics Letters, 103*(23), 233108. doi:10.1063/1.4841535

Lin, Q., Li, Y., & Yang, M. (2012). Tin oxide/graphene composite fabricated via a hydrothermal method for gas sensors working at room temperature. *Sensors and Actuators. B, Chemical, 173*, 139–147. doi:10.1016/j.snb.2012.06.055

Lin, Y., Kan, K., Song, W., Zhang, G., Dang, L., Xie, Y., & Shi, K. et al. (2015). Controllable synthesis of Co_3O_4/polyethyleneimine-carbon nanotubes nanocomposites for CO and NH_3 gas sensing at room temperature. *Journal of Alloys and Compounds, 639*, 187–196. doi:10.1016/j.jallcom.2015.03.139

Liu, F., Chu, X., Dong, Y., Zhang, W., Sun, W., & Shen, L. (2013). Acetone gas sensors based on graphene-$ZnFe_2O_4$ composite prepared by solvothermal method. *Sensors and Actuators. B, Chemical, 188*, 469–474. doi:10.1016/j.snb.2013.06.065

Liu, S., Wang, Z., Zhang, Y., Zhang, C., & Zhang, T. (2015). High performance room temperature NO_2 sensors based on reduced graphene oxide-multiwalled carbon nanotubes-tin oxide nanoparticles hybrids. *Sensors and Actuators. B, Chemical, 211*, 318–324. doi:10.1016/j.snb.2015.01.127

Llobet, E. (2013). Gas sensors using carbon nanomaterials: A review. *Sensors and Actuators. B, Chemical, 179*, 32–45. doi:10.1016/j.snb.2012.11.014

Llobet, E., Vilanova, X., Brezmes, J., López, D., & Correig, X. (2001). Electrical equivalent models of semiconductor gas sensors using PSpice. *Sensors and Actuators. B, Chemical, 77*(1-2), 275–280. doi:10.1016/S0925-4005(01)00710-9

Mao, S., Lu, G., & Chen, J. (2014). Nanocarbon-based gas sensors: Progress and challenges. *Journal of Materials Chemistry A, 2*(16), 5573–5579. doi:10.1039/c3ta13823b

Mendoza, F., Hernández, D. M., Makarov, V., Febus, E., Weine, B. R., & Morell, G. (2014). Room temperature gas sensors based on tin dioxide-carbon nanotubes composite films. *Sensors and Actuators. B, Chemical, 190*, 227–233. doi:10.1016/j.snb.2013.08.050

Meng, F.-L., Guo, Z., & Huang, X.-J. (2015). Graphene-based hybrids for chemiresistive gas sensors. *Trends in Analytical Chemistry, 68*, 37–47. doi:10.1016/j.trac.2015.02.008

Ong, K. G., Zeng, K., & Grimes, C. A. (2002). A wireless, passive carbon nanotube-based gas sensor. *IEEE Sensors Journal, 2*(2), 82–88. doi:10.1109/JSEN.2002.1000247

Peng, N., Zhang, Q., Lee, Y. C., Tan, O. K., & Marzani, N. (2008). Gate modulation in carbon nanotube field effect transistors-based NH_3 gas sensors. *Sensors and Actuators. B, Chemical, 132*(1), 191–195. doi:10.1016/j.snb.2008.01.025

Penza, M., Cassano, G., Rossi, R., Alvisi, M., Rizzo, A., Signore, M. A., & Giorgi, R. et al. (2007b). Enhancement of sensitivity in gas chemiresistors based on carbon nanotube surface functionalized with noble metal (Au, Pt) nanoclusters. *Applied Physics Letters, 90*(17), 173123. doi:10.1063/1.2722207

Penza, M., Cassano, G., Rossi, R., Rizzo, A., Signore, M. A., Alvisi, M., & Giorgi, R. et al. (2007a). Effect of growth catalysts on gas sensitivity in carbon nanotube film based chemiresistive sensors. *Applied Physics Letters*, *90*(10), 103101. doi:10.1063/1.2456258

Penza, M., Rossi, R., Alvisi, M., Cassano, G., Signore, M. A., Serra, E., & Giorgi, R. (2008). Pt- and Pd-nanoclusters functionalized carbon nanotubes networked films for sub-ppm gas sensors. *Sensors and Actuators. B, Chemical*, *135*(1), 289–297. doi:10.1016/j.snb.2008.08.024

Penza, M., Rossi, R., Alvisi, M., Signore, M. A., Serra, E., Paolesse, R., & Di Natale, C. et al. (2010). Metalloporphyrins-modified carbon nanotubes networked films-based chemical sensors for enhanced gas sensitivity. *Sensors and Actuators. B, Chemical*, *144*(2), 387–394. doi:10.1016/j.snb.2008.12.060

Peres, N. M. R., Castro-Neto, A. H., & Guinea, F. (2006). Conductance quantization in mesoscopic graphene. *Physical Review B: Condensed Matter and Materials Physics*, *73*(19), 195411. doi:10.1103/PhysRevB.73.195411

Picaud, F., Girardet, C., & Rao, A. M. (2009). A comparative study of single- and multiwalled carbon nanotube sensitivity to ammonia. *Journal of Applied Physics*, *105*(1), 014315. doi:10.1063/1.3060242

Pike, A., & Gardner, J. W. (1997). Thermal modelling and characterisation of micropower chemoresistive silicon sensors. *Sensors and Actuators. B, Chemical*, *45*(1), 19–26. doi:10.1016/S0925-4005(97)00261-X

Pu, H. H., Rhim, S. H., Gajdardziksa-Josifovska, M., Hirschmugl, C. J., Weinert, M., & Chen, J. H. (2014). A statistical thermodynamics model for monolayer gas adsorption on graphene-based materials: Implications for gas sensing applications. *RSC Advances*, *4*(88), 47481–47487. doi:10.1039/C4RA08124B

Rao, C. N. R., & Voggu, R. (2010). Charge-transfer with graphene and nanotubes. *Materials Today*, *13*(9), 34–40. doi:10.1016/S1369-7021(10)70163-2

Rigoni, F., Drera, G., Pagliara, S., Goldoni, A., & Sangaletti, L. (2014). High sensitivity, moisture selective, ammonia gas sensors based on single-walled carbon nanotubes functionalized with indium tin oxide nanoparticles. *Carbon*, *80*, 356–363. doi:10.1016/j.carbon.2014.08.074

Rumyantsev, S., Liu, G., Shur, M. S., Potyrailo, R. A., & Balandi, A. A. (2012). Selective gas sensing with a single pristine graphene transistor. *Nano Letters*, *12*(5), 2294–2298. doi:10.1021/nl3001293 PMID:22506589

Rushi, A. D., Datta, K. P., Ghosh, P. S., Mulchandani, A., & Shirsat, M. D. (2014). Selective discrimination among benzene, toluene, and xylene: Probing metalloporphyrin-functionalized single-walled carbon nanotube-based field effect transistors. *The Journal of Physical Chemistry C*, *118*(41), 24034–24041. doi:10.1021/jp504657c

Sadrzadeh, A., Farajian, A. A., & Yakobson, B. I. (2008). Electron transport of nanotube-based gas sensors: An *ab initio* study. *Applied Physics Letters*, *92*(2), 022103. doi:10.1063/1.2829609

Salehi-Khojin, A., Field, C. R., Yeom, J., & Masel, R. I. (2010). Sensitivity of nanotube chemical sensors at the onset of Poole-Frenkel conduction. *Applied Physics Letters*, *96*(16), 163110. doi:10.1063/1.3379291

Schedin, F., Geim, A. K., Morozov, S. V., Hill, E. W., Blake, P., Katsnelson, M. I., & Novoselov, K. S. (2007). Detection of individual gas molecules adsorbed on graphene. *Nature Materials*, *6*(9), 652–655. doi:10.1038/nmat1967 PMID:17660825

Selvaraj, K., Kumar, S., & Lakshmanan, R. (2014). Analytical expression for concentration and sensitivity of a thin film semiconductor gas sensor. *Ain Shams Engineering Journal*, *5*(3), 885–893. doi:10.1016/j.asej.2014.01.004

Shao, L., Chen, G., Ye, H., Niu, H., Wu, Y., Zhu, Y., & Ding, B. (2014). Sulfur dioxide molecule sensors based on zigzag graphene nanoribbons with and without Cr dopant. *Physics Letters. [Part A]*, *378*(7-8), 667–671. doi:10.1016/j.physleta.2013.12.042

Smith, A. D., Elgammal, K., Niklaus, F., Delin, A., Fischer, A. C., Vaziri, S., & Lemme, M. C. et al. (2015). Resistive graphene humidity sensors with rapid and direct electrical readout. *Nanoscale*, *7*(45), 19099–19109. doi:10.1039/C5NR06038A PMID:26523705

Some, S., Xu, Y., Kim, Y., Yoon, Y., Qin, H., Kulkarni, A., & Lee, H. et al. (2013). Highly sensitive and selective gas sensor using hydrophilic and hydrophobic graphenes. *Scientific Reports*, *3*, 1868. doi:10.1038/srep01868 PMID:23736838

Terranova, M. L., Orlanducci, S., & Rossi, M. (2012). *Carbon nanomaterials for gas adsorption*. Boca Raton, FL: Pan Stanford Publishing.

Umadevi, D., & Sastry, G. N. (2014). Feasibility of carbon nanomaterials as gas sensors: A computational study. *Current Science*, *106*(9), 1224–1234.

Vargas-Bernal, R. (2015). Advances in computational modeling of electronic devices based on graphene. *IEEE Journal on Emerging and Selected Topics in Circuits and Systems*, *5*(1), 109–116. doi:10.1109/JETCAS.2015.2398219

Vargas-Bernal, R., & Herrera-Pérez, G. (2012). Carbon nanotube- and graphene based devices, circuits and sensors for VLSI design. In E. Tlelo-Cuautle & S. Tan (Eds.), *VLSI Design* (pp. 41–66). Rijeka, Croatia: Intech; doi:10.5772/38743

Vargas-Bernal, R., & Herrera-Pérez, G. (2014). Importance of the nanostructured ceramic materials on gas sensing. In M. Aliofkhazraei (Ed.), Nanotechnology for Optics and Sensors (pp. 1-34). Manchester, UK: One Central Press.

Varghese, S. S., Lonkar, S., Singh, K. K., Swaminathan, S., & Abdala, A. (2015b). Recent advances in graphene based gas sensors. *Sensors and Actuators. B, Chemical*, *218*, 160–183. doi:10.1016/j.snb.2015.04.062

Varguese, S. S., Varghese, S. H., Swaminathan, S., Singh, K. K., & Mittal, V. (2015a). Two-dimensional materials for sensing: Graphene and beyond. *Electronics*, *4*(3), 651–687. doi:10.3390/electronics4030651

Wang, Y., & Yeow, J. T. W. (2009). A review of carbon nanotubes-based gas sensors. *Journal of Sensors*, *493904*. doi:10.1155/2009/493904

Wongchoosuk, C., Wisitsoraat, A., Tuantranont, A., & Kerdcharoen, T. (2010). Portable electronic nose based on carbon nanotube-SnO_2 gas sensors and its application for detection of methanol contamination in whiskeys. *Sensors and Actuators. B, Chemical, 147*(2), 392–399. doi:10.1016/j.snb.2010.03.072

Wu, Z., Chen, X., Zhu, S., Zhou, Z., Yao, Y., Quan, W., & Liu, B. (2013). Enhanced sensitivity of ammonia sensor using graphene/polyaniline nanocomposite. *Sensors and Actuators. B, Chemical, 178*, 485–493. doi:10.1016/j.snb.2013.01.014

Wu, Z.-S., Zhou, G., Yin, L. C., Ren, W., Li, F., & Cheng, H.-M. (2012). Graphene/metal oxide composite electrode materials for energy storage. *Nano Energy, 1*(1), 107–131. doi:10.1016/j.nanoen.2011.11.001

Yang, A., Tao, X., Wang, R., Lee, S., & Surya, C. (2007). Room temperature gas sensing properties of SnO_2/multiwall-carbon-nanotube composite nanofibers. *Applied Physics Letters, 91*(13), 133110. doi:10.1063/1.2783479

Yang, Y., Brenner, K., & Murali, R. (2012). The influence of atmosphere on electrical transport in graphene. *Carbon, 50*(5), 1727–1733. doi:10.1016/j.carbon.2011.12.008

Yao, Y., Chen, X., Guo, H., Wu, Z., & Li, X. (2012). Humidity sensing behaviors of graphene oxide-silicon bi-layer flexible structure. *Sensors and Actuators. B, Chemical, 161*(1), 1053–1058. doi:10.1016/j.snb.2011.12.007

Yi, J., Lee, J. M., & Park, W. I. (2011). Vertically aligned ZnO nanorods and graphene hybrid architectures for high-sensitive flexible gas sensors. *Sensors and Actuators. B, Chemical, 155*(1), 264–269. doi:10.1016/j.snb.2010.12.033

Zanolli, Z., Leghrib, R., Felten, A., Pireaux, J.-J., Llobet, E., & Charlier, J.-C. (2011). Gas sensing with Au-decorated carbon nanotubes. *ACS Nano, 5*(6), 4592–4599. doi:10.1021/nn200294h PMID:21553864

Zhang, J., Hu, P., Zhang, R., Wang, X., Yang, B., Cao, W., & O'Neill, W. et al. (2012). Soft-lithographic processed soluble micropatterns of reduced graphene oxide for wafer-scale thin film transistors and gas sensors. *Journal of Materials Chemistry, 22*(2), 714–718. doi:10.1039/C1JM14071J

Zhang, T., Mubeen, S., Myung, N. V., & Deshusses, M. A. (2008). Recent progress in carbon nanotube-based gas sensors. *Nanotechnology, 19*(33), 332001. doi:10.1088/0957-4484/19/33/332001 PMID:21730614

Zhou, Y., Jiang, Y., Xie, G., Du, X., & Tai, H. (2014). Gas sensors based on multiple-walled carbon nanotubes-polyethylene oxide films for toluene vapor detection. *Sensors and Actuators. B, Chemical, 191*, 24–30. doi:10.1016/j.snb.2013.09.079

ADDITIONAL READING

Battie, Y., Ducloux, O., Thobois, P., Dorval, N., Laurent, J. S., Attal-Trétout, B., & Loiseau, A. (2011). Gas sensors based on thick films of semi-conducting single walled carbon nanotubes. *Carbon, 49*(11), 3544–3552. doi:10.1016/j.carbon.2011.04.054

Chung, M. G., Kim, D. H., Lee, H. M., Kim, T., Choi, J. H., Seo, D. K., & Kim, Y. H. et al. (2012). Highly sensitive NO_2 gas sensor based on ozone treated graphene. *Sensors and Actuators. B, Chemical*, *166-167*, 172–176. doi:10.1016/j.snb.2012.02.036

Dan, Y., Lu, Y., Kybert, N. J., Luo, Z., & Johnson, A. T. C. (2009). Intrinsic response of graphene vapor sensors. *Nano Letters*, *9*(4), 1472–1475. doi:10.1021/nl8033637 PMID:19267449

Dhall, S., Jaggi, N., & Nathawat, R. (2013). Functionalized multiwalled carbon nanotubes based hydrogen gas sensor. *Sensors and Actuators. A, Physical*, *201*, 321–327. doi:10.1016/j.sna.2013.07.018

Dua, V., Surwade, S. P., Ammu, S., Agnihotra, S. R., Jain, S., Roberts, K. E., & Manohar, S. K. et al. (2010). All-organic vapor sensor using inkjet-printed reduced graphene oxide. *Angewandte Chemie International Edition*, *49*(12), 2154–2157. doi:10.1002/anie.200905089 PMID:20187049

Dude, I., Jiménez, D., Fedorov, G., Boyd, A., Gayduchenko, I., Paranjape, M., & Barbara, P. (2015). Understanding the electrical response and sensing mechanism of carbon-nanotube-based gas sensors. *Carbon*, *87*, 330–337. doi:10.1016/j.carbon.2015.01.060

Gautam, M., & Jayatissa, A. H. (2011). Gas sensing properties of graphene synthesized by chemical vapor deposition. *Materials Science and Engineering C*, *31*(7), 1405–1411. doi:10.1016/j.msec.2011.05.008

Gautam, M., & Jayatissa, A. H. (2012). Ammonia gas sensing behavior of graphene surface decorated with gold nanoparticles. *Solid-State Electronics*, *78*, 159–165. doi:10.1016/j.sse.2012.05.059

Hwang, S., Lim, J., Park, H. G., Kim, W. K., Kim, D.-H., Song, I. S., & Jun, S. C. et al. (2012). Chemical vapor sensing properties of graphene based on geometrical evaluation. *Current Applied Physics*, *12*(4), 1017–1022. doi:10.1016/j.cap.2011.12.021

Jaaniso, R., Kahro, T., Kozlova, J., Aarik, J., Aarik, L., Alles, H., & Sammelselg, V. et al. (2014). Temperature induced Inversion of oxygen response in CVD graphene on SiO_2. *Sensors and Actuators. B, Chemical*, *190*, 1006–1013. doi:10.1016/j.snb.2013.09.068

Jeong, H. Y., Lee, D.-S., Choi, H. K., Lee, D. H., Kim, J.-E., Lee, J. Y., & Choi, S.-Y. et al. (2010). Flexible room-temperature NO_2 gas sensors based on carbon nanotubes/reduced graphene hybrid films. *Applied Physics Letters*, *96*(21), 213105. doi:10.1063/1.3432446

Jung, D., Han, M., & Lee, G. S. (2014). Gas sensor using a multi-walled carbon nanotube sheet to detect hydrogen molecules. *Sensors and Actuators. A, Physical*, *211*, 51–54. doi:10.1016/j.sna.2014.03.005

Kang, I.-S., So, H.-M., Bang, G.-S., Kwak, J.-H., Lee, J.-O., & Ahn, C. W. (2012). Recovery improvement of graphene-based gas sensors functionalized with nanoscale heterojunctions. *Applied Physics Letters*, *101*(12), 123504. doi:10.1063/1.4753974

Ko, G., Kim, H.-Y., Ahn, J., Park, Y.-M., Lee, K.-Y., & Kim, J. (2010). Graphene-based nitrogen dioxide gas sensors. *Current Applied Physics*, *10*(4), 1002–1004. doi:10.1016/j.cap.2009.12.024

Lee, C., Ahn, J., Lee, K. B., Kim, D., & Kim, J. (2012). Graphene-based flexible NO_2 chemical sensors. *Thin Solid Films*, *520*(16), 5459–5462. doi:10.1016/j.tsf.2012.03.095

Lin, X., Ni, J., & Fang, C. (2013). Adsorption capacity of H_2O, NH_3, CO, and NO_2 on the pristine graphene. *Journal of Applied Physics*, *113*(3), 034306. doi:10.1063/1.4776239 PMID:23405036

Lu, G., Ocola, L. E., & Chen, J. (2009). Gas detection using low-temperature reduced graphene oxide sheets. *Applied Physics Letters*, *94*(8), 083111. doi:10.1063/1.3086896

Lu, G., Ocola, L. E., & Chen, J. (2009). Reduced graphene oxide for room-temperature gas sensors. *Nanotechnology*, *20*(44), 445502. doi:10.1088/0957-4484/20/44/445502 PMID:19809107

Maeng, S., Moon, S., Kim, S., Lee, H.-Y., Park, S.-J., Kwak, J.-H., & Hong, S. et al. (2008). Highly sensitive NO_2 sensor array based on undecorated single-walled carbon nanotube monolayer junctions. *Applied Physics Letters*, *93*(11), 113111. doi:10.1063/1.2982428

Ndiaye, A. L., Varenne, C., Bonnet, P., Petit, É., Spinelle, L., Brunet, J., & Lauron, B. et al. (2012). Elaboration of single wall carbon nanotubes-based gas sensors: Evaluating the bundling effect on the sensor performance. *Thin Solid Films*, *520*(13), 4465–4469. doi:10.1016/j.tsf.2012.02.071

Nomani, M. W. K., Shishir, R., Qazi, M., Diwan, D., Shields, V. B., Spencer, M. G., & Koley, G. et al. (2010). Highly sensitive and selective detection of NO_2 using Epitaxial graphene on 6H-SiC. *Sensors and Actuators. B, Chemical*, *150*(1), 301–307. doi:10.1016/j.snb.2010.06.069 PMID:20161619

Pearce, R., Iakimov, T., Andersson, M., Hultman, L., Spetz, A. L., & Yakimova, R. (2011). Epitaxially grown graphene based gas sensors for ultra sensitive NO_2 detection. *Sensors and Actuators. B, Chemical*, *155*(2), 451–455. doi:10.1016/j.snb.2010.12.046

Rao, S. S., Stesmans, A., Keunen, K., Kosynkin, D. V., Higginbotham, A., & Tour, J. M. (2011). Unzipped graphene nanoribbons as sensitive O_2 sensors: Electron spin resonance probing and dissociation kinetics. *Applied Physics Letters*, *98*(8), 083116. doi:10.1063/1.3559229

Singh, G., Choudhary, A., Haranath, D., Joshi, A. G., Singh, N., Singh, S., & Pasricha, R. (2012). ZnO decorated luminescent graphene as a potential gas sensor at room temperature. *Carbon*, *50*(2), 385–394. doi:10.1016/j.carbon.2011.08.050

Soldano, C., Mahmood, A., & Dujardin, E. (2010). Production, properties and potential of graphene. *Carbon*, *48*(8), 2127–2150. doi:10.1016/j.carbon.2010.01.058

Wang, J., Singh, B., Maeng, S., Joh, H.-I., & Kim, G.-H. (2013). Assembly of thermally reduced graphene oxide nanostrucures by alternating current dielectrophoresis as hydrogen-gas sensors. *Applied Physics Letters*, *103*(8), 083112. doi:10.1063/1.4819378

Wu, W., Liu, Z., Jauregui, L. A., Yu, Q., Pillai, R., Cao, H., & Pei, S.-S. et al. (2010). Wafer-scale synthesis of graphene by chemical vapor deposition and its application in hydrogen sensing. *Sensors and Actuators. B, Chemical*, *150*(1), 296–300. doi:10.1016/j.snb.2010.06.070

Xie, H., Sheng, C., Chen, X., Wang, X., Li, Z., & Zhou, J. (2012). Multi-wall carbon nanotube gas sensors modified with amino-group to detect low concentration of formaldehyde. *Sensors and Actuators. B, Chemical*, *168*, 34–38. doi:10.1016/j.snb.2011.12.112

Yavari, F., Castillo, E., Gullapalli, H., Ajayan, P. M., & Koratkar, N. (2012). High sensitivity detection of NO_2 and NH_3 in air using chemical vapor deposition grown graphene. *Applied Physics Letters*, *100*(20), 203120. doi:10.1063/1.4720074

Yoon, H. J., Jun, D. H., Yang, J. H., Zhou, Z., Yang, S. S., & Cheng, M. M.-C. (2011). Carbon dioxide gas sensor using a graphene sheet. *Sensors and Actuators. B, Chemical*, *157*(1), 310–313. doi:10.1016/j.snb.2011.03.035

KEY TERMS AND DEFINITIONS

Carbon Nanomaterials: Nanostructures of carbon such as fullerenes, carbon nanotubes, nanofibers and graphene with unique physicochemical properties with multiple technological applications.

Carbon Nanotubes: Allotropes of carbon with a cylindrical nanostructure of length-to-diameter of up to 132,000,000:1, which have unusual properties and valuable for nanotechnology, electronics, optics and other fields of materials science and technology.

Gas Sensor: A device that detects and/or quantifies the presence (qualitative) or the concentration of gases (quantitative) in a specific volume, regularly using the air as the reference environment.

Graphene: A two-dimensional, crystalline allotrope of carbon, whose atoms are densely packed in a hexagonal pattern to atomic-scale under a regular sp^2-bonded system, composed by a layer of graphite with a thickness of one-atom.

Operating Temperature: The temperature at which the maximum sensitivity of the sensor can be achieved.

Recovery Time: The time that requires the output signal of a sensor, to return to its initial value without gas, after that a change of concentration is applied to the sensor, starting from a certain value to zero.

Response Time: The time required by the sensing material in a sensor, to react to one step concentration, and to modify the output from zero to a certain concentration value.

Selectivity: The quality that determines whether a sensor can react selectively to a certain group of gases or even specifically to a single gas.

Sensitivity: The change in measured output signal per gas concentration unit, that is, the slope of its calibration curve.

Stability: The ability of a gas sensor to keep reproducible performance in a specific period of time of parameters such as sensitivity, selectivity, amplitude and form of response, and recovery time.

Chapter 8
Development of Gas Sensor Model for Detection of NO_2 Molecules Adsorbed on Defect-Free and Defective Graphene

Meisam Rahmani
Universiti Teknologi Malaysia, Malaysia

Hediyeh Karimi
Swinburne University of Technology, Australia

Komeil Rahmani
Islamic Azad University, Iran

Elnaz Akbari
Universiti Teknologi Malaysia, Malaysia

Mohammad Javad Kiani
Islamic Azad University, Iran

Mohammad Taghi Ahmadi
Universiti Teknologi Malaysia, Malaysia

Razali Ismail
Universiti Teknologi Malaysia, Malaysia

ABSTRACT

A wide popularity has been generated by graphene as a result of fundamental scientific interest in nanomaterials. Graphene-based nanostructure then possess a wide range of special physical uniqueness which can be used in many types of applications including some categories of sensors like optical, magnetic, electronic field, strain and mass sensors as well as field-effect, electrochemical and piezoelectric gas sensors. Graphene is believed to be a fantastic sensor material because of its single atomic layer of graphite with surface.

INTRODUCTION

A wide popularity has been generated by graphene as a result of fundamental scientific interest in nanomaterials (Zheng, 2008). Graphene-based nanostructure then possess a wide range of special physical uniqueness which can be used in many types of applications including some categories of sensors like optical, magnetic, electronic field, strain and mass sensors as well as field-effect, electrochemical and piezoelectric gas sensors (Hill, 2011; Huang, 2011). As shown in Figure 1, graphene is believed to be

DOI: 10.4018/978-1-5225-0736-9.ch008

Figure 1. Graphene as a fantastic sensor material

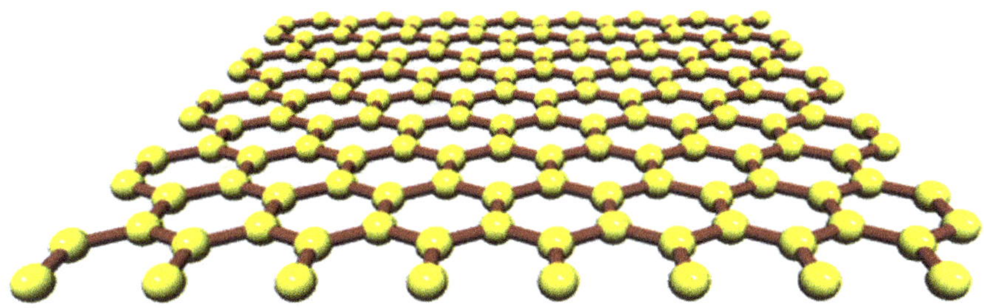

a fantastic sensor material because of its single atomic layer of graphite with surface. This assumption is because of the ease with which the electronic features of the graphene can be adjusted by the directly interaction between each atom in the structure and the sensing environment. Therefore, the interaction between the surface dopants and absorbents can be maximized. It is evident to date that graphene-based sensors can be achieved with best sensor-performance out of graphene (Hill, 2011; Huang, 2011; Ko et al., 2010).

Theoretical study on the adsorption of gas molecules on monolayer graphene has proved that the sensitivity can be improved by doping in carbon nanostructures (Anupama, 2009). This does not only increase the sensitivity of the adsorption process of graphene, but also lowers nonspecific binding and elevates sensitivity for the desired analysis (Ratinac, 2010; Miao, 2011; Dankerl, 2010). Furthermore, a few of the fictionalization method evolved for other set of gas sensors, particularly for single-walled carbon nanotubes, will likely get applications in the graphene-based gas sensors of the future (Yong-Hui et al., 2009). There is a broad knowledge of physics and chemistry behind NO_2 physisorptions or chemisorptions on doped graphene; in addition, magnetic pairing between adsorbed paramagnetic molecules is critical for its applications in electronics (Dong, 2009). Graphene is sensitive to the adsorption of NO_2 because of its transport properties and the system displays n-type semiconducting property after NO_2 adsorption (Tang, 2011; Chen, 2011). Based on quantum transport calculations, NO_2 molecules can be differentiated from another gas molecules by the graphene-based sensors (Anupama, 2009; Zhang, 2009). The extraordinary mobility of carriers in graphene has been used to explain its high sensitivity, which provides extremely low noise sensing at room temperature (Jesse, et al., 2009). Figure 2 indicates the schematic diagram of gas sensing used in our study. Small-width graphene field effect transistor (FET) in the structure of NO_2 sensor with the assumption of ballistic carrier transportation in channel is supposed. A major strength of adopting graphene as a channel material is due to its strong capability to control the electrostatics and hence expected to reduce the short channel effects that rely on the device electrostatics (Dankerl, 2010; Dong, 2010; Wang, 2010; Cohen-Karni, 2010).

The promise of defect-free and defective graphene in generating nanoscale NO_2 gas sensor for environmental monitoring is investigated in this chapter. This study will theoretically inspect the adsorption of NO_2 gas on N-doped graphene. Furthermore, band structure and density of states (DOS) of defect-free and defective graphene based on the tight-binding method will be carried out to find out the adsorption process. An open-shell NO_2 has a significant effect on modifying the DOS of graphene (Pramanik, 2011). Relying on this study, graphene-based sensor is also adopted to derive the current-voltage characteristic at different concentrations of NO_2 gas.

Figure 2. Graphene sensor exposed to NO_2 gas

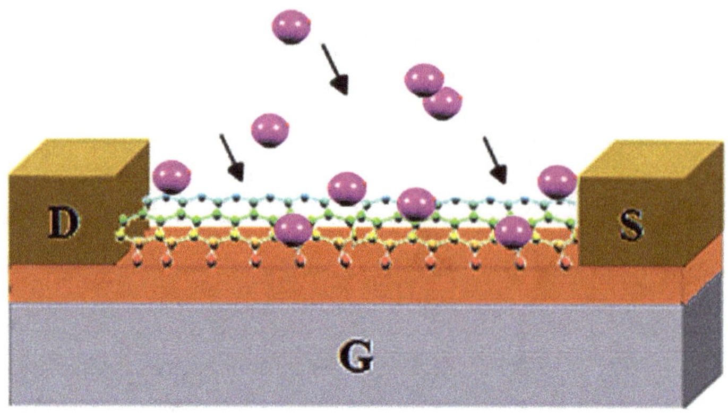

Proposed Model of Defect-Free Graphene for NO_2 Gas Sensor

The computational details of our calculations and later the presentation of results for electronic band structure, DOS and current-voltage properties of graphene-based sensor at different concentrations of NO_2 gas will be discussed in following sections. There will be a suggestion of physical model for the alteration of graphene's DOS during gas adsorption. A detailed study of the adsorption of NO_2 gas on graphene is also presented based on DOS. We study the bandwidth of nitrogen dioxide acceptor bands with the graphene bands through the band structure calculations. Where the linear band dispersion relation at the Brillion zone is given attention, it might be of interest to commence considering small-width graphene from the band structure of graphene. Majority of small-width graphene electronics features are explained with the band energy near the K and K' points. Hence, the energy band dispersion relation in graphene is given by (Johari, 2010)

$$E(\vec{k}) = \pm t\sqrt{1 + 4\cos\left(\frac{k_x 3a_{c-c}}{2}\right)\cos\left(\frac{k_y \sqrt{3}a_{c-c}}{2}\right) + 4\cos^2\left(\frac{k_y \sqrt{3}a_{c-c}}{2}\right)} \tag{1}$$

where $a_{c-c} = 1.42\,\overset{0}{\mathrm{A}}$, $t = 2.7\,eV$ is the tight-binding energy for nearest neighbor C-C atoms and $k_{x,y,z}$ are the wave vector components. Due to the approximation of the small-width graphene band structure, the relationship of energy-wave vector, *E(k)* can be gained as (Ahmadi, 2010)

$$E(\vec{k}) = \pm\frac{3a_{c-c}t}{2}\sqrt{k_x^2 + \beta^2} \tag{2}$$

where the quantized wave vector β is defined as

$$\beta = \frac{2\pi}{\sqrt{3}a_{c-c}}\left(\frac{p_i}{N+1} - \frac{2}{3}\right) \tag{3}$$

where p_i is the sub-band index and N is the number of dimer lines which determines the width of the ribbon. The band energy can be calculated as

$$E = \pm \frac{E_g}{2}\sqrt{1 + \frac{k_x^2}{\beta^2}} \tag{4}$$

The equation 4 is not in parabolic type and the Taylor expansion can be utilized to formulate the parabolic relation between energy and wave vector for semiconducting small-width graphene. Therefore the band energy in the low energy boundary ($k = 0$) is modeled as

$$E(k) \approx \frac{E_g}{2} + \frac{E_g}{4\beta^2}k_x^2 \tag{5}$$

$$E(k) \approx \frac{E_g}{2} + \frac{\hbar^2}{2m^*}k_x^2 \tag{6}$$

where m^* is the effective mass of graphene and $\hbar$ is reduced Planck constant or Dirac constant. As shown in Figure 3, the band structure of small-width graphene near the minimum energy in the E-k relationship is parabolic. The Taylor expansion results to Fermi-Dirac integral which leads to the parabolic band structure. In the parabolic area, it is adequate to utilize the Fermi-Dirac integral for carrier concentration calculation. However, in the non-parabolic region, it will result in a dissimilar type of Fermi integral (Ahmadi, 2010).

Figure 3. The energy band structure of small-width graphene

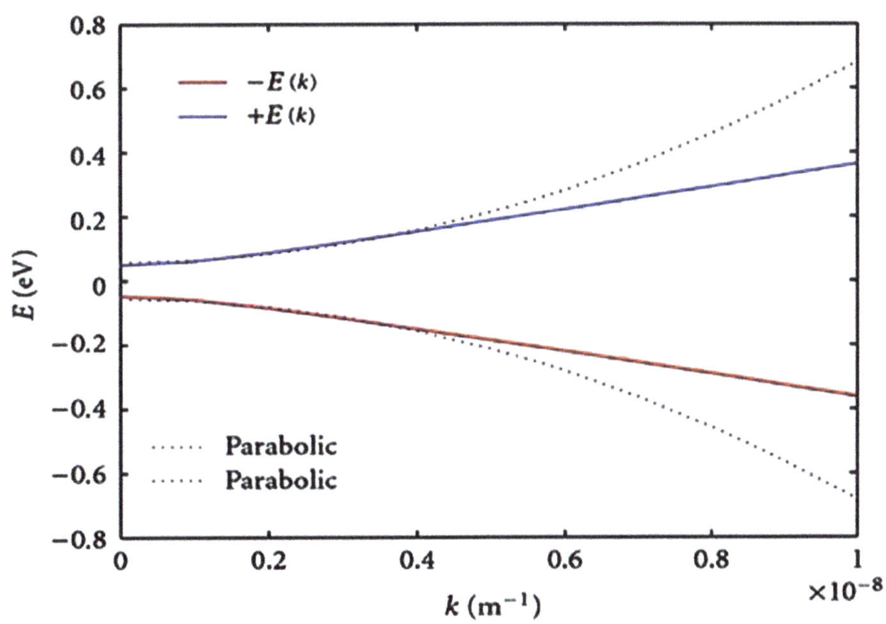

The DOS as a fundamental parameter which shows number of energy states for parabolic limit of one-dimensional small-width graphene band energy is defined as (Johari, 2010; Rahmani, 2011)

$$DOS_{graphene}(E) = \frac{\Delta n}{l\Delta E} = \frac{1}{2\pi}\left(\frac{2m^*}{h^2}\right)^{\frac{1}{2}}\left(E - \frac{E_g}{2}\right)^{-\frac{1}{2}} \tag{7}$$

where L is the length of the ribbon and h is Planck's constant. According to the quantum confinement effect in parabolic band structure of semiconductor graphene, conductance is defined as a function of Fermi-Dirac integral which is based on Maxwell approximation in non-degenerate region (Peres, 2006). Conductance of small-width graphene at Dirac point indicates minimum conductance at charge neutrality point which depends on temperature. The conductance model of small-width graphene can be gained by (Ahmadi, 2010)

$$G = \frac{3q^2\left(3\pi a^3 t^3 k_B T\right)^{\frac{1}{2}}}{hl}\left[\int_0^{+\infty} \frac{x^{-\frac{1}{2}}}{\left(\frac{1}{1+e^{x-\eta}}\right)}dx + \int_0^{+\infty} \frac{x^{-\frac{1}{2}}}{\left(\frac{1}{1+e^{x+\eta}}\right)}dx\right] \tag{8}$$

where $x = \frac{E - Eg}{k_B T}$, normalized Fermi energy is $\eta = \frac{E_F - Eg}{k_B T}$, q is the electron charge, T(E) is the transmission probability, k_B is the Boltzmann coefficient and L is the length of small-width graphene channel. The length of channel has a strong influence on conductivity function. In order to distinguish the role of degenerate and non-degenerate limits, presenting Fermi-Dirac integral shape of conductance is used as

$$G = \frac{3q^2\left(3\pi a^3 t^3 k_B T\right)^{\frac{1}{2}}}{hl}\left[\Im_{-\frac{1}{2}}(\eta) + \Im_{-\frac{1}{2}}(-\eta)\right] \tag{9}$$

There is usually a consequence similarity to the time for the hole to be injected into the semiconductor if the flow of electron from the valence band of the semiconductor into the metal is taken for granted. The drain current consists of both a component-thermal and tunnel, ensuring the intrusion of the conventional current into the drain terminal and leaving the source terminal. Normally, no current exists through the oxide to the gate terminal. For small values of V_{DS}, the channel region has the features of a resistor. Furthermore, the relationship between current and conductance (Neamen, 2003) can be obtainedby model of small-width graphene conductance as

$$I_D = \frac{3q^2(3\pi a^3 t^3 k_B T)^{\frac{1}{2}}}{hl}\left[\Im_{-\frac{1}{2}}(\eta) + \Im_{-\frac{1}{2}}(-\eta)\right](V_{gs} - V_t) \tag{10}$$

where V_{gs} is gate-source voltage and V_t is threshold voltage. Graphene is an excellent low noise material though in the limit of no carriers and with little extra electrons which causes an obvious alteration in the carrier concentration (Hill, 2011). Based on this property, a new analytically model for NO_2 sensor is proposed in this chapter. In presented model, ($\propto$) is the factor of NO_2concentration and in the non-saturation region, NO_2 concentration is simulated as a function of gate-source voltage V_{gs}~F_{NO2}.

$$V_{gs(gas\ concentration)} = \frac{\alpha}{F} V_{gs(without\ gas\ concentration)} \tag{11}$$

where F is the NO_2 concentration. The current-voltage relation for small-width graphene based n-channel MOSFET device can be modified as

$$I_D = \frac{3q^2\left(3\pi a^3 t^3 k_B T\right)^{\frac{1}{2}}}{hl}\left[\Im_{-\frac{1}{2}}(\eta) + \Im_{-\frac{1}{2}}(-\eta)\right]\left(\frac{\alpha}{F} V_{gs(without\ gas\ concentration)} - V_t\right) \tag{12}$$

The adsorbed molecules of NO_2 on small-width graphene surface by iteration technique is modeled as

$$\alpha = aF + b \tag{13}$$

where different concentrations ofNO_2 molecules are assumed in form of (a,b) parameters. Parameters obtained from extracted data indicate a=75, b=-1490. Based on the presented model for NO_2 sensor using nano-structured small-width graphene layer, the current-voltage response as a major characteristic can be achieved as

$$I_D = \frac{3q^2\left(3\pi a^3 t^3 k_B T\right)^{\frac{1}{2}}}{hl}\left[\Im_{-\frac{1}{2}}(\eta) + \Im_{-\frac{1}{2}}(-\eta)\right]\left(\frac{75F - 1490}{F} V_{gs(without\ gas\ concentration)} - V_t\right) \tag{14}$$

Graphene-based devices are adopted to attain the current-voltage characteristic of graphene-based sensors at different concentrations of nitrogen oxide gas. Coupled with the calculations, we also consider any dependence on electrical response for the changing of NO_2 gas concentration.

Proposed Model of Defective Graphene

The controllable defects in GNR mainly include the adatoms, substitution, disorder and Stone Wales (SW) defects (Zeng, 2011). It has been predicted that chemically modified SW defects reveal local out-of-plane dislocation of the carbon atoms. As shown in Figure 4, SW defects involve an in-plane 90° rotation of two carbon atoms with respect to the midpoint of the bond. In this transformation, four hexagons are changed into two heptagons and two pentagons. In fact, the SW defects cause vertical displacement of hundreds of atoms in a typical simulation cell (Ma, 2009).

Figure 4. Atomic structure of (a) Pristine (b) Flat SW defect monolayer graphene
Ma, 2009.

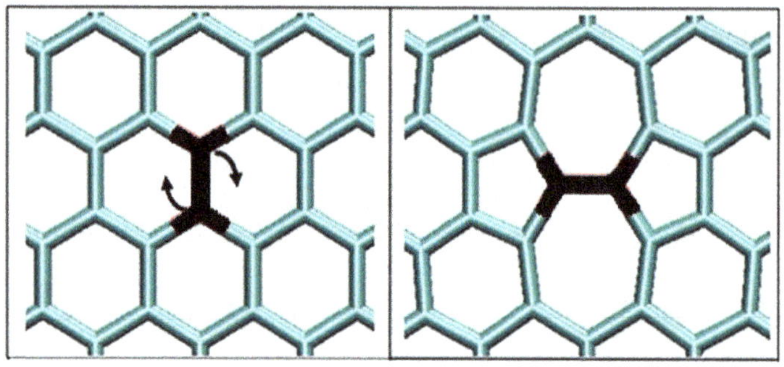

The influence of SW defects on the properties of GNRs has been extensively investigated. Until now, formation of several kinds of defects on graphene sheet has been experimentally studied including defects arising because of pure rearrangement, or reconstruction caused by removal of one, two or multiple carbon atoms from the graphitic lattice (Sena, 2013). When introducing a SW defect into GNR, the energy band structure is obtained (Rodrigues, 2011) as

$$\left(\frac{\varepsilon}{t}\right)^2 = \left|\left(1 - e^{-iq_\sigma}\xi_A^\sigma\right)\right|^2 \tag{15}$$

Low energy sates correspond to $\left(\frac{\varepsilon}{t}\right)^2 << 1$, that $\left|\xi_A^-\right| \approx 1$ around $k_a = \pm\frac{\pi}{3}$, while $\left|\xi_A^+\right| \geq \sqrt{2}$ for all k_a in the F.B.Z. Therefore, the $\sigma = -$ modes are the low energy bulk states when k is near the Dirac value and propagating states of the $\sigma = +$ modes have energy of order t. The corresponding eigenvalues are denoted by $\xi_A^\pm$ and $\xi_B^\pm$, respectively. These quantities depend on the value of the longitudinal momentum k and are gained by (Rodrigues, 2011)

$$\xi_A^+ = -2\cos\left(\frac{ka}{2}\right)e^{i\frac{ka}{2}} \tag{16a}$$

$$\xi_A^- = 2i\sin\left(\frac{ka}{2}\right)e^{i\frac{ka}{2}} \tag{16b}$$

$$\xi_B^+ = -2\cos\left(\frac{ka}{2}\right)e^{-i\frac{ka}{2}} = \left(\xi_A^+\right)^* \tag{16c}$$

$$\xi_B^- = -2i\sin\left(\frac{ka}{2}\right)e^{-i\frac{ka}{2}} = \left(\xi_A^-\right)^* \tag{16d}$$

The Equation (15) can be written as

$$\left(\frac{\varepsilon}{t}\right)^2 = \left(1 - e^{-iq_\sigma}\xi_A^\sigma\right)\left(1 - e^{+iq_\sigma}\left(\xi_A^\sigma\right)^*\right) \tag{17}$$

According to the Equations (16a-16d) and (17), the energy band structure of defective monolayer graphene for negative state ξ_A^- can be approximated as

$$\left(\frac{\varepsilon}{t}\right) = \left(2a^2k^2 - 2aqk + 1\right)^{\frac{1}{2}} \tag{18}$$

The Equation (18) can be written as

$$2a^2k^2 - 2aqk + 1 - E^2 = 0 \tag{19}$$

where $k = \dfrac{q \pm \sqrt{q^2 - 2\left(1 - E^2\right)}}{2a}$. The DOS of defective monolayer graphene band energy is modeled as

$$DOS_{Defective\ graphene}(E) = \frac{\left[\frac{1}{2}\left(q \pm \sqrt{q^2 - 2\left(1 - E^2\right)}\right)^2 - q\left(q \pm \sqrt{q^2 - 2\left(1 - E^2\right)}\right) + 1\right]^{\frac{1}{2}}}{\pm 2a\pi\sqrt{q^2 - 2\left(1 - E^2\right)}} \tag{20}$$

In the presence of parabolic approximation, the statistical behavior of GNR carriers can be explained with Fermi-Dirac integrals. Well-known carrier concentration of defective monolayer graphene in a band can be modeled by integrating the Fermi-Dirac distribution function over energy band as

$$dn_{Defective\ graphene} = \frac{\begin{bmatrix}\frac{1}{2}\left(q \pm \sqrt{q^2 - 2(1 - (x + E_{co})^2(k_BT)^2)}\right)^2 \\ -q\left(q \pm \sqrt{q^2 - 2(1 - (x + E_{co})^2(k_BT)^2)}\right) + 1\end{bmatrix}^{\frac{1}{2}}(k_BT)dx}{\pm 2a\pi\sqrt{q^2 - 2(1 - (x + E_{co})^2(k_BT)^2)}(1 + e^{x - \eta})} \tag{21}$$

where $x = \frac{E - E_c}{k_B T}$, normalized Fermi energy $\eta = \frac{-E_c + E_f}{k_B T}$, $E_{co} = \frac{E_c}{k_B T}$, k_B is the Boltzmann constant, E_f is the 1D Fermi energy and T is the temperature. According to the tight-binding method, Figures 5 and 6 show the energy band structure of defect-free and defective monolayer infinite graphene with 60 atoms in each unit cell. As shown in Figure 6, after the SW defect is applied, the band gap in the band structure of monolayer infinite graphene is created.

Comparison between the energy band structure of defect-free and defective monolayer infinite graphene with 60 atoms in each unit cell is indicated in Figure 7. As shown in Figure 7, the slope of energy levels in the defective graphene is decreased.

Results and Discussion

For the analysis to be more quantitative, a model is presented based on the presence of electrons and holes in graphene and electrons in the NO_2 acceptor states. Electrons and holes in graphene have almost same and rather high mobility of 5000 cm^2/Vs (Wehling et al., 2008). By making use of graphene-based gas detector, the single NO_2 gas molecule can be derived due to the alternation in the DOS and as a result that this material can easily be doped by several adsorbents. Whether from doping or defects, stronger adsorption generally leads to more significant changes in DOS of graphene, as are needed for active sensors. Calculating the kind and the concentration of charge carriers is at the spirit of modern electronics. With band structure measurements, the bandwidth of the NO_2 acceptor with the graphene bands is studied. The outcome of the proposed model for energy band structure of small-width graphene is that the measured DOS function with molecules adsorption is elaborated in Figure 8.

Figure 5. The energy band structure of defect-free monolayer infinite graphene with 60 atoms in each unit cell

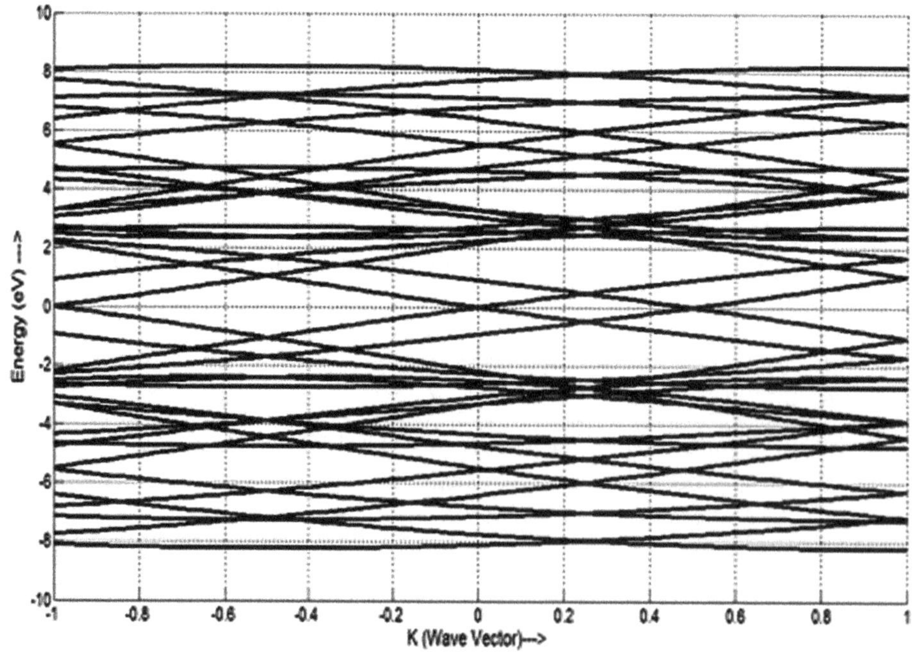

Figure 6. The energy band structure of defective monolayer infinite graphene with 60 atoms in each unit cell

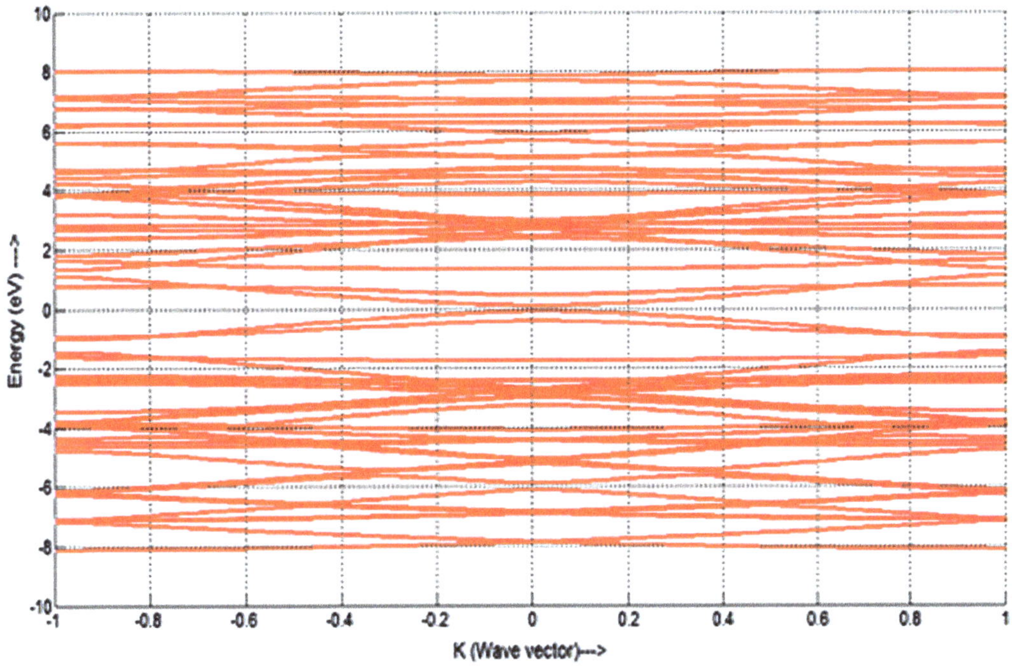

Figure 7. Comparison between the energy band structure of defect-free and defective monolayer infinite graphene with 60 atoms in each unit cell

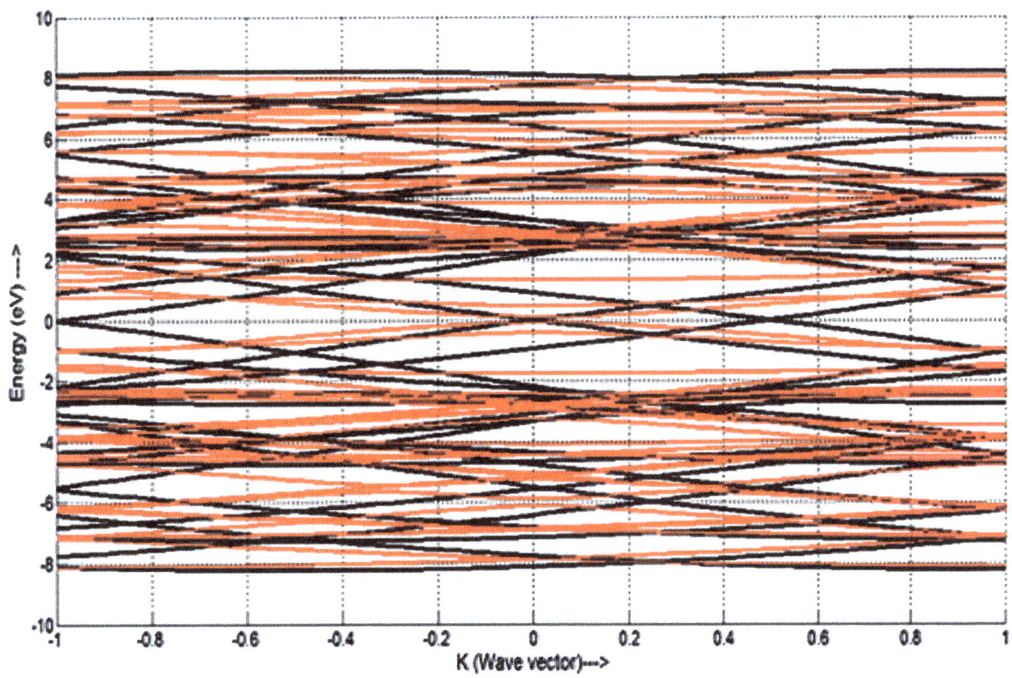

Figure 8. Spin-polarized DOS of the small-width graphene super cells with adsorbed NO_2 (alpha is the factor of NO_2 concentration)

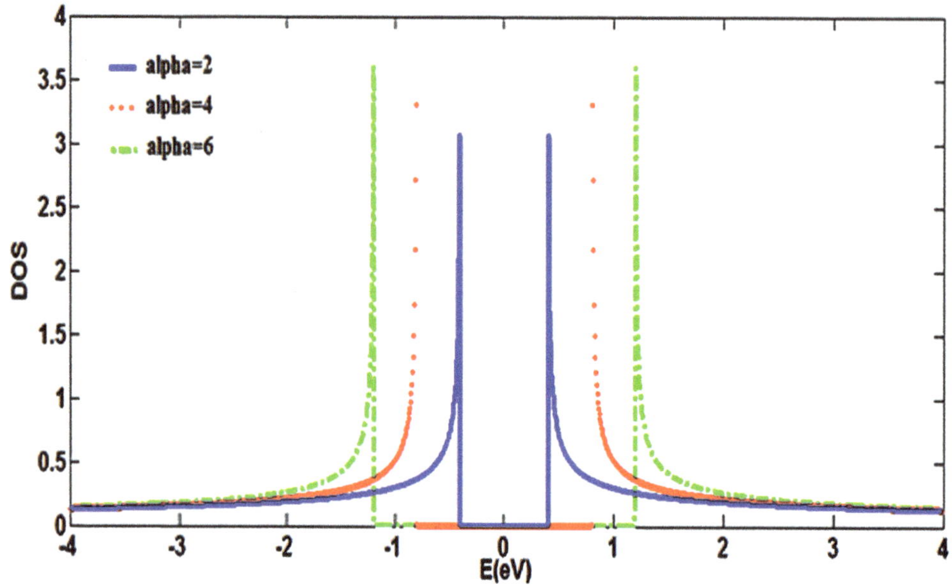

By combining gate voltage and factor of NO_2 concentration, the density of electrons or holes of graphene based FET device is locally changed which permits the variety of currently available semiconductor-based devices (Wehling et al., 2008). In other words, the energy band will be increased with escalating NO_2 doping. As indicated in figure 8, the molecular orbitals of NO_2 correspond to flat bands and manifest themselves as peaks in the DOS. The adsorbent orientation is virtually independent of the energies of these peaks. However, figure 8 shows the potentiality of doping effects as a result of adsorbed dimmers with the use of DOS. Several related electronic instabilities and ferromagnetism are due to the flat bands close to the Fermi level. The occupancy of flat impurity bands close to the Dirac point are created and controlled by NO_2 molecules adsorbed on small-width graphene as much as result to exchange scattering. It is remarkable that the DOS is influenced by the increased number of carriers in the channel of graphene based FET device. Due to this fact, DOS function can be known as a good index for NO_2 detection. As the gas concentration changes, the number of adsorbed gas molecules increases which result in changing of DOS and the sensor will be able of attracting more molecules. The potentiality of graphene can lead to strongly spin-polarized impurity states and result in a possible pathway to high temperature magnetic order in graphene which makes it necessary for future attention (Wehling et al., 2008). By optimization of the sensor configuration and by means of a monolayer graphene, the sensitivity can be more improved. Depending on the ambient situations, NO_2 gas molecules were absorbed/desorbed on the surface of graphene. Gas cylinders with concentrations of 100 ppm, 500 ppm and 1% NO_2 gas have been procured for these experiments (Ko et al., 2010). A typical gas sensing feature of the graphene samples for various concentrations of NO2 (100ppm and 500ppm) at room temperature are indicated in Figures 9 and 10. The current-voltage characteristic of graphene-based sensor is obtained at mentioned amounts of NO_2 concentrations.

Figure 9. The current-voltage characteristic of proposed model and NO_2 gas sensor when gas concentration is 100 ppm

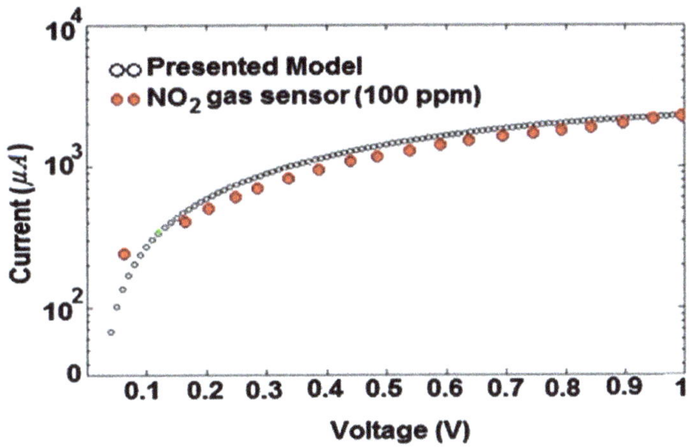

Figure 10. The current-voltage characteristic of proposed model and NO_2 gas sensor when gas concentration is 500 ppm

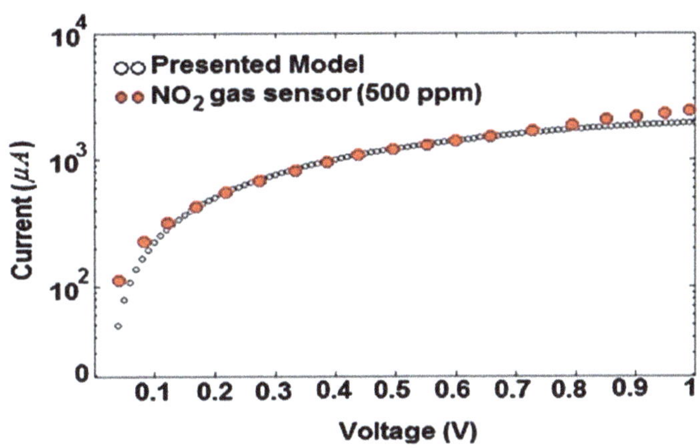

However, current versus voltage characteristic indicates linear ohmic behavior for low concentration of NO_2 exposure. Where the NO_2 molecules operate as an acceptor, an escalation in current can be associated to the charge transfer between NO_2 molecules and graphene. This technique is also known as chemical doping by gas molecules. The sensitivity can be presented in these figures, indicating the response of small-width graphene-based gas sensor under 100ppm and 500ppm NO_2 gas. Clearly, when value of concentration is F= 500 ppm, it is notable that, the proposed mode is nearer to the experimental data (Ko et al., 2010); in the same manner, it can be compared to other extracted data's as well. I-V characteristic of proposed model in comparison with the experimental results is described in Figure 11. As shown in this figure, the performance of NO2 sensor-based graphene is evaluated for two different concentrations of complementary NO_2 (100 ppm and 500 ppm).

Figure 11. Comparison between the transfer characteristic of presented model with experimental extracted data for two different concentrations of complementary NO_2
Ko et al., 2010.

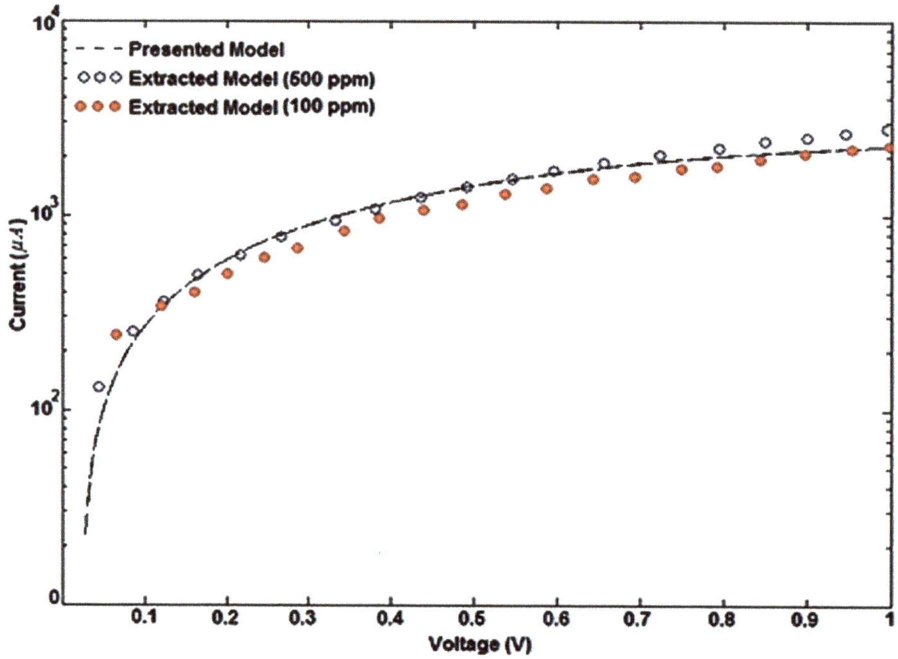

It clearly gives an illustration of the fact that there is a good agreement between proposed model and experimental extracted data (Ko et al., 2010). Adsorption of gas molecules on the surface of graphene leads to an increase in the charge carrier concentrations which can be adopted to fabricate sensitive gas sensor. As a result, there is no mark of worsening in the sensitivity of NO_2 detection. According to these results, graphene is a great potential material for ultra-sensitive gas sensor by the optimization of structure.

CONCLUSION

In this chapter, the promise of defect-free and defective graphene in generating nanoscale NO_2 gas sensor for environmental monitoring is investigated. Summarily, this research has carried out a first-principles measurement to investigate the adsorption geometries and electronic structure of small-width graphene with NO_2 gas molecule adsorption. There is a demonstration of the applicability of small-width graphene-based gas sensor by exploiting the charge transfer mechanism between NO_2 molecules and graphene, which the selectivity also has been tested under NO_2 gas ambient. When the ambient was changed, the novel small-width graphene-based NO_2 sensor exhibited perfect reproducibility. It has been discovered that NO_2 molecule adsorption can substantially impact the DOS and current-voltage characteristic of graphene. For an in-depth knowledge of the electronic features of intrinsic and doped graphene systems, we find the electronic DOS for the defect-free and defective graphene and also its dependence on the factor of NO_2 concentration. Coupled with the calculations, we also investigated the I-V characteristic of small-width graphene-based sensor at various concentrations of NO_2 gas. The characteristic of current-voltage indicates linear region of low concentration for NO_2 exposure. Based on simulated results, we conclude I-V characteristic can be controlled by varying NO_2 concentration through the NO_2 factor; in addition the proposed model is strongly approximate with experimental extracted data.

REFERENCES

Ahmadi, M. T., Johari, Z., Amin, N. A., Fallahpour, A. H., & Ismail, R. (2010). Graphene nanoribbon conductance model in parabolic band structure. *Journal of Nanomaterials*, 4. doi:10.1155/2010/753738

Anupama, G., Late Dattatray, J., Panchakarla, L. S., Govindaraj, A. & Rao, C. N. R. (2009). NO_2 and humidity sensing characteristics of few-layer graphene. *Physics, Chemical Physics,* 560064.

Chen, Y., Gao, B., Zhao, J.-X., Cai, Q.-H., & Fu, H.-G. (2011). Si-doped graphene: An ideal sensor for NO- or NO2-detection and metal-free catalyst for N2O-reduction. *Journal of Molecular Modeling*, *18*(5), 2043–2054. doi:10.1007/s00894-011-1226-x PMID:21881853

Cohen-Karni, T., Qing, Q., Li, Q., Fang, Y., & Lieber, C. M. (2010). Graphene and nanowire transistors for cellular interfaces and electrical recording. *Nano Letters*, *10*(3), 1098–1102. doi:10.1021/nl1002608 PMID:20136098

Dankerl, M., Hauf, M. V., Lippert, A., Hess, L. H., Birner, S., Sharp, I. D., & Garrido, J. A. et al. (2010). Graphene solution-gated field-effect transistor array for sensing applications. *Advanced Functional Materials*, *20*(18), 3117–3124. doi:10.1002/adfm.201000724

Dong, L. X., & Chen, Q. (2010). Properties, synthesis, and characterization of graphene. *Frontiers of Materials Science in China*, *4*(1), 45–51. doi:10.1007/s11706-010-0014-3

Dong, X., Fu, D., Fang, W., Shi, Y., Chen, P., & Li, L.-J. (2009). Doping single-layer graphene with aromatic molecules. *Small*, *5*(12), 1422–1426. doi:10.1002/smll.200801711 PMID:19296561

Hill, E. W., Vijayaragahvan, A., & Novoselov, K. (2011). Graphene sensors. *IEEE Sensors Journal*, *11*(12), 3161–3170. doi:10.1109/JSEN.2011.2167608

Huang, Y., Dong, X., Liu, Y., Li, J., & Chen, P. (2011). Graphene-based biosensors for detection of bacteria and their metabolic activities. *Journal of Materials Chemistry*, *21*(33), 12358–12362. doi:10.1039/c1jm11436k

Jesse, D. F., Matthew, J. A., Vincent, C. T., Yang, Y., Richard, B. K., & Bruce, H. W. (2009). Practical chemical sensors from chemically derived graphene. *ACS Nano*, *3*(2), 301–306. doi:10.1021/nn800593m PMID:19236064

Johari, Z., Ahmadi, M. T., Desmond Chang, Y. C., Amin, N. A., & Ismail, R. (2010). Modelling of graphene nanoribbon fermi energy. *Journal of Nanomaterials*, 6. doi:10.1155/2010/909347

Ko, G., Kim, H., Ahn, J., Park, Y.-M., Lee, K.-Y., & Kim, J. (2010). Graphene-based nitrogen dioxide gas sensors. *Current Applied Physics*, *10*(4), 1002–1004. doi:10.1016/j.cap.2009.12.024

Ma, J., Alfè, D., Michaelides, A., & Wang, E. (2009). Stone-Wales defects in graphene and other planar sp(2)-bonded materials. *Physical Review B: Condensed Matter and Materials Physics*, *80*(3), 033407. doi:10.1103/PhysRevB.80.033407

Miao, Z., Yun-Hao, L., Yong-Qing, C., Chun, Z., & Yuan-Ping, F. (2011). Adsorption of gas molecules on transition metal embedded graphene: A search for high-performance graphene-based catalysts and gas sensors. *Nanotechnology*, *22*(38), 385502. doi:10.1088/0957-4484/22/38/385502 PMID:21869463

Neamen, D. A. (2011). *Semiconductor physics and devices*. University of New Mexico.

Peres, N. M. R., Neto, A. H. C., & Guinea, F. (2006). Conductance quantization in mesoscopic graphene. *Physical Review B: Condensed Matter and Materials Physics*, *73*(19), 195411–195419. doi:10.1103/PhysRevB.73.195411

Pramanik, A., & Seok Kang, H. (2011). Density functional theory study of O_2 and NO adsorption on heteroatom-doped graphenes including the van der waals interaction. *PhysChemComm*, *115*, 10971–10978.

Rahmani, M., Ahmadi, M. T., Ghadity, M. H., Anwar, S., & Ismail, R. (2011). the effect of applied voltage on the carrier effective mass in ABA trilayer graphene nanoribbon. *Journal of Computational and Theoretical Nanoscience*, *9*, 1–4.

Ratinac, K., Wenrong, R., Ringer, Y., Simon, P., & Filip, B. (2010). Toward ubiquitous environmental gas sensors capitalizing on the promise of graphene. *Environmental Science & Technology*, *44*(4), 1167–1176. doi:10.1021/es902659d PMID:20099803

Rodrigues, J. N. B., Gonçalves, P. A. D., Rodrigues, N. F. G., Lopes dos Santos, J. M. B., & Peres, N. M. R. (2011). Zigzag graphene nanoribbon edge reconstruction with Stone-Wales defects. *Physical Review B: Condensed Matter and Materials Physics*, *84*(15), 155435. doi:10.1103/PhysRevB.84.155435

Sena, D., Thapab, R., & Chattopadhyaya, K. K. (2013). Small Pd cluster adsorbed double vacancy defect graphene sheet for hydrogen storage: A first-principles study. *International Journal of Hydrogen Energy*, *38*(7), 3041–3049. doi:10.1016/j.ijhydene.2012.12.113

Tang, S., & Cao, Z. (2011). Adsorption of nitrogen oxides on graphene and graphene oxides: Insights from density functional calculations. *Phys*, *134*, 044710. PMID:21280788

Wang, L. J., Cao, G., Tu, T., Li, H. O., Zhou, C., Hao, X.-J., & Guo, G.-P. et al. (2010). A graphene quantum dot with a single electron transistor as an integrated charge sensor. *Applied Physics Letters*, *97*(26), 262113–262115. doi:10.1063/1.3533021

Wehling, T. O., Novoselov, K. S., Morozov, S. V., Vdovin, E. E., Katsnelson, M. I., Geim, A. K., & Lichtenstein, A. I. (2008). Molecular doping of graphene. *Nano Letters*, *8*(1), 173–177. doi:10.1021/nl072364w PMID:18085811

Yong-Hui, Z., Ya-Bin, C., Kai-Ge, Z., Cai-Hong, L., Jing, Z., Hao-Li, Z., & Yang, P. (2009). Improving gas sensing properties of graphene by introducing dopants and defects: A first-principles study. *Nanotechnolog*, *20*(18), 185504–185512. doi:10.1088/0957-4484/20/18/185504 PMID:19420616

Yu, S. S., Zheng, W. T., & Jiang, Q. (2008). Oxidation of graphene nanoribbon by molecular oxygen. *IEEE Transactions on Nanotechnology*, *7*(5), 628–635. doi:10.1109/TNANO.2008.2002267

Zeng, H., Zhao, J., Wei, J. W., & Hu, H. F. (2011). Effect of N doping and Stone-Wales defects on the electronic properties of graphene nanoribbons. *The European Physical Journal B*, *79*(3), 335–340. doi:10.1140/epjb/e2010-10725-4

Zhang, J., Wang, S. R., Wang, Y. M., & Wu, S. H. (2009). NO_2 sensing performance of SnO_2 hollow-sphere sensor. *Sensors and Actuators. B, Chemical*, *135*(2), 610–617. doi:10.1016/j.snb.2008.09.026

Chapter 9
Modeling of Sensing Layer of Surface Acoustic-Wave-Based Gas Sensors

Hediyeh Karimi
Universiti Teknologi Malaysia, Malaysia

Ali Cheloee Darabi
Iran University Science and Technology, Iran

Rasoul Rahmani
Swinburne University of Technology, Australia

Meisam Rahmani
Universiti Teknologi Malaysia, Malaysia

Elnaz Akbari
Universiti Teknologi Malaysia, Malaysia

Mohammad Taghi Ahmadi
Universiti Teknologi Malaysia, Malaysia

Saba Anbari
Universiti Teknologi Malaysia, Malaysia

ABSTRACT

Industrial activities have polluted the atmosphere very rapidly in these days. There are a various varieties of air pollutants having a strong effect on human health as well as on climate and specially environment, such as nitrogen dioxide (NO_2), oxides of carbon (CO_x) and hydrocarbons. The World Health Organization (WHO) estimates that each year about 4.6million people die directly from causes of air pollution that will be a serious threat to human health. Therefore, there is a growing demand towards highly sensitive, cheap, low consumption, user-friendly devices which could monitor the quality of air indoor and outdoor areas for protecting human health. It is proved to be an efficient and economically feasible alternative for measuring different gas concentrations.

INTRODUCTION

Industrial activities have polluted the atmosphere very rapidly in these days (Michigan). There are a various varieties of air pollutants having a strong effect on human health as well as on climate and specially environment, such as nitrogen dioxide (NO_2), oxides of carbon (CO_x) and hydrocarbons(page). The World Health Organization (WHO) estimates that each year about 4.6million people die directly from causes of air pollution that will be a serious threat to human health (UK). Therefore, there is a growing demand

DOI: 10.4018/978-1-5225-0736-9.ch009

towards highly sensitive, cheap, low consumption, user-friendly devices which could monitor the quality of air indoor and outdoor areas for protecting human health (Organization, 2012). It is proved to be an efficient and economically feasible alternative for measuring different gas concentrations.

ACOUSTIC WAVE DEVICES

Acoustic wave devices are used as filters, delay-lines. There are also various applications for acoustic devices, such as medical (chemical sensors), automotive (torque and pressure sensors) and industrial and commercial applications (Drafts, 2001). Acoustic wave sensors specifications are defined as being small, inexpensive, very sensitive and easily designed for various required measurements.

Surface Acoustic Waves Devices

SAW devices have been developed as high sensitive gravimetric sensors due to their strong surface acoustic energy confinement (Schweyer, Andle, McAllister, French, & Vetelino, 1996). Surface acoustic wave devices have been used for about 40 years. In 1887, Lord Rayleigh who discovered surface acoustic propagation mode described its longitudinal and a vertical shear component which has the ability to contact with any media (Morgan, 1985).

Each SAW device is composed of two sets of interdigital transducers (IDTs) called input and output transducers which are patterned on a piezoelectric substrate. The IDTs are made of thin metal electrodes. In 1960s, White and Voltmer demonstrated that by depositing IDTs metal electrodes on piezoelectric substrates, surface acoustic waves could be generated (Strutt, 1885). Acoustic waves are generated when voltage changes through metal interdigital transducers on the input side. The generated acoustic wave travels through the substrate to reach the other side of output IDTs. The output IDT will detect the generated waves and convert the acoustic wave energy back into electrical energy. This region is called delay line where a voltage generates an output interdigital transducer by mechanical displacement (White & Voltmer, 1965).

SAW Sensors

Surface acoustic waves (SAW) have been used for chemical gas sensing purposes for about 20 years (Michigan). One of the distinctive advantages of SAW devices is that they are so sensitive to mass and conductivity changes. Mass loading mechanism is one of the commonly used ones. Change in the acoustic wave velocity is caused by a mass increase due to the molecules absorbed on the structure surface. Typically used mainly for liquid sensing, mass adsorption mechanism has been also employed for gas sensing, e.g. for volatile organic compounds.

For first time, the SAW technology was used as a sensor for pressure sensing in 1975(Nicolay et al., 2009; Reeder, Cullen, & Gilden, 1975). SAW temperature sensors have good resolution (about one-thousandth of a degree). A SAW device with a sensitive layer of polyaniline nanofibres for H_2 sensing, which could be operated at room temperature, has been reported in 2005 (He et al., 2010). Recently, (Penza, Antolini, & Antisari, 2004) developed a highly sensitive layered SAW device for H_2, NH_3 and NO_2 sensing using carbon nanotubes on $ZnO/LiTaO_3$ obtaining high sensitivity at room temperature.

SAW Gas Sensor Structure

Employing a SAW device satisfies the rigorous requirements necessary for a modern gas sensing device (e.g. easy-to-use, portable, cheap, stable, low power consumption) to monitor the quality of air and to provide safety protection. A Surface Acoustic Wave (SAW) gas sensor consists of a piezoelectric crystal substrate on which metal electrodes (interdigital transducers - IDTs) are fabricated by photolithography techniques. For specific crystal orientations in many piezoelectric substrates, propagation of surface acoustic waves is confined to the surface within one or two wavelengths (Comini, 2006). This confinement of energy makes them highly sensitive to surface perturbations. The sensitive gas layer is deposited upon the substrate material. In this research, Polyisobutylene (PIB) is chosen to be sensitive to the targeted gas analyst Dichloromethane (DCM). The interaction of the gas molecules with the sensitive PIB layer affects the propagation of the acoustic wave between the transmitting and receiving transducers. This could be measured as an incremental change in frequency of oscillation when the SAW sensor is connected to an RF amplifier in an oscillator configuration (El Gowini & Moussa, 2010). When PIB sensing layer exposes to DCM gas, a change in the mass of the PIB layer occurs. This will result in a change of its operational frequency, which could be correlated to the gas concentration. The sensitivity of a SAW sensor could be further enhanced by employing a nanostructured thin film as a gas sensitive layer due to the increased surface-to-volume ratio for gas molecules in such layer (Comini, 2006).

Piezoelectric Materials

All acoustic wave devices usually composed of a piezoelectric substrate which has the ability to produce acoustic waves by applying mechanical stress. Brothers, Pierre and Paul-Jacques Curie in 1880 were the first who discovered piezoelectricity (Ballato, 1996). Different kinds of piezoelectric materials could be used as a substrate for acoustic wave devices to produce acoustic waves. The most widely utilized substrate crystals are: quartz, lithium niobate ($LiNbO_3$), lithium tantalite ($LiTaO_3$), zinc oxide and bismuth germanium oxide. Their application is based on their pros and cons that are dependence on attenuation and velocity (Morgan. D, 1991). They have various temperature sensitive and piezoelectric coupling coefficients. For temperature stable devices, the ST quartz is suitable and used for most of these devices. the rang of wave velocity is 1500 m/s to 4800 m/s that is a function of the substrate material. This is 105 times as low as the electromagnetic wave velocity and makes it possible to production of a small size delay line of a significant delay.

Some famous materials used for this purpose are Lithium Niobate ($LiNbO_3$) and Lithium Tantalite ($LiTaO_3$). They are used base on their own advantages and disadvantages, which are attenuation and velocity (Morgan, D., 1991).

Interdigital Transducer (IDTS)

Figure 1 illustrates the basic SAW device that consists of two interdigital transducers (IDTs) on a piezoelectric substrate such as quartz. Each IDT is utilized in order to covert electrical signals to vice versa and acoustic wave as well IDT are made of interleaved metal electrodes. An IDT is a bidirectional transducer and send equal energy out on both sides of the electrodes. The minimum value of an IDT is roughly 6 dB and also the frequency range of Saw devices is between 10 MHz to several GHz. The IDTs and uniformly spaced electrodes are similar in lengths, ratio of electrons, width and spacing (Ahmadi,

Figure 1. Configuration of SAW gas sensor

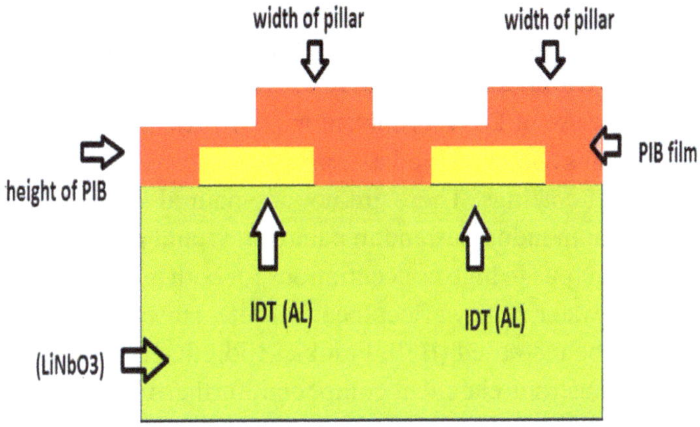

Hassani, Korman, Rahaman, & Zaghloul, 2004; Drafts, 2001). The frequency bandwidth of a SAW device is defined by the number of electrodes. Some specifications of electrons should be to match the IDT input resistance such as: lengths and their number, and matching networks at the electrical ports, the centre frequency f_0, load resistance RL and the generator resistance *R*g. Generally, the minimum loss due to IDTs is 12 dB. The wavelength equivalent to the centre frequency is 2*d* (the distance between the electrodes of the same polarity). The IDT`s geometry and the substrate type are used in order to determine the centre frequency and the bandwidth (Hribšek & Tošić).

Proposed Structure for Modeling

The SAW configuration considered in this research is composed of $LiNbO_3$ as piezoelectric substrate, PIB film as gas sensing layer and two IDTS made up of aluminum as seen in Figure 1.

Lithium Niobate (LiNbO3)

The authors have chosen YX $LiNbO_3$ substrates, which support SH-SAW wave propagation. Appropriate selection of the most suitable substrate orientation and cut leads to a minimal thermal coefficient of delay (TCD) and a maximum electromechanical coupling coefficient (K_2)(Atashbar, Bazuin, Simpeh, & Krishnamurthy, 2005). The K_2 of piezoelectric substrate is one of the most important parameters relating to SAW propagation and is a measure of efficiency of a substrate in converting an applied electric energy into mechanical energy associated with a SAW (Campbell, 1998).

Lithium niobatehas big temperature coefficient (TCD) (roughly 90 ppm/^{o}C) and also it has high electro-acoustic coupling constant. Therefore for temperature maximum 200°C, Lithium niobateis suitable in order to temperature sensors.

Takayanagi et al. (Takayanagi, Yamanouchi, & Shibayama, 1970) has shown that the surface wave propagating along the X-axis of Y-cut $LiNbO_3$ reduces to a SH-type surface wave at a certain rotation angle. As $LiNbO_3$ belongs to the same class of crystals as $LiTaO_3$, it was suggested that this type of wave could occur in this substrate as well (Nakamura, Kazumi, & Shimizu, 1977).

Dichloromethane (DCM)

Dichloromethane (DCM) which is used in this study as a targeted gas is a colorless, un-flammable clear liquid with a sweet odor which is evaporated easily. DCM is known as a very important Volatile Organic Compound (VOCS). The major use of DCM is solvent in paint removers. Furthermore, it is flammable in contact with alkane propellants and employed as a processing solvent in the manufacture of steroids, antibiotics, vitamins, and tablet coatings. There are not any natural sources for producing DCM. It is emitted from its use as a solvent in industries and in paint sprays and paint strippers to the air. It should be concerned that it can be dangerous in high concentrations for both human and environmental health. If DCM is emitted on the surface water, it may affect local wildlife. However, because of its rapid evaporation, the harmful effect would be increased (Bell, Franks, Hildreth, & Melius, 1991).

DCM is also known as a dangerous chemical compound, furthermore it is proven to have a harmful effect on the nervous system, blood, eyes and skin, and in long-term use, and DCM could cause cancer (Bois & Smith, 1995).

nti-spatter sprays used in welding, containing DCM gas, are popular choices due to their cheap cost and effective way they could be used. Inside pressure vessels, still dangerous anti-spatter is commonly used because non-dichloromethane product is not much effective in welding. If DCM is used continuously, the employers must provide protective clothing and eye protection because of above mentioned damages to eyes and skin. The important noteworthy point is that ventilation must be adequate and levels of dichloromethane emission in the air should be controlled to below the maximum exposure limit of 100 parts per million. Being exposed to DCM gas may cause dizziness, drowsiness, headache and unconsciousness, redness of eyes, dry skin and death (DHS, 1989).

PIB

One of professional polymer is PIB that can be attained by the cationic polymerization of isobutylene (IB). PIB has distinctive properties such as: desperately low permeability, excellent thermal and oxidative stability, ozone and chemical resistance, high hysteresis loops, and tack. Also PIBs because of low- and medium-molecular-weight can modify the viscosity that is used as fuel and lubricating oil additives (Lee, Joo, Huh, & Lee, 2006; Sayago et al., 2011).

PIB is also known as a non-polar and non-conductive material which used as a target sensitive layer in this study (Sauviat & Cohen-Addad, 1981; Wibawa, Khoiroh, Afrizal, & Suki, 2010). According to PIB structure as shown in Figure 2, it can satisfy our major requirement of a sensor coating because of its properties.

PIB is known as a non volatile compound which allows the diffusitive vapors into and out of the films (Ho, Lindgren, Rawlinson, McGrath, & Wright, 2003; Lee et al., 2006; Rapp, Rupp, & Ieee, 2007). When PIB expose to DCM gas, PIB sensing layer adsorbed DCM molecules and the mass loading occur which result in reducing the resonant frequency (Grate, Patrash, Abraham, & Du, 1996), demonstrated the volume increase of the polymer film on vapor adsorption on SAW sensor. As discussed above, mass loading and elastic effect are assumed to be the sensing mechanism of the developed sensor.

Figure 2. PIB structure

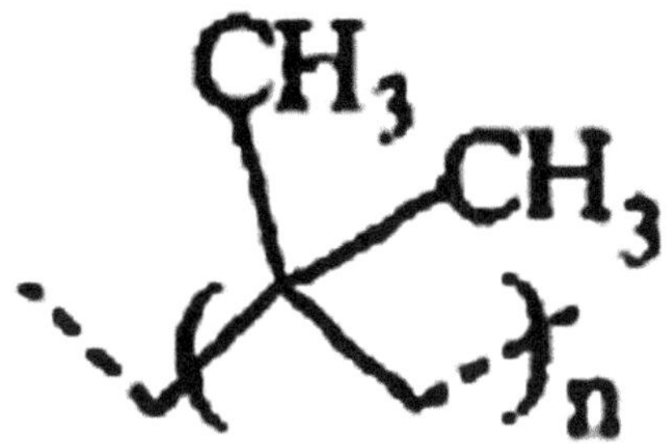

PIB

OPERATION PRINCIPLE OF SURFACE ACOUSTIC WAVE BASED SENSORS

This section is dedicated to analyses of the SAW gas sensors. Section 3 describes the operation principles of SAW devices (Rao & Zhang, 2006). Operation of a SAW sensor is based on detection of small changes in its acoustic wave propagation characteristics (e.g. change of the acoustic velocity or attenuation) caused by perturbations on the active surface of the device (A. J. Ricco & Martin, 1991). These perturbations could be affected by many external factors; each of which could cause a potential sensor response (Ballantine Jr et al., 1996). The following general Equation 1 illustrates the possible external perturbations of the acoustic wave phase velocity due to

T - Temperature,
E - Electric field,
ε - Permittivity,
δ - Electrical conductivity,
m – Mass, and
p - Pressure (A. Ricco, Martin, & Zipperian, 1985).

V_P is the phase velocity and V_0 is the unperturbed velocity:

$$\frac{\Delta V}{V_0} \cong \frac{1}{V_P}\left[\frac{\partial v}{\partial T}\Delta T + \frac{\partial v}{\partial E}\Delta E + \frac{\partial v}{\partial \varepsilon}\Delta \varepsilon + \frac{\partial v}{\partial \delta} + \frac{\partial v}{\partial p} + \ldots\right] \quad (1)$$

Changes in the parameters shown in Equation 3.1 could cause a change on the surface of the device producing a measurable shift of the SAW mode phase velocity. Since SAW devices use a piezoelectric substrate for generating and detecting acoustic waves, the parameters involved in a sensor application could be either mechanical or electrical and it is possible to obtain expressions for $\Delta V / V$ in different cases using theory of perturbation and the appropriate boundary conditions (e.g. mechanical or electrical) (Auld, 1973). In terms of a chemical sensor in the early stages of development in the 1980s, the emphasis was on accumulation of surface mass since the first SAW device utilized thin polymer, metallic or metal oxide layer (d'AMico, Palma, & Verona, 1982).

SAW Sensing Mechanism

A SAW based gas sensor consists usually of a delay line device with a gas sensitive layer deposited on the active surface between the input and output IDTs (Avramescu et al., 2009; Elmazria, 2008; Wang, Lim, Lee, & Yang, 2009). The sensing mechanism involves the chemical interaction between the analytic gas and the gas sensitive layer producing a shift in the SAW mode phase velocity, due to perturbations described by Equation 2.

In the case of a SAW device, this operates due to the mass loading of the gas sensing layer. When the molecules adsorbed on to the surface, the interactions between the gas sensitive layer and the gas analyte result in either change in the mass density or conductivity on the device surface via mass loading accordingly, the velocity and attenuation of the propagating wave is changed by this surface interaction. There are two common methods used for measuring the change in the SAW velocity, the phase and frequency measurement methods (Galipeau, Story, Vetelino, & Mileham, 1997). In this work the frequency measurement technique has been employed by measuring oscillation frequency of the device. The change in the SAW phase velocity Δv could be related to the change in the device oscillation frequency or frequency shift Δf by the following relationship:

$$\frac{\Delta f}{f_0} = \frac{\Delta v}{v_0} \tag{2}$$

where f_0 is the oscillation frequency and v_0 is the SAW velocity (A. J. Ricco & Martin, 1991). In this way, the measured frequency shift Δf could be correlated to the concentrations of the gas analyte. More details regarding the frequency measurement technique and a description of the measurement set-up are included in following sections. For the proposed layered SAW structure, in this research, the sensitive layer covers the entire surface of the device. In this way, the sensitive layer interaction area is increased. A SAW propagating in a piezoelectric material generates a layer of bound charge, which accompanies the mechanical wave. When SAW propagates through the device, the charge carriers in the surface film redistribute to compensate for the layer of bound charge generated by the propagating wave (Ballantine Jr et al., 1996).

DESIGN AND SIMULATION OF SAW GAS SENSOR

In this section, the proposed SAW sensor for gas sensing application is shown as Figure 3 where it consists of an IDTS etched onto a piezoelectric $LiNbO_3$ substrate and covered with a thin PIB sensing gas layer. The mass of the PIB film increases as PIB selectively adsorbs CH_2Cl_2 in air. This causes a shift in resonance to a slightly lower frequency.

Figure 3. Conceptual view of the SAW gas sensor, showing the IDT electrodes (in black), the thin PIB film (light gray), and the $LiNbO_3$ substrate (dark gray)
Tigli & Zaghloul, 2007.

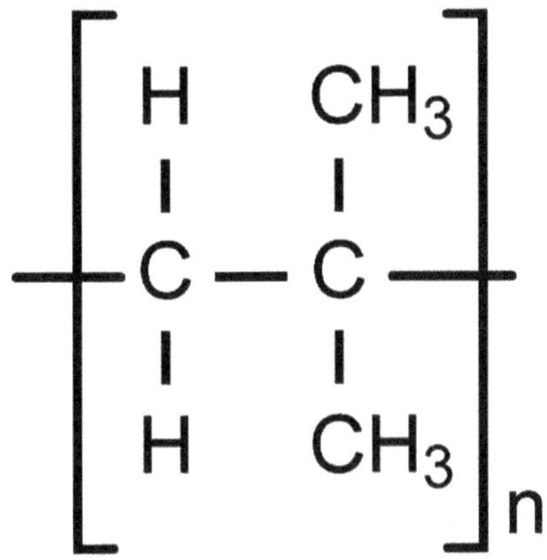

Figure 3 shows a conceptual view of the gas sensor. In this model, IDTs used in SAW devices may have hundreds of identical electrodes, and each electrode could be about 100 times longer than is wide (Wen, He, Li, Liu, & Yong, 2007). Therefore, the edge effects can be neglected and the model geometry is reduced to the periodic unit cell shown in Figure 3.

The height of this cell does not have to extend all the way to the bottom of the substrate but only a few wavelengths down, so that the SAW has almost died out at the lower boundary. In the model, this boundary is fixed to a zero displacement. Design of the whole SAW sensor involves five steps:

1. Modeling the geometry,
2. Physics settings,
3. Solving the problem,
4. Meshing, and
5. Post processing.

Details of each step will be explained completely in following stages.

Geometry Modeling

In the first step, the whole model geometry has been created by using the CAD tools on draw menu, considering specification of the SAW gas sensor as mentioned in Table 1.

As could be seen from Figure 4, the whole geometry is obtained where width of each electrode is 1μm; therefore, the wavelength was calculated 4μm. A 300 nm PIB sensing gas layer has covered two 1μm wide electrodes on top of the $LiNbO_3$ substrate.

In this model, the aluminum IDT electrodes and the PIB film are added. This causes the lowest SAW mode to split up in two eigensolutions, the lowest one representing a series resonance, where propagating waves interfere constructively, the other one a parallel ("anti-") resonance, where they interfere destructively. These two frequencies constitute the edges of the stop band, within which no waves could propagate through the IDT.

Table 1. Specification of SAW geometry

Width (μm)	Height (μm)	Base: Corner x (μm)	Base: Corner Y (μm)
4	22	0	-22
1	0.2	0.5	0
1	0.2	2.5	0
4	0.3	0	0

Figure 4. Model geometry

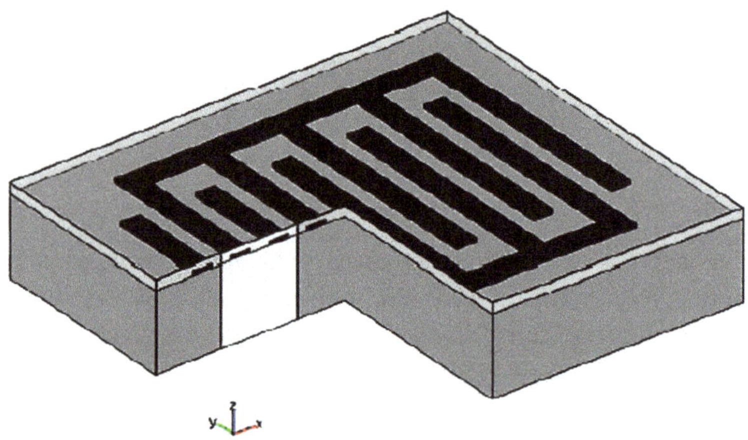

Physics Settings

After drawing is done, the next step is to specify the model physics. For investigating the reaction of SAW gas sensor to apply different physical parameters, the physical modeling is set up which is entering all the descriptions and the settings for the physics and equations in the model.

Normally, material properties are defined in the Sub domain Settings dialog box constraints or loads Table 2, to achieve a deeper understanding of the underlying physics of an application mode, its Equation system settings dialog boxes have been studied. It could be understandable how the software translates the material properties and boundary conditions into the Coefficient form, General form, or Weak form partial differential equations. They also list variables available for coupling and analysis purposes.

The amount of DCM which is adsorbed in PIB film is represented as a small increase in the density of the PIB film. Finally, the sensor is exposed to different concentrations of DCM gas changing from 100 to 800 ppm of DCM. The "partial density" of DCM in the PIB film is then calculated as:

$$\rho_{DCM,PIB} = KMc \quad (3)$$

where K is the partition coefficient for DCM, c: concentration in the air and M is the Molar mass. Other information such as PIB density, $LiNbO_3$ properties, Poisson ratio (0.48) and Young modulus (10GPa) have been obtained from (Schweyer et al., 1996). Poisson ratios considered in this modeling

Table 2. Constants and descriptions

Name	Expression	Description
P	101.325[kPa]	Air pressure
T	25[degC]	Air temperature
R	8.3145[Pa*m^3/(k*mol)]	Gas constant
c_DCM_air	100e-6*p/(R*T)	DCM concentration
M_DCM	84.93[g/mol]	Molar mass of DCM
K	10^1.4821	PIB/air partition constant for DCM
rho_DCM_PIB	K*M_DCM*c_DCM_air	Mass concentration of DCM in PIB
rho _PIB	0.918[g/cm^3]	Density of PIB
E_PIB	10[GPa]	Young 'modulus of PIB
nu_PIB	0.48	Poisson's ratio of PIB
Eps_PIB	2.2	Relative permittivity of PIB

represent a rather rubbery material. We would face multitude of eigen-modes located inside the film, but for this special case it has been focused on SAW sensor only. It should be mentioned that any other effects of DCM adsorption on the material except density have been neglected.

Solving a Model

The analysis type sets up various equations for the application mode. The Solver Parameters dialog box is opened and eigen frequencies solver from the Solver list was selected.

Post Processing and Visualization

The solver returns six eigen solutions with eigen frequencies in the vicinity of 850 MHz at 869.8 MHz, two of them are similar within the numerical accuracy. These show the shape and the frequency for a SAW with wavelength equal to the width of the geometry. Figure 5 shows the SAW as it propagates along the surface of the piezoelectric substrate. The frequency corresponding to a 4μm wavelength computes to 870 MHz, giving a phase velocity of 3479 m/s.

RESULTS AND ANALYSIS

Performance of the developed gas sensors has been analyzed from different numbers of pillars and different concentrations of applied DCM gas. The pillars here represent the surface roughness of the sensing layer. The performances of the developed saw sensors have been investigated by examining (frequency shift-t-Δf) for each saw sensors in both conditions (Ballantine Jr et al., 1996; Wohltjen, 1984). The frequency before gas exposure was measured and recorded, then the gas sensor has been exposed to DCM gas, the frequency was measured again and the frequency shift was obtained by considering the frequency difference before and after gas exposure. As it could be seen from Figure 6, shift frequency before and after gas exposure have been obtained and analyzed.

Figure 5. SAW wave propagates through the substrate: the color scale shows the magnitude of the displacements.

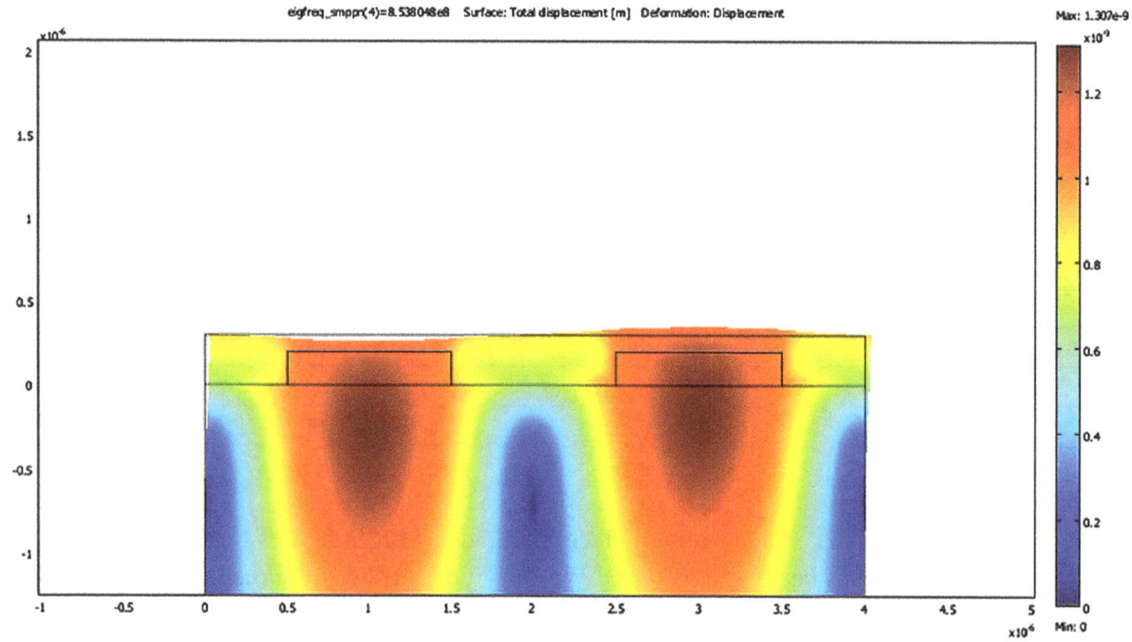

For obtaining the modeling of novel gas sensitive layer of SAW sensor, different configuration of sensing layer was designed, as the following figure shows (Figure 7). The saw sensor has been initially modeled without any roughness of sensing layer and the frequency shift was analyzed as the reference. Then by increasing the number of pillars (surface roughness properties) from 0 to 10 pillars, (Figure 7), different configurations were obtained as observed in the following figure.

As shown in Figure 8, width of each electrode is 1μm, the space between two electrodes is 1μm, the height of PIB sensing layer is 300 nm, and the height of piezoelectric is 22μm. All these constants are constant during this modeling process, the only parameter that will be changed is the width of pillars.

As it could be seen from Table 3, as the width of each pillar is decreased, the total surface area ratio increases; i.e. roughness of the sensing layer has been increased.

Figure 7 depicts 0 pillars SAW gas sensor whose total surface area is 8.6×10^{-6}μm.The developed SAW gas sensor was exposed to different concentrations of DCM gas from 100 to 800 ppm. As shown in Table 4 and Figure 9, frequency before and after gas exposure was measured and recorded for 0 pillars.

Figure 10 and Table 5 show the frequency before and after gas exposure for 4 pillars SAW gas sensor, while it was exposed to DCM gas.

In the next level, the SAW sensor with 10 pillars (the maximum investigated one) is simulated for DCM detection. As seen in Figure 11 and Table 6, the frequency before and after gas exposure was measured for 10 pillars.

Figure 12 illustrates the change of frequency-shift based on changing DCM concentration. It is obviously seen that, by increasing the number of pillars, the shift-frequency has increased. It means that the bigger the frequency-shift, the higher percentages of gas molecular adsorbed.

Figure 6. Frequency-shift before and after gas exposure

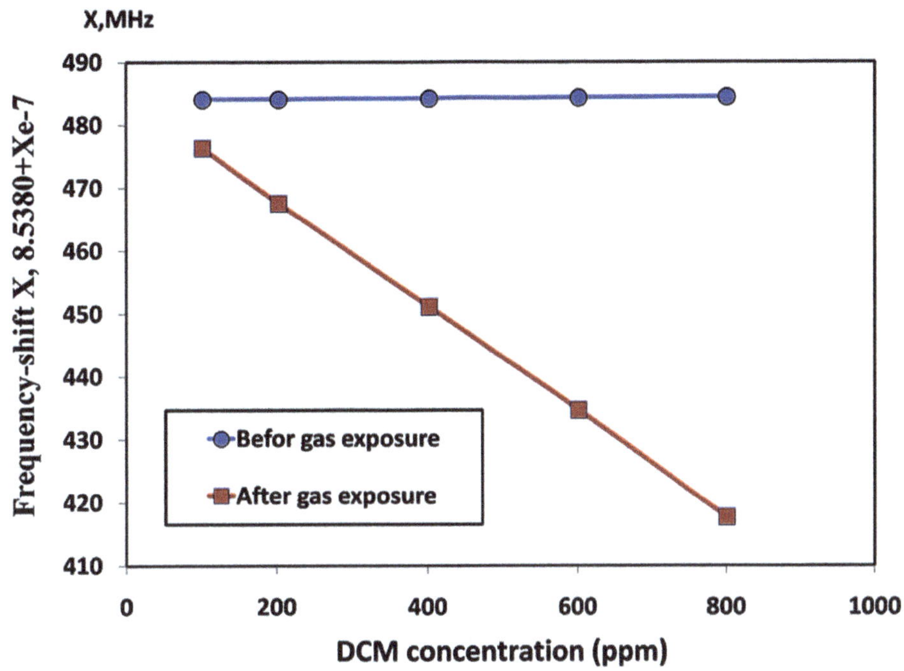

Figure 7. Different configurations of surface roughness for SAW sensor

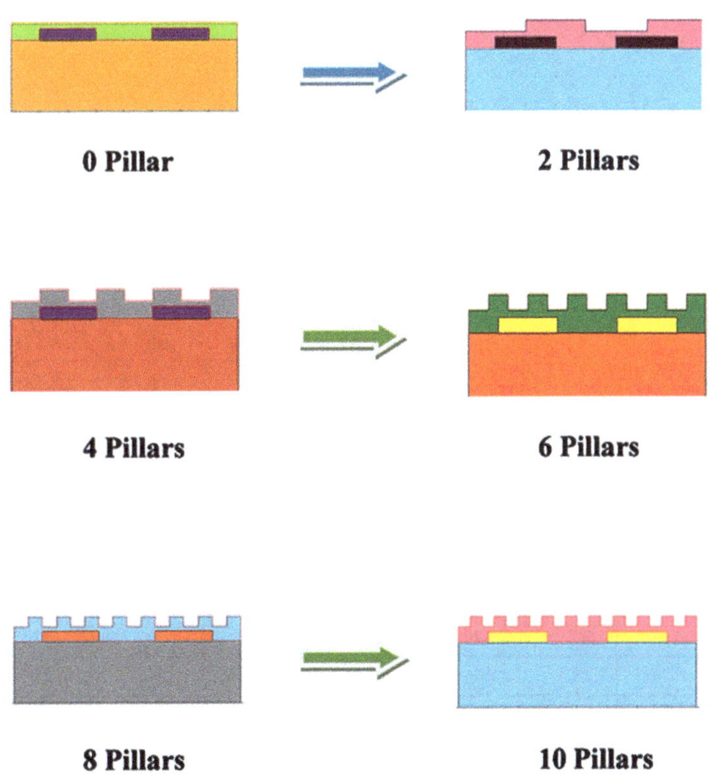

Figure 8. Simulation of SAW gas sensor

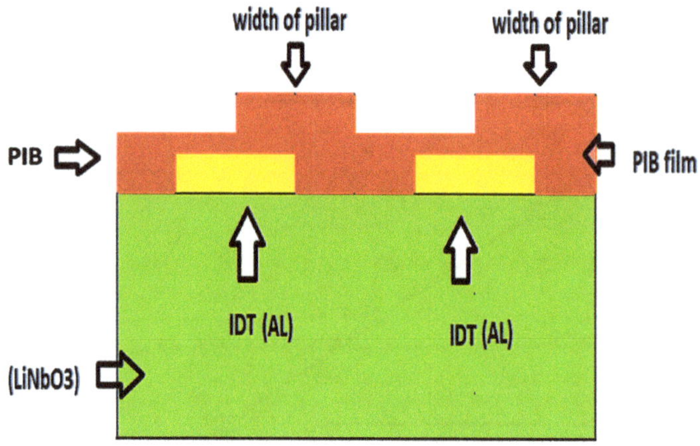

Table 3. Total surface area

Pillars	Width of Each Pillar (μm)	Surface Area (μm)
0-pillar	0	8.6E-6
2-pillars	1.0E-6	9.4E-6
4-pillars	5.0E-7	10.2E-6
6-pillars	0.333E-7	11.0E-6
8-pillars	0.25E-7	11.8E-6
10-pillars	0.2E-7	12.8E-6

Table 4. Frequency before and after gas exposure for 0 pillars

DCM Concentration (ppm)	Frequency Before Gas Exposure (Hz)	Frequency After Gas Exposure (Hz)
100	8.538047575430210E8	8.538047575430210E8
200	8.538048406688360E8	8.538046744170670E8
400	8.538048406688360 E8	8.538045081647340E8
600	8.538048406688340E8	8.538043419118490E8
800	8.538048406688380E8	8.538041756584140E8

Table 5. Frequency before and after gas exposure for 4 pillars

DCM Concentration (ppm)	Frequency Before Gas Exposure (Hz)	Frequency After Gas Exposure (Hz)
100	8.479269811802510E8	8.479268019728E8
200	8.479269811802470E8	8.479266227641E8
400	8.479269811802500E8	8.479262643434E8
600	8.479269811802510E8	8.479259059179E8
800	8.479269811802470E8	8.479255474878E8

Figure 9. Frequency shift of 0 pillars

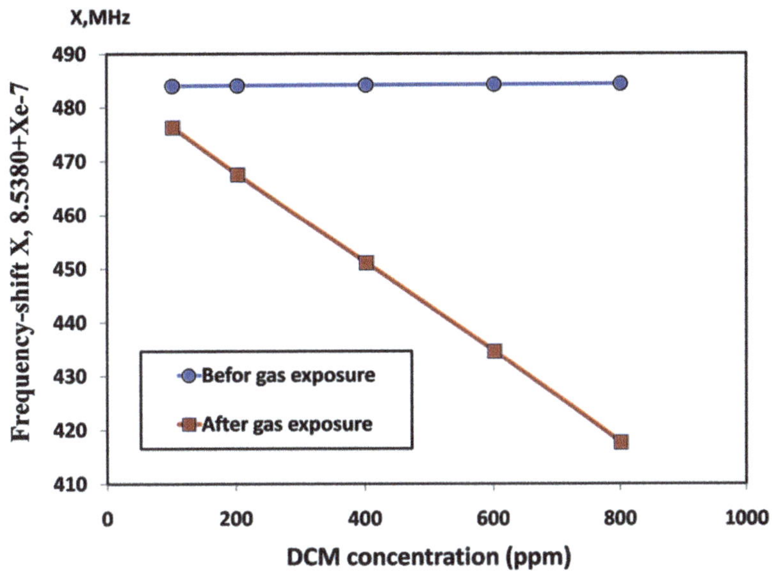

Figure 10. Frequency shift of 4 pillars

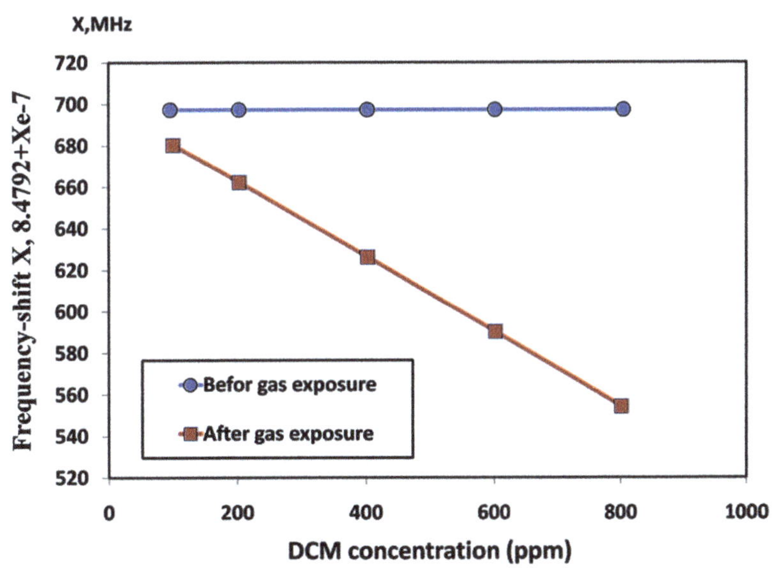

As an important finding for this research, one could state that the higher number of pillars, the better performance will be achieved for one concentration. In other words, for a specific gas concentration, the sensor with a higher amount of pillars on the sensing layer, the more roughness, will give us a better and more sensitive gas sensor which is capable of adsorbing a higher amount of gas on its surface.

Figure 11. Frequency shift for 10 pillars

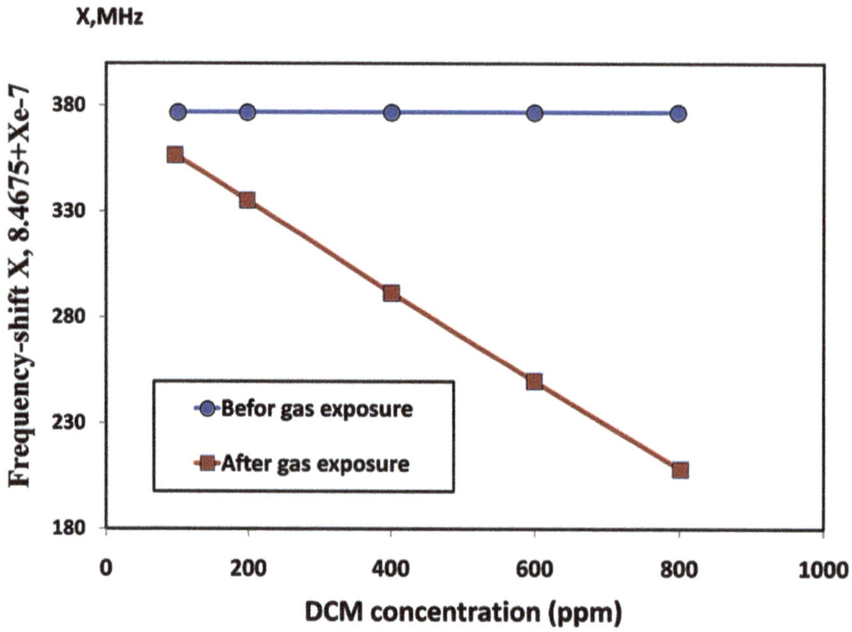

Table 6. Frequency before and after gas exposure for 10 pillars

DCM Concentration (ppm)	Frequency Before Gas Exposure (Hz)	Frequency After Gas Exposure (Hz)
100	8.467537735005930E8	8.467535627030780E8
200	8.467537735605970E8×10^8	8.467533518433580E8
400	8.467537735605940E8	8.467529301173070E8
600	8.467537735605960E8	8.467525083824360E8
800	8.467537735605900E8	8.467520866387500E8

Moreover, the highest frequency shift was accounted for 10 pillars; whereas, 0 pillar has the lowest frequency-shift. There is a significant increase in performance of gas sensor as the number of pillars is increased. On the other hand, to get a deep understanding of sensing layer roughness on, the effect of increasing surface roughness of sensing layer on performance of gas sensor is depicted in Figure 13.

It is apparently seen that in low concentrations, for instance 100 ppm, increasing the number of pillars enhances the performance of gas sensor. The sensing layer has the ability to absorb even very tiny small molecular of DCM up to 200 ppm. In nanostructures, to increase the ability of polymer for adsorption of gases, one could use nanostructure of sensing layer. It can be concluded that in high concentration of DCM gas like 800 ppm, by just increasing the number of pillars from 0 to 10, the sensor performance will be enhanced by three times which is considered as a noticeable result.

Figure 12. All frequency shifts versus DCM concentration variations

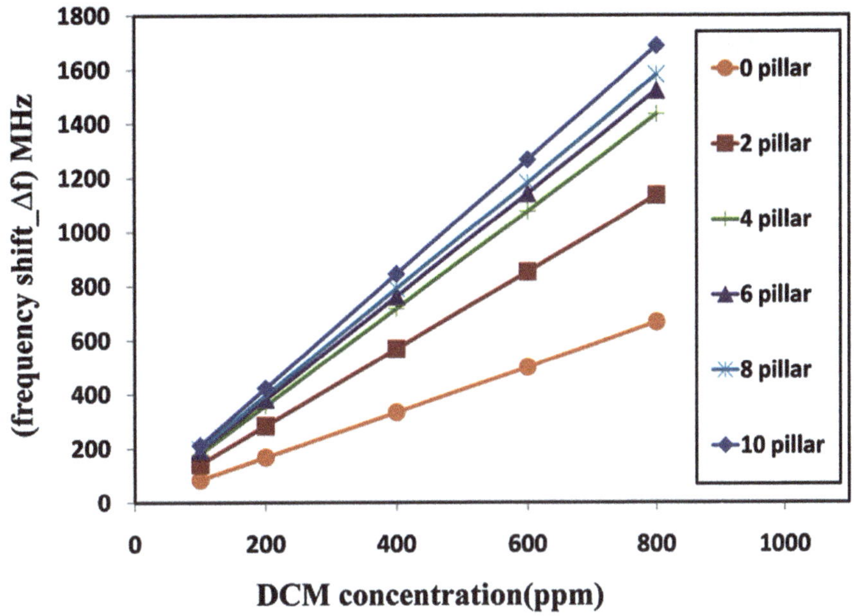

Figure 13. Summary of all pillars

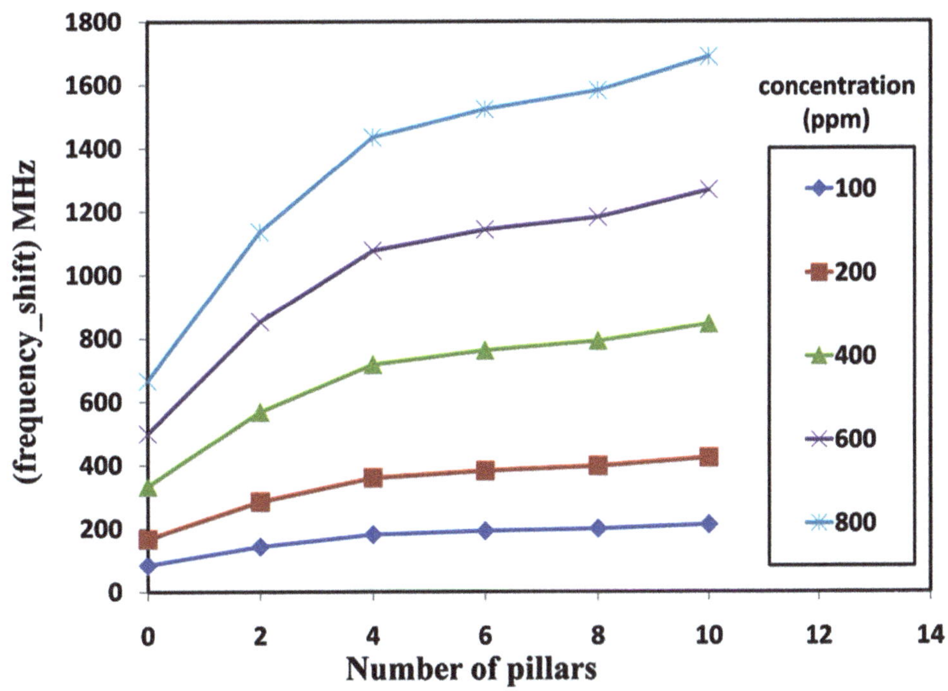

CONCLUSION

In conclusion, models of the sensing layer of SAW device have been carried out. Furthermore, different surface roughness of the sensing layer has been represented by number of pillars on the sensing layer and, finally, surface properties and concentration of the SAW device sensing layer based on frequency shift was analyzed. In conclusion, it is apparently shown that, by increasing the number of pillars (surface roughness), the SAW gas performance could be enhanced. Before gas exposure to all different types of surface roughness, the frequency shift remained unchanged.

All models show linear changes in frequency shift after gas exposure. It could be concluded that, the bigger the frequency shift the more gaseous molecular adsorbed to the sensitive layer of the SAW device. Furthermore, the surface roughness increases the effectiveness of absorption of gaseous molecules.

REFERENCES

Ahmadi, S., Hassani, F., Korman, C., Rahaman, M., & Zaghloul, M. (2004). *Characterization of multi- and single-layer structure SAW sensor [gas sensor]*. Paper presented at the Sensors, 2004.

Atashbar, M. Z., Bazuin, B. J., Simpeh, M., & Krishnamurthy, S. (2005). 3D FE simulation of H(2) SAW gas sensor. *Sensors and Actuators. B, Chemical, 111*, 213–218. doi:10.1016/j.snb.2005.06.054

Auld, B. A. (1973). *Acoustic fields and waves in solids*. Рипол Классик.

Avramescu, V., Bostan, C., Serban, B., Georgescu, I., Costea, S., & Varachiu, N. (2009). Surface acoustic wave devices and their sensing capabilities Cas. *2009 International Semiconductor Conference*. doi:10.1109/SMICND.2009.5336615

Ballantine, D. S., Jr., White, R. M., Martin, S. J., Ricco, A. J., Zellers, E. T., Frye, G. C., & Wohltjen, H. (1996). Acoustic wave sensors: Theory, design, & physico-chemical applications. Academic Press.

Ballato, A. (1996). *Piezoelectricity: History and new thrusts*. Paper presented at the Ultrasonics Symposium. doi:10.1109/ULTSYM.1996.584046

Bell, B. P., Franks, P., Hildreth, N., & Melius, J. (1991). Methylene chloride exposure and birthweight in Monroe County, New York. *Environmental Research, 55*(1), 31–39. doi:10.1016/S0013-9351(05)80138-0 PMID:1855488

Bois, FY, & Smith, AE. (1995). *Evaluation of variability in predictions of cancer risk from occupational exposure to methylene chloride as a consequence of variability in model input parameters*. OSHA.

Campbell, C. (1998). *Surface acoustic wave devices for mobile and wireless communications*. Academic press.

Comini, E. (2006). Metal oxide nano-crystals for gas sensing. *Analytica Chimica Acta, 568*(1), 28–40. doi:10.1016/j.aca.2005.10.069 PMID:17761243 d'AMico, A., Palma, A., & Verona, E. (1982). *Hydrogen sensor using a palladium coated surface acoustic wave delay-line*. Paper presented at the 1982 Ultrasonics Symposium.

DHS. (1989). *Proposed Identification of Methylene Chloride as a Toxic Air Contaminant. Part B- Health Effects of Methylene Chloride*. Berkeley, CA: DHS.

Drafts, B. (2001). Acoustic wave technology sensors. *IEEE Transactions on Microwave Theory and Techniques*, *49*(4), 795–802. doi:10.1109/22.915466

El Gowini, M. M., & Moussa, W. A. (2010). A finite element model of a MEMS-based surface acoustic wave hydrogen sensor. *Sensors (Basel, Switzerland)*, *10*(2), 1232–1250. doi:10.3390/s100201232 PMID:22205865

Elmazria, O. (2008). *Surface acoustic wave devices for wireless sensing applications: Principle, technology and market*. Academic Press.

Galipeau, D. W., Story, P. R., Vetelino, K. A., & Mileham, R. D. (1997). Surface acoustic wave microsensors and applications. *Smart Materials and Structures*, *6*(6), 658–667. doi:10.1088/0964-1726/6/6/002

Grate, J. W., Patrash, S. J., Abraham, M. H., & Du, C. M. (1996). Selective vapor sorption by polymers and cavitands on acoustic wave sensors: Is this molecular recognition? *Analytical Chemistry*, *68*(5), 913–917. doi:10.1021/ac950518z PMID:21619189

He, X., Arsat, R., Sadek, A. Z., Wlodarski, W., Kalantar-zadeh, K., & Li, J. (2010). Electrospun PVP fibers and gas sensing properties of PVP/36 YX LiTaO 3 SAW device. *Sensors and Actuators. B, Chemical*, *145*(2), 674–679. doi:10.1016/j.snb.2010.01.013

Ho, C. K., Lindgren, E. R., Rawlinson, K. S., McGrath, L. K., & Wright, J. L. (2003). Development of a surface acoustic wave sensor for in-situ monitoring of volatile organic compounds. *Sensors (Basel, Switzerland)*, *3*(7), 236–247. doi:10.3390/s30700236

Hribšek, M., & Tošić, D. (n.d.). *Analysis and modelling of surface acoustic wave chemical vapour sensors*. Academic Press.

Lee, J. H., Joo, B. S., Huh, J. S., & Lee, D. D. (2006). Fabrication of SAW sensor for detecting chemical agent. *Rare Metal Materials and Engineering*, *35*, 433–435.

Morgan, D. (1991). Surface-Wave Devices for Signal Processing. Amsterdam: Academic Press.

Morgan, D. P. (1985). *Surface-wave devices for signal processing* (Vol. 19). Elsevier Publishing Company.

Nakamura, K., Kazumi, M., & Shimizu, H. (1977). *SH-type and Rayleigh-type surface waves on rotated Y-cut LiTaO 3*. Paper presented at the Ultrasonics Symposium. doi:10.1109/ULTSYM.1977.196951

Nicolay, P., Elmazria, O., Sarry, F., Bouvot, L., Kambara, H., Singh, K. J., & Alnot, P. (2009). *SAW pressure sensor for vacuum control applications*. Paper presented at the Frequency Control Symposium, 2009 Joint with the 22nd European Frequency and Time forum. doi:10.1109/FREQ.2009.5168151

Penza, M., Antolini, F., & Antisari, M. (2004). Carbon nanotubes as SAW chemical sensors materials. *Sensors and Actuators. B, Chemical*, *100*(1), 47–59. doi:10.1016/j.snb.2003.12.019

Rao, Y. L., & Zhang, G. (2006). Enhancing the sensitivity of SAW sensors with nanostructures. *Current Nanoscience*, *2*(4), 311–318. doi:10.2174/157341306778699374

Rapp, M., & Rupp, S. (2007). New sensing materials for SAW sensors using polymer-silicate composites. *IEEE Sensors,* (pp. 155-157).

Reeder, T. M., Cullen, D. E., & Gilden, M. (1975). *SAW oscillator pressure sensors.* Paper presented at the 1975 Ultrasonics Symposium. doi:10.1109/ULTSYM.1975.196515

Ricco, A. J., & Martin, S. J. (1991). Thin metal film characterization and chemical sensors: Monitoring electronic conductivity, mass loading and mechanical properties with surface acoustic wave devices. *Thin Solid Films*, *206*(1), 94–101. doi:10.1016/0040-6090(91)90399-I

Ricco, A. J., Martin, S. J., & Zipperian, T. E. (1985). Surface acoustic wave gas sensor based on film conductivity changes. *Sensors and Actuators*, *8*(4), 319–333. doi:10.1016/0250-6874(85)80031-7

Sauviat, M., & Cohen-Addad, J. P. (1981). Polyisobutylene: Fractional precipitation from binary solvent induced by non-solvent addition. *Polymer*, *22*(4), 461–464. doi:10.1016/0032-3861(81)90160-9

Sayago, I., Fernandez, M. J., Fontecha, J. L., Horrillo, M. C., Vera, C., Obieta, I., & Bustero, I. (2011). Surface acoustic wave gas sensors based on polyisobutylene and carbon nanotube composites. *Sensors and Actuators. B, Chemical*, *156*(1), 1–5. doi:10.1016/j.snb.2011.04.047

Schweyer, M. G., Andle, J. C., McAllister, D. J., French, L. A., & Vetelino, J. F. (1996). *An acoustic plate mode sensor for aqueous mercury.* Paper presented at the Ultrasonics Symposium. doi:10.1109/ULTSYM.1996.583991

Strutt, J. W. (1885). On waves propagated along the plane surface of an elastic solid. *Proceedings of the London Mathematical Society*, *17*(1), 4–11.

Takayanagi, A., Yamanouchi, K., & Shibayama, K. (1970). Piezoelectric leaky surface wave in LiNbO3. *Applied Physics Letters*, *17*(5), 225–227. doi:10.1063/1.1653375

Tigli, O., & Zaghloul, M. E. (2007). A novel SAW device in CMOS: Design, modeling, and fabrication. *Sensors Journal, IEEE*, *7*(2), 219–227. doi:10.1109/JSEN.2006.883771

Wang, W., Lim, C., Lee, K., & Yang, S. (2009). Wireless surface acoustic wave chemical sensor for simultaneous measurement of CO(2) and humidity. *Journal of Micro-Nanolithography Mems and Moems*, *8*(3). doi: 03130610.1117/1.3158610

Wen, W., He, S., Li, S., Liu, M., & Yong, P. (2007). Enhanced sensitivity of SAW gas sensor coated molecularly imprinted polymer incorporating high frequency stability oscillator. *Sensors and Actuators. B, Chemical*, *125*(2), 422–427. doi:10.1016/j.snb.2007.02.037

White, R. M., & Voltmer, F. W. (1965). Direct piezoelectric coupling to surface elastic waves. *Applied Physics Letters*, *7*(12), 314–316. doi:10.1063/1.1754276

Wibawa, G., Khoiroh, I., Afrizal, D., & Suki, G. (2010). Solubilities of Dichloromethane, Diethyl Ether, Ethyl Acetate, and Nitrobenzene in three polymers using the Piezoelectric Quartz Sorption method. *Journal of Chemical & Engineering Data*, *55*(12), 5581–5586. doi:10.1021/je100389d

Wohltjen, H. (1984). Mechanism of operation and design considerations for surface acoustic wave device vapour sensors. *Sensors and Actuators*, *5*(4), 307–325. doi:10.1016/0250-6874(84)85014-3

World Health Organization. (2012). *Estimated deaths and DA-LYs attributable to selected environmental risk factors*. WHO.

Chapter 10
Optimization of Current-Voltage Characteristics of Graphene-Based Biosensors

Hediyeh Karimi
UniversitiTeknologi Malaysia, Malaysia

Elnaz Akbari
UniversitiTeknologi Malaysia, Malaysia

Rasoul Rahmani
Swinburne Universty of Technology, Australia

Meisam Rahmani
Universiti Teknologi Malaysia, Malaysia

Mohammad Taghi Ahamdi
Universiti Teknologi Malaysia, Malaysia

ABSTRACT

The aim of this project is to study and develop graphene-based DNA sensor model for detection of DNA hybridization application. This includes modeling and simulation of carrieconcentration, conductance, and current-voltage characteristics of graphene-based sensors on the field effect transistor (FET) platform. The main challenge is to validate the developed modelwith the experimental data,sincegraphene is considered as a new emerging material and research is still rapidly taking place with fabrication effort reported so far. In this research, first, numerical model is developed which shows the dependency of current-voltage characteristics on the DNA concentration factor. The iteration method is used for developing the numerical model. The proposed model is simulated utilizing MATLAB software to validate with experimental data of DNA hybridization. The I_d-V_g *characteristic of the proposed numerical model is depicted for different concentrations of DNA molecules and compared with experimental data for the verification purpose. After determining the accuracy of the models, particle swarm optimization (PSO) technique is used to minimize the error of the numerical model.Then, optimization results are shown. Overally, the accuracy of more than 98% represents an overall error of less than 2\% which is quite acceptable for the optimized numerical model.*

DOI: 10.4018/978-1-5225-0736-9.ch010

GRAPHENE

Graphene has been the subject of much interest even though it has been a relatively short perio since the discovery of it. The reason for that being the idiosyncratic physical and chemical characteristics like exceptional strength (Novoselov et al., 2004), and its great thermal conductivity plus biocompatibility that it possess (Novoselov, Jiang, et al., 2005). Graphene and nanotubes have been utilized as an essential element in several electronic device e.g. biosensors. There has been major evidence indicating biosensing applications shown in graphene and its by-products (Y. Shi, Dong, Chen, Wang, & Li, 2009). Due to the biocompatibility and extreme environmental distress affectability of thin graphene plates, it provides an essential biosensing application for them. Meaning they are greatly sensitive to variables such as electronic doping (Collins, Bradley, Ishigami, & Zettl, 2000; Kong et al., 2000; Y. B. Shi et al., 2006; Snow, Perkins, Houser, Badescu, & Reinecke, 2005) and molecule adsorption (X. Dong et al., 2009; Mohanty & Berry, 2008; Y. Shi et al., 2009; Wehling et al., 2008). In addition it is considerable that, because of easier contact, the surface structure of graphene (Displayed in Figure 1) in comparison to other carbon derivatives ensures a higher adsorption of DNA(Zheng et al., 2003).This resulted in the selection of grapheme as the DNA detection sensory framework in this study. With the help of the study and molecular analysis of nucleic acids, almost 400 and increasing genetic conditions are diagnosable now (Mastrangelo, 1999).

Graphene-Based Fet Structure

Plus the rewarding experiments on electronic peripherals such as field effect transistor (FETs) (Nilsson, Neto, Guinea, & Peres, 2007) made possible with the application of thin grapheme plates (Novoselov, Geim, et al., 2005)nano microlithographic fabrication (L. X. Dong & Chen, 2010). This suggests and justifies the call for action and the necessity of further DNA detection studies (Drummond, Hill, & Barton, 2003) as well as understanding the high sensitivity of the transport carriers in graphene plates and the conductance to environmental distress (Huang, Dong, Liu, Li, & Chen, 2011). In the process of observing the conductance variations of fabricated device base carbon materials electrical and label free DNA detection hybridization was possible to attain (Star et al., 2006). Graphene conductive characteristic as a channel in the FETs was considered as a seminal electronic features in the material (Dankerl et al., 2010). Presently the attention has moved towards been a focus on the exploratory capabilities in single-layer graphene (Abergel, Apalkov, Berashevich, Ziegler, & Chakraborty, 2010); Besides, single-

Figure 1. Monolayer graphene structure with one atom thickness

electron transistor base on monolayer graphene was discovered in an experiment as well (Cohen-Karni, Qing, Li, Fang, & Lieber, 2010).

Current-Voltage Characteristics (I-V) of Graphene-Based FET Structure

As depicted in Figure 2, several factors determine the position of the Dirac point such as: the variation between the work function of the gate and the graphene, the type and density of the charges, and any doping of the graphene.

The effects of graphene width on the I-V characteristics were studied by Choudhury et al. [44] assuming a Schottky-barrier FET. The outcome of the work suggested that the current is larger for graphene with higher width and vice versa. Yan et al. [45] worked on the GFET I-V characteristics by implementing first-principles transport calculations. The work carried out showed that GFET can achieve a performance level with other FETmade of single-walled carbon nanotubes (SWNT). It is claimed that graphene could show p-type behavior in these structures because of residual hydroxyl groups.

SCHEMATIC OF GRAPHENE-BASED DNA SENSOR STRUCTURE

Among the bonuses in the incorporation of graphene as an FETs' channel material is a great electrostatics control capability. Therefore, The reduction of short channel effect is anticipated (Choi, Choi, & Hong). Large graphene film has incorporated to be DNA detection template constructed of multi-layer and

Figure 2. Current-voltage characteristics of graphene-based FETs

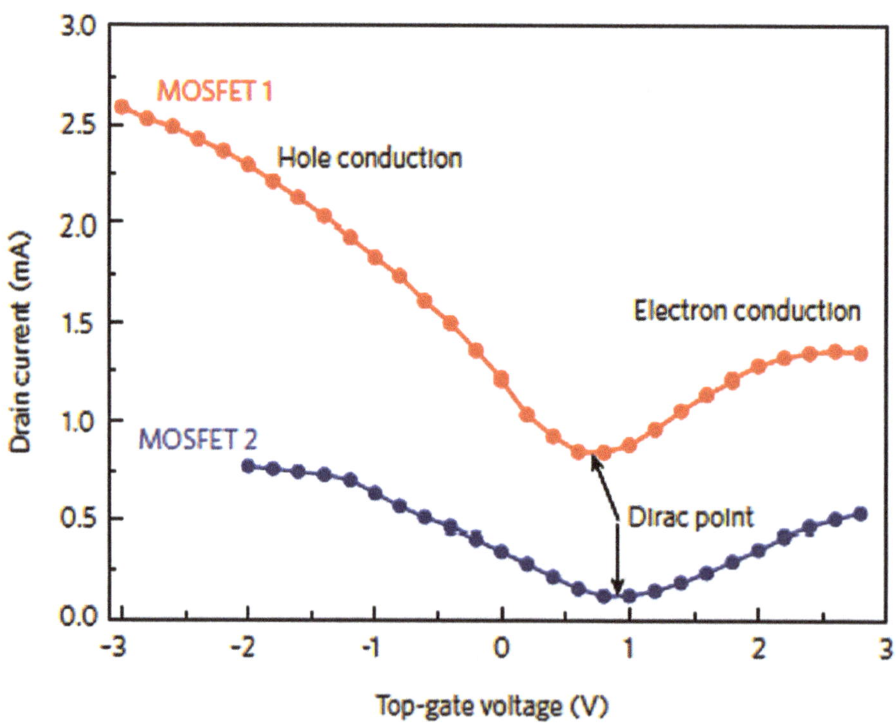

single-layer geaphene with a ratio of 40 to 60 percent. These graphene films are made using the chemical vapor deposition (CVD) method (Reina et al., 2008; Reina et al., 2009) . As a notable fact, transference of the graphene film from Nickel to glass substrate results in the possibility of the fabrication FET transistor based on large graphene (Di, Liu, Yu, & Zhu, 2009). The configuration of the Liquid-gated FETs is possible with the incorporation of graphene films (Figure 3) which have the DNA molecule detection sensitivity of 0.01nM (X. Dong, Y. Shi, W. Huang, P. Chen, & L.-J. Li, 2010). Another one of the abilities in this device is the detection of target DNA hybridization to the probe –DNAs that are pre-fixated on graphene.

The result of producing a DNA detection solution chamber was achieved through incorporating silver paint as source and drain electrodes and Polydimethylsiloxanes (PDMS) as an insulator.

Sensing Mechanism

The $(\pi - \pi)$ reactions among the carbon plane and the DNA molecules strongly affects the device conductivity of device by the means of different mechanisms like electronic doping (Star et al., 2006) and electrostatic gating (Fu & Li, 2010; Johnson Jr, Gelperin, & Staii, 2011). Both the DNA and graphene are negative in charges, hence the considerable role of buffer in single stranded DNA (ssDNA) hybridizing sensing of graphene (TEL-VERED, Willner, & Willner, 2010). Phosphate buffer saline (PBS) or Sodium Chloride solution has recurrently been incorporated in the form of a buffer for charge screening (Chen et al., 2008; X. Dong, Y. Shi, W. Huang, P. Chen, & L. J. Li, 2010) The outcome of which is DNA adsorption due to both Van Der Waals and electrical attraction (X. Dong et al., 2010; W. Wang & He, 2009). Meaning, the detection process could be defined as transference of the electron between DNA and graphene films (Wen et al., 2010). A detection mechanism is depicted in Figure 4 To clearly explain the transference of electrons from DNA molecules to grapheme nnaomaterial during attachment.

Here in this research, the concentration of DNA in the form of a gate voltage is replicated and DNA sensing factor $(\propto)$ is suggested. Eventually the collation study between suggested model and the data attained from the experiments is reported.

Figure 3. Graphene based FET structure for DNA detection

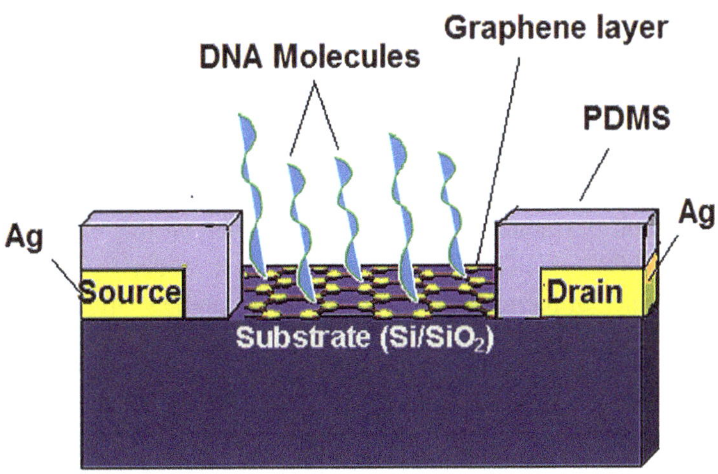

Figure 4. Detection mechanism

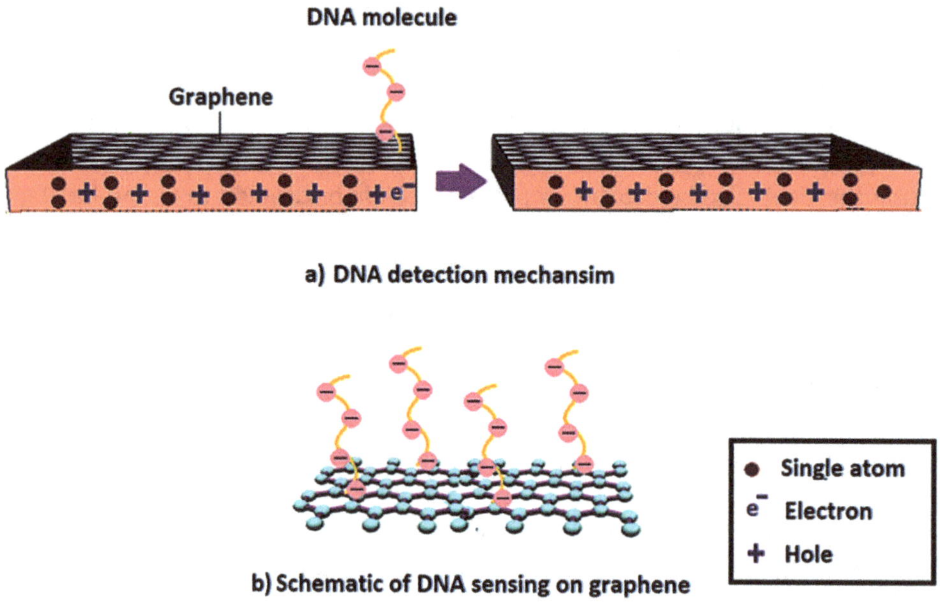

NUMERICAL MODELING OF GRAPHENE-BASED DNA SENSOR

In order to obtain a model for graphene-based DNA sensor, first we need to choose a physical measurable parameter as a sensing element which can be capacitance, resistance, conductance, carrier concentration, velocity of carrier in the cahnnel and etc. In this study, conductance is chosen as a sensing parameter due to ease of comparision with existing experimental work as well as graphene highe sensitivity to conductance change. It is noteworthy to mention that graphene channel in FETs assumed to have a ballistic carrier transportation due to high carrier mobility reported (Ahmadi, Johari, Amin, Fallahpour, & Ismail, 2010). Ahmadi.et al calculated the conductance of graphene based FET structures by considering number of available modes (Datta, 2002, 2005)

$$G = \frac{2q^2}{h}\int_{-\infty}^{+\infty} dE M\left(E\right) T\left(E\right)\left(-\frac{df}{dE}\right) \qquad (1)$$

where q is the electron charge, h is Planck's constant, E is the energy band structure, T (E) is the transmission probability, M (E) is the number of modes and f is the Fermi-Dirac distribution function. In other words, T is the probability of one injected electron at one end that can be transmit to other end is considered equal to one (T(E) =1) (Datta, 2002). Number of modes in corporation with Landauer Formula indicates conductance of monolayer graphene which can be written as

$$M\left(E\right) = \frac{3at}{2L}\left(\frac{4E}{3at} - 2\beta^2\right)^{1/2} \qquad (2)$$

where L is the length of graphene channel, $a_{c-c} = 1.42\,\overset{0}{\mathrm{A}}$, $t = 2.7\,eV$ is the tight-binding energy for nearest neighbor C-C atoms and the quantized wave vector β is defined as $\beta = \frac{2\pi}{\sqrt{3}a_{c-c}}\left(\frac{P_i}{N+1} - \frac{2}{3}\right)$ where p_i is the subband index and N is the number of dimer lines which determines the width of the ribbon. It can be stated that the length of the channel has a strong influence on the conductivity function(Peres, Neto, & Guinea, 2006). To consider the effect of temperature on graphene conductance, the boundary of integral changes as follows

$$G = \frac{3q^2\left(3\pi a^3 t^3 k_B T\right)^{\frac{1}{2}}}{hL} \times \left[\int_{-\infty}^{+\infty} \frac{x^{-\frac{1}{2}}}{\left(\frac{1}{1+e^{x-\eta}}\right)} dx + \int_{-\infty}^{+\infty} \frac{x^{-\frac{1}{2}}}{\left(\frac{1}{1+e^{x+\eta}}\right)} dx\right] \tag{3}$$

This equation can numerically be solved by employing the partial integration method in the simplified form, where $x = \left(E - E_g\right)/k_B T$ and the normalized Fermi energy is $\eta = \left(E_F - E_g\right)/k_B T$. Thus the general conductance model of single layer graphene can be obtained similar to that of silicon reported by Gunlycke(Gunlycke, Areshkin, & White, 2007). According to the current-voltage characteristic of graphene based FETs devices, the performance of DNA sensor can be evaluated through this equation (Passlack, 2008). By assuming the source and substrate terminals are held in ground potential, the channel region has the characteristics of the resistor in small voltage between source and drain (V_{DS}). Moreover, the relationship between current and conductance can be replaced by Fermi-Dirac integral shape of general conductance model of single layer graphene as

$$I_d = \frac{3q^2\left(3\pi a^3 t^3 k_{BT}\right)^{\frac{1}{2}}}{hL}\left[\mathcal{I}_{-\frac{1}{2}}(\eta) + \mathcal{I}_{-\frac{1}{2}}(-\eta)\right]\left(V_{gs} - V_t\right) \tag{4}$$

where V_{gs} is the gate-source voltage and V_t is the threshold voltage. As shown in Figure 5, the performance of DNA sensor base on graphene nanostructured is evaluated by the current-voltage characteristic. Before the probe DNA has been added, pure PBS (40 μL) has been added in the chamber to measure the transfer curve, i.e., drain current I_d versus gate voltage (Xiaochen Dong et al., 2010).

There is a favorable agreement between proposed model for DNA sensor based graphene and experimental result which is extracted from ref (Xiaochen Dong et al., 2010).

CURRENT-VOLTAGE MOLDING

In the next step we need to analyze the behavior of graphene based FET with presence of DNA molecules. For this purpose, DNA molecules with different concentration in 40 μL PBS buffer have been added in the chamber to allow the saturate attachment to graphene surfaces. In addition, in each step the

Figure 5. Current-Voltage model of (starred line) graphene with extracted experimental data (circle line)

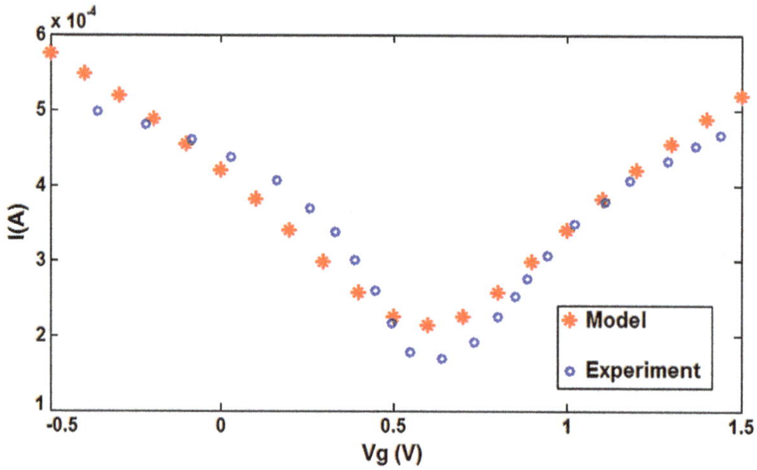

current-voltage was measured and recorded. As depicted in Figure 6, by immobilization of probe DNAs on the graphene surface, the $V_{g, min}$ is dramatically left-shifted. Due to this fact, it can beconcluded that $V_{g, min}$ is so sensitive to the immobilization of probe DNA and hybridization of complementary target DNAs. For supporting this, the gate voltage shifted leftwards owing the fact that DNA molecules n-dopes the graphene film (Liu, Robinson, Sun, & Dai, 2008; Mohanty & Berry, 2008). On the other hand, the conductivity of the graphene based FET devices is influenced by the increased number of carriers in the channel. The bond between graphene and nucleotides is a kind of non-electrostatic interaction that studied experimentally and theoretically (Liu et al., 2008; Manohar et al., 2008; L.-J. Wang et al., 2010). In total, $V_{g, min}$ shift can be known as a good indicator for DNA detection. As shown in Figure 6, the concentration of DNAs are increasing from 0.01nM to 500 nM, more DNA molecules will be adsorbed and the sensor will be capable of attracting more molecules which has led to change in the V_{gmin} on the device(Xiaochen Dong et al., 2010).

By analyzing the behavior of graphene in presense of DNA molecules, now the model can be proposed ispiring the dependence of gate voltage onDNA concentration variations. Ultimately, gate voltage can be modeled by DNA concentration (follow), in other words, V_g is assumed as a function of DNA concentration V_{gs}~F_{DNA}.

$$V_{gs(DNA)} = \frac{\propto}{F} V_{gs(withoutDNA)} \tag{5}$$

where DNA sensing factor ($\propto$) is suggested and F is DNA concentration.

$$I_d = \frac{3q^2\left(3\pi a^3 t^3 k_{BT}\right)^{\frac{1}{2}}}{hL}\left[\mathcal{I}_{-\frac{1}{2}}\left(\eta\right)+\mathcal{I}_{-\frac{1}{2}}\left(-\eta\right)\right]\left(\frac{\propto}{F} V_{gs(withoutDNA)} - V_t\right) \tag{6}$$

Figure 6. Transfer characteristic of graphene FET for DNA concentration changing from 0.01 to 500 nM

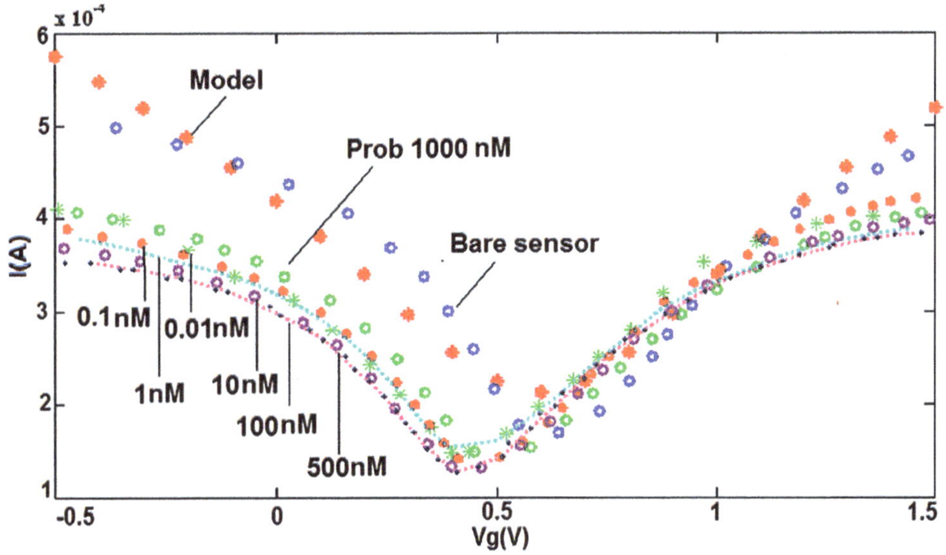

Different concentration of DNA molecules is presented in form of (F) parameter. Thus the DNAmolecules adsorbed on graphene surface by iteration method is modeled as

$$\alpha = aF^2 + bF + c \tag{7}$$

From extracted data, (a, b, c) parameters are calculated where a=13, b=50, c=4070. Eventually, according to the proposed model of DNA sensor using nanostructured graphene layer, the current-voltage characteristic can be modified as

$$I_d = \frac{3q^2 \left(3\pi a^3 t^3 k_{BT}\right)^{\frac{1}{2}}}{hL} \left[\mathcal{I}_{-\frac{1}{2}}(\eta) + \mathcal{I}_{-\frac{1}{2}}(-\eta)\right] \left(\frac{13F^2 + 50F + 4070}{F} V_{gs(withoutDNA)} - V_t\right) \tag{8}$$

Results of Current-Voltage Molding

To avoid from graphs overlapping, each figure with specific concentration has been considered separately in Figure 7 and 8. Also comparison study between our models in detail shows acceptable agreement with reported data as shown in Figures 7 and 8. When DNA molecules n-dopes the graphene layer, $V_{g, min}$ has beenleft-shifted. In both experimental and theoretical data, apparently the shift in $V_{g, min}$ rosewith the increasing concentration of the complementary DNA.

It is apparently shown that, by varying DNA concentration through the DNA factor, I-V characteristic curve can be controlled. Furthermore, the proposed model is strongly approximate with experimental data.

According to Figure 8 (a, b), the amount of $V_{g, min}$ shift remains unchanged by increasing the DNA concentration from 10nM to 500nM likely holding the fact that the number of DNA molecules is limited

Figure 7. I-V characteristic of proposed alpha model with experimental data for different DNA concentrations: (a) 0.01nM; (b) 0.1nM; (c) 10nM

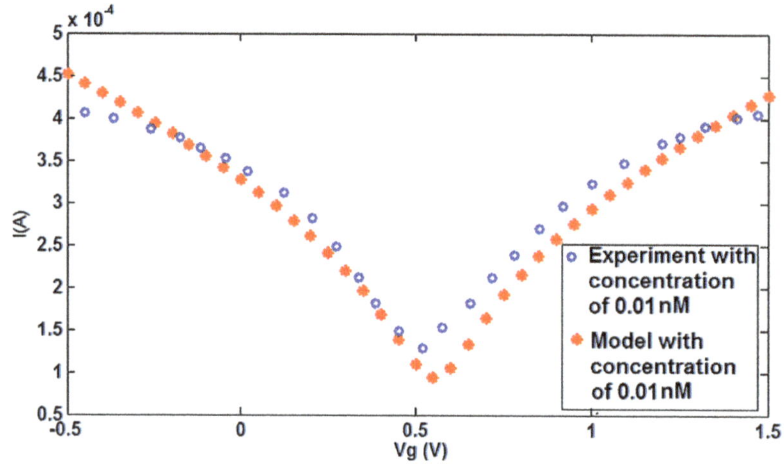

(a)

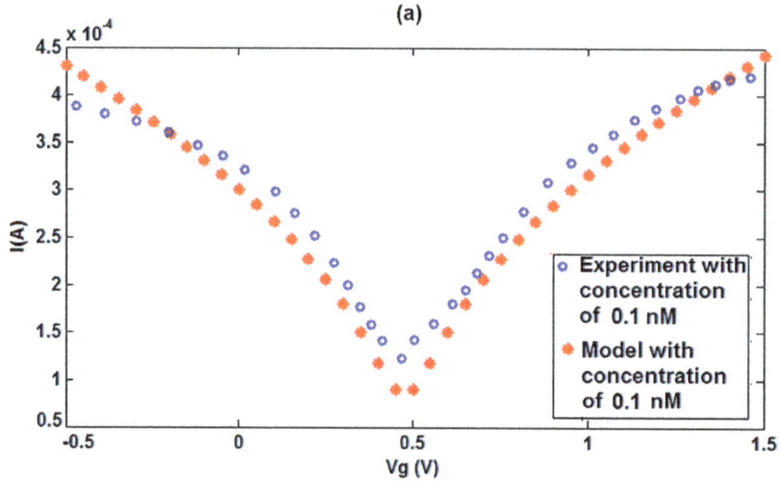

(b)

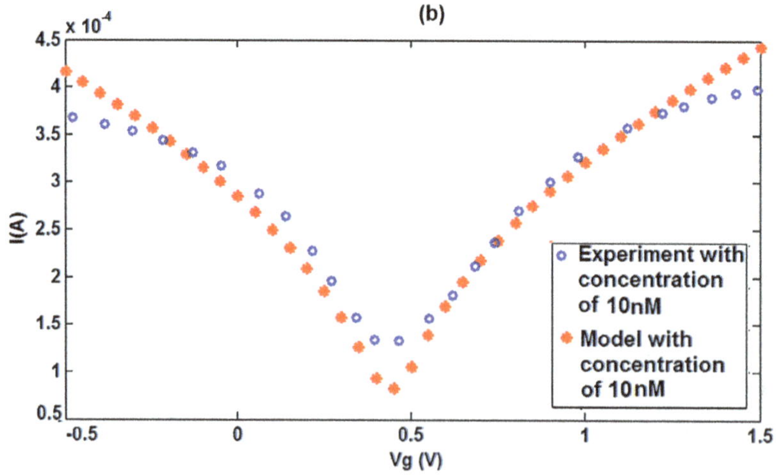

(c)

Figure 8. I-V characteristic of proposed alpha model with experimental data: (a) with concentration of 100nM; (b) with concentration of 500nM

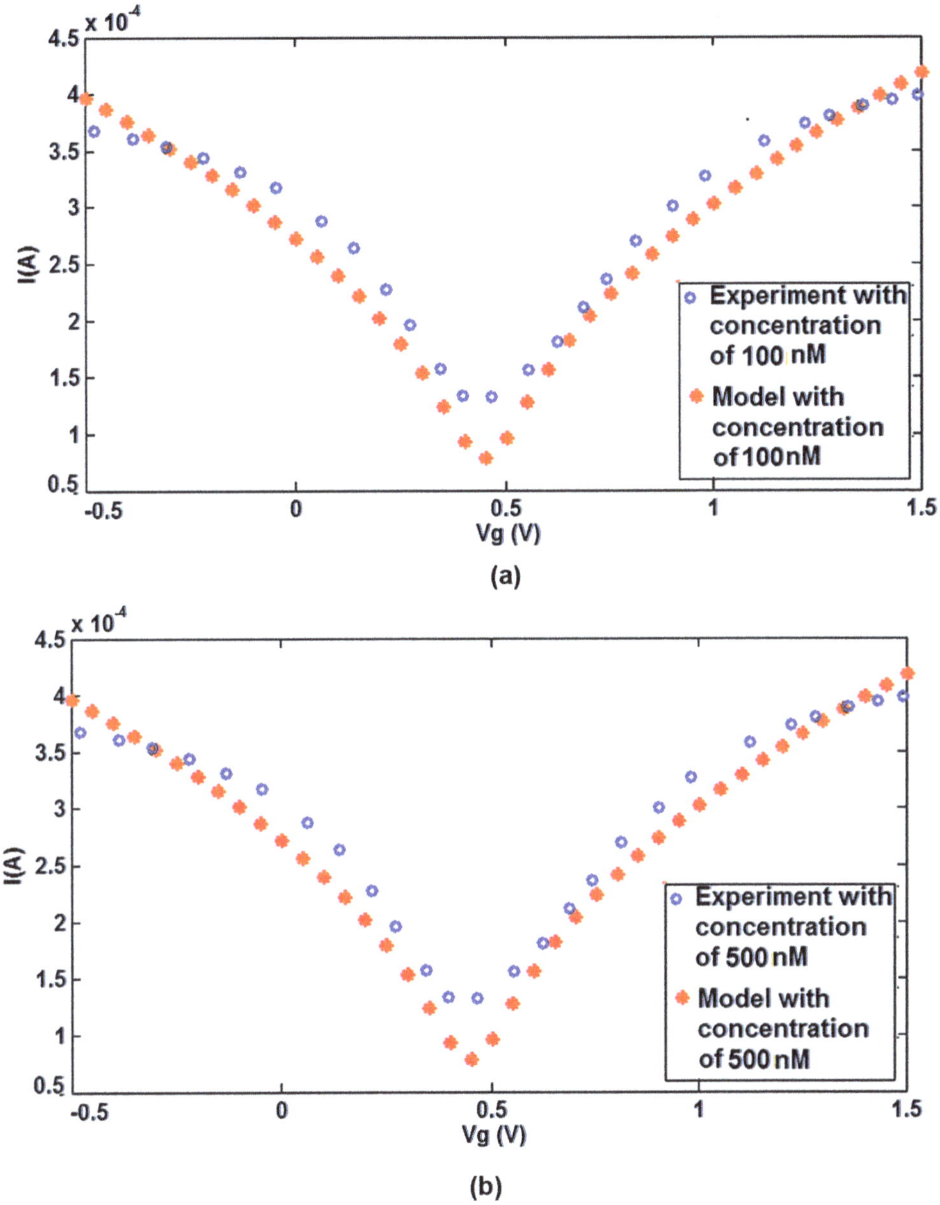

and the graphene surface has become saturated.The electron transfer from DNA molecules (electron rich) to graphene likely have been explored experimentally (Xiaochen Dong et al., 2010). Also as shown in Figure 8 (a, b) each diagram depicts specific concentration of DNA molecules. For example when concentration value is F=100 nM, it is noteworthy that, the model is closer to the blue line; in the same manner we can compare other concentrations as well (for more information see the appendix).

OPTIMIZATION OF CURRENT-VOLTAGE CHARECTERISTICS OF GRAPHENE BASED DNA SENSOR

Due to the fact that the figures of of the parameters *A, B* and *C* in Equation (8) are calculated by the iteration method which is a trial and error based technique, the need to incorporate a methodological approach for attaining a feasible and precise model for utilization in individual aspects of the graphene-based DNA sensor does exist. In order to achieve this, an evolutionary algorithm (EA) known as particle swarm optimization (PSO) has been utilized in importing the mathematical model displayed shown in Equation (8). The PSO method is conventionally incorporated in solving varying problems, such as fine materials, medical science, control theory and energy. Notable characteristics that contribute to the popularity of PSO in the research circles include its speed, the ability of refraining from entrapment in the local optima, plus the great and variable capability of utilization in numerous distinct enhancement situations(Boeringer & Werner, 2004; Clerc, 2006; Das, Abraham, & Konar, 2008; R. Eberhart, Shi, Y., 2001).

As a result, due to fact that individual cases commonly possess a restricted area of interaction with the environment the initiation of another particle will not alter the swarm as a whole(Y. Wang, Lv, Zhu, & Ma, 2010). Therefore the change of the scale in the swarm would less complicate depending on the scale of the research space. This pliability would result in the adaptability of the swarm in different circumstances. The proper swarm algorithm will have the ability to utilize and study a new domain, such as the considerable gap huge gap in the search space. Regarding the wholeness, if a few faulty cases exist in the swarm, the passivity of them not influencing the global optimum in the swarm exists. Another major upside would economic feasibility in terms of production of affordable of convenient sized and low cost components in relation to the computation and memory per individual.

One of the great benefits of the particle swarm optimization technique in comparison to global optimization methods like the simulated annealing would be the existence of numerous agents, that will result in the resilience of particle swarm technique to the local optima points of the search space. Meaning, this will facilitate the algorithm in avowing entrapment in the local optima points.

Particle Swarm Optimization (PSO)

The PSO algorithm is majorly built upon particle population motion also known as a 'Swarm 'which is named after the motion of groups of fish and birds in the nature(del Valle, Venayagamoorthy, Mohagheghi, Hernandez, & Harley, 2008; R. C. Eberhart, Shi, & Kennedy, 2001; Poli, Kennedy, & Blackwell, 2007). The assessment of the search space would be executed in a gradual manner.Some factors need to be taken into the account in each step, due to the fact that they will prove to be crucial in the calculation of the following step. Equations (9) and (10) show the velocity and location of i^{th}particle at k^{th}step.

$$V_i^{k+1} = W \times V_i^k + c_1 \times r_1 \left(Gbest_i^k - X_i^k \right) + c_2 \times r_2 \left(Pbest_i^k - X_i^k \right) \tag{9}$$

$$X_i^{k+1} = X_i^k + V_i^k \tag{10}$$

i= 1, 2,..., *nop* (number of particles);

k=1, 2, ..., k_{max} (maximum iteration number)

where,

K is the iteration number;
i is particle number;
W is referred to inertia weight coefficient which is normally chosen between 0.5 and 1.2;
c1 and *c2* are acceleration constants;
r1 and *r2* random are random values between 0 and 1;
V_i^k is the velocity of particle i at iteration k and X_i^k indicates the position of particle i at iteration k.

Other than the aforementioned factors which are necessary to account for a respective particle, social factors leading the swarm to the area contain the global optima must be noted. "Personal best" and "Global best" are two factors utilized in all the PSO variations. Generally they are mentioned as "*Pbest*" and "*Gbest*". The *Pbest* is considereed as the best location that each particle happens to be in the runtime, while the *Gbest* is the best global optimum observed so far in the swarm. It is displayed in Figure 9 with a simple illustration of movement of a typical particle.

Implementing Particle Swarm Optimization in DNA Sensor Model

Swarm is the particles which are moving and giving solutions for solving the problem. The motion of the particles happens in the territory of the proposed space and they respectively pose as a solution for the problem. The parameters to be optimized in the DNA sensor model are A, B, and C coefficients as in Equation (8). Figure 10 illustrates a three dimensional search space of the problem. If $\widehat{A}$, $\widehat{B}$ and $\widehat{C}$ are the vectors which return A, B and C, respectively; then particle i flies in the search space to find the best possible values for the respective parameters. Vector V_i is the movement velocity vector of the particle *i*which is obtained from Equation (4).

In the optimization process, each location represented by the particles is a possible solution. The search space consists of arrays and matrices. For example, X[i][j] is a matrix of location of the particles in the proposed algorithm while i indicates the particle and j is the index for the optimizing parameter.

Figure 9. A simple diagram for movement of a sample particle in PSO

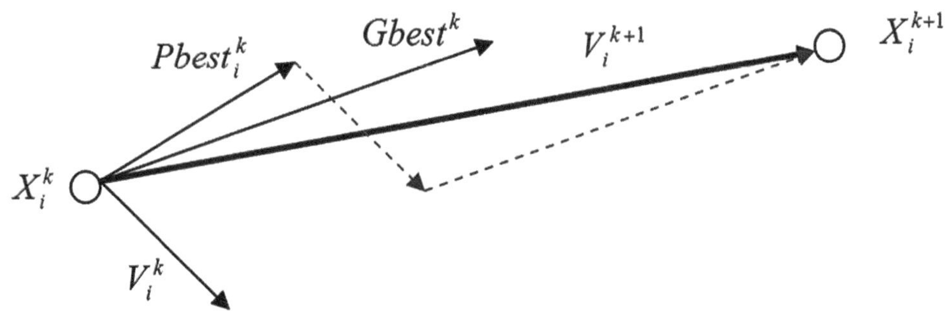

Figure 10. The search pace of the problem

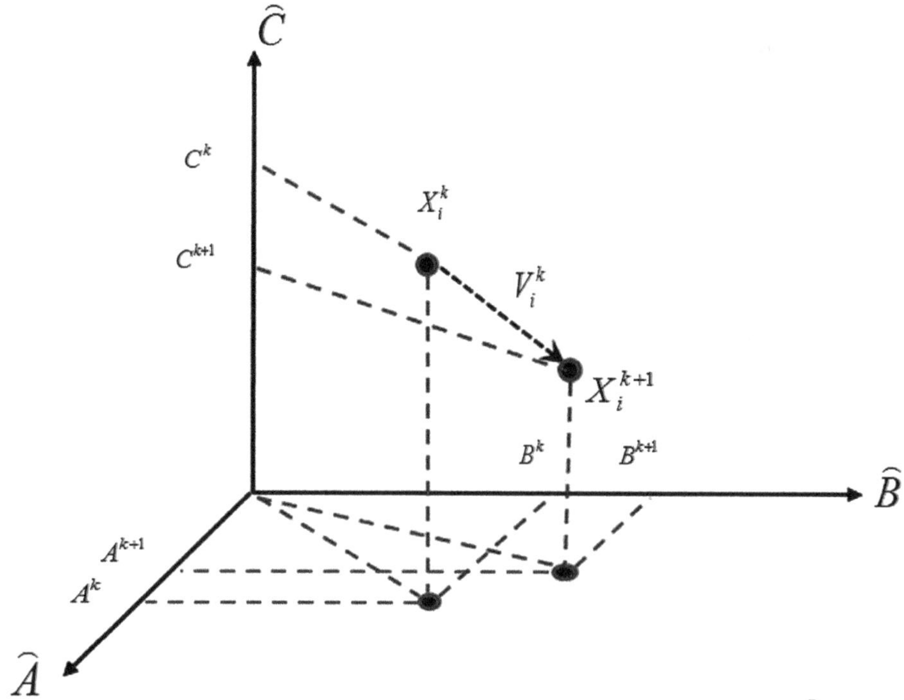

In Figure 10, the dimension of the X[i][j] matrix will be 200×3, if we consider the number of particles to be equal to 200.

To assess the solutions suggested by the particles, it is necessary to have a fitness function. This function needs to have the ability of determining the better more productive solution with consideration of solutions attained by the particles at each replication. Usually the function is designed to have the slightest possible value at an optimum point. Here in this study, there is also the need to achieve the lowest error tolerance in the experimental data in comparison to research. Therefore the function is suggested as follows:

$$Fitness\ function = \psi_i = \sum_{k=1}^{max}\left(\hat{I}_i(k) - I(k)\right)^2 \tag{11}$$

where

$\hat{I}_i(k)$ represents the modelled current waveform for particle *i*,

$I(k)$ is the experimental current of the DNA sensor, and

ψ_i is the fitness value of the ith particle.

Based on the fitness function chosen, the best solution will propose the most fitted curve for the current waveform which is desired for a suitable DNA sensor model. The parameters of the PSO algorithm are chosen based on the values tabulated in Table 1, and the strategy of the optimization is shown in a flowchart as in Figure 5.

Optimization Results for DNA Sensor Model

After the optimization process, the best values obtained for the parameters A, B and C are tabulated in Table 2. These values are the best among the 20 runs of the optimization algorithm, and the lowest fitness value is the indicator of choosing the best parameters. The lowest obtained fitness value is also shown in the table which determines how good the modelled waveform fits the experimental curve.

The train for the optimizing parameters is performed using the experimental curve of the bare DNA sensor. Figure 11 illustrates the experimental waveform and the optimized model for the probe DNA sensor.

To display the concurrence propety of the optimization process that incorporates the PSO technique, the concurrence profile of the best fitness value displayed in Figure 12 that is built upon the fitness value in comparison to iteration. The graph displays the fact that the algorithm could converge to the optimized values after around 60 recurrences which show a valid convergence pace of the suggested method.

Using the values attained for A, B and C factors, several DNA sensors concentrations are replicated and studied in comparison with the related waveforms from the experiments. Meaning the absolute percentage error (MAPE) index is utilized to evaluate the quality of the simulated waveforms (see Equation 7).

$$MAPE = \frac{1}{n}\sum_{k=1}^{n}\left|\frac{\hat{I}(k) - I(k)}{I(k)}\right| \qquad (12)$$

Table 3 displays the permission of the suggested numerical system for six distinguished concentrations of the DNA sensor covering a range from 0.01 to 500 nM, prior to and subsequent to optimization. The lowest accuracy attained is 90.10\% for the concentration of 0.01nM meanwhile the highest accuracy is 99.41\%, a byproduct of to the concentration of 100 nM. Altogether, the precision of more than 98\% displays an comprehensive tolerance of less than 2\% that is quite valids for the optimized numerical model.

Table 1. The parameters chosen for the PSO algorithm

Number of Particles (nop)	Number of Dimension (nod)	C1	C2	Maximum Iteration	Inertia Weight (W)
200	3	2	2	300	0.4<W<1.2

Table 2. The best values of the optimizing parameters over the 20 runs

The Best Fitness Value Obtained	Optimized Value for A	Optimized Value for B	Optimized Value for C
4.484e-07	1.0865e10	9.9662e9	-5.921e3

Figure 11. The experimental and optimized model waveforms for the bare DNA sensor

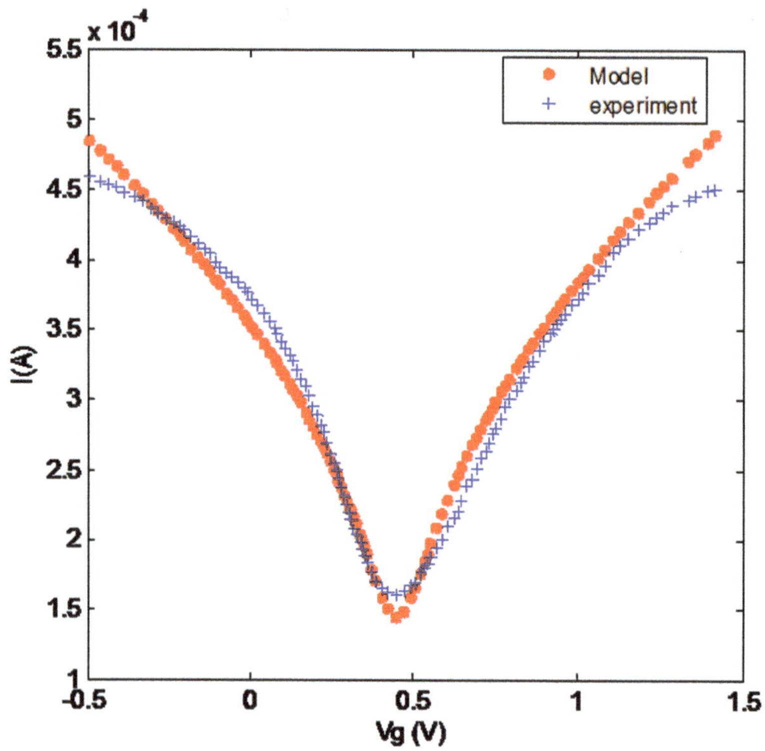

Figure 12. The convergence profile of the optimization process using PSO technique

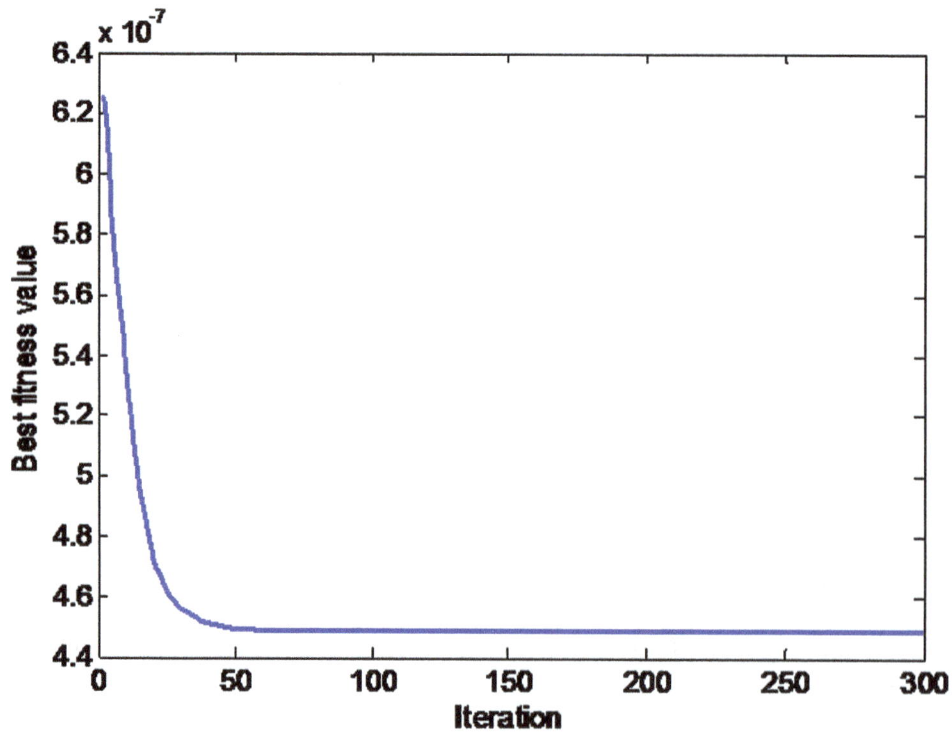

Table 3. The MAPE value in respect to the different concentrations (F)

Concentration F (nM)	MAPE Value (%)	Accuracy Based on MAPE (%)
F=0.01	4.58	95.42
F=0.1	3.52	96.48
F=1	3.33	96.67
F=10	3.39	96.61
F=100	3.52	96.48
F=500	3.94	96.06

Furthermore, the optimized model compared with experimental data in Figure 13 for various concentrations. It needs to be mentioned that the middle part of the graph is very important, since the voltage used in experiment is normally between 0 and 1 volts. The optimized model fits even better around the mentioned voltage range.

CONCLUSION

Due to its unique C–C bonding, 1D structure, large surface–to–volume ratio, high conductivity and biocompatibility, graphene has remarkable potential for bio-sensing material applications. DNA sensors based on graphene have a similar structure to traditionalmetal-oxide-semiconductor FETs (MOSFETs) which its gate is exposed to theelectrolyte solution containing DNA molecules. In these sensors, DNA adsorptionoccurringon a graphene surface may change the I-V characteristicsof the transistors due to the intrinsic negative charge of DNA. In this research, DNA sensing factor which was presented in our previous work, is optimized using particle swarm optimization (PSO) technique. The optimization process is performed based on finding the optimized values for fitting parameters in a current-voltage characteristic of DNA sensor model to predict graphene doping behavior in GFET devices. Based on the results, an accuracy of more than 95% is obtained for the optimized model comparing with the experimental data available. It can be concluded that the optimized graphene based DNA sensor model with the accuracy obtained, is now suitable and reliable for being employed in its different applications such as diagnosis of genetic and pathogenic deseases.

Figure 13. The optimized model and experimental waveforms for different concentrations of the DNA sensor: (a) F=0.01 nM; (b) F=0.1 nM; (c) F=1 nM; (d) F=10 nM; (e) F=100 nM; and (f) F=500 nM

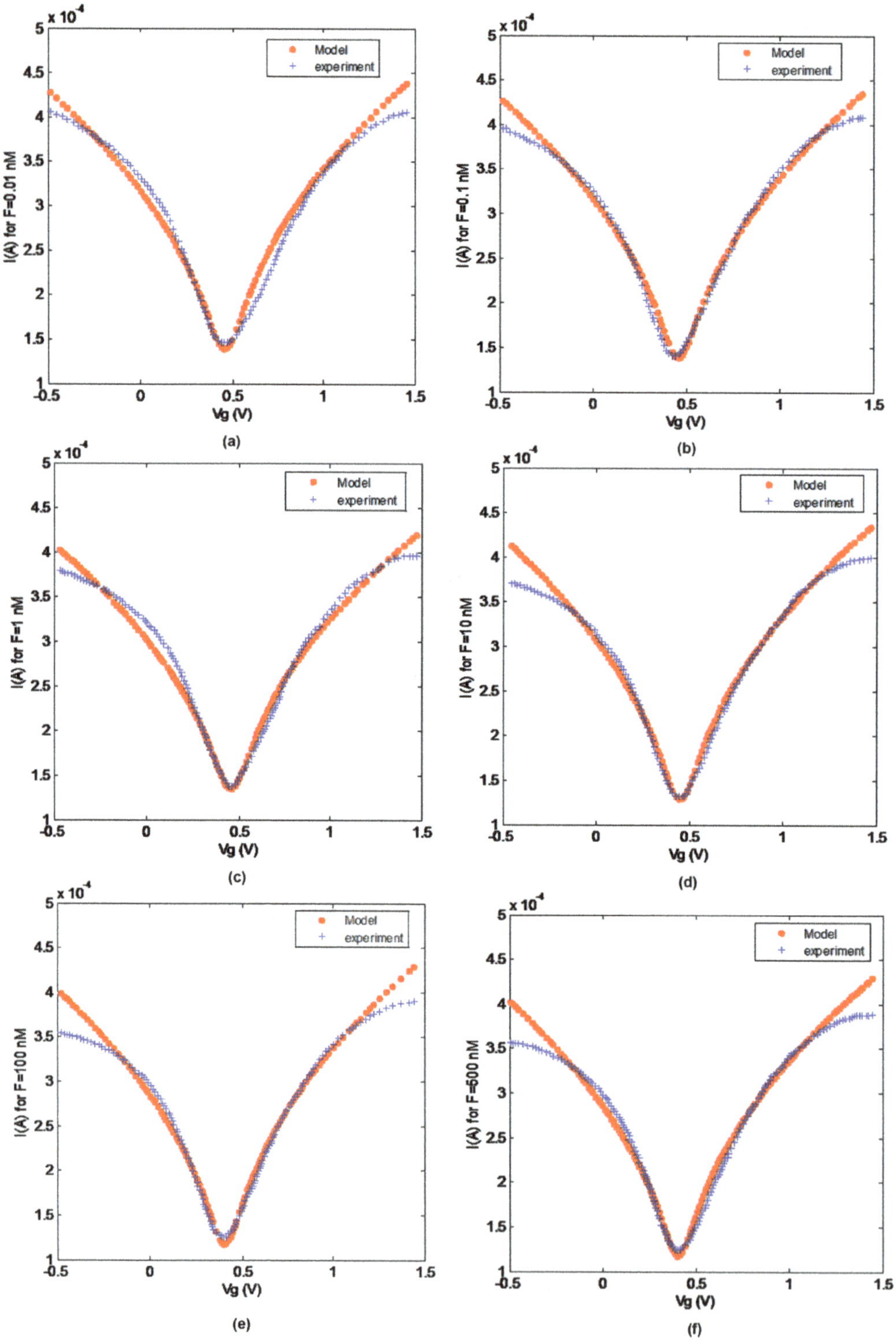

REFERENCES

Abergel, D. S. L., Apalkov, V., Berashevich, J., Ziegler, K., & Chakraborty, T. (2010). Properties of graphene: A theoretical perspective. *Advances in Physics*, *59*(4), 261–482. doi:10.1080/00018732.2010.487978

Ahmadi, M. T., Johari, Z., Amin, N. A., Fallahpour, A. H., & Ismail, R. (2010). Graphene nanoribbon conductance model in parabolic band structure. *Journal of Nanomaterials*, *2010*, 12. doi:10.1155/2010/753738

Boeringer, D. W., & Werner, D. H. (2004). Particle swarm optimization versus genetic algorithms for phased array synthesis. *Antennas and Propagation. IEEE Transactions on*, *52*(3), 771–779. doi:10.1109/TAP.2004.825102

Chen, J. H., Jang, C., Adam, S., Fuhrer, M. S., Williams, E. D., & Ishigami, M. (2008). Charged-impurity scattering in graphene. *Nature Physics*, *4*(5), 377–381. doi:10.1038/nphys935

Clerc, M. (2006). *Particle swarm optimization*. Wiley-ISTE. doi:10.1002/9780470612163

Cohen-Karni, T., Qing, Q., Li, Q., Fang, Y., & Lieber, C. M. (2010). Graphene and nanowire transistors for cellular interfaces and electrical recording. *Nano Letters*, *10*(3), 1098–1102. doi:10.1021/nl1002608 PMID:20136098

Collins, P. G., Bradley, K., Ishigami, M., & Zettl, A. (2000). Extreme oxygen sensitivity of electronic properties of carbon nanotubes. *Science*, *287*(5459), 1801–1804. doi:10.1126/science.287.5459.1801 PMID:10710305

Dankerl, M., Hauf, M. V., Lippert, A., Hess, L. H., Birner, S., Sharp, I. D., & Garrido, J. A. et al. (2010). Graphene solution-gated field-effect transistor array for sensing applications. *Advanced Functional Materials*, *20*(18), 3117–3124. doi:10.1002/adfm.201000724

Das, S., Abraham, A., & Konar, A. (2008). Particle swarm optimization and differential evolution algorithms: Technical analysis, applications and hybridization perspectives. *Advances of Computational Intelligence in Industrial Systems*, *116*, 1–38. doi:10.1007/978-3-540-78297-1_1

Datta, S. (2002). *Electronic transport in mesoscopic systems*. Cambridge, UK: Cambridge University Press.

Datta, S. (2005). *Quantum transport: Atom to transistor*. New York, US: Cambridge University Press. doi:10.1017/CBO9781139164313

del Valle, Y., Venayagamoorthy, G. K., Mohagheghi, S., Hernandez, J. C., & Harley, R. G. (2008). Particle swarm optimization: Basic concepts, variants and applications in power systems. *Evolutionary Computation. IEEE Transactions on*, *12*(2), 171–195.

Di, C.-A., Liu, Y., Yu, G., & Zhu, D. (2009). Interface engineering: An effective approach toward high-performance organic field-effect transistors. *Accounts of Chemical Research*, *42*(10), 1573–1583. doi:10.1021/ar9000873 PMID:19645474

Dong, L. X., & Chen, Q. (2010). Properties, synthesis, and characterization of graphene. *Frontiers of Materials Science in China*, *4*(1), 45–51. doi:10.1007/s11706-010-0014-3

Dong, X., Fu, D., Fang, W., Shi, Y., Chen, P., & Li, L. J. (2009). Doping single-layer graphene with aromatic molecules. *Small*, *5*(12), 1422–1426. doi:10.1002/smll.200801711 PMID:19296561

Dong, X., Shi, Y., Huang, W., Chen, P., & Li, L. J. (2010). Electrical detection of DNA hybridization with single-base specificity using transistors based on cvd-grown graphene sheets. *Advanced Materials*, *22*(14), 1649–1653. doi:10.1002/adma.200903645 PMID:20496398

Dong, X., Shi, Y., Huang, W., Chen, P., & Li, L-J. (2010). Electrical detection of DNA hybridization with single-base specificity using transistors based on cvd-grown graphene sheets. *Advanced Materials*, *22*(14). doi: 10.1002/adma.200903645

Drummond, T. G., Hill, M. G., & Barton, J. K. (2003). Electrochemical DNA sensors. *Nature Biotechnology*, *21*(10), 1192–1199. doi:10.1038/nbt873 PMID:14520405

Eberhart, R., & Shi, Y. (2001). *Particle swarm optimization: Developments, applications and resources*. Paper presented at the Evolutionary Computation. doi:10.1109/CEC.2001.934374

Eberhart, R. C., Shi, Y., & Kennedy, J. (2001). *Swarm intelligence*. Indianapolis, IN: Elsevier.

Fu, D., & Li, L.J. (2010). Label-free electrical detection of DNA hybridization using carbon nanotubes and graphene. *Nano Reviews, 1*.

Gunlycke, D., Areshkin, D. A., & White, C. T. (2007). Semiconducting graphene nanostrips with edge disorder. *Applied Physics Letters*, *90*(14), 142104. doi:10.1063/1.2718515

Huang, Y., Dong, X., Liu, Y., Li, L.-J., & Chen, P. (2011). Graphene-based biosensors for detection of bacteria and their metabolic activities. *Journal of Materials Chemistry*, *21*(33), 12358–12362. doi:10.1039/c1jm11436k

Johnson, A.T., Jr., Gelperin, A., & Staii, C. (2011). *Single walled carbon nanotubes functionally adsorbed to biopolymers for use as chemical sensors: Google Patents*.

Kong, J., Franklin, N. R., Zhou, C., Chapline, M. G., Peng, S., Cho, K., & Dai, H. (2000). Nanotube molecular wires as chemical sensors. *Science*, *287*(5453), 622–625. doi:10.1126/science.287.5453.622 PMID:10649989

Liu, Z., Robinson, J. T., Sun, X., & Dai, H. (2008). PEGylated Nanographene Oxide for Delivery of Water-Insoluble Cancer Drugs. *Journal of the American Chemical Society*, *130*(33), 10876–10877. doi:10.1021/ja803688x PMID:18661992

Manohar, S., Mantz, A. R., Bancroft, K. E., Hui, C. Y., Jagota, A., & Vezenov, D. V. (2008). Peeling single-stranded DNA from graphite surface to determine oligonucleotide binding energy by force spectroscopy. *Nano Letters*, *8*(12), 4365–4372. doi:10.1021/nl8022143 PMID:19368004

Mastrangelo, C. H. (1999). DNA analysis systems on a chip. *Adv. Sci. Technol*, *26*, 465–476.

Mohanty, N., & Berry, V. (2008). Graphene-Based Single-Bacterium Resolution Biodevice and DNA Transistor: Interfacing Graphene Derivatives with Nanoscale and Microscale Biocomponents. *Nano Letters*, *8*(12), 4469–4476. doi:10.1021/nl802412n PMID:19367973

Nilsson, J., Castro Neto, A. H., Guinea, F., & Peres, N. M. R. (2007). Transmission through a biased graphene bilayer barrier. *Physical Review B: Condensed Matter and Materials Physics*, *76*(16), 165416–165410. doi:10.1103/PhysRevB.76.165416

Novoselov, K. S., Geim, A. K., Morozov, S. V., Jiang, D., Katsnelson, M. I., Grigorieva, I. V., & Firsov, A. A. et al. (2005). Two-dimensional gas of massless Dirac fermions in graphene. *Nature*, *438*(7065), 197–200. doi:10.1038/nature04233 PMID:16281030

Novoselov, K. S., Geim, A. K., Morozov, S. V., Jiang, D., Zhang, Y., Dubonos, S. V., & Firsov, A. A. et al. (2004). Electric Field Effect in Atomically Thin Carbon Films. *Science*, *306*(5696), 666–669. doi:10.1126/science.1102896 PMID:15499015

Novoselov, K. S., Jiang, D., Schedin, F., Booth, T. J., Khotkevich, V. V., Morozov, S. V., & Geim, A. K. (2005). Two-dimensional atomic crystals. *Proceedings of the National Academy of Sciences of the United States of America*, *102*(30), 10451–10453. doi:10.1073/pnas.0502848102 PMID:16027370

Passlack, M. (2008). III-V metal-oxide-semiconductor technology. *2008 IEEE 20th International Conference on Indium Phosphide and Related Materials (IPRM).*

Peres, N. M. R., Neto, A. H. C., & Guinea, F. (2006). Conductance quantization in mesoscopic graphene. *Physical Review B: Condensed Matter and Materials Physics*, *73*(19), 195411. doi:10.1103/PhysRevB.73.195411

Poli, R., Kennedy, J., & Blackwell, T. (2007). Particle swarm optimization. *Swarm Intelligence*, *1*(1), 33–57. doi:10.1007/s11721-007-0002-0

Reina, A., Jia, X., Ho, J., Nezich, D., Son, H., Bulovic, V., & Kong, J. et al. (2008). Large area, few-layer graphene films on arbitrary substrates by chemical vapor deposition. *Nano Letters*, *9*(1), 30–35. doi:10.1021/nl801827v PMID:19046078

Reina, A., Thiele, S., Jia, X., Bhaviripudi, S., Dresselhaus, M. S., Schaefer, J. A., & Kong, J. (2009). Growth of large-area single-and bi-layer graphene by controlled carbon precipitation on polycrystalline Ni surfaces. *Nano Research*, *2*(6), 509–516. doi:10.1007/s12274-009-9059-y

Shi, Y., Dong, X., Chen, P., Wang, J., & Li, L. J. (2009). Effective doping of single-layer graphene from underlying SiO_ {2} substrates. *Physical Review B: Condensed Matter and Materials Physics*, *79*(11), 115402. doi:10.1103/PhysRevB.79.115402

Shi, Y. B., Xiang, J. J., Feng, Q. H., Hu, Z. P., Zhang, H. Q., & Guo, J. Y. (2006). Binary channel SAW mustard gas sensor based on PdPc(0.3)PANI(0.7)hybrid sensitive film. In J. Tan (Ed.), *4th International Symposium on Instrumentation Science and Technology* (Vol. 48, pp. 292-297).

Snow, E. S., Perkins, F. K., Houser, E. J., Badescu, S. C., & Reinecke, T. L. (2005). Chemical detection with a single-walled carbon nanotube capacitor. *Science*, *307*(5717), 1942–1945. doi:10.1126/science.1109128 PMID:15790850

Star, A., Tu, E., Niemann, J., Gabriel, J. C. P., Joiner, C. S., & Valcke, C. (2006). Label-free detection of DNA hybridization using carbon nanotube network field-effect transistors. *Proceedings of the National Academy of Sciences of the United States of America*, *103*(4), 921–926. doi:10.1073/pnas.0504146103 PMID:16418278

Tel-Vered, R., Willner, B., & Willner, I. (2010). Biohybrid Electrochemical Devices. *Electrochemistry of Functional Supramolecular Systems, 6*, 333.

Wang, L-J., Cao, G., Tu, T., Li, H-O., Zhou, C., Hao, X-J., . . . Guo, G-P. (2010). A graphene quantum dot with a single electron transistor as an integrated charge sensor. *Applied Physics Letters, 97*(26). doi: 262113 10.1063/1.3533021

Wang, W., & He, S. (2009). Theoretical analysis on response mechanism of polymer-coated chemical sensor based Love wave in viscoelastic media. *Sensors and Actuators. B, Chemical*, *138*(2), 432–440. doi:10.1016/j.snb.2009.03.006

Wang, Y., Lv, J., Zhu, L., & Ma, Y. (2010). Crystal structure prediction via particle-swarm optimization. *Physical Review B: Condensed Matter and Materials Physics*, *82*(9), 094116. doi:10.1103/PhysRevB.82.094116

Wehling, T. O., Novoselov, K. S., Morozov, S. V., Vdovin, E. E., Katsnelson, M. I., Geim, A. K., & Lichtenstein, A. I. (2008). Molecular doping of graphene. *Nano Letters*, *8*(1), 173–177. doi:10.1021/nl072364w PMID:18085811

Wen, Y., Xing, F., He, S., Song, S., Wang, L., Long, Y., & Fan, C. et al. (2010). A graphene-based fluorescent nanoprobe for silver (I) ions detection by using graphene oxide and a silver-specific oligonucleotide. *Chemical Communications*, *46*(15), 2596–2598. doi:10.1039/b924832c PMID:20449319

Zheng, M., Jagota, A., Semke, E. D., Diner, B. A., McLean, R. S., Lustig, S. R., & Tassi, N. G. et al. (2003). DNA-assisted dispersion and separation of carbon nanotubes. *Nature Materials*, *2*(5), 338–342. doi:10.1038/nmat877 PMID:12692536

Chapter 11
Graphene Based-Biosensor:
Graphene Based Electrolyte Gated Graphene Field Effect Transistor

Mohammad Javad Kiani
Islamic Azad University, Iran

M. H. Shahrokh Abadi
Hakim Sabzevari University, Iran

Meisam Rahmani
Universiti Teknologi Malaysia, Malaysia

Mohammad Taghi Ahmadi
Urmia University, Iran

F. K. Che Harun
Universiti Teknologi Malaysia, Malaysia

Karamollah Bagherifard
Islamic Azad University, Iran

ABSTRACT

Because of unique electrical properties of graphene, it has been employed in many applications, such as batteries, energy storage devices and biosensors. In this chapter modelling of bilayer graphene nanoribbon (BGNR) sensor is in our focus. Based on the presented model BGNR quantum capacitance variation effect by the prostate specific antigen (PSA) injected electrons into the FET channel as a sensing mechanism is considered. Also carrier movement in BGNR as another modelling parameter is suggested. PSA adsorption and local pH value of injecting carriers on the surface of player BGNR is modelled. Carrier concentration as a function of control parameters (f, p) is predicted. Furthermore, changes in charged lipid membrane properties can be electrically detected by graphene based electrolyte gated Graphene Field Effect Transistor (GFET). In this chapter, monolayer graphene-based GFET with a focus on conductance variation occurred by membrane electric charges and thickness is studied. Monolayer graphene conductance as an electrical detection platform which is tuned by neutral, negative and positive electric charged membrane together with membrane thickness is suggested. Electric charge and thickness of the lipid bilayer (Q_{LP} and L_{LP}) as a function of carrier density are proposed and the control parameters are defined. Finally, the proposed analytical model is compared with experimental data which indicates good overall agreement.

DOI: 10.4018/978-1-5225-0736-9.ch011

INTRODUCTION

Background of Graphene Based-Biosensor

Graphene is a monolayer of sp^2 bonded carbon atom, this sp^2 bond makes the graphene structure apparently looks like honeycomb crystal as shown in Figure 1. Graphene is called the mother of graphite (many layers of graphene), because it can act as basic building block of these allotropes(Enoki, Kobayashi, & Fukui, 2007). Graphene is theoretically discovered back in the 1940s, however at that time graphene (a 2D layer crystal) are believed thermodynamically too unstable to be produced in real world(Wang & Hu, 2012). After Andre Geim and Konstanstin Novoselov successful production of graphene by scotch tape in 2004 the research attention has been moved forward on graphene strongly(Wang & Hu, 2012).

There are few ways to produce graphene other than mechanical exfoliation, the common ways are Epitaxial grown, Reduced Graphene Oxide and Chemical Vaporization Deposition (CVD) (J.-L. Chen & Yan, 2010). The main reason to focus on graphene production is its promising electrical property and applications (Liu, Tian, Wang, & Sun, 2011). Intrinsic graphene actually has no band gap, which is unfawerable but we can produce a tune-able band gap of graphene. Firstly is to make it bilayer, then induced E-field to the bilayer or by doping the graphene chemically (M. J. A. Kiani, M T. Rahmani, M. Harun, FK Che, 2013; Mak, Lui, Shan, & Heinz, 2009).

Different layers of graphene indicate different electrical properties. In this part two diverse stacking structures (AA and AB) for bilayer graphene nanoribbons (BGNs) with armchair edge are studied (Barbier, Vasilopoulos, Peeters, & Pereira, 2009). The AA-stacked configuration is metallic, but the AB-stacked which is in our focus behaves like a semiconductor material (with a band gap of 0.02 eV). In the AB structure of bilayer graphene with hexagonal carbon lattice the atoms where located on the top layer of BGNs called A1 and B1, whereas atoms on bottom layer of BGN are A2 and B2as shown in Figure 2 (Fiori & Iannaccone, 2009).

By applying an exact electric field on BGNs can open a gap on its band energy(Mousavi, Ahmadi, & Sadeghi, 2011; Sadeghi, Ahmadi, Ishak, Mousavi, & Ismail, 2011). In addition, value of band gap energy by applying voltage can be controlled. Theoretical study on BGNs still is a novel and velocity characteristic based on band structure are unexplored yet.

Figure 1. Monolayer graphene structure with one atom thickness

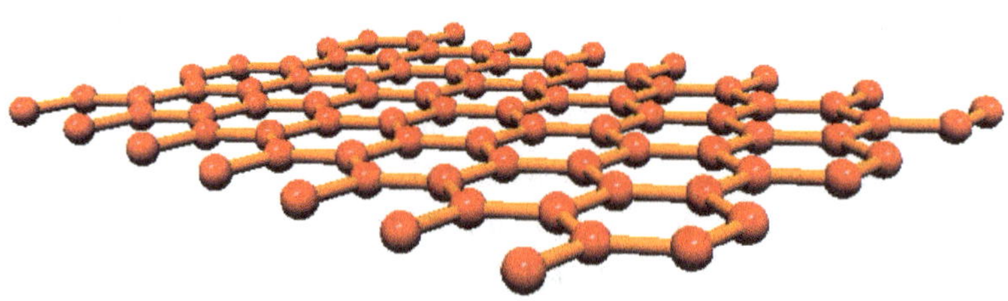

Figure 2. Schematic of AB bilayer graphene lattice configuration

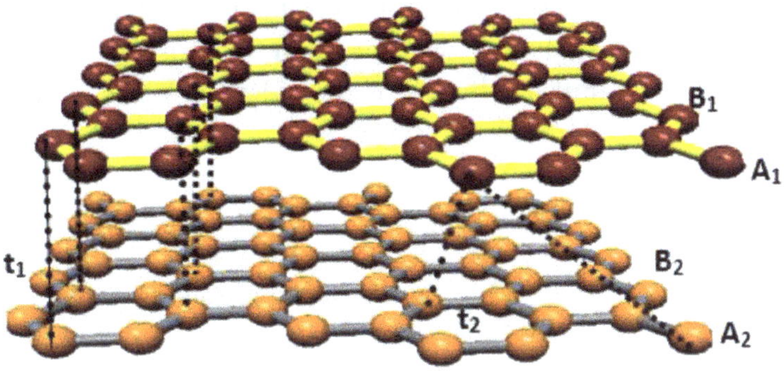

GRAPHENE BASED-BIOSENSOR

PSA Sensor

Graphene has been applied in biological systems, such as detection of DNA and metal ion, protein and pathogen, design of cell/bacterial nanodevice and drug delivery carrier (Bertini, Lee, Luchinat, Piccioli, & Poggi, 2001; Lin, Liu, Tang, & Li, 2011; Ohno, Maehashi, Yamashiro, & Matsumoto, 2009; Qiu, Qu, Dong, Ai, & Han, 2011). Graphene-modified electrode would facilitate electron transfer and provide a non-cytotoxic, larger surface for biomolecules immobilization because of its amazing electron transport property and high surface area (Baby, Aravind, Arockiadoss, Rakhi, & Ramaprabhu, 2010; Liao, Lin, Macosko, & Haynes, 2011). Graphene-based sensors have the potential to detect several types of molecules and ions by logical, physical and chemical modification (Myung, Solanki, & Kim, 2011; Xie, Chen, Du, & Lin, 2011). Cancer marker has been used as a great medical instrument in the diagnosis, monitoring and disease prediction (Li & Yang, 2011).

In order to detect cancer markers, general methods have been employed such as surface plasma resonance, microcantilevers, enzyme linked immunosorbent assay (Salavagione, Díez-Pascual, & Lázaro, 2014). The clinical utility of protein biomarker requires the ability to measure very low concentration proteins to diagnose health and disease(Bin, Sargent, & Kelley, 2010). The early diagnosis of many diseases such as, cancer and HIV is vital. Since many of methods have some limitation such as ultra low detection, large detection ranges, extremely expensive and complex to realize. It is essential to provide effective approaches for novel biosensors with nanomaterials, such as nanoparticle, carbon nanotube and silicon nanowire which have superior performance (Abdelwahab, Koh, Noh, & Shim, 2010; Du, Zou, & Shin, 2010; B. Zhang & Cui, 2011)). For example, through the binding of these biomolecules to the surface of nanostructure and based on electrical altering, carbon based nanomaterials with high sensitivity and selectivity applied to detect small amounts of target biomolecules (Myung & Solanki, 2011; B. Zhang & Cui, 2011).

Therefore, CNTs and graphene based FETs biosensors have attracted much attention for detection of biomarker protein with high sensitivity (Wei, Frank, Chang, & Wong, 2011). Because of the limitations of CNTs, such as variations in electrical properties of CNT-based devices and the limited surface area of CNTs and important characteristics of the graphene use of graphene in FET based biosensors is becoming more and more attractive (Du & Zou, 2010). Nevertheless, there arent any reports on the improvement of graphene FET-based biosensors and their potential as biosensors has not been completely explored(Du & Zou, 2010).Consequently, it is essential to develop nanoscopic graphene based biosensors and important criterion for biosensors because to small in size, simple in device structure, allow label-free detection and real-time monitoring of biomarkers(F. Chen, Qing, Xia, & Tao, 2010; B. Zhang & Cui, 2011).

Structure of PSA sensor

As shown in Figure 3 detection method of biosensor is demonstrated schematically. Based on the assumed detection method bilayer graphene nanoribbons (BGNs) on polyethylene terephthalate (PET) substrate as a sensor has been suggested.

In addition low-cost, flexible and high sensitive cancer marker for real detection of prostate specific antigen(PSA) has been synthesized (B. Zhang & Cui, 2011). Also conductivity variation by gas molecule adsorption on graphene has been reported (H. Mousavi, 2011).

BGNs Velocity Model of Biosensor Channel

The velocity characteristic of a bilayer graphene (BG) in nanoscale regime is applied to study of BGN electrical properties such as conductance, mobility and also to compare graphene based system by another device like MOSFET. In the tight-binding energy method, the biased energy of BGNs has been stated as:

$$E_k(V) = \pm\sqrt{\in_k^2 + \frac{t_\perp^2}{2} + \frac{V^2}{4} + \sqrt{\left(t_\perp^2 + V^2\right)\in_k^2 + \frac{t_\perp^4}{4}}} \tag{1}$$

where

Figure 3. Schematic illustration of bilayer graphene cancer sensor on a flexible PET substrate

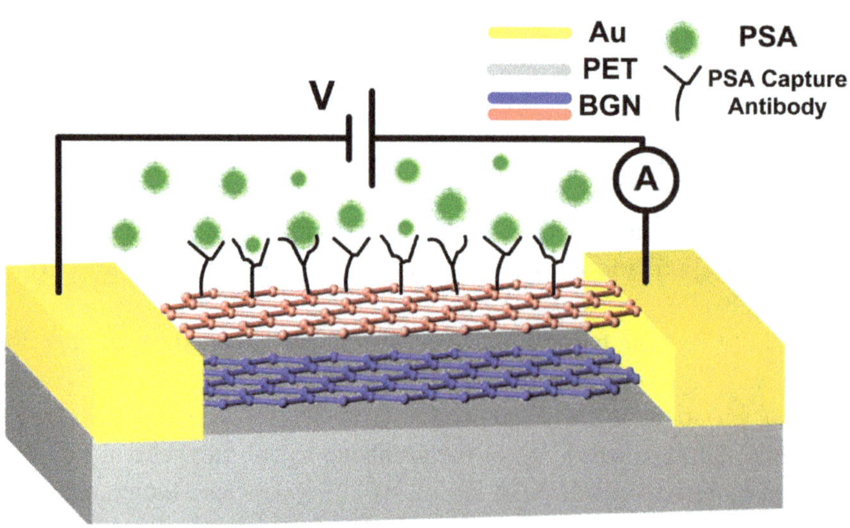

$$\in_k^2 = \left(\frac{V^4}{4} + V^2 \frac{t_\perp^2}{2}\right) \Big/ \left(V^2 + t_\perp^2\right)$$

$$\beta = \upsilon_F^4 \Big/ V_{t_\perp}^2 ,$$

$$\alpha = \left(V \big/ t_\perp^2\right) \upsilon_F^2 ,$$

V is applied voltage,

$\upsilon_F = 3ta_{c-c} \big/ 2$ is Fermi velocity and

$t\perp_1$ is an interlayer hopping (the experimental value for bulk graphite is approximately equal to 0.39 eV).

In-plan hopping t is about equal to 3.1 (eV). Taylor expansion has been used to simplify the electronic dispersion of BG near to the K point in the form of Mexican hat as shown in Figure 4:

$$E(k) \approx (V\ /\ 2) - \alpha k^2 + \beta k^4 \tag{2}$$

Density of state (DOS) can illustrate existing energy states in any materials that can be considered in the monolayer graphene, BG and multilayer graphene naoribbons as a fundamental parameters as well. By the gradient of DOS near the Dirac point, electrical property of materials from metal to semi-conductor will be changed, so, density of state of BGN on the first subband energy can be obtained as .

$$D(E) = \frac{1}{2\pi\left(4\beta k^3 - 2\,\alpha k\right)} \tag{3}$$

Figure 4. The energy band structure of gated bilayer graphene near the Fermi level (Mexican-hat)

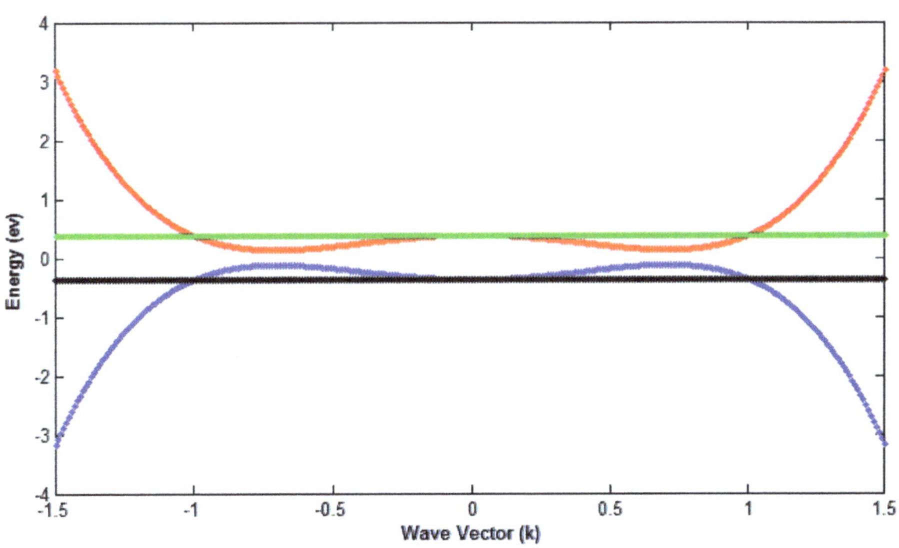

The average velocity of the carriers can be calculated as a accumulative velocity of all carriers divided by number of the carriers, in other words velocity in a band can be obtained by integrating the distribution function over energy band. Based on the presented DOS the velocity is written as:

$$\langle v\rangle = \frac{\int_0^\infty vDos(E)\cdot f(E)dE}{\int_0^\infty Dos(E)\cdot f(E)dE} \tag{4}$$

and

$$f(E) = 1/\left[(1+\exp\left(E-E_f/k_BT\right)\right],$$

where f(E) is the Fermi-Dirac distribution function, E_f is the one-dimensional Fermi energy, T is the temperature and k_B is Boltzmann constant. By respect to the Equation (2) and (4) biased BGN average velocity is modified:

$$\langle v\rangle = \frac{\sqrt{\frac{2}{m^*}}\int_0^\eta \frac{\sqrt{x+V/2}\left[\frac{\alpha\pm\sqrt{\alpha^2+4x\beta}}{2\beta}\right]^{-\frac{1}{2}}}{\left[\sqrt{\alpha^2/(2\beta)+2xk_BT}\right]\left[1+\exp\left(x-\eta\right)\right]}dx}{\int_0^\eta \frac{\left[\frac{\alpha\pm\sqrt{\alpha^2+4x\beta}}{2\beta}\right]^{-\frac{1}{2}}}{\left[\sqrt{\alpha^2/(2\beta)+2xk_BT}\right]\left[1+\exp\left(x-\eta\right)\right]}dx} \tag{5}$$

where

m* is effective mass of carriers in BGN in the form of $m^* = \frac{\hbar^2}{48\beta n^2\pi^2 - 2\alpha l^2}$,
$\eta = (Ef - V/2)/kBT$ is the normalize Fermi energy, and
$x = (E - V/2)/kB$.

As shown in Figure 5 based on this model bilayer graphene nanoribbon carrier velocity is a function of normalized Fermi energy. Also normalized Fermi energy in general is a function of temperature in other words carrier velocity as a function of temperature can be defined. In general degeneracy effect on the velocity will affect carrier movement in the form of Fermi velocity or thermal velocity based on the degeneracy limit.

Figure 5. The velocity for the first subband of BG

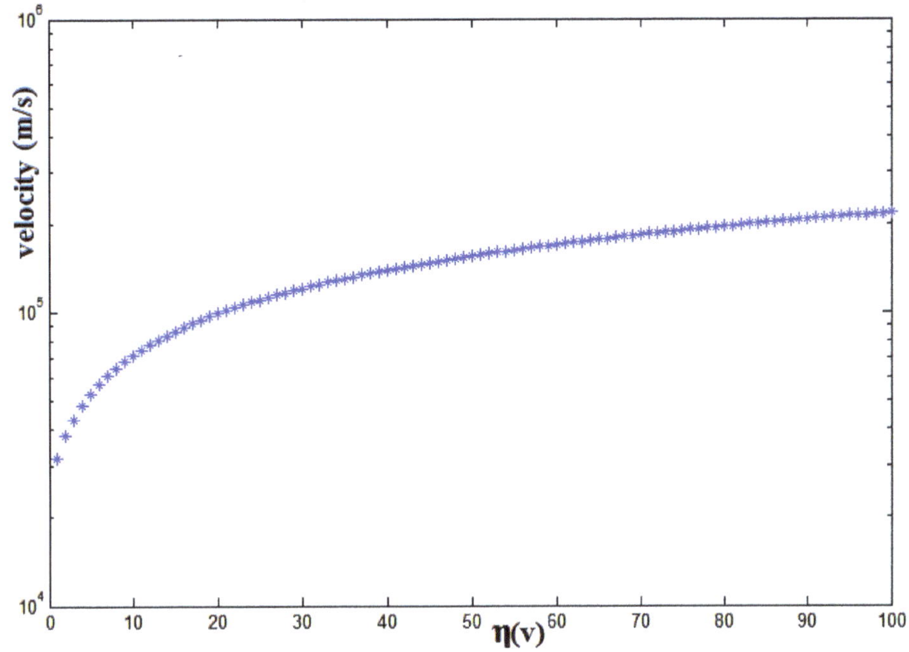

Non-degenerate approximation have been illustrated with the distance more than $3K_BT$ from either valence or conduction band edge in the form of band gap near the Fermi level(Ahmadi, Ismail, Tan, & Arora, 2008). In semiconductors, non-degenerate regime is situated within a band, less than $3K_BT$ far away from the conduction and valence band which normalized Fermi energy less than -3 indicate this approximation. Therefore, in the non-degenerate regime, $E - E_F > 3k_{BT}$ then we can write, $f(E) = \exp\left(\frac{E_F - E}{k_{BT}}\right)$. Thus, Fermi integral in non-degenerate regime can be modified in the form of exponential function therefore carrier velocity is written as:

$$\langle v \rangle_{nd} = \frac{\sqrt{\frac{2}{m^*}} \int_0^\eta \frac{\sqrt{x+V/2}\left[\frac{\alpha \pm \sqrt{\alpha^2+4x\beta}}{2\beta}\right]^{-\frac{1}{2}}}{\left[\sqrt{\alpha^2/(2\beta)+2xk_BT}\right]\exp\left(x-\eta\right)}dx}{\int_0^\eta \frac{\left[\frac{\alpha \pm \sqrt{\alpha^2+4x\beta}}{2\beta}\right]^{-\frac{1}{2}}}{\left[\sqrt{\alpha^2/(2\beta)+2xk_BT}\right]\exp\left(x-\eta\right)}dx} \qquad (6)$$

Based on the suggested model carrier velocity in bilayer graphene as a function of thermal energy is reported similar to the conventional materials as well. In other words temperature can affect the velocity of the carriers in this regime . As shown in Figure 6 normalized Fermi energy less than zero shows non-degenerate approximation.

On the other hand in the degenerate BG, $E - E_F < 3k_{BT}$, probability function equal to one have been reported $F(E)$=1 (Arora, Tan, Saad, & Ismail, 2007). Therefore because of the very small magnitude of x-η in this regime exp (x-η) in the comparison of 1 can be neglected which means carrier velocity can be modified as:

$$\langle v \rangle_d = \frac{\sqrt{\frac{2}{m^*}} \int_0^{\eta} \frac{\sqrt{x + V/2} \left[\frac{\alpha \pm \sqrt{\alpha^2 + 4x\beta}}{2\beta} \right]^{-\frac{1}{2}}}{\left[\sqrt{\alpha^2 / (2\beta) + 2xk_B T} \right]} dx}{\int_0^{\eta} \frac{\left[\frac{\alpha \pm \sqrt{\alpha^2 + 4x\beta}}{2\beta} \right]^{-\frac{1}{2}}}{\left[\sqrt{\alpha^2 / (2\beta) + 2xk_B T} \right]} dx} \tag{7}$$

Ultimately, carrier velocity in both degenerate and degenerate regimes is shown in Figure 7.

Figure 6. The comparison of the carrier velocity(v) and non-degenerate, (v_{nd})

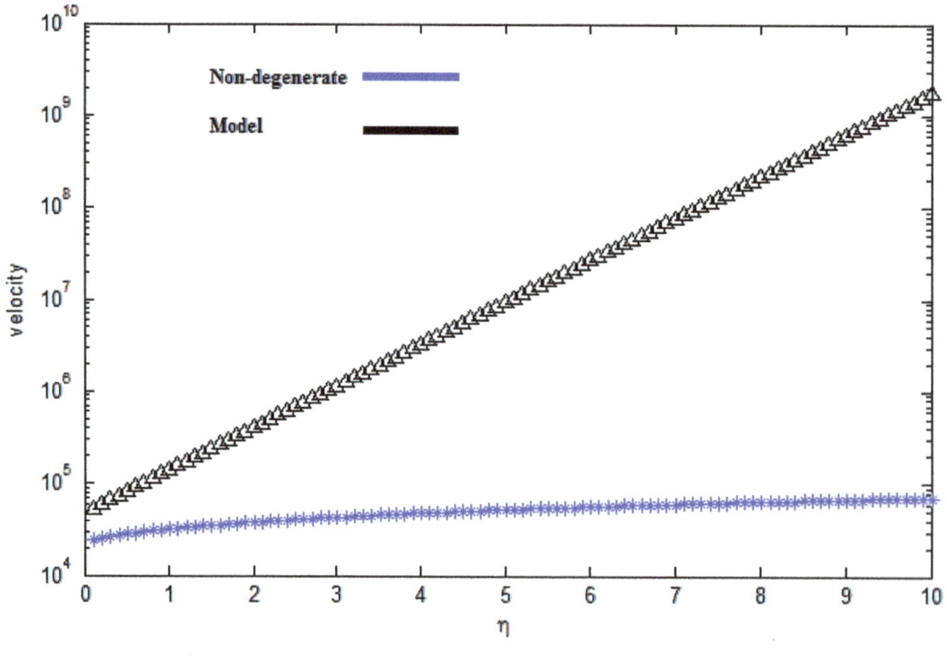

Figure 7. The comparison of the carrier velocity(v), degenerate(v_d) and non-degenerate, (v_{nd})

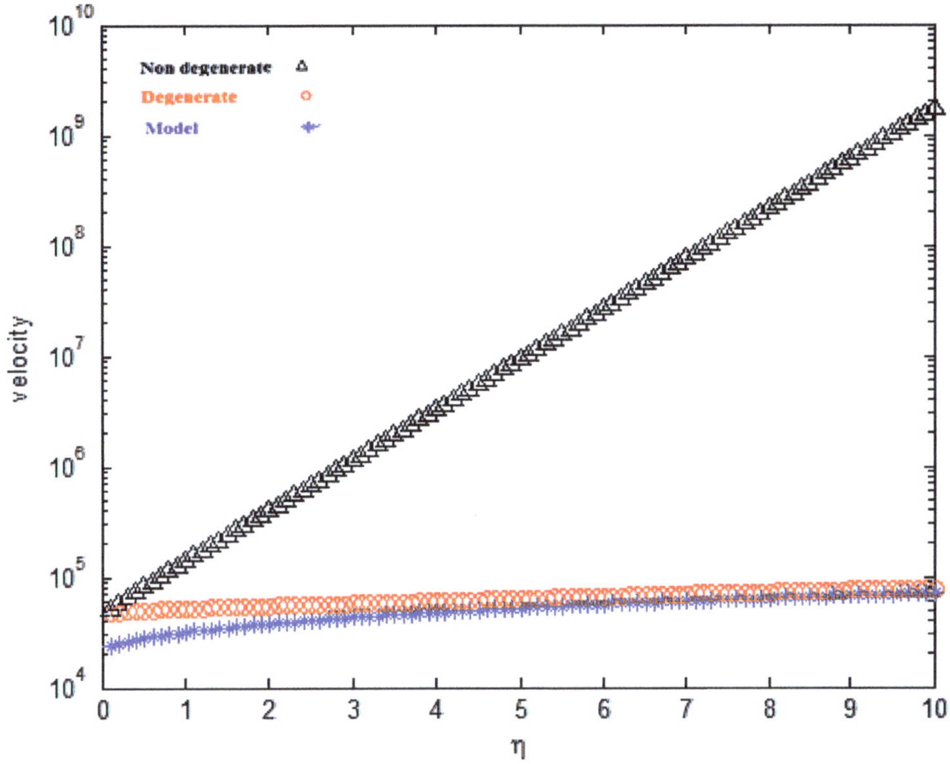

Consequently, that curve demonstrates which carrier velocity in the range of small amount of normalized Fermi energy is behaves like thermal velocity which is equally non-degenerate carrier velocity. On the other hand normalized Fermi energy with amount more than 3 indicates Fermi velocity that illustrates degenerate carrier velocity as well.

BGNs Velocity Model of Biosensor Channel

In order to estimate the detectable prostate specific antigen (PSA) on graphene based biosensor the quantum capacitance characteristic of a bilayer graphene (BG) in nanoscale is applied. So, Carrier concentration in a subband can be obtained by integrating the distribution function over energy band(S. M. A. Mousavi, M. T. Webb, J. F. Sadeghi, H. Nilghaz, A. Amin, A. Johari, Z. Ismail, R., 2011).

$$n = \int_0^\infty \left(2\pi \left(4\beta \left[\frac{\alpha \pm \sqrt{\alpha^2 - 4\left(E - V/2\right)\beta}}{2\beta} \right]^{\frac{3}{2}} - 2\alpha \left[\frac{\alpha \pm \sqrt{\alpha^2 - 4\left(E - V/2\right)\beta}}{2\beta} \right]^{\frac{1}{2}} \times \left(1 + \eta\right) \right] \right)^{-1} dE \tag{8}$$

where $\eta = e^{\frac{E-E_F}{K_B T}}$ is normalized Fermi energy. In conventional metal oxide field effect transistor (MOS-FETs), the electrostatic (classical) capacitance is dominant compare to the quantum capacitance. But as the devices model is scaled down into nano-size technology, the quantum capacitance is comparable with the classical capacitance. Nowadays, the application of the quantum capacitance has been taken considered for nanoscale devices especially in the carbon-based material. In order to model bilayer graphene quantum capacitance we can write:

$$c_q = e^2 \frac{\partial n}{\partial E} \tag{9}$$

Consequently the quantum capacitance model by carrier concentration modelling, will be modified as

$$C_q = e^2 \left(2\pi \left(4\beta \left[\frac{\alpha \pm \sqrt{\alpha^2 - 4\left(E - V/2\right)\beta}}{2\beta} \right]^{\frac{3}{2}} - 2\alpha \left[\frac{\alpha \pm \sqrt{\alpha^2 - 4\left(E - V/2\right)\beta}}{2\beta} \right]^{\frac{1}{2}} \right) \times \left(1 + \eta\right) \right)^{-1} \tag{10}$$

As shown in Figure 8 derived from this modelling of the bilayer graphene nanoribbon, quantum capacitance is a function of normalized Fermi energy while normalized Fermi energy is a function of temperature.

Figure 8. The carrier velocity for the first subband of bilayer graphene

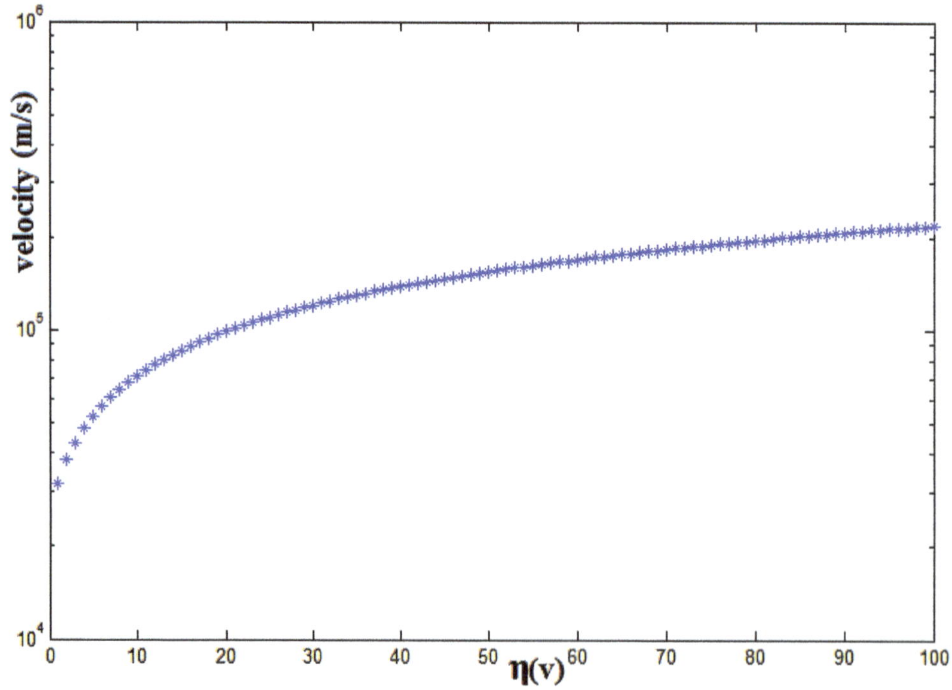

Non-degenerate approximation have been illustrated with the distance more than $3K_BT$ from either the conduction or valance band edge in the form of band gap near the Fermi level. In semiconductors, non-degenerate regime is situated within a band less than $3K_BT$ far away from the conductance and the valence band.normalized Fermi energy is less than -3indicates of non-degenerate approximation. So, Fermi integral in non-degenerate regime is obtained by exponential function as:

$$C_{qnd} = e^2 \left(2\pi \left(4\beta \left[\frac{\alpha \pm \sqrt{\alpha^2 - 4\left(E - V/2\right)\beta}}{2\beta} \right]^{\frac{3}{2}} - 2\alpha \left[\frac{\alpha \pm \sqrt{\alpha^2 - 4\left(E - V/2\right)\beta}}{2\beta} \right]^{\frac{1}{2}} \right) \times \eta \right)^{-1} \tag{11}$$

In the nano-scale devices degenerate regime will play a main role in the quantum capacitance study. For the degenerate regime of BG, $E - E_F < 3k_{BT}$, therefore, η=1.So we can write

$$C_d = e^2 \left(2\pi \left(4\beta \left[\frac{\alpha \pm \sqrt{\alpha^2 - 4\left(E - V/2\right)\beta}}{2\beta} \right]^{\frac{3}{2}} - 4\alpha \left[\frac{\alpha \pm \sqrt{\alpha^2 - 4\left(E - V/2\right)\beta}}{2\beta} \right]^{\frac{1}{2}} \right) \right)^{-1} \tag{12}$$

In other words, in this regime η, can be neglected because of the very small magnitude of η.

As shown in Figure 9 it is discovered that temperature can affect quantum capacitance in non-degenerate regime for the reason that non-degenerate quantum capacitance is a function of thermal energy(η) while the increases quantum capacitance rate independent of temperature in degenerate approximation .

Ultimately, the comparison of quantum capacitance, degenerate approximation quantum capacitance and non-degenerate regime quantum capacitance is shown in Figure 10. Subsequently, that curve illustrates which carrier concentration for less than zero amount of changing energy than the Fermi energy divided thermal energy(η) is equally non-degenerate concentration and temperature is dependent variable while for an amount more than zero of changing energy than the Fermi energy divided thermal energy(η) is equal degenerate quantum capacitance and quantum capacitance is independent of the temperature.

Proposed Model for PSA Effect on BGNs Based Biosensor

In order to model the effect of PSA adsorption by graphene we have monitored capacitance variation as

$$C_q = e^2 \frac{\partial n}{\partial E} + e^2 \frac{\partial n'}{\partial E} \tag{13}$$

It has been demonstrated that the conductance of biosensor depends on the concentration of PSA and different local pH values. Carrier concentration as a function of PSA concentration and different local pH values is considered, which are presented in the form of PSA and PH factors (f,p). The carriers injected from the PSA and localized pH value into the bilayer graphene is added to the conventional carrier concentration as:

Figure 9. The comparison of the quantum capacitance, degenerate(c_{qd}) and non-degenerate, (c_{qnd})

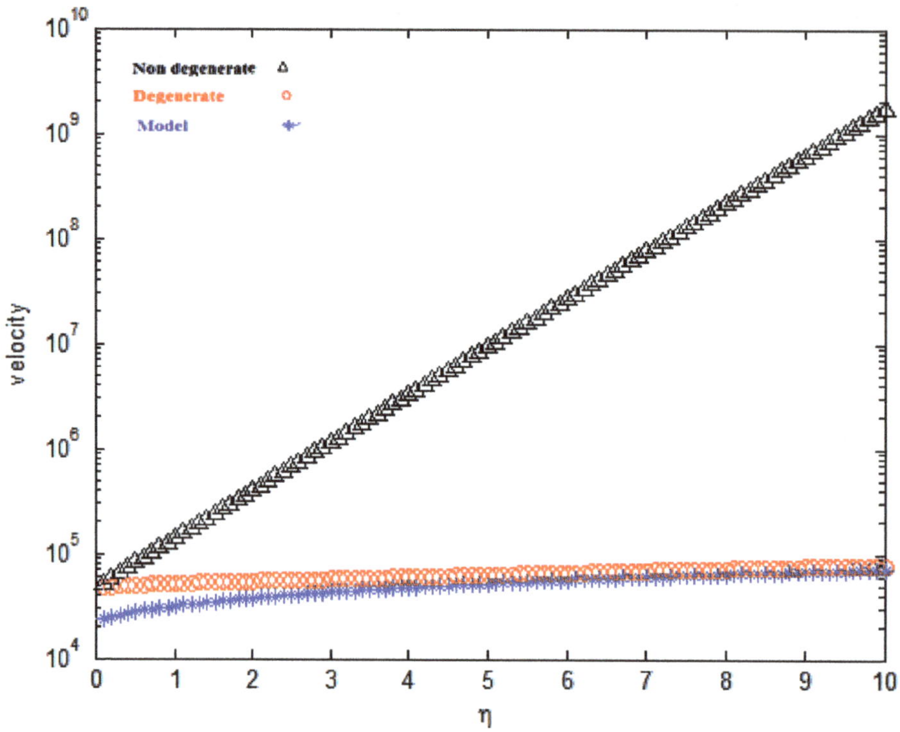

Figure 10. Quantum capacitance in general, non-degenerate and degenerate regime as a function of Fermi energy

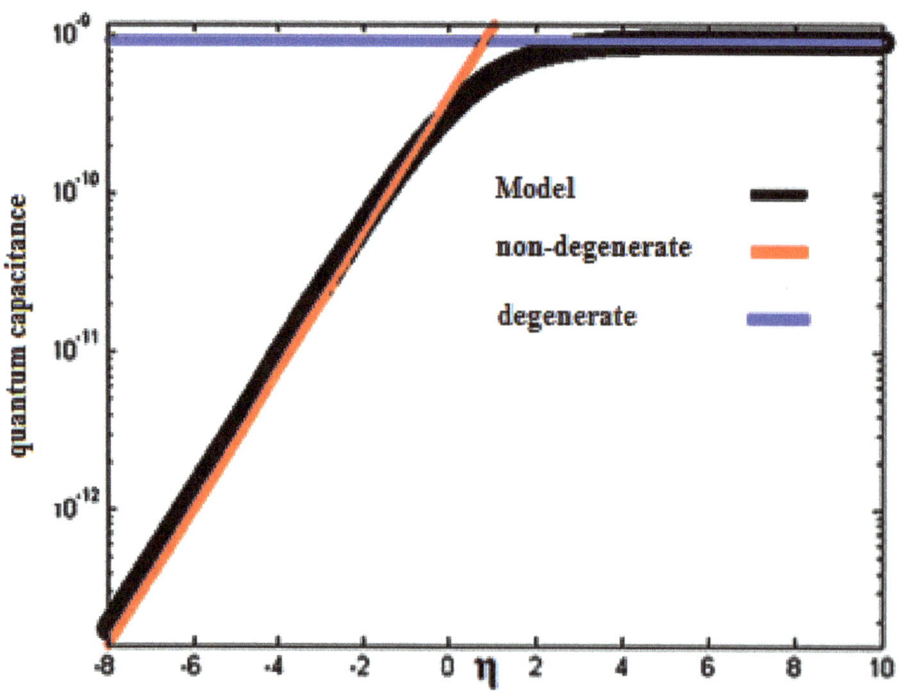

$$n' = f_{PSA} + P_{pH} \tag{14}$$

So the capacitance based on the proposed model of BG carriers and adsorbing carriers (PSA) by the surface of the biosensor is modified as:

$$C_q = e^2\left(2\pi\left(4\beta\left[\frac{\alpha \pm \sqrt{\alpha^2 - 4\left(E - V/2\right)\beta}}{2\beta}\right]^{\frac{3}{2}} - 2\alpha\left[\frac{\alpha \pm \sqrt{\alpha^2 - 4\left(E - V/2\right)\beta}}{2\beta}\right]^{\frac{1}{2}}\right) \times \left(1+\eta\right)\right)^{-1} + e^2 f\frac{\partial F}{\partial t}\frac{\partial t}{\partial E} + e^2 p\frac{\partial P_H}{\partial t}\frac{\partial t}{\partial E} \tag{15}$$

where ∂F is the PSA concentration variation and ∂P_H is different local pH values, therefore the current voltage characteristic can be modified by proposing capacitance model as.

$$I = \left(e^2\left(2\pi\left(4\beta\left[\frac{\alpha \pm \sqrt{\alpha^2 - 4\left(E - V/2\right)\beta}}{2\beta}\right]^{\frac{3}{2}} - 2\alpha\left[\frac{\alpha \pm \sqrt{\alpha^2 - 4\left(E - V/2\right)\beta}}{2\beta}\right]^{\frac{1}{2}}\right) \times \left(1+\eta\right)\right)^{-1} + e^2 f\frac{\partial F}{\partial t}\frac{\partial t}{\partial E} + e^2 f P_H\right) \times V_{ds}\mu_e E \tag{16}$$

This Equation (16) can be numerically solved for different potential. Thus the proposed quantum capacitance model of the performance of graphene-based PSA sensor in the nanostructured region by the I-V characteristic is evaluated in Figure 11(a).

As shown in Figure 11(b) current - voltage characteristic of graphene based PSA biosensor by the quantum capacitance effect is plotted. Based on the experimental data it is explored that Sensor I-V characteristic can be controlled by PSA concentration (as an antibodies parameter), therefore the concentration coefficient by iteration method is estimated as:

$$f = f_1 Ln(F) + f_2 \tag{17}$$

where f_1=1.84×10^{-2} and f_2=0.3991.

As shown in Figure 12, after incubating bio sensor in different concentrations of PSA and labeled by HRP the normalized conductance of grouping not only raises with the increasing of PSA concentration but also conductance of the bilayer graphene changes with the different local pH values. Thus the pH effect on the carrier concentration in the same manner with PSA concentration by iteration method is modeled as:

Figure 11. (a) The curve of I-V model of graphene-based PSA sensor; (b) comparison of current than voltage based on quantum capacitance modeling and experimental data for PH=5 that shows a promising agreement between theoretical models and experiments
B. C. Zhang, Tianhong, 2011.

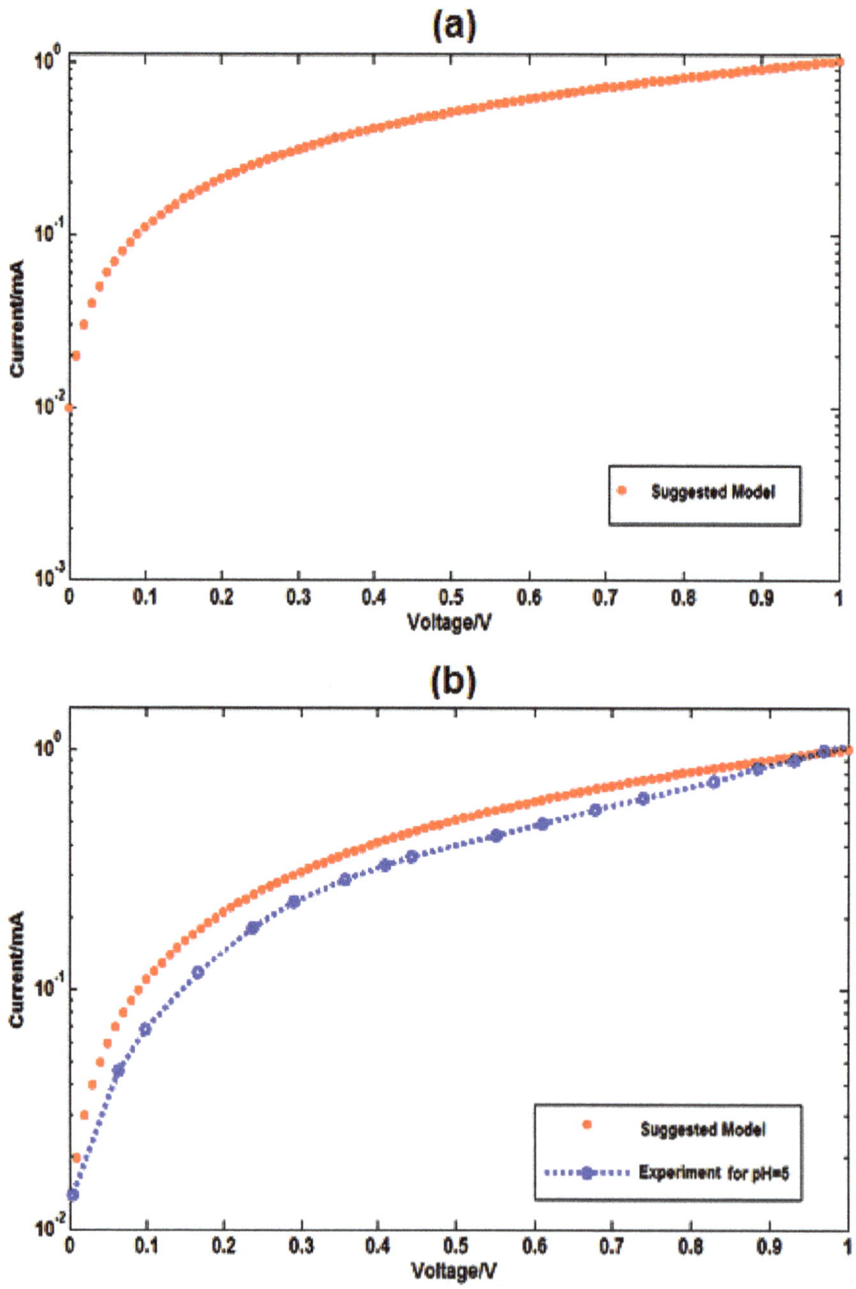

$$p = p_1 P_H^{\ 2} - p_2 P_H + p_3 \tag{18}$$

where p_1=170.31, p_2=15.063 and p_3=5.3251. Also the suggested PSA detection modelling can be employed for almost any target antigen with a known antibody.

$$I = \begin{pmatrix} e^2 \left(2\pi \left(4\beta \left[\frac{\alpha \pm \sqrt{\alpha^2 - 4\left(E - V/2\right)\beta}}{2\beta} \right]^{\frac{3}{2}} - 2\alpha \left[\frac{\alpha \pm \sqrt{\alpha^2 - 4\left(E - V/2\right)\beta}}{2\beta} \right]^{\frac{1}{2}} \right) \times (1+\eta) \right) \\ + e^2 \left(1.84 \times 10^{-2} Ln(F) + 0.3991\right) \frac{\partial F}{\partial t} \frac{\partial t}{\partial E} + e^2 \left(170.31 P_H^{\ 2} - 15.063 P_H + 5.3251\right) P_H \end{pmatrix} \times V_{ds} \mu_e E \tag{19}$$

As shown in Figure 13 based on the presented model it is notable that by raising the concentration (as an example for pH value is P=5) model is closer to the experimental data (black line) in the same manner we can compare other experimental datas as well.

Current-voltage characteristic altering by carrier concentration and local pH values, also based on the supposed model sensor I-V characteristic is controlled by PSA concentration parameter and PH parameter.

Figure 12. Comparison of current than voltage based on quantum capacitance modeling and experimental data for PH=6 that shows a promising agreement between theoretical model and experimental

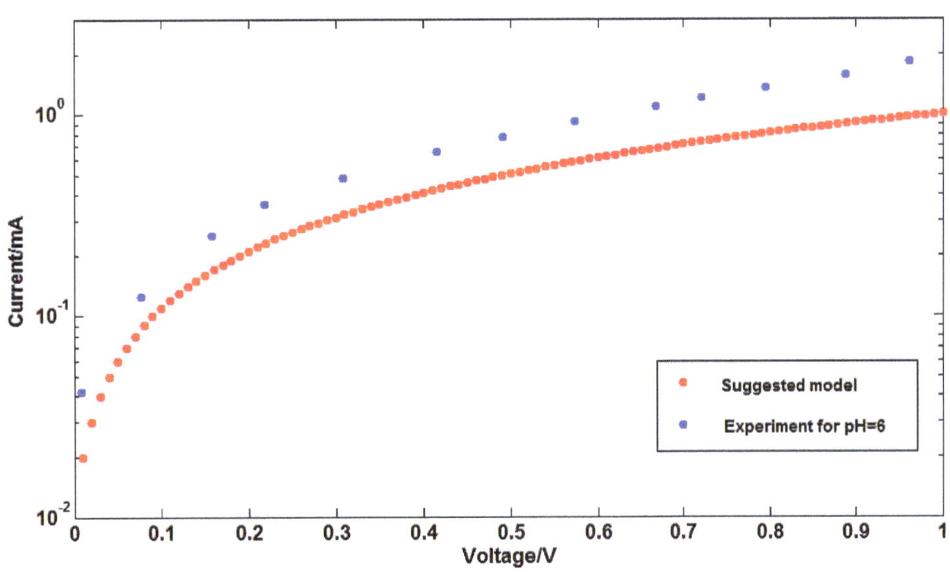

Figure 13. Comparison of current than voltage based on quantum capacitance modeling and experimental data for different PH

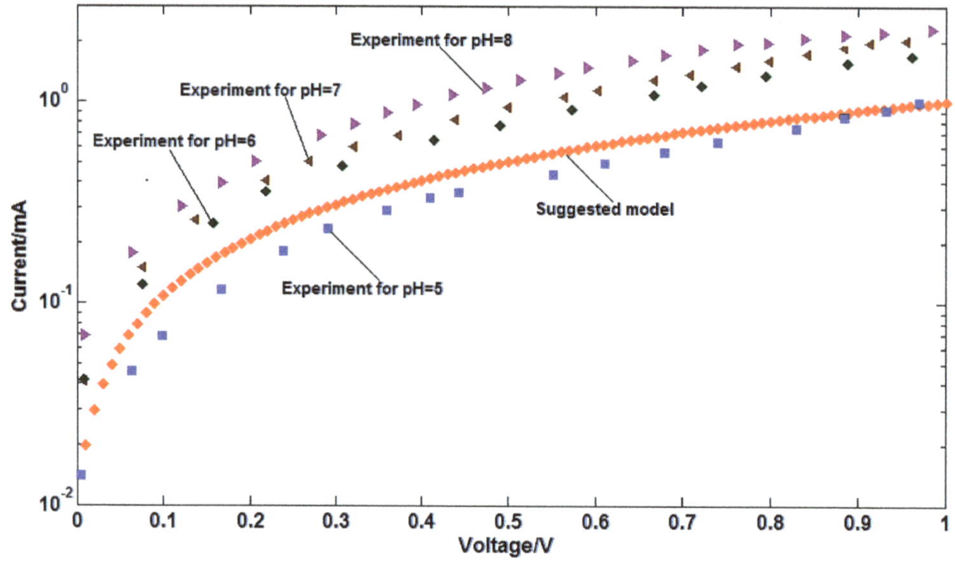

LIPID BILAYER USING ELECTROLYTE GATED GRAPHENE-FET

Detection Mechanism of Charged Lipid Bilayer Using Electrolyte Gated Graphene-FET

Graphene has attractive electrical properties such as ability to continuously tune the charge carriers from holes to electron, high mobility and high carrier velocity (M. J. Kiani, Ahmadi, Akbari, Karimi, & Che Harun, 2013; M. J. Kiani, Ahmadi, & Harun, 2013). Charge carrier (electron) in graphene can be explained by electron propagation through honeycomb lattice of graphene that after that electrons lose their effective mass, which yield a quasi-particle called "Dirac-fermions" (Y. B. Zhang, Brar, Girit, Zettl, & Crommie, 2009). This Dirac fermions particle is hard to imagine because it has no known analogies. They can illustrate by a combination of both Dirac and Schrödinger equations. In addition graphene stipulate current effectively, precision and faster than any other metal on biosensors same as biomimetic membrane-coated graphene biosensor. Several types of animal cells and plant are surrounded with two-layer covering, which called phospholipid bilayer .

As shown in Figure 14 the molecules that make up the phospholipid bilayer called phospholipids, which organize themselves in two corresponding layers, shaping a covering that can only be infiltrated by definite kinds of substances . This gives the cell an apparent frontier, and keeps useless materials out . Although the phospholipid bilayer frequency works well, it can be injured, and some kinds of superfluous materials can evade it. Two ends in phospholipids exist, the first one is hydrophilic, that attract the water, and the second is hydrophobic, that resist the water . While the inside of cells is typically water, and the region outside of cells is generally water, these molecules organize themselves into two sheets, with the hydrophilic ends of each layer pointing externals, and the hydrophobic parts pointing inwards. While they are fats or, lipids they are not crashed by water, and are firm enough not to allow large molecules go through without the assist of another material.

Some smaller molecules, such as carbon dioxide and oxygen, can go through without difficulty on their own, but the larger molecules such as water, sodium, or magnesium, cannot pass easily (Leonenko, Cramb, Amrein, & Finot, 2008). In the interior of the membrane is also liquid, and lets proteins, cholesterol, sphingolipids, or, and sterols to converge it. The role of sphingolipids is protecting the outside of the cell, and the role of sterols and cholesterols is stabilizing the phospholipid bilayer in plant and animal cells, respectively.

Although this is critical for cells to have enough constancy, a large amount of the cholesterol can make them inflexible, which is hazardous especially if they are part of a vein that should be flexible to allow blood flow . The proteins are used to transfer materials in or out of the cell throughout the bilayer and supply places for definite materials to attach to the exterior of the cell . Supplying the structure to a cell is one of the major aims of the phospholipid bilayer, which it performs because of the natural arrangement of the hydrophilic and hydrophobic ends of the phospholipids, and with becoming stable sterols and cholesterol. Its other aim is to control the kinds of materials that can go into the cell or attach with it, which it does in a number of ways using proteins (Wang & Hu, 2012). The kinds of protein that expand from the top of the membrane, can be used to recognize the cell or to make a place for specific materials to attach to it.

Also some types of proteins can shape tunnels or channels or for definite substances to go through. Some channels are always open for certain types of molecules, while others need energy to open and close like gates (Weiss, Sakai, Ghebremariam, Ni, & Matile, 1997). This kind of transportation is active transport, and can be worked in both way, to bring substances in and out of the cell. Its generally used with materials like calcium, potassium, and sodium (Shrivastava & Sansom, 2000). Charged lipid bilayer adsorbing on the surface can adapt the electronic properties of graphene adapt. An electrolyte-gated biomimetic membrane-graphene transistor can be used to monitore electerically the bio-recognition events which guide to changes in membrane uprightness . Graphene can sense electrically the bactericidal motion of antimicrobial peptides based on a multipart interaction of ionic screening effect and biomolecular doping. Graphene based FET structure can be used in the sensing biological events in the precens of electrical parameters variation. The observed transfers of the Dirac point along with the indication of

Figure 14. Structure of phospholipid bilayer

lipid charges is a significance of charge-impurity potential made by the lipid membranes and clearly express that the exciting lipid membranes adapt the electronic properties of graphene considerably. Assuming an equivalent division of exciting lipids in the two leaflets since graphene is an electrically neutral substrate, the concentration of charged pollutions in the lipid membranes can be approximated from the surface area connected to a lipid head group. By means of charged lipid bilayer adsorbing on the membrane surface, the conductance of graphene can be adapted and replicated. Biorecognition actions which cause modifies in membrane integrity can be considered electrically, by an electrolyte-gated biomimetic membrane-graphene biosensor (GFET). In the current chapter the effect of charged lipid membranes on the conductance of graphene-based GFET is estimated regarding to significant shift on the Dirac point in G-V_g characteristic of graphene-based biosensor.

Structure of Charged Lipid Bilayer Using Electrolyte Gated Graphene-FET

In this model, the thickness and the type of coated charge as a function of gate voltage is simulated and control parameters are suggested. Subsequently, to get a greater insight into the role of both thickness and the type of lipid bilayer, GFET modelling is employed to obtain relevance between conductance versus the voltage of liquid gate, where two electrodes of sensor as shown in Figure 15 are considered as the source and drain contacts.

Subsequently, to get a greater insight into the role of both thickness and the type of lipid bilayer, GFET modelling is employed to obtain relevance between conductance versus the voltage of liquid gate, where two electrodes of sensor as shown in Figure 15 are considered as the source and drain contacts.

Proposed Model of Charged Lipid Bilayer Using Electrolyte Gated Graphene-FET

Monolayer graphene in electrolyte-gated biomimetic membrane garaphene transistor with ballistic channel is supposed to monitor the changes in membrane integrity. High carrier mobility reported by experiments in the graphene owing to assume totally ballistic carrier transportation in the graphene, which leads to higher transmission probability. Higher transmission probability causes higher carrier conductance from source to drain that can be provided by the Boltzmann transport equation as.

Figure 15. Schematics of the proposed structure and the electrical circuit of the electrolyte gated graphene-FET for charged lipid bilayer detection

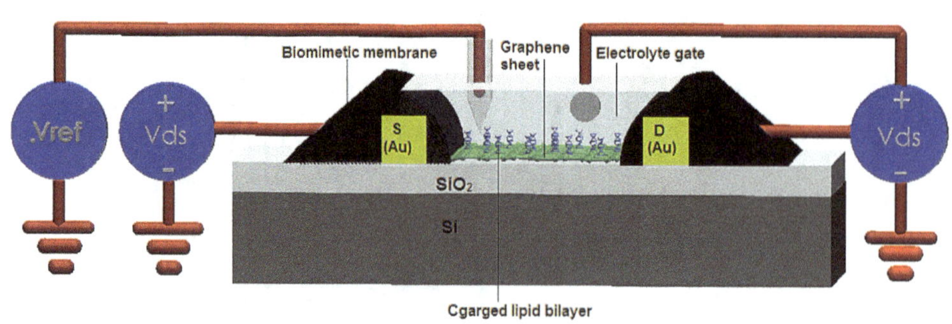

$$G = \frac{2q^2}{h} \int_{-\infty}^{+\infty} dEM(E)T(E)\left(-\frac{df}{dE}\right) \tag{20}$$

where

q is the electron charge,
h is Plancks constant,
E is the energy band structure,
$M(E)$ is the number of modes,
f is the Fermi-Dirac distribution function, and
$T(E)$ is the transmission probability.

On the other hand, because of ballistic transport T as a possibility of one inserted electron at one end that can be convey to other end is considered equivalent to one (T(E) =1). Number of modes in corporation with Landauer Formula with respect to conductance of monolayer graphene can be written as.

$$M(E) = \frac{\alpha - 3\beta k^2}{1} \tag{21}$$

where

l is the length of graphene channel,
k is the wave vector, and
$\alpha = \frac{\Delta v_F}{\sqrt{2t}}, \quad \beta = \frac{v_{F^3}}{\Delta\sqrt{2t}}$.

It can be affirmed that the length of channel has a strong influence on conductivity function. Taking into consideration the effect of temperature on graphene conductance, the boundary of integral is changed. This equation can numerically be solved by employing the partial integration method.

$$G = \frac{3q^2\left(3\pi a^3 t^3 k_B T\right)^{\frac{1}{2}}}{hl} \times \left[\int_0^{+\infty} \frac{X^{-1/2}}{\left(1/\left(1+e^{X-\eta}\right)\right)} dX + \int_0^{+\infty} \frac{X^{-1/2}}{\left(1/\left(1+e^{X+\eta}\right)\right)} dX\right] \tag{22}$$

where $x = \left(E - E_g\right)/k_B T$ and the normalized Fermi energy is $\eta = \left(E_F - E_g\right)/k_B T$. Thus the general conductance model of single layer graphene can be gained similar to that of silicon reported by Gunlycke. According to the conductance-gate voltage characteristic of graphene based electrolyte gated Graphene Field Effect Transistor (GFET) devices, the performance of biomimetic membrane-coated graphene biosensor can be estimated through this equation. By assuming the source and substrate terminals are detained in ground potential, the channel region has the characteristics of the resistor in small voltage between source and drain (V_{DS}). As shown in Figure 16, the performance of lipid bilayer-based sensor base on graphene nanostructured is assessd by the conductance characteristic. Before the electrolyte solution has been added, pure water as water-gated ambipolar was added in the membrane to measure the transfer curve. There is a constructive agreement between proposed model for lipid bilayer-based biosensor and experimental result which is extracted from ref .

As depicted in Figure 17, by applying the gate voltage to the biomimetic membrane, it is clearly stated that the conductance of GFET based graphene shows amipolar behavior. The doping states of graphene is monitored by the $V_{g,min}$ to measure the smallest amount conductance of the graphene layer which is identified from transfer characteristic curve. In total, $V_{g,min}$ (at Dirac point) shift can be known as a good indicator for lipid bilayer modulated and measurement. Nevertheless, the magnitude of voltage shift from both positive and negative lipids is quite comparable, when this shift is measured from the position of conductivity minimum of bare graphene. As shown in Figure 17, The changes in membrane electric charge can be detected electrically. The conductivity cure is changed when the electric charges are changing for biomimetic membrane-coated graphene gated. So, more electrically charged molecules will be adsorbed and the sensor will be capable of attracting more molecules which has led to change in the Vg, min on the device and the value of hole density can be estimated to decrease. Negatively exciting membrane demonstrates an enhance in conductivity smallest amount value and a positive change in Dirac point comparative to that of exposed graphene.

This is because of an enhanced in remaining pollution charges persuaded by the negatively charged membrane. A detection charged lipid bilayer can be obtained based on a detectable Dirac point shift. In light of this fact, the main objective of the current section is to present a new model for biomimetic membrane-coated graphene biosensor. In this model, the thickness and the type of coated charge as a function of gate voltage is simulated and control parameters are suggested. Subsequently, to get a greater insight into the role of both thickness and the type of lipid bilayer, GFET modelling is employed to obtain relevance between conductance versus the voltage of liquid gate, where two electrodes of sensor as shown in Figure 18 are considered as the source and drain contacts.

Figure 16. Comparison between bipolar transfer curve of conductance model (blue line) with experimental extracted data (red line) for neutral membrane

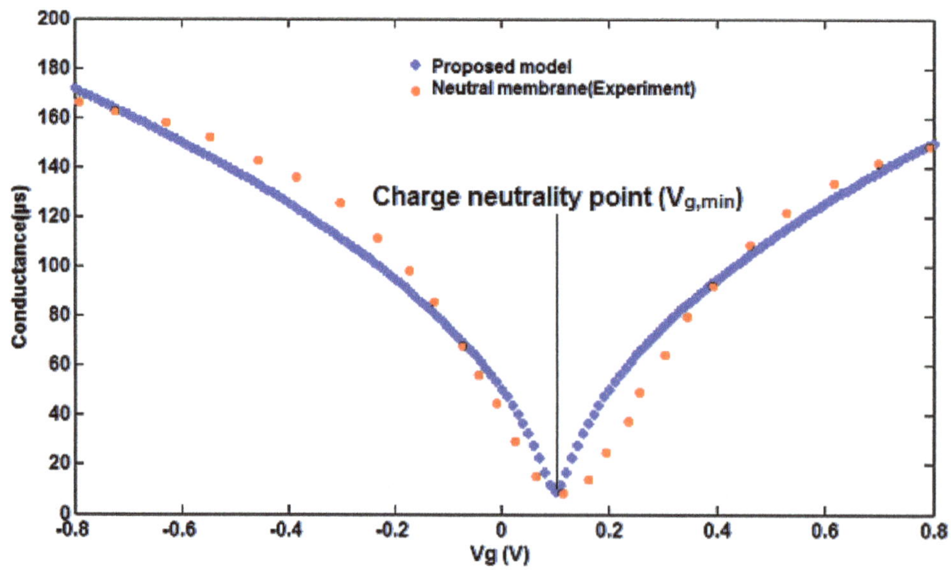

Figure 17. The comparison between GFET-conductance model with extracted experimental data for graphene coated with negatively charged, positively charged and neutral POPC membranes
Fu, 2011.

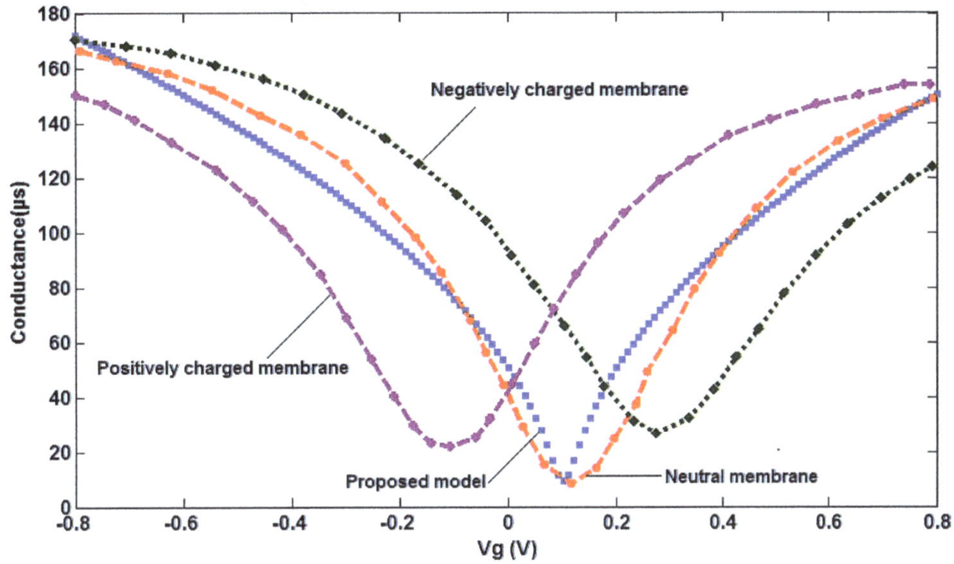

Figure 18. Schematics of the proposed structure and the electrical circuit of the electrolyte gated graphene-FET for charged lipid bilayer detection

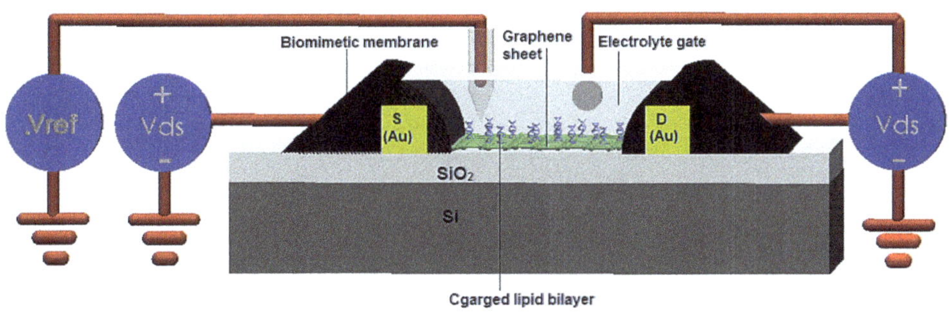

The conductance of GFET channel not only is dependent on the operation voltage on the source-drain channel and graphene organization, but also it depends on the biomimetic membranes of diverse surface charges were accumulated on graphene and charged lipid bilayer adsorbing by the graphene surface. The conductivity of the graphene based-GFET device is influenced by the charge carrier density changing in the channel. As it is shown in Figure 19, because of the membrane thinning effect, the conductance of the channel is altered.

Different ions can be adsorbed by changes in membrane electric charge and membrane thickness, subsequently the sensor will be capable of attracting the ions in the solution which has caused to transform in the conductance of the graphene based-biosensor. Dependent upon the channel conductance in the biomimetic membrane-coated graphene biosensor it is concluded that GLP is a function of electric charge and thickness. where GLP is the channel conductance after adding lipid bilayer.

Figure 19. Schematic of lipid bilayer-adsorption processes by surface area of single layer graphene

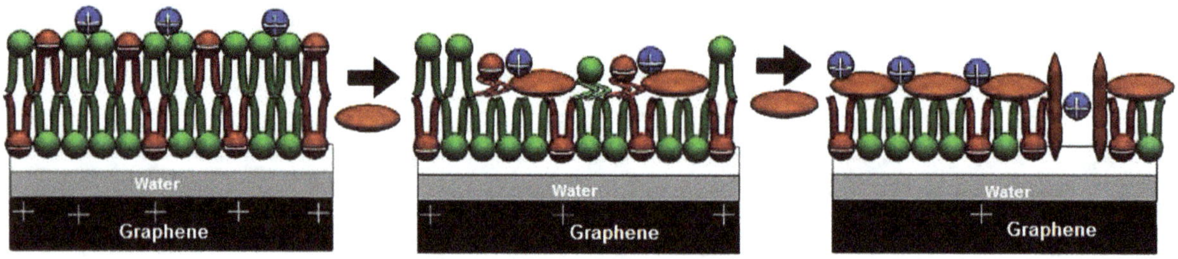

The focus of the present chapter is to demonstrate a new model for GFET to measure changes in membrane electric charge and membrane thickness, in other words the conductance of GFET device as a function of different electric charge and thickness is simulated and electric charge factor (α) and thickness factor (β) are suggested. Subsequently, for more understanding of the role of lipid bilayer, FET modelling is employed to obtain an equation between the conductance than electric charge and thickness, where the suggested structure of GFET is shown in Figure 19. This means that, GLP is supposed as a function of electric charge (QLP) as follows.

$$G_{LP} = G_{Neutral} + \alpha Q_{LP} \tag{23}$$

where electric charge factor (α) is assumed, GLP is the channel conductance of graphene with biomimetic membranes of different surface charges, Q_{LP} is the electrical charge of the membrane. Consequently, the supposed conductance model of graphene base GFET channel can be written as.

$$G_{LP} = \left(\frac{3q^2 \left(3\pi a^3 t^3 k_{BT}\right)^{\frac{1}{2}}}{hl} \left[\Im_{\frac{-1}{2}}(\eta) + \Im_{\frac{-1}{2}}(-\eta) \right] \right) + \left(\alpha Q_{LP}\right) \tag{24}$$

It is apparently shown in Figure 20 (a, b) that each diagram depicts the specific electric charge. For example when graphene coated with negative charge, it is noteworthy that, the model is closer to the experimental data; in the same manner we can compare graphene coated with the positive charge as well. It is apparently shown that, by varying electric charge trough the electric charge factor, G-V_g characteristic curve can be controlled.

Furthermore, the proposed model is strongly supported with experimental data. According to Figure 21, the amount of $V_{g, min}$ shift remains changed by changing the electric charge from negatively membrane to positively membrane likely holding the fact that the changing electric charge has an effective role in Dirac point. It is evident that, the graphene channel will be doped to n-type region with negatively charged membrane, whereas it changes to hole doping under positively charged membrane.

By increasing the membrane thickness on the graphene surface, the $V_{g,min}$ is dramatically left-shifted. It can therefore be concluded that $V_{g,min}$ is very sensitive to the electric charge and the thickness of the membrane. To support this, the gate voltage shifted leftwards owing to the fact that the graphene will

Figure 20. The comparison between graphene conductance model with extracted experimental data for different electric charge
Fu, 2011.

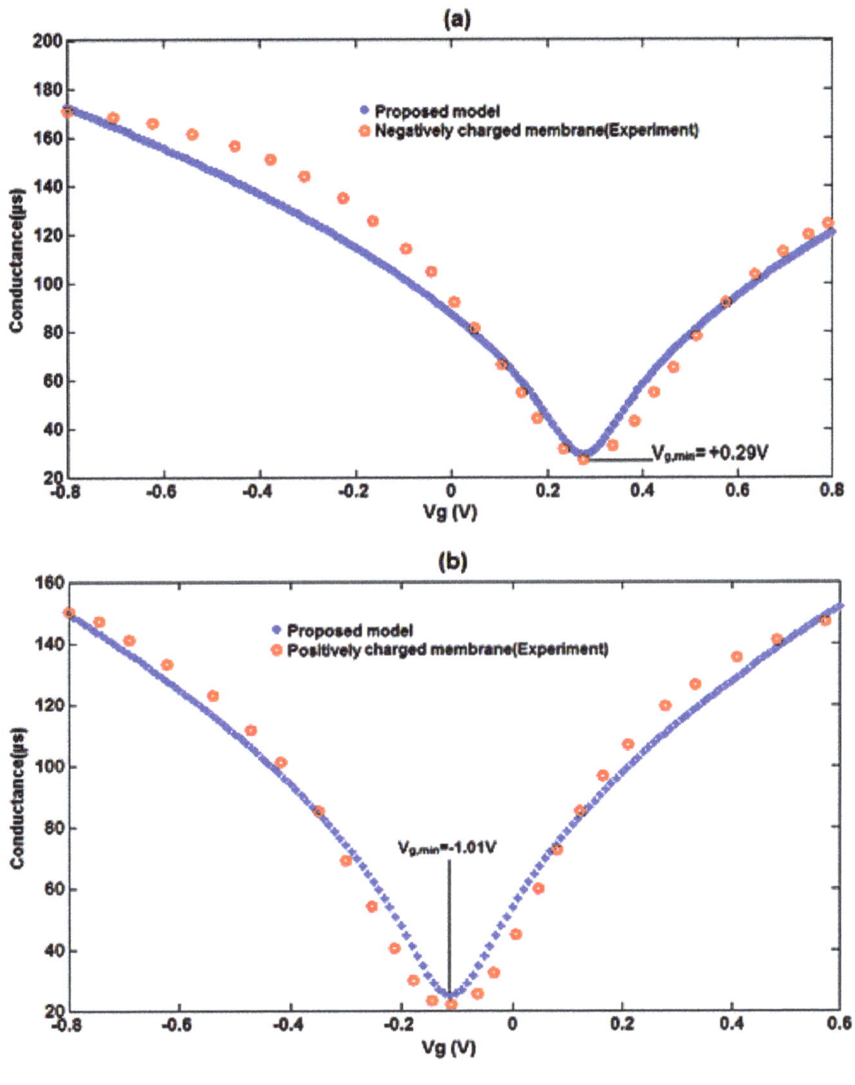

be n-doped by the high membrane thickness. On the other hand, the increased number of carriers in the channel influences the conductivity of the graphene-based FET devices.

In other words, the $V_{g,min}$ will be shifted leftwards and the extent of the shift increases with the increasing thickness of the membrane from 0.01 nM to 10 μM, as stated in Ang, Jaiswal, Lim, & Wang (2010). In order to verify the proposed model, the effect of membrane thickness will be assumed and G_{LP} is modified as a function of electric charge (Q_{LP}) and membrane thickness as follows:

$$G_{LP} = G_{Neutral} + \left(\alpha Q_{LP} + \beta L_{LP}\right) \tag{25}$$

Figure 21. Extracted experimental data for membrane thickness effect of biomimetic membrane-coated graphene biosensor

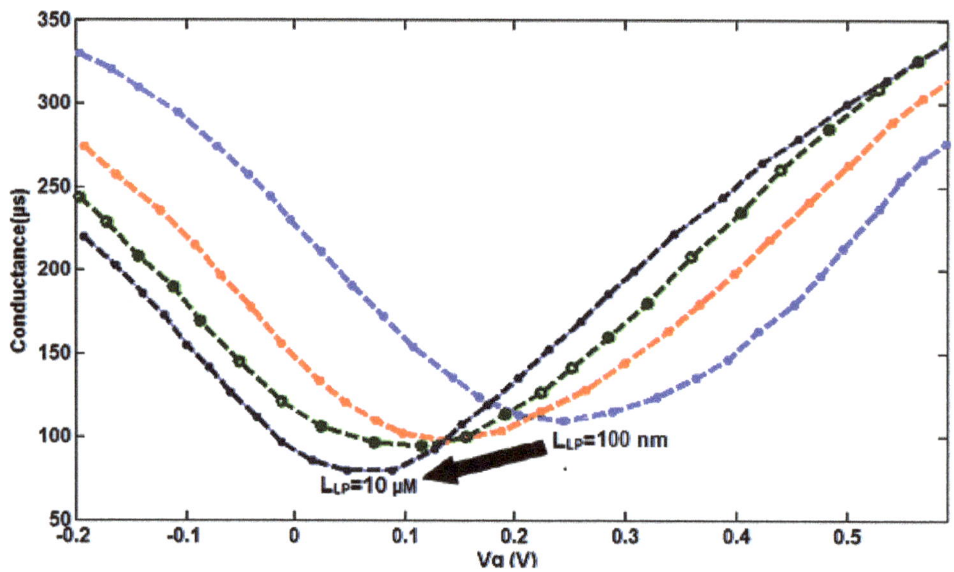

where (β) and L_{LP} are the thickness parameter and thickness of adsorbed lipid bilayer respectively. In the non-saturation region, the GFET conductance model is involved as a purpose of gate electrical energy and the perfect conductance–voltage related to the graphene channel of the GFET device which leads to the modified conductance as:

$$G_{LP} = \left(\frac{3q^2 \left(3\pi a^3 t^3 k_{BT}\right)^{\frac{1}{2}}}{hl} \left[\Im_{\frac{-1}{2}}(\eta) + \Im_{\frac{-1}{2}}(-\eta) \right] \right) \left(\frac{\alpha_1 1e^{-\alpha_2 \cdot pH}}{pH} V_{gs}(whitout\ electrolyte\ gate) - V \right)_t \quad (26)$$

In Figure 22, all the theoretical G_{LP}–V_g characteristics of graphene-based GFET with L_{LP}=10 µM are plotted. Comparing Figures 20 and 22, it can be seen that the biomimetic membrane-coated graphene biosensor model according to the suggested parameters (α and β) indicates the same trends as those reported by Ang, Jaiswal, & Lim (2010). In both the experimental and theoretical data, there is a clear shift in $V_{g,min}$ with increasing membrane thickness. The comparison of the experimental data depicted with the theoretical data in Figure 22 shows that a 10 µM membrane thickness caused a 10 mV shift in $V_{g,min.}$

In the suggested model, different charged lipid bilayer and membrane thickness are demonstrated in the form of GLP and LLP parameters respectively to an agreement with the reported data which is shown in Table 1 . The Vg,min did not shift further at higher thickness membrane due to saturation current density of injected carrier concentration by charged lipid bilayer.

According to the saturation region of the presented conductance model and $g_{m,min}$ belong to graphene-based biosensor; the control parameter with respect to the iteration method is suggested as:

Figure 22. G-Vg characteristic of proposed conductance model with experimental data, for 10 μM membrane thickness

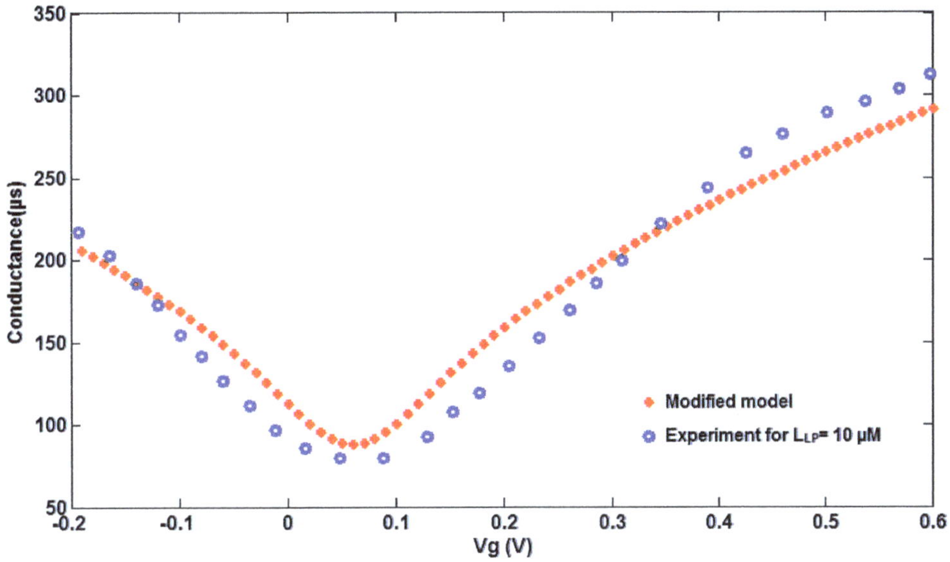

Table 1a. Different Q_{LP} values with $V_{g,min}$ changes

Q_{LP}	$V_{g,min}$(V)
Neutral	0.11
Negatively	0.29
Positively	-1.1

Table 1b. Different L_{LP} values with $V_{g,min}$ changes

L_{LP}	$V_{g,min}$ (V)
10 nm	0.24
0.1 μm	0.135
1 μm	0.09
10 μm	0.045

$$\beta = l_1 e^{-l_2 L_{LP}} \tag{27}$$

where l1=0.4157, l2=-0.543.In addition α for neutral, negatively and positively membrane is assumed 0, 1 and -1 respectively. Consequently, the justified model for the interaction of charged impurity and the consequence of charged lipid membranes in a biomimetic membrane-coated graphene biosensor is proposed as.

$$G_{LP} = \left(\alpha Q_{LP} + L_{LP} l_1 e^{-l_2 L_{LP}}\right) + \left(\frac{3q^2 \left(3\pi a^3 t^3 k_{BT}\right)^{\frac{1}{2}}}{hl}\left[\Im_{\frac{-1}{2}}(\eta) + \Im_{\frac{-1}{2}}(-\eta)\right]\right) \tag{28}$$

The proposed model in coupled with experimental data is shown in this work to confirm that the conductivity of the graphene-based biosensor is changed by electric charge and the membrane thickness of lipid bilayer. In a nutshell, electrolyte-gated graphene field-effect transistor structure was used

Chemical vapor deposited (CVD) as the electrical transduction stage because of the high electrical conductivity, optical transparency, and its large area, give the likelihood of manufacturing a dual-mode optical and electrical detection system for detecting the changes of membrane properties. Based on what has been discussed, one could firmly claim that by changes of charged lipid membranes and biomimetic membranes of different thickness charges, significant shift in Vg,min of the ambipolar FET in response is occurred due to device electronic on both n-doping and p-doping materials.

CONCLUSION

The emerging potentials of nanostructured graphene based biosensors in highly sensitive and readily detected of single-base polymorphorism or mutation that is thought as the key to diagnosis of genetic diseases and realization of personalized medicine have been demonstrated. In this research current voltage characteristic of a bilayer graphene (BG) in terms of capacitance model is applied to evaluate the detectable prostate specific antigen (PSA) on the graphene based biosensors. PSA adsorption and local pH value effects on the carrier concentration are monitored and injected carriers as a function of control parameters (f, p) are presented We have concluded that sensor characteristics altering by PSA concentration and local pH values, which means of the proposed model sensor can be controlled. In addition, in a lipid bilayer-based biosensor, the graphene carrier concentration as a function of lipid bilayer can be modeled. In this research, the total conductance of graphene as a function of electric charge (Q_{LP}) and thickness of the adsorbed lipid bilayer (L_{LP}) is presented. Dramatic decrease on the minimum conductance related gate voltage ($Vg_{, min}$) by both changing the electrical charge from negative to positive and decreasing of lipid thickness have been reported. In the presented model the $Vg_{, min}$ variation based on the adapted experimental data as an electrical detection platform is considered and the sensor control parameters are defined. Finally, good agreement between proposed model and experimental data is reported.

REFERENCES

Abdelwahab, A. A., Koh, W. C. A., Noh, H.-B., & Shim, Y.-B. (2010). A selective nitric oxide nanocomposite biosensor based on direct electron transfer of microperoxidase: Removal of interferences by co-immobilized enzymes. *Biosensors & Bioelectronics*, *26*(3), 1080–1086. doi:10.1016/j.bios.2010.08.070 PMID:20864332

Ahmadi, M. T., Ismail, R., Tan, M. L. P., & Arora, V. K. (2008). The ultimate ballistic drift velocity in carbon nanotubes. *Journal of Nanomaterials*, *2008*, 1–8. doi:10.1155/2008/769250

Ang, P. K., Jaiswal, M., Lim, C. H. Y. X., Wang, Y., Sankaran, J., Li, A., & Loh, K. P. et al. (2010). A bioelectronic platform using a graphene– lipid bilayer interface. *ACS Nano*, *4*(12), 7387–7394. doi:10.1021/nn1022582 PMID:21067155

Ang, P. K., Jaiswal, M., Lim, C. H. Y. X., Wang, Y., Sankaran, J., Li, A., & Loh, K. P. et al. (2010). A bioelectronic platform using a graphene– lipid bilayer interface. *ACS Nano*, *4*(12), 7387–7394. doi:10.1021/nn1022582 PMID:21067155

Arora, V. K., Tan, M. L. P., Saad, I., & Ismail, R. (2007). Ballistic quantum transport in a nanoscale metal-oxide-semiconductor field effect transistor. *Applied Physics Letters*, 91.

Baby, T. T., Aravind, S. S. J., Arockiadoss, T., Rakhi, R. B., & Ramaprabhu, S. (2010). Metal decorated graphene nanosheets as immobilization matrix for amperometric glucose biosensor. *Sensors and Actuators. B, Chemical*, *145*(1), 71–77. doi:10.1016/j.snb.2009.11.022

Barbier, M., Vasilopoulos, P., Peeters, F. M., & Pereira, J. M. (2009). Bilayer graphene with single and multiple electrostatic barriers: Band structure and transmission. *Physical Review B: Condensed Matter and Materials Physics*, *79*(15), 155402. doi:10.1103/PhysRevB.79.155402

Bertini, I., Lee, Y. M., Luchinat, C., Piccioli, M., & Poggi, L. (2001). Locating the metal ion in calcium-binding proteins by using cerium(III) as a probe. *ChemBioChem*, *2*(7-8), 550–558. doi:10.1002/1439-7633(20010803)2:7/8<550::AID-CBIC550>3.0.CO;2-T PMID:11828488

Bin, X., Sargent, E. H., & Kelley, S. O. (2010). Nanostructuring of sensors determines the efficiency of biomolecular capture. *Analytical Chemistry*, *82*(14), 5928–5931. doi:10.1021/ac101164n PMID:20568723

Chen, F., Qing, Q., Xia, J., & Tao, N. (2010). Graphene field-effect transistors: Electrochemical gating, interfacial capacitance, and biosensing applications. *Chemistry, an Asian Journal*, *5*(10), 2144–2153. doi:10.1002/asia.201000252 PMID:20715049

Chen, J.-L., & Yan, X.-P. (2010). A dehydration and stabilizer-free approach to production of stable water dispersions of graphene nanosheets. *Journal of Materials Chemistry*, *20*(21), 4328–4332. doi:10.1039/c000177e

Du, D., Zou, Z., Shin, Y., Wang, J., Wu, H., Engelhard, M. H., & Lin, Y. et al. (2010). Sensitive immunosensor for cancer biomarker based on dual signal amplification strategy of graphene sheets and multienzyme functionalized carbon nanospheres. *Analytical Chemistry*, *82*(7), 2989–2995. doi:10.1021/ac100036p PMID:20201502

Du, D., Zou, Z., Shin, Y., Wang, J., Wu, H., Engelhard, M. H., & Lin, Y. et al. (2010). Sensitive immunosensor for cancer biomarker based on dual signal amplification strategy of graphene sheets and multienzyme functionalized carbon nanospheres. *Analytical Chemistry*, *82*(7), 2989–2995. doi:10.1021/ac100036p PMID:20201502

Enoki, T., Kobayashi, Y., & Fukui, K.-I. (2007). Electronic structures of graphene edges and nanographene. *International Reviews in Physical Chemistry*, *26*(4), 609–645. doi:10.1080/01442350701611991

Fiori, G., & Iannaccone, G. (2009). Performance analysis of graphene bilayer transistors through tight-binding simulations. *Iwce-13: 2009 13th International Workshop on Computational Electronics*, (pp. 301-304).

Fu, W. N., Knopfmacher, Tarasov, Weiss, Calame, & Schonenberger, C. (2011). Graphene transistors are insensitive to pH changes in solution. *Nano Letters, 11*(9), 3597-3600.

Kiani, M. J., Ahmadi, M., Akbari, E., Karimi, H., & Che Harun, F. (2013). Graphene nanoribbon based gas sensor. *Key Engineering Materials*, *553*, 7–11. doi:10.4028/www.scientific.net/KEM.553.7

Kiani, M. J., Ahmadi, M., & Harun, F. (2013). Quantum capacitance effect on bilayer graphene nanoribbon based nanoscale transistors. *Journal of Nanoengineering and Nanomanufacturing*, *3*(2), 138–141. doi:10.1166/jnan.2013.1119

Kiani, M. J. A., Rahmani, Harun, & Che. (2013). Degeneracy effect on carrier transport in bilayer graphene nanoribbon. *International Journal of Nano Devices, Sensors and Systems (IJ-Nano), 2*(1).

Leonenko, Z., Cramb, D. T., Amrein, M., & Finot, E. (2008). Atomic force microscopy: Interaction forces measured in phospholipid monolayers, bilayers and cell membranes. *Applied Scanning Probe Methods*, *IX*, 207–234.

Li, T., Yang, M., & Li, H. (2011). Label-free electrochemical detection of cancer marker based on graphene–cobalt hexacyanoferrate nanocomposite. *Journal of Electroanalytical Chemistry*, *655*(1), 50–55. doi:10.1016/j.jelechem.2011.02.009

Liao, K.-H., Lin, Y.-S., Macosko, C. W., & Haynes, C. L. (2011). Cytotoxicity of graphene oxide and graphene in human erythrocytes and skin fibroblasts. *ACS Applied Materials & Interfaces*, *3*(7), 2607–2615. doi:10.1021/am200428v PMID:21650218

Lin, L., Liu, Y., Tang, L., & Li, J. (2011). Electrochemical DNA sensor by the assembly of graphene and DNA-conjugated gold nanoparticles with silver enhancement strategy. *Analyst (London)*, *136*(22), 4732–4737. doi:10.1039/c1an15610a PMID:21952074

Liu, S., Tian, J., Wang, L., & Sun, X. (2011). A method for the production of reduced graphene oxide using benzylamine as a reducing and stabilizing agent and its subsequent decoration with Ag nanoparticles for enzymeless hydrogen peroxide detection. *Carbon*, *49*(10), 3158–3164. doi:10.1016/j.carbon.2011.03.036

Mak, K. F., Lui, C. H., Shan, J., & Heinz, T. F. (2009). Observation of an electric-field-induced band gap in bilayer graphene by infrared spectroscopy. *Physical Review Letters*, *102*(25), 256405. doi:10.1103/PhysRevLett.102.256405 PMID:19659105

Mousavi, H. (2011). Graphene as Gas Sensors. *Communications in Theoretical Physics*, *56*(2), 373–376. doi:10.1088/0253-6102/56/2/30

Mousavi, S. M., Ahmadi, M. T., Sadeghi, H., Nilghaz, A., Amin, A., Johari, Z., & Ismail, R. (2011). Bilayer graphene nanoribbon carrier statistic in degenerate and non degenerate limit. *Journal of Computational and Theoretical Nanoscience*, *8*(10), 2029–2032. doi:10.1166/jctn.2011.1921

Mousavi, S. M. A., Webb, M. T., Sadeghi, J. F., Nilghaz, H., Amin, A., Johari, A., & Ismail, Z., R. (2011). Bilayer graphene nanoribbon carrier statistics in the degenerate regime. In N. W. J. F. V. P. Barsoum (Ed.), *Proceedings of the Fourth Global Conference on Power Control and Optimization* (Vol. 1337, pp. 180-183). doi:10.1063/1.3592463

Myung, S., Solanki, A., Kim, C., Park, J., Kim, K. S., & Lee, K.-B. (2011). Graphene-encapsulated nanoparticle-based biosensor for the selective detection of cancer biomarkers. *Advanced Materials*, *23*(19), 2221–2225. doi:10.1002/adma.201100014 PMID:21469221

Myung, S., Solanki, A., Kim, C., Park, J., Kim, K. S., & Lee, K.-B. (2011). Graphene-encapsulated nanoparticle-based biosensor for the selective detection of cancer biomarkers. *Advanced Materials*, *23*(19), 2221–2225. doi:10.1002/adma.201100014 PMID:21469221

Ohno, Y., Maehashi, K., Yamashiro, Y., & Matsumoto, K. (2009). Electrolyte-gated graphene field-effect transistors for detecting pH protein adsorption. *Nano Letters*, *9*(9), 3318–3322. doi:10.1021/nl901596m PMID:19637913

Qiu, Y., Qu, X., Dong, J., Ai, S., & Han, R. (2011). Electrochemical detection of DNA damage induced by acrylamide and its metabolite at the graphene-ionic liquid-Nafion modified pyrolytic graphite electrode. *Journal of Hazardous Materials*, *190*(1-3), 480–485. doi:10.1016/j.jhazmat.2011.03.071 PMID:21497017

Sadeghi, H., Ahmadi, M., Ishak, B., Mousavi, S., & Ismail, R. (2011). Ballistic conductance model of bilayer graphene nanoribbon (BGN). *Journal of Computational and Theoretical Nanoscience*, *8*(10), 1993–1998. doi:10.1166/jctn.2011.1915

Salavagione, H. J., Díez-Pascual, A. M., Lázaro, E., Vera, S., & Gómez-Fatou, M. A. (2014). Chemical sensors based on polymer composites with carbon nanotubes and graphene: The role of the polymer. *Journal of Materials Chemistry A*, *2*(35), 14289–14328. doi:10.1039/C4TA02159B

Shrivastava, I. H., & Sansom, M. S. (2000). Simulations of ion permeation through a potassium channel: Molecular dynamics of KcsA in a phospholipid bilayer. *Biophysical Journal*, *78*(2), 557–570. doi:10.1016/S0006-3495(00)76616-1 PMID:10653771

Wang, X.-N., & Hu, P.-A. (2012). Carbon nanomaterials: Controlled growth and field-effect transistor biosensors. *Frontiers of Materials Science*, *6*(1), 26–46. doi:10.1007/s11706-012-0160-x

Wei, L., Frank, D. J., Chang, L., & Wong, H. S. P. (2011). A fully anlytical model for carbon nanotube FETs including quantum capacitances and electrostatics. Boston, MA: Academic Press.

Weiss, L. A., Sakai, N., Ghebremariam, B., Ni, C., & Matile, S. (1997). Rigid rod-shaped polyols: Functional nonpeptide models for transmembrane proton channels. *Journal of the American Chemical Society*, *119*(50), 12142–12149. doi:10.1021/ja973126d

Xie, Y., Chen, A., Du, D., & Lin, Y. (2011). Graphene-based immunosensor for electrochemical quantification of phosphorylated p53 (S15). *Analytica Chimica Acta*, *699*(1), 44–48. doi:10.1016/j.aca.2011.05.010 PMID:21704756

Zhang, B., & Cui, T. (2011). An ultrasensitive and low-cost graphene sensor based on layer-by-layer nano self-assembly. *Applied Physics Letters*, *98*(7), 073116. doi:10.1063/1.3557504

Zhang, B. C., & Cui, T. (2011). An ultrasensitive and low-cost graphene sensor based on layer-by-layer nano self-assembly. *Applied Physics Letters*, *98*(7), 073116. doi:10.1063/1.3557504

Zhang, Y. B., Brar, V. W., Girit, C., Zettl, A., & Crommie, M. F. (2009). Origin of spatial charge inhomogeneity in graphene. *Nature Physics*, *5*(10), 722–726. doi:10.1038/nphys1365

Chapter 12
Graphene and CNT Field Effect Transistors Based Biosensor Models

Ali Hosseingholi Pourasl
Universiti Teknologi Malaysia, Malaysia

Meisam Rahmani
Universiti Teknologi Malaysia, Malaysia

Mohammad Taghi Ahmadi
Urmia University, Iran

Razali Ismail
Universiti Teknologi Malaysia, Malaysia

Michael Loong Pengl Tan
Universiti Teknologi Malaysia, Malaysia

ABSTRACT

In this chapter, novel ideas of graphene and CNT based electrical biosensors are provided. A liquid gated graphene field effect transistor (LG-GFET) based biosensor model is analytically developed for electrical detection of Escherichia coli (E. coli) bacteria. E. coli absorption effects on the graphene surface in the form of conductance variation is considered. Moreover, the current-voltage characteristic in terms of conductance model is applied to evaluate the performance of the biosensor model. Furthermore, the CNT-FET platform is employed for modeling biosensor in order to detect Glucose. For diagnosing and monitoring the blood glucose level, glucose oxidase (GOx) based enzyme sensors have been immensely used. According to the proposed CNT-FET structure, charge based analytical modeling approach is used. The charge-based carrier velocity model is implemented to study electrical characteristics of CNT-FET. In the presented model, the gate voltage is considered as a function of glucose concentration. Finally, the both of presented models are compared with published experimental data.

INTRODUCTION

Silicon based technologies are approaching their scaling limits. To overcome the limitations of silicon planner devices, new device designs are being explored to replace the existing planar device technology. Hence, smaller dimensions, less power consumption, low energy delay product, very large computing power and high density as well as high speed is demanding for future electronic technology. Nanoelec-

DOI: 10.4018/978-1-5225-0736-9.ch012

tronic, is a new filed which is opining novel horizons for electronic device industry. Nevertheless, the downscaling procedure of semiconductor device dimensions lead to some drawbacks such as gate leakage current and short channel effect, like in the conventional metal oxide semiconductor field effect transistors (MOSFETs). Therefore, devices based on novel nanomaterials like graphene nanoribbon field effect transistors (GNR-FETs), carbon nanotube based field effect transistors (CNT-FET), nanowire field effect transistors (NW-FETs) and nanoscrolls field effect transistors (NS-FETs) are promising candidates to overcome planner devise constrains.

MODERN SENSOR TECHNOLOGY

In the early 1990s (since the foundation of MOSFET), miniaturization and integration of very small functional units, open a new horizon for future technologies, from micro down to nano and molecular scales. Sensor play a very important role in most products and systems. The development of sensors could not keep pace with the advancements in microelectronic devices and components. Therefore, sensor technology is in restructuring stage for miniaturization and integration in signal processing system (Meixner, 2008). This fact causes technologies that permit manufacturing of sensors and the related electronics with low costs, to be given high importance. In the past two decades, numerous research works has been done in sensor technology area. Among those researches, biosensors have attracted enormous attentions because of their amazing selectivity and sensitivities (Kirsch, Siltanen, Zhou, Revzin, & Simonian, 2013; Kriz, Ramström, & Mosbach, 1997).

Research and development in this field is broad and involves areas such as physical chemistry, electrochemistry, bioreactor science and biochemistry. Majority of the focus was on amperometric biosensors and colorimetric paper enzyme strips. But, nowadays more focus is being given to the research of the different transducer types and their application to biosensors (Chaplin & Bucke, 1990; Taneja, 2012). Biosensors application are growing at an annual rate of 60% where the health-care industries, food quality appraisals and environmental monitoring mainly contribute to this growth. The approximate value of the world analytical market is about 12 billion dollars per year which the share of medical area is nearly a third of total market. Of this total market, only approximately 0.1% accounts for the use of biosensors, hence exhibiting the vast potential growth in this area (Taneja, 2012).

In spite of remarkable advancement in the field of electronics and biosensors in recent years, this area is in the initial steps. Some crucial issues such as selecting ideal biomaterial and improvement of sensing techniques based on electronic detection approaches are getting a lot of attentions (Dankerl et al., 2009; Fromherz, Offenhausser, Vetter, & Weis, 1991; Härtl et al., 2004; Stern et al., 2007; Voelker & Fromherz, 2005; Yang et al., 2002). Because of some reasons such as industrial, technical and technological maturities, silicon is the most popular and commonly used material in this field (Fromherz et al., 1991; Stern et al., 2007; Voelker & Fromherz, 2005). But, because of inherent limitations of silicon, novel nanomaterials are called upon to challenge silicon dominance. Silicon suffers low chemical stability and biocompatibility. Furthermore, silicon platform based biosensors have very complex design and fabrication process. In addition, electronic devices based on silicon have relatively high noise which effect on device sensitivity (Dankerl et al., 2009; Härtl et al., 2004; Yang et al., 2002).

But, superior electrical properties and low noise allow carbon based material such as graphene and carbon nanotubes to be excellent candidates for sensing applications (Galib, 2012; Geim & Novoselov, 2007). One of the most important benefits of these materials is their large surface to volume ratio, which makes them very efficient to sense adsorbed molecules such as organic cells and bacteria (Choi, Lahiri, Seelaboyina, & Kang, 2010; Geim & Novoselov, 2007; J. Wang, 2005). Therefore, developing the use of graphene and carbon nanotubes for medical diagnostics applications is currently very popular research area.

CNT

Offering unique structure, carbon nanotubes are raised as great candidates for future electronic device technology. They show excellent electrical, mechanical and chemical properties such as long mean free path, high carrier mobility, and potential of carrying high current density. Figure 1 shows a single wall carbon nanotube (SWNT) which is made by rolling up a graphene sheet. If the diameter of the tube is less than de-Broglie wave length it can be modeled as one dimensional.

CNT Structure and Properties

The structural properties of the carbon nanotubes are dependent on the diameter and helicity of the formation of graphitic rings, which electrically determines whether, it is semiconducting or metallic. As shown in the Figure 1, the diameter and helicity of a defect-free SWCNT is described by the chiral vector, which is defined as $\vec{C}_h = n\vec{a}_1 + m\vec{a}_2$, where $\vec{a}_1$ and $\vec{a}_2$ are the graphite lattice unit cell base vectors and n and m are the integer coefficients. Based on the chiral vector, graphene sheet can be rolled up to form armchair or zigzag cylinders which can be either metallic or semiconducting (Sangtarash et al., 2012). CNTs can be in the form of single layer or single wall and multilayer or multiwall tubes with two or more concentric cylinders (Martel, Schmidt, Shea, Hertel, & Avouris, 1998; Pantano, Parks, & Boyce, 2004). By rolling graphene in different directions, three main types of CNTs, namely armchair, zigzag and chiral can be made (M. Dresselhaus, G. Dresselhaus, & Saito, 1995), which are demonstrated in Figure 2.

Figure 1. A SWNT structure with length higher than de-Broglie wave length and diameter less than de-Broglie wave length

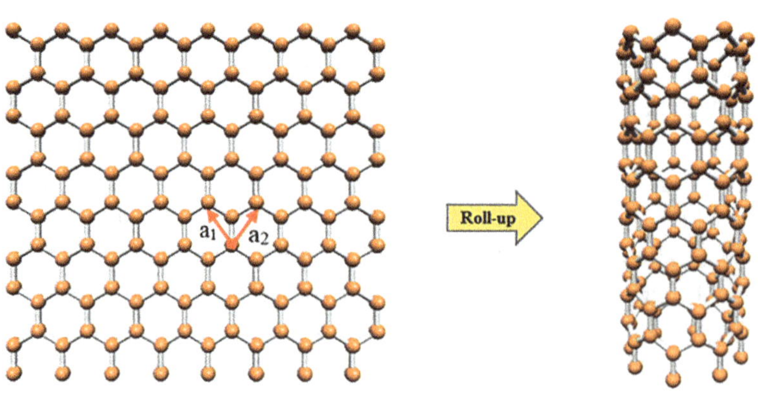

Figure 2. Carbon nanotube configurations presenting armchair, zigzag, and chiral structure

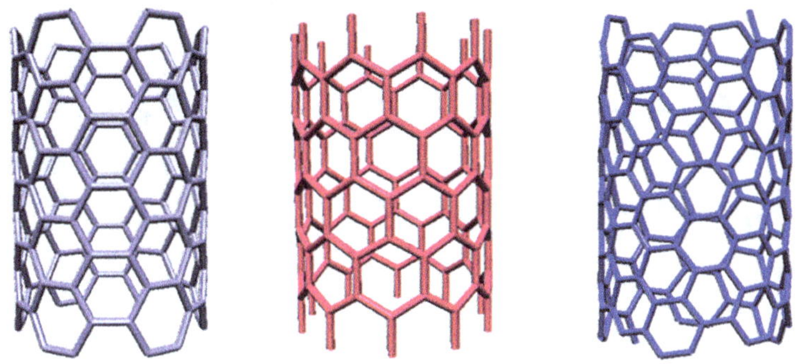

MONOLAYER GRAPHENE STRUCTURE

As another derivative of the graphite and the main substitute for silicon, researchers are increasingly using graphene in transistors and different settings. Scientists consider it as a suitable material for implementing in transistors and integrated circuits (Avouris & Xia, 2012; Y. M. Lin et al., 2011). Graphene is a monolayer of graphite in which carbon atoms are strongly packed into a two-dimensional honeycomb lattice structure, which is a fundamental building block for carbon materials (Z. F. Wang, Zheng, Shi, & Chen, 2009). By wrapping into zero dimensional (0D), it can be converted to fullerenes, rolling into one-dimensional (1D), can make carbon nanotubes, cutting in the form of (1D), narrow strips and ribbons can be made or it can be stacked in the form of three-dimensional (3D) graphite as shown in Figure 3.

Using micromechanical cleavage technique, for the first time graphene was extracted from graphite (Novoselov et al., 2004). This method eases the fabrication of graphene with high crystal quality and in addition lead to a lot of research activities and experimental works (Katsnelson, 2007). Inherent graphene is categorized in zero gap or semi-metal semiconductor group and its exceptional electronic properties provide a surprisingly high opacity for one atom thick layer, with a remarkable *2.3%* low white light adsorption ratio (Kuzmenko, Heumen, Carbone, & Marel, 2008). As electrical characterization, graphene showed notably high electron mobility of $15,000\ cm^2V^{-1}s^{-1}$ at room temperature (Novoselov et al., 2004). Furthermore, its corresponding resistivity reported as $10^{-6}\ \Omega\text{-}cm$, less than silver resistivity which has the lowest substance resistivity known at room temperature (Novoselov et al., 2004).

Structural and Electrical Properties

Various properties of graphene make it as a strong nominee to replace silicon in next generation of nanoscale electronic devices (Geim, 2009; X. Huang, Qi, Boey, & Zhang, 2012; X. Huang, Zeng, Fan, Liu, & Zhang, 2012; X. S. Li et al., 2009). For example, electrons in graphene behave like zero rest mass and have a ballistic transport (travel with light speed), but electrons travel over the silicon in a sequences of collision which leads to generation of heat and limits the speed and dimension of silicon transistor. Ballistic transport in graphene allows designing transistors with low power consumption and faster switching.

Figure 3. Derivations of carbon based materials with different dimensions: a) 2D Graphene sheet b) 1D Graphene nanoribbon c) 1D Carbon nanotube d) 0D Fullerene e) 3D Graphite

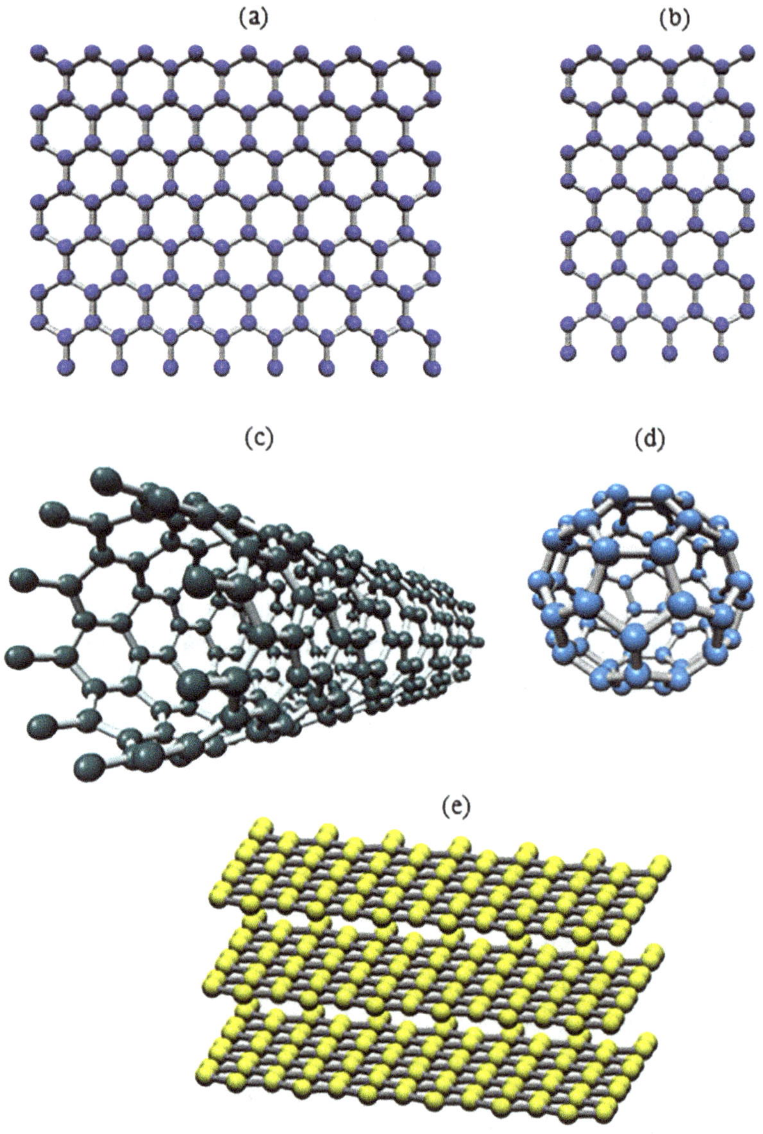

Furthermore, depending on the edge shape and width graphene nanoribbons can be either semiconducting or metallic. This property enables much flexibility in designing and fabricating interconnections and several electronic devices including electric switches and transistors. Recent experiments have illustrated that researchers can use chemical stripping or plasma etching e-beam lithography technique to manage and control shape of the graphene nanoribbons (X. Li, Wang, Zhang, Lee, & Dai, 2008; Schedin et al., 2007). Furthermore, Ci et al. (Ci et al., 2008) have proposed a new approach to cut graphene and extract ribbons with perfect atomic edges. Using these approaches, there is no need for metal wires for interconnections, and resolves the problem of high electric resistances happening when metal wires are attached to nanotubes.

Graphene Band Structure

There are two atoms per graphene's unit cell which yields two equivalent points in Brillion zone, where band crossing occurs at *K* and *K'* (Dirac points). The energy of electron is dependent on the wave vector linearly, close to these crossing points (Novoselov et al., 2005). In graphene, charge carriers are explained by a massless Dirac spectrum, unlike nonrelativistic quantum particles which are explained by the typical Schrodinger equation. In the special crystal of graphene, there is a couple of carbon sublattices, A and B, per unit cell. The intersection of their energy band close to the edges of the Brillouin zone leads to a conical energy spectrum. As shown in Figure 4, because of linear spectrum, it can be expected that different from conventional semiconductors and metals, quasi-particle behaves in a different way, where the spectrum of energy can be estimated by a parabolic dispersion relation (Geim & Novoselov, 2007; Katsnelson, 2007).

Graphene Nanoribbon

Similar to carbon nanotube, Graphene nanoribbon (GNR) is a one-dimensional (1D) nanomaterial, which is an unrolled carbon nanotube. Two-dimensional graphene is a zero band gap material which makes it inappropriate to be used as a channel in a transistor. Zero band gap make graphene to show semi-metal manner (Han, Özyilmaz, Zhang, & Kim, 2007). Quantum confinement of graphene leads to formation of energy band gap between valence band and conduction band in graphene nanoribbons which make it suitable to be implemented as a channel in field effect transistors. Graphene nanoribbon can be produced from graphene by techniques such as chemical stripping, e-beam lithography and plasma etching (X. Li et al., 2008; Schedin et al., 2007).

Based on the recent studies on the graphene nanoribbon FET, channel conductivity of the graphene was found difficult to change and control under ambient conditions (Fiori & Iannaccone, 2007). The crucial issue in this regard is to find an approach to fabricate nanoribbon with extremely small width with appropriate band gap in room temperature. Based on the theoretical studies it has been found that there is an inverse relation between nanoribbon width and energy band gap (Son, Cohen, & Louie, 2006). To reach minimum band gap in room temperature, the maximum allowed width of the ribbon has been

Figure 4. Graphene band structure
As cited in Recher & Trauzettel, 2011.

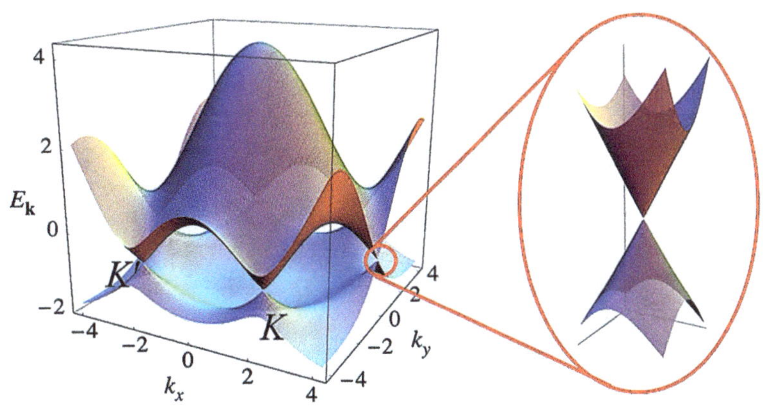

obtained to be 10 nm (X. R. Wang et al., 2008). Graphene nanoribbon characteristics to a large extent depends on edge and width of the ribbon. Figure 5 illustrates the schematics of the graphene nanoribbon with width less than de-Broglie wave length (λ_D >10nm) which is formed from graphene sheet.

Graphene Nanoribbon Geometries

The structure of graphene nanoribbon with a width less than de-Broglie wavelength was demonstrated in Figure 5. It was mentioned that in order to implement graphene in transistor efficiently, the formation of band gap is required. To this end, graphene should be cut in narrow strips or nano ribbons. The quantum confinement that form standing waves along the chiral vector leads to band gap formation in GNR (Chen, Lin, Rooks, & Avouris, 2007; Han et al., 2007). Theoretical studies indicated that, the GNR band gap formation, is dependent to the chirality and width of the ribbons and according to geometry and edge structure GNR shows different carrier transport properties (Sols, Guinea, & Neto, 2007; Son et al., 2006). Graphene nanoribbons can be formed in either zigzag or armchair arrangements as shown in Figure 6.

Similar to carbon nanotubes, graphene nanoribbon shows semiconducting and metallic electrical properties. The zigzag GNR can carry a spin current, that make it to behave as a semi-metal material. In contrast, depending on the width of the ribbons armchair GNR can exhibit either semiconducting or metallic properties (Berger et al., 2004). The width of GNR can be varied by the number of dimmer lines, N, given as $W_{GNR} = \left(N-1\right)\left(3^{1/2}/2\right)a_{C-C}$, where a_{C-C}=0.142nm is the lattice constant (Lundstrom & Guo, 2006). The electrical characteristics of GNR is affected by the value of N so that armchair GNR shows semiconducting properties when N = 3p or 3p + 1, where *p* is an integer. It also behaves as a quasi-metallic when N=3p+2 (Datta, 2005).

Figure 5. Schematic of GNR with width less than de-Broglie wavelength

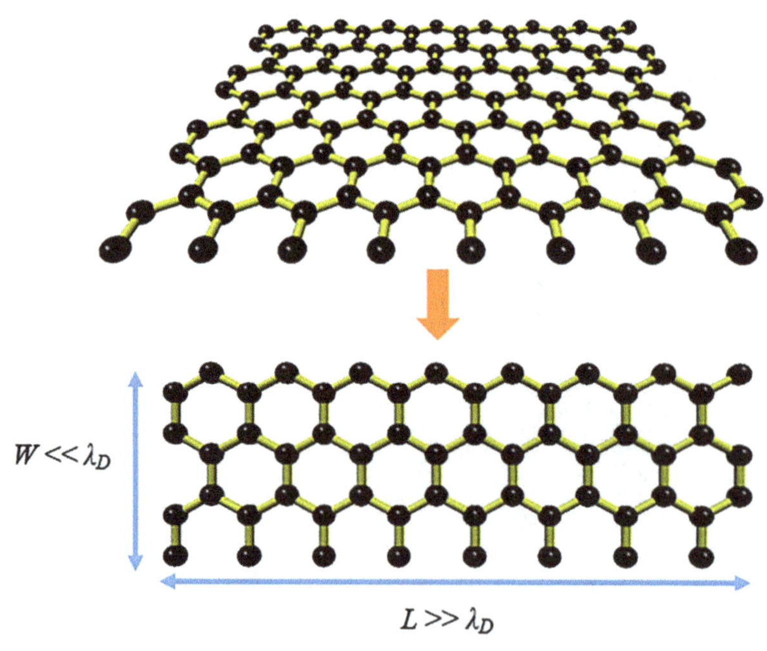

Figure 6. Atomic graphene structure: a) armchair graphene; b) zigzag graphene

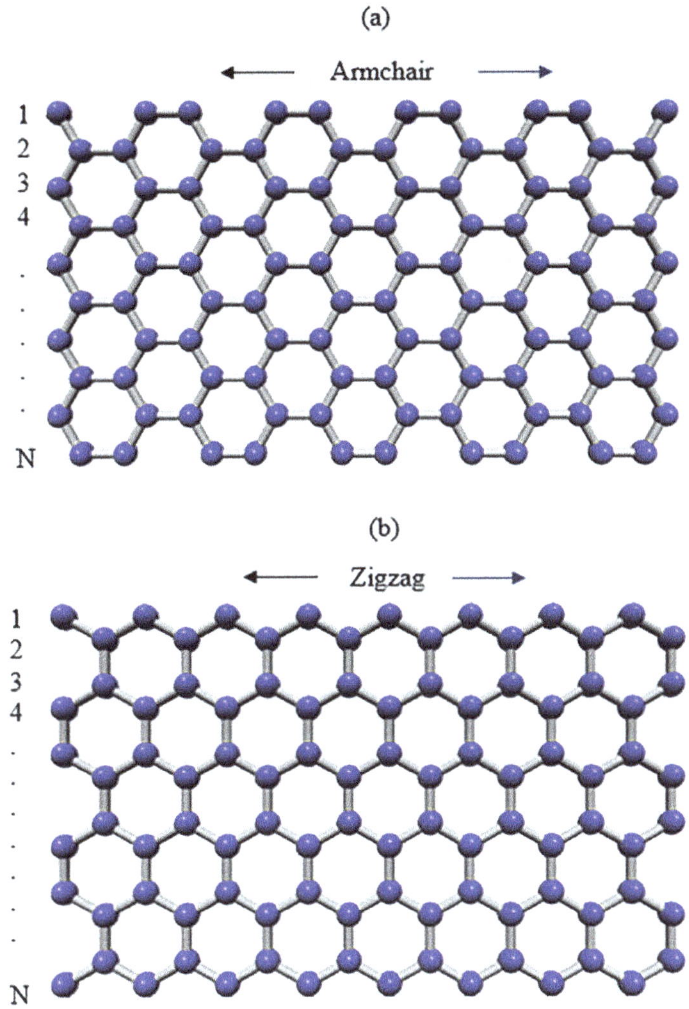

BIOSENSORS

Biosensors are analytical devices that contain biological responsive material embedded in transducers that convert biological response into an electrical signal. There is an ever-increasing demand for biosensors in various fields as their applications are slowly being discovered. Some of these areas include medicine, biotechnology and environmental control (e.g. for detection of greenhouse gases and pollutants in water). The main characteristic of biosensors that responsible for this high demand application is the receptor. It offers shape-specific recognition, hence making the sensor highly sensitive and specific in nature. One of the main advantages that biosensor has over other analytical detection methods such as liquid and gas chromatography is that they can be used in situations where the molecule to be measured is in micro concentrations (Byfield & Abuknesha, 1994).

The illustration of a biosensors consists of three main components: a very sensitive bioelement, a detecting elements, and a signal processing element. The bioelement can be, micro-organisms, living cells, or enzymes (Held, Schuhmann, Jahreis, & Schmidt, 2002; Vianello, Ragusa, Cambria, & Rigo,

2006), which detects the target. The detecting element can be employed for monitoring the changes of impedance (Lillie, Payne, & Vadgama, 2001), electric current and potential (Andle & Vetelino, 1994), electromagnetic radiation (Haes, Chang, Klein, & Duyne, 2005), and optical density (Choudhury, Shinar, & Shinar, 2004), among others. The bio-element interfaces directly to a signal transducer (detecting element) which converts the variation of the target analyte into a quantitative response. Various compositions of a bio-element matched to a signal transducer lead to different applications. Figure 7 shows a schematic diagram of the concept on a biosensor device.

Operational Principle of Electronic Nanosensors

Due to adsorption of the molecules and biomolecules on the graphene and CNT surfaces, their electrical conductivity changes. This variation in the presence of molecules is the basic operational principle of the CNT and graphene based gas or biosensors (Besteman, Lee, Wiertz, Heering, & Dekker, 2003; C. Lee, Wei, Kysar, & Hone, 2008). The variation of conductivity can be explained in the form of variation of carrier concentration of graphene and CNT due to the molecules which are absorbed to their surfaces and act as a donor or acceptor. Moreover, graphene and CNT possess some interesting properties which aid to increase their sensitivity even to single atom. For instance, they have large surface to volume ratio so that expose all surface to the target analyte. They have very high conductivity with low Johnson noise, therefore, a small variation in carrier concentration can yield a notable change of conductance (C. Li, Thostenson, & Chou, 2008; Schedin et al., 2007). Furthermore, they have less crystal defects which yields low noise level due to thermal switching (Novoselov et al., 2007; Qi et al., 2003). Also, four-probe measurements can be applied on single crystal graphene device with low resistance Ohmic contacts (C. Chen et al., 2007; Fowler et al., 2009).

Enzymatic Detection

Due to the importance of use of biosensors in some applications including environmental monitoring, bioprocess monitoring, clinical diagnostics and food freshness, they have received increasing attentions as an important research area(Malhotra, Chaubey, & Singh, 2006). The type of materials which are used in the biosensor electrodes governs the performance of the enzyme electrode. Enzymatically produced redox reactions are moved by freely diffusing natural or artificial redox mediators through a shuttle mechanism (Schuhmann, 1995).

Figure 7. Basic block diagram of a biosensor

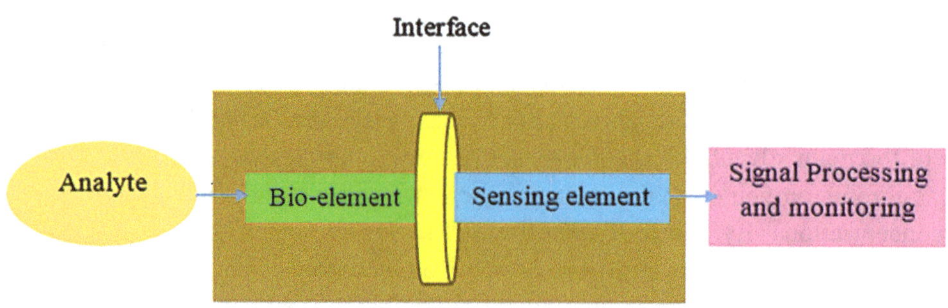

The classical enzyme electrodes can be constructed by integrating the electrode surfaces of gold, platinum, glassy carbon or carbon past (Moyo, Okonkwo, & Agyei, 2012). To select electrode material, some main factors such as hardness and electrical conductivity should be taken into account (Sarma, Vatsyayan, Goswami, & Minteer, 2009). Enzymes can be combined by electrodes through three main approaches namely; covalent binding on the electrode, physical absorption or cross linking with the polymer. In order to have interface between the biocatalytical layer and the transducer to make well organized and efficient electron transfer between enzyme and electrode, it is necessary to develop amperometric, conductometric or potentiometric enzyme electrodes. The enzyme function in these biosensors is to provide selectivity considering its biological affinity for a specific molecule of substrate that is of biological origin (M. Husain & Q. Husain, 2007; Sarma et al., 2009).

To develop the performance and functionality of biosensors it is needed to evaluate the activity of the surface of the modified electrodes which is done by reduction or oxidation of electroactive species on the surface of modified electrodes. Nevertheless, electroactive species direct oxidation including thiocholine at classical electrodes is low, disposed to interfering challenges from other electroactive species and moreover demands high over-potential with the additional electrode fouling problem (Arduini et al., 2009; Ndlovu, Arotiba, Krause, & Mamba, 2010; White, Lawrence, Davis, & Compton, 2002).

Field Effect Transistor Based Biosensor

Biosensors based on the field effect transistor have grabbed great attention among various electrical sensing device architectures. This is due to the interface between target analyte and FET which directly convert the biological interactions between molecules and transistor channel to measurable electrical responses (Patolsky, Timko, et al., 2006; Patolsky, Zheng, & Lieber, 2006). The composition of a FET sensor is similar to common three electrode transistor, so that drain and source electrodes bridge semiconductor channel and the gate modulates conductance of the channel. The schematics of a common FET based sensor is illustrated in the Figure 8. In the FET nanosensor architecture a nanomaterial is implemented as a channel and works as a sensing element of the device. The channel can be made of different nanomaterials such as nanowires, carbon nanotubes, graphene and graphene nanoribbons (Fiori & Iannaccone, 2007; Stern, Vacic, & Reed, 2008).

Conversely, the channel is buried insides substrate in conventional complementary metal oxide semiconductor transistors as shown in the Figure 9. The direct interface with the environment in FET based sensors yields a better control over the surface charge. This indicates that the FET based sensors are more sensitive to their environment: Direct electron transfer to channel surface, modulates surface potential and effects on the electrical properties of the channel. In other words, biological reactions which happen at the surface of channel could result to variation of potential of the semiconductor channel. The potential variation of the channel modulates channel conductance, capacitance, and carrier concentration; and changes the current-voltage characteristics of the biosensor (Liu & Guo, 2012).

Furthermore, to make sensitivity and selectivity toward a particular target, a specific functionalization group (also called probe or receptor) is anchored to the semiconductor channel surface as depicted in Figure 9. This recognition group is selected to detect its target molecule (analyte) with high affinity and specify (Curreli et al., 2008). In the following sections, we are going to analytically model and investigate the working principles and sensing mechanisms of the graphene and carbon nanotube based biosensors.

Figure 8. Schematic of a FET based nano-biosensor structure

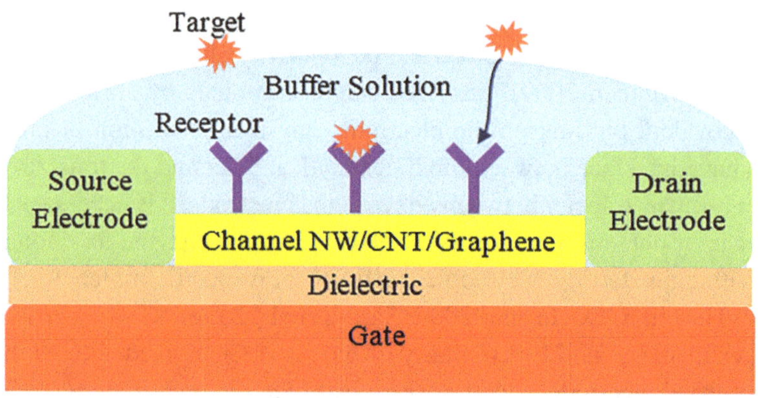

Figure 9. Metal oxide semiconductor field effect transistor (MOSFET) structure

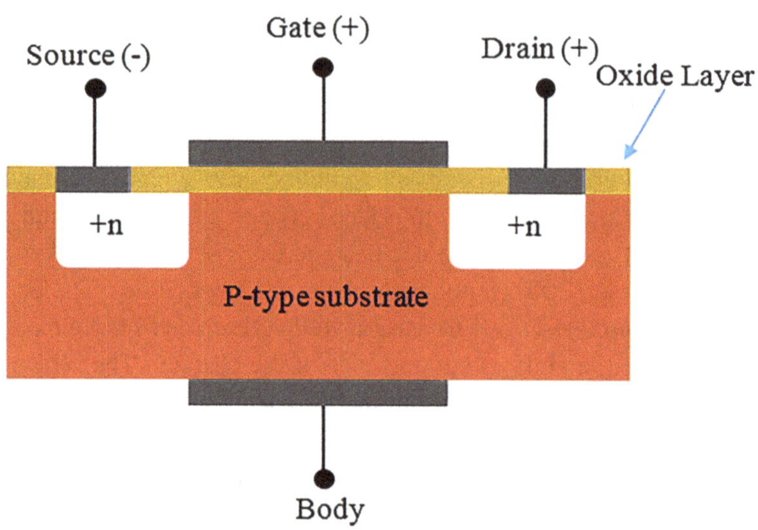

CNT Based Sensors

Carbon nanotubes (CNTs) with superior electrical performance are essential in designing modern biosensors (Abdelwahab, Koh, Noh, & Shim, 2010; Alwarappan, Boyapalle, Kumar, Li, & Mohapatra, 2012; Du et al., 2010). CNT-based biosensors have an economical production process, rapid response, high sensitivity, and good selectivity and are easily available in the market. The variety of CNTs potential and practical applications such as chemical sensors (J. Wang, 2005; J. Wang & Musameh, 2003; Zhang, Smith, & Gorski, 2004), and electronic devices has been reported so far. CNT show high electron transfer property when applied as an electrode in biosensors which aid to act as a good solid support for immobilization of molecules. The electronic properties of CNT implies that it can act as a good mediator in reactions and transfer electrons with electroactive species. Therefore, CNTs have been immensely used in nanosensors as an electrode material. CNT presents better behaviour than its traditionally used electrode material counterparts with high conductivity chemical stability (J. Wang & Musameh, 2003).

CNT Biosensor Model for Glucose Sensing

Glucose Sensing Mechanism

Due to their excellent mechanical stability, high conductivity, and antifouling properties, CNTs have been widely employed for GOx immobilization in biosensors (Sotiropoulou, Gavalas, Vamvakaki, & Chaniotakis, 2003). Moreover, the CNT platform provides a more appropriate environment for immobilized GOx and therefore provides a quick shuttling of electrons with the surface of an electrode (Sotiropoulou et al., 2003; Y.Q. Zhang et al., 2004). To explain the sensing mechanism, immobilization methods need to be described to explain the sensing mechanism of a biosensor. Immobilization means to integrate a biocatalyst with a matrix that it is not soluble in aqueous media. A wide variety of approaches can be applied for immobilization of enzymes or cells on the variety of natural and synthetic supports, that both immobilization approach and support are dependent to the type of enzyme and substrate (Guisan, 2006; Y. Q. Zhang et al., 2004). Enzymes are very instable and sensitive to their environment [19]. In the case that there is no special precaution, some common approaches such as deactivation on a adsorption and chemical or thermal inactivation have been usual (Chaniotakis, 2004; Skoog, West, & Holler, 1988).

The important techniques that maintain the enzyme activity of immobilization are encapsulation, covalent immobilization and site-specific mutagenesis (Grieshaber, MacKenzie, Vörös, & Reimhult, 2008; Sotiropoulou et al., 2003). In other words, application of the new materials will affect the sensing mechanism. Because of high surface-volume ratio, CNTs show good performance (Cao et al., 2013) when are used in biosensors (Park et al., 2012).

The CNT application on glucose detection, which is in our focus, is experimentally reported (D. Lee & Cui, 2010). Based on the reported method, multilayer poly diallyldimethylammonium chloride / Glucose oxidase (PDDA/GOx) has been assembled in the ultrasonic bath for glucose detection. The fabrication process of the single wall carbon nanotube (SWCNT) (1–2*nm* in diameter, 50*μm* in length) (D. Lee & Cui, 2009a, 2009b) based electrochemical glucose biosensors using glucose oxidase (GOx) has been reported (D. Lee & Cui, 2010) as shown in the simplified form in Figures 10 and 11. Figure 10 shows the assembly of (PDDA/SWCNT) on the Polyethylene terephthalate polyester flexible substrate, and extra (PDDA/GOx) biomolecular assembly is depicted in Figure 11.

To produce stable negative charges, GOx is dissolved into a phosphate-buffered saline (PBS) of a concentration of 1*mg/mL*. The phosphate monobasic (NaH_2PO_4) and dibasic (Na_2HPO_4) are employed as a standard pH buffer solution. The standard D-glucose solutions have been used in the glucose concentration test (D. Lee & Cui, 2010) and the results is shown in terms of drain current versus drain voltage (*I-V*) characteristics.

Proposed Model

Figure 11 shows the structure of the SWCNT FET with PET polyester as a back-gate and Chromium (Cr) or Aurum (Au) as the source and drain. A SWCNT is employed as a channel to connect the source and drain. Based on the proposed structure, two main modeling approaches in the carbon nanotube field-effect transistor (CNTFET) analytical modeling can be utilized. The first approach is derived from the charge-based framework, and the second modeling approach is a non-charge based analytical model using the surface-potential-based analysis method. The charge-based carrier velocity model is implemented in

Figure 10. The schematic fabrication process of glucose sensor
Adapted from D. Lee & Cui, (2010).

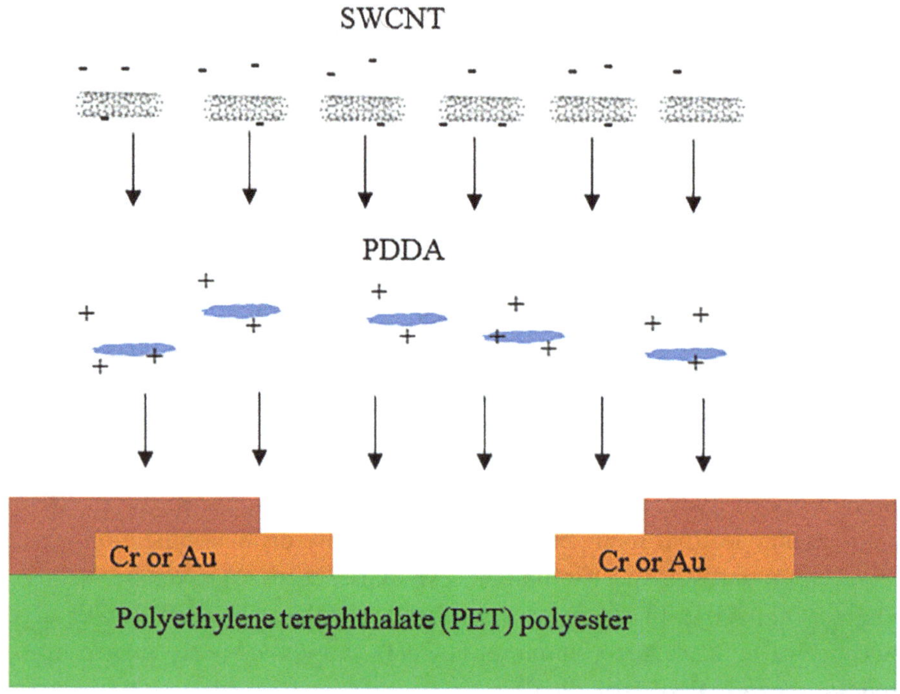

Figure 11. Proposed combinations of metal electrodes and SWCNTs in the form of FET

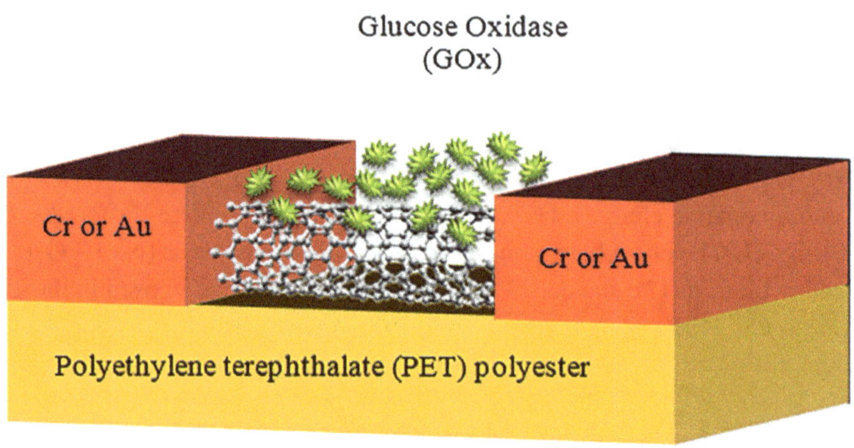

this work. The drift velocity of carrier in the presence of an applied electric field (Ahmadi, Tan, Ismail, & Arora, 2009) is given as

$$v_D = \frac{\mu_0 E}{1 + \frac{E}{E_c}} \tag{1}$$

where

μ is the mobility of the carriers,
E is the electric field, and
E_c is the critical electric field under high applied bias.

From Equation (1), the drain current as a function of gate voltage and drain voltage is obtained as

$$I_D = {}^2\frac{\left[2V_{GT}V_D - V_D^2\right]}{1+\frac{V_D}{V_c}} \tag{2}$$

where

$\beta = \mu C_G / (2L)$,
$V_{GT} = V_G - V_T$ and critical applied voltage as $V_c = \left(v_{sat}/¼\right)L$, where
v_{sat} is the saturation velocity,
V_G is the gate to source voltage,
V_T is the threshold voltage (Chek, Tan, Ahmadi, Ismail, & Arora, 2010),
C_G is the gate capacitance per unit length, and
L is the effective channel length (Ahmadi, Karamdel, Ismail, Dee, & Majlis, 2008).

The value of $\mu_{ef} = 60000 cm^2 / V.s$ is obtained (Anantram & Leonard, 2006). Because of the unknown nature of the quantum emission, it is not considered in this calculation. According to the geometry of the CNTFET that is proposed in the Figure 11, the gate capacitance (C_G) can be defined as

$$C_G = \frac{C_E C_Q}{C_E + C_Q} \tag{3}$$

where C_E and C_Q are the electrostatic gate coupling capacitance of the gate oxide and the quantum capacitance of the gated SWCNT, respectively (Anantram & Leonard, 2006). Figure 12 shows the *I-V* characteristics of a bare SWCNT FET for different gate voltages without any PBS and glucose concentration that is based on Equation (2).

The electrostatic gate coupling capacitance C_E for Figure 11 is given by

$$C_E = \frac{2\pi\varepsilon}{\ln\left(4H_{PET}/d\right)}L \tag{4}$$

where

Figure 12. I-V characteristic of single wall CNT-FET based on the proposed model for different gate voltage (V)

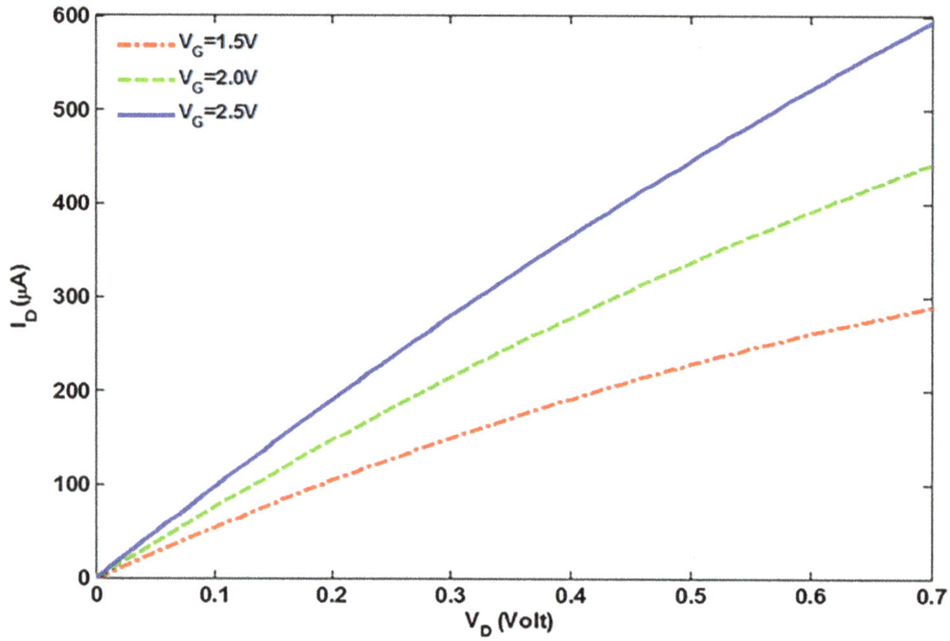

H_{PET} is the Polyethylene terephthalate (PET) polyester thickness,
d is the diameter of CNT,
$\varepsilon=3.3\varepsilon_0$ is the dielectric permittivity of PET.

The existence of the quantum capacitance is due to the displacement of the electron wave function at the CNT insulator interface. The C_Q relates to the electron Fermi velocity (v_F) in the form of $C_Q = 2\ e/v_F$, where $v_F \approx 10^6 m/s$ (Y. M. Lin et al., 2005). Numerically, the quantum capacitance is 76.5 $aF/\mu m$ and shows that both the electrostatic and quantum capacitance have a high impact on the CNT characteristics (Heller, Kong, Williams, Dekker, & Lemay, 2006; Ilani, Donev, Kindermann, & McEuen, 2006). At the saturation velocity, the electric field is very severe at the early stage of current saturation at the drain end of channel.

In this research, the effect of glucose concentration, F_g on the *I-V* characteristics of the CNTFET is studied. First, PBS should be added to the solution so that glucose is detectable and the change in *I-V* can be observed. The *I-V* change is due to the carrier concentration gradient of the injected carriers from the PBS to the channel and vice versa. The channel carrier concentration can be modeled in the function of gate voltage variations as

$$V_{GS1\,(with\ PBS)} = V_{GS\left(without\ PBS\right)} + V_{PBS} \tag{5}$$

where

$V_{GS\ 1}$ is the gate voltage in the presence of PBS,
V_{PBS} is voltage due to the interaction of PBS with CNT in the solution, and
$V_{GS1\ (without\ PBS)}$ indicates the gate voltage in a bare channel.

The effect of PBS in the *I-V* characteristics is modeled as

$$I_D = \frac{\left[2\left(V_{GS(without\ PBS)} + V_{PBS} - V_T\right)V_D - V_D^2\right]}{1 + \frac{V_D}{V_c}} \quad (6)$$

Before glucose and PBS is added, $V_{GS(without\ PBS)}$ is set to be $1.5V$. The V_{PBS} is found to $0.6V$ when PBS concentration, $F_{PBS} = 1mg/mL$ is added into the solution. By using Equation (5) and Equation (6), the presented model provides a good consensus between the model and the experimental data as shown in Figure 13.

In the glucose sensing mechanism reported in Ref. (D. Lee & Cui, 2010), *β-D* glucose oxidizes to D-glucono-δ-lactone and H_2O_2 as a result of the catalyst reaction of GOx. The hydrolyzation of D-glucose-δ-lactone and the electrooxidation of H_2O_2 under an applied gate voltage produces two hydrogen ions and two electrons which contributes to the additional carrier concentration in the SWCNT channel. On the whole, the glucose sensing mechanism can be summarized as followed

Figure 13. Comparison of the I-V simulation output and the experimental data: PBS concentration F_{PBS} = 1 mg/mL, $V_{GS(without\ PBS)}$ = 1.5V and V_{PBS} = 0.6V
D. Lee & Cui, 2010.

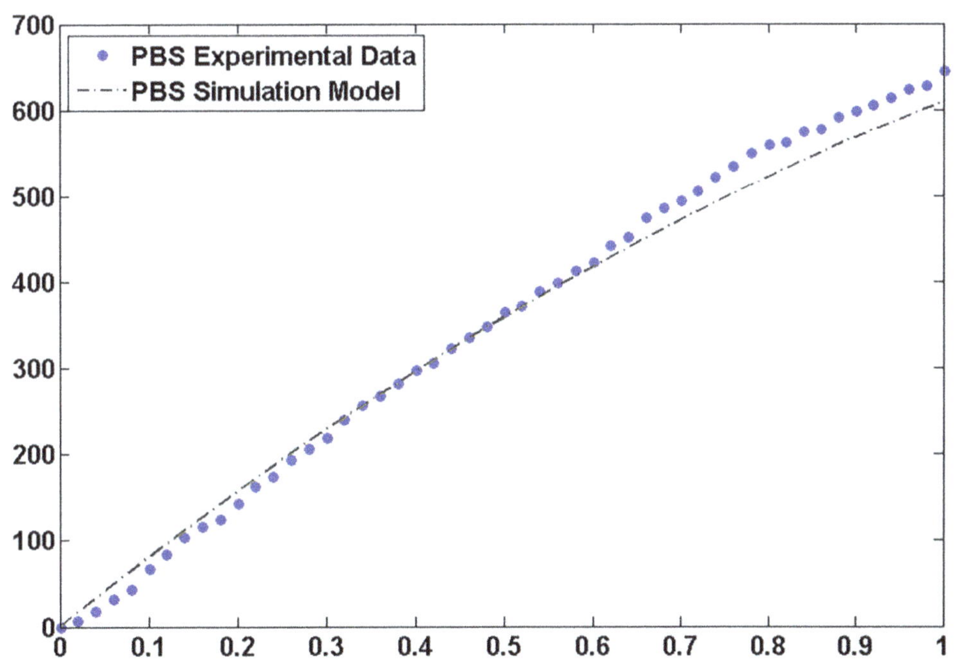

β-D-glucose + O_2 with GOx enzyme → D-glucono-δ-lactone + H_2O_2 (7)

D-glucono-δ- lactone + H_2O → D-gluconate$^-$ + H^+ (8)

$H_2O_2 \rightarrow O_2 + 2H^+ + 2e^-$ (9)

The variation of the proximal ionic deposition and the direct electron transfer to the electrode surface modify the electrical conductance of the SWCNT. The direct electron transfer leads to a variation of the drain current in the SWCNT FET. Therefore, Equation (10) that incorporates the gate voltage change due to the additional electrons from the glucose interaction with PBS is given by

$$V_{GS2(with\ PBS\ and\ Glucose\ Detection)} = V_{GS(without\ PBS)} + V_{PBS} + V_{Glucose} \quad (10)$$

By incorporating Equation (10), Equation (6) then becomes

$$I_D = \frac{\left[2\left(V_{GS(without\ PBS)} + V_{PBS} + V_{Glucose} - V_T\right)V_D - V_D^2\right]}{1 + \frac{V_D}{V_c}} \quad (11)$$

$V_{Glucose}$ is the glucose-based controlling parameter that highlights the effects of glucose concentration, F_g against gate voltages. In the proposed model, Equation (12) is obtained by analyzing the rise I_D with gate voltages versus glucose concentration. Based on the iteration method demonstrated in Rahmani et al. (2013), the concentration control parameter as a function of glucose concentration in a piecewise exponential models is expressed by

$$V_{Glucose}(F_g) = \begin{cases} 0V & if\ F_g = 0mM \\ 1.42V - \exp\left(-0.1 \times F_g\right) & if\ F_g > 0mM \end{cases} \quad (12)$$

In other words, the *I-V* characteristics of the biosensor can also be controlled by changing glucose concentration. To evaluate the proposed model, the drain voltage is varied from 0 to 0.7 *V* similar to the measurement work, and F_g is changed from the range of $2mM$ to $50mM$ (D. Lee & Cui, 2010).

Having proposed glucose biosensor model, now it is time to employ developed to monitor biosensor functionality. The biosensor manner changes according to the Equation (11), which was extracted in previous chapter. The glucose concentration controlling equation (Equation 12) is applied into Equation (11). By means of changing F_g (concentration of the glucose) with respect to the value of the concentration in the paper, Glucose concentration can be controlled using this equation. Then the simulated results will be compared with experimental data to verify proposed model.

Glucose Sensing and Accuracy of Sensor Model

To evaluate the presented model, the drain voltage with a step of $20mV$ in the same range of experimental work is changed from range of 0 to $0.7V$ and V_G was turned from the range of 0 to $2V$ with a step of $0.3V$ (D. Lee & Cui, 2010). Based on the experimental data, it is found that *I-V* characteristic of the biosensor can be controlled by changing glucose concentration as well.

By increasing the glucose concentration in multiple steps from 2–50*mM*, a fairly good consensus between our simulation model and experimental data particularly in the linear region is illustrated in Figures 14-20. The results show the accuracy of our predictive model against the measurement data of the glucose biosensor for various glucose concentrations up to 50*mM*. It is observed that the current in the CNTFET increases exponentially with glucose concentration.

From Figures 14-20, the glucose sensor model shows a sensitivity of 18.75 *A/mM* on a linear range of 2–10*mM* at $V_D = 0.7V$. The high sensitivity is due to the additional electron per glucose molecule from the oxidation of hydrogen peroxide (H_2O_2) and the high quality of polymer substrate that are able to sustain immobilized GOx (D. Lee & Cui, 2010). It is shown that by increasing the concentration of glucose, the current in CNTFET increases. It is also evident that gate voltage increases with higher glucose concentrations. Table 1 shows the results for the Root Mean Square (RMS) errors (absolute and normalized) when the glucose is varied from 2*mM* to 50m. The normalized RMS error is given by the absolute RMS divided by the mean of actual data. It also revealed that the corresponding average RMS errors do not exceed 13%. The discrepancy between simulation and experimental data is due to the onset of saturation effects of the drain current at higher gate voltages and glucose concentration where enzyme reactions are limited.

Finally the accuracy of presented biosensor model is veryfied by glucose concentration in range of 50*mM* as shown in Figure 20. Based on the experimental data that are simulated in this paper, it can be seen that by increasing the concentration of the glucose, the current voltage characteristics of the biosensor changes. Based on the Figure 14, when the glucose concentration is 2*Mm*, by increasing the gate source voltage (V_{GS}) from 0 to 0.7 (*V*), the current varies between the range of 0 to around $600\mu A$. Hereafter, by increased the glucose concentration into 4*Mm* for the same range of voltage in Figure 15, the current changes between 0 to near $700\mu A$. In continue in the Figures 16-20, by increasing glucose concentration to the 6*Mm* and above the current increases. It is driven variation of carrier concentration of the channel. When the glucose concentration increases, the number of electrons injected to the CNT channel increase that leads to growth of the carrier concentration.

Therefor carrier concentration of the CNT channel has a direct relation with glucose concentration. In means that by increasing concentration of the glucose number of electrons shuttling to the surface of the electrode and the channel carrier concentration increase that lead to increase the current and vice versa. Based on the Figure 20, the glucose concentration of 50 *Mm* is the highest concentration and it can be seen it has the maximum current value that is around $860\mu A$. It means that there are most numbers of electrons in the CNT channel for 50*Mm* glucose. It is evident that for different glucose concentrations, the proposed model can be used properly.

Figure 14. An I-V comparison of the simulated output and the measured data for glucose concentration, F_g = 2mM: the other parameters used in the simulation data are $V_{GS(without\ PBS)}$ = 1.5V and V_{PBS} = 0.6V. D. Lee & Cui, 2010.

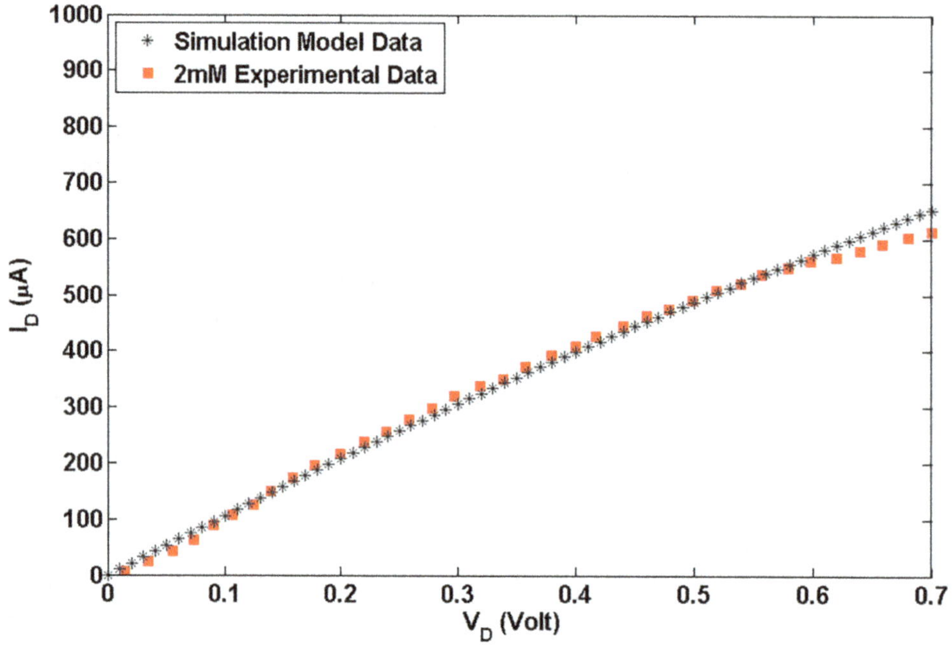

Figure 15. An I-V comparison of the simulated output and the measured data for glucose concentration, F_g = 4mM: the other parameters used in the simulation data are $V_{GS(without\ PBS)}$ = 1.5V and V_{PBS} = 0.6V. D. Lee & Cui, 2010.

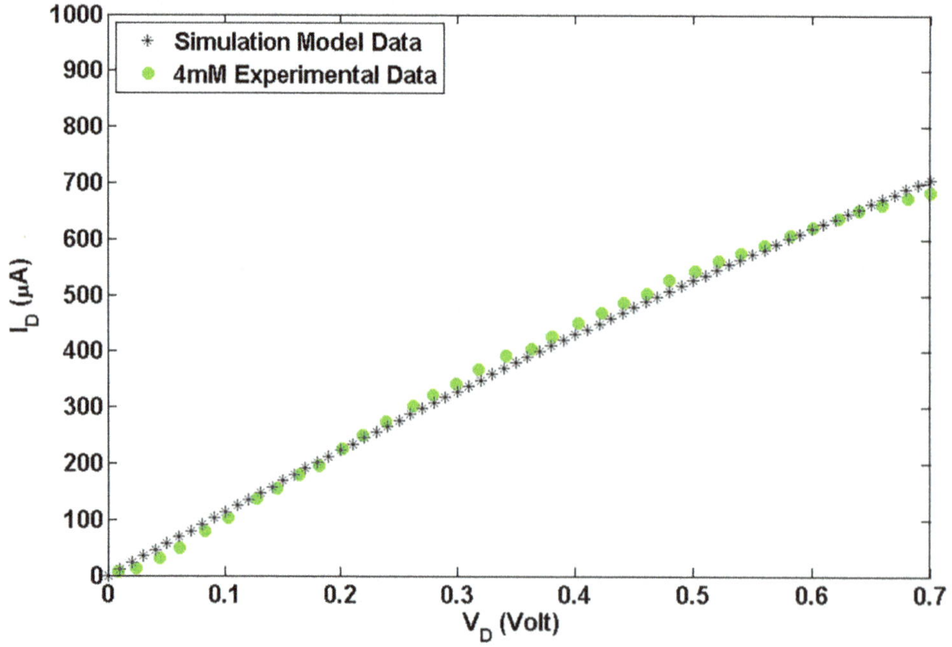

Figure 16. An I-V comparison of the simulated output and the measured data for glucose concentration, F_g = 6mM: the other parameters used in the simulation data are $V_{GS(without\ PBS)}$ = 1.5V and V_{PBS} = 0.6V. D. Lee & Cui, 2010.

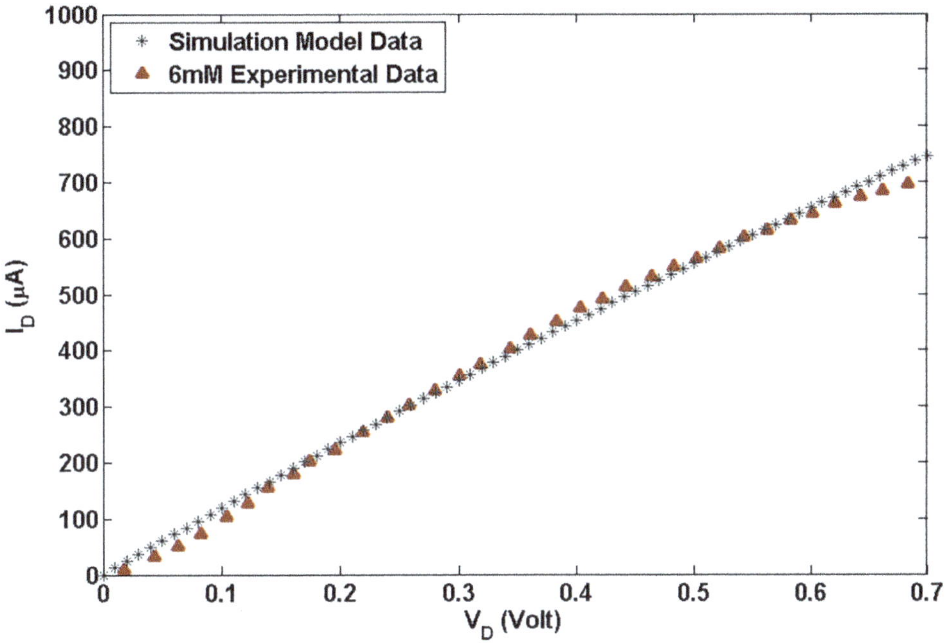

Figure 17. An I-V comparison of the simulated output and the measured data for glucose concentration, F_g = 8mM: the other parameters used in the simulation data are $V_{GS(without\ PBS)}$ = 1.5V and V_{PBS} = 0.6V. D. Lee & Cui, 2010.

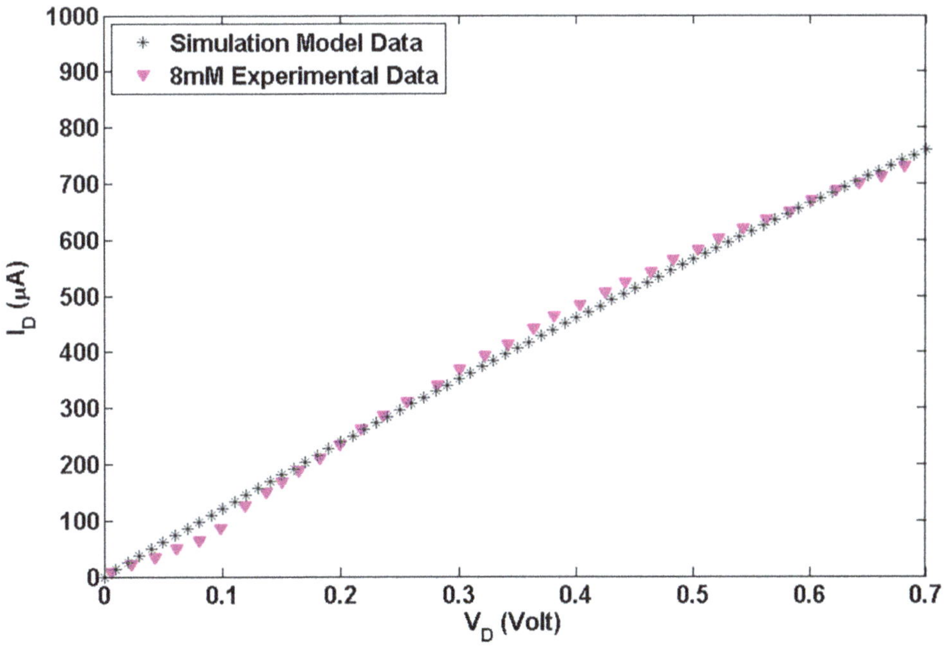

Figure 18. An I-V comparison of the simulated output and the measured data or glucose concentration, F_g = 10mM: the other parameters used in the simulation data are $V_{GS(without\ PBS)}$ = 1.5V and V_{PBS} = 0.6V. D. Lee & Cui, 2010.

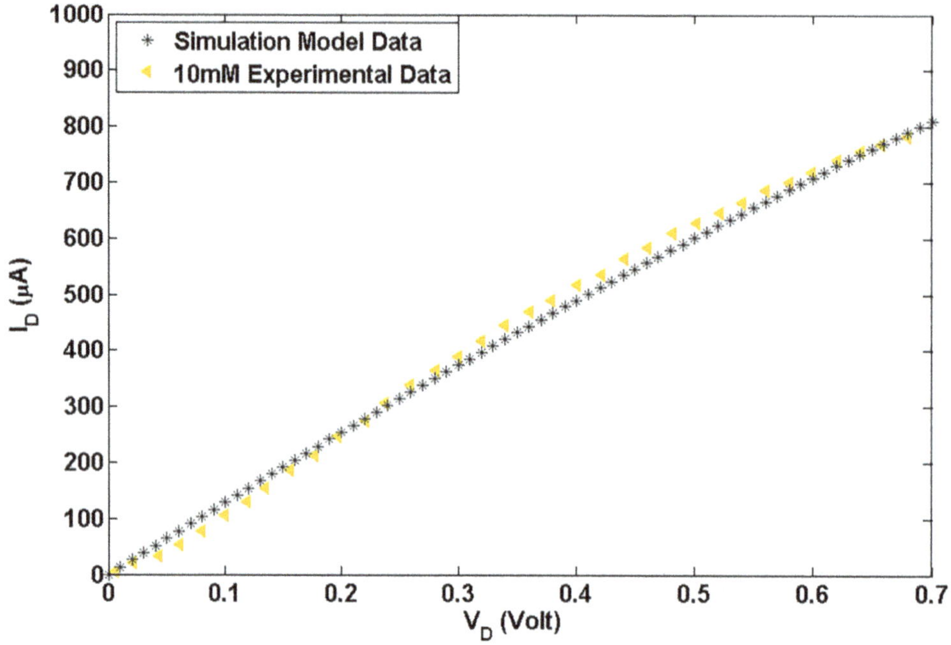

Figure 19. An I-V comparison of the simulated output and the measured data for glucose concentration, F_g = 20mM: the other parameters used in the simulation data are $V_{GS(without\ PBS)}$ = 1.5V and V_{PBS} = 0.6V. D. Lee & Cui, 2010.

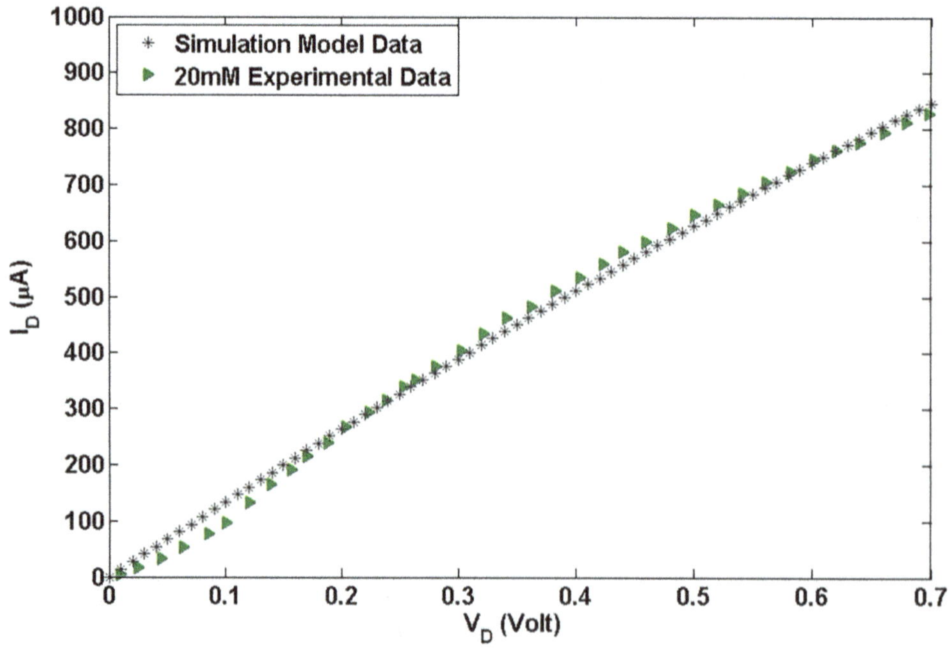

Figure 20. An I-V comparison of the simulated output and the measured data for glucose concentration, F_g = 50mM: the other parameters used in the simulation data are $V_{GS(without\ PBS)}$ = 1.5V and V_{PBS} = 0.6V. D. Lee & Cui, 2010.

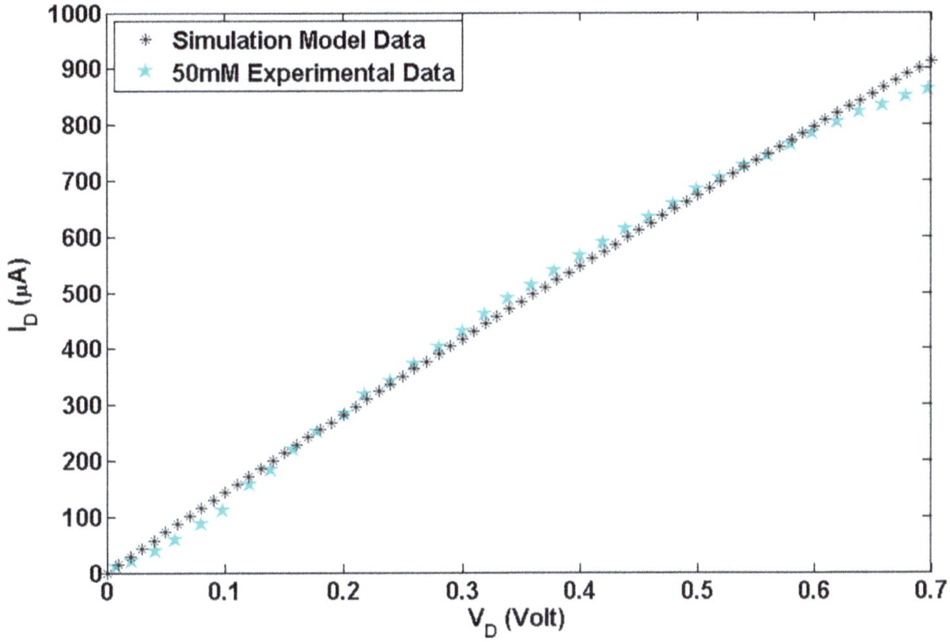

Table 1. Average RMS errors in IV comparison to the experimental results of the simulation model and the measured data

Glucose	Absolute RMS Error	Normalized RMS Error
0mM with PBS	19.24	5.66%
2mM	57.55	12.22%
4mM	49.05	9.75%
6mM	59.47	11.23%
8mM	53.99	9.80%
10mM	55.60	9.53%
20mM	69.18	11.17%
50mM	75.07	11.60%

GRAPHENE BASED SENSORS

There have been many studies on graphene for sensing application, since the first report on the graphene sensing by Schedin et al. (2007). Graphene showed excellent results for sensing applications in expose to NO2, NH3, H2O and CO (Schedin et al., 2007). It was demonstrated that after exposure to the analyte, sensing properties of graphene can be recovered by illumination to UV. Moreover, based on the reported work, the mobility of the carriers cannot be affected by chemical doping of the graphene, even in high concentrations.

Graphene Biosensor Modeling for Escherichia Coli Bacteria Detection

Escherichia Coli Bacteria Sensing Mechanism

The development of biosensors using nanomaterials like graphene has progressed recently (Putzbach & Ronkainen, 2013). Using graphene for medical diagnostics applications is currently an active research area. The E. coli bacteria is the most common cause of food poisoning, and shortens the shelf life of food products. Some E. coli serotypes cause serious food poisoning in their hosts and occasionally are responsible for product recall due to food poisoning(Abdel-Hamid, Ivnitski, Atanasov, & Wilkins, 1999; Clermont, Bonacorsi, & Bingen, 2000).Among the pathogenic bacteria, E. coli O157:H7 is known as one of the most noxious food poisoning pathogens which can cause food-borne disease and even death (Rijal & Mutharasan, 2013; Y. X. Wang, Ping, Ye, Wu, & Ying, 2013). Hence, the demand for rapid, simple and cost effective techniques for the specific and sensitive determination of E. coli O157:H7 is gaining mass appeal. (Akhavan & Ghaderi, 2012; S. Huang et al., 2008; H. Lin, Lu, Ge, Cai, & Grimes, 2010; Y. Wang, Ye, Si, & Ying, 2011). Y. Huang et al. (2011) have fabricated a liquid gated graphene field effect transistor (LG-GFET) to sense E. coli and monitor its metabolic activity. As shown in Figure 21, in order to specifically sense E. coli, anti-E. coli antibodies have been immobilized onto a graphene film through linker molecules (1-pyrenebutanoic acid succinimidyl ester) whose pyrene group from one side links to the surface of graphene via solid pi-pi junctions, and from the other side, the succinimidyl ester group makes covalent junctions with amino groups on the antibody. In order to prevent nonspecific binding, ethanolamine has been employed to quench the unreacted succinimidyl esters on the linker molecules, and Tween 20 has been used to passivate the uncoated graphene surface. Then, on top of the functionalized graphene film, the E. coli bacterium has been attached. Figure 22 is presented to understand the biological reactions occurring in the sensing mechanism of the E. coli which indicates the covalent bonding of linker with Fab region through amino group and Fc region of antibody with linker through COOH.

Narrow width graphene FET is assumed in the biosensor structure, and ballistic transport for the carriers in the channel is considered. The width of narrow-width graphene is comparable to the wavelength of free electrons (or de-Broglie wave length, λ_D=10 nm,) and can be varied up to 100nm. By increasing the channel width, the resistivity of the channel decreases so that the current increases. Furthermore, a larger channel leads to smaller band gap which causes it to be very sensitive and suitable for application in sensors (Ouyang, Yoon, & Guo, 2007). The surface to thickness ratio of graphene is very high. In other words, graphene has got a very large surface area. This allows it to effectively absorb molecules and to be used as a sensitive sensor. Furthermore, the graphene band gap is very small which means that the small variation in the graphene carrier concentration will result in a large change in graphene conductance. Therefore, the aim of this study is to develop a LG-GFET-based biosensor model for E. coli detection, and validate the theoretical predictions with the experimental results.

Proposed Model

It has been reported that the gradual increase in the number of E. coli molecules trapped by the antibodies on the graphene films leads to an increase in graphene conductance (Y. Huang, Dong, Liu, Li, & Chen, 2011). Hence, the graphene conductance as a main sensing parameter is considered, and its variation in the sensing area in the presence of bacteria is analytically modeled. To calculate graphene

Figure 21. Schematics of liquid gated graphene FET biosensor for detection of E. coli bacteria
Adapted from (Y. Huang et al., 2011).

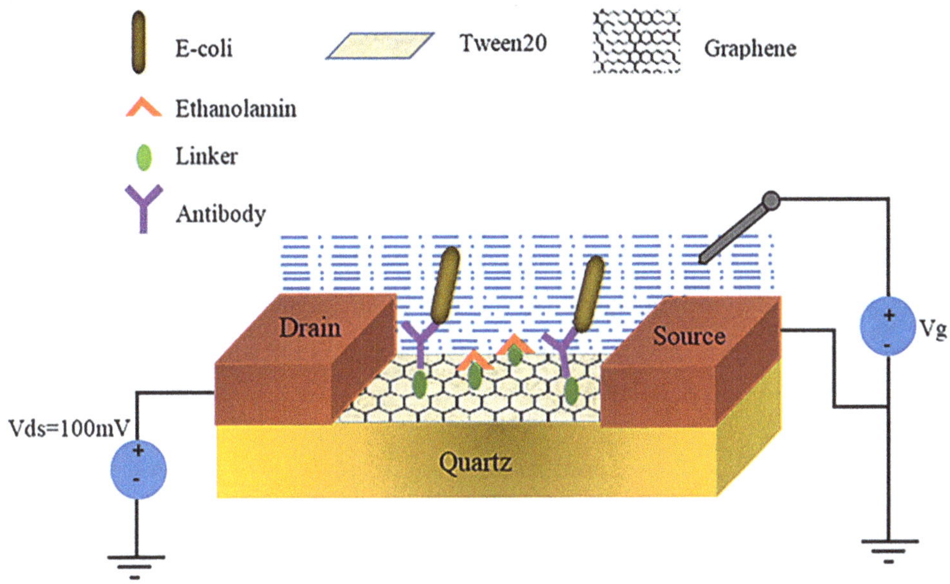

Figure 22. Schematics of binding of linker with antibody: a) binding of Fc region of antibody with linker; b) binding of Fc region of antibody with linker through COOH

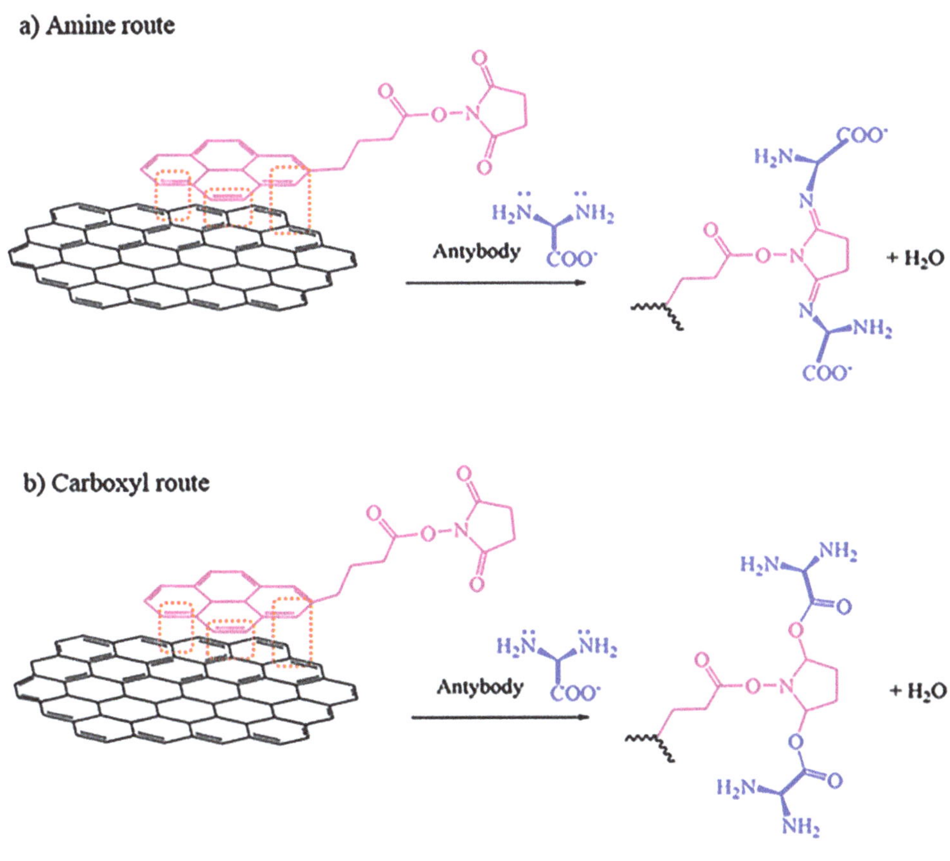

conductance, the $E(\vec{k})$ relation of the narrow width graphene with Taylor expansion has been employed (Ahmadi, Johari, Amin, Fallahpour, & Ismail, 2010; Berger et al., 2006; Karimi, Yusof, Rahmani, Hosseinpour, & Ahmadi, 2014) as

$$E(\vec{k}) = \pm \frac{3ta_{C-C}}{2}\sqrt{k_x^2 + \beta^2} \tag{13}$$

where (±) signs indicate the conductance band and valence band, $a_{c-c} = 1.42\overset{0}{A}$ is the carbon-carbon bond length, t=2.7eV is the tight-binding overlap energy for the nearest neighbor C-C atoms, $\vec{k}_x$ is the wave vector along the x direction, and $\vec{\beta}$ is the quantized wave vector which has been reported as (Ahmadi, Johari, Amin, Fallahpour, & Ismail, 2010)

$$\beta = \frac{2\pi}{a_{c-c}\sqrt{3}}\left(\frac{p_i}{N+1} - \frac{2}{3}\right) \tag{14}$$

where N represents the number of dimmer lines which determines the width of the graphene, and p_i is the sub-band index. Since the relation between energy and wave vector is not parabolic, applying Taylor expansion leads to the following parabolic relation (Ahmadi, Ismail, Tan, & Arora, 2008; Datta, 1997),

$$E(k) \approx \frac{E_g}{2} + \frac{E_g}{4\beta^2}k_x^2 \tag{15}$$

This can be compared by the conventional parabolic dispersion relation Equation (16) as well.

$$E \approx \frac{E_g}{2} + \frac{\hbar^2 k_x^2}{2m^*} \tag{16}$$

where

m^* indicates the effective mass of graphene, and
$\hbar$ is reduced Plank constant.

At a given energy, the actual modes number *M(E)* which depends on the sub-band location can be obtained based on the wave vector. By taking the derivation of the wave vector, *k*, over the energy *E(dk/dE)*, the number of modes, *M(E)*, can be written as

$$M(E) = \frac{\Delta E}{\Delta k \cdot L} = \frac{3ta_{c-c}}{2L}\left(\frac{4E}{3ta_{c-c}} - 2\beta^2\right)^{1/2} \tag{17}$$

where L indicates the length of the narrow width graphene. The number of conduction channels can be obtained by considering spin and degeneracy, which in the presence of the Landauer formula, the conductance can be expressed as

$$G = \frac{2q^2}{h}\int_{-\infty}^{+\infty} dEM(E)T(E)\left(-\frac{df}{dE}\right) \tag{18}$$

where

q is the electronic charge,
h is the Planck's constant and
f is the Fermi-Dirac distribution function.

The transmission probability, $T(E)$, is average probability of injected electrons at one end that will be transmitted to the other end. Due to the completely ballistic carrier transportation in graphene, $T(E)$ is approximately equal to one (Naeemi & Meindl, 2007). With the replacement of the mode numbers (number of sub-bands) as well as Fermi-Dirac distribution function, the graphene length depended conductance is given by

$$G = \frac{3q^2}{h}\times\frac{a_{C-C}t}{L}\left(3a_{C-C}t\right)^{1/2}\int_{-\infty}^{+\infty}\left(E-\frac{2\beta^2}{3a_{C-C}t}\right)^{1/2} d\left(-\frac{1}{1+e^{(E-E_F)/k_BT}}\right) \tag{19}$$

where

E_F is the Fermi energy level,
k_B is the Boltzmann constant and
T is the temperature.

The energy and number of modes are employed in the conductance calculation. Subsequently variation of conductance of narrow width graphene in the presence of biological molecules is utilized to immobilize and sense E. coli bacteria. It has been distinguished that for a variety of materials (molecules), its conductance value differs. For sensing E. coli, it has been immobilized by means of linker molecules therefore the graphene conductance is changed by binding other molecules to it (Y. Huang, Dong, Liu, Li, & Chen, 2011). Hence the variation on total conductance can be supposed as a conductance of bare graphene minus variations of conductance after binding different molecules (minus sign is because of G reduction) which can be explained by the reduction in carrier density.

$$G_T = G - G_P \tag{20}$$

where

G_T is the total conductance, and
G_p is the variation of conductance in the presence of other molecules.

Based on the results presented by Y. Huang et al. (2011), the G_p is a function of the molecule type bonded to the graphene therefore varies for different molecules, so it is modeled as:

$$G_p = \lambda \times P \tag{21}$$

where *P* illustrates the type of the molecule bonded to the graphene and indicates the $\pi - \pi$ junction overlap energy with a unit of *e.V*, and *λ (1/Ω e.V)* is defined as a molecule type sensing parameter which varies by bonded molecule nature, therefore the related *I-V* characteristic is in the form of:

$$I_{ds(\mathrm{P})} = (G - \lambda \times P) \times V_{ds(\mathrm{P})} \tag{22}$$

where $I_{ds(p)}$ and $V_{ds(p)}$ are the drain–source current and voltage corresponding to the different types of the bonded molecules. In other words, while molecules bind to the graphene the graphene conductance in the presence of the bonded molecule is changed.

Results and Discussion

By employing the model described in the preceding section, the response of graphene conductance to the binding of different molecules to the graphene surface is plotted in Figure 23.

In this work, the response of the presented model by molecule variation is modeled and molecule parameter to describe the minimum conductance shift is defined therefore, this phenomenon is incorporated into sensor modeling. In order to model a biosensor, the effect of E. coli on the conductance of the narrow width graphene in the form of carrier variation is considered. Thus, the total conductance is assumed to be the bare graphene conductance (*G*) minus the conductance variation in the presence of E. coli.

$$G_T = G - G_{E.c} \tag{23}$$

Here, G_T is total conductance of graphene and the $G_{E.c}$ indicates the conductance of graphene affected by E. coli. Since the conductance of graphene is a function of E. coli concentration based on the reported experiment by Y. Huang et al. (2011) which verifies, by changing E. coli concentration, the $G_{E.c}$ and total conductance G_T has been change. Thus, the $G_{E.c}$ can be written as a function of E. coli concentration as

$$G_{E.c} = \alpha \times F_{E.c} \tag{24}$$

where $F_{E.c}$*(cfu/mL)* is the concentration of E. coli, and α is defined as the concentration control parameter and its unit is given by *cfu/mL.Ω*.

The total concentration is modified

Figure 23. Response to graphene conductance variation for different bonded molecules

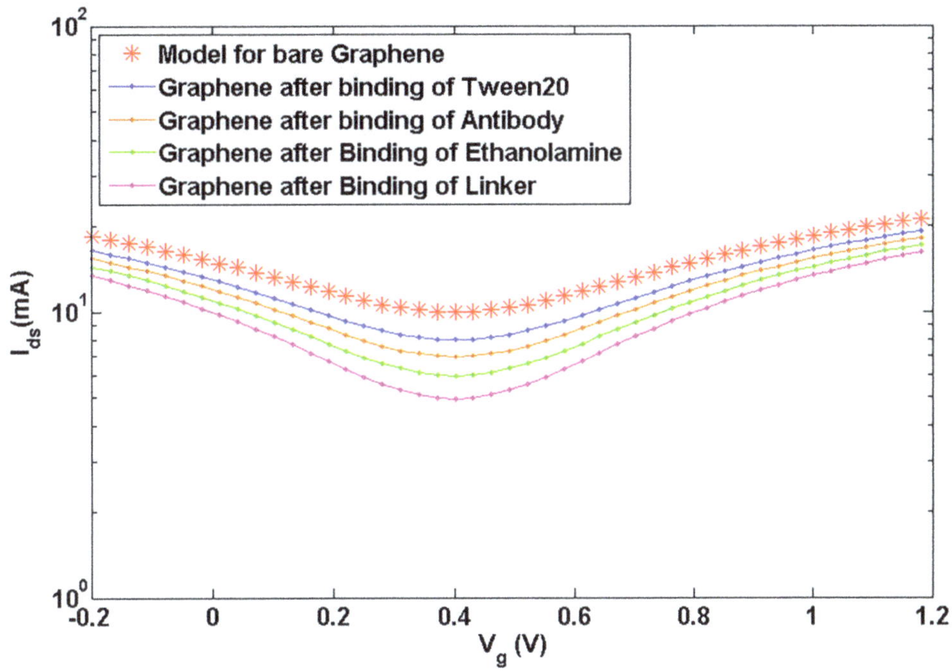

$$G_T = G - \left(\alpha \times F_{E.c}\right) \tag{25}$$

Consequently, the *I-V* model of the biosensor which is a function of total conductance is obtained by

$$I_{ds(\mathrm{E.c})} = (G - \alpha \times F_{E.c}) \times V_{ds(\mathrm{E.c})} \tag{26}$$

Figures 24-28 show the I_{ds} versus V_{ds} of the LG-GFET biosensor in the presence of E. coli, for concentrations between 0 to 10^5*cfu/mL* when the solution-gated voltage is zero *(V_g=0V)*. It can be seen that by increasing the concentration of E. coli, the current increases. In other words, as the concentration increases, the number of E. coli bacteria caught by linker molecules increases, which leads to the rise in current.

Charge transfer is involved within the sensing mechanism of graphene FET based biosensor. This happens during the E. coli molecules interaction with electrode surface which leads to modify the channel media conductivity through this interaction. It is likely that this phenomenon occur as a result of interaction between graphene surface and bacteria molecule which via linker molecules is attached to the graphene. Based on the reported experimental work by Y. Huang et al. (2011), for the sensing mechanism of E. coli, hydrogen peroxide (H_2O_2) is released as a result of chemical interaction between the linker molecules and E. coli then catalyze under applied voltage and releases electron. The direct electron transfer to LG-GFET leads to the variation of drain source current.

Figure 24. Comparison of I_{ds} versus V_{ds} of LG-GFET biosensor model with experimental data for E. coli concentration, 10 cfu/mL
Y. Huang, Dong, Liu, Li, & Chen, 2011.

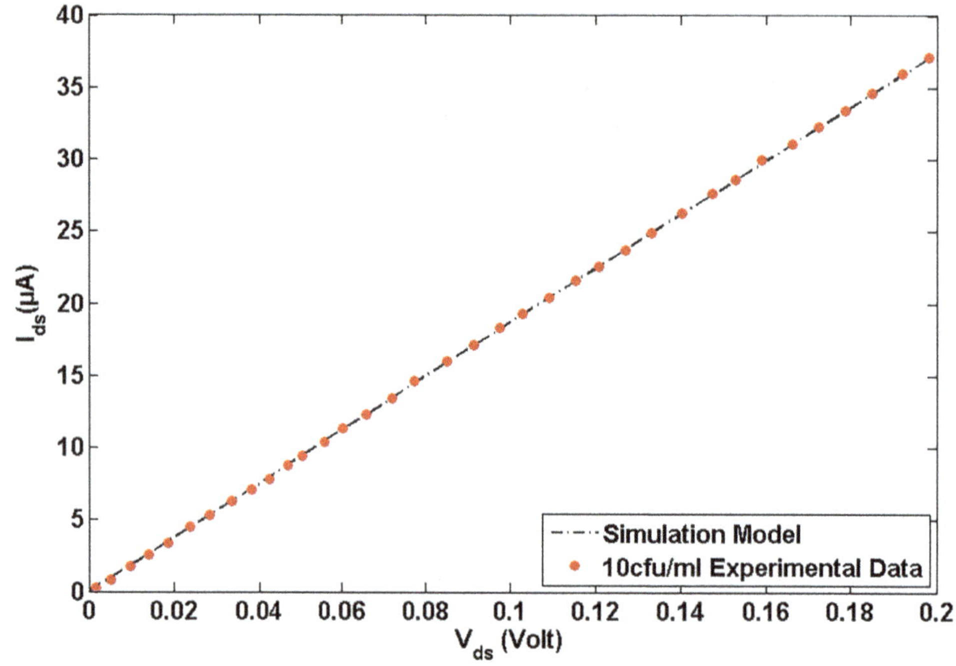

Figure 25. Comparison of I_{ds} versus V_{ds} of LG-GFET biosensor model with experimental data for E. coli concentration, 100 cfu/mL
Y. Huang, Dong, Liu, Li, & Chen, 2011.

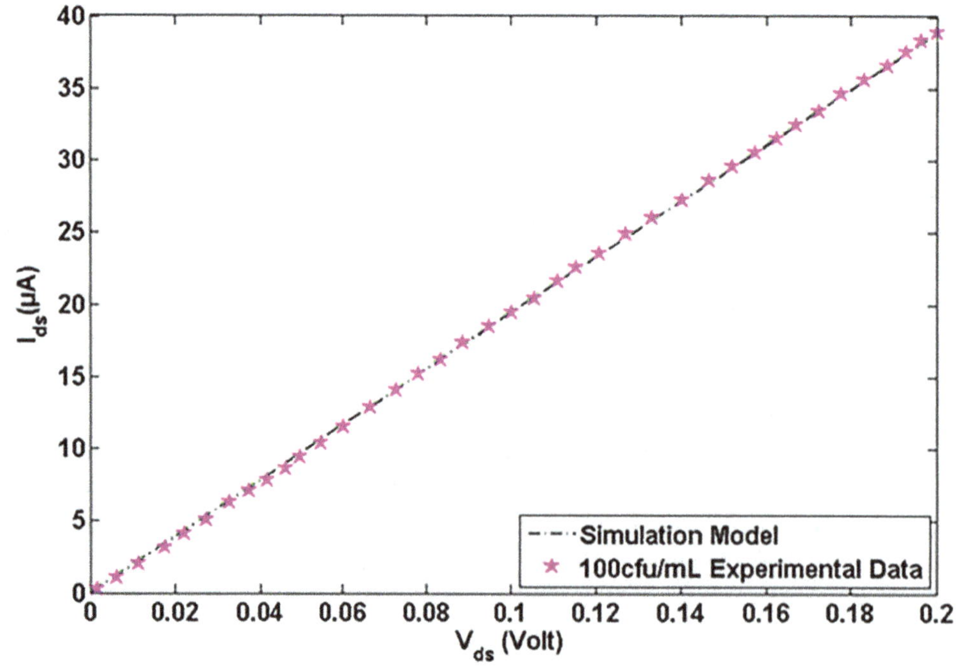

Figure 26. Comparison of I_{ds} versus V_{ds} of LG-GFET biosensor model with experimental data for E. coli concentration, 1000 cfu/mL
Y. Huang, Dong, Liu, Li, & Chen, 2011.

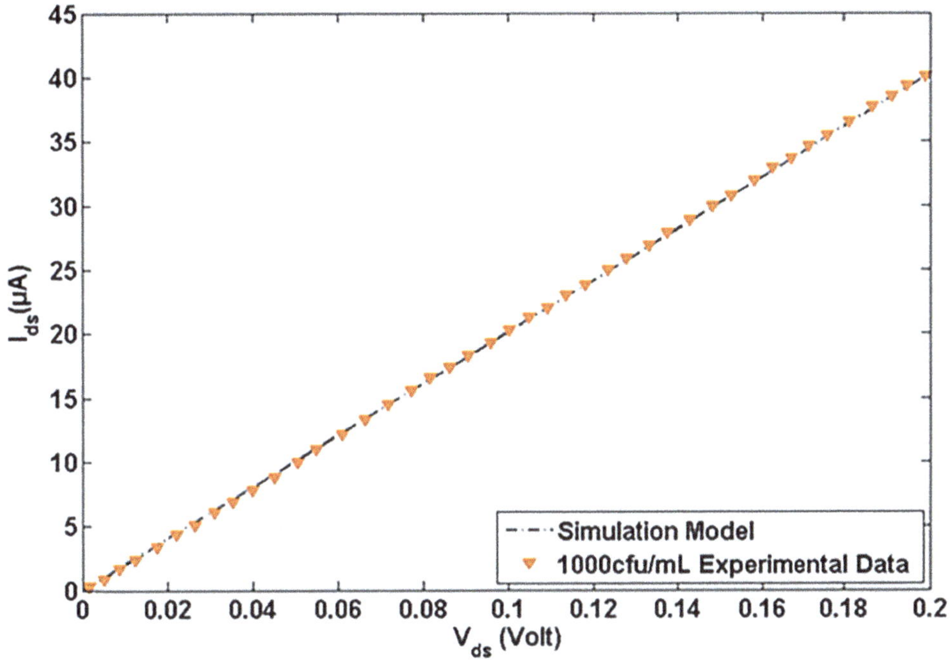

Figure 27. Comparison of I_{ds} versus V_{ds} of LG-GFET biosensor model with experimental data for E. coli concentration, 10000 cfu/mL
Y. Huang, Dong, Liu, Li, & Chen, 2011.

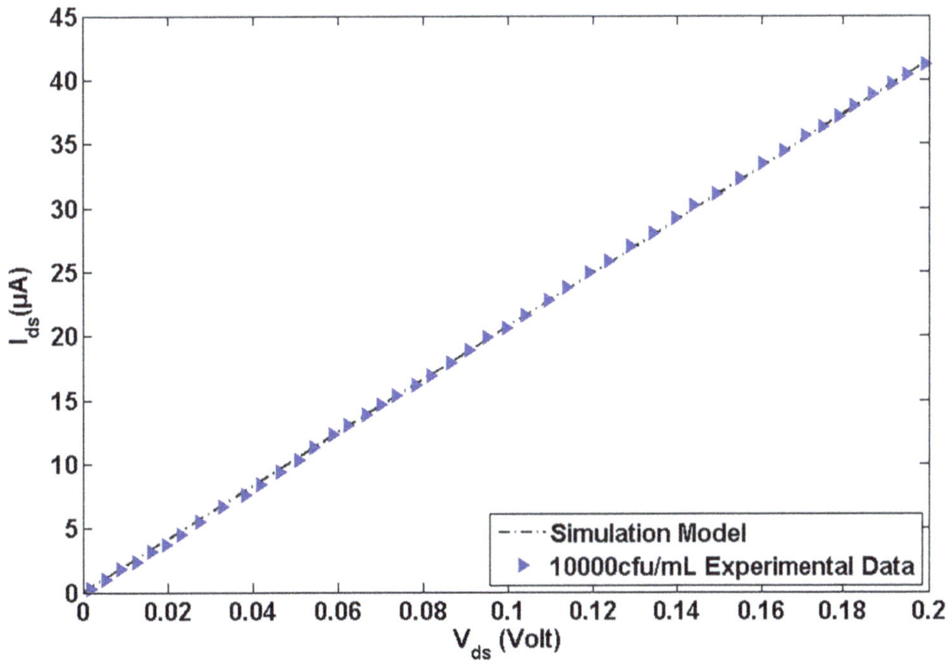

Figure 28. Comparison of I_{ds} versus V_{ds} of LG-GFET biosensor model with experimental data for E. coli concentration, 100000 cfu/mL
Y. Huang, Dong, Liu, Li, & Chen, 2011.

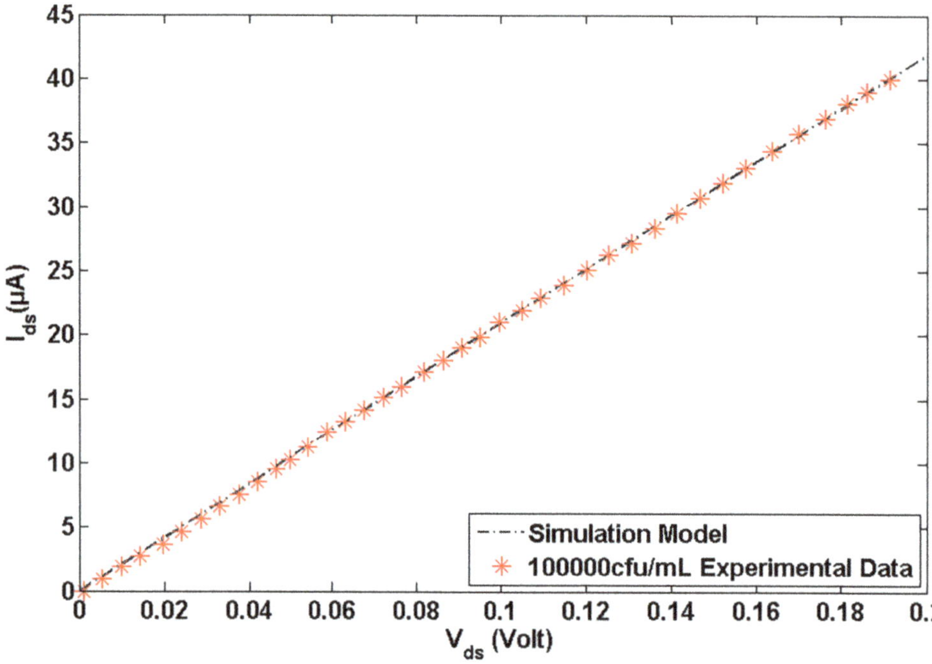

It needs to be made clear that since the graphene FET operates in the *p-type* region, the device conductance increases by means of increased hole density induced by the highly negatively charged bacterial wall which is different from the functionalizing group materials.

By increasing the concentration of E. coli, the conductance rises which is induced by increasing the carrier concentration, and leads to the elevation of biosensor current. In other words the rise of current can be explained by the adsorption of E. coli molecules to the graphene surface which lead to charge transfer between bacteria and electrode.

According to Y. Huang et al. (2011), the functionalized graphene devices, for 30 min duration, have been incubated with various concentrations of E. coli and thoroughly rinsed with PBS solution. Their electrical characteristic has been monitored by measuring the I_{ds}-V_{ds} value whereas the solution gate voltage V_g was held at $0V$. The figures above present obvious illustration of the fact that a good consensus between the presented model and extracted data can be seen. In the developed model, α is introduced as bacteria concentration control parameter. In our study the analytical model showed that, the rate of changes in graphene conductivity is dependent on the concentration which is introduced by the F parameter. The optimized equation for α versus E. coli concentration based on the mathematical methods leads to logarithmic relation given as

$$\alpha = A \times Ln\left(F\right) - B \quad (27)$$

where parameters A and B are calculated to be A=1.56896916 $1/\Omega$ and B=14.673495 $cfu/mL.\Omega$. The proposed sensing parameter can be implemented as a mediator between the concentrations of bacteria and biosensor to control current-voltage characteristics. The *I-V* characteristics of the graphene device in exposure to E. coli showed that the proposed sensing parameter, α, can be applied for accurate detection of E. coli.

CONCLUSION

Graphene and CNTs as carbon allotropes illustrate the amazing mechanical, chemical, and electrical properties that are preferable for use in biosensors. It is crucial to improve the biosensors' functionality for use in medical applications and food industry. In this chapter the SWCNT-FET based biosensor was analytically modeled for glucose detection. To validate the proposed model, a comparative study between the model and the experimental data from other research work was prepared, and good consensus was observed. Based on the modeling results it was shown that the current of the biosensor is a function of glucose concentration and therefore can be utilized for a wide process variation such as length and diameter of nanotube, capacitance of PET polymer, and PBS voltage. The glucose sensing parameters with gate voltages were also defined in exponential piecewise function. Based on a good consensus between the analytical model and the measured data, the predictive model can provide a fairly accurate simulation based on the change in glucose concentration.

Furthermore, the LG-GFET biosensor model was proposed for electrical detection of E. coli O157:H7bacteria. Chemical reactions between the graphene surface and E. coli trapped by linker molecules led to the release of electrons and modification in the channel conductance. The process of increasing E. coli concentration by the number of E. coli molecules trapped on the graphene surface in the form of conductance variation was modeled, and relevant controlling parameters are explored. The E. coli sensing parameter also was defined by using a logarithmic piecewise equation. Moreover, the current-voltage characteristics in terms of a conductance model were applied and the performance of the biosensor was investigated. Based on the good agreement between the presented model and experimental data, the biosensor predictive model which can provide an accurate simulation based on the biomolecule concentration, was offered.

REFERENCES

Abdel-Hamid, I., Ivnitski, D., Atanasov, P., & Wilkins, E. (1999). Flow-through immunofiltration assay system for rapid detection of E. coli O157: H7. *Biosensors & Bioelectronics*, *14*(3), 309–316. doi:10.1016/S0956-5663(99)00004-4 PMID:10230031

Abdelwahab, A. A., Koh, W. C. A., Noh, H.-B., & Shim, Y.-B. (2010). A selective nitric oxide nanocomposite biosensor based on direct electron transfer of microperoxidase: Removal of interferences by co-immobilized enzymes. *Biosensors & Bioelectronics*, *26*(3), 1080–1086. doi:10.1016/j.bios.2010.08.070 PMID:20864332

Ahmadi, M. T., Ismail, R., Tan, M. L., & Arora, V. K. (2008). The ultimate ballistic drift velocity in carbon nanotubes. *Journal of Nanomaterials*.

Ahmadi, M. T., Johari, Z., Amin, N. A., Fallahpour, A. H., & Ismail, R. (2010). Graphene nanoribbon conductance model in parabolic band structure. *Journal of Nanomaterials*.

Ahmadi, M. T., Johari, Z., Amin, N. A., Fallahpour, A. H., & Ismail, R. (2010). Graphene nanoribbon conductance model in parabolic band structure. *Journal of Nanomaterials*. doi:10.1155/2010/753738

Ahmadi, M. T., Karamdel, J., Ismail, R., Dee, C., & Majlis, B. Y. (2008). *Modelling of the current-voltage characteristics of a carbon nano tube field effect transistor.* Paper presented at the Semiconductor Electronics, 2008. ICSE 2008. IEEE International Conference on. doi:10.1109/SMELEC.2008.4770391

Ahmadi, M. T., Tan, M. L. P., Ismail, R., & Arora, V. K. (2009). The high-field drift velocity in degenerately-doped silicon nanowires. *International Journal of Nanotechnology*, *6*(7-8), 601–617. doi:10.1504/IJNT.2009.025299

Akhavan, O., & Ghaderi, E. (2012). Escherichia coli bacteria reduce graphene oxide to bactericidal graphene in a self-limiting manner. *Carbon*, *50*(5), 1853–1860. doi:10.1016/j.carbon.2011.12.035

Alwarappan, S., Boyapalle, S., Kumar, A., Li, C.-Z., & Mohapatra, S. (2012). Comparative study of single-, few-, and multilayered graphene toward enzyme conjugation and electrochemical response. *The Journal of Physical Chemistry C*, *116*(11), 6556–6559. doi:10.1021/jp211201b

Anantram, M., & Leonard, F. (2006). Physics of carbon nanotube electronic devices. *Reports on Progress in Physics*, *69*(3), 507–561. doi:10.1088/0034-4885/69/3/R01

Andle, J., & Vetelino, J. (1994). Acoustic wave biosensors. *Sensors and Actuators. A, Physical*, *44*(3), 167–176. doi:10.1016/0924-4247(94)00801-9

Arduini, F., Cassisi, A., Amine, A., Ricci, F., Moscone, D., & Palleschi, G. (2009). Electrocatalytic oxidation of thiocholine at chemically modified cobalt hexacyanoferrate screen-printed electrodes. *Journal of Electroanalytical Chemistry*, *626*(1), 66–74. doi:10.1016/j.jelechem.2008.11.003

Avouris, P., & Xia, F. (2012). Graphene applications in electronics and photonics. *MRS Bulletin*, *37*(12), 1225–1234. doi:10.1557/mrs.2012.206

Berger, C., Song, Z., Li, T., Li, X., Ogbazghi, A. Y., Feng, R., & First, P. N. (2004). Ultrathin epitaxial graphite: 2D electron gas properties and a route toward graphene-based nanoelectronics. *The Journal of Physical Chemistry B*, *108*(52), 19912–19916. doi:10.1021/jp040650f

Berger, C., Song, Z. M., Li, X. B., Wu, X. S., Brown, N., Naud, C., & de Heer, W. A. (2006). Electronic confinement and coherence in patterned epitaxial graphene. *Science*, *312*(5777), 1191–1196. doi:10.1126/science.1125925 PMID:16614173

Besteman, K., Lee, J.-O., Wiertz, F. G., Heering, H. A., & Dekker, C. (2003). Enzyme-coated carbon nanotubes as single-molecule biosensors. *Nano Letters*, *3*(6), 727–730. doi:10.1021/nl034139u

Byfield, M., & Abuknesha, R. (1994). Biochemical aspects of biosensors. *Biosensors & Bioelectronics*, *9*(4), 373–399. doi:10.1016/0956-5663(94)80038-3 PMID:8068232

Cao, Q., Han, S.-, Tulevski, G. S., Zhu, Y., Lu, D. D., & Haensch, W. (2013). Arrays of single-walled carbon nanotubes with full surface coverage for high-performance electronics. *Nature Nanotechnology*, *8*(3), 180–186. doi:10.1038/nnano.2012.257 PMID:23353673

Chaniotakis, N. A. (2004). Enzyme stabilization strategies based on electrolytes and polyelectrolytes for biosensor applications. *Analytical and Bioanalytical Chemistry*, *378*(1), 89–95. doi:10.1007/s00216-003-2188-3 PMID:13680062

Chaplin, M. F., & Bucke, C. (1990). *Enzyme technology*. CUP Archive.

Chek, D. C., Tan, M. L., Ahmadi, M. T., Ismail, R., & Arora, V. K. (2010). Analytical modeling of high performance single-walled carbon nanotube field-effect-transistor. *Microelectronics Journal*, *41*(9), 579–584. doi:10.1016/j.mejo.2010.05.008

Chen, C., Liu, L., Lu, Y., Kong, E. S.-W., Zhang, Y., Sheng, X., & Ding, H. (2007). A method for creating reliable and low-resistance contacts between carbon nanotubes and microelectrodes. *Carbon*, *45*(2), 436–442. doi:10.1016/j.carbon.2006.08.021

Chen, Z., Lin, Y.-M., Rooks, M. J., & Avouris, P. (2007). Graphene nano-ribbon electronics. *Physica E, Low-Dimensional Systems and Nanostructures*, *40*(2), 228–232. doi:10.1016/j.physe.2007.06.020

Choi, W., Lahiri, I., Seelaboyina, R., & Kang, Y. S. (2010). Synthesis of graphene and its applications: A review. *Critical Reviews in Solid State and Material Sciences*, *35*(1), 52–71. doi:10.1080/10408430903505036

Choudhury, B., Shinar, R., & Shinar, J. (2004). Glucose biosensors based on organic light-emitting devices structurally integrated with a luminescent sensing element. *Journal of Applied Physics*, *96*(5), 2949–2954. doi:10.1063/1.1778477

Ci, L., Xu, Z., Wang, L., Gao, W., Ding, F., Kelly, K. F., & Ajayan, P. M. (2008). Controlled nanocutting of graphene. *Nano Research*, *1*(2), 116–122. doi:10.1007/s12274-008-8020-9

Clermont, O., Bonacorsi, S., & Bingen, E. (2000). Rapid and simple determination of the Escherichia coli phylogenetic group. *Applied and Environmental Microbiology*, *66*(10), 4555–4558. doi:10.1128/AEM.66.10.4555-4558.2000 PMID:11010916

Curreli, M., Zhang, R., Ishikawa, F. N., Chang, H.-K., Cote, R. J., Zhou, C., & Thompson, M. E. (2008). Real-Time, Label-Free Detection of Biological Entities Using Nanowire-Based FETs. *IEEE Transactions on Nanotechnology*, *7*(6), 651–667. doi:10.1109/TNANO.2008.2006165

Dankerl, M., Eick, S., Hofmann, B., Hauf, M., Ingebrandt, S., Offenhäusser, A., & Garrido, J. A. (2009). Diamond transistor array for extracellular recording from electrogenic cells. *Advanced Functional Materials*, *19*(18), 2915–2923. doi:10.1002/adfm.200900590

Datta, S. (1997). *Electronic transport in mesoscopic systems*. Cambridge University Press.

Datta, S. (2005). *Quantum transport: Atom to transistor*. Cambridge University Press. doi:10.1017/CBO9781139164313

Dresselhaus, M., Dresselhaus, G., & Saito, R. (1995). Physics of carbon nanotubes. *Carbon*, *33*(7), 883–891. doi:10.1016/0008-6223(95)00017-8

Du, D., Zou, Z., Shin, Y., Wang, J., Wu, H., Engelhard, M. H., & Lin, Y. (2010). Sensitive immunosensor for cancer biomarker based on dual signal amplification strategy of graphene sheets and multienzyme functionalized carbon nanospheres. *Analytical Chemistry*, *82*(7), 2989–2995. doi:10.1021/ac100036p PMID:20201502

Fiori, G., & Iannaccone, G. (2007). Simulation of graphene nanoribbon field-effect transistors. *Electron Device Letters, IEEE*, *28*(8), 760–762. doi:10.1109/LED.2007.901680

Fowler, J. D., Allen, M. J., Tung, V. C., Yang, Y., Kaner, R. B., & Weiller, B. H. (2009). Practical chemical sensors from chemically derived graphene. *ACS Nano*, *3*(2), 301–306. doi:10.1021/nn800593m PMID:19236064

Fromherz, P., Offenhausser, A., Vetter, T., & Weis, J. (1991). A neuron-silicon junction: A Retzius cell of the leech on an insulated-gate field-effect transistor. *Science*, *252*(5010), 1290–1293. doi:10.1126/science.1925540 PMID:1925540

Galib, O. (2012). *Graphene grown by chemical vapor deposition on evaporated copper thin films*. Academic Press.

Geim, A. K. (2009). Graphene: Status and prospects. *Science*, *324*(5934), 1530–1534. doi:10.1126/science.1158877 PMID:19541989

Geim, A. K., & Novoselov, K. S. (2007). The rise of graphene. *Nature Materials*, *6*(3), 183–191. doi:10.1038/nmat1849 PMID:17330084

Grieshaber, D., MacKenzie, R., Vörös, J., & Reimhult, E. (2008). Electrochemical biosensors-Sensor principles and architectures. *Sensors (Basel, Switzerland)*, *8*(3), 1400–1458. doi:10.3390/s8031400

Guisan, J. M. (2006). *Immobilization of enzymes and cells* (Vol. 22). Springer. doi:10.1007/978-1-59745-053-9_1

Haes, A. J., Chang, L., Klein, W. L., & Van Duyne, R. P. (2005). Detection of a biomarker for Alzheimer's disease from synthetic and clinical samples using a nanoscale optical biosensor. *Journal of the American Chemical Society*, *127*(7), 2264–2271. doi:10.1021/ja044087q PMID:15713105

Han, M. Y., Özyilmaz, B., Zhang, Y., & Kim, P. (2007). Energy band-gap engineering of graphene nanoribbons. *Physical Review Letters*, *98*(20), 206805. doi:10.1103/PhysRevLett.98.206805 PMID:17677729

Härtl, A., Schmich, E., Garrido, J. A., Hernando, J., Catharino, S. C., Walter, S., & Stutzmann, M. (2004). Protein-modified nanocrystalline diamond thin films for biosensor applications. *Nature Materials*, *3*(10), 736–742. doi:10.1038/nmat1204 PMID:15359341

Held, M., Schuhmann, W., Jahreis, K., & Schmidt, H.-L. (2002). Microbial biosensor array with transport mutants of Escherichia coli K12 for the simultaneous determination of mono-and disaccharides. *Biosensors & Bioelectronics*, *17*(11), 1089–1094. doi:10.1016/S0956-5663(02)00103-3 PMID:12392960

Heller, I., Kong, J., Williams, K. A., Dekker, C., & Lemay, S. G. (2006). Electrochemistry at single-walled carbon nanotubes: The role of band structure and quantum capacitance. *Journal of the American Chemical Society*, *128*(22), 7353–7359. doi:10.1021/ja061212k PMID:16734491

Huang, S., Pang, P., Xiao, X., He, L., Cai, Q., & Grimes, C. A. (2008). A wireless, remote-query sensor for real-time detection of Escherichia coli O157: H7 concentrations. *Sensors and Actuators. B, Chemical, 131*(2), 489–495. doi:10.1016/j.snb.2007.12.036

Huang, X., Qi, X., Boey, F., & Zhang, H. (2012). Graphene-based composites. *Chemical Society Reviews, 41*(2), 666–686. doi:10.1039/C1CS15078B PMID:21796314

Huang, X., Zeng, Z., Fan, Z., Liu, J., & Zhang, H. (2012). Graphene-based electrodes. *Advanced Materials, 24*(45), 5979–6004. doi:10.1002/adma.201201587 PMID:22927209

Huang, Y., Dong, X., Liu, Y., Li, L.-J., & Chen, P. (2011). Graphene-based biosensors for detection of bacteria and their metabolic activities. *Journal of Materials Chemistry, 21*(33), 12358–12362. doi:10.1039/c1jm11436k

Husain, M., & Husain, Q. (2007). Applications of redox mediators in the treatment of organic pollutants by using oxidoreductive enzymes: A review. *Critical Reviews in Environmental Science and Technology, 38*(1), 1–42. doi:10.1080/10643380701501213

Ilani, S., Donev, L. A., Kindermann, M., & McEuen, P. L. (2006). Measurement of the quantum capacitance of interacting electrons in carbon nanotubes. *Nature Physics, 2*(10), 687–691. doi:10.1038/nphys412

Karimi, H., Yusof, R., Rahmani, R., Hosseinpour, H., & Ahmadi, M. T. (2014). Development of solution-gated graphene transistor model for biosensors. *Nanoscale Research Letters, 9.*

Katsnelson, M. I. (2007). Graphene: Carbon in two dimensions. *Materials Today, 10*(1), 20–27. doi:10.1016/S1369-7021(06)71788-6

Kirsch, J., Siltanen, C., Zhou, Q., Revzin, A., & Simonian, A. (2013). Biosensor technology: Recent advances in threat agent detection and medicine. *Chemical Society Reviews, 42*(22), 8733–8768. doi:10.1039/c3cs60141b PMID:23852443

Kriz, D., Ramström, O., & Mosbach, K. (1997). Peer Reviewed: Molecular Imprinting: New Possibilities for Sensor Technology. *Analytical Chemistry, 69*(11), 345A–349A. doi:10.1021/ac971657e

Kuzmenko, A., Van Heumen, E., Carbone, F., & Van Der Marel, D. (2008). Universal optical conductance of graphite. *Physical Review Letters, 100*(11), 117401. doi:10.1103/PhysRevLett.100.117401 PMID:18517825

Lee, C., Wei, X., Kysar, J. W., & Hone, J. (2008). Measurement of the elastic properties and intrinsic strength of monolayer graphene. *Science, 321*(5887), 385–388. doi:10.1126/science.1157996 PMID:18635798

Lee, D., & Cui, T. (2009a). Layer-by-layer self-assembled single-walled carbon nanotubes based ion-sensitive conductometric glucose biosensors. *Sensors Journal, IEEE, 9*(4), 449–456. doi:10.1109/JSEN.2009.2014414

Lee, D., & Cui, T. (2009b). pH-dependent conductance behaviors of layer-by-layer self-assembled carboxylated carbon nanotube multilayer thin-film sensors. *Journal of Vacuum Science & Technology B Microelectronics and Nanometer Structures, 27*(2), 842. doi:10.1116/1.3002386

Lee, D., & Cui, T. (2010). Low-cost, transparent, and flexible single-walled carbon nanotube nanocomposite based ion-sensitive field-effect transistors for pH/glucose sensing. *Biosensors & Bioelectronics*, *25*(10), 2259–2264. doi:10.1016/j.bios.2010.03.003 PMID:20417088

Li, C., Thostenson, E. T., & Chou, T.-W. (2008). Sensors and actuators based on carbon nanotubes and their composites: A review. *Composites Science and Technology*, *68*(6), 1227–1249. doi:10.1016/j.compscitech.2008.01.006

Li, X., Wang, X., Zhang, L., Lee, S., & Dai, H. (2008). Chemically derived, ultrasmooth graphene nanoribbon semiconductors. *Science*, *319*(5867), 1229–1232. doi:10.1126/science.1150878 PMID:18218865

Li, X. S., Cai, W. W., Jung, I. H., An, J. H., Yang, D. X., Velamakanni, A., & Ruoff, R. S. (2009). Synthesis, characterization, and properties of large-area graphene films. *Graphene and Emerging Materials for Post-Cmos Applications*, *19*(5), 41–52. doi:10.1149/1.3119526

Lillie, G., Payne, P., & Vadgama, P. (2001). Electrochemical impedance spectroscopy as a platform for reagentless bioaffinity sensing. *Sensors and Actuators. B, Chemical*, *78*(1), 249–256. doi:10.1016/S0925-4005(01)00821-8

Lin, H., Lu, Q., Ge, S., Cai, Q., & Grimes, C. A. (2010). Detection of pathogen Escherichia coli O157:H7 with a wireless magnetoelastic-sensing device amplified by using chitosan-modified magnetic nanoparticles. *Sensors and Actuators. B, Chemical*, *147*(1), 343–349. doi:10.1016/j.snb.2010.03.011

Lin, Y.-M., Appenzeller, A., Chen, Z., Chen, Z.-G., Cheng, H.-M., & Avouris, P. (2005). *Demonstration of a high performance 40-nm-gate carbon nanotube field-effect transistor.* Paper presented at the Device Research Conference Digest.

Lin, Y.-M., Valdes-Garcia, A., Han, S.-J., Farmer, D. B., Meric, I., Sun, Y., & Avouris, P. (2011). Wafer-scale graphene integrated circuit. *Science*, *332*(6035), 1294–1297. doi:10.1126/science.1204428 PMID:21659599

Liu, S., & Guo, X. (2012). Carbon nanomaterials field-effect-transistor-based biosensors. *NPG Asia Materials*, *4*(8), e23. doi:10.1038/am.2012.42

Lundstrom, M., & Guo, J. (2006). *Nanoscale transistors: Device physics, modeling and simulation.* Springer Science & Business Media.

Malhotra, B. D., Chaubey, A., & Singh, S. (2006). Prospects of conducting polymers in biosensors. *Analytica Chimica Acta*, *578*(1), 59–74. doi:10.1016/j.aca.2006.04.055 PMID:17723695

Martel, R., Schmidt, T., Shea, H., Hertel, T., & Avouris, P. (1998). Single-and multi-wall carbon nanotube field-effect transistors. *Applied Physics Letters*, *73*(17), 2447–2449. doi:10.1063/1.122477

Meixner, H. (2008). *Sensors, micro-and nanosensor technology: Trends in sensor markets* (Vol. 8). John Wiley & Sons.

Moyo, M., Okonkwo, J. O., & Agyei, N. M. (2012). Recent advances in polymeric materials used as electron mediators and immobilizing matrices in developing enzyme electrodes. *Sensors (Basel, Switzerland)*, *12*(1), 923–953. doi:10.3390/s120100923 PMID:22368503

Naeemi, A., & Meindl, J. D. (2007). Conductance modeling for graphene nanoribbon (GNR) interconnects. *IEEE Electron Device Letters*, *28*(5), 428–431. doi:10.1109/LED.2007.895452

Ndlovu, T., Arotiba, O. A., Krause, R. W., & Mamba, B. B. (2010). Electrochemical detection of o-nitrophenol on a poly (propyleneimine)-gold nanocomposite modified glassy carbon electrode. *International Journal of Electrochemical Science*, *5*, 1179–1186.

Novoselov, K., Geim, A. K., Morozov, S., Jiang, D., Grigorieva, M. K. I., Dubonos, S., & Firsov, A. (2005). Two-dimensional gas of massless Dirac fermions in graphene. *Nature*, *438*(7065), 197–200. doi:10.1038/nature04233 PMID:16281030

Novoselov, K. S., Geim, A. K., Morozov, S., Jiang, D., Zhang, Y., Dubonos, S., & Firsov, A. (2004). Electric field effect in atomically thin carbon films. *Science*, *306*(5696), 666–669. doi:10.1126/science.1102896 PMID:15499015

Novoselov, K. S., Jiang, Z., Zhang, Y., Morozov, S., Stormer, H., Zeitler, U., & Geim, A. (2007). Room-temperature quantum Hall effect in graphene. *Science*, *315*(5817), 1379–1379. doi:10.1126/science.1137201 PMID:17303717

Ouyang, Y., Yoon, Y., & Guo, J. (2007). Scaling behaviors of graphene nanoribbon FETs: A three-dimensional quantum simulation study. *Electron Devices. IEEE Transactions on*, *54*(9), 2223–2231. doi:10.1109/TED.2007.902692

Pantano, A., Parks, D. M., & Boyce, M. C. (2004). Mechanics of deformation of single-and multi-wall carbon nanotubes. *Journal of the Mechanics and Physics of Solids*, *52*(4), 789–821. doi:10.1016/j.jmps.2003.08.004

Park, H., Afzali, A., Han, S.-J., Tulevski, G. S., Franklin, A. D., Tersoff, J., & Haensch, W. (2012). High-density integration of carbon nanotubes via chemical self-assembly. *Nature Nanotechnology*, *7*(12), 787–791. doi:10.1038/nnano.2012.189 PMID:23103933

Patolsky, F., Timko, B. P., Yu, G., Fang, Y., Greytak, A. B., Zheng, G., & Lieber, C. M. (2006). Detection, stimulation, and inhibition of neuronal signals with high-density nanowire transistor arrays. *Science*, *313*(5790), 1100–1104. doi:10.1126/science.1128640 PMID:16931757

Patolsky, F., Zheng, G., & Lieber, C. M. (2006). Fabrication of silicon nanowire devices for ultrasensitive, label-free, real-time detection of biological and chemical species. *Nature Protocols*, *1*(4), 1711–1724. doi:10.1038/nprot.2006.227 PMID:17487154

Putzbach, W., & Ronkainen, N. J. (2013). Immobilization techniques in the fabrication of nanomaterial-based electrochemical biosensors: A review. *Sensors (Basel, Switzerland)*, *13*(4), 4811–4840. doi:10.3390/s130404811 PMID:23580051

Qi, P., Vermesh, O., Grecu, M., Javey, A., Wang, Q., Dai, H., & Cho, K. (2003). Toward large arrays of multiplex functionalized carbon nanotube sensors for highly sensitive and selective molecular detection. *Nano Letters*, *3*(3), 347–351. doi:10.1021/nl034010k

Rahmani, M., Ahmadi, M., Karimi, H., Kiani, M., Akbari, E., & Ismail, R. (2013). Analytical modeling of monolayer graphene-based NO2 sensor. *Sensor Letters*, *11*(2), 270–275. doi:10.1166/sl.2013.2742

Recher, P., & Trauzettel, B. (2011). A defect controls transport in graphene. *Physics*, *4*, 25. doi:10.1103/Physics.4.25

Rijal, K., & Mutharasan, R. (2013). A method for DNA-based detection of E. coli O157:H7 in a proteinous background using piezoelectric-excited cantilever sensors. *Analyst (London)*, *138*(10), 2943–2950. doi:10.1039/c3an36814a PMID:23535914

Sangtarash, S., Sadeghi, H., Ahmadi, M., Ghadiry, M., Anwar, S., & Ismail, R. (2012). Schottky current in carbon nanotube-metal contact. *Journal of Computational and Theoretical Nanoscience*, *9*(10), 1554–1557. doi:10.1166/jctn.2012.2243

Sarma, A. K., Vatsyayan, P., Goswami, P., & Minteer, S. D. (2009). Recent advances in material science for developing enzyme electrodes. *Biosensors & Bioelectronics*, *24*(8), 2313–2322. doi:10.1016/j.bios.2008.09.026 PMID:19022645

Schedin, F., Geim, A., Morozov, S., Hill, E., Blake, P., Katsnelson, M., & Novoselov, K. (2007). Detection of individual gas molecules adsorbed on graphene. *Nature Materials*, *6*(9), 652–655. doi:10.1038/nmat1967 PMID:17660825

Schuhmann, W. (1995). Conducting polymer based amperometric enzyme electrodes. *Mikrochimica Acta*, *121*(1-4), 1–29. doi:10.1007/BF01248237

Skoog, D. A., West, D. M., & Holler, F. J. (1988). *Fundamentals of analytical chemistry*. Academic Press.

Sols, F., Guinea, F., & Neto, A. C. (2007). Coulomb blockade in graphene nanoribbons. *Physical Review Letters*, *99*(16), 166803. doi:10.1103/PhysRevLett.99.166803 PMID:17995278

Son, Y.-W., Cohen, M. L., & Louie, S. G. (2006). Energy gaps in graphene nanoribbons. *Physical Review Letters*, *97*(21), 216803. doi:10.1103/PhysRevLett.97.216803 PMID:17155765

Sotiropoulou, S., Gavalas, V., Vamvakaki, V., & Chaniotakis, N. (2003). Novel carbon materials in biosensor systems. *Biosensors & Bioelectronics*, *18*(2), 211–215. doi:10.1016/S0956-5663(02)00183-5 PMID:12485767

Stern, E., Klemic, J. F., Routenberg, D. A., Wyrembak, P. N., Turner-Evans, D. B., Hamilton, A. D., & Reed, M. A. (2007). Label-free immunodetection with CMOS-compatible semiconducting nanowires. *Nature*, *445*(7127), 519–522. doi:10.1038/nature05498 PMID:17268465

Stern, E., Vacic, A., & Reed, M. A. (2008). Semiconducting nanowire field-effect transistor biomolecular sensors. *IEEE Transactions on Electron Devices*, *55*(11), 3119–3130. doi:10.1109/TED.2008.2005168

Taneja, L. (2012). A fractal analysis of the detection of glucose molecules on biosensor surfaces. (1515330 M.S.). The University of Mississippi. Retrieved from ProQuest Dissertations & Theses Full Text database.

Vianello, F., Ragusa, S., Cambria, M. T., & Rigo, A. (2006). A high sensitivity amperometric biosensor using laccase as biorecognition element. *Biosensors & Bioelectronics*, *21*(11), 2155–2160. doi:10.1016/j.bios.2005.10.006 PMID:16293408

Voelker, M., & Fromherz, P. (2005). Signal transmission from individual mammalian nerve cell to field-effect transistor. *Small*, *1*(2), 206–210. doi:10.1002/smll.200400077 PMID:17193431

Wang, J. (2005). Carbon-nanotube based electrochemical biosensors: A review. *Electroanalysis*, *17*(1), 7–14. doi:10.1002/elan.200403113

Wang, J., & Musameh, M. (2003). Carbon nanotube/teflon composite electrochemical sensors and biosensors. *Analytical Chemistry*, *75*(9), 2075–2079. doi:10.1021/ac030007+ PMID:12720343

Wang, X. R., Ouyang, Y. J., Li, X. L., Wang, H. L., Guo, J., & Dai, H. J. (2008). Room-temperature all-semiconducting sub-10-nm graphene nanoribbon field-effect transistors. *Physical Review Letters*, *100*(20).

Wang, Y., Ye, Z., Si, C., & Ying, Y. (2011). Subtractive inhibition assay for the detection of E. coli O157: H7 using surface plasmon resonance. *Sensors (Basel, Switzerland)*, *11*(3), 2728–2739. doi:10.3390/s110302728 PMID:22163763

Wang, Y. X., Ping, J. F., Ye, Z. Z., Wu, J., & Ying, Y. B. (2013). Impedimetric immunosensor based on gold nanoparticles modified graphene paper for label-free detection of Escherichia coli O157:H7. *Biosensors & Bioelectronics*, *49*, 492–498. doi:10.1016/j.bios.2013.05.061 PMID:23811484

Wang, Z. F., Zheng, H., Shi, Q. W., & Chen, J. (2009). Emerging nanodevice paradigm: Graphene-Based electronics for nanoscale computing. *Acm Journal on Emerging Technologies in Computing Systems*, *5*(1), 1–19. doi:10.1145/1482613.1482616

White, P. C., Lawrence, N. S., Davis, J., & Compton, R. G. (2002). Electrochemical determination of thiols: A perspective. *Electroanalysis*, *14*(2), 89–98. doi:10.1002/1521-4109(200201)14:2<89::AID-ELAN89>3.0.CO;2-Y

Yang, W., Auciello, O., Butler, J. E., Cai, W., Carlisle, J. A., Gerbi, J. E., & Russell, J. N. (2002). DNA-modified nanocrystalline diamond thin-films as stable, biologically active substrates. *Nature Materials*, *1*(4), 253–257. doi:10.1038/nmat779 PMID:12618788

Zhang, M., Smith, A., & Gorski, W. (2004). Carbon nanotube-chitosan system for electrochemical sensing based on dehydrogenase enzymes. *Analytical Chemistry*, *76*(17), 5045–5050. doi:10.1021/ac049519u PMID:15373440

Zhang, Y.-Q., Tao, M.-L., Shen, W.-D., Zhou, Y.-Z., Ding, Y., Ma, Y., & Zhou, W.-L. (2004). Immobilization of L-asparaginase on the microparticles of the natural silk sericin protein and its characters. *Biomaterials*, *25*(17), 3751–3759. doi:10.1016/j.biomaterials.2003.10.019 PMID:15020151

Chapter 13
Carbon Materials Based Ion Sensitive Field Effect Transistor (ISFET):
The Emerging Potentials of Nanostructured Carbon-Based ISFET with High Sensitivity

Mohammad Javad Kiani
Islamic Azad University, Iran

Mohammad Taghi Ahmadi
Urmia University, Iran

M. H. Shahrokh Abadi
Hakim Sabzevari University, Iran

F. K. Che Harun
Universiti Teknologi Malaysia, Malaysia

Meisam Rahmani
Universiti Teknologi Malaysia, Malaysia

S.N. Hedayat
Urmia University, Iran

S.H. Yaghoobian
Islamic Azad University, Iran

ABSTRACT

Graphene and SWCNT-based Ion Sensitive FET (ISFET) as a novel material with organic nature and ionic liquid gate is intrinsically sensitive to pH changes. pH is an important factor in enzymes stabilities which can affect the enzymatic reaction and broaden the number of enzyme applications. More accurate and consistent results of enzymes must be optimized to realize their full potential as catalysts accordingly. In this chapter, an appropriate structure to ISFET device is designed for the purpose of electrical measurement of different pH buffer solutions. Electrical detection model of each pH value is suggested using conductance modelling of monolayer graphene. In addition, ISFET based on nanostructured SWCNT is studied for the purpose of electrical detection of hydrogen ion concentrations. Electrical detection of hydrogen ion concentrations by modelling the conductance of SWCNT sheets is proposed. pH buffer as a function of gate voltage is assumed and sensing factor is defined. Finally, the proposed new approach improving the analytical model is compared with experimental data and shows good overall agreement.

DOI: 10.4018/978-1-5225-0736-9.ch013

INTRODUCTION

Carbon Materials as a Based-Biosensor

Graphene is a 2 of sp^2 bonded carbon atom, which makes its structure apparently looks like honeycomb crystal as seen in Figure 1 (Enoki, Kobayashi, & Fukui, 2007; Fang, Leiber, Xie, & Xiong). It is called the mother of graphite (many layers of graphene) form because it can act as all these allotropes basic building block.

Graphene is theoretically discovered back in the 1940s, but at that time graphene (a 2D layer crystal) are believed thermodynamically too unstable to be produced in real world (Martoccia, Björck, & Schlepütz, 2010). However, Andre Geim and Konstanstin Novoselov successful to produce graphene by just using scotch tape in 2004 (Geim & Novoselov, 2007). This method is called Mechanical Exfoliation because it mechanically exfoliates layers of graphene from graphite.

Nowadays, there are few ways how to produce graphene other than mechanical exfoliation, the common ways are Epitaxial grown, Reduced Graphene Oxide and Chemical Vaporization Deposition (CVD). CVD is considered as the most promising method to commercialize graphene, because its low cost and readily accessible techniques for growing in a large area and high quality Graphene.

Intrinsic graphene actually has no band gap, which are bad, we can produce a tune-able band gap of graphene by, at first make it bilayer graphene, and then induced E-field to the bilayer of the graphene or by doping the Graphene chemically. Besides graphene, carbon can make many other forms, this variance of form is called allotropes which each allotropes have its own properties. Another kind of carbon allotropes such as carbon nanotube and graphene nanoscrolls is cylindrical and spherical. Graphene is attracting interest majored in electrical, physical, chemical and even biology since of its unique properties (M. J. Kiani, Ahmadi, Abadi, & Rahmani, 2013). Because of physical and electrical properties of graphene it really suit to make this material as Field Effect Transistor.

Also, Carbon nanotubes (CNTs) can be imagined as a sheet of carbon atoms turned up into a pipe with a diameter of approximately ten of nanometres. Two major kinds of CNTs, is exist, namely, Single-walled (SWCNTs) and multi-walled carbon nanotubes (MWCNTs), the latter being shaped by numerous concentric layers of turned graphene (Figure 2). Particularly, a high feature proportion describes SW-CNTs. Furthermore, their multipurpose physicochemical aspects facilitate the noncovalent and covalent beginning of several biosensing and biomedicine function appropriate entities.

Therefore development of their distinctive thermal, optical, electrical, and spectroscopic possessions in a biological framework is expected to defer great progress in the treatment of disease and discovery biomolecules such as antigen–antibody, cells, DNA, and other biomolecules.

Figure 1. Monolayer graphene atoms arrangement with only one atom thickness

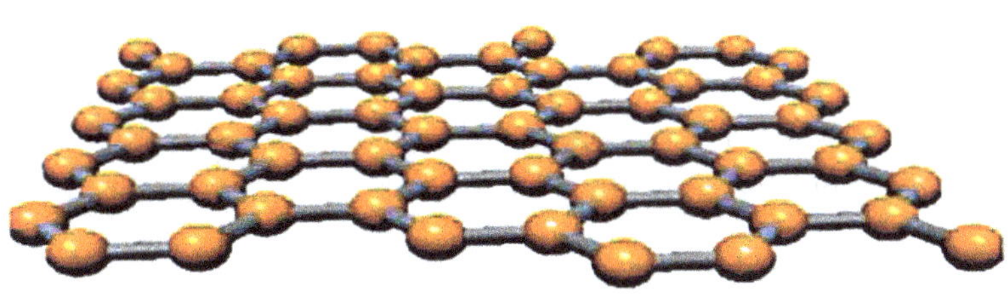

Figure 2. Wrapping of graphene sheet to form SWCNT

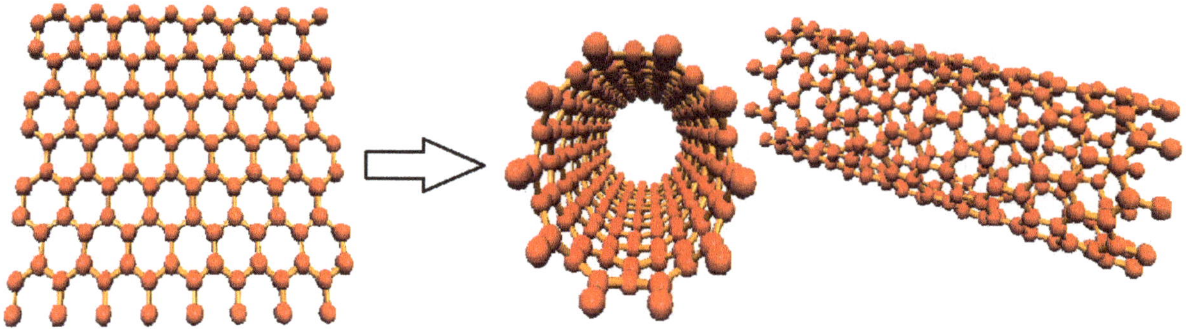

BASIC CONCEPTS OF GRAPHENE BASED-ISFET

Nowadays, Ion-Sensitive FETs (ISFETs) has caught much attention due to their advantages such as small in size and the possibilities for mass production (Gotoh, 1989; M. J. Kiani, Ahmadi, Akbari, Karimi, & Che Harun, 2013; Pham, 1994). Their short and consistent response times are very favorable to the electronics industry (Dongjin, 2012; Schlesinger, 1996). ISFETs introduced new features such as the integration of data processing and compensation circuits in the similar circuit for this type of sensors (W.-Y. L. Chung, Yeong-Tsair Pijanowska, Dorota G.Yang, Chung-Huang Wang, Ming-Chia Krzyskow, Alfred Torbicz, Wladyslaw, 2006; M. J. A. Kiani, Mohammad Taghi Karimi Feiz Abadi, Hediyeh Rahmani, Meisam Hashim, Amin, 2013; S. C. Chen Yan-Kuin, 1986; Shepherd, 2005). By altering the gate material, depositing layers of selective membrane or a bio-recognition element onto the gate, variance of selectivity can be achieved (Kal, 2007; Voigt, 1997).

After the process of depositing, the sensors now are called Chemically Sensitive FET (CHEMFET) (Cobben, 1992; Reinhoudt, 1990). Initially, heterogeneous membranes of silver halides and membranes based on polyvinyl chloride (PVC) have been used for ISFET (Cadogan, 1992; Guth, 2009). Due to poor adherence between PVC base membrane and ISFET surface and inconsistent results, the scientist explores for a new membrane (Cadogan, 1992; Jiménez, 1997). That is where Photocured polymers, which are compatible with the proposed photolithography techniques (Bratov, 1995; Jiménez, 1997). It had the properties of a higher adherence string of the salinized ISFET gate's surface (Kuang, 2012). In order to expand ion selective membranes, numerous polymers such as polysiloxanes, polyurethanes, and different methacrylate derived polymers have been reported (Cecilia, 2009; Seymour, 1966). These new polymers show promising results regarding consistency and longer stability compared to PVC membranes. In addition, almost all effective ionic based ISFETs were developed by clinical analyses and environmental (W.-Y. C. Chung, Febus Reidj G.Szu, Harold Pijanowska, Dorota G.Dawgul, Marek Torbicz, Wladislaw Grabiec, Piotr B.Jarosewicz, Bohdan Chiang, Jung-Lung Chang, Kuo-Chung Cheng, Cheanyeh Ho, Wei-Po, 2009). Recently, microelectronic advances have been exploited and applied to improve ISFET fabrication (Haigang Yang, 2005; Kim, 2009). Because of the electrolyte ionic properties, electrical parts of ISFET cannot have contact with liquid, only the gate area is open (Martinoia, 2004).

Due to its organic nature, gate material for ISFET is intrinsically sensitive to pH changes (Bousse, 1984; Steinhoff, 2003). On the other hand, all enzymes are sensitive to pH changes, but extremely high or low pH values can make these enzymes to lose their activity (Jamasb, 1997; Pijanowska & Torbicz, 1997). PH is also a main factor in enzymes stabilities (Couto, 2006). Each enzyme includes suitable or optimal pH stability range (Couto, 2006; Pijanowska & Torbicz, 1997). Apart from temperature and pH, Ionic strength can also affect the enzymatic reaction (Pokhrel, Joo, Kim, & Yoo, 2012). For more accurate and consistent results, each of these physical and chemical parameters must be considered and optimized accordingly (Morgenshtein, 2004).

Structure of Graphene as an ISFET Device

ISFETs can be based on many materials as their detectors such as membrane and graphene (Chen, 2012). Because of physical and electrical properties of graphene, it can be applied as a sensing material in the structure of FET (Chen, 2012). On the other hand, there is not any information on the development and modelling of ion sensitivity FET, and their potential as ISFET has not been totally studied yet. The reaction between solution with different pH values and surface of graphene has a notable effect on the conductivity of graphene (Zhao, 2012). This means that, the detection mechanism by adsorbing the hydrogen ions from solution to carbon-based materials can be clarified as shown in Figure 2. In other words, based on the electron transfer between ion solutions and graphene surface, analytical model of reaction between buffer solution of different PH and graphene is presented.

Figure 3 illustrates the detection mechanism of solution with different pH using ISFET device is exhibited, so monolayer graphene on silicon oxide and silicon substrate with a deposited epoxy layer (Epotek 302-3M, Epoxy Technology) as an ISFET membrane is proposed. In this chapter, PH solution as a gate voltage is replicated due to the carrier injected to channel from it, and pH as sensing parameter (Ƥ) is suggested. Finally, the presented model is compared with experimental data for purposes of validation of our new model.

Figure 3. Schematics of the proposed structure and the electrical circuit of the graphene based-ISFET for PH detection

Graphene Nanoribbon (GNR) Conductance Model

The remarkable two-dimensional material called graphene is a semi-metal that has applications related to transistor because it has metallic characteristics. Nevertheless, it is possible to utilize lateral confinement to achieve band gap energy. In order to reach this goal, graphene nanoribbon (as shown in Figure 4) which is a standard lithography technique may be employed to make the energy from patterned graphene into narrow graphene stripes. Carrier transport in GNRs is of one of the most important aspects to take care of in order to manipulate its excellent transport properties for electronic applications especially the mobility of charge carriers, which is essential to characterize the device performance in terms of the switching speed. This excellent transport property therefore makes graphene appear in the frontline for advanced applications in nanoelectronics where it may be used as the channel or the device active region conducting the carriers.

This study is crucial since numerous researches has been done theoretically and experimentally to explore the conductance of the Carbon nanotubes, CNT (M. T. Ahmadi, Ismail, Tan, & Arora, 2008). The counterpart of GNR whereas less were carried out for GNR(Mohammad Taghi Ahmadi & Johari, 2010). Besides, the effective mobility is usually extracted using the well-known Matthiessen rule (Gnani, Gnudi, Reggiani, & Baccarani, 2010). Thus, far only a small portion of works has been done to find out the effective mobility based on the conductance approach or known as Drude model. For example used the conductance model in order to investigate the effective mobility in nanowire FETs. Also employed the same approach in order to evaluate the silicon nanowire FETs effective mobility in the presence of scatterings. The lack of study in this conductance-based effective mobility leads to the motivation of this work, aiming to solve the analytical effective mobility of GNRs based on the conductance approach.

Figure 4. Structure of monolayer GNR

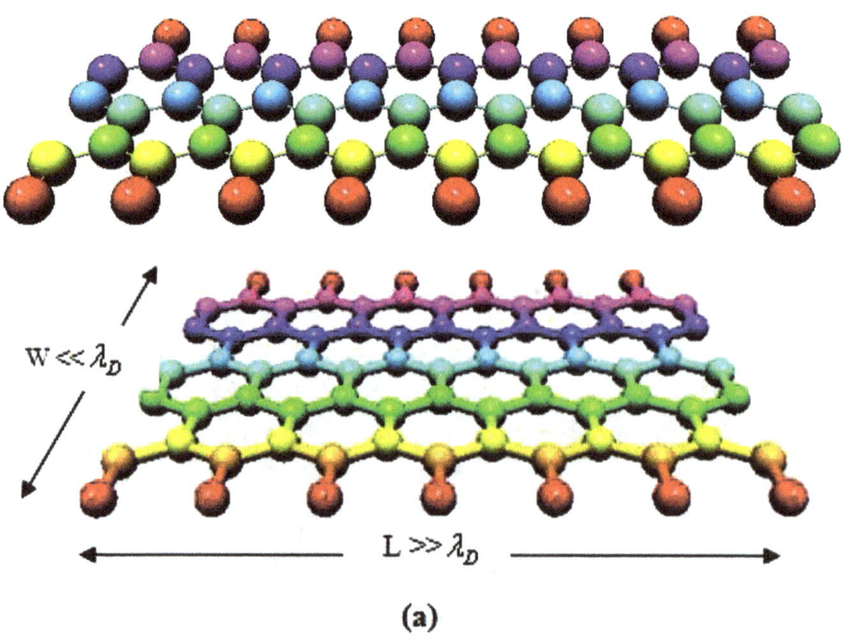

The graphene nanoribbon (GNR) channel is supposed to be completely ballistic for one dimensional (1D) monolayer ISFET to pH sensing since high carrier mobility reported from experiments in the graphene. The important quantity that define the efficiency of electron transport is conductance, G which is proportional to the transmission probability of carriers from one electrode (source) to another (drain) according to the Landauer formula(Mohammad Taghi Ahmadi & Johari, 2010).

$$G = \frac{2e^2}{h} T \tag{1}$$

where e is the electron charge, h is the Planck constant, and T is the total transmission probability (sum over all possible transmission channels). By applying the Taylor expansion on graphene band energy near the Fermi point, the $E\,(k)$ relation of the GNR is obtained as.

$$E\left(\vec{k}\right) = \pm \frac{3ta}{2} \sqrt{k_x^2 + \beta^2} \tag{2}$$

where k_x is the wave vector along the length of the nanoribbon and β is quantized wave vector given by

$$\beta = \frac{2\pi}{a\sqrt{3}} \left(\frac{p_i}{N+1} - \frac{2}{3} \right) \tag{3}$$

here p_i is the sub band index and N is the number of dimer lines which determine the width of the ribbon. The energy band gap, E_g can be assumed as

$$E_g = 3ta\beta \tag{4}$$

where $a = 0.142nm$ is the lattice constant and $t = 2.7eV$ is the tight binding energy. Therefore the modified energy relation is

$$E = \pm \frac{E_g}{2} \sqrt{\left(1 + \frac{k_x^2}{\beta^2} \right)} \tag{5}$$

Equation (5) indicates that relationship between energy and wave vector is not parabolic. However, square root approximation incorporate with Taylor expansion leads to parabolic relation between energy and wave vector.

$$E(k) \approx \frac{E_g}{2} + \frac{E_g}{4\beta^2} k_x^2 \tag{6}$$

So

$$E \approx \frac{E_g}{2} + \frac{\hbar^2 k_x^2}{2m^*} \tag{7}$$

where m^* is the effective mass of GNR. In parabolic part of the band energy the wave vector can be extracted as

$$k = \sqrt{\frac{4E}{3at} - 2\beta^2} \tag{8}$$

Based on this wave vector, number of actual modes *M (E)* at a given energy, which is dependent on the sub bands location, can be calculated. If the related energy includes the bottom of the conduction band then parabolic approximation of band diagram can be used then the mode density *M (E)* increases with energy. However, in the valence band, any information related to the subbands is more difficult to obtain because of the coupled multiple bands are increasing and difficult dispersion relations are needed. By taking the derivatives of wave vector *k* over the energy *E (dk/dE)* of Equation (8), the number of the mode, *M(E)* is written as

$$M(E) = \frac{\Delta E}{\Delta k \cdot L} = \frac{9at}{2L}\left(\frac{4E}{9at} - 2\beta^2\right)^{1/2} \tag{9}$$

where *l* is the length of ISFET channel. Now taking into consideration of spin degeneracy, the number of conducting channels can be finalized as

$$M(E) = 2\frac{\Delta E}{\Delta k \cdot L} = \frac{9at}{L}\left(\frac{4E}{9at} - 2\beta^2\right)^{1/2} \tag{10}$$

A region of lowest G with respect to gate voltage in bulk graphene is calculated and matches to the minimum conductivity at the charge neutrality point, $Vg = V_{\text{Dirac}}$. That is a fundamental constant proportional to the Planck's constant and electron charge given by the Equation (1).

In fact, levels of up spin and down spin in the small channels naturally with same energy as a degenerate level results the minimum conductance two times larger than this amount, which is equal to $2G_0$. In the bad contact, the measured conductance is always lower than this value. Based on Landauer formula, the conductance on large channel follows the Ohmic scaling law but in the smaller two possible corrections need to apply on this law, firstly interface resistance which independent of the length. Secondly, conductance related to the width nonlinearly which depends on the number of the modes in the conductor that is quantized parameter, in the Landauer formula both of these features are corporate and thus the conductance is

$$G = \frac{2q^2}{h}\int_{-\infty}^{+\infty} dEM(E)T(E)\left(-\frac{df}{dE}\right) \tag{11}$$

where T is average probability of injected electron at one end will transmit to the other end, and in our ballistic channel this parameter is equal to one. Replacement by the number of sub bands (mode numbers) in corporate with Fermi – Dirac distribution function conductance is related to the length of nanoribon as well.

$$G = \frac{2q^2}{h}\frac{9at}{L}\left(\frac{9at}{4}\right)^{1/2}\int_{-\infty}^{+\infty}\left(E-\frac{9at\beta^2}{2}\right)^{1/2} d\left(-\frac{1}{1+e^{\frac{E-E_F}{k_BT}}}\right) \tag{12}$$

Temperature effect on nanoribbon conductance can be seen by changing the boundary of integral as follow

$$G = \frac{6q^2}{hL}\left(\pi atk_BT\right)^{1/2}\left[\int_0^{+\infty}\frac{x^{-1/2}}{\left(\frac{1}{1+e^{x-\eta}}\right)}dx + \int_0^{+\infty}\frac{x^{-1/2}}{\left(\frac{1}{1+e^{x+\eta}}\right)}dx\right] \tag{13}$$

In order to simplify the conductance equation, we assumed x= $(E-E_g/K_BT)$ and $\eta = (E_F-E_g)/K_BT$ as normalized Fermi energy. Consequently, the supposed conductance model of graphene base ISFET channel can be written as

$$G = \frac{3q^2\left(3\pi a^3t^3k_{BT}\right)^{\frac{1}{2}}}{hl}\left[\Im_{\frac{-1}{2}}(\eta)+\Im_{\frac{-1}{2}}(-\eta)\right] \tag{14}$$

This equation can be numerically solved for different gate voltages. Thus, the proposed conductance model of the performance of graphene-based ISFET in nanostructured region by the conductance–voltage characteristic is evaluated in Figure 5.

An applying gate voltage between 0.2 to 0.7 V monitors a bipolar characteristic of FET device monitored since Fermi energy can be controlled by gate voltage. Based on this characteristic, it is notable that the graphene can be continuously dropped from the p-doped to the n-doped region by the controllable gate voltage. The minimum conductance is observed at the transition point between electron and hole doping. This conjunction point called the charge-neutrality point (CNP) (Moriconi, 2011).

Figure 5. A bipolar transfer curve of conductance model of graphene-based ISFET

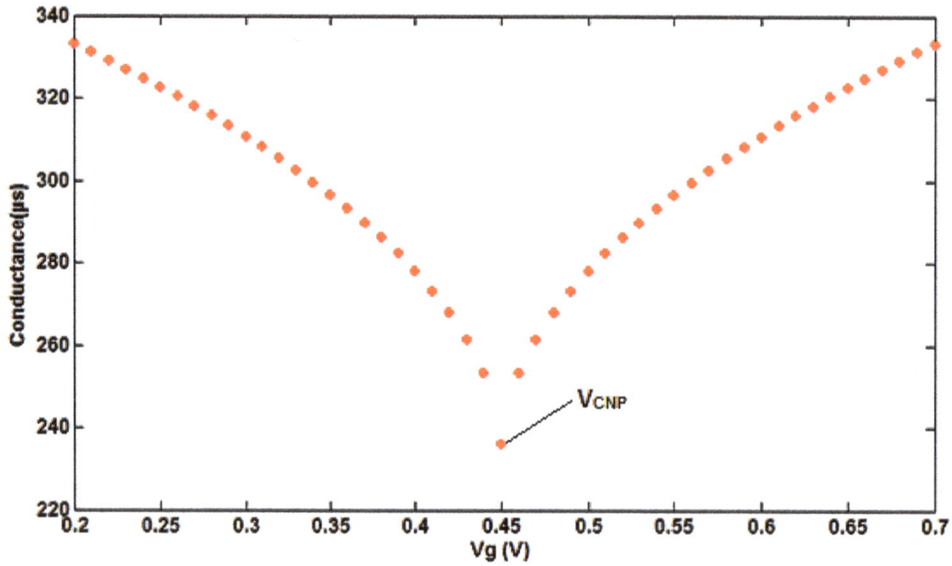

ISFET Model Based on GNR

The conductance of ISFET channel not only is dependent on the graphene structure and operation voltage on the source-drain channel, but also it depends on the electrolyte environment and ion concentration in solution (Bonanni, 2012; W. N. Fu, Cornelia Knopfmacher, Oren Tarasov, Alexey Weiss, Markus Calame, Michel Schönenberger, Christian, 2011). It has been demonstrated that different pH values can affect the ISFET conductance (W. N. Fu, Cornelia Knopfmacher, Oren Tarasov, Alexey Weiss, Markus Calame, Michel Schönenberger, Christian, 2011).

Before the Hydrogen ion concentration has been changed in the solution, natural solution (Pure water) with PH=7 buffer was added in electro-active membrane to measure the dependence of conductance versus gate voltage. There is a favorable agreement between the proposed model for PH sensing based on graphene and experimental data for non-ionic solution (PH=7) which is extracted from Ref (W. N. Fu, Cornelia Knopfmacher, Oren Tarasov, Alexey Weiss, Markus Calame, Michel Schönenberger, Christian, 2011)as can be seen in Figure 6.

The conductivity of the graphene based-ISFET devices is influenced by the number of carriers changing in the channel. ISFET based-graphene with high sensitivity applied to detect the different PH values based on conductance altering (W. N. Fu, Cornelia Knopfmacher, Oren Tarasov, Alexey Weiss, Markus Calame, Michel Schönenberger, Christian, 2011). As can be seen in Figure 7, the conductance of the channel changes due to the binding of the Hydrogen ions in the solution to the surface of ISFET channel. When the PH value of the solution is rising from 5 to 10, less hydrogen ions will be adsorbed and the sensor will be capable of attracting less ions which has led to changes in the conductance of the graphene based-ISFET as shown in Figure 8.

Dependent upon the source-drain conductance of the ISFET device we can write

$$G_{withPH} \approx PH \tag{15}$$

Figure 6. Electrical source-drain conductance versus gate voltage of the graphene-based ISFET for both model and experimental data with non-ionic solution
W. N. Fu, Cornelia Knopfmacher, Oren Tarasov, Alexey Weiss, Markus Calame, Michel Schönenberger, Christian, 2011.

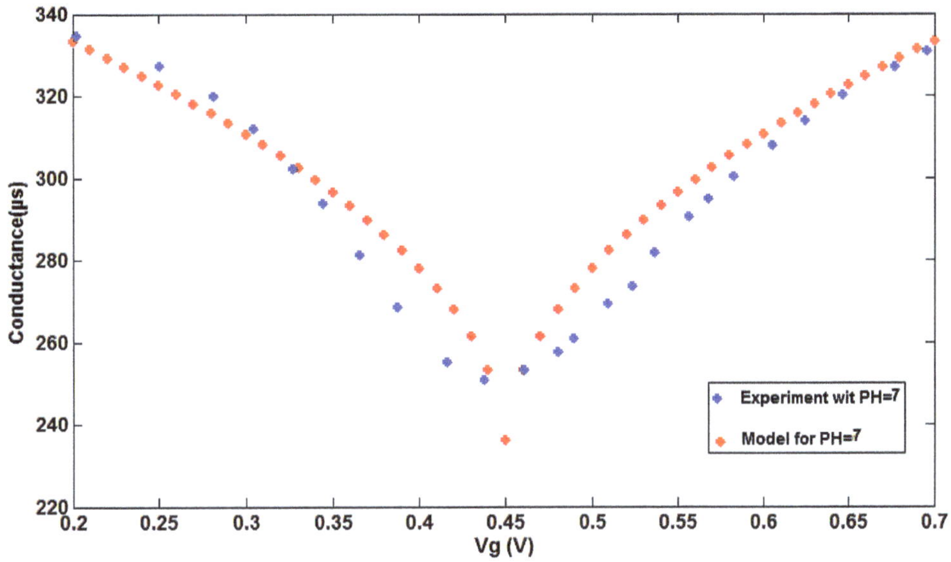

The focus of this paper is to present a new model for ISFET to measure PH changes, in other words, the conductance of ISFET device as a function of different PH values is simulated and PH factor (Ƥ) is suggested. Subsequently, for more understanding of role of hydrogen ion concentration, FET modelling is employed to obtain an equation between the conductance than PH of a solution, where the suggested structure of ISFET is shown in Figure 3 with source and drain as contacts. Ultimately, pH of a solution (follow) can model different PH values. This means that, G_{withPH} is supposed as a function of PH values.

Figure 7. Schematic of hydrogen ion-adsorption processes by surface area of single layer graphene

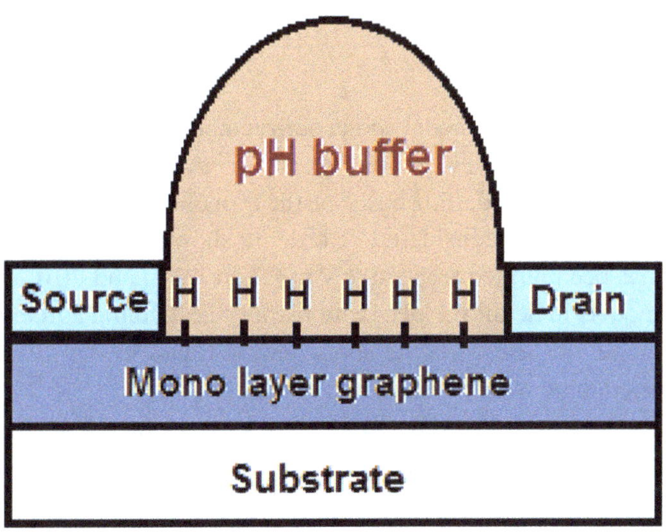

Figure 8. The comparison between graphene conductance model with extracted experimental data for different pH values
W. N. Fu, Cornelia Knopfmacher, Oren Tarasov, Alexey Weiss, Markus Calame, Michel Schönenberger, Christian, 2011.

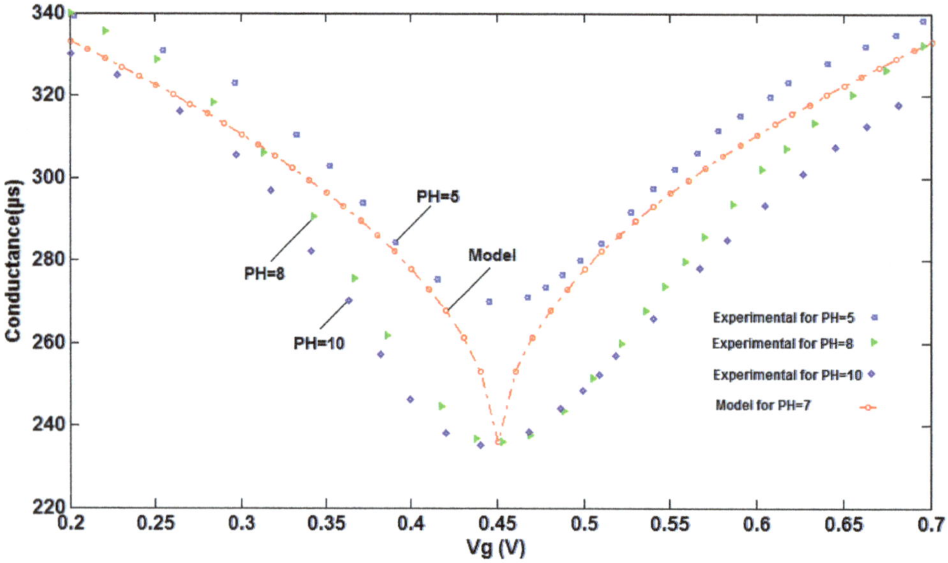

$$G_{withPH} = \frac{\mathcal{P}}{P_H} G_{withoutPH} \tag{16}$$

where pH sensing factor (Ƥ) is assumed and P_H is pH values. In the non-saturation region, the ISFET conductance model is involved as a function of gate voltage and the ideal conductance–voltage relation to the graphene channel of the ISFET device from Equations 14 and 16 is modified as.

$$G_{withPH} = \frac{\mathcal{P}}{P_H}\left(\frac{3q^2(3\pi a^3 t^3 k_{BT})^{\frac{1}{2}}}{hl}\left[\Im_{\frac{-1}{2}}(\eta) + \Im_{\frac{-1}{2}}(-\eta)\right]\right) \tag{17}$$

So, the G–Vg characteristics of both model and experimental of graphene based-ISFET for changing the pH level in solution from 6 to 7 are plotted in Figure 9.

By comparing suggested ISFET modelling based on the proposed parameter model with experimental data in Figure 9, similar trends can be considered. In order to show all figures without overlapping, each figure for per PH value has been plotted respectively in Figures 9(a) and 9(b). In addition, comparison between observed new models in detail per pH, demonstrates acceptable agreement with experimental data. In the suggested model, different PH values is demonstrated in the form of Ƥ parameter to an agreement with the reported data, which is shown in Table 1.

Therefore, based on the iteration method by Table (1), the electro-active ions absorbed by the surface of the ISFET channel as a pH sensing factor (Ƥ) can be suggested by the following equations as

Figure 9. G-V_g characteristic of proposed conductance model with experimental data for solution with (a) pH=5 and (b) pH=6

W. N. Fu, Cornelia Knopfmacher, Oren Tarasov, Alexey Weiss, Markus Calame, Michel Schönenberger, Christian, 2011.

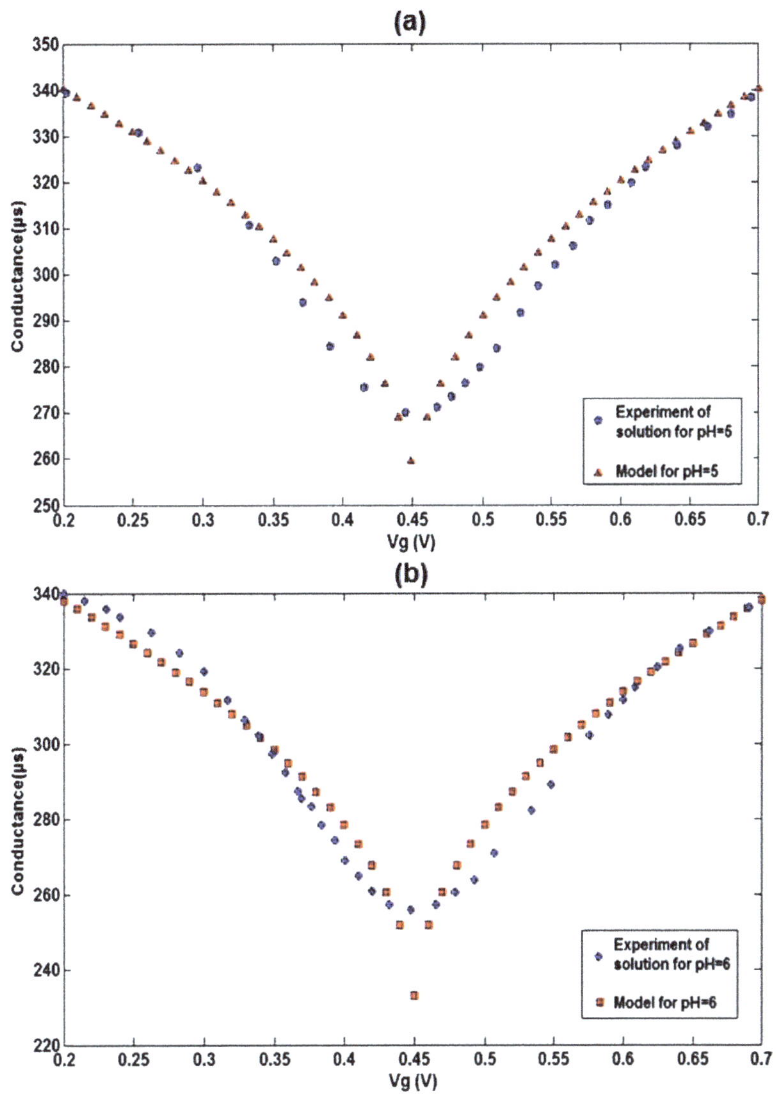

Table 1. Different PH values with P parameter

The Value of P Parameter	pH Values
0.039105	5
0.035142	6
0.034918	7
0.034662	8
0.034437	9
0.034209	10

$$P = \alpha_1 P_H^{\beta_1} \tag{18}$$

and

$$P = \alpha_2 e^{\beta_2 (P_H)} \tag{19}$$

According to the saturation region of the supposed conductance model belong to ISFET device; Equation 20 is acceptable for both saturation behavior and experimental data from (W. N. Fu, Cornelia Knopfmacher, Oren Tarasov, Alexey Weiss, Markus Calame, Michel Schönenberger, Christian, 2011).

$$P = \alpha Ln\left(P_H\right) + \beta \tag{20}$$

From extracted data, α and β parameters are calculated where α = 2.7318 and β = 4.5044. Consequently, based on the proposed model of ISFET device, the conductance versus gate voltage is modified as

$$G_{withPH} = \frac{\alpha Ln\left(P_H\right) + \beta}{P_H} \left(\frac{3q^2 \left(3\pi a^3 t^3 k_{BT}\right)^{\frac{1}{2}}}{hl} \left[\Im_{\frac{-1}{2}}(\eta) + \Im_{\frac{-1}{2}}(-\eta) \right] \right) \tag{21}$$

As can be seen in Figure 10, theoretical G–Vg characteristic of graphene based-ISFET for pH changes from 8 to 10 are plotted.

It is evident that, G–Vg characteristic curve can be controlled by the pH factor (Ƥ) and also the proposed model of ISFET conductance is closely matched with experimental data. In both reported data and theoretical data, the declining of ISFET conductance is noticeable when pH level increases. In addition, the conductance curve is almost symmetric near $V_{CNP,}$ while at the large carrier concentration about 350-400 μS, a saturation behavior is depicted. Comparing both experimental data and theoretical data depicted in Figure 10, it is revealed that when the concentration of hydrogen ion changes from PH=5 to PH=6, the ISFET conductance decreases about 5μs. Moreover, as it is shown in Figures 10(a), (b) and (c), each graph depicts a particular value of PH. For example, when the PH value is PH=8, it is notable that the model is closer to the blue line (experimental data) and also in the different pH values we can compare other ion concentrations as well.

An innovative analytic of matching models using the different values in experimental data has been presented in this work to verify that the conductivity of the graphene based-ISFET is moved down vertically at higher pH values. Briefly, Ion Sensitive FET structure was used with monolayer graphene prepared from CVD grown in large-sized on pieces of p-doped Si covered with a 300 nm as substrate to measure PH changes. According to what was previously mentioned, one could claim that by pH changes in electro-active membrane, significant vertical shift in conductance of the graphene (G_{withPH}) occurred due to ions adsorption on the surface area of monolayer graphene sheet of ISFET channel. In addition, it is notable that the temperature remains constant (about 25°C in solution) in the suggested model because the temperature can have an effect on the behavior of sensing parameter as well.

Figure 10. G-V_g characteristic of proposed conductance model with experimental data for solution with (a) pH=8, (b) pH=9 and (c) pH=10

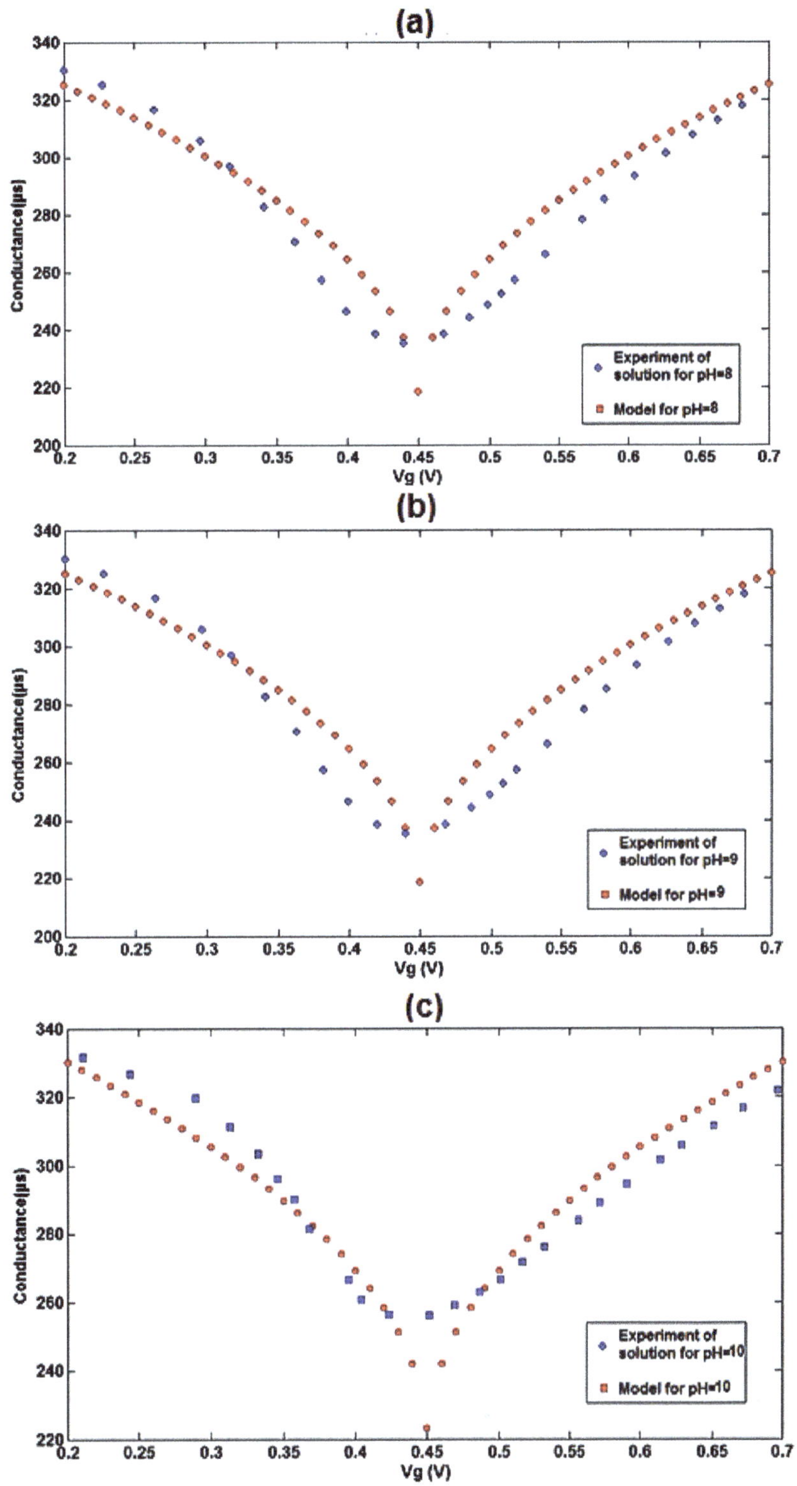

SWCNT APPLICATION IN ISFET DEVICE

SWCNT Base Biosensor

The non-covalent approach by means of electrostatic interface, Van der Waals force, or π–π stacking is a trimming of the CNTs are open as an outcome of oxidation treatment, for biomolecules smaller proteins can be inserted into the tubular practicable immobilization method. Principally, it is hopeful for improving the dispersal proteins of CNTs without devastating of the nanotube construction. In general, this way can be carried out by physical entrapment or adsorption. For biological functions, the development of solvability of CNTs in organic or aqueous flushes is a main undertaking. Enormous attempts have dedicated to investigate gainful approaches to functionalize CNTs for connection to biomolecules as identification factors. In general, this process can be carried out by covalent fictionalization and noncovalent approach. For instance, with carbon nanotube based ISFET, the conductance of the tool is perceptive to pH. In the same way, by means of diarylethylenes as the bridge gives tools that can fluctuability change between nonconjugated and conjugated conditions. The molecular bridge can carry out the double task of transport electrical current and sensing/identification through biological occurrences such as DNA hybridization and protein/substrate binding.

Structure and Detection Mechanism of SWCNT Based-pH Sensor

The homogeneous and transferable SWCNT-based bendable ISFET sensor has been established to be a adaptable biological and chemical sensor, which creates in vivo function capable because of biocompatible character of polymer substrate. The assemblage of nanomaterials on cut-price, clear, and bendable substrate via LbL bottom-up structure is gainful and proposes new concept of industrialized procedure. In this construction electrical contacts were attractive on top of the nanotubes from dissolved Cr-Au films and a accidental network of CNT is attractive into the tool that included interdigitated electrodes with 10µm separation, as demonstrated in Figure 11.

Figure 11. SWCNT based-ISFET structure for pH sensing

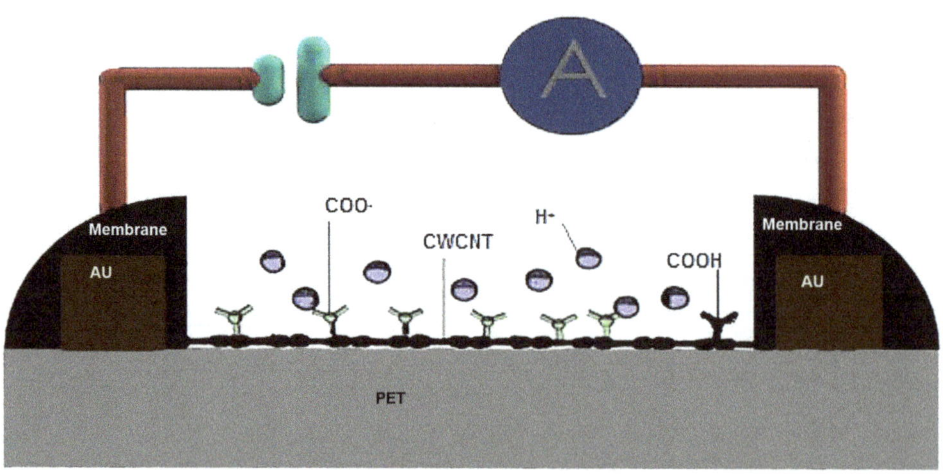

In fact, when π electrons in one-dimensional CNTs are delocalized on the surfaces, the transporters carry prospectively on their surface (D. Fu & Li, 2010). Consequently, the ecological perturbations happening in surrounding area of carbon surfaces have a strapping impact on the conductance of CNTs. In the current paper, the interface between resolution with diverse pH and CNTs cause the considerable modifying in conductance of NTNFET tool throughout electrostatic gating method. In the previous studies, proposed recognition methods are electrostatic gating, transporter mobility modifying, and Schottky obstruction results. For supporting this idea, electrostatic gating is influenced by adsorbed charge type has been approved to describe the conductance amend of carbon nanotube transistors (Ahn, Tsen, Kim, Park, & Park, 2007). In contrast, the conductivity of the NTNFET tools effect the improved number of transporters in the canal. Interface of CNTs with hydrogen ions concentration is due to nucleic acid base π-stacking on the nanotube surface which causes the hydrophilic molecular part indicating to the exterior and a steady hybrid with individual CNTs can be shaped by covering around them through the fragrant interfaces between CNTs sidewalls and nucleotide bases (MacDonald & Laurenzi, 2005). In the present chapter, the model of ISFET tools is planned as detectors of pH sensing. The suggested model imitates the behaviour of ISFET tool to discriminate between wild-type (wt) and mutant (mut) alleles which are deemed as TCT ATG AT**G** ATG AGA GT and TCT ATG AT**C** ATG AGA GT (respectively). Diverse pH rates as a function of entrance voltage is replicated in the meantime pH-sensing aspect is proposed. In the end evaluation study between offered model and experimental data is reported.

I-V Characteristic of Perfectly Symmetric SWCNT

Towards model the band structure of CNT it was started from the modelling of single layer graphene band construction, the energy distribution relation, and obtaining it by means of the Taylor series development near the Fermi indicates (M.T. Ahmadi, Johari, Amin, Mousavi, & Ismail, 2010).

$$E(k) = \pm \frac{t3a_{c-c}}{2}\sqrt{\left(\frac{2}{3d}\right)^2 + k_X^2} \tag{22}$$

where $a_{C\text{-}C} = 1.42$ Å is Carbon-Carbon(C-C) bond length, t=2.7 (eV) is the nearest neighbour C-C firm required extend beyond energy, d is the diameter of the CNT and the (±) signals are connected to the conductance and valence bands. It can be deduced that the first band gap energy can be written as

$$E_G = 2a_{c-c}t \,/\, d = (0.8eV) \,/\, d(nm) .$$

Additionally, due to the parabolic band structure near k=0 points, Equation (22) can be provided by.

$$E(k) = \frac{E_G}{2}\sqrt{1 + \left(\frac{3k_x d}{2}\right)^2} \tag{23}$$

By means of the square root estimate, the parabolic structure of band gap can be assessed by that of silicon nanowires (SNWs).

$$E \approx \frac{E_G}{2} + \frac{\hbar^2 k_x^2}{2m^*} \tag{24}$$

where m^* is efficient group for CNT that depends on the diameter of the tube, $\hbar$ is the concentrated Plank stable; k_x is signal vector module lengthways of the nanotubes which can be extorted in the parabolic part of the band energy.

$$K_X = \sqrt{\frac{4E}{3a_{c-c}t} - \frac{8}{9d^2}} \tag{25}$$

Since the sub bands position has a strapping impact on the number of definite manners, M (E) at a given energy, the parabolic estimate of the band diagram can be employed when the connected energy consists of the foundation of the transference band. In other words, the mode concentration M (E) enhances with energy (Mohammad Taghi Ahmadi, Johari, & Amin, 2010). Taking into consideration the spin dissoluteness the number of transmission canals can be described as.

$$M(E) = 2\frac{\Delta E}{\Delta k \cdot L} = \frac{t3a_{c-c}}{L}\left(\frac{4E}{3a_{c-c}t} - \frac{8}{9d^2}\right)^{\frac{1}{2}} \tag{26}$$

where L is the canal length. The conductance is influenced by large channels, which make it able of subsequent the ohmic scaling law rooted in Landauer method, is regarded because of two aspects. The first one is the boundary confrontation which is autonomous of the length and the second one is because of the fact that the relation of conductance and the width is nonlinear and depends on the number of the forms in the performer. The forms in the performer, though, is quantized restriction in Landauer formula where both of these characteristics are community (M.T. Ahmadi, Johari, Amin, Fallahpour, & Ismail, 2010).

$$G = \frac{2q^2}{h}\int_{-\infty}^{+\infty} dEM(E)T(E)\left(-\frac{df}{dE}\right) \tag{27}$$

where

q is electron charge,
h is Plank stable and
T is the broadcast likelihood of an introduced electron throughout the canal which is estimated to (*T*(*E*) =1) in the ballistic channels (Datta, 2002).

Because of the fact that the term of $\frac{df}{dE}$ is significant only near the Fermi energy the number of definite methods at the Fermi energy is two. Plugging Equation (26) into Equation (27) and in view of the Fermi–Dirac circulation function, conductance can be voluntarily gained as a function of the length of nanoribbon which has a tough impact on conductivity (Peres, Neto, & Guinea, 2006). Shifting the frontier of essential as pursues, Equation (28) can be stated as.

$$G = \frac{2q^2}{h}\frac{3a_{c-c}t}{L}\left(\frac{4}{3a_{c-c}t}\right)\int_{-\infty}^{+\infty}\left(E - \frac{2a_{c-c}t}{3d^2}\right)^{\frac{1}{2}} d\left(-\frac{1}{1+e^{(E-E_F)/K_BT}}\right) \tag{28}$$

Replacing the boundary of integral as follow, SWCNT conductance model is modified as

$$G = \frac{4q^2}{hL}\left(3a_{c-c}t\pi K_BT\right)^{\frac{1}{2}}\left[\int_{-\infty}^{+\infty}\left(\frac{X}{\left(\frac{1}{1+e^{x-\eta}}\right)}\right)^{-\frac{1}{2}} dx + \int_{-\infty}^{+\infty}\left(\frac{X}{\left(\frac{1}{1+e^{x+\eta}}\right)}\right)^{-\frac{1}{2}} dx\right] \tag{29}$$

where $X = (E - E_g) / K_BT$ and the standardized Fermi energy is $\eta = (E_F - E_g) / K_BT$. This equation can mathematically be resolved by utilizing the inequitable combination technique. The Fermi-Dirac allocation function has diverse shapes in disintegrate and non-disintegrate conditions which are qualified by ($\eta\rangle\rangle 0$)and ($\eta\langle\langle 0$) respectively(Dingle & Dingle, 1973) (Zaharah, Mohammad Taghi, Desmond Chang Yih, N Aziziah, & Razali, 2010). In the non-disintegrate condition there are few electrons in the transmission band and the boundary of the transmission group is far over the Fermi energy evaluated to K_{BT}, so the Fermi-Dirac essential can be estimated by Maxwell-Boltzmann allocation aspect of η (E)=exp(η). In disintegrate condition, conversely, the attentiveness of electrons in the transmission group surpasses the concentration of condition and the Fermi energy reclines inside the conductance group and Fermi-Dirac function can be estimated as η (E)=1. Therefore, the overall conductance model of carbon nano tube supported pH biosensor can be gained comparable to that of silicon described by Gunlycke (Gunlycke, Areshkin, & White, 2007).

$$G = \frac{4q^2}{hL}\left(3a_{c-c}t\pi K_BT\right)^{\frac{1}{2}}\left[\Im_{\frac{-1}{2}}(\eta) + \Im_{\frac{-1}{2}}(-\eta)\right] \tag{30}$$

where $\Im_{\frac{-1}{2}}(\eta)$ is the Fermi-Dirac integral of order (-1/2).

Consistent with the present-voltage feature of SWCNT based ISFET tools, the presentation of pH sensor can be accessed through this equation (Passlack, 2008). The canal area has the features of the resistor in small voltage between drain (V_{DS})and source With assuming the source and substrate terminals are supposed in ground potential. Furthermore, the relationship between channel current and conductance can be substituted by common conductance model of single layer grapheme.

$$G_{LP} = \left(\frac{3q^2 \left(3\pi a^3 t^3 k_{BT}\right)^{\frac{1}{2}}}{hl} \left[\Im_{\frac{-1}{2}}(\eta) + \Im_{\frac{-1}{2}}(-\eta) \right] \right) \left(V_{gs} - V_t\right) \tag{31}$$

As it is shown in Figure 12, the present-voltage feature assesses the presentation of pH sensor base on SWCNT nanostructured. Buffer explanation pH 7 MSDS was inserted in the chamber to calculate the remove camber, i.e., ditch present I_d opposed to gate voltage, before the pH value of electrolyte membrane has been added, (Dong, Shi, Huang, Chen, & Li, 2010).

There is an affirmative conformity between planned model for pH sensor based SWCNT and experimental result which is removed from ref (Lee & Cui, 2010).

Proposed Model for SWCNT Based-ISFET

Diverse rates of pH buffer have been inserted in the chamber to permit the inundate attachment to SWCNT surfaces. As it is represented in Figure 13, by affecting the gate voltage to the electrolyte membrane, it is obviously declared that the conductance of FET based graphene demonstrates different behaviour and after that the I-V feature of ISFET tool will be converted. The difference conditions of SWCNT have been observed by determine the present opposed to voltage which is recognized from transport feature curve.

Altogether, Vg can be recognized as a good sign of pH sensing. As it is shown in Figure 13, when the rates of pH decline from 9 to 5, more H+ ions are adsorbed and the sensor will be able to attract more molecules in the same method used to modify the Vg on the tool. In light of this fact, the current paper has focused on showing a new model for pH sensor. In this model, the pH buffer as a function of gate voltage is replicated and pH aspect ($\propto$) is recommended. Consequently, in order to get a superior insight into the role of pH buffer,ISFET modelling is utilized to gain a significant concept for I-V characteristic of SWCNT based-sensor. To achieve this purpose, two electrodes of sensor, as it is demon-

Figure 12. (a) I-V characteristic model of SWCNT; (b) comparison between carbon nanotube current models with experimental data for Buffer solution pH 7
Lee & Cui, 2010.

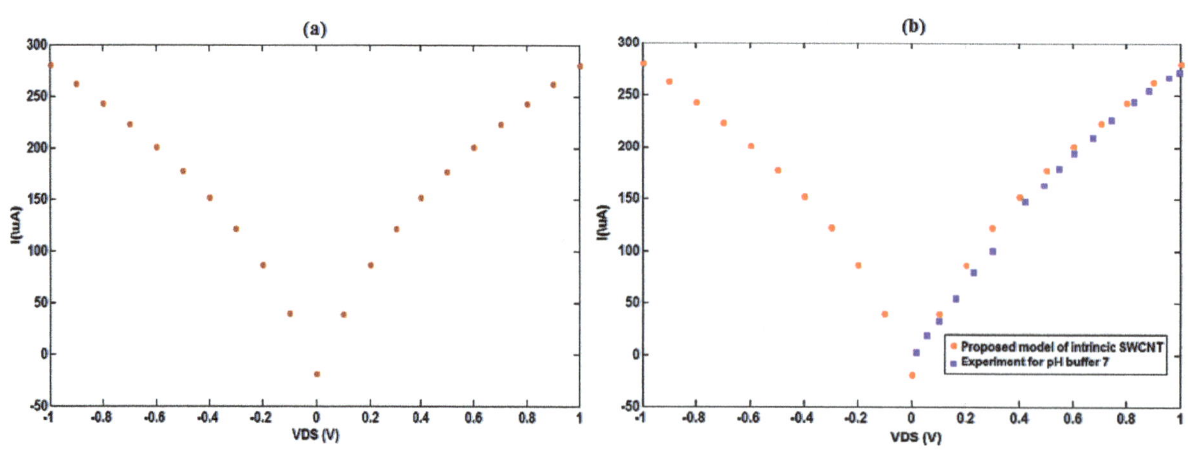

Figure 13. The comparison between SWCNT current model with extracted experimental data for different pH value from pH 5 to pH9 which shows considerable changes in current of SWCNT-based sensor for different pH buffer
Lee & Cui, 2010.

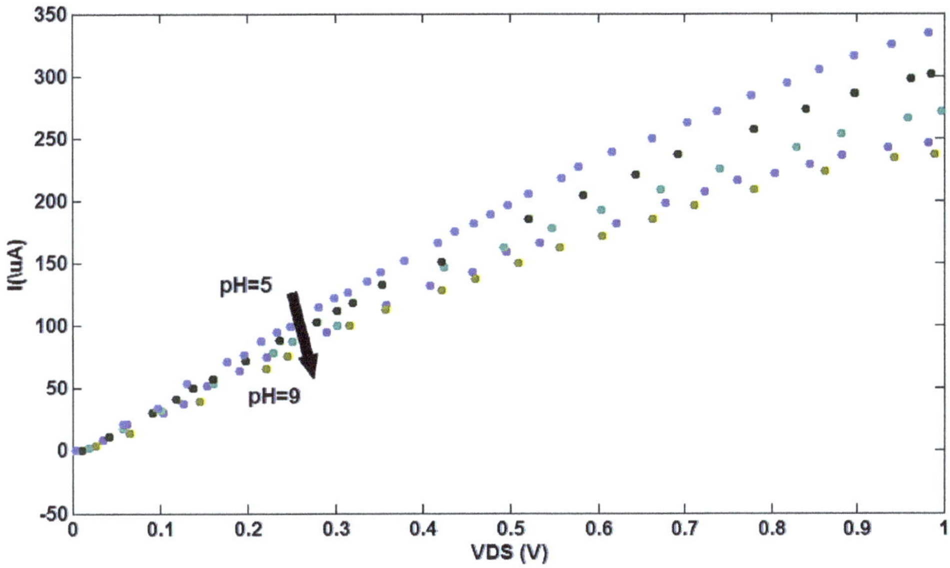

strated in figure 3 are deemed as source and drain contacts. Finally, gate voltage is modelled by pH (following), in other words, Vg is supposed as a function of pH (Vgs~pH).

$$V_{gs}(with\ electrolyte\ gate) = \frac{\alpha}{pH} V_{gs}(without\ electrolyte\ gate) \tag{32}$$

where pH sensing feature ($\propto$) is proposed and pH is the diverse rate of pH buffer. In the nano-saturation area, pH model is utilized as a function of gate voltage and the current-voltage relation for the n-channel MOSFET is personalized as.

$$G_{LP} = \left(\frac{3q^2 \left(3\pi a^3 t^3 k_{BT}\right)^{\frac{1}{2}}}{hl} \left[\Im_{\frac{-1}{2}}(\eta) + \Im_{\frac{-1}{2}}(-\eta) \right] \right) \left(\frac{\alpha}{pH} V_{gs}(whitout\ electrolyte\ gate) - V_t \right) \tag{33}$$

The shape of (pH) parameter offers different rates of pH buffer. Therefore, the hydrogen ion concentration adsorbed on SWCNT surface by iteration method is modelled as.

$$\alpha = \alpha_1 e^{-\alpha_2 \cdot pH} \tag{34}$$

From experiment data, (α_1 and α_2) parameters are estimated where α_1=364.41 and α_2=0.088. Ultimately, consistent with the planned model of pH sensor by means of nanostructured nanotube, the current-voltage changes can be customized as.

$$G_{LP} = \left(\frac{3q^2\left(3\pi a^3 t^3 k_{BT}\right)^{\frac{1}{2}}}{hl} \left[\Im_{\frac{-1}{2}}(\eta) + \Im_{\frac{-1}{2}}(-\eta) \right] \right) \left(\frac{\alpha_1 1 e^{-\alpha_2 \cdot pH}}{pH} V_{gs}(whitout\ electrolyte\ gate) - V_t \right) \quad (35)$$

Each diagram in Figure 14 and Figure 15 shows definite pH buffer of electrolyte membrane.

Figure 14. Current versus drain voltage curves for both proposed model and experimental data after adding the solution with pH 5(a) and pH 6(b) respectively
Lee & Cui, 2010.

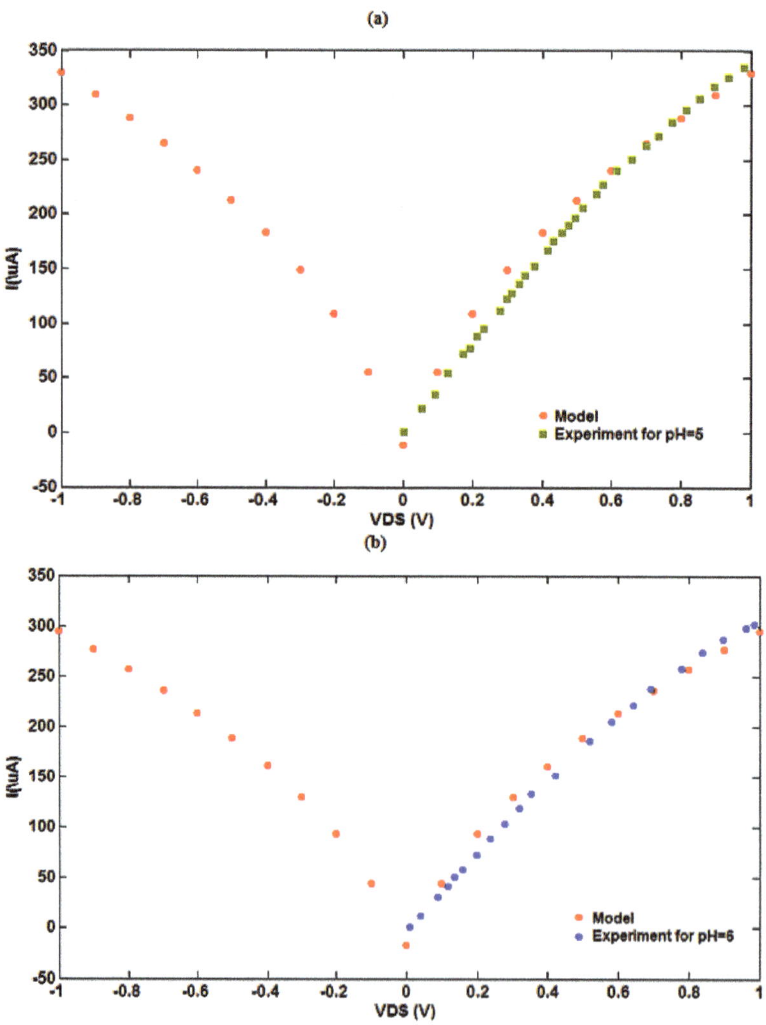

Figure 15. G-V_{DS} characteristic of CNT proposed model with experimental data for solution with pH 8 (a) and pH9(b)
Lee & Cui, 2010.

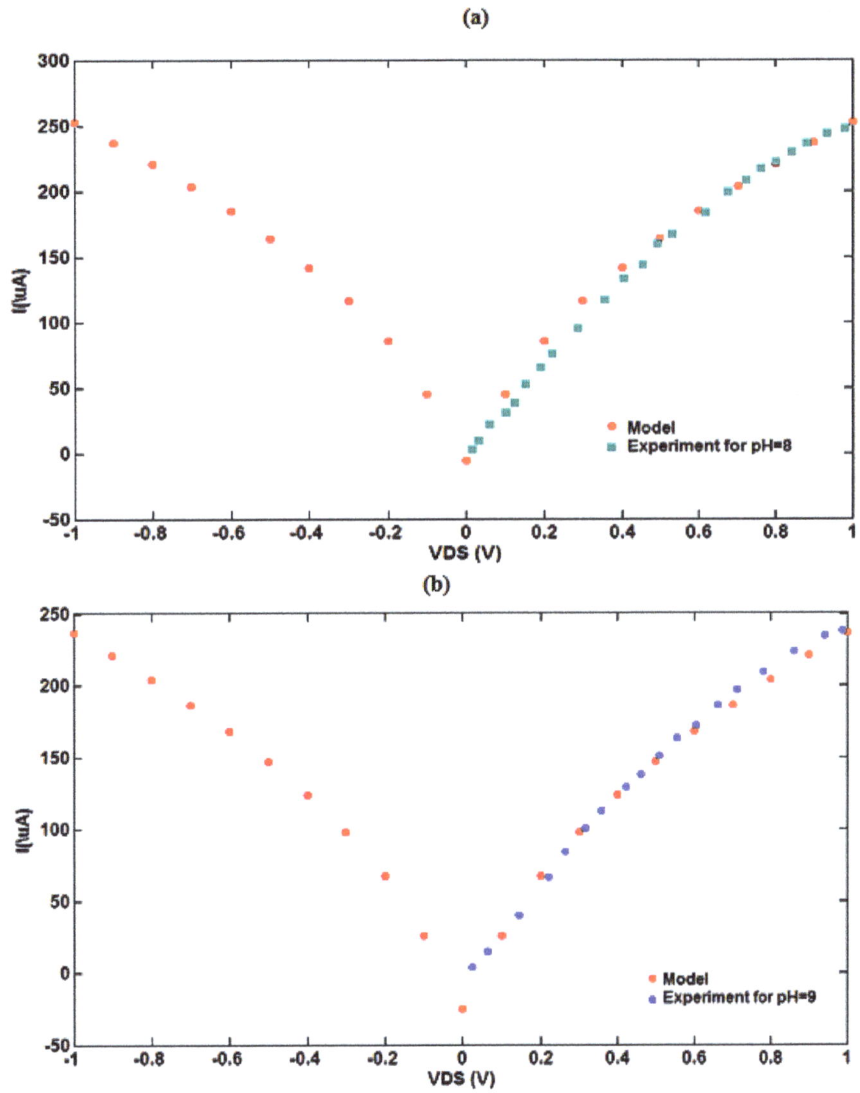

For instance, when pH buffer enhances from 5 to 7, it is remarkable that, the model is closer to the blue line and the conductance of nanotube will be reduced about 35µs; in the same way we can evaluate other experimental data as well. It is actually shown that, I-V characteristic curve can be managed by altering pH buffer through the pH feature. In addition, the planned model is powerfully estimated with experimental data (Lee & Cui, 2010).

According to Figure 15, the quantity of conductance will be saturated by rising the pH buffer from 8 to 9, probably holding the fact that the concentration of hydrogen ions is inadequate and the sensor transfer cure has been saturated. According to what discussed above, one could strongly assert that the current versus gate voltage of SWCNT based-ISFET device with different pH values can be displayed by ion concentration of the solution. In addition, the current of its channel assumed as a function of PH levels, which can be controlled by a control parameter (α). The number of carriers changing in the channel influences the conductance of the CNT-based FET devices. FET-based CNT with high sensitivity was applied to detect the pH changes, based on the conductance variations. The performance of CNT-based biosensor for pH=5 to pH=9 is evaluated and the analytical results of the proposed model for pH sensor with appropriate parameters are compared with the experimental which indicates a good agreement. It is evident in the figures that the points calculated and obtained from our model agreeably overlap the measurement data.

CONCLUSION

The emerging potentials of nanostructured carbon based-ISFET with high sensitivity and readily detection have been applied to electrochemical catalysis through PH sensing. The conductance of an ISFET device with different pH values can be displayed by ion concentration of the solution. In this research, the conductance of graphene and SWCNT assume as a function of PH levels $G_{withPH} \approx PH$ which shows pH factor. Measurements show decreasing conductivity when the pH value of electrolyte is increased. Specially, in V_{CNP}, the changed conductance values are clearly depicted. The suggested model verifies the reported experimental data as well. In other words, based on the good agreement between the presented analytical model and experimental data, Ƥ can be employed as a PH factor to predict graphene and SWCNT behavior in carbon materials based-ISFET.

REFERENCES

Ahmadi, M. T., Ismail, R., Tan, M. L. P., & Arora, V. K. (2008). The ultimate ballistic drift velocity in carbon nanotubes. *Journal of Nanomaterials*, *2008*, 1–8. doi:10.1155/2008/769250

Ahmadi, M. T., Johari, Z., & Amin, N. A. (2010). *Carbon nanotube conductance model in parabolic band structure*. Paper presented at the Semiconductor Electronics (ICSE), 2010 IEEE International Conference on. doi:10.1109/SMELEC.2010.5549582

Ahmadi, M. T., Johari, Z., Amin, N. A., Fallahpour, A. H., & Ismail, R. (2010). Graphene nanoribbon conductance model in parabolic band structure. *Journal of Nanomaterials*, *2010*, 12. doi:10.1155/2010/753738

Ahmadi, M. T., Johari, Z., Amin, N. A., Fallahpour, A. H., & Ismail, R. (2010). Graphene nanoribbon conductance model in parabolic band structure. *Journal of Nanomaterials*, *2010*, 12. doi:10.1155/2010/753738

Ahmadi, M. T., Johari, Z., Amin, N. A., Mousavi, S. M., & Ismail, R. (2010). *Carbon nanotube conductance model in parabolic band structure*. Academic Press.

Ahn, Y. H., Tsen, A. W., Kim, B., Park, Y. W., & Park, J. (2007). Photocurrent imaging of p-n junctions in ambipolar carbon nanotube transistors. *Nano Letters*, *7*(11), 3320–3323. doi:10.1021/nl071536m PMID:17939725

Bonanni, A. L., Loo, A. H., & Pumera, M. (2012). Graphene for impedimetric biosensing. *Trac-Trends in Analytical Chemistry*, *37*, 12–21. doi:10.1016/j.trac.2012.02.011

Bousse, L. B., & Bergveld, P. (1984). The role of buried OH sites in the response mechanism of inorganic-gate pH-sensitive ISFETs. *Sensors and Actuators*, *6*(1), 65–78. doi:10.1016/0250-6874(84)80028-1

Bratov, A. M. o., Dominguez, & , & Bartroli, . (1995). Photocurable polymers applied as encapsulating materials for ISFET production. *Sensors and Actuators. B, Chemical*, *25*(1–3), 823–825. doi:10.1016/0925-4005(95)85183-6

Cadogan, A. G., Lewenstam, Ivaska, Diamond, & Dermot. (1992). All-solid-state sodium-selective electrode based on a calixarene ionophore in a poly(vinyl chloride) membrane with a polypyrrole solid contact. *Analytical Chemistry, 64*(21), 2496-2501.

Cecilia, J.-J. J., Antoni, & Baldi. (2009). ISFET Based Microsensors for Environmental Monitoring. *Sensors (Basel, Switzerland)*, *10*(1). PMID:22315527

Chen, S. Z., Zhang, Z.-B., Ma, L., Ahlberg, P., Gao, X., Qiu, Z., & Zhang, S.-L. et al. (2012). A graphene field-effect capacitor sensor in electrolyte. *Applied Physics Letters*, *101*(15), 154106–154105. doi:10.1063/1.4759147

Chen Yan-Kuin, S. C. (1951). The fabrication and characterisation of ion-sensitive field-effect transistors with a silicon dioxide gate. *Journal of Physics. D, Applied Physics*, *19*(10).

Chung, W.-Y. C., Szu, Pijanowska, Dawgul, Torbicz, Grabiec, … Wei-Po. (2009). *ISFET electronic tongue system for environmental multi-ion sensing with independent component analysis signal processing*. doi: 10.1117/12.821635

Chung, W.-Y. L., Lin, Y.-T., Pijanowska, D. G., Yang, C.-H., Wang, M.-C., Krzyskow, A., & Torbicz, W. (2006). New ISFET interface circuit design with temperature compensation. *Microelectronics Journal*, *37*(10), 1105–1114. doi:10.1016/j.mejo.2006.05.001

Cobben, P. L. H. M. E., Bomer, Bergveld, Verboom, & Reinhoudt. (1992). Transduction of selective recognition of heavy metal ions by chemically modified field effect transistors (CHEMFETs). *Journal of the American Chemical Society, 114*(26), 10573-10582.

Couto, S. R. M., & Sanromán, D. (2006). Optimum stability conditions of pH and temperature for ligninase and manganese-dependent peroxidase from Phanerochaete chrysosporium. Application to in vitro decolorization of Poly R-478 by MnP. *World Journal of Microbiology & Biotechnology*, *22*(6), 607–612. doi:10.1007/s11274-005-9078-0

Datta, S. (2002). *Electronic transport in mesoscopic systems*. Cambridge, UK: Cambridge University Press.

Dingle, R. B., & Dingle, R. (1973). *Asymptotic expansions: Their derivation and interpretation*. Academic Press London.

Dong, X., Shi, Y., Huang, W., Chen, P., & Li, L.-J. (2010). Electrical Detection of DNA Hybridization with Single-Base Specificity Using Transistors Based on CVD-Grown Graphene Sheets. *Advanced Materials*, *22*(14), 1649–1653. doi:10.1002/adma.200903645 PMID:20496398

Dongjin, L. T. (2012). An electric detection of immunoglobulin G in the enzyme-linked immunosorbent assay using an indium oxide nanoparticle ion-sensitive field-effect transistor. *Journal of Micromechanics and Microengineering*, *22*(1), 015009. doi:10.1088/0960-1317/22/1/015009

Enoki, T., Kobayashi, Y., & Fukui, K.-I. (2007). Electronic structures of graphene edges and nanographene. *International Reviews in Physical Chemistry*, *26*(4), 609–645. doi:10.1080/01442350701611991

Fu, D., & Li, L. J. (2010). Label-free electrical detection of DNA hybridization using carbon nanotubes and graphene. *Nano Reviews, 1*.

Fu, W. N., Knopfmacher, Tarasov, Weiss, Calame, & Schonenberger. (2011). Graphene Transistors Are Insensitive to pH changes in Solution. *Nano Letters, 11*(9), 3597-3600.

Geim, A. K., & Novoselov, K. S. (2007). The rise of graphene. *Nature Materials*, *6*(3), 183–191. doi:10.1038/nmat1849 PMID:17330084

Gnani, E., Gnudi, A., Reggiani, S., & Baccarani, G. (2010). Effective Mobility in Nanowire FETs under quasi-ballistic conditions. *IEEE Transactions on Electron Devices*, *57*(1), 336–344. doi:10.1109/TED.2009.2035545

Gotoh, M. S., Suzuki, M., Kubo, I., Tamiya, E., & Karube, I. (1989). Immuno-FET sensor. *Journal of Molecular Catalysis*, *53*(3), 285–292. doi:10.1016/0304-5102(89)80063-X

Gunlycke, D., Areshkin, D., & White, C. (2007). Semiconducting graphene nanostrips with edge disorder. *Applied Physics Letters*, *90*(14), 142104. doi:10.1063/1.2718515

Guth, U. G. F., Decker, M., Oelßner, W., & Vonau, W. (2009). Solid-state reference electrodes for potentiometric sensors. *Journal of Solid State Electrochemistry*, *13*(1), 27–39. doi:10.1007/s10008-008-0574-7

Haigang Yang, H. S., Han, Wei, Lin, Xia, & Zhong. (2005). *A pH-ISFET based micro sensor system on chip using standard CMOS technology*. Paper presented at the Fifth International Workshop on System-on-Chip for Real-Time Applications. doi:10.1109/IWSOC.2005.21

Jamasb, S. C., & Smith, S. D., R. L. (1997). *Correction of instability in ion-selective field effect transistors (ISFETs) for accurate continuous monitoring of pH.* Paper presented at the Engineering in Medicine and Biology Society.

Jiménez, C. B., & Bartroli, J. (1997). Development of an ion-sensitive field effect transistor based on PVC membrane technology with improved long-term stability. *Electroanalysis*, *9*(4), 316–319. doi:10.1002/elan.1140090411

Kal, S. B. P. V. (2007). *Design and modeling of ISFET for pH sensing*. Paper presented at the TENCON 2007 - 2007 IEEE Region 10 Conference.

Kiani, M. J., Ahmadi, M., Akbari, E., Karimi, H., & Che Harun, F. (2013). Graphene nanoribbon based gas sensor. *Key Engineering Materials*, *553*, 7–11. doi:10.4028/www.scientific.net/KEM.553.7

Kiani, M. J., Ahmadi, M. T., Abadi, H. K. F., & Rahmani, M. (2013). Analytical modelling of monolayer graphene-based ion-sensitive FET to pH changes. *Nanoscale Research Letters*, *8*(1), 1–9. doi:10.1186/1556-276X-8-173 PMID:23590751

Kiani, M. J. A., Ahmadi, M., Karimi Feiz Abadi, H., Rahmani, M., Hashim, A., & Che harun, F. (2013). Analytical modelling of monolayer graphene-based ion-sensitive FET to pH changes. *Nanoscale Research Letters*, *8*(1), 173. doi:10.1186/1556-276X-8-173 PMID:23590751

Kim, S. J. (2009). Flexible alcohol vapor sensors using multiple spray-coated SWNTs on PES substrates. *Journal of the Korean Physical Society*, *54*(5), 1779–1783. doi:10.3938/jkps.54.1779

Kuang, B. M., & Hafiz, S. (2012). Sensing soil properties in the laboratory, in situ, and on-line: A Review. In L. S. Donald (Ed.), Advances in Agronomy (Vol. 114, pp. 155-223). Academic Press.

Lee, D., & Cui, T. (2010). Low-cost, transparent, and flexible single-walled carbon nanotube nanocomposite based ion-sensitive field-effect transistors for pH/glucose sensing. *Biosensors & Bioelectronics*, *25*(10), 2259–2264. doi:10.1016/j.bios.2010.03.003 PMID:20417088

MacDonald, R. A., Laurenzi, B. F., Viswanathan, G., Ajayan, P. M., & Stegemann, J. P. (2005). Collagen–carbon nanotube composite materials as scaffolds in tissue engineering. *Journal of Biomedical Materials Research. Part A*, *74*(3), 489–496. doi:10.1002/jbm.a.30386 PMID:15973695

Martinoia, S. M., & Massobrio, P. (2004). ISFET–neuron junction: Circuit models and extracellular signal simulations. *Biosensors & Bioelectronics*, *19*(11), 1487–1496. doi:10.1016/j.bios.2003.12.003 PMID:15093221

Martoccia, D., Björck, M., Schlepütz, C., Brugger, T., Pauli, S. A., Patterson, B. D., & Willmott, P. R. et al. (2010). Graphene on Ru (0001): A corrugated and chiral structure. *New Journal of Physics*, *12*(4), 043028. doi:10.1088/1367-2630/12/4/043028

Morgenshtein, A. D. U., Jakobson, C. G., & Nemirovsky, Y. (2004). Wheatstone-Bridge readout interface for ISFET/REFET applications. *Sensors and Actuators. B, Chemical*, *98*(1), 10–10. doi:10.1016/j.snb.2003.07.017

Moriconi, L. N. D., & Niemeyer, D. (2011). Graphene conductivity near the charge neutral point. *Physical Review B: Condensed Matter and Materials Physics*, *84*(19), 193401. doi:10.1103/PhysRevB.84.193401

Passlack, M. (2008). III-V Metal-oxide-semiconductor technology. *2008 IEEE 20th International Conference on Indium Phosphide and Related Materials (IPRM).*

Peres, N., Neto, A. H. C., & Guinea, F. (2006). Conductance quantization in mesoscopic graphene. *Physical Review B: Condensed Matter and Materials Physics*, *73*(19), 195411. doi:10.1103/PhysRevB.73.195411

Pham, M. T. H., Kunath, S., Kurth, C., & Köhler, E. (1994). Backside membrane structures for ISFETs applied in miniature analysis systems. *Sensors and Actuators. B, Chemical*, *19*(1–3), 333–335. doi:10.1016/0925-4005(93)00985-8

Pijanowska, D., & Torbicz, W. (1997). *Simple method of enzyme immobilization for pH-ISFET-based urea biosensors*. doi: 10.1117/12.266713

Pokhrel, S., Joo, J. C., Kim, Y. H., & Yoo, Y. J. (2012). Rational design of a Bacillus circulans xylanase by introducing charged residue to shift the pH optimum. *Process Biochemistry*, *47*(12), 2487–2493. doi:10.1016/j.procbio.2012.10.011

Reinhoudt, D. N. S., & Ernst, J. R. (1990). The transduction of host-guest interactions into electronic signals by molecular systems. *Advanced Materials*, *2*(1), 23–32. doi:10.1002/adma.19900020105

Schlesinger, R. B. M., Becht, R., Dosenbach, S., Hoffmann, W., & Ache, H. J. (1996). ISFETs with sputtered sodium alumino-silicate glass membranes. *Fresenius' Journal of Analytical Chemistry*, *354*(7-8), 852–856. PMID:15048401

Seymour, R. B. (1966). Plastics. *Industrial & Engineering Chemistry, 58*(8), 61-73.

Shepherd, L. T., & Toumazou, C. (2005). Weak Inversion ISFETs for ultra-low power biochemical sensing and real-time analysis. *Sensors and Actuators. B, Chemical*, *107*(1), 468–473. doi:10.1016/j.snb.2004.11.006

Steinhoff, G. H. M., Schaff, W. J., Eastman, L. F., Stutzmann, M., & Eickhoff, M. (2003). pH response of GaN surfaces and its application for pH-sensitive field-effect transistors. *Applied Physics Letters*, *83*(1), 177–179. doi:10.1063/1.1589188

Voigt, H. S. F., Lange, T., Kullick, T., & Ferretti, R. (1997). Diamond-like carbon-gate pH-ISFET. *Sensors and Actuators. B, Chemical*, *44*(1–3), 441–445. doi:10.1016/S0925-4005(97)00236-0

Zaharah, J., Mohammad Taghi, A., Desmond Chang Yih, C., & Aziziah, N. (2010). A., & Razali, I. (2010). Modelling of Graphene Nanoribbon Fermi Energy. *Journal of Nanomaterials*.

Zhao, Y. S., Song, X., Song, Q., & Yin, Z. (2012). A facile route to the synthesis copper oxide/reduced graphene oxide nanocomposites and electrochemical detection of catechol organic pollutant. *CrystEngComm*, *14*(20), 6710–6719. doi:10.1039/c2ce25509j

Chapter 14
Surface Plasmon Resonance-Based Sensor Modeling

Bahar Meshginqalam
Urmia University, Iran

Mohammad Taghi Ahmadi
Urmia University, Iran

Hamid Toloue Ajili Tousi
Malaysia-Japan International Institute of Technology (MJIIT), Malaysia

Arash Sabatyan
Urmia University, Iran

Anthony Centeno
Malaysia-Japan International Institute of Technology (MJIIT), Malaysia

ABSTRACT

Exceptional optical and electrical characteristics of graphene based materials attract significant interest of the researchers to develop sensing center of surface Plasmon resonance (SPR) based sensors by graphene application. On the other hand refractive index calculation of graphene based structures is necessary for SPR sensor analysis. In this chapter first of all a new method for refractive index investigation of some graphene based structures are introduced and then the effect of carrier density variant in the form of conductance gradient on graphene based SPR sensor response is modeled. The molecular properties such as electro-negativity, molecular mass, effective group number and effective outer shell factor of the molecule are engaged. In addition each factor effect in the cumulative carrier variation is explored analytically. The refractive index shift equation based on these factors is defined and related coefficients are proposed. Finally a semi-empirical model for interpretation of changes in SPR curve is suggested and tested for some organic molecules.

INTRODUCTION

Graphene with a single atomic layer of carbon is an ideal nominee on sensor application because of high surface-to-volume ratio. Two dimensional (2D) honeycomb lattice of graphene has exclusive optoelectronic properties with enormous application potential on future nanoscale devices (Costamagna & Dobry, 2011; Luican et al., 2011). Its interesting optical properties lead to low-cost and accurate optical devices as well. Moreover, band energy as a major factor which plays a significant role on carrier transport has

DOI: 10.4018/978-1-5225-0736-9.ch014

been explored widely (Abergel, Apalkov, Berashevich, Ziegler, & Chakraborty, 2010; Adonkin, Gorelov, Dyakin, Karpov, & Prikhodko, 1994; Akdim & Pachtert, 2011; Elias et al., 2009), moreover different stacking of graphene sheets leads to different electronic properties (Craciun et al., 2009) which cause to different optical features. On the other hand, increasing industrial and clinical demand on graphene based optical devices such as Surface Plasmon Resonance (SPR) sensors requires its optical properties to be explored (Acik & Chabal, 2011; Afzali, Bol, Kasry, & Tulevski; Huang, Dong, Liu, Li, & Chen, 2011; Phan & Viet, 2012). As the first step, Bilayer Graphene Nanoribbon (BGN) which consists of two Bernal AB stacked layers and Trilayer Graphene Nanoribbon (TGN) with three layers of graphene and a tunable band gap as two examples of graphene based structures are considered.

In this chapter, firstly optical properties of BGN and TGN in the presence of applied voltage for different incident wavelength are explored. BGN and TGN dielectric constants and refractive indices based on their conductance are theoretically modeled and obtained results are numerically simulated. Additionally, the applied bias effect on BGN and TGN optical parameters are discussed based on the presented model. It is notable that, obtained results can be used in SPR modeling. Furthermore, SPR based sensors as an attractive configuration which operates by sensing the interactions between sensing element and the sensor metallic surface. From sensor point of view, its sensitivity is one of the main features, so to overcome the sensitivity confinement, new technological developments in device and material characteristic are needed (Zuppella, Tosatto, Corso, Zuccon, & Pelizzo, 2013). Moreover, it should be mentioned that the graphene and Graphene Oxide Sheets (GOS) based SPR structures sensitivity have been approved by (Maharana, Jha, & Palei, 2014; Pradeep Kumar, Triranjita, & Rajan, 2014; Wu, Chu, Koh, & Li, 2010), in addition, the GOS sensor chip is 3.7 times more sensitive than the graphene based chip as reported by Stebunov, et al (Stebunov, Aftenieva, Arsenin, & Volkov, 2015). These advantages motivate researchers to work on analytical prediction model of GOS film-based SPR sensor. So as the second step in this chapter, the analytically modeled response of GOS film-based SPR sensor which is covered by organic molecules is investigated.

OPTICAL PROPERTIES OF GRAPHENE BASED STRUCTURES

Optical Properties of Bilayer Graphene Nanoribbon

BGN with unique optical and electrical properties has been studied in different nanoscale fields (Rahmani, Ahmadi, Ismail, & Ghadiry, 2013; H. Sadeghi et al., 2011). In contrast with graphene nanoribbon, BGN consists of two Bernal AB stacked layers which is shown in Figure 1. Optical properties of BGN, which is necessary for SPR sensor applications (Acik & Chabal, 2011; Afzali et al.; Huang et al., 2011; Huang et al., 2010; Phan & Viet, 2012), in the presence of applied voltage for different incident wavelength based on conductance can be modeled. Conductance is one of the main parameters which need to be discovered then the optical properties of BGN can be derived from conductance in specific condition.

First step to analyze of BGN conductance is started with parabolic band energy approximation. For the proposed BGN, the tight-binding technique is adopted in order to calculate the energy band structure of BGN with AB stacking (Hatef, SeyedMahdi, Meisam, MohammadTaghi, & Razali, 2012):

Figure 1. The schematic of BGN (AB-stacked configuration)

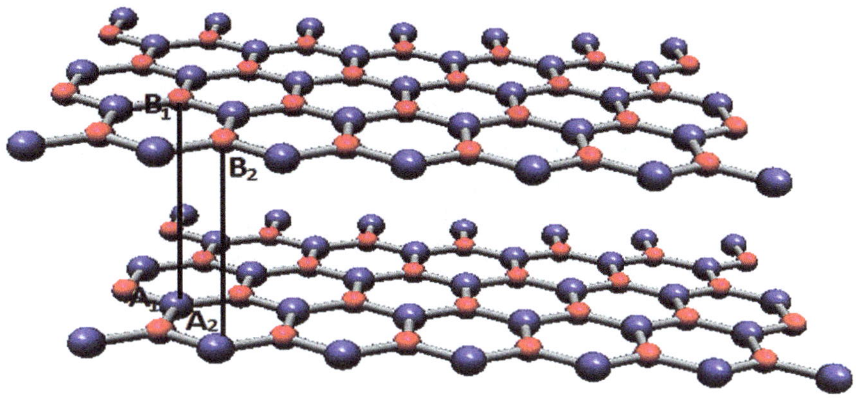

$$E_{k(V)} = \pm\sqrt{\varepsilon_k^2 + \frac{t_\perp^2}{2} + \frac{V^2}{4} \pm \sqrt{\left(t_\perp^2 + V^2\right)\varepsilon_k^2 + \frac{t_\perp^2}{4}}}, \tag{1}$$

where V is the applied voltage, $t_\perp$ is the interlayer hopping energy, ε_k^2 is derived from the below equation which is clear that it is depends on applied voltage and interlayer hopping energy.

$$\varepsilon_k^2 = \frac{\left(\frac{V^4}{4} + \frac{V^2 t_\perp^2}{2}\right)}{\left(V^2 + t_\perp^2\right)}, \tag{2}$$

Energy equation (Equation (1)) can be exposed as:

$$E_{(k)} \approx \Delta - \alpha k^2 + \beta k^4, \tag{3}$$

where $\Delta = V/2$, $k = 2\pi n/\lambda$ is the wave vector, $\alpha = \left(\frac{V}{t_\perp^2}\right) v_F^2$, $\beta = \frac{v_F^4}{V t_\perp^2}$. α and β coefficients depend on the value of Fermi velocity. Fermi velocity is defined by $v_F = \sqrt{3}\gamma_0 a/2\hbar \cong 10^6$ m/s (Avetisyan, Partoens, & Peeters, 2010) where $\gamma_0 \approx 3.12eV$ (Mak, Shan, & Heinz, 2010) is the hopping between π orbitals located at nearest neighbor atoms and also we indicate the interlayer hopping energy as $t_\perp \approx 0.1\gamma_0$ and (a) is the lattice spacing which is equal to $1.4\,\mathring{A}$. In addition, based on conductance definition ($G = I/V$), the Boltzmann transport equation can be written as the Landauer formula which illustrates conductance with Ohmic behavior and parabolic band energy directs to number of mode calculation. However, in the smaller length the interface resistance and width effect in the form of number of modes are considered as:

$$G = \frac{2q^2}{h}\int_{-\infty}^{+\infty} M(E)T(E)\left(-\frac{df}{dE}\right)dE \tag{4}$$

where q is the electron charge, T is transmission probability which is equal to one in ballistic limit, h is the Planck's constant, M is the number of modes and f is the Fermi–Dirac distribution function. Furthermore, 1D quantum confinement effect demonstrates temperature-dependent conductance with minimum value near the Dirac point that proves Fermi-Dirac integral based method on conductance expression. The general mathematical model of conductance for BGN, which can be numerically solved is written as (Hatef et al., 2012).

$$G_{BGN} = -\frac{4q^2}{hl}\int_{-v_g}^{+v_g}\left(\frac{1}{1+e^{\left(x-\frac{(v_g-v)q}{k_BT}\right)}}\right)\left(\frac{3}{4}\sqrt{\frac{a+\sqrt{a^2+4\beta x k_BT}}{2\beta\left(a^2+4\beta x k_BT\right)}}-\frac{a}{8\beta\left(a^2+4\beta x k_BT\right)\sqrt{\frac{a+\sqrt{a^2+4\beta x k_BT}}{2\beta}}}\right)dx \tag{5}$$

where V is the biased voltage, v_g is the gate voltage, k_B is the Boltzmann constant, T is temperature and $x = \left(E-\Delta\right)/k_BT$. Moreover, Fermi-Dirac general integral forms conductance needs to be recognized by the role of degeneracy. An acceptable agreement with previous experimental result achieved by comparing with common form of conductance based on the presented model in the ballistic limit which is depicted in Figure 2 (Hamid. Toloue A.T, 2015).

The conductivity equation is applied to explore the BGN optical parameters such as complex refractive index. Together with Ampere's current law, conductance is employed to calculate the dielectric function of BGN.

Figure 2. BGN conductance model in comparison with experimental data
Ahmadi, Johari, Amin, Fallahpour, & Ismail, 2010.

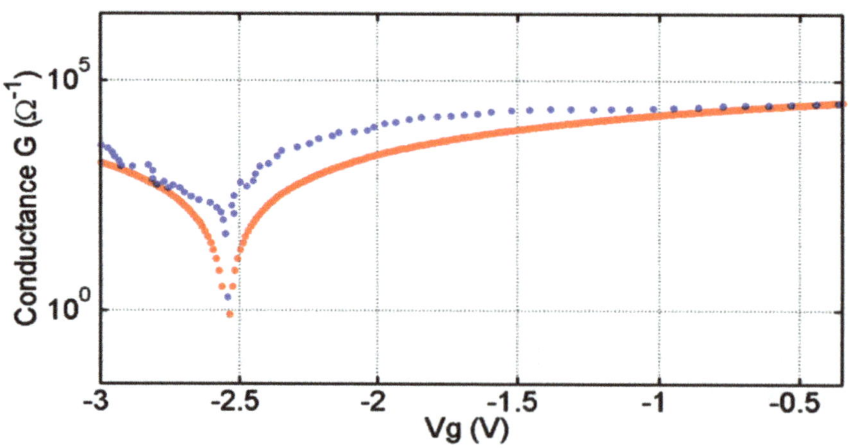

$$J = \frac{G}{\delta} E' \tag{6}$$

where the thickness of the graphene layer is δ and the electric field vector $E' = E_0' e^{+i\omega t}$. By applying the conductance equation, the current density per cross section area is modified as:

$$J = -\frac{4q^2 E}{\delta h l} \int_{-v_g}^{+v_g} \left(\frac{1}{1 + e^{\left(x - \frac{(v_g - v)q}{k_B T} \right)}} \right) \left(\frac{3}{4} \sqrt{\frac{a + \sqrt{a^2 + 4\beta x k_B T}}{2\beta \left(a^2 + 4\beta x k_B T \right)}} - \frac{a}{8\beta \left(a^2 + 4\beta x k_B T \right) \sqrt{\frac{a + \sqrt{a^2 + 4\beta x k_B T}}{2\beta}}} \right) dx \tag{7}$$

Since optical conductivity represents the AC conductivity, which related to electrons oscillation about their equilibrium positions, in low biases it can be equal to the Drude DC conductivity which related to the electrons movement over an arbitrary distance in response to the DC field (Singh, 2012). Therefore, in this condition optical conductivity can replace with electrical conductivity. It is clear that Ampere's current law can relate the conductance with the optical property as:

$$\nabla \times H = J + \frac{\partial D_0}{\partial \tau} = \frac{\partial D}{\partial \tau} \tag{8}$$

where $D_0 = \varepsilon_0 E'$ and $D = \varepsilon_0 \varepsilon_{BGN} E'$ are the displacement fields of vacuum and BGN, respectively. In addition, the conductance is employed which leads to the modified current equation as:

$$\frac{G}{\delta} E_0 e^{+i\omega t} + i\omega \varepsilon_0 E_0 e^{+i\omega t} = i\omega \varepsilon_0 \varepsilon_{BGN} E_0 e^{+i\omega t} \tag{9}$$

where

ε_0 is the vacuum permittivity,
ω is the frequency, and
ε_{BGN} is the BGN dielectric constant.

From this we can extract the relative dielectric constant that presents conductance resembling trends in dielectric constant expression.

$$\varepsilon_{BGN} = \frac{G}{i\delta\omega\varepsilon_0} + 1 \tag{10}$$

By considering the wide application of degenerate regime in nanoscale devices, Equation (5) was analytically solved (Hatef et al., 2012). In degenerate limit, number of carriers increases rapidly and probability of occupied energy levels is equal to one, therefore the dielectric function as a function of BGN geometry is:

$$\varepsilon_{BGN} = \frac{G_{BGN}^{D}}{i\delta\omega\varepsilon_0} + 1 = \frac{q^2}{\sqrt{2} ihl\varepsilon_0 \omega K_B T\delta} \left(\sqrt{\left(\frac{a + \sqrt{a^2 + 4\beta x k_B T}}{\beta} \right)^3} - \frac{\alpha}{\beta} \sqrt{\frac{a + \sqrt{a^2 + 4\beta x k_B T}}{\beta}} \right)_{-v_g}^{v_g} + 1 \quad (11)$$

In non-equilibrium conditions, conductance as a function of applied voltage has been reported (Amin et al., 2011; Dellabetta & Gilbert, 2010; Hatef Sadeghi, Ahmadi, Mousavi, & Ismail, 2012). Analytical model illustrates same performance on dielectric function as shown in Figure 3. In other word, dielectric function is changed by applied voltage together with wavelength variation.

By increasing the applied voltage, the imaginary part of dielectric constant of BGN decreases (absorption increased as absolute value increased) but the real part remains constant. The conductance per atomic layer in mass less Dirac fermium band structure is a universal constant that directs to atomic layer dependent refractive index, in agreement with the theoretical anticipation on atomic density as (Min, Cho, Mason, Lee, & Kim, 2011; Velasco et al., 2012). The degenerate condition plays dominant rule as the BGN shrinks to the nanoscale regime and universal conductance can be used in calculation of BGN optical properties.

Figure 3. The effect of non-equilibrium condition on the dielectric function with different wavelengths that shows bias dependence optical parameters on BNG

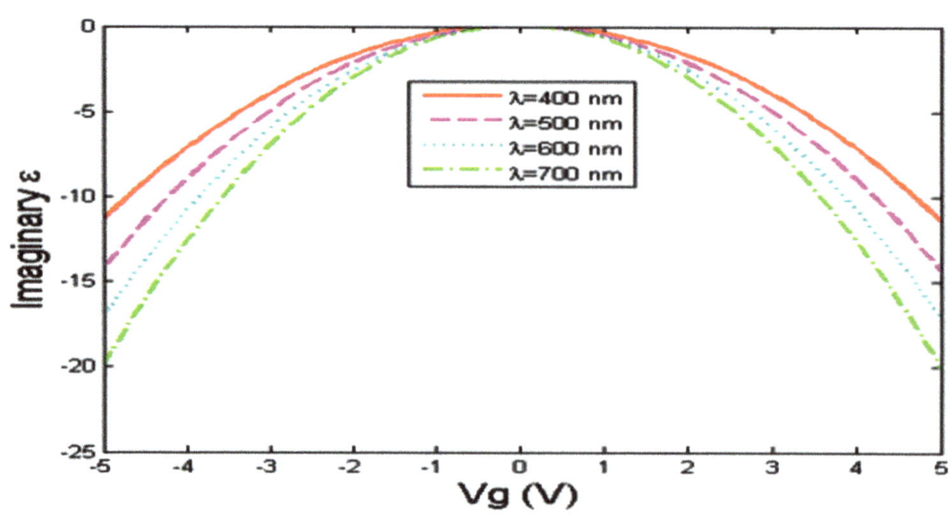

$$G = -\frac{q^2}{\sqrt{2hlK_BT}}\left(\sqrt{\left(\frac{a+\sqrt{a^2+4\beta xk_BT}}{\beta}\right)^3} - \frac{\alpha}{\beta}\sqrt{\frac{a+\sqrt{a^2+4\beta xk_BT}}{\beta}}\right)_{-v_g}^{v_g} \tag{12}$$

$$\approx \frac{q^2}{4\hbar} \approx 6.08\times10^{-5}(\Omega^{-1})$$

The normalization will affect the universal conductance, therefore it will be independent of unit system consequently the refractive index is simplified as:

$$n_{BGN} = \sqrt{\varepsilon_{BGN}} = \sqrt{\frac{G_{BGN}^D}{i\delta\omega\varepsilon_0}+1} \tag{13}$$

Subsequently, the effect of applied bias on refractive index is illustrated in Figure 4.

Additionally, the refractive index as a function of wavelength is explored:

$$n_{BGN} = \sqrt{\frac{G_{BGN}^D}{i2\pi c\delta\varepsilon_0}\lambda+1} \tag{14}$$

The real part of refractive index accounts for refraction while the imaginary part leads to absorption. The real part of n is increased by increasing the applied voltage and wave length but for the imaginary part of n it is vice versa (E.G. Bittle and J.W. Brill, 2010). Additionally, it is confirmed that real part of modeled refractive index is increased, while the imaginary part of it is decreased by increasing wave length for different voltages as illustrated in Figure 5.

Consequently, refractive index is increased by growth of applied bias on BGN. It can be concluded that real and imaginary parts of refractive index can be engineered by controlled applied voltage. Possibility of BGN application under controlled bias as a gold replacement on optical sensors needs to be explored in future work.

Figure 4. (a) Applied bias effect on real part of refractive index; b) applied bias effect on imaginary part of refractive index both for different wave lengths

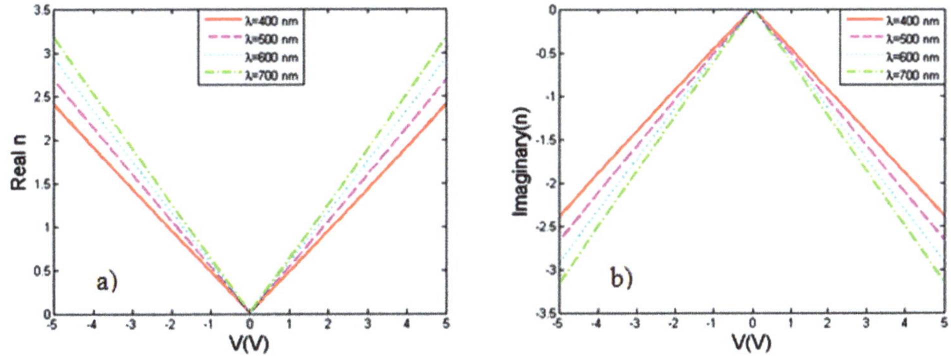

Figure 5. (a) The wave length increment effect on real part of refractive index; (b) the wave length decrement effect on imaginary part of refractive index in degenerate condition with applied bias

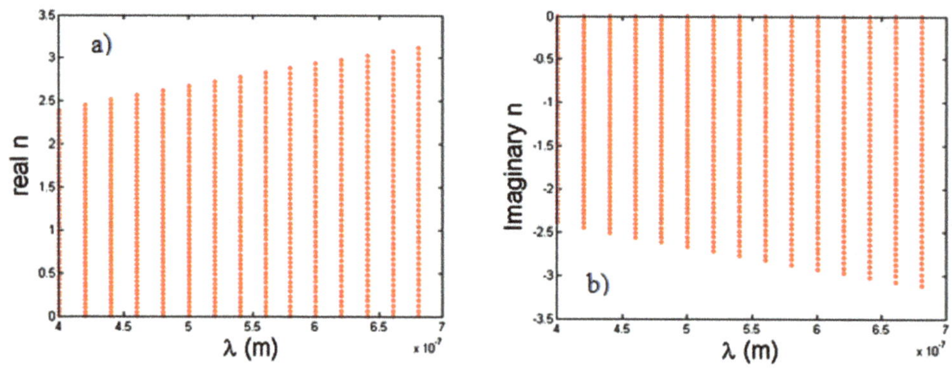

Optical Properties of Trilayer Graphene Nanoribbon

Recently, experimental attention has turned toward TGN because of its tunable band gap and high carrier mobility which are critical for industrial applications (Craciun et al., 2009; Sutter, Hybertsen, Sadowski, & Sutter, 2009). The quantum confinement effect in TGN is assumed in two directions which means that its width and thickness are less than the de Broglie wavelength (10 nm) so it will act as a 1D material. The symmetry between the three layers in TGN structure will be broken by applying the perpendicular external electrostatic fields as shown in Figure 6.

Figure 6. ABA stacking TGN as a 1D material with quantum confinement effect on two directions

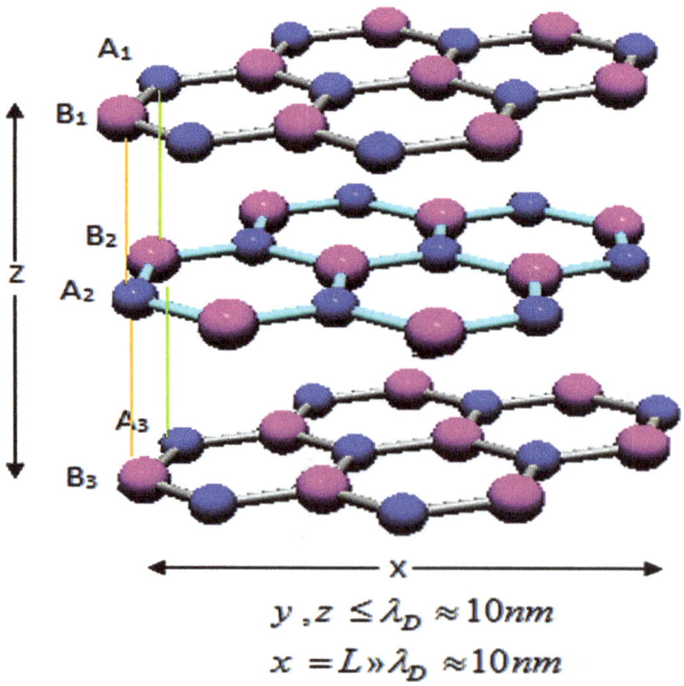

The sensitivity enhancement effect of graphene layer in optical sensors like surface plasmon resonance sensors is proved in several papers (Salihoglu, Balci, & Kocabas, 2012; Simsek, 2013), and investigation on optical properties of multilayer graphene are necessary to graphene based optical sensor simulations. Furthermore, the response of ABA-stacked TGN to an external electric field is different from that of mono or BGN. Although it causes to opening a gap in BGN, manages the magnitude of overlap in TGN. So study of tunable optical properties of TGN will be essential for optical sensor and sensitivity enhancement applications. Additionally, TGN electrical conductance as a fundamental parameter on optical property analysis needs to be discovered. Conductance analyses are started with energy equation. Based on the tight-binding method, the energy band structure of ABA-stacked TGN in the form of polynomial has been reported. On the other hand, the perturbation theory in the limit of $v_F\left|k\right| \ll V \ll t_p$ leads to the electronic band structure of TGN (Mak et al., 2010) as:

$$E(k) = \alpha\left|k\right| - \beta\left|k\right|^3 \tag{15}$$

where

$$\alpha = \frac{v_f V}{t_\perp \sqrt{2}},$$

$$\beta = \frac{v_f^{\,3}}{t_\perp \sqrt{2} V},$$

K is the wave vector in the x direction,

$t_\perp$ is the hopping energy,

v_f is the Fermi velocity, and

V is the applied voltage. The upper layer is at potential V/2, the lower layer is at potential –V/2, and the middle layer is at zero potential.

The external electric field can change the amount of overlap in band structure of TGN. This variation is a unique property of TGN that had not been previously found in other semi-metallic systems (Craciun et al., 2009). In conclusion, the conductance of TGN based on Landauer formula can be calculated as (Hatef Sadeghi et al., 2012):

$$G = \frac{2\alpha q^2}{Lh}\int_{-\infty}^{+\infty}\left[\left(-\frac{d}{dE}\left(\frac{1}{1+e^{\frac{E-E_F}{k_BT}}}\right)\right)\right]dE + \frac{-6\beta q^2}{Lh}\int_{-\infty}^{+\infty}k^2\left[\left(-\frac{d}{dE}\left(\frac{1}{1+e^{\frac{E-E_F}{k_BT}}}\right)\right)\right]dE \tag{16}$$

Consequently, the conductance can be assumed in the form of $G = N_1G_1 + N_2G_2$, where $N_1 = 2\alpha q^2 / Lh$ and $N_2 = -6\beta q^2 / Lh$. Since G_1 is an odd function, its value is zero. So the conductance can be modified in the form of:

$$G_2 = \int_{-\infty}^{+\infty} \left[\left(\left(-\frac{E}{2\beta} + \sqrt{\left(\frac{-\alpha}{3\beta}\right)^3 + \left(\frac{E}{2\beta}\right)^2} \right)^{1/3} + \left(-\frac{E}{2\beta} - \sqrt{\left(\frac{-\alpha}{3\beta}\right)^3 + \left(\frac{E}{2\beta}\right)^2} \right)^{1/3} \right)^2 \left(-\frac{d}{dE} \left(\frac{1}{1 + e^{\frac{E-E_F}{k_B T}}} \right) \right) \right] dE \quad (17)$$

The partial integration method and simplification direct to numerical solution. Thus, the general conductance model of TGN is obtained as (Hatef Sadeghi et al., 2012):

$$G = N_2 G_2 = -N_2 \int_{-V}^{+V} \left\{ \begin{aligned} &\left(\frac{2k_B T}{1 + e^{x-\eta}} \right) \times \left(\frac{ -\frac{k_B T}{2\beta} - \frac{x k_B T + \Delta}{4\beta^2 \sqrt{-\frac{\alpha^3}{27\beta^3} + \frac{E^2}{4\beta^2}}} }{ 3\left(-\frac{x k_B T + \Delta}{2\beta} - \sqrt{-\frac{\alpha^3}{27\beta^3} + \frac{(x k_B T + \Delta)^2}{4\beta^2}} \right)^{2/3} } + \frac{ -\frac{k_B T}{2\beta} + \frac{x k_B T + \Delta}{4\beta^2 \sqrt{-\frac{\alpha^3}{27\beta^3} + \frac{E^2}{4\beta^2}}} }{ 3\left(-\frac{x k_B T + \Delta}{2\beta} + \sqrt{-\frac{\alpha^3}{27\beta^3} + \frac{(x k_B T + \Delta)^2}{4\beta^2}} \right)^{2/3} } \right) \\ &\times \left(\left(-\frac{x k_B T + \Delta}{2\beta} - \sqrt{-\frac{\alpha^3}{27\beta^3} + \frac{(x k_B T + \Delta)^2}{4\beta^2}} \right)^{1/3} + \left(-\frac{x k_B T + \Delta}{2\beta} + \sqrt{-\frac{\alpha^3}{27\beta^3} + \frac{(x k_B T + \Delta)^2}{4\beta^2}} \right)^{1/3} \right) \end{aligned} \right\} dx \quad (18)$$

where $x = (E - \Delta) / k_B T$ and $\eta = (E_F - \Delta) / k_B T$, η is normalized Fermi energy and $\Delta = \frac{qV}{2}$. By considering the formula as $\eta = (E_F - \Delta) / k_B T$ where $\Delta = \frac{qV}{2}$ the linear relation between η and applied voltage will be cleared so changing the gate voltage not only varies the band structure of TGN but also modifies the Fermi energy level (η). Also, the conductivity of TGN increases by increasing the magnitude of the gate voltage which is clearly shown. It is notable that conductance based model shows acceptable agreement with published data presented by (Ahmadi et al., 2010). Moreover, in degenerate limit the Fermi energy distribution function is estimated to be one. Since degenerate condition is dominant in nano-scale devices, we should consider this condition. General conductance model in degenerate condition can be written as:

$$G_D = -N_2 \int_{-V}^{+V} \times \left(\frac{-\frac{1}{\beta} - \frac{E}{2\beta^2 \sqrt{-\frac{\alpha^3}{27\beta^3} + \frac{E^2}{4\beta^2}}}}{3\left(-\frac{E}{2\beta} - \sqrt{-\frac{\alpha^3}{27\beta^3} + \frac{E^2}{4\beta^2}}\right)^{2/3}} + \frac{-\frac{1}{\beta} + \frac{E}{2\beta^2 \sqrt{-\frac{\alpha^3}{27\beta^3} + \frac{E^2}{4\beta^2}}}}{3\left(-\frac{E}{2\beta} + \sqrt{-\frac{\alpha^3}{27\beta^3} + \frac{E^2}{4\beta^2}}\right)^{2/3}} \right)$$
$$\times \left(\left(-\frac{E}{2\beta} - \sqrt{-\frac{\alpha^3}{27\beta^3} + \frac{E^2}{4\beta^2}}\right)^{1/3} + \left(-\frac{E}{2\beta} + \sqrt{-\frac{\alpha^3}{27\beta^3} + \frac{E^2}{4\beta^2}}\right)^{1/3} \right) dE \tag{19}$$

So after solving analytically, the general conductance model in degenerate limit is reported as:

$$G_D = N_2 \frac{\left(-\frac{9E}{\beta} - \sqrt{\frac{81E^2 - 12\alpha^3}{\beta^3}}\right)^{2/3} + \left(-\frac{9E}{\beta} + \sqrt{\frac{81E^2 - 12\alpha^3}{\beta^3}}\right)^{2/3}}{6.87} \tag{20}$$

The degenerate limit conductivity equation is used to find the TGN optical parameters such as complex dielectric function. Applying Equation (20) and doing the same proses as mentioned in section 1 lead to the dielectric constant of TGN which depends on applied voltage, frequency and thickness of graphene layers. The degenerate conductance is used to calculate the dielectric constant of TGN in this limit as:

$$\varepsilon_{TGND} = \frac{\left(N_2 \frac{\left(-\frac{9E}{\beta} - \sqrt{\frac{81E^2 - 12\alpha^3}{\beta^3}}\right)^{2/3} + \left(-\frac{9E}{\beta} + \sqrt{\frac{81E^2 - 12\alpha^3}{\beta^3}}\right)^{2/3}}{6.87} \right)}{i\omega\varepsilon_0\delta} + 1 \tag{21}$$

By considering the relation between frequency ω and wave length λ, changes in TGN dielectric constant as a function of applied voltage for different wave lengths are simulated as illustrated in Figure 7.

By increasing the applied voltage, the imaginary part of dielectric constant of TGN decreases (absorption increased as absolute value increased) but the real part remains constant. In the case of solids, the complex dielectric function is closely connected to band structure. The main quantity that characterizes the electronic structure of any crystalline material is the probability of photon absorption, which is directly related to the imaginary part of the optical dielectric function (Bishop, 2001).

Figure 7. The effect of degenerate condition on dielectric constant as a function of applied voltage for different wave lengths

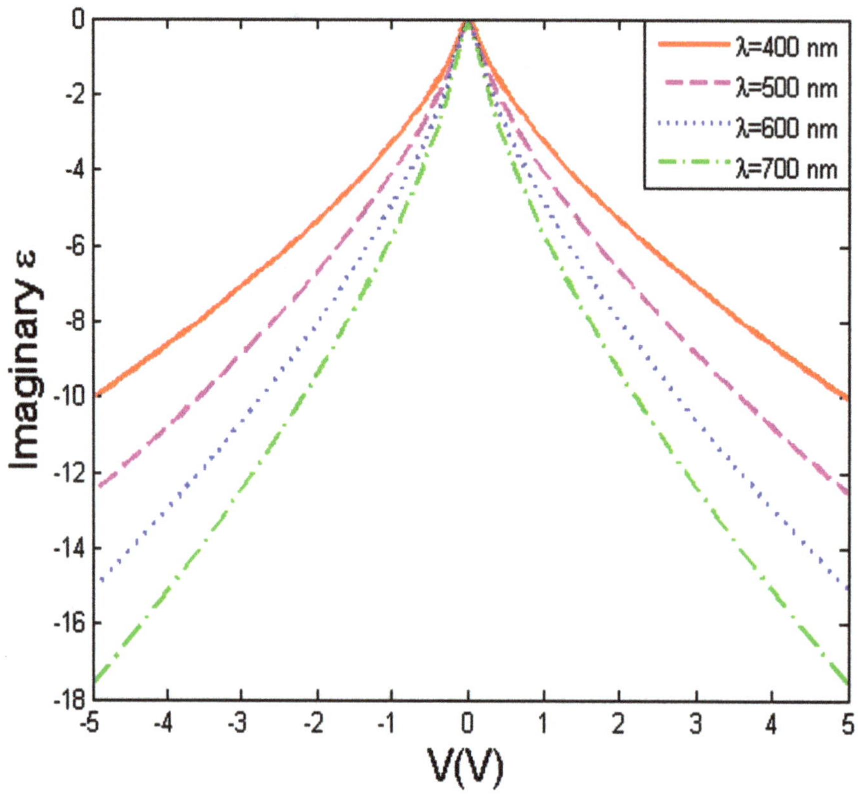

$$n = \sqrt{\frac{\left(N_2 \frac{\left(-\frac{9E}{\beta} - \sqrt{\frac{81E^2 - 12\alpha^3}{\beta^3}} \right)^{2/3} + \left(-\frac{9E}{\beta} + \sqrt{\frac{81E^2 - 12\alpha^3}{\beta^3}} \right)^{2/3}}{6.87} \right)}{i\omega\varepsilon_0\delta} + 1} \tag{22}$$

Also, the applied voltage affects the relation of dielectric constant; consequently refractive index will be affected too. From Equation 22 the refractive index as a function of wave length can be written as:

$$n = \sqrt{\left(\frac{G_D\lambda}{i2\pi c\varepsilon_0\delta} \right) + 1} \tag{23}$$

The refractive index as a function of applied bias for different wave lengths is simulated as shown in Figure 8.

Figure 8. (a) Applied bias effect on real part of refractive index; (b) Applied bias effect on imaginary part of refractive index both for different wave lengths
Hamid. Toloue A.T, 2015.

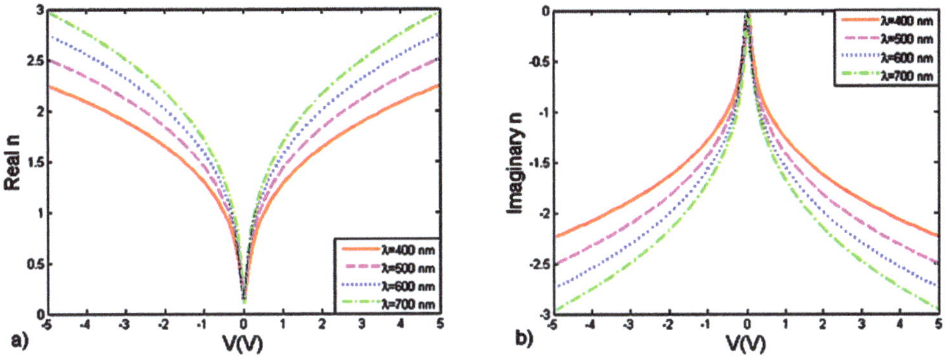

The real part of refractive index accounts for refraction and increased by increasing the applied voltage and wave length while for the imaginary part which leads to absorption it is vice versa (Meshginqalam, Toloue, Ahmadi, & Sabatyan, 2016). Additionally, it is confirmed that real part of modeled refractive index is increased, while the imaginary part of it, is decreased by increasing wave length for different voltages as depicted in Figure 9.

Consequently, raising the applied bias on TGN leads to increment in the real part and decrement in the imaginary part of refractive index, which means by controlling the applied voltage, both parts of refractive index can be engineered.

GRAPHENE BASED SURFACE PLASMON RESONANCE SENSOR

SPR based sensors are one of the attractive configurations which operate by sensing the interactions between sensing element and the sensor metallic surface. This mechanism leads to the local gradient of refractive index by changes in molecule concentrations and chemical reactions. The effects of sensing

Figure 9. (a) The wave length increment effect on real part of refractive index; (b) The wave length decrement effect on imaginary part of refractive index in degenerate condition with applied bias

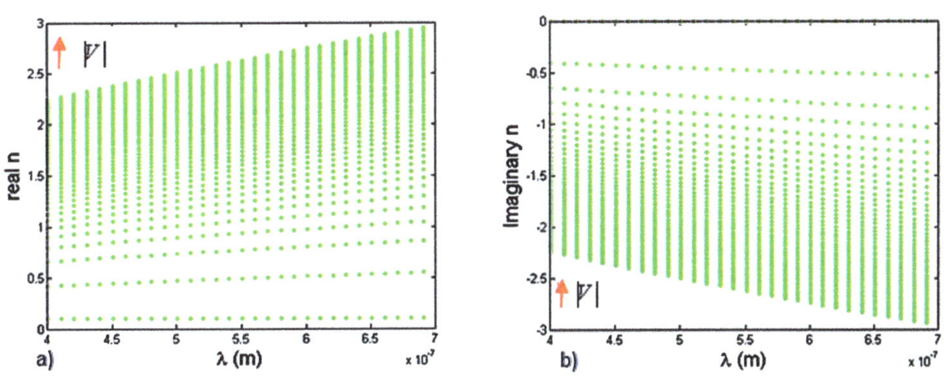

element concentrations and chemical reactions can be optically measured by the attenuated total reflection method (Choi, Kim, & Byun, 2011). In fact, a minimum reflectivity occurs when the momentum of the incident p-polarized light in the surface plane matches that of the SPP mode and no reflection appears. The sensitivity is an important feature for a sensor, so to overcome the sensitivity confinement, new technics are needed (Zuppella et al., 2013). To vanquish the poor adsorption of molecules on to the SPR sensor surface, different solutions are considered. One of these solutions is to use Bio-molecular Recognition Elements (BREs) such as nanoparticles (Homola & Dostálek, 2006) furthermore the fine control of the geometrical and optical properties of these nanostructures is a favored research topic. As an example, Structures based on Au/graphene have been studied recently (Cheon et al., 2012) indicating the improvement in the biomolecular affinity and decrement in contamination and oxidation by using the graphene (Wu et al., 2010). In addition, GOSs as a new class of two-dimensional carbon nanostructures with promising electrical properties, have received significant attention in recent years (Bao & Loh, 2012; Jablan, Soljacic, & Buljan, 2013). Also, GOSs comprise notable optical (Eda & Chhowalla, 2010; Johari & Shenoy, 2011; Shang et al., 2012) and biosensing (Guo & Dong, 2011; Hu et al., 2011; Lee, Kuo, Chiu, & Ieee, 2012; Liu, Fu, Yuan, Li, & Deng, 2010; Salihoglu et al., 2012; Wang, Zhou, Zhang, Boey, & Zhang, 2009) properties and are expected to have a wide range of applications. Moreover, GOSs contain oxygen at their surfaces and the band gap of a GOS can be controlled by modification of oxidation (Shukla & Saxena, 2011). Furthermore, GOS band gap and dielectric constant can be increased by modifying the functional groups on its surface (Chiu, Huang, Lai, & Liu, 2014), also it should be mentioned that the graphene and GOS based SPR structures sensitivity have been approved by (Maharana et al., 2014; Pradeep Kumar et al., 2014; Wu et al., 2010), moreover the GOS sensor chip is 3.7 times more sensitive than the graphene based chip as reported by Stebunov, et al (Stebunov et al., 2015). These advantages motivate researchers to work on analytical prediction model of GOS film-based SPR sensor. In this section, the analytically modeled response of GOS film-based SPR sensor which is covered by organic molecules is investigated. The common configuration of graphene oxide based SPR sensor indicates a layer of GOS on top of the gold surface, where the TiO_2 layer (5nm) acts as an adhesion layer makes the gold film mechanically stable, therefore the sensor chip can be integrated in a more complex device (Ovchinnikov & Shevchenko, 2013). This structure is shown in Figure 10.

Figure 10. Schematic figure of graphene based SPR system where GOS surface covered by organic molecules

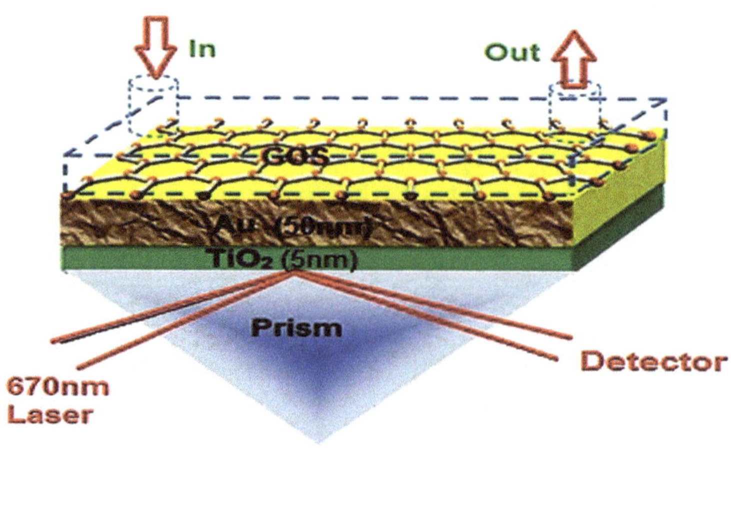

After the light illumination on the prism, an evanescent wave penetrates through the gold film leads to excitation of the plasmons which causes to a minimum reflectivity appearance in the reflected light. Material sensing leads to variation in this minimum reflectivity. The plot of totally reflected intensity versus angle of incidence is called SPR curve. In order to form this curve, we use the model based on transverse-magnetic wave for N-layer system that has been reported by (Wu et al., 2010):

$$R = \left| \frac{\left(M_{11} + M_{11}q_N\right)q_1 - \left(M_{21} + M_{22}q_N\right)}{\left(M_{11} + M_{11}q_N\right)q_1 + \left(M_{21} + M_{22}q_N\right)} \right|^2 \quad (24)$$

where

$$M_{ij} = \left(\prod_{k=2}^{N-1} M_k \right)_{ij}, i \cdot j = 1,2,$$

M_k is the transfer matrix,

$$q_k = \frac{\sqrt{\varepsilon_k - n_1^2 \sin^2\theta}}{\varepsilon_k} \text{ and}$$

$$\beta_k = d_k \left(\frac{2\pi}{\lambda} \right) \sqrt{\varepsilon_k - n_1^2 \sin^2\theta}.$$

In the proposed configuration, the first layer is prism with refractive index $n_1 = n_{prism}$ as shown in Figure 10. The k_{th} layer has a thickness of d_k and the local dielectric function $\varepsilon_{(\lambda_0)}$ or the refractive index $n_{(\lambda_0)}$. These parameters for Ti, Au, graphene oxide and BK7 prism are,

$$n_{Ti} = 2.36 + i3.47\left(d = 5nm\right),$$

$$n_{Au} = 0.197 + i3.67\left(d = 50nm\right),$$

$$n_{graphene\ oxide} = 2.1 + i0.56$$

and

$$n_1 = 1.52$$

respectively. In addition, the employed wavelength is equal to 670 nm. Graphene with narrow width illustrates small band gap which means by small gradient on its carrier concentration the electrical conductivity will changed dramatically. Graphene has been employed in the form of reduced graphene oxide in the SPR system because of its positive effect on adsorption efficiency and its deterrence effect on Au layer (Zhang et al., 2014). This modification by graphene oxide helps to wires surface plasmons to the visible range also provides a tunable propagation and excitation platform (Balapanuru et al., 2010). Applying a GOS film on top of the Au layer will affect the SPR curve because of its oxygen functional group. The material property effect on conductance can lead to refractive index variation (Chang, Huang, & Lin, 2009). The conductance variation is modeled due to the molecular properties of the material, such as outer shell electrons effect, molecular mass, effective group number and effective outer shell factor. Constant parameters such as a, b, c and d are explored based on the effects of these factors on the refractive index. The refractive index shift caused by each molecule is considered in the form of effective parameters that is addressed as refractive index shift equation (Toloue, Ahmadi, Meshginqalam, Centeno, & Sabatyan, 2015).

$$\left|Xa + Yb + Zc + Vd\right| = n' \tag{25}$$

where X, Y, Z and V indicate outer shell electrons, molecular mass, effective group number and effective outer shell factors, respectively. (Outer shell electrons effect (X) is related to the electronegativity factor where the effective outer shell factor (V) is related to the number of dangling bonds of a molecule which is zero in the case of molecules without any dangling bonds, moreover the parameter "Z" denoted the effective group number which is the group number of the element except carbon and hydrogen in the desired molecule, so it is zero for the molecules consist of only carbon and hydrogen atoms). Also, the molecular shape effect for determination of the "V" parameter values can be viewed in appendix 1. The modified model of reflected intensity or SPR response also supports the experimental work (Valentini, Carbone, & Palleschi, 2013; Wan, Wang, Wu, & Zhag, 2011) as illustrated in Figure 11.

The electro-negativity differences between all components of the sensing molecule are considered as outer shell electrons effect. In order to find the constant parameters, this equation is solved for four different sensing elements. The Mannose, Lactose, Poly Ethylene Imine (PEI) and Poly Sodium 4-Styrenesulfonate (PSS) as sensing elements have been used and the effective parameters have been reported (Meshginqalam, Toloue, Ahmadi, Sabatyan, et al., 2016). For the Mannose molecule ($C_6H_{12}O_6$) the electronegativity factor is equal to 2.48, molecular mass is equal to 180 and the effective group number because of the functionalized oxygen is 6 and the related value of the parameter "V" is zero, so the refractive index shift equation can be written as:

$$2.48a + 180b + 6c = 0.0015 \tag{26}$$

where 0.0015 indicates refractive index shift because of mannose molecule (Subramanian et al., 2014). In the same manner, the set of refractive index shift equation for the rest of the sensing elements is considered as:

Lactose:

$$2.48a + 342b + 6c = 0.0030 \tag{27}$$

Poly Ethylene Imine (PEI):

$$1.33a + 1800b + 5c = 0.0045 \tag{28}$$

Figure 11. The modeled SPR curve in the presence of GOS and without GOS in comparison with experimental data

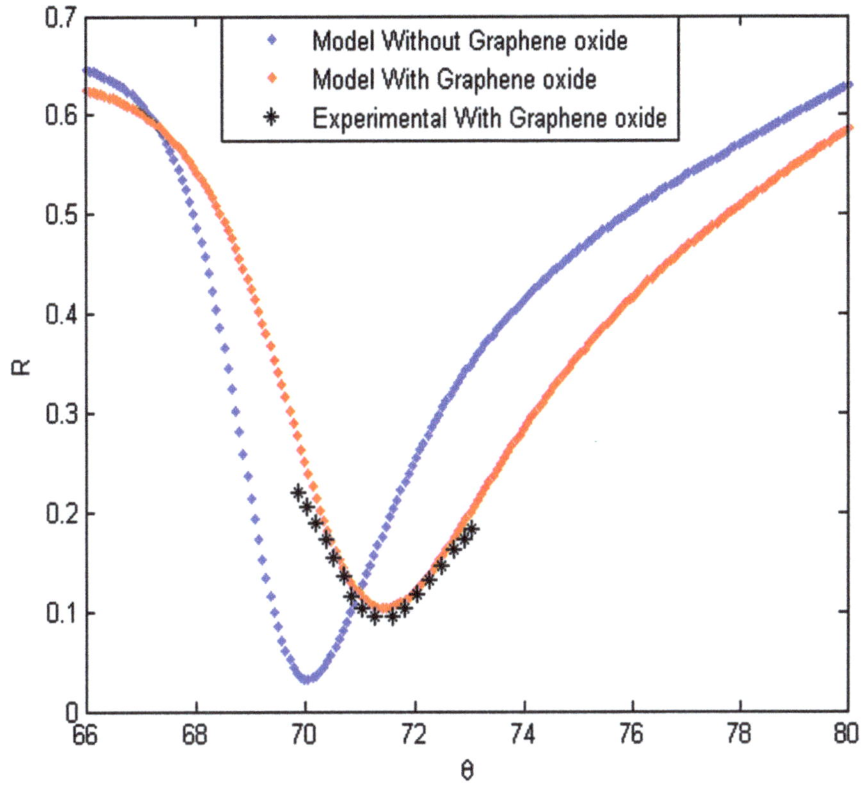

Poly Sodium 4-Styrenesulfonate (PSS):

$$3.78a + 70000b + 6c + 2d = 0.0060 \quad (29)$$

Eventually the constant coefficients can be obtained by simultaneously solving these four equations.

$$a = 0.016327\,,$$

$$b = 0.000009\,,$$

$$c = -0.00677\,,$$

$$d = -0.33160\,.$$

The modeled constant coefficients that are calculated by simultaneous solution of the set of equations cause to fantastic agreement between the presented model and experimental data as shown in Figures 12, 13, 14 and 15.

Figure 12. The refractive index with magnitude of 0.0015 is shifted because of mannose attendance in graphene layer

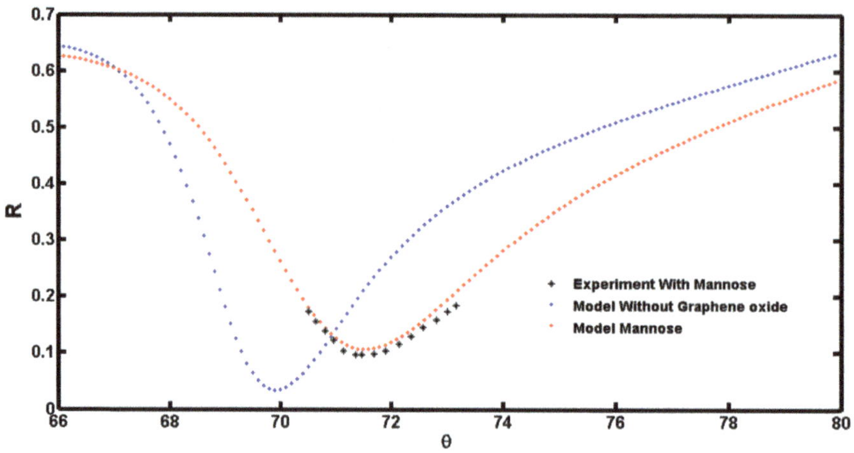

Figure 13. Lactose molecules effect on the refractive index in comparison with modeled parameter

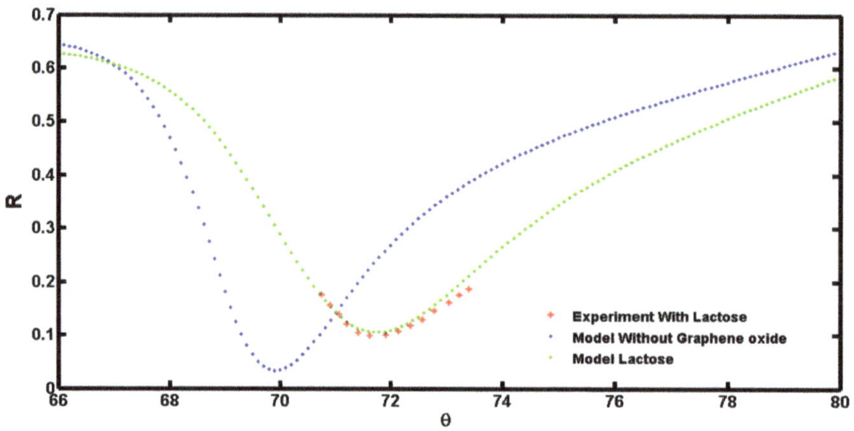

Figure 14. PEI polymer shifts the refractive index more than mannose and lactose however presented model predicts its behavior splendidly

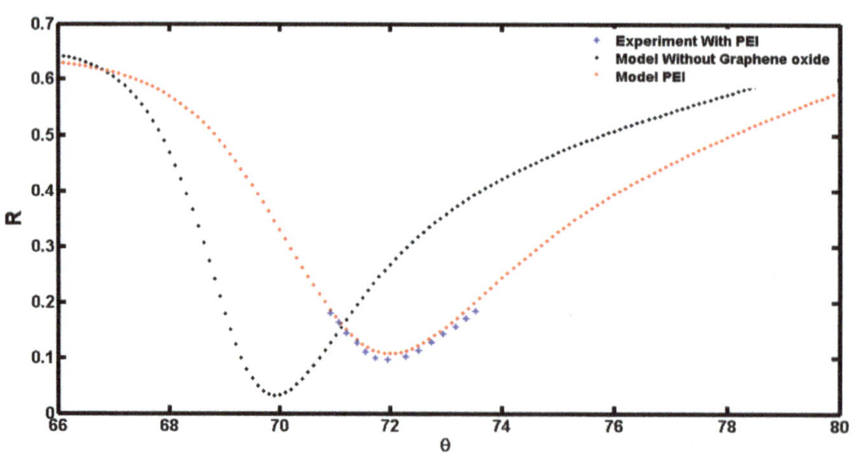

Figure 15. PSS polymer molecule effect in the SPR system evaluated and superior agreement with model is observed

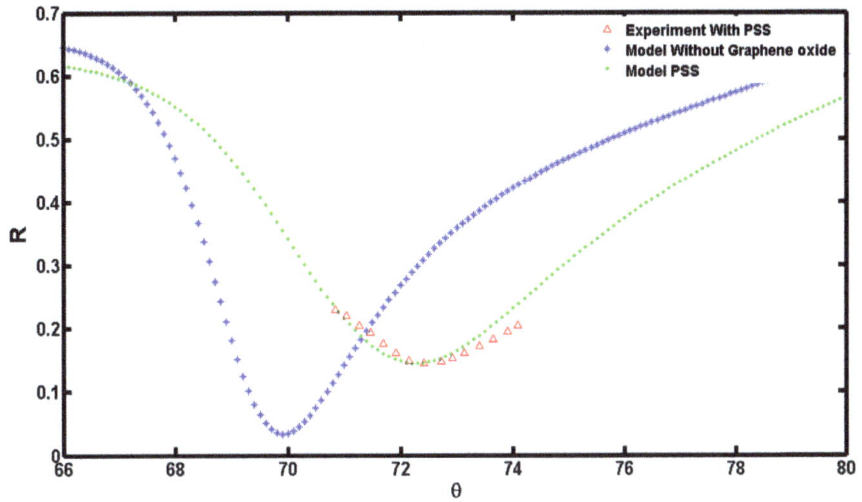

Finally to validate the all processes presented models simultaneously compared with experimental data as depicted in Figure 16. From the proposed method good agreement with experiment is publicized which means it can be improved into the prediction of any molecule detection by graphene based SPR sensor.

The calculated constant coefficients can be evaluated by generalization of the refractive index shift equation for each desired material like organic molecules. Furthermore, the refractive index values are

Figure 16. The model in comparison with the experimental data extracted from Subramanian 2014 with acceptable agreement
Subramanian et al., 2014; Toloue. H, 2015.

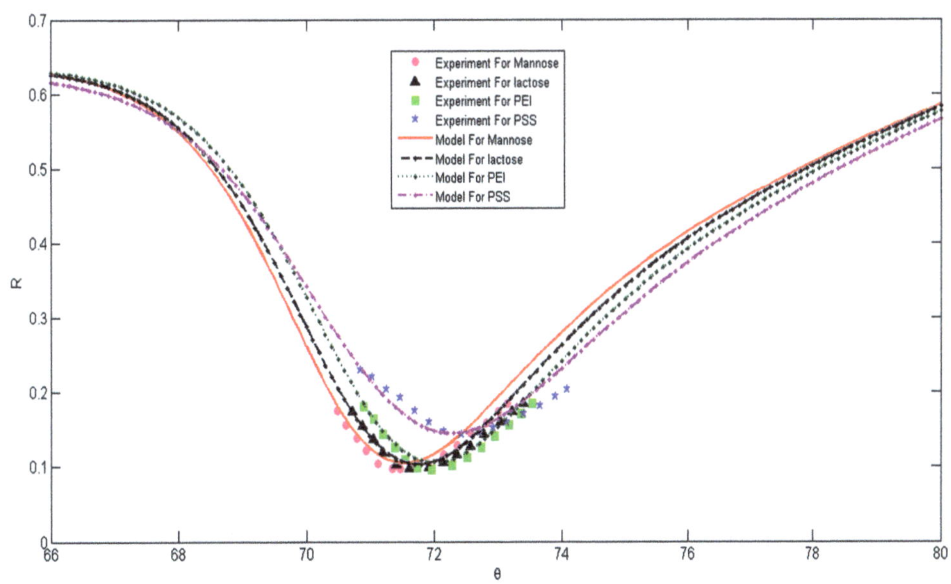

calculated theoretically form resonance angles of SPR curves of Figure 16 as well and the differences with experimental values in order of 10^{-4} are reported in appendix 2.

Hydrocarbons or organic compounds consist of carbon and hydrogen, furthermore Alkanes are considered as the simplest organic molecules, which have the general formula C_nH_{2n+2}. It is worth mentioned that for all materials which have only carbon and hydrogen in their structures, the electronegativity factor is 0.35. In addition, molecular mass for Alkanes is calculated by

$$M_1 = C_m n + H_m \left(2n + 2\right)$$

where n is the number of carbon atoms in the special organic structure, C_m and H_m are the atomic masses of carbon and hydrogen respectively, which guides to the refractive index shift equation as:

$$\left|0.35a + M_1 b\right| = n_1' \quad (30)$$

where a and b are replaced by above values then n_1' as the shift in sensed refractive index is obtained. It is important to emphasis that "c" and "d" are the constant coefficients with non-zero values calculated in the text so only "Z" or "V" parameters can be equal to zero which cause to neglecting the related terms. For example for methane molecule (CH_4), molecular mass will be equal to 16 so the shift value will be 0.0059 and resulted SPR curve is simulated as shown in Figure 17.

Moreover, an Alkene is a hydrocarbon that contains at least one carbon–carbon double bond and it has the general formula of C_nH_{2n} that represent the molecular mass as

Figure 17. Methane molecules effect on the SPR curve

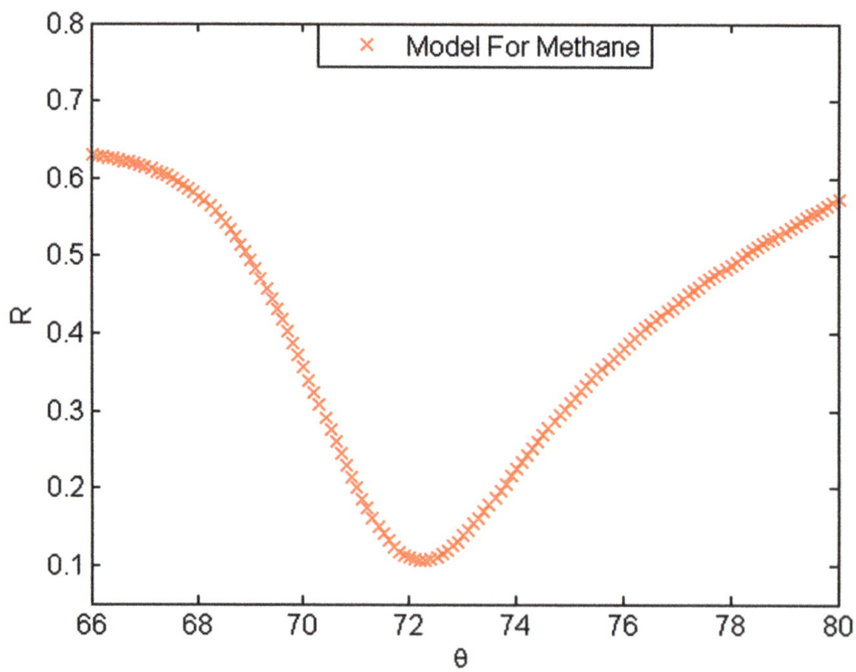

$$M_2 = C_m n + H_m (2n).$$

Consequently the refractive index shift equation for Alkenes is reported as:

$$\left|0.35a + M_2 b\right| = n_2' \tag{31}$$

For ethene (C_2H_4) as the simplest Alkene, molecular mass will be equal to 28 so the shift value will be equal to 0.0060.

This shift value is considered to form the SPR response for ethene as illustrated in Figure 18. Alcohol as an organic compound is the next sensing element in which the hydroxyl group (-OH) is bound to a saturated carbon atom. Its general formula is $C_nH_{2n+1}OH$ so the related electronegativity factor is 2.48, the effective group number is 6 and its molecular mass is calculated by

$$M_3 = C_m n + H_m (2n + 2) + O_m,$$

where O_m is the atomic mass of oxygen, therefore the refractive index shift equation for Alcohol takes the form of:

$$\left|2.48a + M_3 b + 6c\right| = n_3' \tag{32}$$

Figure 18. Ethene molecules effect on the refractive index and SPR curve

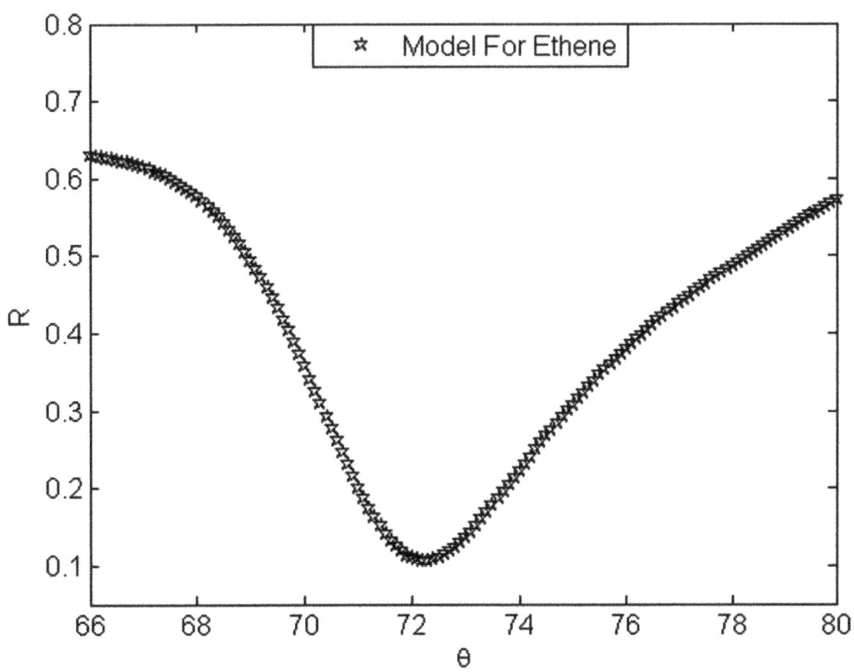

The methanol as a simplest member of the saturated straight chain alcohols is considered, its molecular mass is equal to 32, and therefore the shift value of 0.0001 is obtained which leads to its SPR curve as illustrated in Figure 19.

In the next step, we examine our proposed model for Ethers with general formula of R–O–R', so the electronegativity factor of 1.24 and molecular mass as

$$M_4 = 2\left(C_m n + H_m\left(2n+1\right)\right) + O_m$$

are considered and diethyl ether (CH_3-CH_2-O-CH_2-CH_3) with molecular mass of 74 is employed as an example furthermore the refraction index shift of 0.0197 is resulted as shown in Figure 20.

An Alkyl with the general formula of C_nH_{2n+1} is typically a part of a larger molecule. Methyl Bromide (CH_3Br) as an example of this family is considered and its modeled results are summarized in Table 1 where related SPR curve is mentioned in Figure 21.

In an Alkyne there is at least one triple bond between two carbon atoms and its general chemical formula have been mentioned As C_nH_{2n-2}, therefore the calculation results for propyne as an example of Alkynes are indicated in table 1 furthermore the simulation result for SPR curve is indicated in Figure 22.

Finally Amines are considered in this stage, related results for the methylamineas a primary amine are declared in Table 1 (Meshginqalam, Toloue, Ahmadi, Sabatyan, et al., 2016) and Figure 23.

Eventually the SPR graphs for organic molecules based on the proposed model are plotted together as shown in Figure 24. SPR technique relies on the principle that in order to satisfy the resonance condition, any changes on the dielectric sensing surface will cause a shift in the angle of reflectivity (Daghestani & Day, 2010) so a change in refractive index at the surface of the GOS film will cause an angle shift. Since

Figure 19. The modeled SPR response for methanol sensing

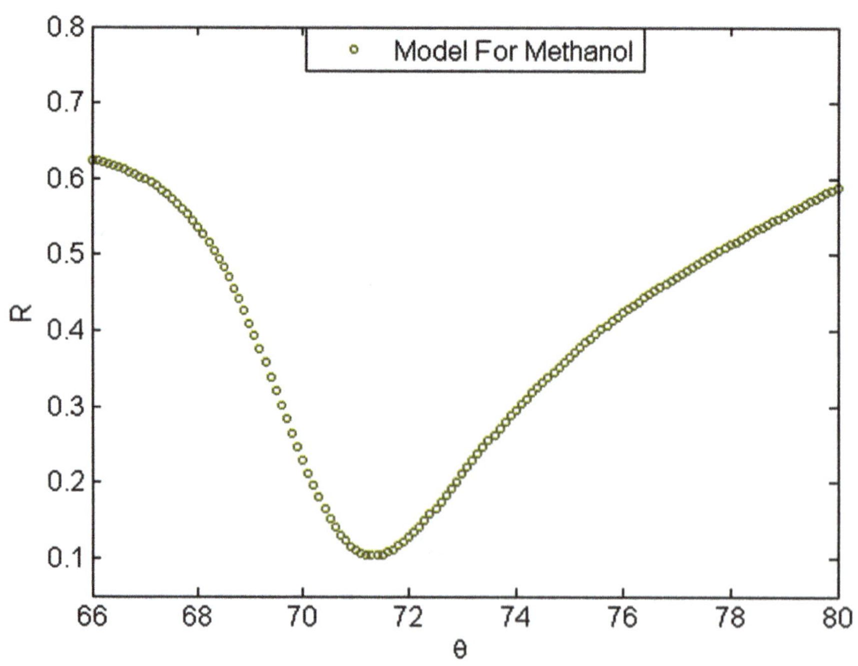

Figure 20. Diethyl Ether molecules' effect on the SPR curve

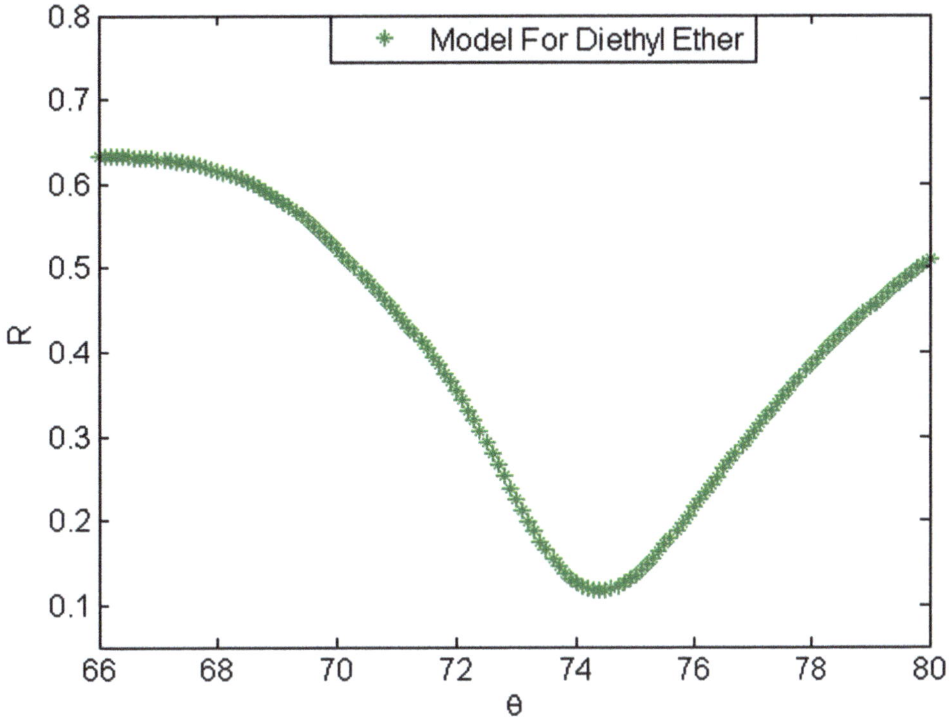

Figure 21. The modeled SPR response for methyl bromide

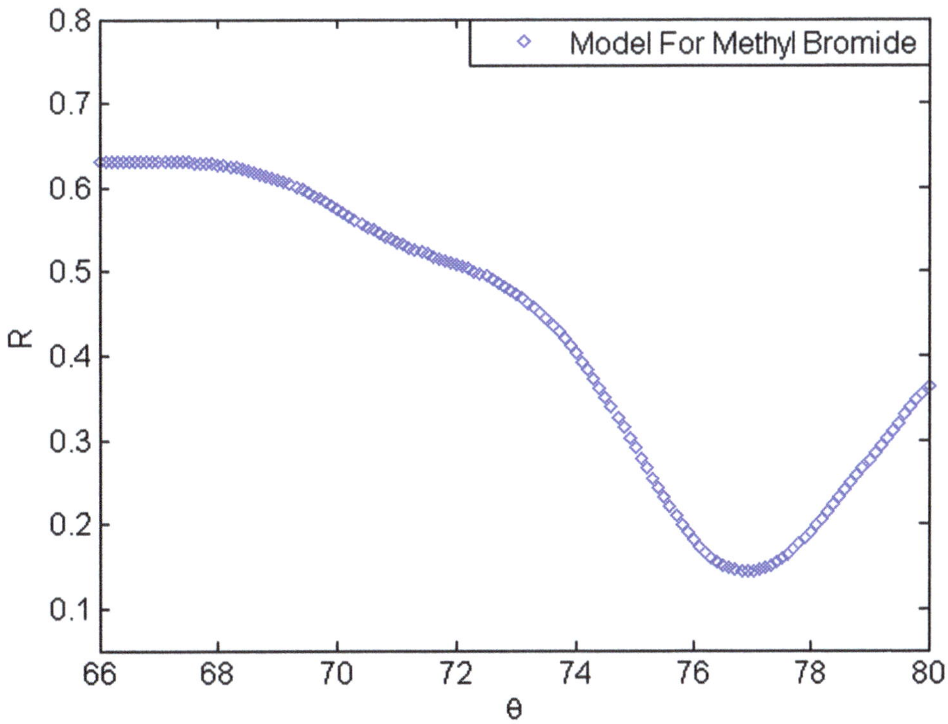

Table 1. The effect of molecular properties on refractive index shift for different organic molecules

Sensing Material	Chemical Formula	Electronegativity Factor	Molecular Mass	Refraction Index Shift Equation	Example	Refraction Index Shift Value
Alkane	C_nH_{2n+2}	0.35	$M_1 = C_m n + H_m(2n+2)$	$0.35a + M_1 b = n_1'$	Methane 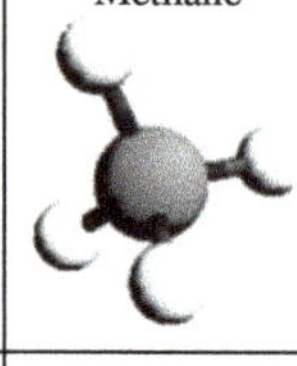	0.0059
Alkene	C_nH_{2n}	0.35	$M_2 = C_m n + H_m(2n)$	$0.35a + M_2 b = n_2'$	Ethene 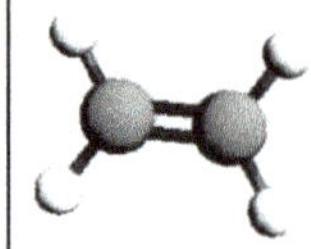	0.0060
Alcohol	$C_nH_{2n+1}OH$	2.48	$M_3 = C_m n + H_m(2n+2) + Om$	$2.48a + M_3 b + 6c = n_3'$	Methanol 	0.0001
Ethers	R–O–R'	1.24	$M_4 = Om + 2(C_m n + H_m(2n+1))$	$1.24a + M_4 b + 6c = n_4'$	Diethyl Ether 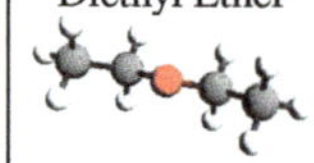	0.0197
Alkyl	C_nH_{2n+1}	0.76	$M_5 = C_m n + Br_m + H_m(2n+1)$	$0.76a + M_5 b + 7c = n_5'$	Methyl Bromide 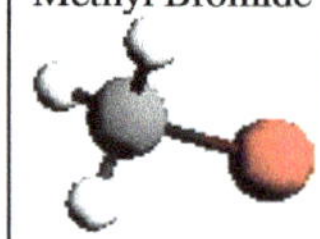	0.0341
Alkyne	C_nH_{2n-2}	0.35	$M_6 = C_m n + H_m(2n-2)$	$0.35a + M_6 b = n_6'$	Propyne 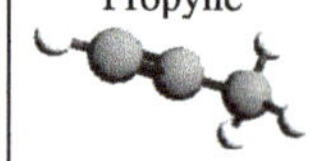	0.0061
Amine	$C_nH_{2n+1}NH_2$	1.68	$M_7 = C_m n + N_m + H_m(2n+3)$	$1.68a + M_7 b + 5c = n_7'$	Methylamine 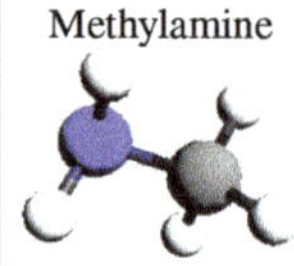	0.0062

Figure 22. .Propyne molecules effect on the SPR sensogram

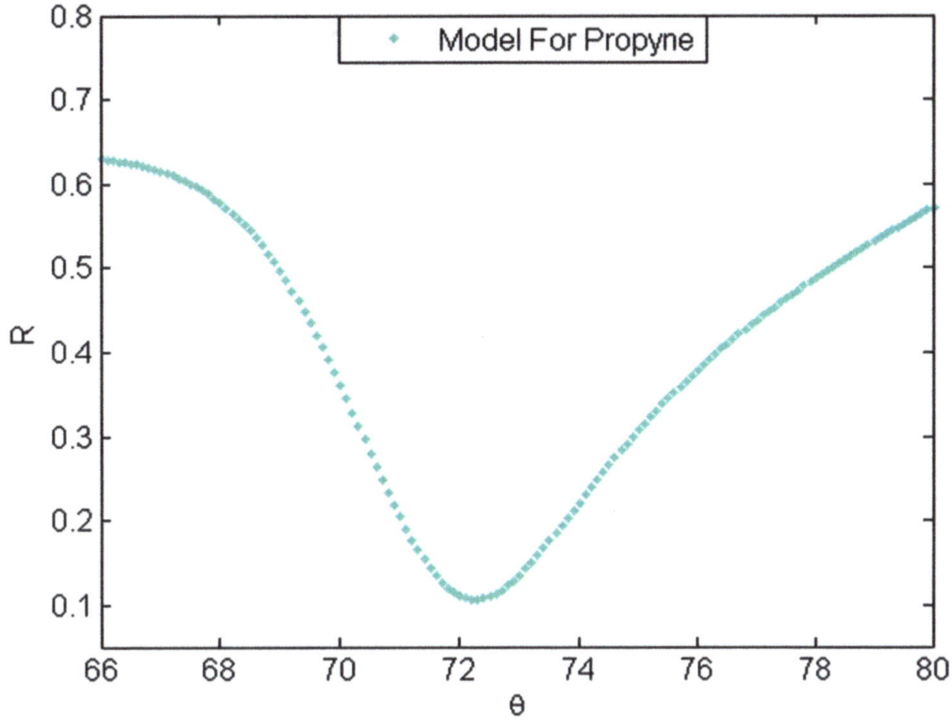

Figure 23. The modeled SPR response for methylamine

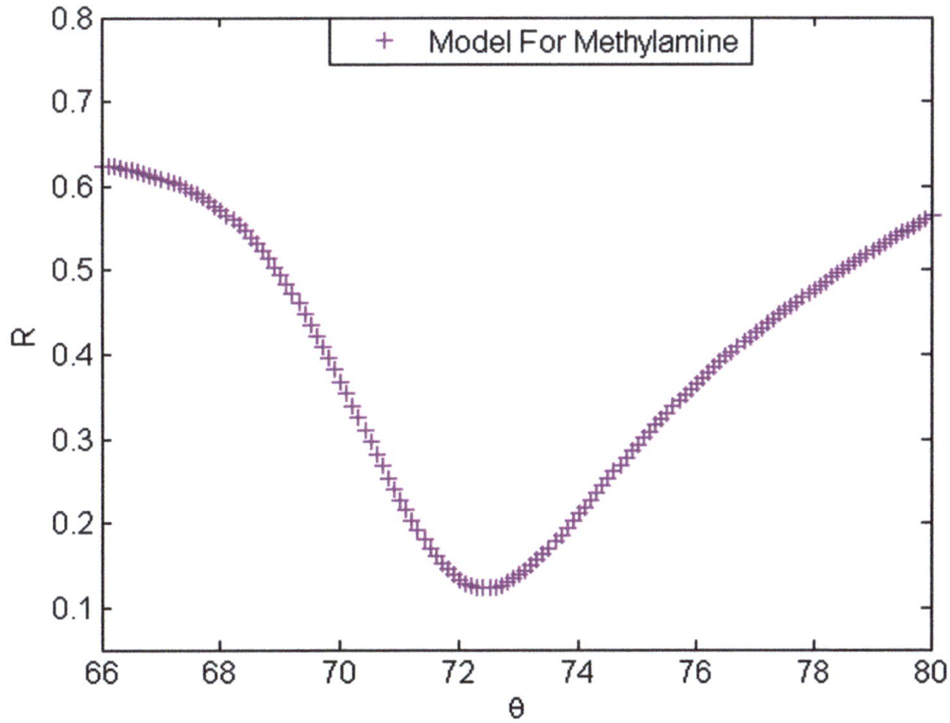

Figure 24. SPR response prediction model for organic molecules

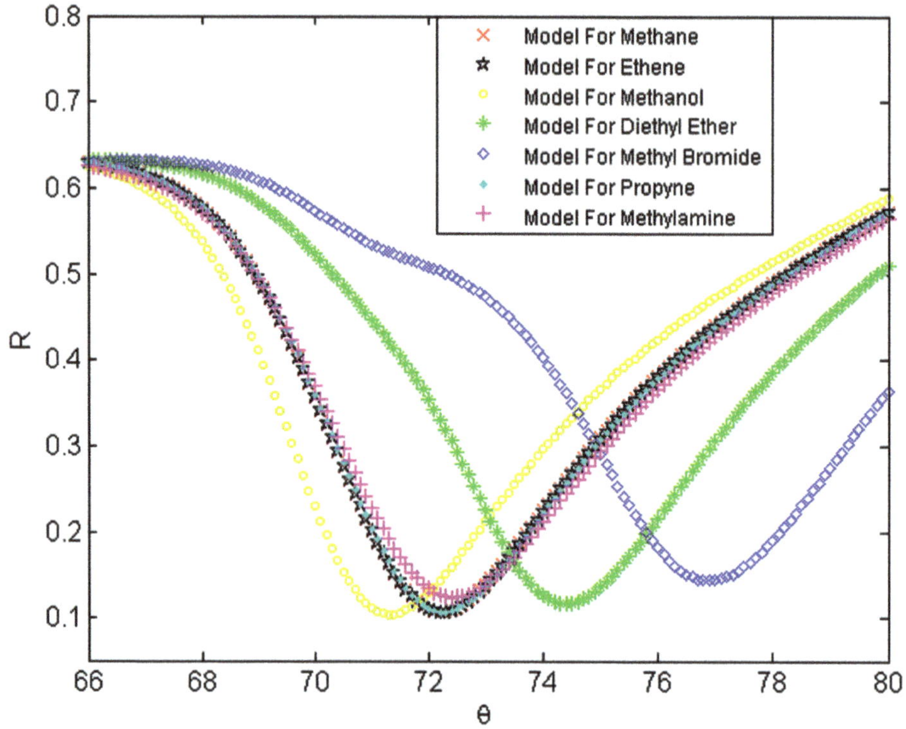

the refractive index of the sensing medium increases, the SPR angle shifts to bigger angles. Figure 24 indicates that the incident angle for methanol with 0.0001 changing in refractive index is smaller than that is of diethyl ether with 0.0197 refractive index change.

Sensitivity is an important parameter of a SPR sensing system. If the refractive index of the sensing layer is altered by δn_s and the corresponding shift in the resonance angle is $\delta\theta_{SPR}$, then the sensitivity (S) of a SPR sensor is defined as the change in resonance angle per unit change in refractive index of the sensing region at the sensor surface (Maharana et al., 2014).

$$S = \frac{\delta\theta_{SPR}}{\delta n_s} \tag{33}$$

Table 2 shows values of θ_{SPR} after sensing molecules adsorption, refractive index shift induced by sensing materials and corresponding sensitivity values. It is clear from Figure 11 that θ_0 (Without Graphene Oxide Layer) = 70.06 deg. and θ_0 (With Graphene Oxide Layer) =71.46 deg.

Therefore applying a GOS layer over the gold layer in the proposed model increases the sensitivity by more than 7%.

The variation of minimum reflectivity with refractive index of sensing medium is indicated in Figure 25. By considering the reflectance relation with respect to refractive index (Woan), it can be seen that the reflectivity increased with increasing the refractive index of the analyte (Sadrolhosseini).

Table 2. Sensitivity calculations for proposed model of SPR system

Sensing Material	δn_s	θ_{SPR} Without GOS	$\delta\theta_{SPR}$ Without GOS	θ_{SPR} With GOS	$\delta\theta_{SPR}$ With GOS	S Without GOS	S With GOS	S Enhancement
Diethyl Ether	0.01972	72.8	2.74	74.4	2.94	138.9	149.1	7.3%
Methyl Bromide	0.03414	75.1	5.04	76.9	5.44	147.6	159.3	7.9%
Propyne	0.00608	70.8	.74	72.3	0.84	121.7	138.2	13.5%

CONCLUSION

Electrical and optical property investigation plays critical role on future nano-electronic devices, graphene based materials demonstrate outstanding optical, mechanical and electrical properties which make them promising materials for future optoelectronic devices. Each multilayer graphene arrangement behaves like a new material and different electronic properties are the result of stacking variations of graphene sheets so a variety of optical properties will be obtained from these electronic features. In addition, optical features investigation is essential to performance modeling and simulation of optical sensors like surface plasmon resonance based sensors. Furthermore, achieving the tunable dielectric constant and refractive index in BGN and TGN make it easy to use these materials in different sensor systems with different optical values. In the presented chapter, optical conductance of BGN and TGN in the calculation of dielectric constant and refractive index are employed. The manageable dielectric constants by

Figure 25. Reflectivity with induced refractive index shift

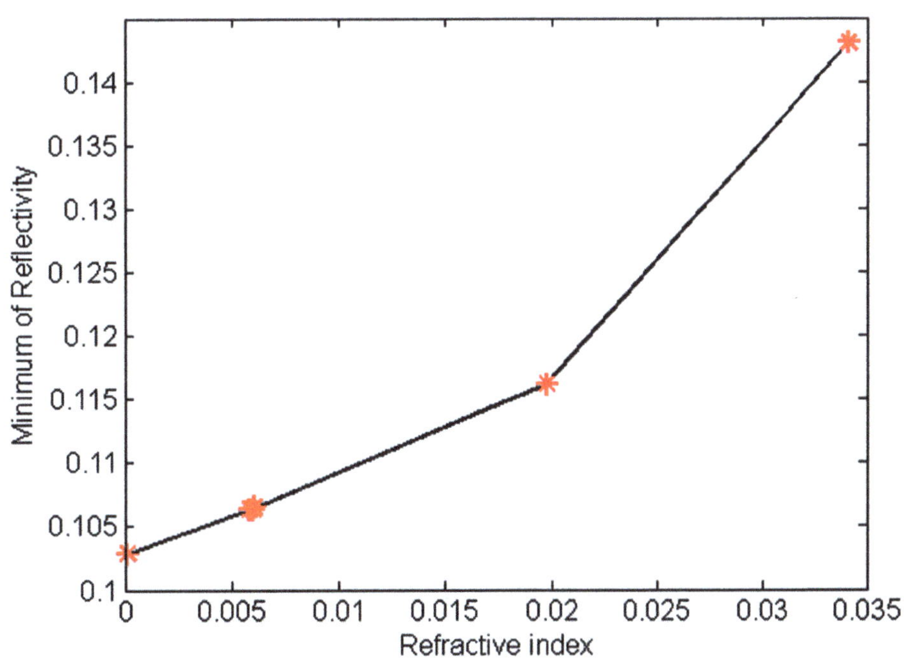

applied voltage and incident wave length are pointed out. Consequently, applied voltage and wave length dependent tunable refractive indices are reported based on the conductance model in the degenerate condition. Therefore, the possibility of BGN and TGN applications as a gold or silver replacement under controlled bias in surface plasmon resonance based sensors is recommended.

Furthermore, oxygen functionalized structure in graphene oxide lead to a small band gap which makes it so sensitive to the environmental materials so it can be placed on top of the Au layer in common SPR sensor to improving the sensitivity of the sensor. In the presented chapter, the carrier density variations effect on graphene based SPR sensor response is modeled. In addition in the presence of different organic molecules, the refractive index shift is formulized and molecular properties of each sensing material such as electro negativity, molecular mass and effective group number are considered. Based on these parameters, sets of equations are simultaneously analyzed and the related coefficients are reported. Finally, a semi-empirical model for interpretation of changes in SPR curve is suggested and tested for some organic molecules.

REFERENCES

Abergel, D. S. L., Apalkov, V., Berashevich, J., Ziegler, K., & Chakraborty, T. (2010). Properties of graphene: A theoretical perspective. *Advances in Physics*, *59*(4), 261–482. doi:10.1080/00018732.2010.487978

Acik, M., & Chabal, Y. J. (2011). Nature of graphene edges: A review. *Japanese Journal of Applied Physics, 50*(7). doi: 07010110.1143/jjap.50.070101

Adonkin, V. T., Gorelov, B. M., Dyakin, V. V., Karpov, I. I., & Prikhodko, G. P. (1994). Effect of intercalation and exfoliation on positron-annihilation lifetimes in graphites. *Journal of Physics and Chemistry of Solids*, *55*(5), 443–446. doi:10.1016/0022-3697(94)90171-6

Ahmadi, M. T., Johari, Z., Amin, N. A., Fallahpour, A. H., & Ismail, R. (2010). Graphene nanoribbon conductance model in parabolic band structure. *Journal of Nanomaterials*. doi: 75373810.1155/2010/753738

Akdim, B., & Pachtert, R. (2011). Switching behavior of carbon chains bridging graphene nanoribbons: Effects of uniaxial strain. *ACS Nano*, *5*(3), 1769–1774. doi:10.1021/nn102403j PMID:21344875

Amin, N. A., Ahmadi, M. T., Johari, Z., Webb, J. F., Mousavi, S. M., & Ismail, R. (2011). Drift velocity and mobility of a graphene nanoribbon in a high magnitude electric field. In N. W. J. F. V. P. Barsoum (Ed.), *Proceedings of the Fourth Global Conference on Power Control and Optimization* (Vol. 1337, pp. 177-179). doi:10.1063/1.3592462

Avetisyan, A. A., Partoens, B., & Peeters, F. M. (2010). Stacking order dependent electric field tuning of the band gap in graphene multilayers. *Physical Review B: Condensed Matter and Materials Physics*, *81*(11), 115432. doi:10.1103/PhysRevB.81.115432

Balapanuru, J., Yang, J.-X., Xiao, S., Bao, Q., Jahan, M., Polavarapu, L., & Loh, K. P. et al. (2010). A Graphene oxide–organic dye ionic complex with dna-sensing and optical-limiting properties. *Angewandte Chemie International Edition*, *49*(37), 6549–6553. doi:10.1002/anie.201001004 PMID:20669201

Bao, Q., & Loh, K. P. (2012). Graphene photonics, plasmonics, and broadband optoelectronic devices. *ACS Nano*, *6*(5), 3677–3694. doi:10.1021/nn300989g PMID:22512399

Bishop, C. (2001). *The relationship between loss, conductivity, and dielectric constant*. Academic Press.

Bittle, E. G., Brill, J. W., & Anthony, J. E. (2010). Electro-optic measurement of carrier mobility in an organic thin-film transistor. *Applied Physics Letters*, *97*(1), 013302. doi:10.1063/1.3460526

Chang, P.-Y., Huang, W.-M., & Lin, H.-H. (2009). Impurity-induced conductance anomaly in zigzag carbon nanotubes. In P. Kes, & R. Jochemsen (Eds.), *25th International Conference on Low Temperature Physics* (Vol. 150). doi:10.1088/1742-6596/150/2/022007

Cheon, S., Kihm, K. D., Park, J. S., Lee, J. S., Lee, B. J., Kim, H., & Hong, B. H. (2012). How to optically count graphene layers. *Optics Letters*, *37*(18), 3765–3767. doi:10.1364/OL.37.003765 PMID:23041852

Chiu, N.-F., Huang, T.-Y., Lai, H.-C., & Liu, K.-C. (2014). Graphene oxide-based SPR biosensor chip for immunoassay applications. *Nanoscale Research Letters*, *9*(1), 1–7. doi:10.1186/1556-276X-9-445 PMID:25232298

Choi, S. H., Kim, Y. L., & Byun, K. M. (2011). Graphene-on-silver substrates for sensitive surface plasmon resonance imaging biosensors. *Optics Express*, *19*(2), 458–466. doi:10.1364/OE.19.000458 PMID:21263585

Costamagna, S., & Dobry, A. (2011). From graphene sheets to graphene nanoribbons: Dimensional crossover signals in structural thermal fluctuations. *Physical Review B, 83*(23). doi: 23340110.1103/PhysRevB.83.233401

Craciun, M. F., Russo, S., Yamamoto, M., Oostinga, J. B., Morpurgo, A. F., & Thrucha, S. (2009). Trilayer graphene is a semimetal with a gate-tunable band overlap. *Nature Nanotechnology*, *4*(6), 383–388. doi:10.1038/nnano.2009.89 PMID:19498401

Daghestani, H. N., & Day, B. W. (2010). Theory and applications of surface plasmon resonance, resonant mirror, resonant waveguide grating, and dual polarization interferometry biosensors. *Sensors (Basel, Switzerland)*, *10*(11), 9630–9646. doi:10.3390/s101109630 PMID:22163431

Dellabetta, B., & Gilbert, M. J. (2010). *The effect of disorder in superfluid graphene bilayers*. Academic Press.

Eda, G., & Chhowalla, M. (2010). Chemically derived graphene oxide: Towards large-area thin-film electronics and optoelectronics. *Advanced Materials*, *22*(22), 2392–2415. doi:10.1002/adma.200903689 PMID:20432408

Elias, D. C., Nair, R. R., Mohiuddin, T. M. G., Morozov, S. V., Blake, P., Halsall, M. P., & Novoselov, K. S. et al. (2009). Control of graphene's properties by reversible hydrogenation: Evidence for graphane. *Science*, *323*(5914), 610–613. doi:10.1126/science.1167130 PMID:19179524

Guo, S., & Dong, S. (2011). Graphene and its derivative-based sensing materials for analytical devices. *Journal of Materials Chemistry*, *21*(46), 18503–18516. doi:10.1039/c1jm13228h

Hamid, T., Jameil, Taghi, Ahmadi, & Centeno. (2015). Investigation on optical and electrical properties of bilayer graphene. *Diyala Journal of Engineering Sciences*, 583-545.

Hatef, S., SeyedMahdi, M., Meisam, R., MohammadTaghi, A., & Razali, I. (2012). Bilayer graphene nanoribbon transport model. In Advanced Nanoelectronics (pp. 203-222). CRC Press.

Homola, J. I., & Dostálek, J. (2006). *Surface plasmon resonance based sensors*. Academic Press.

Hu, Y., Li, F., Bai, X., Li, D., Hua, S., Wang, K., & Niu, L. (2011). Label-free electrochemical impedance sensing of DNA hybridization based on functionalized graphene sheets. *Chemical Communications*, *47*(6), 1743–1745. doi:10.1039/C0CC04514D PMID:21125081

Huang, Y., Dong, X., Liu, Y., Li, L.-J., & Chen, P. (2011). Graphene-based biosensors for detection of bacteria and their metabolic activities. *Journal of Materials Chemistry*, *21*(33), 12358–12362. doi:10.1039/c1jm11436k

Huang, Y., Dong, X., Shi, Y., Li, C. M., Li, L.-J., & Chen, P. (2010). Nanoelectronic biosensors based on CVD grown graphene. *Nanoscale*, *2*(8), 1485–1488. doi:10.1039/c0nr00142b PMID:20820739

Jablan, M., Soljacic, M., & Buljan, H. (2013). Plasmons in graphene: Fundamental properties and potential applications. *Proceedings of the IEEE*, *101*(7), 1689–1704. doi:10.1109/JPROC.2013.2260115

Johari, P., & Shenoy, V. B. (2011). Modulating optical properties of graphene oxide: Role of prominent functional groups. *ACS Nano*, *5*(9), 7640–7647. doi:10.1021/nn202732t PMID:21875075

Lee, W.-C., Kuo, C.-C., & Chiu, N.-F. (2012). Simple fabrication of glucose biosensor based on graphene-nafion composite by amperometric detections. *2012 IEEE Sensors Proceedings* (pp. 1625-1628).

Liu, J., Fu, S., Yuan, B., Li, Y., & Deng, Z. (2010). Toward a universal "adhesive nanosheet" for the assembly of multiple nanoparticles based on a protein-induced reduction/decoration of graphene oxide. *Journal of the American Chemical Society*, *132*(21), 7279–7281. doi:10.1021/ja100938r PMID:20462190

Luican, A., Li, G. H., Reina, A., Kong, J., Nair, R. R., Novoselov, K. S., & Andrei, E. Y. et al. (2011). Single-layer behavior and its breakdown in twisted graphene layers. *Physical Review Letters*, *106*(12). doi. *Artn*, *126802*. doi:10.1103/Physrevlett.106.126802

Maharana, P. K., Jha, R., & Palei, S. (2014). Sensitivity enhancement by air mediated graphene multilayer based surface plasmon resonance biosensor for near infrared. *Sensors and Actuators. B, Chemical*, *190*, 494–501. doi:10.1016/j.snb.2013.08.089

Mak, K. F., Shan, J., & Heinz, T. F. (2010). Electronic Structure of Few-Layer Graphene: Experimental demonstration of strong dependence on stacking sequence. *Physical Review Letters*, *104*(17), 176404. doi:10.1103/PhysRevLett.104.176404 PMID:20482122

Meshginqalam, B., Toloue, H., Ahmadi, M. T., Sabatyan, A., Centeno, A., & Ismail, R. (2016). Graphene embedded surface plasmon resonance based sensor prediction model. *Optical and Quantum Electronics*, *48*(6), 1–14. doi:10.1007/s11082-016-0597-8

Meshginqalam, B., Toloue, H. A. T., Ahmadi, M. T., & Sabatyan, A. (2016). Engineer-able optical properties of trilayer graphene nanoribbon. *Physica Scripta*, *91*(3), 035802. doi:10.1088/0031-8949/91/3/035802

Min, S. K., Cho, Y., Mason, D. R., Lee, J. Y., & Kim, K. S. (2011). Theoretical design of nanomaterials and nanodevices: Nanolensing, supermagnetoresistance, and ultrafast DNA sequencing. *The Journal of Physical Chemistry C, 115*(33), 16247–16257. doi:10.1021/jp203748r

Ovchinnikov, V., & Shevchenko, A. (2013). Self-organization-based fabrication of stable noble-metal nanostructures on large-area dielectric substrates. *Journal of Chemistry, 10.* doi:10.1155/2013/158431

Phan, A. D., & Viet, N. A. (2012). A new type of optical biosensor from DNA wrapped semiconductor graphene ribbons. *Journal of Applied Physics, 111*(11). doi: 11470310.1063/1.4728196

Pradeep Kumar, M., Triranjita, S., & Rajan, J. (2014). Low index dielectric mediated surface plasmon resonance sensor based on graphene for near infrared measurements. *Journal of Physics. D, Applied Physics, 47*(38), 385102. doi:10.1088/0022-3727/47/38/385102

Rahmani, M., Ahmadi, M. T., Ismail, R., & Ghadiry, M. H. (2013). Performance of bilayer graphene nanoribbon schottky diode in comparison with conventional diodes. *Journal of Computational and Theoretical Nanoscience, 10*(2), 323–327. doi:10.1166/jctn.2013.2699

Sadeghi, H., Ahmadi, M. T., Mousavi, S. M., & Ismail, R. (2012). Channel conductance of ABA stacking trilayer graphene nanoribbon field-effect transistor. *Modern Physics Letters B, 26*(8). doi: 125004710.1142/s0217984912500479

Sadeghi, H., Ahmadi, M. T., Webb, J. F., Mousavi, S. M., Ismail, R., & Ismail, I. (2011). CARRIER statistics model for a bilayer graphene nanoribbon in the nondegenerate regime. In N. W. J. F. V. P. Barsoum (Ed.), *Proceedings of the Fourth Global Conference on Power Control and Optimization* (Vol. 1337, pp. 184-187). doi:10.1063/1.3592464

Sadrolhosseini, A. R. (n.d.). *Surface Plasmon resonance characterization of biodiesel.* (PhD thesis). Universiti Putra Malaysia.

Salihoglu, O., Balci, S., & Kocabas, C. (2012). Plasmon-polaritons on graphene-metal surface and their use in biosensors. *Applied Physics Letters, 100*(21), 213110. doi:10.1063/1.4721453

Shang, J., Ma, L., Li, J., Ai, W., Yu, T., & Gurzadyan, G. G. (2012). The Origin of fluorescence from graphene oxide. *Scientific Reports, 2*, 792. doi:10.1038/srep00792 PMID:23145316

Shukla, S., & Saxena, S. (2011). Spectroscopic investigation of confinement effects on optical properties of graphene oxide. *Applied Physics Letters, 98*(7), 073104. doi:10.1063/1.3555438

Simsek, E. (2013). Improving tuning range and sensitivity of localized spr sensors with graphene. *Photonics Technology Letters, IEEE, 25*(9), 867–870. doi:10.1109/LPT.2013.2253316

Singh, R. J. (2012). *Solid State Physics*. Dorling Kindersley.

Stebunov, Y. V., Aftenieva, O. A., Arsenin, A. V., & Volkov, V. S. (2015). Highly sensitive and selective sensor chips with graphene-oxide linking layer. *ACS Applied Materials & Interfaces, 7*(39), 21727–21734. doi:10.1021/acsami.5b04427 PMID:26358000

Subramanian, P., Barka-Bouaifel, F., Bouckaert, J., Yamakawa, N., Boukherroub, R., & Szunerits, S. (2014). Graphene-coated surface plasmon resonance interfaces for studying the interactions between bacteria and surfaces. *ACS Applied Materials & Interfaces*, *6*(8), 5422–5431. doi:10.1021/am405541z PMID:24433135

Sutter, P., Hybertsen, M. S., Sadowski, J. T., & Sutter, E. (2009). Electronic structure of few-layer epitaxial graphene on Ru(0001). *Nano Letters*, *9*(7), 2654–2660. doi:10.1021/nl901040v PMID:19505134

Tang, Y., Zeng, X., & Liang, J. (2010). Surface plasmon resonance: An introduction to a surface spectroscopy technique. *Journal of Chemical Education*, *87*(7), 742–746. doi:10.1021/ed100186y PMID:21359107

Toloue, H., Ahmadi, M., Meshginqalam, B., Centeno, A., & Sabatyan, A. (2015). Graphene based surface plasmon resonance bio-sensor modeling. *Annual International RIAPA Meeting on Low Dimensional Systems*.

Toloue, H. M. B., Ahmadi, M. T., & Centeno, A. (2015). *Graphene based surface plasmon resonance bio-sensor response modeling*. Nanomeghyas.

Valentini, F., Carbone, M., & Palleschi, G. (2013). Graphene oxide nanoribbons (GNO), reduced graphene nanoribbons (GNR), and multi-layers of oxidized graphene functionalized with ionic liquids (GO-IL) for assembly of miniaturized electrochemical devices. *Analytical and Bioanalytical Chemistry*, *405*(11), 3449–3474. doi:10.1007/s00216-012-6615-1 PMID:23274557

Velasco, J. Jr, Lee, Y., Jing, L., Liu, G., Bao, W., & Lau, C. N. (2012). Quantum transport in double-gated graphene devices. *Solid State Communications*, *152*(15), 1301–1305. doi:10.1016/j.ssc.2012.04.024

Wan, Y., Wang, Y., Wu, J., & Zhag, D. (2011). Graphene oxide sheet-mediated silver enhancement for application to electrochemical biosensors. *Analytical Chemistry*, *83*(3), 648–653. doi:10.1021/ac103047c PMID:21175166

Wang, Z., Zhou, X., Zhang, J., Boey, F., & Zhang, H. (2009). Direct electrochemical reduction of single-layer graphene oxide and subsequent functionalization with glucose oxidase. *The Journal of Physical Chemistry C*, *113*(32), 14071–14075. doi:10.1021/jp906348x

Woan, G. (n.d.). *The Cambridge Handbook of Physics Formulas*. Cambridge University Press.

Wu, L., Chu, H. S., Koh, W. S., & Li, E. P. (2010). Highly sensitive graphene biosensors based on surface plasmon resonance. *Optics Express*, *18*(14), 14395–14400. doi:10.1364/OE.18.014395 PMID:20639924

Zhang, J., Sun, Y., Wu, Q., Gao, Y., Zhang, H., Bai, Y., & Song, D. (2014). Preparation of graphene oxide-based surface plasmon resonance biosensor with Au bipyramid nanoparticles as sensitivity enhancer. *Colloids and Surfaces. B, Biointerfaces*, *116*, 211–218. doi:10.1016/j.colsurfb.2014.01.003 PMID:24480068

Zuppella, P., Tosatto, S., Corso, A. J., Zuccon, S., & Pelizzo, M. G. (2013). Graphene–noble metal bilayers for inverted surface plasmon resonance biosensors. *Journal of Optics*, *15*(5), 055010. doi:10.1088/2040-8978/15/5/055010

APPENDIX 1: MOLECULE SHAPE EFFECT

For the effective outer shell factor determination, the chemical Structure of Mannose, Lactose, PEI and PSS (Subramanian et al., 2014), should be considered. It is clear that only in the case of PSS molecule there are two dangling bonds, so the related value for parameter "V" is 2 for this material and zero for others as indicated in Table 3.

Table 3. The effective outer shell factor determination

Name of the Molecule	Chemical Structure	The Number of Dangling Bands
Mannose	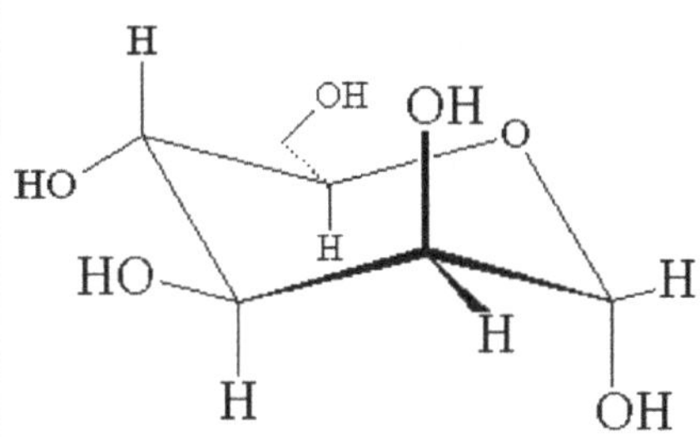	0
Lactose	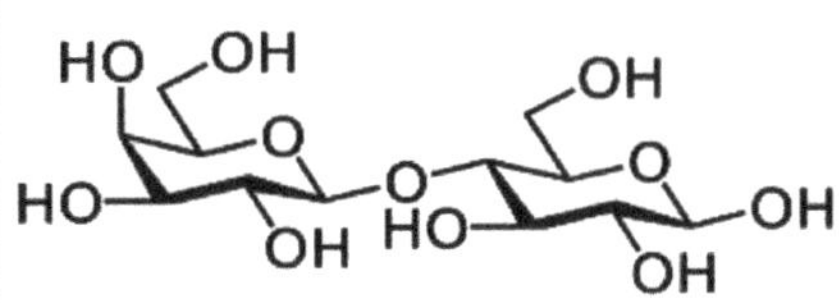	0
PEI	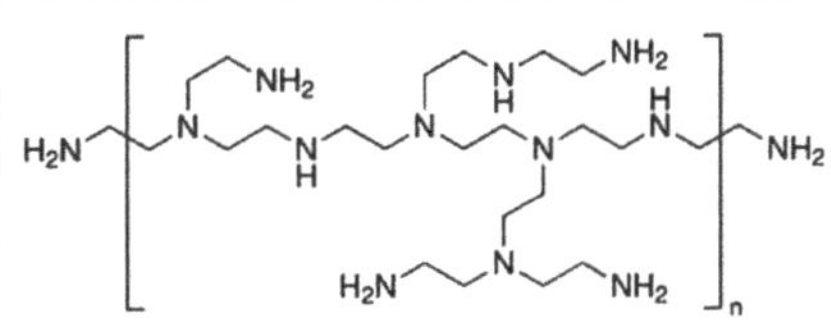	0
PSS	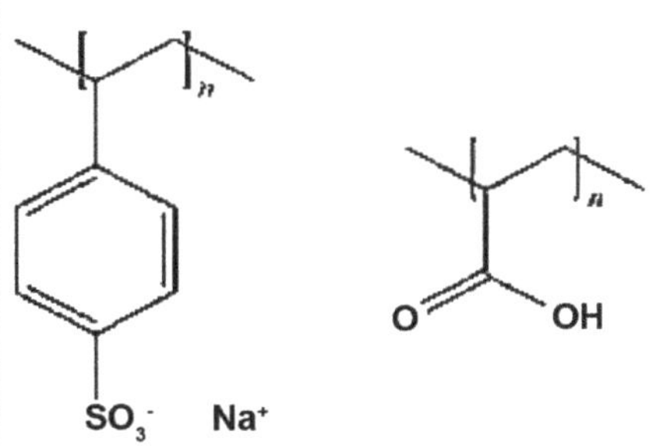	2

APPENDIX 2: REFRACTIVE INDEX CALCULATION FROM SPR CURVE

The refractive index values of different materials in desired SPR configuration can be used for resonance angle calculations as mentioned in (Tang, Zeng, & Liang, 2010). So one may use this relation for refractive index investigation from resonance angle too as shown in Figure 26.

Figure 26. Desired SPR configuration with different layers

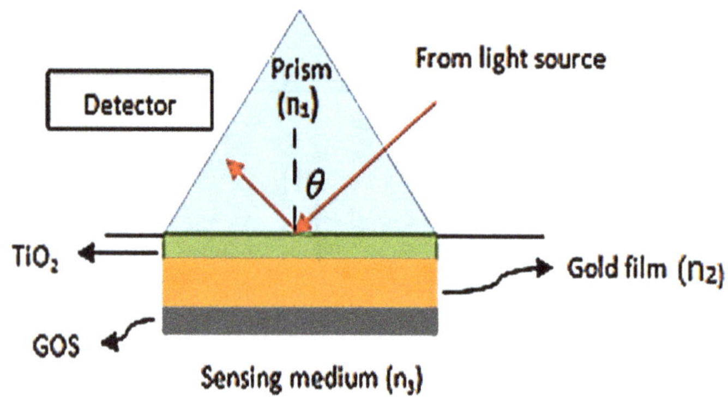

$$\theta_{SPR} = \sin^{-1}\left(\frac{1}{n_1}\sqrt{\frac{n_3^{\,2}n_2^{\,2}}{n_3^{\,2}+n_2^{\,2}}}\right)$$

Table 4 shows refractive index values which are extracted from SPR curves of Figure 16.

Table 4. Refractive index values extracted from SPR curves

Name of the Molecule	Resonance Angle from Theoretical Data	Resonance Angle from Experimental Data	Theoretical Refractive Index (n)	Experimental Refractive Index (n_x)	n-n_x
Mannose	71.5	71.36	1.3423 - 0.0096i	1.3414 - 0.0096i	(9.5522 -0.20451i) 10^{-4}
Lactose	71.8	71.62	1.3444 - 0.0096i	1.3432 - 0.0096i	(12 - 0.0000i) 10^{-4}
PEI	72	71.96	1.3457 - 0.0097i	1.3454 - 0.0097i	(2.6477- o.o570i) 10^{-4}
PSS	72.3	72.42	1.3477 - 0.0097i	1.3484 - 0.0097i	(-7.7744 + 0.16799i) 10^{-4}

Chapter 15
Fast Neuron Detection

Hadi Kasani
University of Mohaghegh Ardabili, Iran

Mohammad Taghi Ahmadi
Urmia University, Iran

Rasoul Khoda-Bakhsh
Urmia University, Iran

Dariush Rezaei Ochbelagh
Amirkabir University of Technology, Iran

ABSTRACT

In many research fields and industry such as nuclear physics, notably nuclear technology, fusion plasma diagnostics, radiotherapy and radiation protection, it is very substantial that measure fast neutron spectra. For example in nuclear reactor primary generated neutrons have energies around 2 MeV that lie fast neutron category. Also particle accelerators and Am-Be neutron source raise fast neutrons. Therefore a review of silicon based fast neutron detection with proton recoil methods is surveyed. Furthermore Carbon nanoparticles (CNPs) with simple and low cost preparation methods with exceptional electrical properties have been used widely in nanoelectronic applications such as radiation sensors. In this chapter, fast neutron detectors using Carbon based semiconductor, back-to-back Schottky diode type, and polyethylene as convertor are developed and the Am-Be fast neutron source is used in experimental measurements.

INTRODUCTION

Kind of Neutron Detectors

Neutrons are detected by nuclear reactions which result in prompt charged particles such as protons, alpha, and so on. Virtually, all types of the neutron detectors include the combination of a target material to carry out neutron to ionizing particle conversion together with one of the charged particle detectors. Because the neutron energy is the important parameter in use of the cross section for neutron interactions in most materials, several techniques have been introduced for neutron detection in various energy ranges.

In selecting for nuclear reactions that could be used in neutron detection, several elements must be considered. For example, the cross section of the reaction must be as large as possible so that efficient detectors with small dimensions can be used. In many applications, fields of gamma rays are also pre-

DOI: 10.4018/978-1-5225-0736-9.ch015

sented with neutrons, so the choice of reaction is involved on the ability of discriminate gamma rays from the neutrons in the detection process (Fazzi, Agosteo, Pola, Varoli, & Zotto, 2003).

The discovery of the neutron is based on the measurements of neutron energy due to scattering neutrons on hydrogen (Chadwick, 1932) or nitrogen (Feather, 1932), and measuring the energy of the recoiled nuclei. Neutron detection has adopted to the development of nuclear industry since 1932 and also in many research fields such as nuclear physics, nuclear technology, fusion plasma diagnostics, radiotherapy and radiation protection, it is very significant that measure fast neutron spectra.

Methods of neutron spectrometry can be categorized into seven groups (Brooks & Klein, 2002):

1. Methods based on measurement energy of the recoiled nuclei, as in the discovery of the neutron.
2. Methods in which the neutron is induced in nuclear reaction and results the charge particles such as alpha particles.
3. Methods based on the measurements of neutron velocity.
4. Threshold methods, so that minimum neutron energy is indicated by the presence of a neutron irradiation effect such as neutron activation, radioactivity, specific gamma-ray energy or a phase transition.
5. Methods based on the determination of neutron energy distribution by unfolding a set of detectors (or detector geometries) which their response to neutrons are differed.
6. Methods in which neutron diffraction effects are observed and
7. Methods based on the measurement of time-distribution of the high-energy neutron slowing downs in a suitable medium.

In this survey we particularly discuss methods belonging to group 1 and we introduce detectors that working phenomena lies in this class.

The most useful method of fast neutron spectrometry is based on elastic scattering of neutrons by hydrogenous materials. The recoil nuclei that result from (n,p) reaction are called recoil protons and consequently, spectrometry method is known as proton recoil method.

One of the proton recoil detector is the 'SP2' counter, investigated by Benjamin et al. (Benjamin, Kemshall, & Redfearn, 1968). A thin thickness spherical stainless steel chamber, 40mm in diameter, filled with hydrogen gas at a pressure range from 10^5 to 10^6 Pa (1–10 atm), which neutrons interact by scattering from a proton in the hydrogen atoms and the obtained proton deposits its energy in the gas medium, and generates primary ionization along its track until it stopped by the gas (or hits the wall). In the nearby to the anode wire, that the electric field is relatively strong, multiplication of the primary charge is started. The main disadvantage of the SP2 counter is to gamma ray sensitivity, which primary photoelectrons from the gamma rays generate secondary electrons near the anode wire.

Working Phenomena of Proton Recoil Neutron Detectors

As be mentioned previously, neutrons are detected by indirectly methods where charged particles are generated by nuclear reactions. The energy of charged particles is depends on the neutron energy.

Elastic scattering of neutrons with nuclei of the convertor materials such as methane or polyethylene is followed by production of recoil protons in hydrogenous material or of alpha particles in ^{4}He filling gas. The maximum energy transfer from the neutron to the recoil nucleus with mass of M is given by:

$$E_{\max} = \frac{4M}{(M+1)^2} E_n \tag{1}$$

where, E_n is the incident neutron energy. The energies of the recoil nuclei are distributed from 0 to E_{max}, where each energy value is generated with the same probability.

Since, on the basis of centre-of-mass coordinate system, the scattering cross section is isotropic, the resulting distribution of proton recoil energy, P(E), for an incident monoenergetic neutron energy E would ideally have the characteristic rectangular shape as shown in Figure 1.

SILICON BASED FAST NEUTRON DETECTORS

Silicon based semiconductor detectors are a suitable devises for detection of energetic charged particles, photons and neutrons. In comparison with gas chamber or scintillation techniques, these detectors have some advances, such as their compactness, small size, low weight, simple operation, low voltage power supply, and high energy resolution. For these reasons, semiconductor detectors have found more applications in physics and radiation protection research fields (Knoll, 2000).

PIN Diode Based Neutron Detectors

Commercial silicon Positive-Intrinsic-Negative (PIN) photodiodes not only are used for visible light detection but also they have been used to detect charged particles (Yamamoto, Hatakeyama, Norimura, & Tsuchiya, 1984). These diodes are such a photodiodes with a large, neutrally doped intrinsic region sandwiched between p-type and n-type semiconductors. The PIN photodiode has many advantages like low cost, simple operation, low weight, low voltage power supply and good energy resolution. Detector is used with 6LiF (for thermal neutrons) and polyethylene (for fast neutrons) convertors. Also the different fast neutron source has been used for characteristics and calibration of the detector.

Figure 1. Ideal response function of proton recoil detector to monoenergetic neutrons

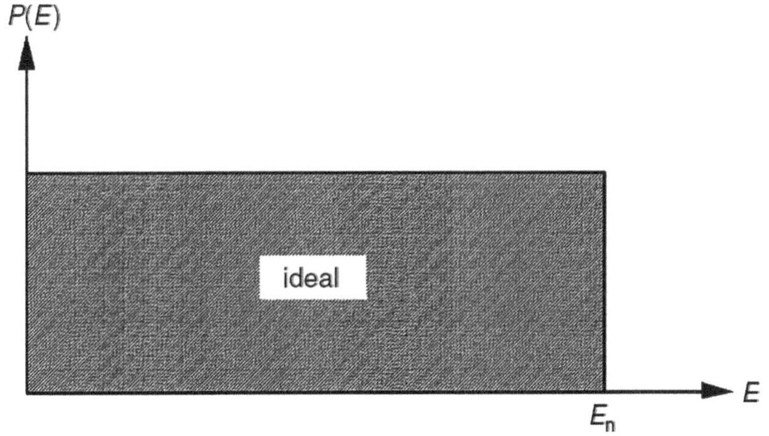

The neutron detectors discussed in this section consists of a polyethylene converter (Voytchev et al., 2003). Neutrons are detected through the energy deposition of proton recoil in depletion region of PIN diode. In this case secondary neutrons generated by 3.7, 4.0 and 5.0 MeV protons impacting a 1-mm-thick beryllium target. A PIN diode with 300 mm thick and a reverse bias voltage of 30 V (totally depleted) was set up in the reverse-injection configuration for the experimental validation. Also the sensitive area of the diode is 3 mm2.

Exploiting the different rise times of charge carrier pulses in the silicon detector is used to discrimination of the pulses from secondary electrons from those due to low-energy recoil-protons. The minimum detectable energy which can be measured with pulse-shape discrimination (PSD) is about 0.9 MeV. The spectrum of energy deposited in the diode with PSD is shown in Figure 2 which detector irradiated with the secondary neutron field generated by 5 MeV protons on a beryllium target (Agosteo et al., 2005).

The maximum detectable energy is about 6 MeV and is obtained by the thickness of the fully depleted layer (300 μm).

In the next type fast neutron detector (Hosono, Sjafruddin, Iguchi, & Nakazawa, 1995), a windowless PIN photodiode with polyethylene convertor and a low noise charge sensitive preamplifier is used. A schematic of setup is shown in Figure 3. The PD detector (Hamamatu photonics S 3590-02, windowless type, sensitive area = 10 × 10 mm2) with a polyethylene convertor (10 × 10 mm^2, thickness 45 pm) is used. It is important to mention that the equivalent capacitance of detector is about 70 pF. The thickness of depletion layer is estimated about 180-200 pm at a reverse voltage of 24 V. On the other hand this is equivalent to the range of about 5 MeV protons. The electron-hole pairs are also created In the N+ layer of the PD, however, they do not generate significant output signals because recombination of electron-hole pairs is occurred in a very short time.

A low noise charge sensitive preamplifier was used to spectroscopy procedure. Figure 4 shows the circuit diagram of preamplifier.

Figure 2. Energy spectrum of secondary neutrons from 5MeV proton impact

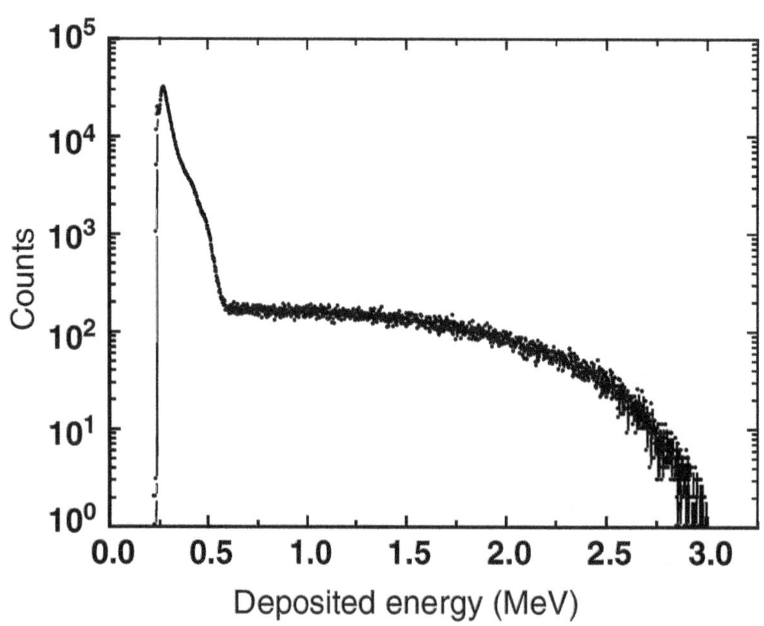

Figure 3. Schematic of experimental setup

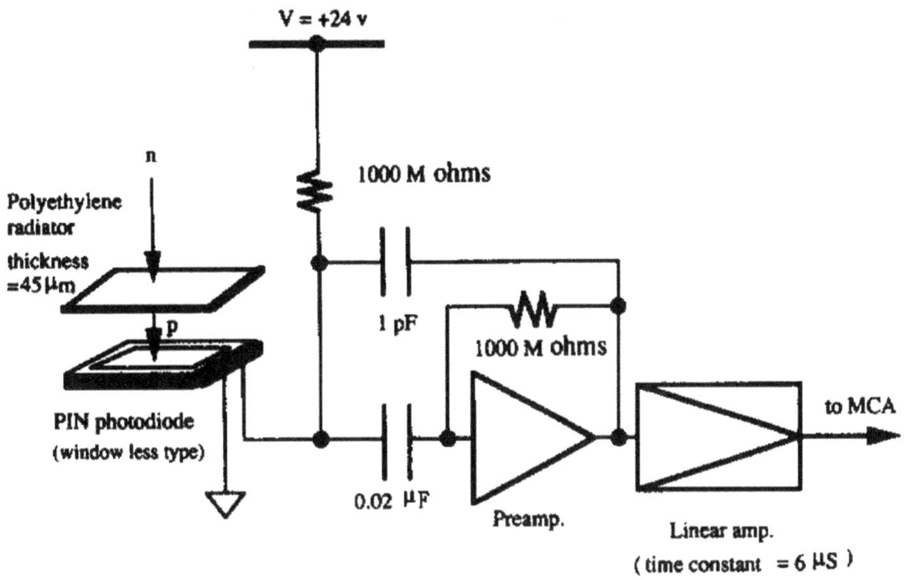

Figure 4. A low noise charge sensitive preamplifier

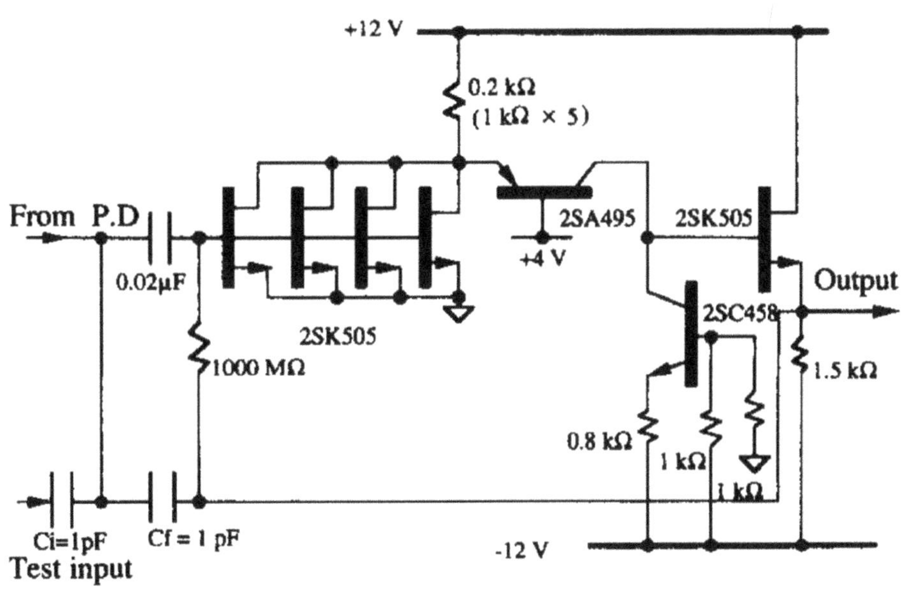

In order to calibration of the detector, the test pulse and the following relation between proton energy Ep and the test pulse is used.

$$E_p = \frac{wC_iV_p}{q} \tag{2}$$

where w is the average energy required for one electron-hole pair generation, Ci is the test input preamplifier capacitance, Vp is the amplitude of the test pulse and q is the basic charge.

Figure 5 shows:

- With polyethylene (a),
- Without polyethylene (b), and
- Background (c) spectrum of 252Cf neutrons is presented.

From Figure 5, it is clear that fast neutrons are included the higher energy region above 1.2 MeV in (a) Recoil proton signals. The energy region lower than 1.2 MeV in (b) is due to γ-ray and neutron from the detector by Si(n,p), (n,γ) reactions. Finally, the efficiency for ^{252}Cf neutrons was obtained as 7.8×10^{-5} (counts/incident neutron)

On the basis of PIN diode, detector system with Am-Be neutron source and ^{6}LiF and polyethylene converters is investigated (Adamiec, Iñiguez, Lorente, Voytchev, & Gallego, 2004). The 4mm thick square polyethylene slab placed in front of the PIN diode is used for the fast neutrons converter. The length of the depletion region is about 200 μm. Also the range of protons in PE and Si were founded from $R = aE^{b}$, where the constants a and b were calculated using SRIM code (Adamiec et al., 2004).

Figure 5. Pulse height distribution for ^{252}Cf neutrons in different conditions

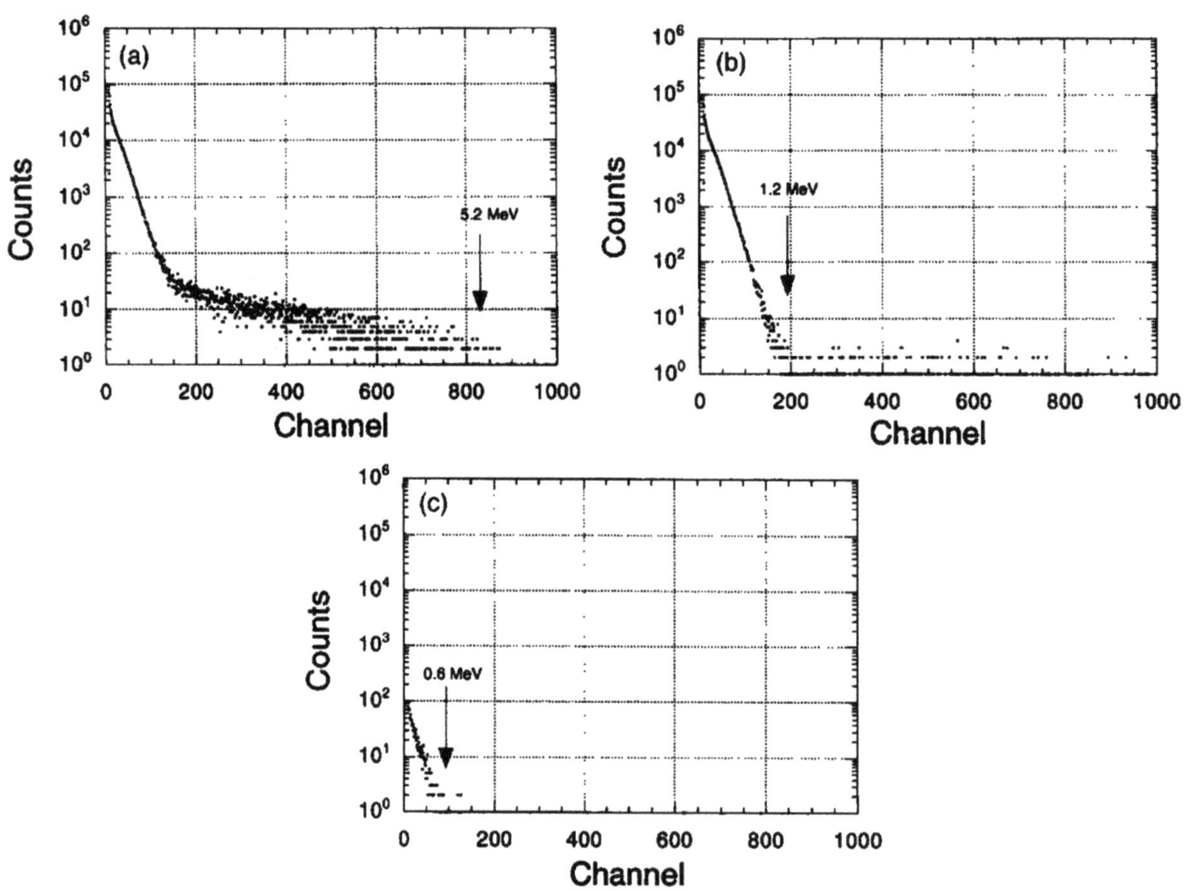

SRIM calculate the stopping power and range of ions in targets using a quantum mechanical behavior of ion-atom collisions.

The energy spectrum of recoil protons was measured and comparison with modeling results (Figure 6). It can be seen that the experimental and simulated energy spectra have the same in shape for higher energies. In the low energy region, the most difference in the two spectra can be observed. The simulated spectrum falls to zero for low energies.

Energy spectrum of recoil protons was measured in different neutron incident angles (Figure 7). Results show that the detector is sensitive to the direction of neutrons enter the PE convertor.

In addition PIN based detectors can be used for dosimetric applications. Therefor the count rate was measured in different distances from the neutron source in the fully depleted condition of detector with 6LiF convertor. A summary of obtained results is represented in the Table 1.

On the other hand since the monoenergetic neutrons impacted to the detector have the continuum spectrum, so the extra energies recorded in the detector (Figure 1). In order to cancel the later energies, an unfolding computer code is required. There are various algorithms for these calculations. Then the neutron energy spectrum can be obtained from the detector data by the deconvolution procedure. The following unfold linear equation gives the response of a neutron spectrum (Zaki Dizaji, Kakavand, & Abbasi Davani, 2014).

Figure 6. (a) Experimental result of measured spectrum; (b) simulated energy spectrum

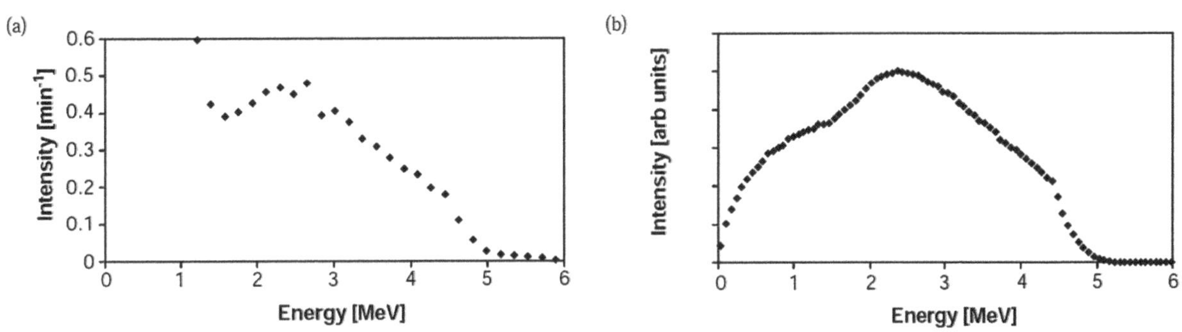

Figure 7. Pulse height distribution for ^{252}Cf neutrons in different conditions

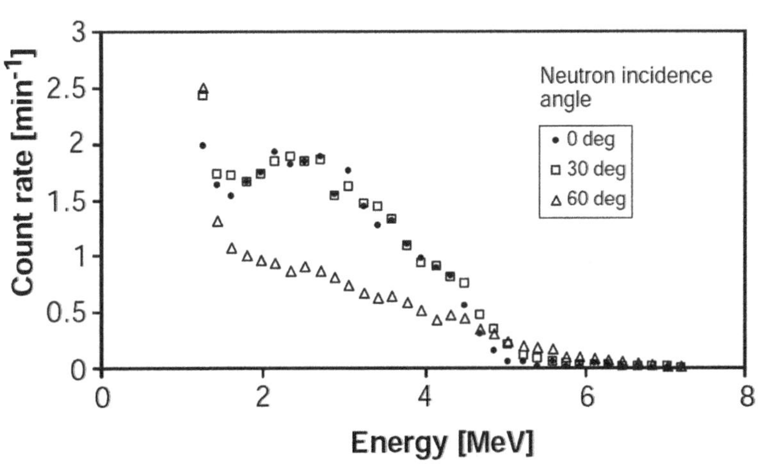

Table 1. measured dose in different distances

Distance [cm]	Count Rate [min-1]	Ambient Dose Equivalentrate H*(10) [μS v h⁻¹]
17	369 ± 17	2673 ± 10
22	259 ± 6	Not measured
35	94.1 ± 1.4	625±3
50	46.8 ± 0.8	3 1 7 ± 2
100	1 0.3 ± 0.3	83 ± I
115	10.2 ± 0.3	64± I
1 50	6.27 ± 0 . 1 5	40 ± I

$$R_i = \int_{E_{min}}^{E_{max}} \sigma_i(E)\Phi(E)dE, i = 1,2,3,...,m \quad (3)$$

where

R_i is the ith detector response,
$\sigma_i(E)$ is the sensitivity of the ith detector to a neutron with the energy of E,
$\Phi(E)$ is the incident neutron fluence, and
m is the number of detectors.

Discrete version of Equation 3 is used in various energy intervals. The detector system had been irradiated by monoenergitic neutrons in 0.5-11 MeV energy range which each energy interval is included the specific detector. Figure 8 shows the unfolded and standard energy spectrum of Am-Be neutron source.

Schottky Diode-Based Neutron Detectors

A Schottky junction refers to a metal-semiconductor contact having a potential barrier. This barrier can be controlling the current conduction as well as its capacitance behavior. Also the depletion layer of a metal-semiconductor contact is generated as well as p-n junction. This region is sensitive to pass of charged particles and they create the electron-hole pairs through the path.

Silicon carbide neutron detectors were introduced in several literatures (Babcock, 1965). For example the samples as schottky diode and p-n junction were investigated (Ruddy, Dulloo, Seidel, Seshadri, & Rowland, 1998). Samples are fabricated by vapor-phase epitaxy onto high-purity 4H-SiC substrate wafers with a nitrogen doping. A schematic of detectors are presented in Figure 9. Two sizes of detectors have diameters of 400 and 200 μm with active areas of 0.217 and 0.066 mm^2, respectively.

First the response of detector is investigated for alpha particles. The clear peak observed in the spectrum is related to the 238Pu alpha particles entering the diode. The energy spectrum of the alpha particles is shown in Figure 10.

Using the full width at half-maximum (FWHM) of peak, the detector resolution is obtained of 5.8%. Also the detector is sensitive to gamma rays. As shown in Figure 11 the gamma and alpha pulses can be discriminate so that the rise time of alpha particles faster than that of gamma pulses.

Figure 8. Energy spectrum of Am-Be neutrons from unfolded and ISO
ISO 8529-1, Reference Neutron Radiations. Part 1: Characteristics and methods of production. International Organization for Standardization ISO8529-1, 2001.

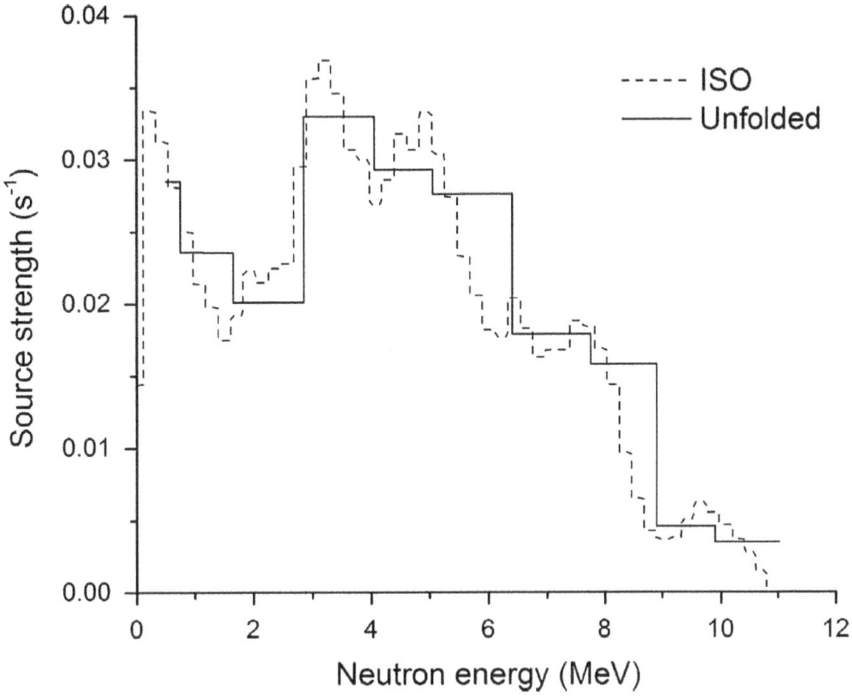

Figure 9. Schematic of fabricated detectors: (a) Schottky and (b) p-n junction diodes [17]

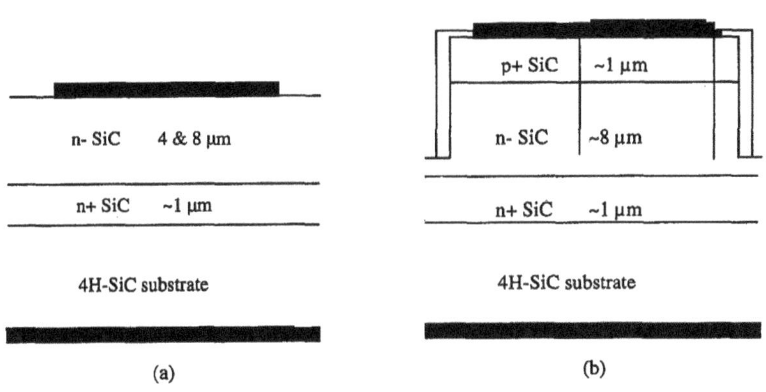

Additionally the large area 4H-SiC Schottky diode with a 6LiF converter was tested as neutron detector (Lo Giudice et al., 2007). Furthermore, the SiC based detectors due to their wide band gap are very suitable for high temperature applications (Bertuccio, Casiraghi, Cetronio, Lanzieri, & Nava, 2004; Dulloo et al., 1999). The operations of SiC detectors were investigated with neutron fluences in the range of 109–1013 cm-2. Energy spectra of alpha and tritium particles, created by 6Li(n,a)3H reaction, were obtained at different applied bias voltages and by interrupting polyethylene moderators of thickness from 10 to 55 mm. Figure 12 shows the spectra from a 20mm2 area detector at different applied bias voltages with neutrons of 1W reactor power.

Figure 10. Response of Shottky detector to alpha particles

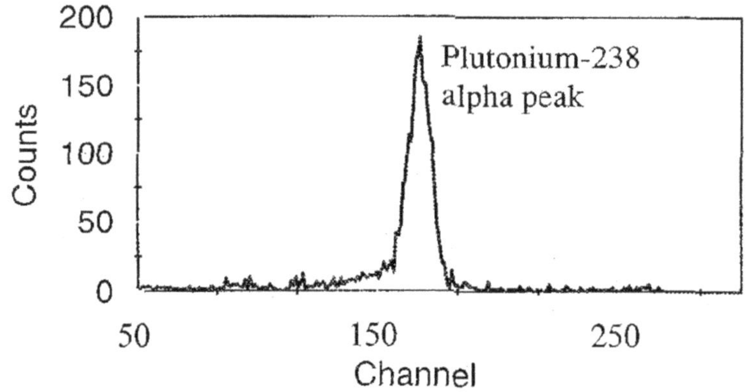

Figure 11. Simultaneously response of Shottky detector to alpha particles and 60Co gamma rays

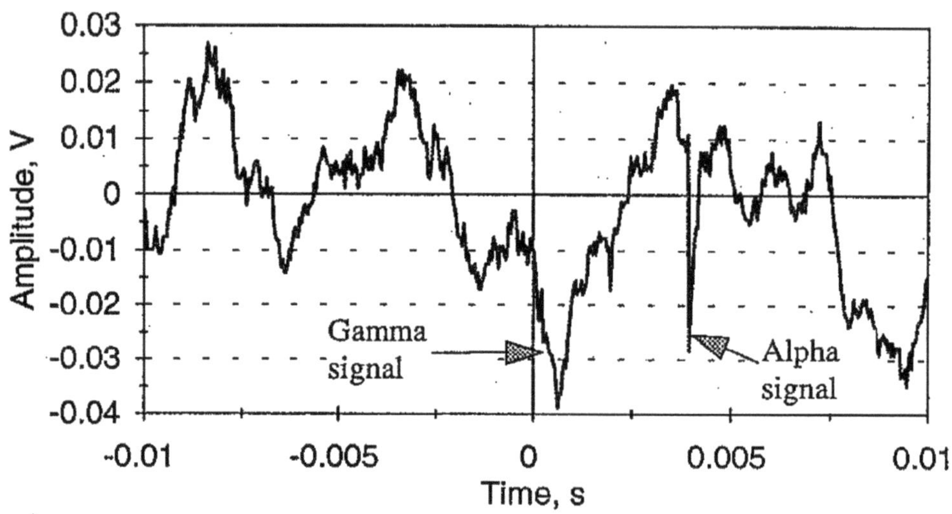

Also the 4H-silicon carbide (4H-SiC) detectors are designed for neutron pulse form measurement (Wu et al., 2015). I-V characteristic curve of the diode is presented in Figure 13.

Furthermore, the response spectrum of the 4H-SiC detector to 241Am-239Pu alpha particles before and after 18 reactor pulses is shown in Figure 14.

CARBON BASED RADIATION DETECTORS

Carbon nanomaterials such as one-dimensional (1D) carbon nanotubes (CNT) or two-dimensional (2D) graphene layers, which have exceptional electrical properties, can be used in electronic devices instead of silicon based devices (Avouris, Chen, & Perebeinos, 2007). According to the definition, carrier mobility determines the electrical conductivity per unit charge so it shows the sensitivity of the device. This feature of semiconductors is very important to use in radiation detection devices. Using the carrier

Figure 12. Energy spectrum of alpha and tritium with Q-value of 4.78MeV of reaction

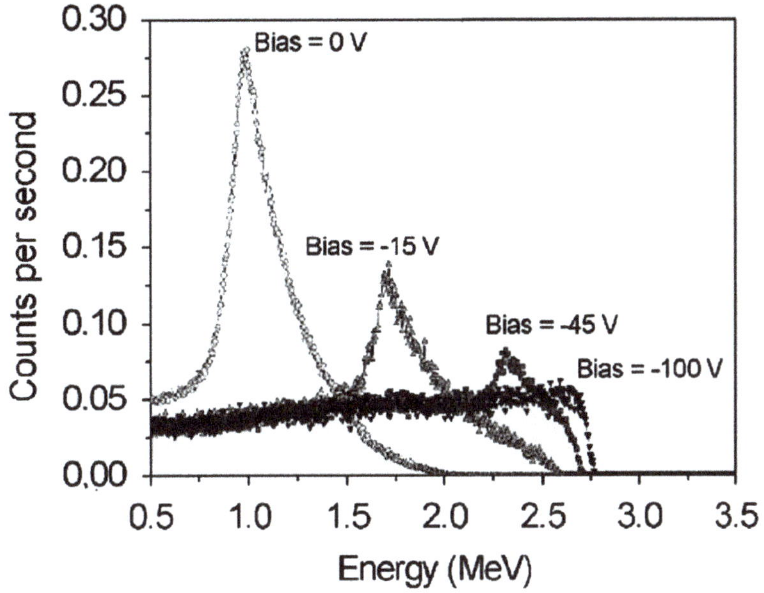

Figure 13. Reverse biased I-V characteristic of the detector

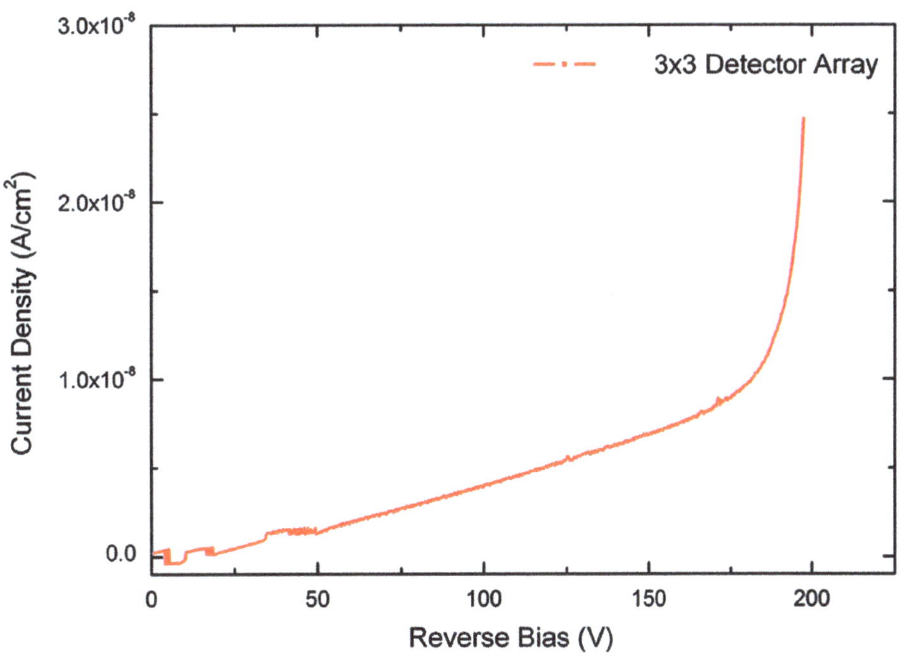

transport characteristics in Carbon based transistors, field-effect mobility is obtained equal to 79,000 cm2/Vs and the intrinsic mobility is estimated over 100,000 cm2/Vs at room temperature (Tameev, Jiménez, Pereshivko, Rychwalski, & Vannikov, 2007) which these values much bigger than for other semiconductors. In this section, the carbon nanoparticle (CNP) based radiation detector, structured with

Figure 14. The response spectrum to ^{241}Am-^{239}Pu alpha particles in two conditions

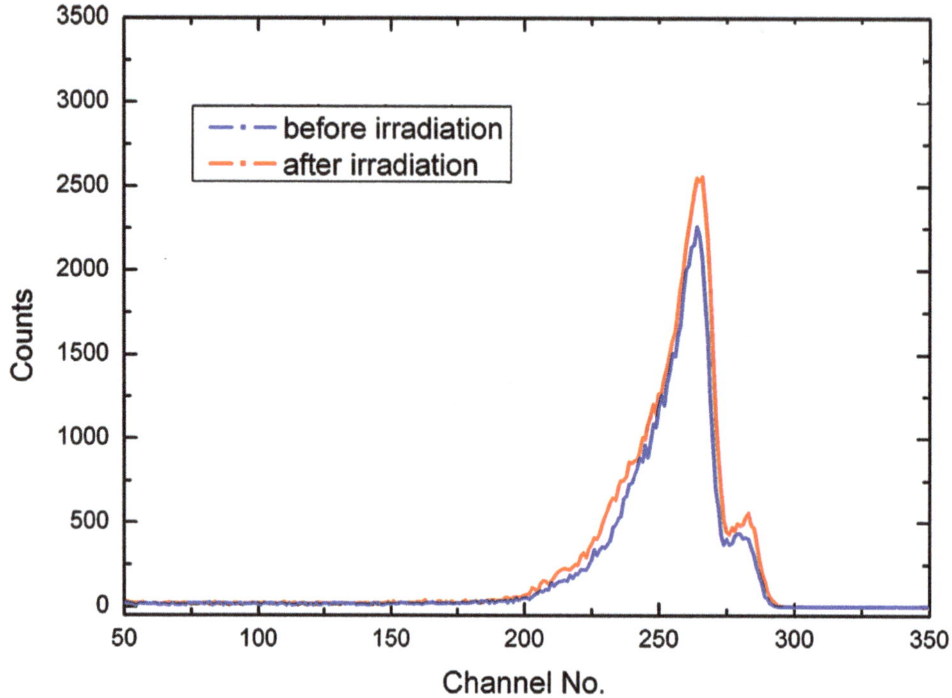

metal-semiconductor-metal (MSM), is introduced which can be used to fast neutron detection. First the detector was tested by beta-rays and the change in electrical properties is considered (Kasani, Taghi Ahmadi, Khoda-bakhsh, RezaeiOchbelagh, & Ismail, 2016).

Experimental Setup and Characteristics

The arc discharge method is one of the CNPs synthesis methods which has been described in details by ref. (Arora & Sharma, 2014). In the presented work the arc discharge method with two Copper electrodes in a fixed frequency is used for CNP synthesis as shown in Figure 15 and 16.

Figure 15. Schematic of the setup used for CNPs growth

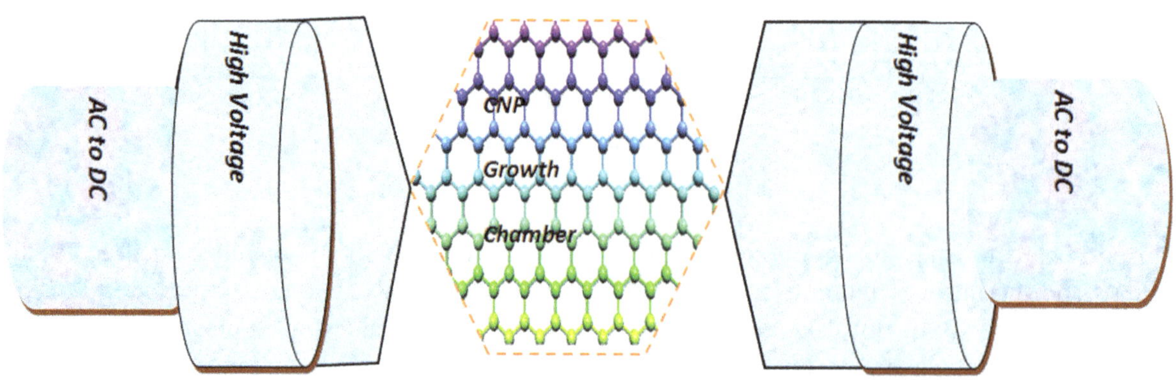

Figure 16. Glass chamber used to CNPs growth

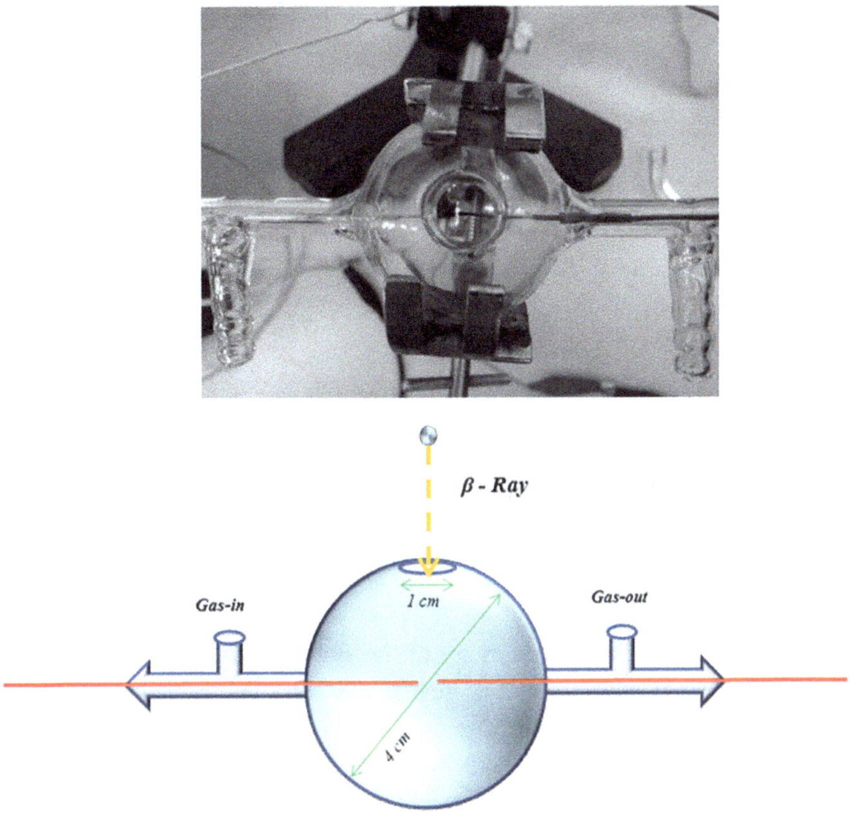

The experimental system was designed in following manner. A spherical glass chamber with inner diameter of 40 mm, was selected as shown in Figure 2. Then in both sides of the sphere, two pipes for input and output of butane (C4H10) gas are connected. However, in order to prevent re-entry air molecules into the chamber, output pipe is connected to a bubbling system. Also the chamber consists of two electrodes which are mounted in the each pipe. Since the stability of CNPs, grown by arc discharge method, depends on electrode geometry (Arora & Sharma, 2014) therefore the conically electrodes are selected. In the presence of gas, high voltage generator is connected to the chamber by electrodes. It is important to mention that gas chamber has atmospheric pressure.

In order to β exposure, circular hole with 1 cm diameter is mounted on the top of the chamber. After growth process, I-V measurements are obtained by a BHP-2064 electrochemical analysis system using two-probe method. In this technique two probes, one as Counting and Reference probe and another as a Working probe, are employed for I-V characteristic measurements. The measured I-V curve is presented in Figure 17.

Scanning electron microscopy (SEM) is one of the most significant tools in nanotechnology for nanostructure characterization. Since SEM is simple and could be carried out routinely, it is been preferred in comparison with other microscopy techniques (Lehman, Terrones, Mansfield, Hurst, & Meunier, 2011; Sydlik, Lee, Walish, Thomas, & Swager, 2013).The morphology of the obtained material was visualized by SEM after separation from the electrodes. Cross section and surface of bulk CNPs are shown in Figure 18a. Also Figure 18b shows the magnified cross-section SEM image of the broken bulk CNPs.

Figure 17. An I-V characteristic curve obtained by BPH system

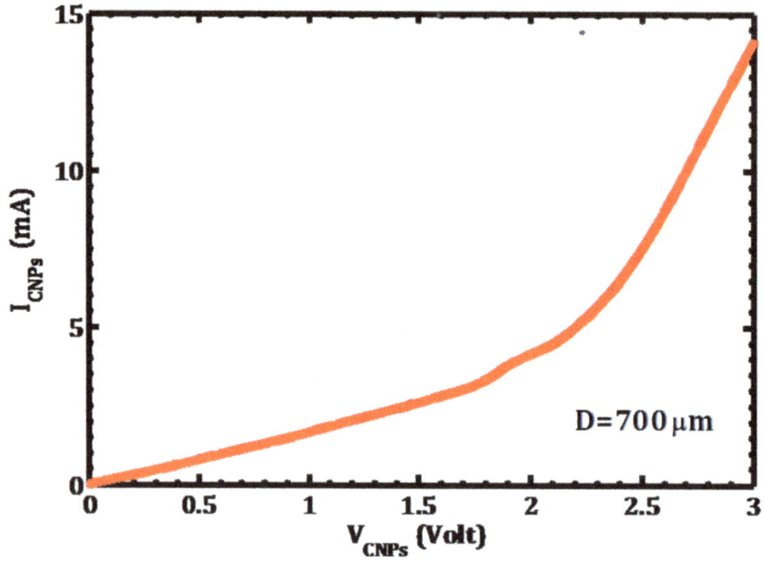

Additionally Figure 5c shows the average diameter of the synthesized cylindrical bulk CNPs of 29 µm. SEM images, taken at different magnifications, reveal the existence of nanostructures grown on the surface of the cylindrical bulk CNPs from the arc discharge as demonstrated in Figure 18d.

Infrared spectroscopy is another analyses technique has been used in nanotechnology science to investigate functionalization. In this method infrared (IR) radiation is exposure on the sample and absorption (or transmittance) is observed, which is a quantized process. It is clear that two molecules don't have exactly the same IR spectrum. CNPs infrared spectra measurements are performed on Shimadzu IR spectrometer in the wavenumbers range of 500–4000 cm^{-1} using KBr pellets. Figure 19 shows the IR spectra of the background, CNPs and MWCNTs. Since the little amount of CNPs uses for analyses the intensity of absorption spectra are very week but they relatively are visible. As shown in Figure 6 the peak at 3443 cm-1 (arrow 1 in the Figure 19) could be ascribed by O–H vibration in the carboxyl groups (Stobinski et al., 2010). Two peaks appear between 2850 and 3000 cm-1 (arrows 2,3 in the Figure 19) due to the H–C stretch modes of H–C=O in the carboxyl group (Zaragoza-Contreras et al., 2009).

Unfortunately the region between 1640-1670 cm-1(arrow 4 in the Figure 19) has big noise amount and existence of C=C and O=C bonds cannot be recognized exactly. The appeared peaks in this region can be attributed to the stretching of the carbon nanotube backbone and carboxyl groups (Goyanes et al., 2007).

Monte Carlo Simulation

Monte Carlo simulation is a method to understand the interaction of charged particles with matter. This method is widely used to model random events and statistics phenomena. In this work, Monte Carlo simulation using MCNPX code shows that the large ratio of β-ray energies are deposited within the Copper electrodes and ionization caused by β electrons in the C_4H_{10} gas, due to its very low density, is negligible (Figure 20). Figure 21 shows the flux of electrons in the chamber so that the mesh is carried out in the y direction as one part and its thickness is selected of 2 mm.

Figure 18. a) SEM image; b) cross-section; c) surface; and d) existence of CNPs on the surface of the synthesized sample

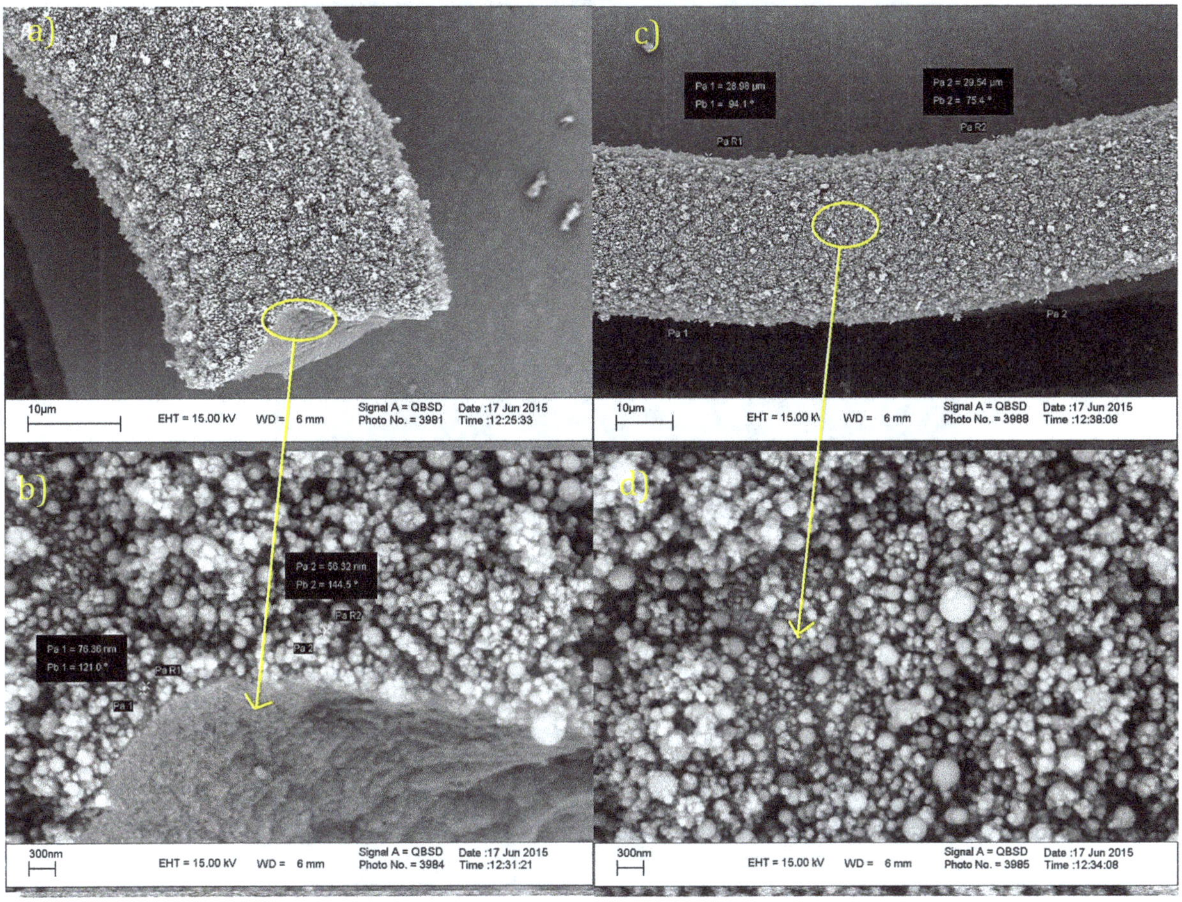

Figure 19. IR spectra of: (a) background; (b) our CNPs; and (c) FT-IR spectrum of MWCNTs
Cunha et al., 2012.

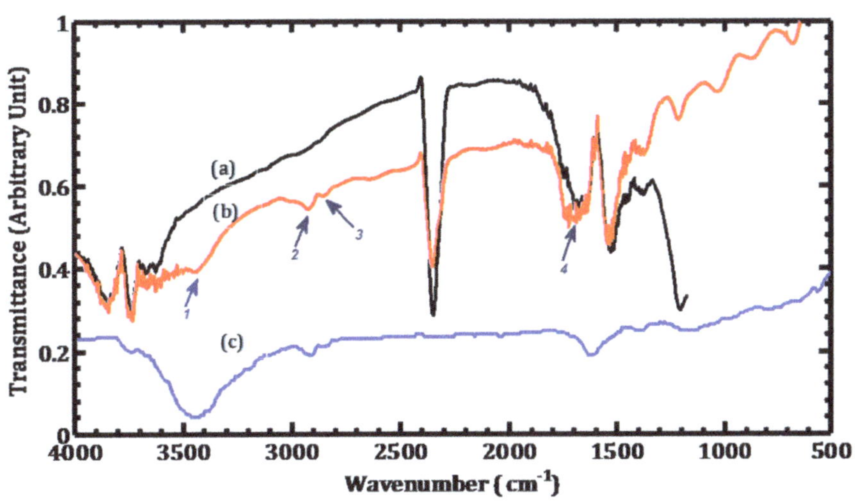

Figure 20. Averaged deposition energy in unit volume

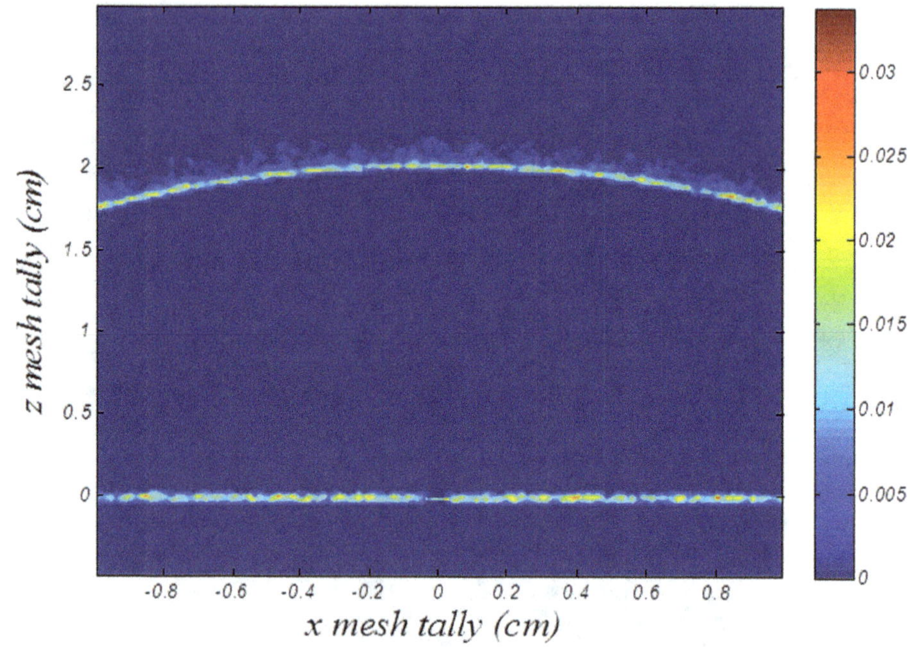

Figure 21. Number of electrons passesper unit area and per unit time (Flux)

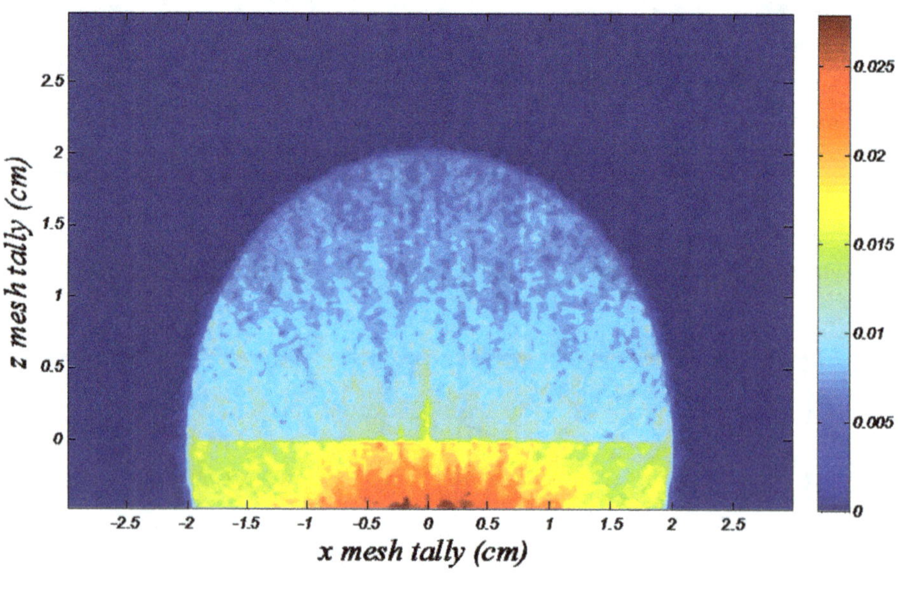

The energy spectrum of beta source which is been simulated is presented in Figure 22. The execution time (CTME) is considered 15 min and rectangular mesh generator is used in Tally Mesh. Also we calculate the flux of the β-rays that pass though circular hole by using the MCNPX which varied by change the distance of CNPs from β source.

Figure 22. Energy spectrum of 90-Sr&90-Y beta source
Prause, 2006.

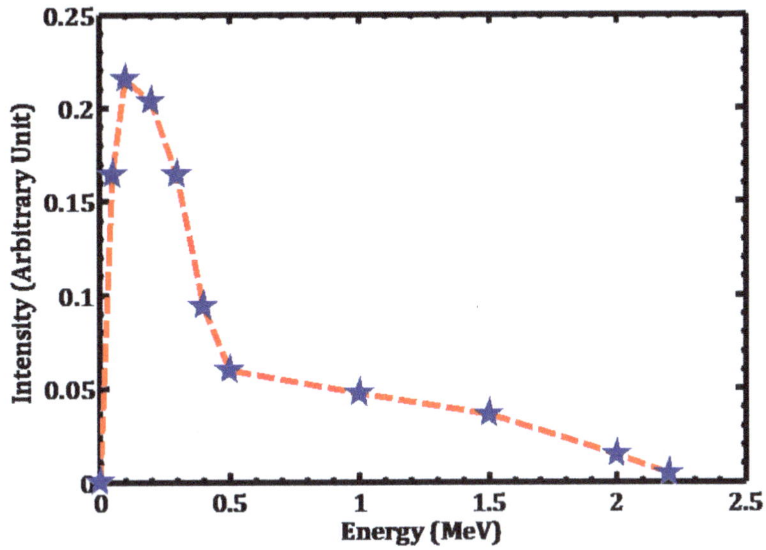

Semiconducting CNPs

Semiconducting CNP has been especially used in high speed nanoelectronics and other various application fields (Tans, Verschueren, & Dekker, 1998). Electrical properties of carbon nanoparticles strongly depend on the geometry of CNPs for example, their band gap varies between 0.4 to 6 eV by the size (Afrin et al., 2012). In this work, the distance between the electrodes has been changed and then semiconducting CNPs are synthesized. It is noted that different distance of electrodes is equivalent to the different length of bulk CNPs which affects their electrical properties by change in the size. Also it is investigated that, the threshold voltage is increased by increasing the bulk CNPs length (D) as shown in Figure 23.

In many literatures P type semiconductor features has been reported for semiconducting CNPs (Derycke, Martel, Appenzeller, & Avouris, 2001). The reason of this fact was not well-known until the year 2000. Recently some expressions have been presented to explain this behavior. For example generation of the functionality due to adsorption of Oxygen in CNPs is one of them that led to acceptor doping behaviour on CNPs(McEuen, Fuhrer, & Park, 2002). However N type CNPs can be generated in controlled conditions by the doping of Nitrogen as donor (Zhou, Kong, Yenilmez, & Dai, 2000).

Some significant effects on P type semiconductors can be introduced by radiation exposure such as generation of electron-hole pairs, recombination of electrons and holes, transport of holes, hole trapping and generation of hole in surface of semiconductors (Pejovi et al., 2013). The fraction of β-rays energy which is converted into electron–hole pairs depends on the property of CNPs. Additionally the mechanism of electron-hole pairs creation shows the extremely efficient electron-hole pair generation in CNP based photodiodes (Gabor, Zhong, Bosnick, Park, & McEuen, 2009). It is important to mention that the generation of electron-hole pairs in depletion region is the basic concept of semiconductor radiation detectors (Knoll, 2000). So electron-hole generation is the main effect of radiation induced current in CNPs additionally it is clear that holes are the majority carriers on P type CNP based devices(Avouris,

Figure 23. I-V characteristic curves in different channel length

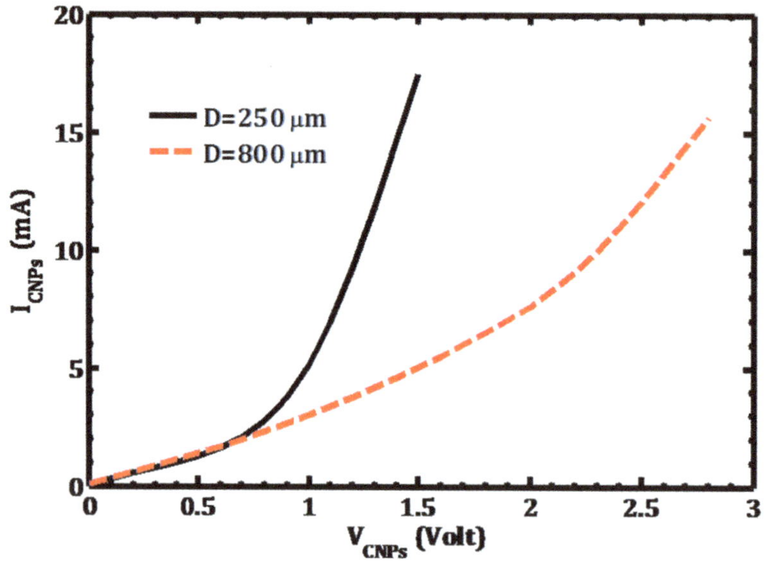

2002). Therefore because of metallic contacts in both sides of CNP the electrical behaviour of CNP in the form of back-to-back Schottky diode in the presence of radiation is analysed as shown in Figure 24.

In order to compare the results from different length of bulk CNPs, experiment is repeated several times and same behavior on the I-V characteristics is observe and sensible current variation is noticed. Also the obtained I-V characteristics are compared (Figures 24 and 25) and CNP characteristic distinction under electrode distance variation is reported. It seems that I-V curves have a saturation region which means that current cannot be increased by voltage after the specific value of applied voltage on

Figure 24. Obtained I-V curves for semiconducting CNPs under irradiation and without irradiation

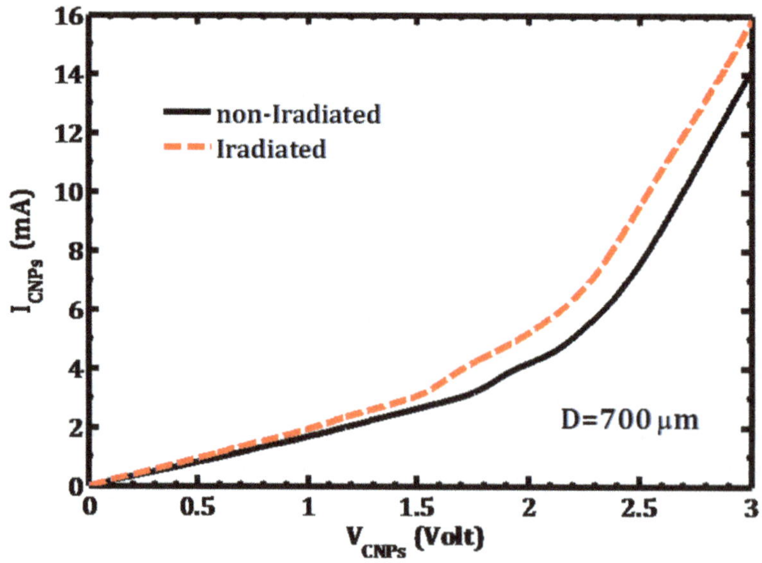

Figure 25. I-V characteristic curves for CNPs under irradiation and without irradiation

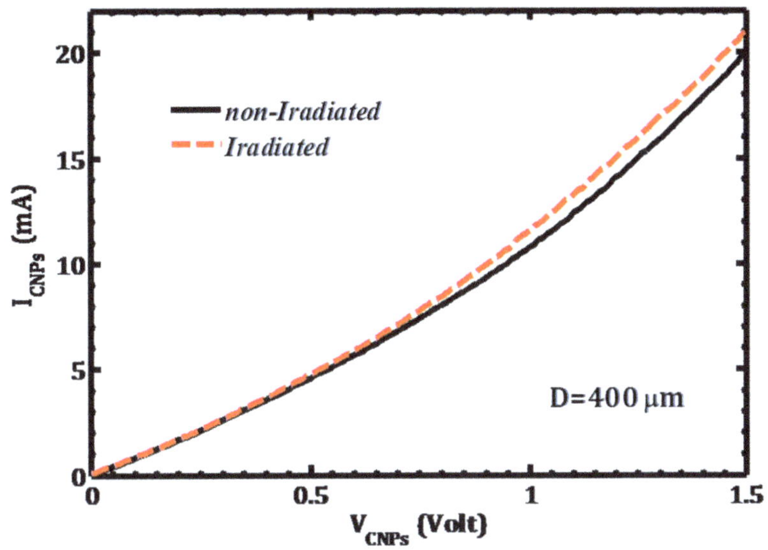

CNPs. That can be understood in the form of depletion region length on MSM systems as well. As we mentioned before the reverse bias is considered as constant and so greater than the forward bias voltage that this is caused to saturate of increasing current in higher voltage with presence of β-rays.

In the next stage the I-V curves in several incident β-rays fluxes (different distance of CNPs from the source) are measured and I-V characteristics are obtained in a fixed applied voltage (V=1.4 Volt,) as shown in Figure 26 for the case by the length of D=400 µm. Also the current of 16.7 mA for all other voltages on non-irradiated condition is achieved.

In the nearest distance from the radiation source (biggest value of flux) the current slightly decreases compared to the non-irradiated conditions (Figure 26). It should be noted that the scattering of electrons also will play an important role in semiconducting CNPs I-V characteristics and it leads to the current reduction in this device as well. Furthermore by flux reduction the current is amplified in the other words the electron-hole pair generation is increased. This process will be continuing until the specific flux and finally the current returns to the non-irradiated value by increasing the distance of CNPs from source.

Polymer/MWCNTs Composite as Neutron Detector

Polyethylene and multi walled Carbon nanotubes (PE/MWCNTs) composite had been fabricated by arc discharge method and it characterized with SEM and IR spectroscopy. Also the I-V curve obtained and investigated. In order the modeling of devise the thermionic emission theorem was handled and the results compared with measurements.

In this case, the MSM system is modelled by two Schottky barriers back-to-back in series with a resistor R, resulting from the CNPs/PE composite as shown in Figure 27.

It is clear that a metal-semiconductor device is described as a diode. While Ohmic contacts have a linear current–voltage (I–V) characteristic and Schottky contacts vice versa. Additionally in the literature Schottky contacts of MSM system are named by back-to-back Schottky diodes (Adenilson et al., 2012). One of the current conduction mechanism is thermionic emission (Sze, 1981) and it is always associated

Figure 26. Variations of current with incident flux

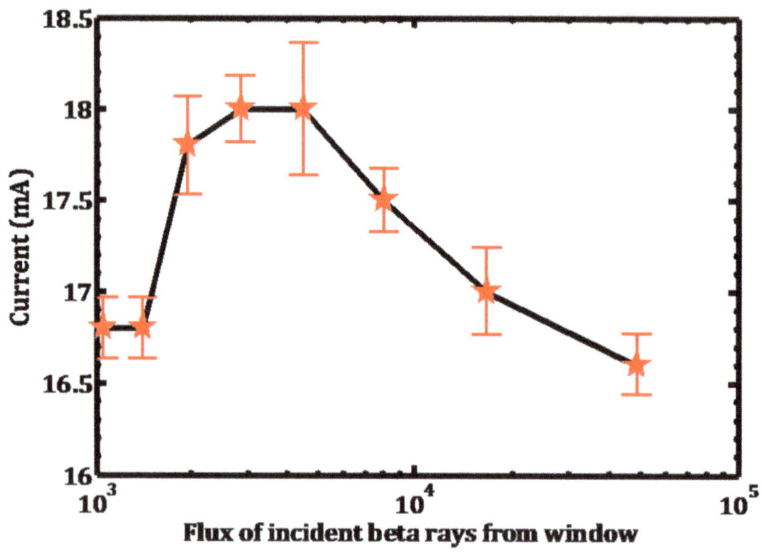

Figure 27. a) Energy band diagram of Back-to-Back Schottky diodes with a resistor and b) Proposed circuit for use to modeling

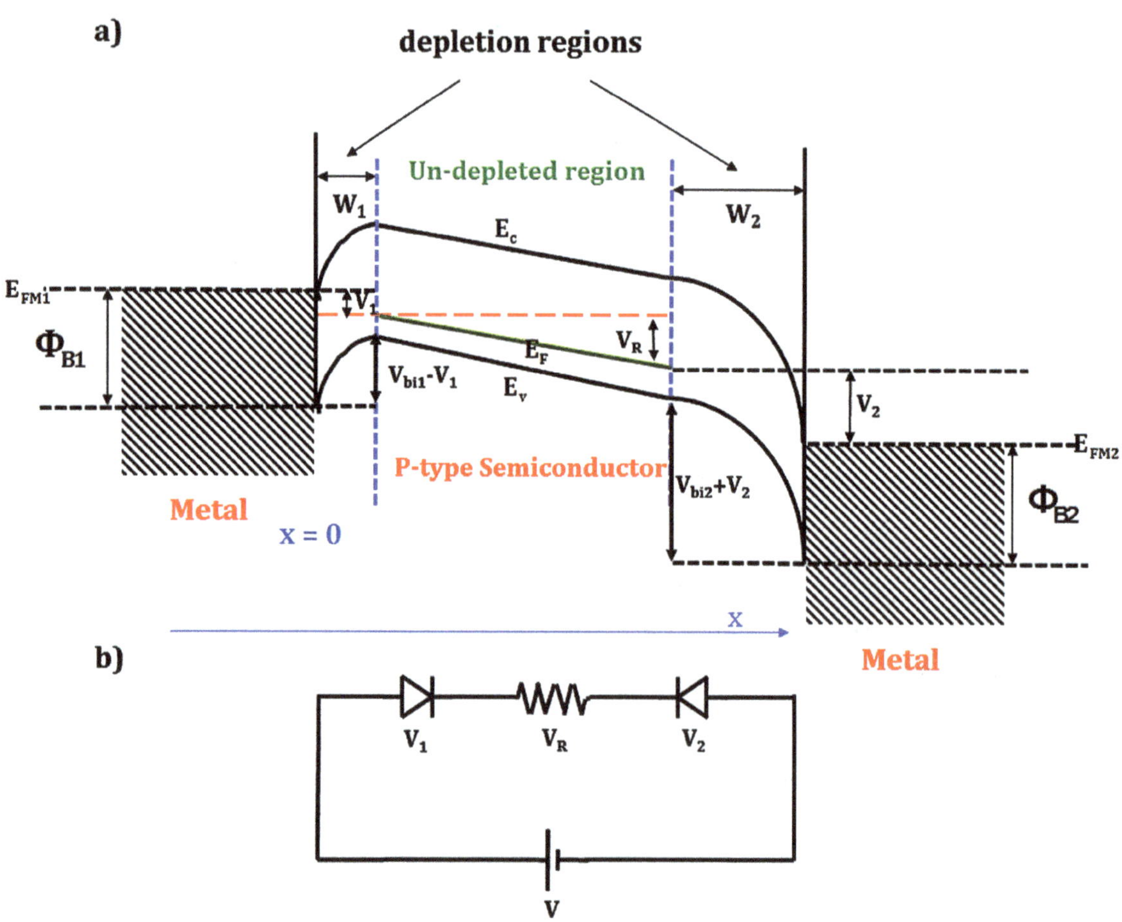

with a potential barrier. Using the thermionic emission theory, the current at the specific temperature T for barriers 1 and 2 are written as (Sze, 1981):

$$I_1 = I_{s1}\left[\exp\left(\frac{qV_1}{k_B T}\right) - 1\right] \tag{1}$$

$$I_2 = -I_{s2}\left[\exp\left(-\frac{qV_2}{k_B T}\right) - 1\right] \tag{2}$$

where

$$I_{s1,s2} = AA^*T^2 \exp\left(\frac{q\Phi_{B1,B2}}{k_B T}\right),$$

A^* is the Richardson constant,
A is the area of contact cross section,
Φ_{B1} and Φ_{B2} are the Schottky barrier heights (Figure 27),
q is electron elementary charge and
kB is the Boltzmann constant.

According to the current continuous theory the total current I across the devise is obtained by I=I1=I2=IR and setting up V=V1+V2+VR. So the I-V relationship of MSM system with two Schottky contacts is given by:

$$I(V) \cong \frac{2I_{s1}I_{s2}\sinh\frac{qV}{2k_B T}}{\left(I_{s1} + \frac{qR}{k_B T}I_{s1}I_{s2}\right)\exp\left(\frac{qV}{2k_B T}\right) + I_{s2}\exp\left(\frac{-qV}{2k_B T}\right)} \tag{3}$$

where, R is the resistance of bulk MWCNTs/PE composite between two identical Cu electrodes. It is important to mention that in the analytical modeling the exp(-qRI/kBT) term is employed by approximation form while the first order of total current I is kept. Also by the voltage dependence of barrier heights which are given by (Adenilson et al., 2012):

$$\Phi_{B1} = \Phi_{B01} + V_1\left(\frac{1}{n} - 1\right) \tag{4a}$$

$$\Phi_{B2} = \Phi_{B02} + V_2\left(\frac{1}{n} - 1\right) \tag{4b}$$

where Φ_{B01} and Φ_{B02} are values of the barrier in an ideal Schottky junction and V1 and V2 are applied voltage on barriers 1 and 2, respectively. As shown in Figure 28 there is a good agreement between measurement and modeling results.

Furthermore, the temperature effect on electrical behaviour, such as I-V characteristic curve, has been studied. As a result an intense variation in I-V characteristic for different temperatures is shown in Figure 5, where the temperature difference is 10 K.

This behavior of MSM device can be described by temperature dependence of ideal Schottky barrier height. For the temperature dependence of barrier height, inhomogeneities are assumed to model the interface of Schottky diode by Gaussian distribution. Therefore, it is convenient to assume the existence of a Gaussian distribution of the barrier heights with mean $\bar{\Phi}_{B01}$ and standard deviation σ_0 over the contact area. So we can write this distribution in the form of (Werner & Güttler, 1991):

$$P(\Phi_{B01}) = \frac{1}{\sigma_0\sqrt{2\pi}} \exp\left(-\frac{\left(\Phi_{B01} - \bar{\Phi}_{B01}\right)^2}{2\sigma_0^2}\right) \quad (5)$$

where, the term of $\frac{1}{\sigma_0\sqrt{2\pi}}$ is the normalizing constant. As a result we obtained for barrier height by (Hidayet & Enise, 2005; von Wenckstern et al., 2006; Werner & Güttler, 1991):

$$\Phi_{B01} = (\Phi_{B02}) = \bar{\Phi}_{B01} - \frac{q\sigma_0^2}{2k_B T} \quad (6)$$

Figure 28. I-V characteristics curve obtained from model and measurement

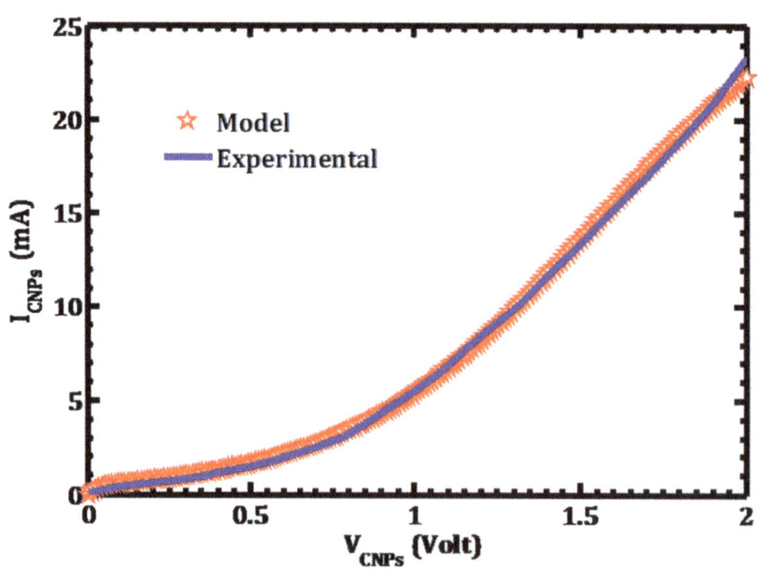

where, the $\overline{\Phi}_{B01}$ is mean value of barrier height and σ_0 is standard deviation of distribution. Also we set to equal for barriers in both contacts. It is important to mention that if the current direction is exchange then the reverse bias V2 would be replaced with the forward bias V1 and so a symmetrical I-V curve is obtained by two identical metals. Additionally another approach for analysis of MSM system has been improved in ref. (Elhadidy, Sikula, & Franc, 2012) that the presented voltage distribution along the MSM system confirmed the our plan in selection of constant reverse voltage and biggest contribution of applied voltage V. For example the reverse bias voltage V2 in Figure 28 is obtained by 1.482 Volts.

We are obtained I-V curve in two temperatures (300 K and 310 K) by applying this equation in modeling of I-V-T measurements as shown in Figure 30. It can be seen that I-V curves in Figure 29 and Figure 30 are similar in two temperatures obtained by experimental and modeling results.

It is important to mention that the curves are also reversible, meaning that by removing the heating source, they will go back to pre-contact condition. Also changes in electrical properties because of the human expiration temperature are obtained experimentally.

Finally, the effect of the On-Line Am-Be neutron irradiation on I-V characteristics (of the described MSM system) had been studied in the different neutron exposure. Results obtained in the different times after the starting of the exposure as shown in Figure 31.

It can be seen that the current pulses related to neutrons are observed in the voltage of 1.8 V. These pulses extracted from proton recoils due to neutron scattering from PE.

CONCLUSION

Conventional semiconductor (such as Si and Ge) based detectors are common devices in radiation detection. The operation of these detectors is similar to the gas-filled detectors. In addition in the case of semiconductor detectors, electron-hole pairs play critical role in signal generation. One of the major

Figure 29. I-V characteristics curve obtained in different temperatures

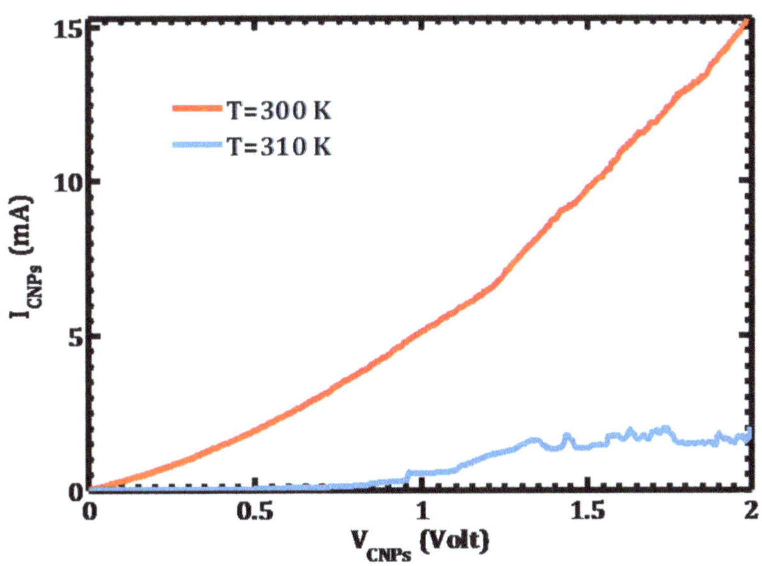

Figure 30. Modelled I-V characteristics curve obtained in different temperatures

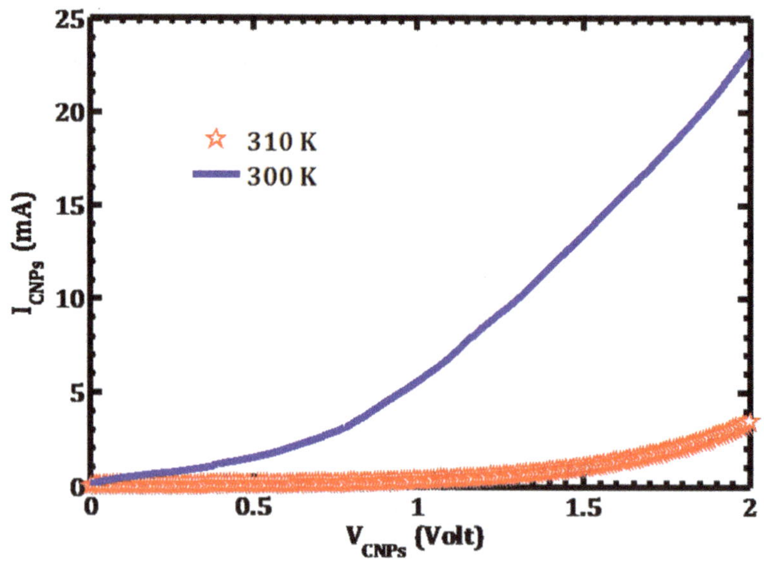

Figure 31. I-V characteristics in different neutron irradiation conditions

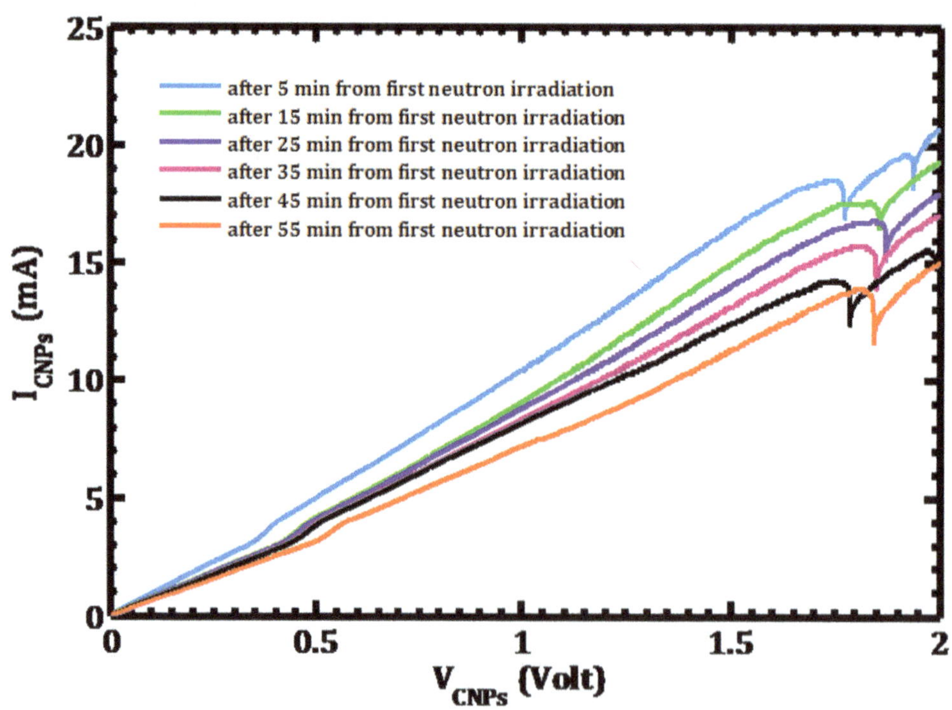

advantages of semiconductor radiation detectors is their high energy resolution. For example high purity germanium HPGe semiconductor detector is used in gamma spectroscopy, but, high cost and difficulties in maintenance make it ineffective device for common applications.

Currently CNPs with outstanding electrical and mechanical properties engages in nanotechnology applications. We demonstrate the effect of β-ray exposure on electrical property of metallic and semiconducting CNPs. In order to characterize, SEM and IR analysis are used for synthesized CNPs. Then the devise responses to β-rays in metallic and semiconducting junctions are investigated. Also thermionic emission concept on our prototype devise modelling is employed. In order to prove that β-rays can deposit energy on the electrodes and CNPs Monte Carlo simulation is taken and also flux of the β-rays is calculated by MCNPX code. In semiconducting CNPs due to generation of electron-hole pairs the current is increased. Also the carbon based nanocomposite has introduced as fast neutron detector. Therefore we have investigated I-V characteristic curves in on-line case. Studies show that electrical characteristics of our prototype device change in on-line measurements.

REFERENCES

Adamiec, G., Iñiguez, M. P., Lorente, A., Voytchev, M., & Gallego, E. (2004). Response of a silicon PIN photodiode to an Am-Be neutron source. *Nuclear Instruments & Methods in Physics Research. Section A, Accelerators, Spectrometers, Detectors and Associated Equipment*, *534*(3), 544–550. doi:10.1016/j.nima.2004.06.145

Adenilson, J. C., Cleber, A. A., Olivia, M. B., Luana, S. A., Eric, P. B., & Edson, R. L. (2012). Back-to-back Schottky diodes: The generalization of the diode theory in analysis and extraction of electrical parameters of nanodevices. *Journal of Physics Condensed Matter*, *24*(22), 225303. doi:10.1088/0953-8984/24/22/225303 PMID:22556197

Afrin, R., Khaliq, J., Islam, M., Gul, I. H., Bhatti, A. S., & Manzoor, U. (2012). Synthesis of multiwalled carbon nanotube-based infrared radiation detector. *Sensors and Actuators. A, Physical*, *187*(0), 73–78. doi:10.1016/j.sna.2012.08.028

Agosteo, S., D'Angelo, G., Fazzi, A., Para, A. F., Pola, A., Ventura, L., & Zotto, P. (2005). Performance of a neutron spectrometer based on a PIN diode. *Radiation Protection Dosimetry*, *116*(1-4), 180–184. doi:10.1093/rpd/nci102 PMID:16604623

Arora, N., & Sharma, N. N. (2014). Arc discharge synthesis of carbon nanotubes: Comprehensive review. *Diamond and Related Materials*, *50*(0), 135–150. doi:10.1016/j.diamond.2014.10.001

Avouris, P. (2002). Molecular electronics with carbon nanotubes. *Accounts of Chemical Research*, *35*(12), 1026–1034. doi:10.1021/ar010152e PMID:12484790

Avouris, P., Chen, Z., & Perebeinos, V. (2007). Carbon-based electronics. *Nat Nano*, *2*(10), 605–615. doi:10.1038/nnano.2007.300 PMID:18654384

Babcock, R. (1965). Radiation Damage in SiC. *IEEE Transactions on Nuclear Science*, *12*(6), 43–47. doi:10.1109/TNS.1965.4323922

Benjamin, P. W., Kemshall, C. D., & Redfearn, J. (1968). A high resolution spherical proportional counter. *Nuclear Instruments and Methods*, *59*(1), 77–85. doi:10.1016/0029-554X(68)90347-9

Bertuccio, G., Casiraghi, R., Cetronio, A., Lanzieri, C., & Nava, F. (2004). Silicon carbide for high resolution X-ray detectors operating up to 100°C. *Nuclear Instruments & Methods in Physics Research. Section A, Accelerators, Spectrometers, Detectors and Associated Equipment*, *522*(3), 413–419. doi:10.1016/j.nima.2003.11.413

Brooks, F. D., & Klein, H. (2002). Neutron spectrometry—historical review and present status. *Nuclear Instruments & Methods in Physics Research. Section A, Accelerators, Spectrometers, Detectors and Associated Equipment*, *476*(1–2), 1–11. doi:10.1016/S0168-9002(01)01378-X

Chadwick, J. (1932). The existence of a neutron. *Proceedings of the Royal Society of London A: Mathematical, Physical and Engineering Sciences, 136*(830), 692-708. doi:10.1098/rspa.1932.0112

Cunha, C., Panseri, S., Iannazzo, D., Piperno, A., Pistone, A., Fazio, M., & Galvagno, S. et al. (2012). Hybrid composites made of multiwalled carbon nanotubes functionalized with Fe 3 O 4 nanoparticles for tissue engineering applications. *Nanotechnology*, *23*(46), 465102. doi:10.1088/0957-4484/23/46/465102 PMID:23093179

Derycke, V., Martel, R., Appenzeller, J., & Avouris, P. (2001). Carbon nanotube inter- and intramolecular logic gates. *Nano Letters*, *1*(9), 453–456. doi:10.1021/nl015606f

Dulloo, A. R., Ruddy, F. H., Seidel, J. G., Davison, C., Flinchbaugh, T., & Daubenspeck, T. (1999). Simultaneous measurement of neutron and gamma-ray radiation levels from a TRIGA reactor core using silicon carbide semiconductor detectors. *IEEE Transactions on Nuclear Science*, *46*(3), 275–279. doi:10.1109/23.775527

Elhadidy, H., Sikula, J., & Franc, J. (2012). Symmetrical current–voltage characteristic of a metal–semiconductor–metal structure of Schottky contacts and parameter retrieval of a CdTe structure. *Semiconductor Science and Technology*, *27*(1), 015006. doi:10.1088/0268-1242/27/1/015006

Fazzi, A., Agosteo, S., Pola, A., Varoli, V., & Zotto, P. (2003). *Pulse discrimination between recoil protons and secondary electrons for a silicon diode based neutron spectrometer.* Paper presented at the Nuclear Science Symposium Conference Record. doi:10.1109/NSSMIC.2003.1351811

Feather, N. (1932). The collisions of neutrons with nitrogen nuclei. *Proceedings of the Royal Society of London A: Mathematical, Physical and Engineering Sciences, 136*(830), 709-727. doi:10.1098/rspa.1932.0113

Gabor, N. M., Zhong, Z., Bosnick, K., Park, J., & McEuen, P. L. (2009). Extremely efficient multiple electron-hole pair generation in carbon nanotube photodiodes. *Science*, *325*(5946), 1367–1371. doi:10.1126/science.1176112 PMID:19745146

Goyanes, S., Rubiolo, G. R., Salazar, A., Jimeno, A., Corcuera, M. A., & Mondragon, I. (2007). Carboxylation treatment of multiwalled carbon nanotubes monitored by infrared and ultraviolet spectroscopies and scanning probe microscopy. *Diamond and Related Materials, 16*(2), 412-417. doi: 10.1016/j.diamond.2006.08.021

Hidayet, C., & Enise, A. (2005). Temperature dependence of electrical parameters of the Au/n-InP Schottky barrier diodes. *Semiconductor Science and Technology*, *20*(6), 625–631. doi:10.1088/0268-1242/20/6/025

Hosono, Y., Sjafruddin, , Iguchi, T., & Nakazawa, M. (1995). Fast neutron detector using PIN-type silicon photodiode. *Nuclear Instruments & Methods in Physics Research. Section A, Accelerators, Spectrometers, Detectors and Associated Equipment, 361*(3), 554–557. doi:10.1016/0168-9002(95)00206-5

ISO 8529-1. (2001). *Reference Neutron Radiations. Part1: Characteristics and methods of production*. International Organization for Standardization ISO8529-1.

Kasani, H., Taghi Ahmadi, M., Khoda-bakhsh, R., RezaeiOchbelagh, D., & Ismail, R. (2016). Influences of Sr-90 beta-ray irradiation on electrical characteristics of carbon nanoparticles. *Journal of Applied Physics, 119*(12), 124510. doi:10.1063/1.4944901

Knoll, G. F. (2000). *Radiation Detection and Measurement* (3rd ed.). Wiley.

Lehman, J. H., Terrones, M., Mansfield, E., Hurst, K. E., & Meunier, V. (2011). Evaluating the characteristics of multiwall carbon nanotubes. *Carbon, 49*(8), 2581–2602. doi:10.1016/j.carbon.2011.03.028

Lo Giudice, A., Fasolo, F., Durisi, E., Manfredotti, C., Vittone, E., Fizzotti, F., & Rosi, G. et al. (2007). Performances of 4H-SiC Schottky diodes as neutron detectors. *Nuclear Instruments & Methods in Physics Research. Section A, Accelerators, Spectrometers, Detectors and Associated Equipment, 583*(1), 177–180. doi:10.1016/j.nima.2007.08.241

McEuen, P. L., Fuhrer, M. S., & Park, H. (2002). Single-walled carbon nanotube electronics. *Nanotechnology. IEEE Transactions on, 1*(1), 78–85. doi:10.1109/TNANO.2002.1005429

(2013). Pejovi, #x107, M., #x107, Ciraj-Bjelac, O., Kova, . . ., G. (2013). Sensitivity of P-Channel MOSFET to X- and Gamma-Ray Irradiation. *International Journal of Photoenergy, 6*. doi:10.1155/2013/158403

Prause, C. A. (2006). *External detection and measurement of inhaled radionuclides using thermoluminescent dosimeters.* (Master's thesis). Texas A&M University.

Ruddy, F. H., Dulloo, A. R., Seidel, J. G., Seshadri, S., & Rowland, L. B. (1998). Development of a silicon carbide radiation detector. *Nuclear Science. IEEE Transactions on, 45*(3), 536–541. doi:10.1109/23.682444

Stobinski, L., Lesiak, B., Kövér, L., Tóth, J., Biniak, S., Trykowski, G., & Judek, J. (2010). Multiwall carbon nanotubes purification and oxidation by nitric acid studied by the FTIR and electron spectroscopy methods. *Journal of Alloys and Compounds, 501*(1), 77–84. doi:10.1016/j.jallcom.2010.04.032

Sydlik, S. A., Lee, J.-H., Walish, J. J., Thomas, E. L., & Swager, T. M. (2013). Epoxy functionalized multi-walled carbon nanotubes for improved adhesives. *Carbon, 59*(0), 109–120. doi:10.1016/j.carbon.2013.02.061

Sze, S. M. (1981). *Physics of Semiconductor Devices*. New York: Wiley.

Tameev, A. R., Jiménez, L. L., Pereshivko, L. Y., Rychwalski, R. W., & Vannikov, A. V. (2007). Charge carrier mobility in films of carbon-nanotube- polymer composites. *Journal of Physics: Conference Series, 61*(1), 1152–1156. doi:10.1088/1742-6596/61/1/228

Tans, S. J., Verschueren, A. R. M., & Dekker, C. (1998). Room-temperature transistor based on a single carbon nanotube. *Nature, 393*(6680), 49–52. doi:10.1038/29954 von Wenckstern, H., Biehne, G., Rahman, R. A., Hochmuth, H., Lorenz, M., & Grundmann, M. (2006). Mean barrier height of Pd Schottky contacts on ZnO thin films. *Applied Physics Letters, 88*(9), 092102. doi:10.1063/1.2180445

Voytchev, M., Iñiguez, M. P., Méndez, R., Mañanes, A., Rodríguez, L. R., & Barquero, R. (2003). Neutron detection with a silicon PIN photodiode and 6LiF converter. *Nuclear Instruments & Methods in Physics Research. Section A, Accelerators, Spectrometers, Detectors and Associated Equipment, 512*(3), 546–552. doi:10.1016/S0168-9002(03)02013-8

Werner, J. H., & Güttler, H. H. (1991). Barrier inhomogeneities at Schottky contacts. *Journal of Applied Physics, 69*(3), 1522-1533. doi:10.1063/1.347243

Wu, J., Li, M., Jiang, Y., Li, J., Zhang, Y., Gao, H., & Lei, J. et al. (2015). Performance of a 4H-SiC Schottky diode as a compact sized detector for neutron pulse form measurements. *Nuclear Instruments & Methods in Physics Research. Section A, Accelerators, Spectrometers, Detectors and Associated Equipment*, *771*, 17–20. doi:10.1016/j.nima.2014.10.032

Yamamoto, H., Hatakeyama, S., Norimura, T., & Tsuchiya, T. (1984). A radiation detector fabricated from silicon photodiode. *Radioisotopes*, *33*(12), 864–866. doi:10.3769/radioisotopes.33.12_864 PMID:6528065

Zaki Dizaji, H., Kakavand, T., & Abbasi Davani, F. (2014). Spectrometry and dosimetry of fast neutrons using pin diode detectors. *Nuclear Instruments & Methods in Physics Research. Section A, Accelerators, Spectrometers, Detectors and Associated Equipment*, *741*, 84–87. doi:10.1016/j.nima.2013.12.018

Zaragoza-Contreras, E. A., Lozano-Rodríguez, E. D., Román-Aguirre, M., Antunez-Flores, W., Hernández-Escobar, C. A., Flores-Gallardo, S. G., & Aguilar-Elguezabal, A. (2009). Evidence of multi-walled carbon nanotube fragmentation induced by sonication during nanotube encapsulation via bulk-suspension polymerization. *Micron (Oxford, England)*, *40*(5–6), 621–627. doi:10.1016/j.micron.2009.02.007 PMID:19299150

Zhou, C., Kong, J., Yenilmez, E., & Dai, H. (2000). Modulated chemical doping of individual carbon nanotubes. *Science*, *290*(5496), 1552–1555. doi:10.1126/science.290.5496.1552 PMID:11090348

Chapter 16
Sensors and Amplifiers:
Sensor Output Signal Amplification Systems

Amir Fathi
Urmia University, Iran

Sarkis Azizian
K. N. Toosi University of Technology, Iran

Nastaran Sharifan
Tehran University, Iran

ABSTRACT

Sensors are electrical-mechanical elements which are the interface between environment and electrical systems. The input of sensors is characteristics of the environment for example temperature, pressure and etc. and their output is a small electric voltage or current. Their job is to convert environment characteristics to an electric voltage or current at their outputs. Since the output current or voltage is very small, it must be amplified in order to be suitable for use in electronic systems. In this chapter we completely explain the design procedure and characteristics of sensor amplifiers. The important parameters of sensor amplifiers are input and output resistance, gain, unity gain bandwidth and etc. One of the most important characteristics of amplifiers is the linearity of amplification in a way that it must have uniformity for all amplitude voltages or currents in all frequencies of the bandwidth. For this purpose, first the operational amplifier is completely discussed, then the linearity of feedback operation will be explained.

INTRODUCTION

The signals which are obtained from the environment by sensors have often small amplitudes and need to be amplified for better analysis and interpretation. This job is done by amplifiers which are transistor based circuits.

DOI: 10.4018/978-1-5225-0736-9.ch016

One of the important evolutions happened in the middle of 20-th century, was the invention of p-n junction diodes which is followed by the invention of bipolar junction transistors. Before that, the diodes were produced from vacuum-tubes and nobody could imagine a good future for these components. Along with the scientific developments, the semiconducting properties of carbon, silicon and germanium were studied and the result was the invention of p-n junction diode composed of P-type and N-type semiconductors. The developments continued and reached to their turning point which led to the introduction of bipolar transistors. With the technology advancements, the mass production of transistors started and an industrial revolution happened. The computers, cell phones and many other devices came to the market and the use of transistors and their amplifying properties continued to grow more and more. With the fabrication of unipolar transistors which are known as FETs and MOSFETs, the technology extended its development. The base of today's amplifying circuits is still upon transistors and their compounds such as Op-Amps, Comparators, and etc. The only difference is that everything is now being fabricated in a small area of a silicon wafer which is a part of big industry called integrated circuit design technology.

Today different ICs are available in markets with very chip prices and each of them is doing a desired job. In order to have a sufficient knowledge about the behavior of different ICs and their comprising circuits, at first step, we have to collect useful information about the diodes, transistors and their amplifying properties.

This chapter is written to give useful information about the semiconducting based amplifiers which are largely used in the design of electronic circuits and devices. At first, the semiconductors are introduced. Then, their derivatives such as diode, BJT and MOSFET with their complete analysis are introduced. Then the idea will be expanded for multistage amplifiers which leads to OpAmps. Along with the introduction of each semiconductor based component, some of the applications of the component in conjunction with the sensors will be discussed.

WHAT IS A SEMICONDUCTOR?

Different States of Materials for Electric Conduction

Based upon the capability of current conduction, the materials are categorized in 3 groups of conductors, insulators and semiconductors.

Conductors

In electronics, a conductor is a kind of material which permits the flow of electrical current along its length. In elements like copper or aluminum which are the basic substance of electric wires, the electrons are the ambulant charged particles. As illustrated in the basic electric circuit of Figure 1, the flow of electrons across the wire which is the conducting medium, enlightens the lamp.

Insulators

An insulator is an object whose internal electric charges even in presence of external electric field cannot flow freely. Thus, its nominal property is its high resistance. In general, the value of resistivity for insulators is much higher than that of semiconductors and conductors.

Figure 1. Basic electrical circuit and flow of electrons in opposite direction of current

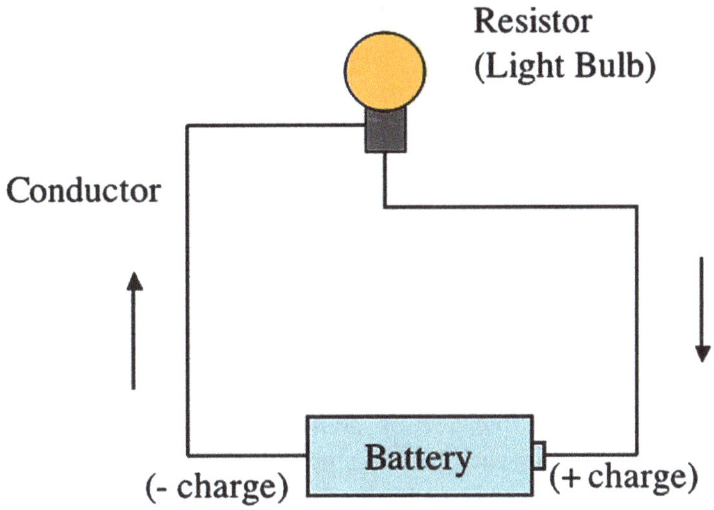

Semiconductors

A semiconductor material treats like an insulator naturally, but in certain conditions such as adding some materials it can show an electrical conductivity value between that of a conductor and an insulator (Mehta, 2008). As shown in Figure 2, because of the unique arrangement of crystal lattice, silicon and germanium are chosen as the most popular elements in the production of semiconducting devices.

Figure 2. Crystal lattice arrangement of silicon atoms

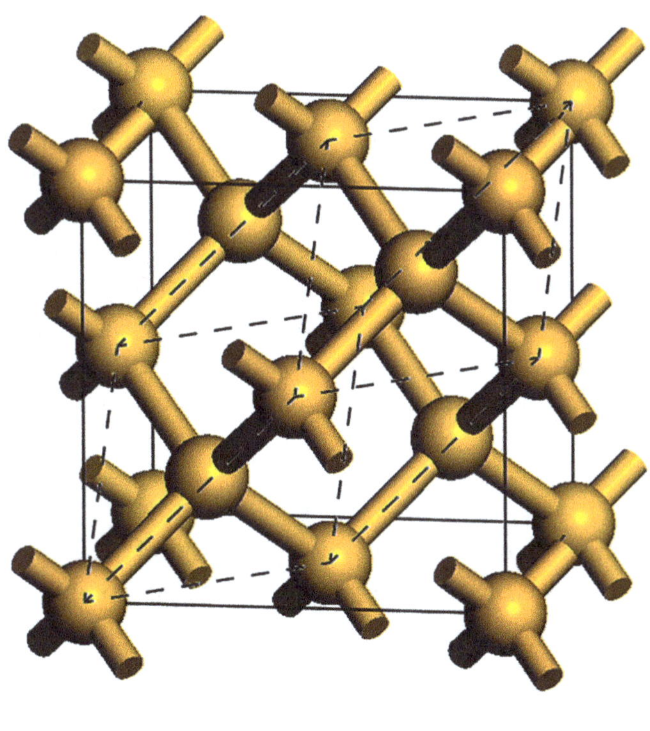

The electrical conductivity of a semiconductor differs with environmental conditions such as temperature, in which unlike that of a metal, the conductivity increases with temperature rise. In addition, semiconductor devices have the interesting property of ferrying current across themselves much easier in one direction than the other. Therefore, their resistance has a variable value respect to the current amount and polarity. The travel of free electrons and holes through a semiconductor which are together recognized as charge carriers makes the current conductivity possible.

In normal conditions, the number of free electrons in single crystal lattice of silicon has small value and actually, it is treated like an insulator. But its electrical characteristics can be changed by supervised addition of impurities which is another interesting property of a semiconductor element. By means of the added impurity atoms to the crystal lattice of a semiconducting material, the number of charge carriers will significantly be increased. This phenomenon which is known as doping, directly affects the resistive behavior of a semiconductor. A doped semiconductor which comprises free holes at most it is called "P-type", and when it includes free electrons at most it is known as "N-type" (Neamen, 2003).

In an innate Si or Ge semiconductor, each nucleus employs its four valence bond electrons to form four covalent bonds with its neighbors as shown in Figure 2. Because there are no more electrons or holes, in a normal state the number of present electrons and holes at any given time are always equal together.

N-Type Semiconductor

When a pentavalent element is injected as an impurity to an innate semiconductor, it is said to be an n-type semiconductor. That's why pentavalent materials such as phosphorus, arsenic and antimony are known as donor impurities (Hook, & Hall, 2001).

For example, consider that phosphorus is to be added to silicon as shown in Figure 3. The process starts with heating silicon up to temperature near 1000º C. Then the phosphorus in gas state flows above the warm, nearly melting silicon. During this situation phosphorus diffuses into the silicon. A phosphorus atom has 5 valence electrons against 4 electrons in valance bond of silicon. Four electrons of each phosphorus atom form 4 covalent bonds with the 4 neighboring silicon atoms whilst the fifth valence electron cannot build any covalent bond formation. Therefore, the extra electron of phosphorus atom won't be associated in the formation of covalent bonds and will be free to move.

It must be mentioned that total electric charge of an N-type semiconductor is neutral. Although an N-type semiconductor has large number of free electrons, just keep it in mind that they are obtained by the pentavalent atoms which are electrically neutral.

As illustrated in Figure 4, if an N-type is set in the path of an external voltage, the free electrons move towards the positive terminal whilst the holes migrate to the negative terminal of applied voltage.

Because in an n-type semiconductor the population of free electrons is more than that of the holes, these electrons are called majority carriers and the holes are considered as minority carriers. To cut a long story short, in an n-type semiconductor the conduction mainly comes from the motion of free electrons.

P-Type Semiconductor

In contrast with N-type semiconductor, when a trivalent impurity is injected as impurity to a pure semiconductor, then it is said to be a P-type semiconductor. In this process, one of the atoms in the semiconductor lattice is substituted by an element with three valence electrons, such as a Group 3 element

Figure 3. Formation of an N-type semiconductor

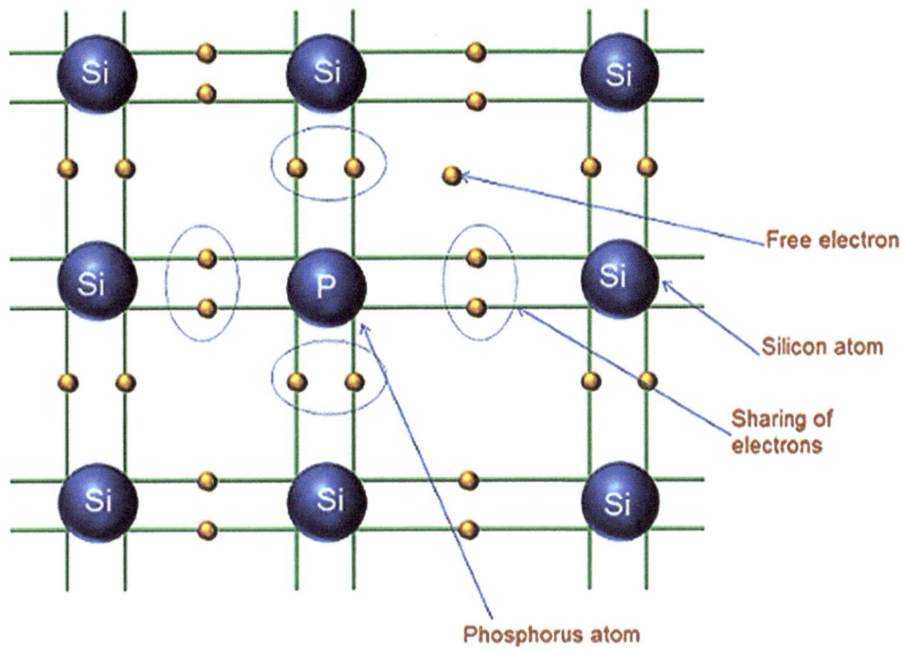

Figure 4. Applying external voltage to N-type semiconductor

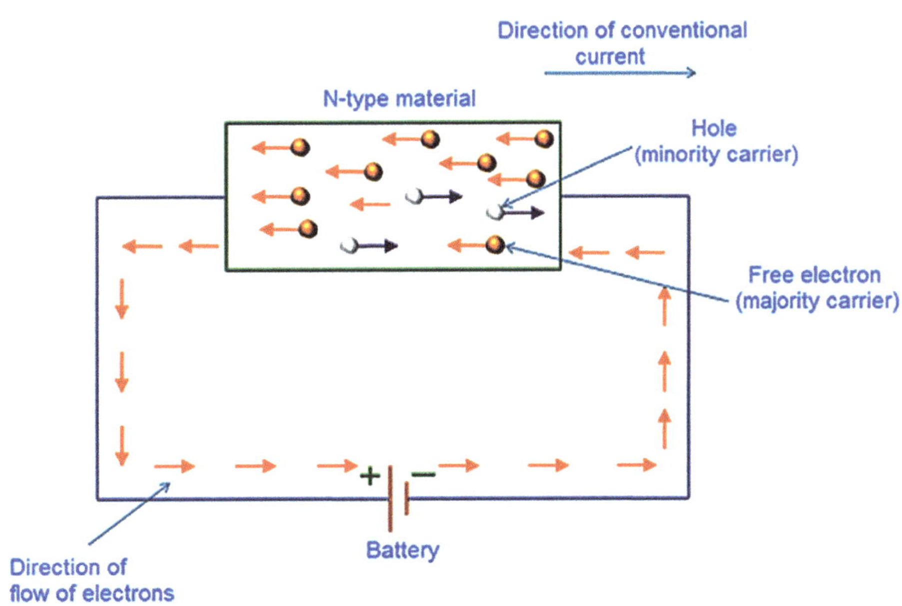

like Boron (B) or Gallium (Ga). As shown in Figure 5, this impurity is only able to share three valence electrons with the lattice, leaving one excess hole. Trivalent materials such as Boron (B), Aluminum (Al), Gallium (G) and Indium (In) because of the capability of accepting free electrons are called acceptor impurity (Hook, & Hall, 2001)

Figure 5. A P-type semiconductor

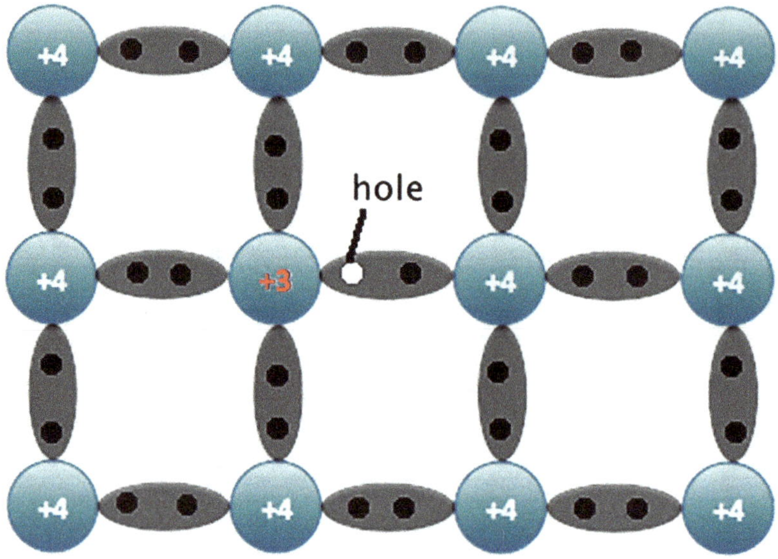

For example, consider that boron is to be added to silicon as shown in Figure 5. Again, the process starts with heating silicon up to temperature near 1000° C. Then the vapor of boron flows above the warm, nearly melting silicon. During this situation boron diffuses into the silicon. A boron atom contains 3 valence electrons whilst the corresponding value for silicon is 4. After constitution of 3 covalent bonds, for the fourth covalent bond, the silicon atom contributes one valence electron, while the boron atom has no valence electron to share. Therefore, the fourth covalent bond remains imperfect with shortage of one electron. The location of this missing electron is called hole. This fact implies that each boron atom can accept one electron to fill the hole. In microscopic view, a small addition of boron impurity provides millions of holes.

Since an acceptor provides extra holes which are assumed to be positively charged, a semiconductor doped with an acceptor element is denominated a P-type semiconductor. The notation P stands for positive. Just like the case for N-type semiconductor, the material as a whole is neutral. In a P-type semiconductor the current is mainly produced and flown by the holes. Thus, the holes are the majority carriers, while the electrons are the minority carriers. It must be mentioned that in fact, the electrons are moving in the crystal lattice to generate the current. But with the consideration of holes as the majority carriers, it is assumed that the holes are moving in crystal.

THE DIODE

Structure and Behavior

As illustrated in Figure 6, in semiconductor physics a diode is a crystalline piece of semiconductor substance consisting of a p-n junction in which the free parts of the junction are terminated with two metal leads. In simple words, with the connection of a P-type semiconductor to an N-type semiconductor, a p-n junction will be obtained which is known as diode in modern electronics (Horowitz, Hill, 1989; Lowe, 2013).

Figure 6. A p-n junction known as diode

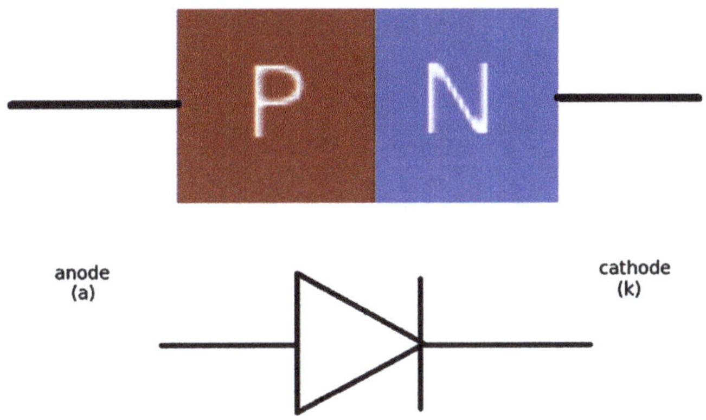

The free electrons and holes cannot attract each other because they belong to the crystal lattice of the corresponding semiconductor. As mentioned before, each material as a whole is neutral and there's no pure charge to produce an electric field in order to make these free particles to imbibe the opposite charge.

An interesting event which occurs in this state is that positive charges (holes) start to move towards the N-type semiconductor and on the other side, the free electrons of N-type semiconductor begin to proceed to the P-type side. This relocation isn't due to the attraction between electrons and holes, but because of the diffusion phenomenon (Neamen, 2003; Mijoe, 2015).

In order to explain the diffusion phenomenon one can say that if there is any agglomeration of the electric charges in a system, this reposition wants to be dislocated in a way such that the agglomeration becomes uniform in the whole system. Therefore, as said in the previous paragraph, the free electrons and holes move towards the opposite side semiconductor to balance the distribution of the free charges of valance bonds which aren't associated in covalent bonds. Because of the diffusion phenomenon which mobilizes electrons and holes, the structure of a p-n junction becomes like the one shown in Figure 7.

This movement and the attraction between opposite charges results in the neutralization of them. Also, the N-type semiconductor because of more positive charges becomes electrically positive, whilst the P-type side as a result of stationary negative charges becomes electrically negative. Therefore, an electric field from N side to P side will be established (Mijoe, 2015).

The diffusion phenomenon will be continued until the force of the generated electric field becomes equal to that of the phenomenon. From this moment, the electric field barricades the transfer of the electrons and holes from one side to another part. The final result is the constitution of a region free of any charges around the connection area of the P-type and N-type semiconductors which is known as the Depletion Region and is illustrated in Figure 7.

Reverse Bias of the Diode

As illustrated in Figure 8, if the positive terminal of a battery will be connected to N-type region of and the negative terminal will be connected to P-type region of the diode, it is said that the diode is reverse biased. At this state, the external electric field which comes from the battery, is at the same direction with the internal electric field of the diode that is originated from the diffusion process. This makes the depletion region grow larger in proportion with the increment of the electric field. With the growth of

Figure 7. Formation of the depletion region in a p-n junction

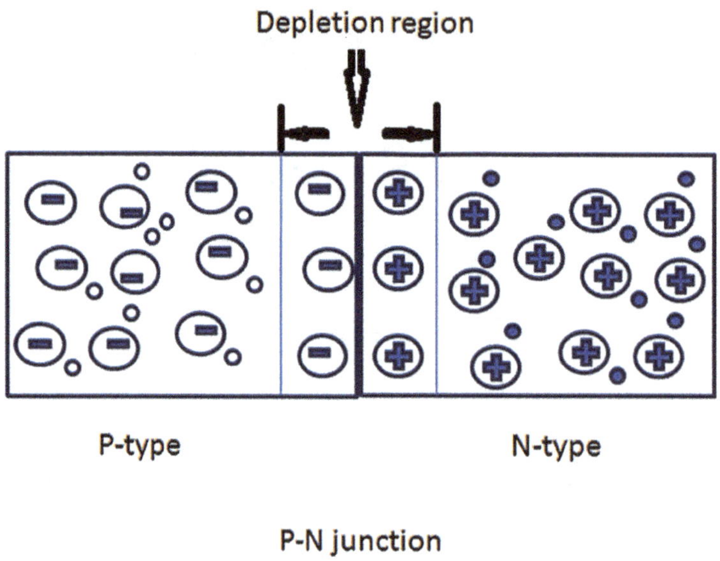

the depletion region, it will block the flow of the electric current through the diode and the value of the current is almost zero (Boylestad, Nashelsky, 2002; Sedra, Smith, 2004).

Sometimes in any region, an electron or a hole might have enough energy to separate from its original atom. If this situation happens in P or N region, the charge will return back to its atom valance bond. But in depletion region the conditions are different. Since there exists a strong electric field, it mobilizes this charged particle and conducts it outside of the deplete region and the existence of these charges causes a very small electric current to be flown across the diode which is known as the reverse saturation current of the diode, I_s.

Figure 8. A reverse biased diode

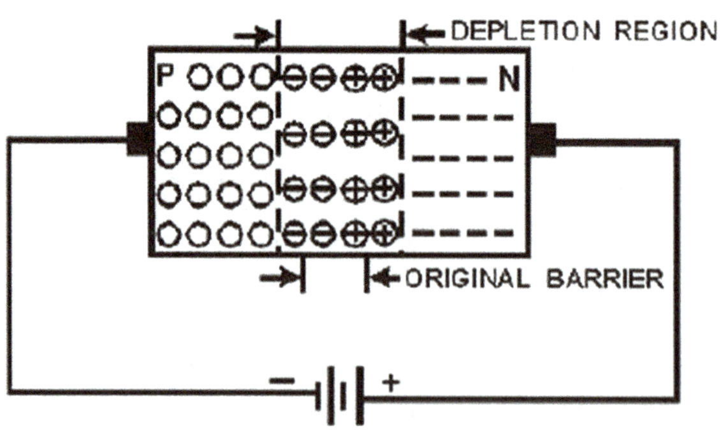

Forward Bias of the Diode

As shown in Figure 9, in forward bias state the positive terminal of the battery is connected to P-type region of and the negative terminal is attached to N-type region of the diode. The opposite direction of the external electric field with internal field of the diode, makes the deplete area more narrow because the terminals of the battery are connected to the regions of the diode from the same polarity. Therefore, the charges are banished towards the center of the junction. The greater the electric field from outside, the narrower the depletion layer which leads to the flow of electric current from P region to N region (Sedra, & Smith, 2004).

The relation between the current and voltage of the diode will be expressed by Shockley equation and is as follows (Popadic, Lorito, & Nanver, 2009):

$$I_D = I_s\left(e^{\frac{V_D}{\eta V_T}} - 1\right) \tag{1}$$

where

I_D is the diode current,
I_s is the reverse saturation current (leakage current),
V_D is the voltage drop across the diode,
η is the process coefficient which is considered as 1 for silicon, and
V_T is the thermal voltage of the diode that is defined by the following equation:

$$V_T = \frac{KT}{q} \tag{2}$$

Figure 9. A forward biased diode and its behavior as a short circuit

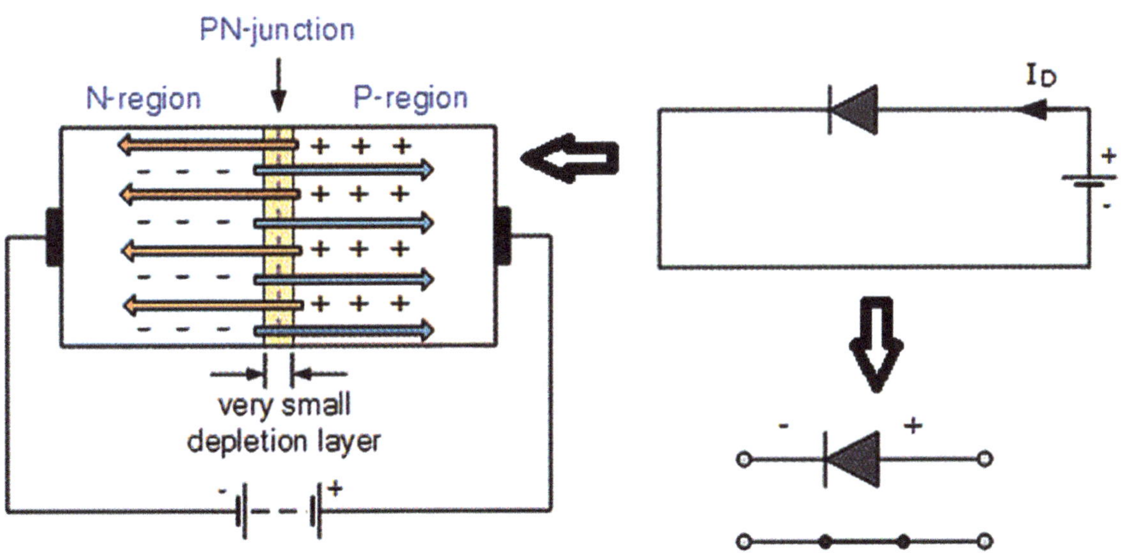

In equation (2),

K is the Boltzmann coefficient,
T is the temperature in Kelvin, and
q is the electric charge of an electron.

In room temperature, V is almost equal to $25mV$. According to Equation (1), and by assuming $V_D \cong 0$ the diode current for silicon diode can be expressed as:

$$\begin{cases} I_D \cong I_s e^{\frac{V_D}{V_T}} \; for \; V_D > 0 \\ I_D = -I_s \; for \; V_D < 0 \end{cases} \tag{3}$$

Diode Fabrication Process

Just as discussed before, it looks like that the diode is being obtained by joining two separate P and N doped semiconductors. But in modern Integrated Circuit (IC) design rules in which millions of diodes and transistors are fabricated in a small silicon wafer, the design process totally differs. Nowadays, in order to build a diode, at first step the silicon wafer is heated to the temperature near 1000° C (Ghandi). Then during the heating process the vapor of an acceptor element like boron is being passed over the heated nearly melting silicon to diffuse into silicon and make a P-type semiconductor which is called P-type Substrate (P-Sub). At the second step, the masking process on the P-Sub will be carried out and the areas in which N-type impurities should be added, will be uncovered. Then, by means of the secondary diffusion process, a pentavalent material with higher density than the previous trivalent will be injected inside the P-Sub to create a well of N-type. At the third step, the second stage will be repeated for a pentavalent element to generate a P-well inside the N-well. At the final stage a layer of SiO_2 will cover the produced p-n junction and the leads should be added to complete the process. The final product which is shown in Figure 10, is a yield known as the monolithic diode which is nowadays fabricated in a mass production process of manufacturing thousands of diodes on a single wafer (Jaeger, 1988).

Figure 10. A monolithic diode after fabrication

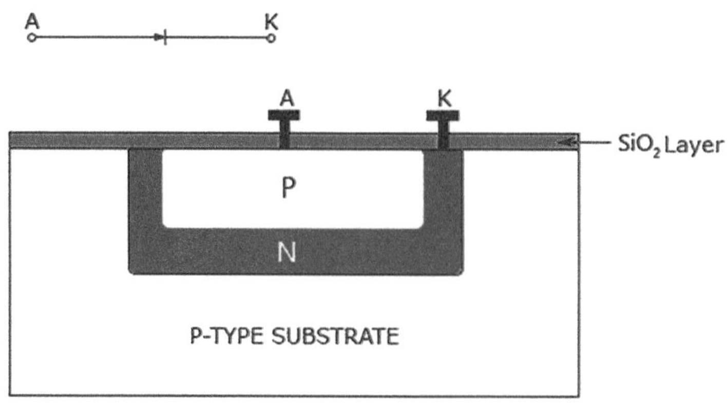

Avalanche Breakdown of Diode

In certain conditions, an interesting event can occur in both of the insulators and semiconductors which may be considered as a kind of electric current multiplication. This effect which is a type of electron avalanche, is called the avalanche breakdown of the diode. It happens when the reverse bias voltage of the diode gets as large as possible to accelerate the minority carriers. Along with the acceleration, the energy level rises (McKay, 1954; Jaeger, Blalock, 2003). When these high energy carriers want to pass the deplete region, they may have a collision with the atoms and break their covalent bonds to free more charged carriers. This process continues and the number of charged particles increases like an avalanche. A large reverse current can be produced by this effect and this is the principle used to produce Zener Diode. But the difference of the action in a zener diode is that the reverse applied electric field itself breaks the covalent bonds to produce the current. This effect is known as the Zener Breakdown Phenomenon.

It must be mentioned that in normal room temperatures, the breakdown voltage of the diode has a large value. From now, we can draw a current-voltage characteristics curve for a diode consisting of three regions: forward bias, reverse bias and breakdown region. The curve is illustrated in Figure 11.

Diodes and Sensors

Besides the discussed details, the diode itself can be used as a sensor. The simplest example in this field is semiconductor temperature sensor. Since the diode voltage drops almost linearly with temperature increment, the p-n junction diode can be the basis of an accurate thermometer for cryogenic temperatures up to about 200°C.

To achieve this, a constant current is kept through the diode while the voltage across it gives an indication of the temperature. The amplitude of the constant current is typically chosen small enough to minimize diode self-heating. As a result, the diode voltage falls linearly with temperature making it necessary to use an operational amplifier conditioning circuit for direct readout. Output calibration

Figure 11. Current-Voltage characteristics of a diode

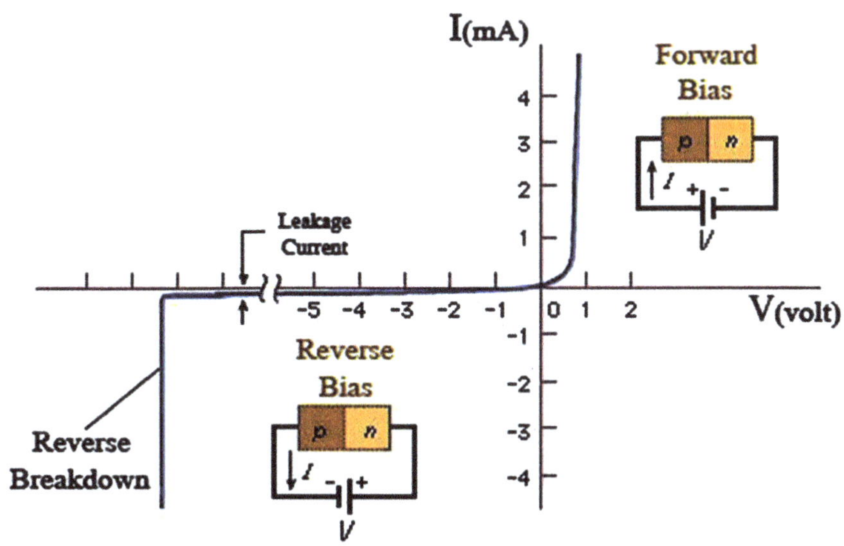

can be achieved by means of a voltage reference. The dependence of diode current to the temperature is obvious if the value of V_T is substituted from equation (2) in (1). Figure 12 shows the effect of two different currents on the diode voltage as it is cooled in a water bath from temperature T_0. In fact, the backward extrapolated segments AB and CD would meet at absolute zero (-273°C). However, impurity and some other effects in the diode will limit circuit usage to about -200°C.

As it can be observed, the voltage difference between two voltages at a given temperature T is directly proportional to the temperature provided in which the currents are switched at the measurement temperature. More details are available at (Ocaya, 2013 /) where the proposed circuit is shown in Figure 13.

Figure 12. Operating principle of the direct RMS readout thermometer

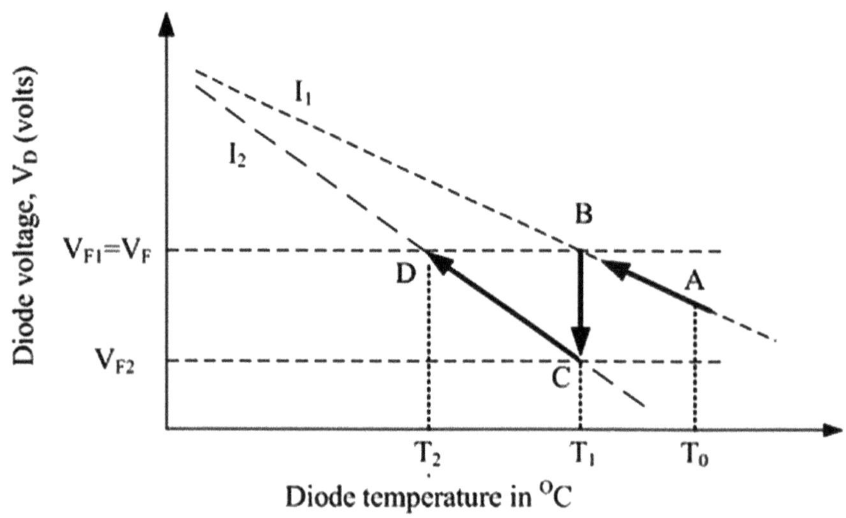

Figure 13. Switched constant-current source temperate sensor

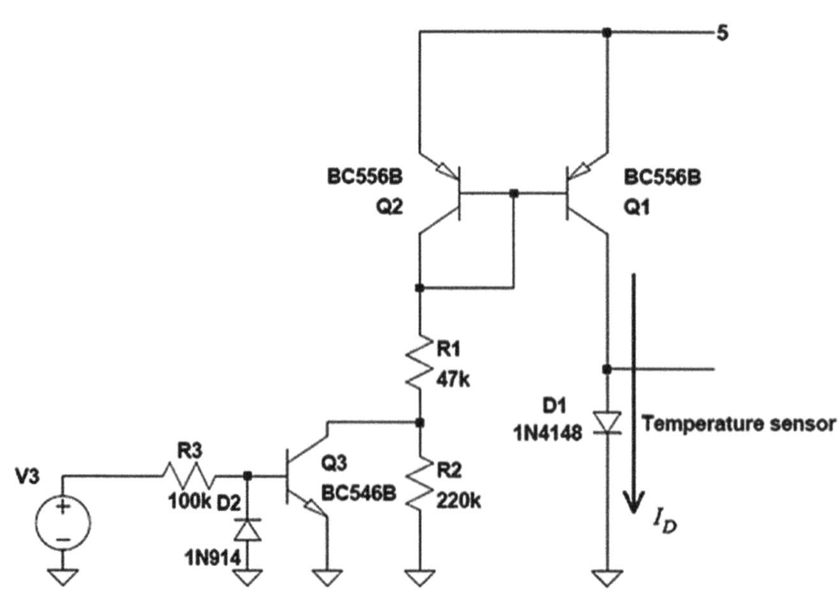

Over recent years, a whole new range of semiconductor temperature sensors is arriving from different manufacturers. However, the most popular ones include *AD590* and the *LM35*.

THE BIPOLAR JUNCTION TRANSISTOR

Structure

With the interlaced junction of three semiconductors in which the middle crystal has an opposite diffused impurity versus the side crystals, a bipolar junction transistor (BJT) will be obtained. The sandwiched region is known as the Base (B) and the surrounding semiconductors are known as the Emitter (E) and the Collector (C). Because of two possible alternate positioning for semiconductor crystals, two structures may be realized for a BJT. These two realizations, according to the arrangement of the P-type and N-type semiconductors are form NPN and PNP transistors as shown in Figure 14 (Boylestad, Nashelsky, 2002; Sedra, Smith, 2004).

In order to distinguish the emitter terminal from collector, an arrow which indicates the base-emitter diode is used and if the arrow in pointing to the base, the transistor is PNP and for the pointing out state the transistor is NPN.

From now, we are concerning on the analysis of NPN transistor. The same method applies on PNP with the exception that the direction of the currents must be changed.

Operating Principle of NPN Transistor

In order to analyze the amplifying behavior of a transistor, let's connect the emitter to zero reference voltage and considering a situation in which the base-emitter diode is forward biased whilst the base-collector diode is reverse biased. In this condition, the transistor is said to be in Active Bias Region where $V_{BE}\geq 0.7$ and $V_{BC}<0$. This configuration is shown in Figure 15 in which V_{BE} is equal to 0.7 volts and V_{CE} is equal to 5 volts (Sedra, & Smith, 2004).

Figure 14. Two types of bipolar transistor

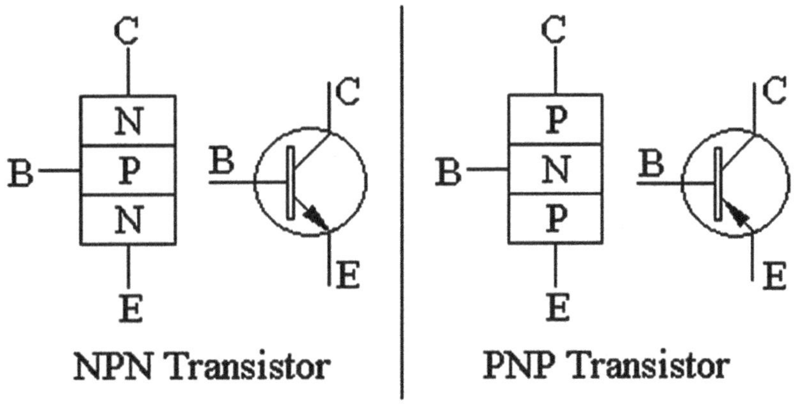

From a macroscopic view, a BJT is a three lead electronic component as shown in Figure 16. By considering the whole transistor as a cut set, we can write:

$$I_E = I_B + I_C \tag{4}$$

In Equation (4), I_E is the emitter current, while I_B and I_C are the base and collector currents, respectively. We can also write:

$$V_{CB} + V_{BE} = V_{CE} \tag{5}$$

in which V_{BE} is the base-emitter voltage, V_{CE} is the collector-emitter voltage and V_{CB} is the collector-base voltage.

From the circuit configuration of Figure 15, we can obtain:

$$V_{CB} = V_{CE} - V_{BE} = 5 - 0.7 = 4.3v \tag{6}$$

which guaranties that the base-collector diode is reverse biased.

Now, we want to study the transistor in microscopic view like the one done for the diode. The microscopic view of an NPN transistor which is biased in active region, is shown in Figure 17.

Figure 15. Biasing an NPN transistor in active region

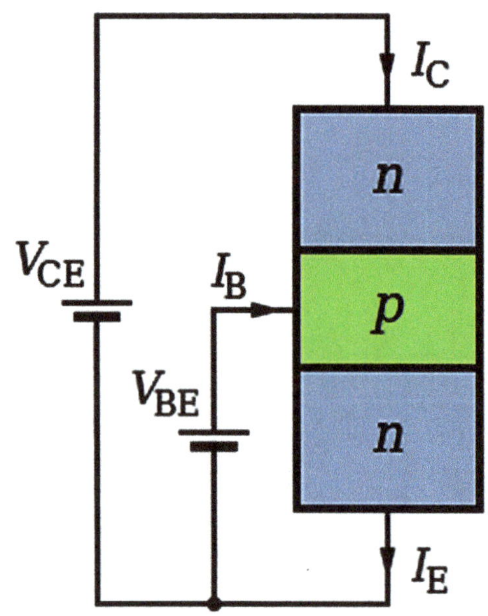

Figure 16. A BJT as a three lead electronic component

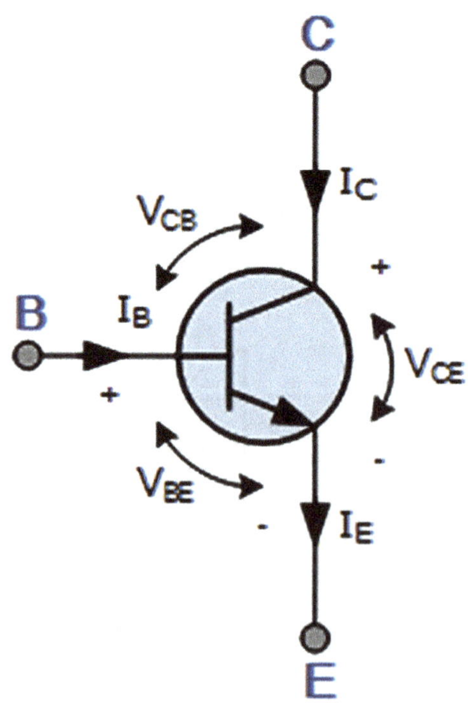

Figure 17. Microscopic view of an NPN transistor in active region

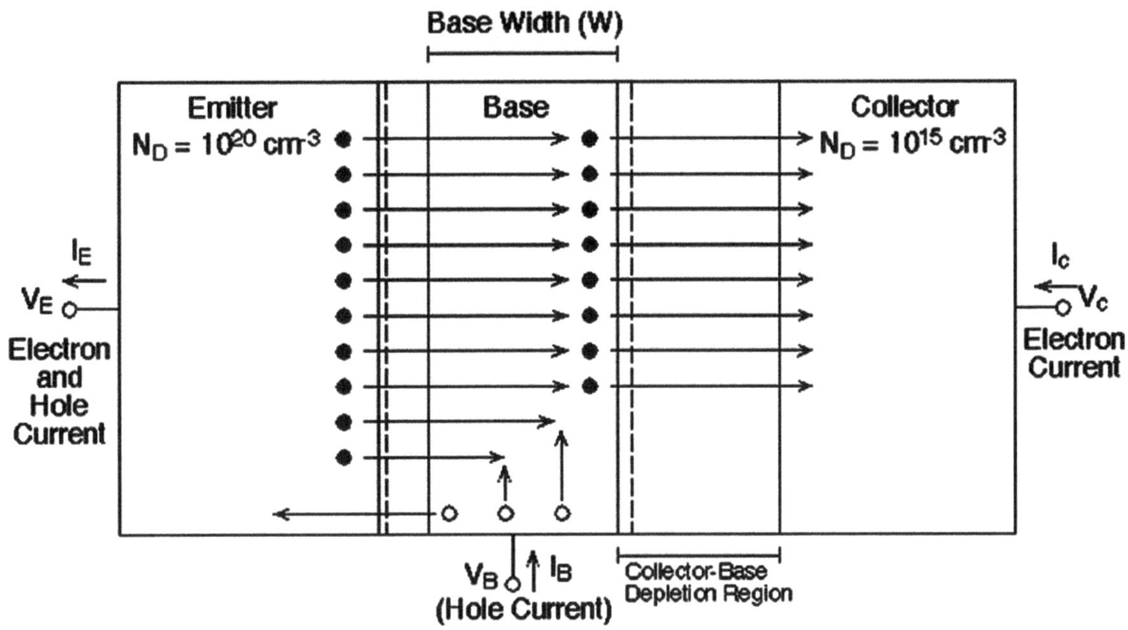

Since the base-emitter diode is forward biased, the internal electric field is decreased enough in which the free electrons of emitter region can easily penetrate to the base region. If these free particles can successfully reach to the collector region, they will face a strong electric field arising from reverse biased base-collector p-n junction. This field can collect the successful electrons and that is why this zone is known as the collector of the transistor.

If all of the injected electrons from the emitter can reach to the collector, then we can write:

$$I_E = I_C \tag{7}$$

But in fact, the base is full of holes and the injected electrons must pass a region full of holes to reach the edge of base and enter to the collector. The possibility of collision between the particles with opposite charges is not equal to zero and as the system is in balance, the combined holes will be replaced by the base to constitute the base current.

Therefore, Equation (7) must be corrected and the equality must be corrected as follows:

$$I_C = \alpha I_E \tag{8}$$

The coefficient α which has a value less than unity, indicates that a fraction of penetrated electrons from emitter will be neutralized by the holes in base and α is the fraction of electrons which successfully reached to the collector zone. This fact is shown in Figure 18 and justifies the Equation (4) (Jaeger, Blalock, 2003; Horowitz, Hill, 2015).

Figure 18. Current components in a BJT

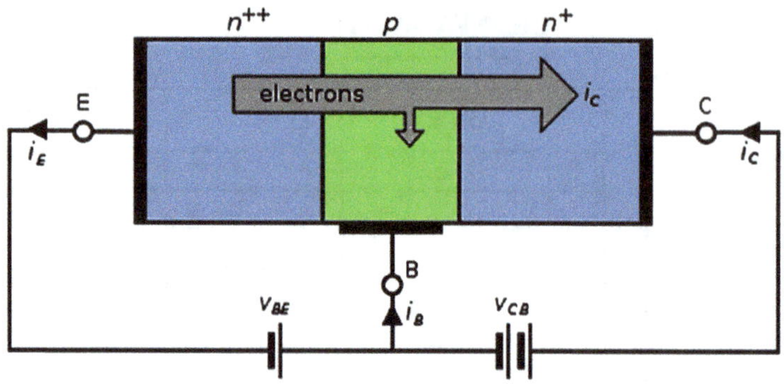

Since we want to reach to the Equation (16-8), the parameter must become equal to unity. In this case, the transistor will be the ideal one we were looking for. To realize this, the base must be narrowed so that the injected electrons of emitter can pass a shorter way to reach the collector. The less the path, the less the combination possibility with the holes in the base.

To simplify the calculations we define a secondary parameter named β just as follows:

$$\beta = \frac{I_C}{I_B} \tag{9}$$

From Equation (8), we can obtain:

$$\alpha = \frac{I_C}{I_E} \tag{10}$$

By substituting (10) in (4), we have:

$$\frac{I_C}{\alpha} = I_C + I_B \tag{11}$$

$$I_C\left(\frac{1}{\alpha} - 1\right) = I_B \tag{12}$$

$$\frac{I_C}{I_B} = \frac{\alpha}{1 - \alpha} \tag{13}$$

By comparing (13) with (9), we obtain:

$$\beta = \frac{\alpha}{1-\alpha} \tag{14}$$

For α=1, as Equation (14) shows, β tends to infinity which is equal with I_B=0. In the detailed explanation for equalizing α with the unity value, the thickness of base was reduced. But a factor that is not considered on the calculation of the base current, was the penetration of the holes from base to emitter as a result of forward biasing.

Because of these injected particles, α will be less than unity and this indicates on the flow of a current from base to emitter where the collector doesn't know anything about it!

To solve this problem, the doping of emitter must be increased to neutralize the base current in forward bias state of base-emitter diode. In NPN transistor where emitter is of N-type semiconductor, if the doping of emitter is selected at least 1000 times more than the P-type base region, α will almost be equal to the unity value.

Since base-emitter is forward biased, we can write:

$$I_E = I_{SE}\left(e^{\frac{V_{BE}}{V_T}} - 1\right) \cong I_{SE}e^{\frac{V_{BE}}{V_T}} \tag{15}$$

For the collector current, we have:

$$I_C = \alpha I_E = \alpha I_{SE}e^{\frac{V_{BE}}{V_T}} \tag{16}$$

I_{SE} is the reverse saturation current of the base-emitter diode, which is a function of temperature and the area of base-emitter p-n junction. By defining $I_S=\alpha I_{SE}$, we obtain:

$$I_C = I_S e^{\frac{V_{BE}}{V_T}} \tag{17}$$

Equation (17) shows that the collector current is an exponential function of base-emitter voltage (Sedra & Smith, 2004; Jaeger & Blalock, 2003; Horowitz & Hill, 2015). Therefore, any small variations in V_{BE}, can significantly change the collector current. This fact is employed in the design of amplification systems. Another fact which must be considered is the dependency of V_{BE} to the temperature changes of the environment. For every centigrade degree increment in temperature, V_{BE} drops by 2.5 mili volts. Figure 19 shows the different values of I_C versus changes of V_{BE} because of temperature variations.

At the beginning of this section, we supposed that the base-emitter diode is forward biased whilst the base-collector diode is reverse biased. For this purpose, the collector voltage is set on 5 volts to strongly ensure the base-collector diode in cut off region. Let's suppose a case in which the collector voltage is adjustable and we start to decrease the value of voltage from 5 to 0.7 volts. Since V_{CE}=0.7, the

Figure 19. I_C–V_{BE} characteristics curvature according to temperature changes

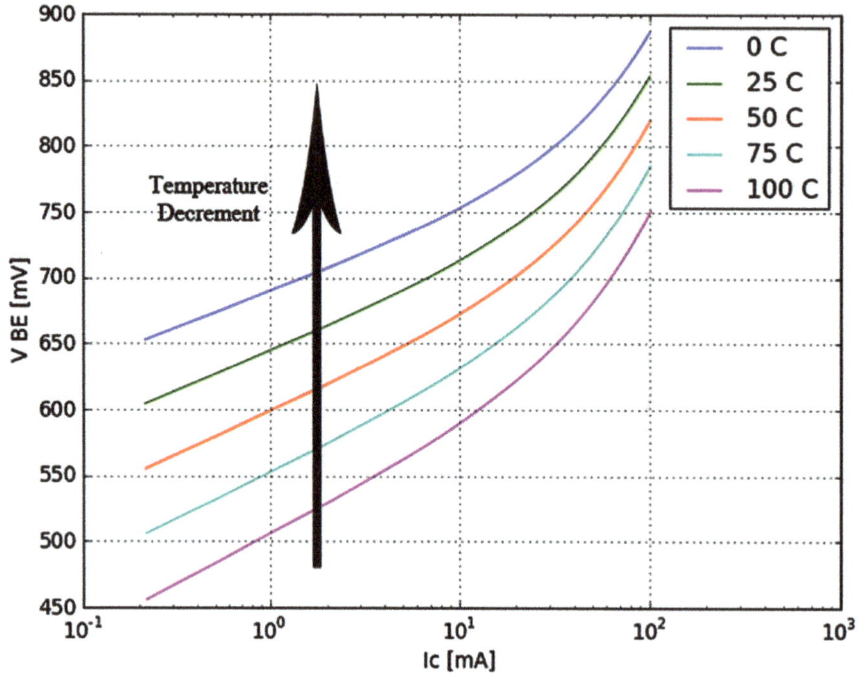

base-collector diode observes a zero volt external field. The free electrons coming from emitter will be collected by built-in field (internal field) of base-collector diode and Equation (17) is still valid.

For more voltage decrement, the base-collector diode is going to be forward biased. But the leakage current from collector to base, is now equal to zero. In V_{CE}=0.2v, the base-collector diode is almost ON and the injected electrons from collector to base which are now the majority carriers, will be increased. As a reason of this situation, the free electrons from emitter even if they are collected by built-in field of the base-collector, because of the diffusion process, will be kicked back. Therefore, the collector itself starts to release free electrons to base. From now on, the transistor loses its amplification property and regardless of any voltage drop on base-collector diode (since this diode is ON), V_{CE} is equal to 0.2 volts. This working region is known as the saturation region. Therefore, the four working regions of the BJT will be defined as follows:

1. Active Region in which the base-emitter diode is forward biased and the base-collector diode is reverse biased. In this region we always have $V_{CE}>0.2v$.
2. Saturation Region in which both diodes are forward biased. In this region we always have $V_{CE}\cong0.2v$ and in most cases for the simplification of the calculations, V_{CE} is considered equal to zero.
3. Cut Off Region where both the diodes are reverse biased.
4. Reverse Active Region where the base-emitter diode is reverse biased and the base-collector diode is forward biased (the roles of emitter and collector are changed). In this region because the doping of emitter is much higher than collector (the reason will be discussed in Early Effect), the parameters α and β will have small values and there won't be any current amplification between base and collector current.

In practice, we never bias the transistor in reverse active region. Also, the bias in saturation region is not the desired one because Equation (9) isn't valid in this region and in fact, we have:

$$I_C < \beta I_B \tag{18}$$

For the collector current in saturation region, we have:

$$I_C = I_S e^{\frac{V_{BE}}{V_T}} - I_{SC} e^{\frac{V_{BC}}{V_T}} \tag{19}$$

$$I_C = \alpha I_{SE} e^{\frac{V_{BE}}{V_T}} - I_{SC} e^{\frac{V_{BC}}{V_T}} \tag{20}$$

In Equations (19) and (20), the second exponential terms are because of forward biased base-collector diode. This term is the result of inequality $I_C<\beta I_B$. For the emitter current, we can write:

$$I_E = I_{SE} e^{\frac{V_{BE}}{V_T}} - \alpha_R I_{SC} e^{\frac{V_{BC}}{V_T}} \tag{21}$$

in which α_R pertains to the sequence of reverse active region of the transistor where the roles of emitter and collector are replaced. For α_R we have:

$$\alpha_R < \alpha \tag{22}$$

and it is because of the smaller value of the collector doping.

If we apply the Kirchhoff's Voltage Law (KVL) to the collector-emitter loop, we'll obtain the BJT's output characteristics function for I_C versus V_{CE} which is a straight line with negative slope. If we draw the characteristics function known as DC load curve at the same plane with the base current or base-emitter voltage, we can easily obtain the transistor's operating point. Figure 20 shows this characteristics curve in which the three operating regions are specified with the consideration of maximum allowable power dissipation limitation.

In Figure 20, we can see that based upon the base current, we can have different operating points. Choosing the best operating point is one of the important principles which must be emphasized in the bias circuit design of a transistor. For example, the point Q_1 is located near the saturation region and Q_3 is located near the cut off region. Any thermal runaway or other factors may replace the location of these points. Therefore, the best operating point must be selected in middle of the output characteristics curve so that we can obtain a good swing of the collector voltage (Jaeger & Blalock, 2003).

As Figure 20 illustrates, in active region the base currents are considered horizontal line (with zero slope). But in practice, the base currents are not constant and the lines contain different slopes as shown in Figure 21. To explain this, one can say that the wide span of charge space is the cause of these inclined lines. For high values of base-collector reverse bias, the deplete region of this diode will be wide which is

Figure 20. I_C–V_{CE} characteristics curvature of an NPN transistor

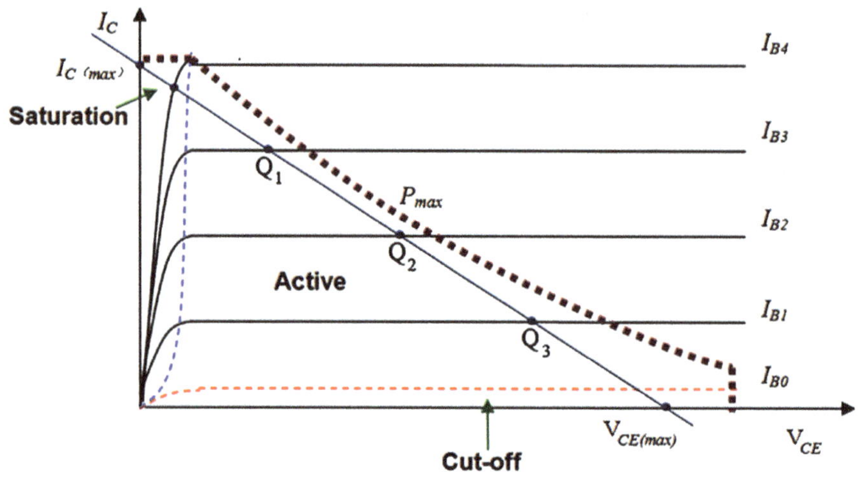

Figure 21. The inclined lines of base current in I_C–V_{CE} characteristics curvature known as the Early Effect

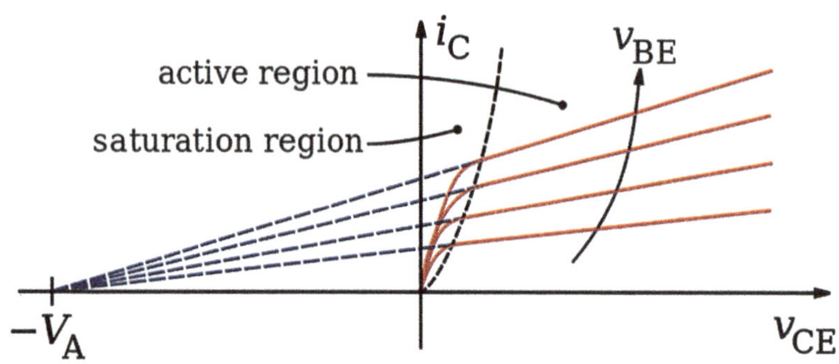

equal to thin base layer when BJT is operating in active region. As a result, the free electrons will travel a shorter distance to reach the collector whilst for low values of reverse bias this path becomes longer which is equal to more neutralization of injected electrons from emitter with the provided holes of base.

As shown in Figure 21, if we continue the lines from left side, we'll observe that in $-V_A$, all of these lines will intercept each other. This is known as the Early Effect and to explain its efficacy in transistor current relation (Jaeger & Blalock, 2003; Alioto & Palumbo, 2005), we have to write:

$$I_C = I_S\left(1+\frac{V_{CE}}{V_A}\right)e^{\frac{V_{BE}}{V_T}} \tag{23}$$

For $V_A \rightarrow \infty$, Equation (23) turns to (17) and the slopes tend to zero which leads to Figure 20. In a BJT, V_A has a great value and is about 100 volts. To reach the ideal form of the transistor, we must get higher values and it is carried out by consolidation of charge space regardless of V_{CE} variations. For this purpose the doping of base must be higher than the one for collector which decreases the charge space around the base. This consideration along with the higher doping of emitter in comparison with base was previously shown in Figure 17 (Hu, 2010). It must be mentioned that the Early Voltage affects β of the transistor as well as I_C with the same coefficient.

BJT Fabrication

With these considerations, the fabrication process of the BJT will be somehow similar to the procedure discussed for the diode with two exceptions. The first one is the higher doping level of the semiconductors in comparison of diode fabrication because the provided reasons for the dosage of impurities and the second one is the repetition of the second and third steps once more for the third n++ layer of emitter (Sze, 1985; Chang & Sze, 1996). The final product with the covering of SiO_2 and the addition of leads is illustrated in Figure 22.

It must be mentioned that a same procedure goes on for the fabrication of the resistors and capacitors on the integrated circuit design processes. The examples of this kind are shown in Figure 23 for resistor and Figure 24 for capacitor design.

Figure 22. A monolithic NPN transistor after fabrication

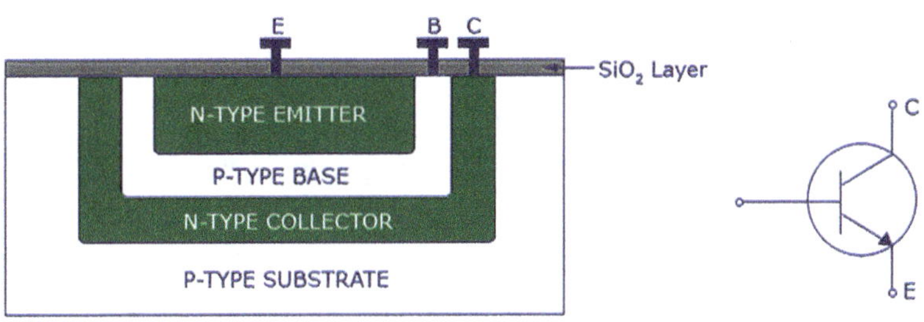

Figure 23. A monolithic resistor

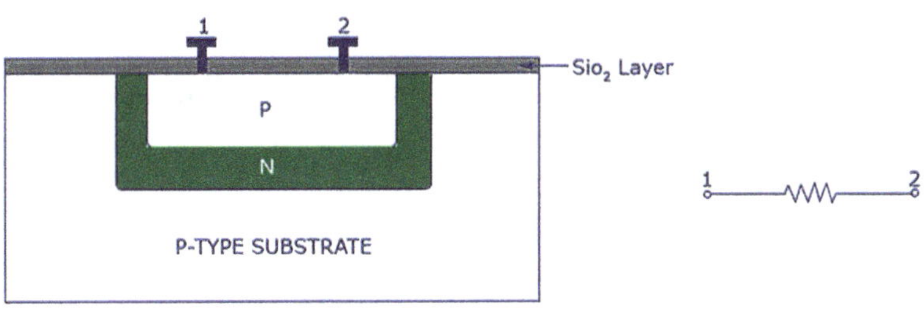

Figure 24. A monolithic diffused capacitor

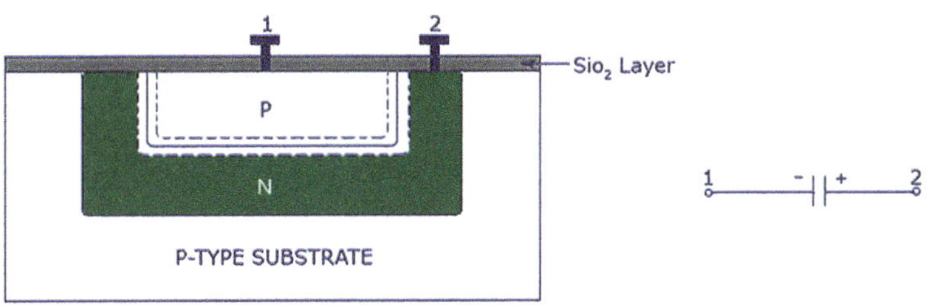

BJT Bias Circuits

As discussed in previous section, choosing the best operating point for a BJT is one of the important principles which must be concerned in the bias circuit design. In discrete circuit design in which the elements are implemented on the Printed Circuit Boards (PCBs), based on the design specifications, the combination of resistors and capacitors and somehow inductors may constitute the biasing circuit. But in IC fabrication technologies, the bias is usually obtained by a bandgap reference circuits or current mirrors. The emphasis in this chapter is on discrete implementation of the bias circuit.

By means of KVL on the collector-emitter loop, the BJT's output characteristics function which is the relation between I_C and V_{CE} will be obtained. The DC current of the collector represents the operating point current, whilst the DC voltage between collector and emitter exhibits the operating point voltage. These two components are known as the Q-point parameters. The AC signal sources of the system don't have any effect on Q-point parameters.

In Figure 20 it was explained that from the intersection of the base current curvature with the output characteristics function, the Q-point parameters will be attained. Therefore, one of the most important factors on the transistor bias is the base current. Other factors such as the collector resistance or the DC supply may also be considered. Since the transistor is to be operating in active region and there is almost a linear relation between I_C and I_B, the main focus is on base current (the leakage current between collector and base coming from reverse biased diode is neglected). According to Figure 20, Q_2 seems to be a good operating point. But in order to have it, at first the base current must be set on desired value. The simplest method for transistor bias is shown in Figure 25.

From Equation (17), one can say that if instead of temperature changes the base-emitter voltage is assumed constant, along with the increment of the temperature the base current will rise. Hence, the collector current will be increased by means of Equation (9). This effect known as thermal runaway, can fully deflect the transistor from its operating point. One of the effective parameters on thermal runaway is the reverse saturation current of the base-collector junction which doubles by each 10º C increment in temperature and it also worsens the situation.

Therefore, the use of bias configuration shown in Figure 25 is completely lapsed because of the discussed thermal runaway effect. It also has no use because two different power supply voltages are needed which is not suitable for transistor bias due to power consumption considerations. In addition, the dependency of I_C to V_{BE} in exponential form makes us difficult to correctly adjust the collector current.

In order to obtain a good DC bias, at first the base current must be stabilized. By means of a resistor in base, this problem can be solved by the configuration of Figure 26.

The most remarkable advantage of this circuit over Figure 25 is the use of one power source. Also, the operating point can easily be shifted by means of the base resistor. For the base current and V_{CE}, we have:

$$I_B = \frac{V_{CC} - V_{BE}}{R_B} \tag{24}$$

$$V_{CE} = V_{CC} - R_C I_C \tag{25}$$

Figure 25. The simplest form of BJT bias
Sedra & Smith, 2004; Jaeger & Blalock, 2003.

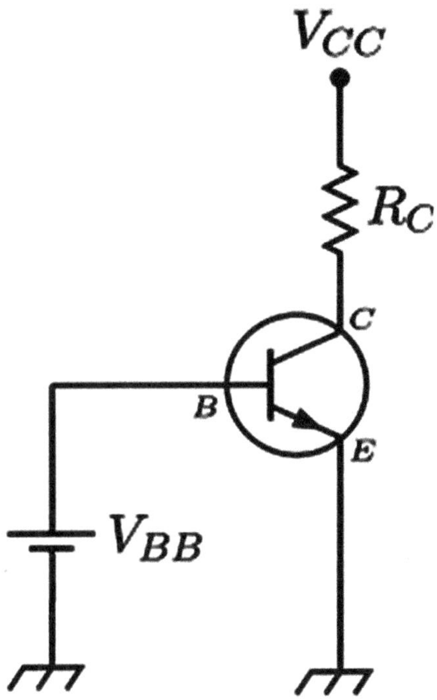

Figure 26. Fixed bias of the BJT

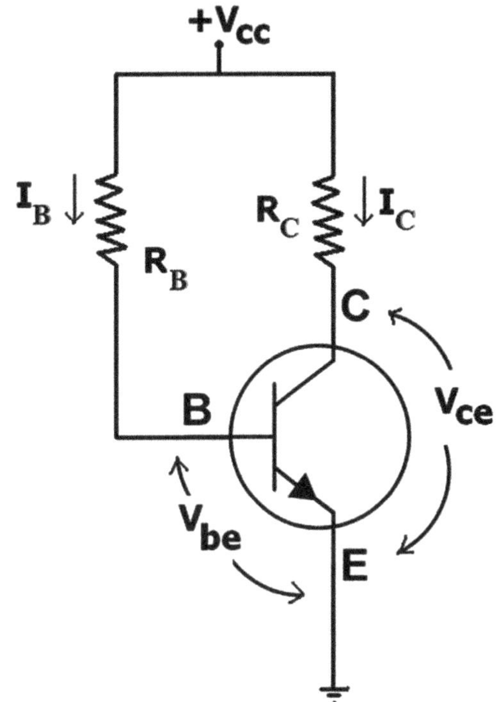

Despite of fixed bias, this circuit still has some drawbacks. The thermal runaway effect still exists and the dependency of I_C to β is another problem, because the range of variations of β is high and it differs from the transistor of a type to another due to fabrication factors.

To solve the thermal drift effect, the circuit of Figure 27 can be employed. Because of a feedback loop, it is known as collector feedback bias.

The existence of negative feedback stabilizes the operating point and removes the effect of thermal runaway. The only difference of this circuit with the configuration of Figure 26 is that the base resistor is attached to the collector instead of being directly connected to the power supply. Any thermal runaway will increase the collector current which itself will cause a higher voltage drop across the collector resistor. If the base-emitter voltage is assumed constant, the reduction of collector voltage will reduce the base current and consequently decrease the collector current by means of Equation (9). For the base current and V_{CE}, we have:

$$I_B = \frac{V_{CC} - V_{BE}}{R_B + (\beta + 1) R_C} \tag{26}$$

$$V_{CE} = V_{CC} - R_C (I_C + I_B) \tag{27}$$

Figure 27. Collector feedback bias of the BJT

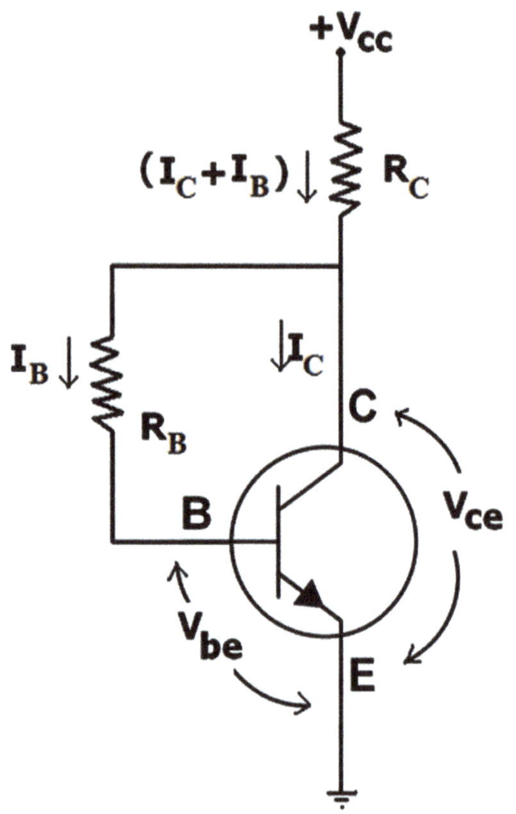

Figure 28. Voltage divider bias of the BJT

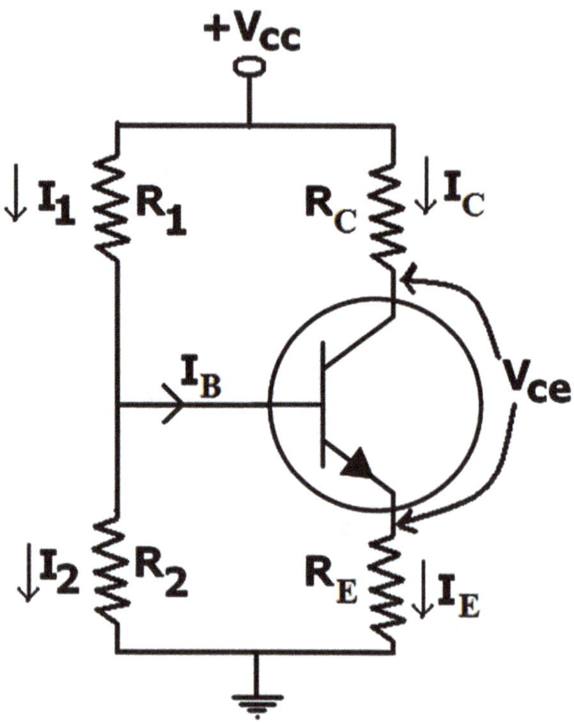

This circuit is mainly used in RF circuit design because of simplicity and good resistance against temperature changes. However, the dependency case of the collector current to the variations of β still is unsolved. To overcome this problem, the voltage divider biasing is suggested which is shown in Figure 28.

The main difference of this configuration against previous circuits is the existence of emitter resistor which reduces the sensitivity of base-emitter voltage changes to the temperature variations by means of a local negative feedback in the loop consisting of base and emitter.

This configuration is also known as the Self Bias Circuit. Instead of a base resistor, two external resistors are employed in which the voltage drop over R_2, determines the base and the emitter bias voltage. The emitter resistor has the responsibility to compensate the thermal runaway of the collector current. For the base voltage we have:

$$V_B = \frac{R_2}{R_1 + R_2} V_{CC} - \frac{R_1 R_2}{R_1 + R_2} I_B \tag{28}$$

If $I_B \ll I_2$ which means $\beta \gg 1$, then we can use the approximation formula below for the base voltage:

$$V_B \cong \frac{R_2}{R_1 + R_2} V_{CC} \tag{29}$$

By means of Equation (28), the emitter current would be calculated as follows:

$$I_E = \frac{V_{CC} - V_{BE}}{\frac{R_B}{\beta + 1} + R_E} \quad (30)$$

For $R_B \ll (\beta+1)R_E$, the above equation can be simplified as follows:

$$I_E \cong \frac{V_{CC} - V_{BE}}{R_E} \quad (31)$$

which implies that the emitter current and respectively, the collector current with good approximation will be free of β coefficient and has a good bias stability. It must be mentioned that the difference of this circuit with collector feedback bias configuration is that in Figure 27, because of the bias considerations, $R_B \ll (\beta+1)R_C$ cannot be achieved.

For practical circuit design, if we set:

$$R_B = 0.1\left(\beta + 1\right)R_E \quad (32)$$

then the self bias circuit will provide the independency of collector current from β along with the specifications discussed.

Large Signal Behavior of BJT and the Swing

By attention to Equation (17), we'll observe that a small variation in base-emitter voltage, can significantly change the collecor current. For example, if the temperature will vary from 25ºC to 65ºC, this increment causes a reduction in V_{BE} value for about $100mV$. If V_T=$25mV$ and the collector current in 25ºC is denoted with I_{C_0}, then according to Equation (17), for the new value of collector current we have:

$$I_{C_1} = I_S e^{\frac{V_{BE_1}}{V_T}} = I_S e^{\frac{V_{BE}-0.1}{V_T}} = I_S e^{\frac{V_{BE}}{V_T}} \times e^{\frac{-0.1}{0.025}} \quad (33)$$

$$I_{C_1} = I_{C_0} \times e^{-4} \quad (34)$$

$$\frac{I_{C_1}}{I_{C_0}} \cong \frac{1}{55} \quad (35)$$

which shows that the collector current is almost 55 times stepped down. As a result, since $100mV$ causes a great change in collector voltage it is a large signal and cannot be considered as small signal amplitude.

From now, we want to separate the small signal behavior of the transistor from its large signal. For this purpose, we want to find the required conditions to analyze the small signal behavior a BJT. To do this, the base-emitter must be divided to its DC and AC components:

$$V_{BE} = V_{BE,DC} + V_{BE,AC} \tag{36}$$

$$V_{BE} = V_{be} + v_{be} \tag{37}$$

in which V_{be} stands for the DC component and v_{be} represents the AC component. The Equation (17) can be rewritten as follows:

$$I_C = I_S e^{\frac{V_{BE}}{V_T}} = I_S e^{\frac{V_{be}+v_{be}}{V_T}} = I_S e^{\frac{V_{BE}}{V_T}} \times e^{\frac{v_{be}}{V_T}} \tag{38}$$

$I_S e^{\frac{V_{BE}}{V_T}}$ is the Q-point parameter denoted as I_{CQ}. Therefore, we can write:

$$I_C = I_{CQ} e^{\frac{v_{be}}{V_T}} \tag{39}$$

The exponential term can be written as its Taylor Series around zero, known as Maclaurin Series shown below:

$$e^x = 1 + \frac{x}{1!} + \frac{x^2}{2!} + \frac{x^3}{3!} + \ldots \tag{40}$$

For small values of x, Equation (40) can be approximated as follows:

$$e^x \cong 1 + x \tag{41}$$

Substituting Equation (41) in (39), results in:

$$I_C = I_{CQ}\left(1 + \frac{v_{be}}{V_T}\right) \tag{42}$$

For the linear behavior of the transistor and in other words $I_C = I_{CQ}$, we must have:

$$\frac{v_{be}}{V_T} \ll 1 \Rightarrow v_{be} \ll V_T \tag{43}$$

$v_{be} \ll V_T$ is the small signal requirement of a bipolar transistor, while there is no resistor in emitter. It is obvious that for the existence of emitter resistor, a voltage division must be done for small signal component of a BJT. But in most cases, this resistor is being bypassed by a capacitor in parallel so that the gain of the transistor won't drop. As a result, the Equation (43) is usually valid and according to it, the amplitude of AC signal must be less than $2.5mV$ (Mohamed, 1966).

Since the base voltage consists of DC and AC components in which the DC component is a result of transistor bias and the AC component comes from AC supply, thus the operating point parameters will contain both parameters. From now, we'll use the Small Signal term instead of AC component because its amplitude is negligible in comparison with DC component known as large signal component.

As shown in Figure 20, the output characteristics curve determines Q-point parameters. Just as defined for base-emitter voltage, the effect of small signal on operating point parameters can be defined with following equations:

$$V_{CE} = V_{ce} + v_{ce} \tag{44}$$

$$I_C = I_c + i_c \tag{45}$$

in which by means of Equation (42), we can write:

$$I_c = I_{CQ}, V_{ce} = V_{CEQ} \tag{46}$$

Because the transistor amplifies the small signal of input node, the output signal must have large amplitude. Here comes the important reason of choosing Q_2 as the operating point in Figure 20, since the output signal needs to play freely around this point without any distortion. Q_1 is near saturation region and Q_3 is close to cutoff region and any relocation of output signal can send transistor to these unwanted regions. These variations of the output signal from top peak to bottom peak is called the output swing.

The best condition is to set the biasing point which is the collector DC voltage in the middle of emitter DC voltage and V_{cc} so that the symmetrical changes of transistor output reach the maximum value as illustrated in Figure 29. Only if the emitter is used as the output node, the collector voltage as the basis of swing is replaced with emitter voltage and V_E must vary between V_C and the zero reference voltage. For more accurate calculations, the saturation voltage which is about 0.2 volts may also be considered.

As illustrated in Figure 29, in order to maximize the symmetric swinging, the positive and negative swings must be equal. To calculate the maximum available symmetric output swing of a BJT, the following steps must be done:

1. The DC load curve must be determined.
2. The DC power sources must be eliminated by setting them to zero because the circuit is biased in desired point. Now, the effect of AC power sources must be considered. This means that a second-

Figure 29. Output signal swing in BJT

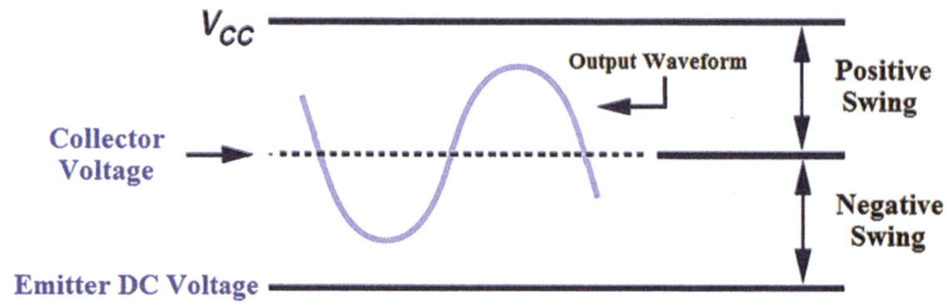

ary load curve should be drawn which is known as AC load curve. To draw the curve, the Q-point is considered as the origin.

3. The DC load curve without any changes and the AC load curve with positive slope is being put in the system of equations. By solving the equations, the maximum available swing will be obtained.

It must be mentioned that there is no need for the equality of positive and negative swing at all. These swings can have different values.

Now for example, we want to calculate the maximum available symmetric swing of the output in the voltage divider circuit of Figure 30 which uses the bypass capacitor in parallel with emitter resistor (Jaeger & Blalock, 2003).

Figure 30. Voltage divider with bypass capacitor

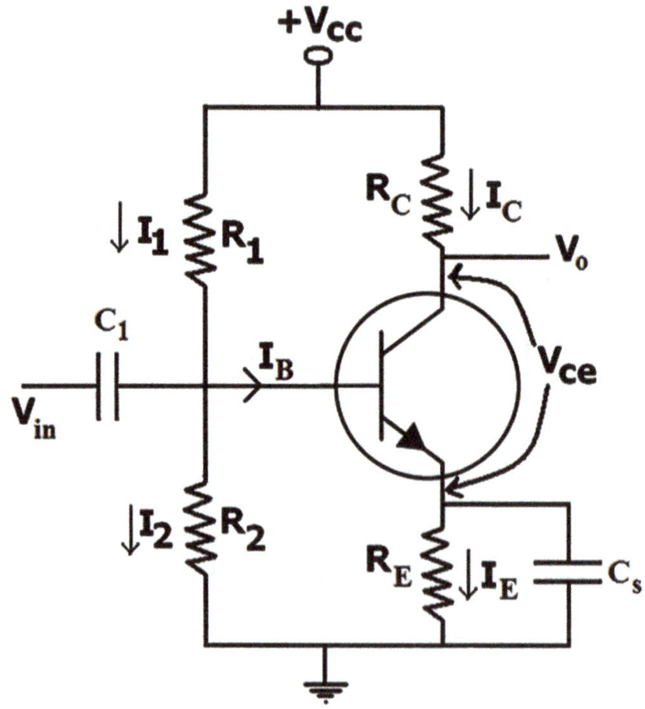

Since in large signal analysis capacitor is open circuited, then for the DC load curve, we have:

$$V_{CEQ} = V_{CC} - R_{dc} I_{CQ} \tag{47}$$

where, $R_{dc}=R_C+R_E$. For the AC load curve, at first the DC power sources must be set to zero. The equivalent circuit is shown in Figure 31 in which $R_B=R_1||R_2$.

For AC sources, C_s bypasses the emitter resistor. Therefore, the AC load curve will be as follows:

$$v_{ce} = -R_{ac} i_c \tag{48}$$

in which, $R_{AC}=R_C$. Since the Q-point parameters satisfy Equation (48), hence we can write:

$$V_{CEQ} = R_{AC} I_{CQ} \tag{49}$$

which is the AC load curve with positive slope. By means of Equations (47) and (49), the maximum available symmetric swing can be calculated. If we consider the effect of saturation voltage, we'll obtain the maximum swing parameters as follows:

$$I_{CQ} = \frac{V_{CC} - V_{CE(sat)}}{R_{ac} + R_{dc}}, V_{CEQ} = \frac{R_{ac}}{R_{ac} + R_{dc}} \left(V_{CC} - V_{CE(sat)} \right) \tag{50}$$

Figure 31. AC equivalent circuit of Figure 30

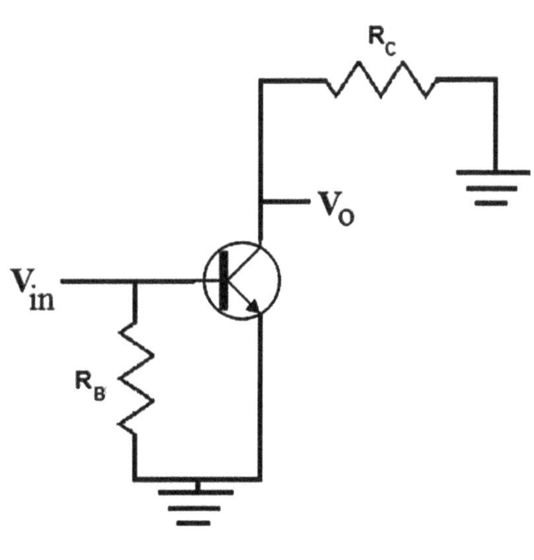

Amplification of BJT

As discussed in previous section, the amplification comes from small signal analysis of BJT. But in order to analyze the amplification, a suitable model must be defined for BJT. A bipolar transistor has three terminals and cannot be modeled with a one port network. A two port model is a suitable choice for this purpose. The two port network shown in Figure 32 can be described by different models depending on which signals are assumed as the inputs and which signals are considered as the outputs.

Since the transistor is a three port component, one of the terminals can be held in common and the two port configuration of Figure 32 would be simplified as shown in Figure 33.

For the configuration of Figure 33 the modeling will be simplified because the T and π models can be defined without any restriction. Based on this structure, 6 different models would be available for a transistor. But since the collector will rarely be used as the signal input (except for avalanche transistors which is out of our discussion) and the base isn't used as the output, there will exist 3 patterns for a BJT. These are:

1. Common Emitter (CE) configuration in which the base is used as the input node and the collector is employed as the output.
2. Common Collector (CC) configuration in which the base is used as the input node and the emitter is employed as the output.
3. Common Base (CB) configuration in which the emitter is used as the input node and the emitter is employed as the output.

Based on these structures, 3 different models will be defined for a BJT. But as the base-emitter diode is forward biased and the base-collector diode is in reverse bias region, some factors are in direct rela-

Figure 32. A two port network

Figure 33. Transistor as two port network

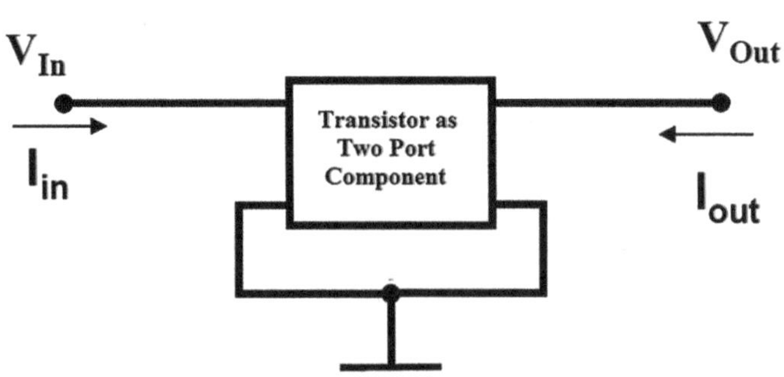

tion with each other. By means of Equation (17), I_C and V_{BE} are related to each other. Also, we know that every increment in base voltage enhances the emitter voltage whilst the collector voltage drops along with the increment of base voltage. We'll define the π model for CE configuration which has no limitations for use in CC and CB structures. If the π models of CC and CB are employed, with proper coefficients they will be transformed to CE model. But the primary question which was about the usage of proper two port parameters for modeling is not answered yet. Each of the models from Z parameters to Y can be used to define a model. But the simplest one will be obtained by means of Hybrid Model parameters. Based on hybrid parameters, the π model of CE configuration is shown in Figure 34 (Sedra & Smith, 2004; Jaeger & Blalock, 2003).

According to the π model of CE, four important parameters must be calculated which are g_m, r_π, r_o and r_μ. The parameters will be obtained as follows.

As the first parameter, we want to calculate the effect of input in output (known as the gain). This parameter is known as a_m and one can say that it is the most important factor which is used in transistor calculations. As shown in Figure 34, if the small signal section of base-emitter voltage is changed by V_π value, then it will cause a change in collector current equal to $g_m V_\pi$. Therefore, we can write:

$$g_m = \frac{\partial I_C}{\partial V_{BE}} = \frac{I_S}{V_T} e^{\frac{V_{BE}}{V_T}} \tag{51}$$

However, $I_S e^{\frac{V_{BE}}{V_T}}$ is the operating point current. Therefore, one can say:

$$g_m = \frac{I_C}{V_T} \tag{52}$$

g_m is known as the transistor transconductance. As the second parameter, we want to calculate r_π. Since it is the input impedance, in a two port network the output must be short circuited to obtain this component. Therefore, we'll obtain:

Figure 34. π model of CE configuration

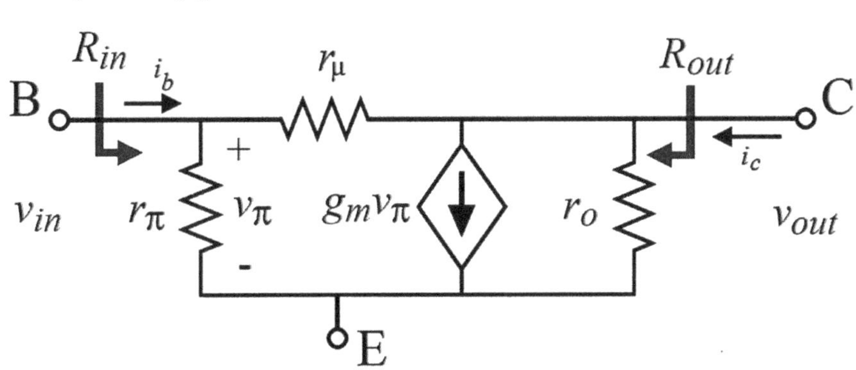

$$r_{\pi} = \frac{\Delta V_{BE}}{\Delta I_B} \tag{53}$$

which means:

$$\frac{1}{r_{\pi}} = \frac{\partial I_B}{\partial V_{BE}} = \frac{1}{\beta}\frac{\partial I_C}{\partial V_{BE}} \tag{54}$$

But from Equation (51) we know that $g_m = \frac{\partial I_C}{\partial V_{BE}}$. Therefore, one can write:

$$r_{\pi} = \frac{\beta}{g_m} \tag{55}$$

As the third parameter, we want to calculate the output impedance the model. If we change V_{CE}, because of the inclined lines in active region of Figure 21, I_C will be varied which was shown in Equation (23). The changes are linear and we can use a resistor in π model denoted as r_o.

To calculate r_o, the input is set to zero. Therefore, V_{π} becomes zero and there will be no Voltage Controlled Current Source (VCCS) in output node. Now, one can say:

$$r_o = \frac{\Delta V_{CE}}{\Delta I_C} \tag{56}$$

which means:

$$\frac{1}{r_o} = \frac{\partial I_C}{\partial V_{CE}} \tag{57}$$

By means of Equation (23), we'll obtain:

$$\frac{\partial I_C}{\partial V_{CE}} = I_S e^{\frac{V_{BE}}{V_T}} \times \frac{1}{V_A} = \underline{\left| I_S \left(1 + \frac{V_{CE}}{V_A}\right) e^{\frac{V_{BE}}{V_T}} \right.} \times \frac{1}{V_A + V_{CE}} \tag{58}$$

The selected term in Equation (58) is the operating point current. Therefore, one can write:

$$r_o = \frac{V_A + V_{CE}}{I_C} \tag{59}$$

Usually $V_{CE} \ll V_A$, which simplifies Equation (59) as follows:

$$r_o = \frac{V_A}{I_C} \tag{60}$$

The final parameter is defined by the effect of output at input. As discussed before, along with changes of V_C, the thickness of base differs which affects I_B. So, the input current has knowledge of the output voltage changes. This effect is modeled with a resistor named r_μ. Just like previous calculations, we can write:

$$\frac{1}{r_\mu} = \frac{\partial I_B}{\partial V_{CE}} \tag{61}$$

which results in:

$$r_\mu = \beta r_o \tag{62}$$

In practice, r_μ is greater than βr_o. To explain this, one can say that the base current is composed two components: the first one which comes from the penetrated electrons of emitter and the second one which is the reflected current from base to emitter. r_μ is only affected by first part and in new IC design technologies the first fraction constitutes only 10% of the total base current. Therefore:

$$r_\mu \gg \beta r_o \tag{63}$$

and we can remove it from the π model and simplify it as shown in Figure 35.

By means of the attained model, we are going to analyze the amplification feature of BJT. The start point for this purpose is CE configuration of Figure 30 which contains the bypass capacitor for emitter resistor. If this amplifier is to drive a load like R_L, then we'll obtain the scheme of Figure 36.

Figure 35. Simplified π model

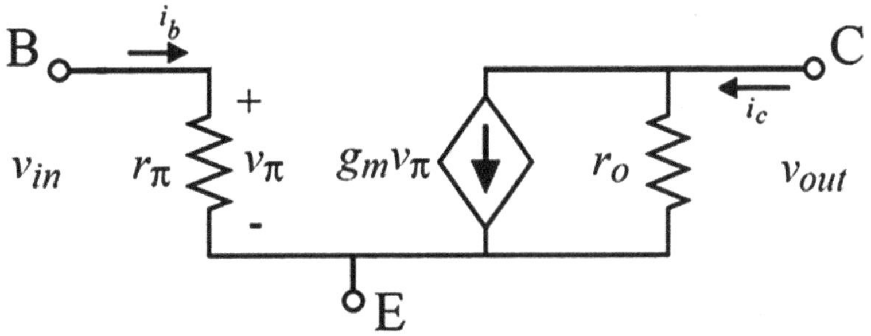

Figure 36. CE configuration with load resistor

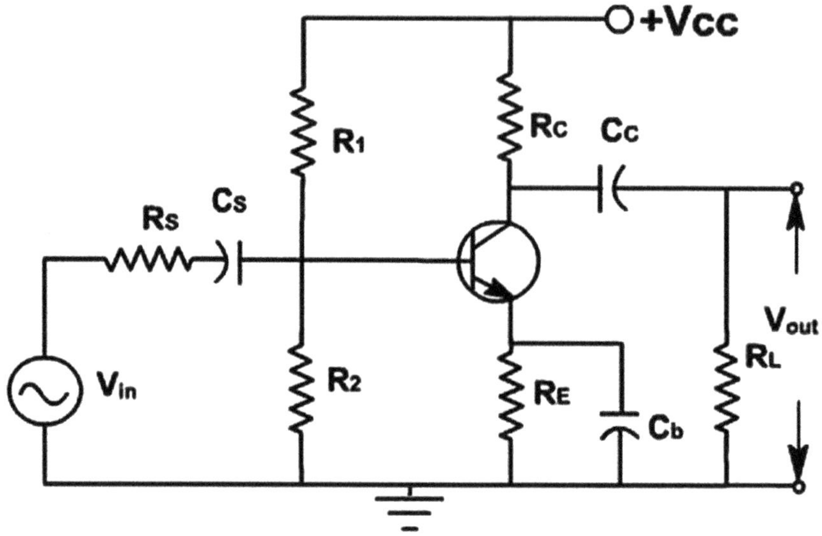

The equivalent circuit with the use of π model is shown Figure 37 which in small signal analysis. For this circuit, four parameters must be calculated. They are:

- Current gain,
- Voltage gain,
- Input impedance, and
- Output impedance.

The voltage gain is defined as the ratio of output signal V_{out} to input signal V_{in}. Therefore, we can write:

$$v_{out} = -i_{out}\left(R_C \parallel R_L \parallel r_o\right) = -g_m V_\pi \left(R_C \parallel R_L \parallel r_o\right) \tag{64}$$

Figure 37. Small signal equivalent circuit for circuit of Figure 36

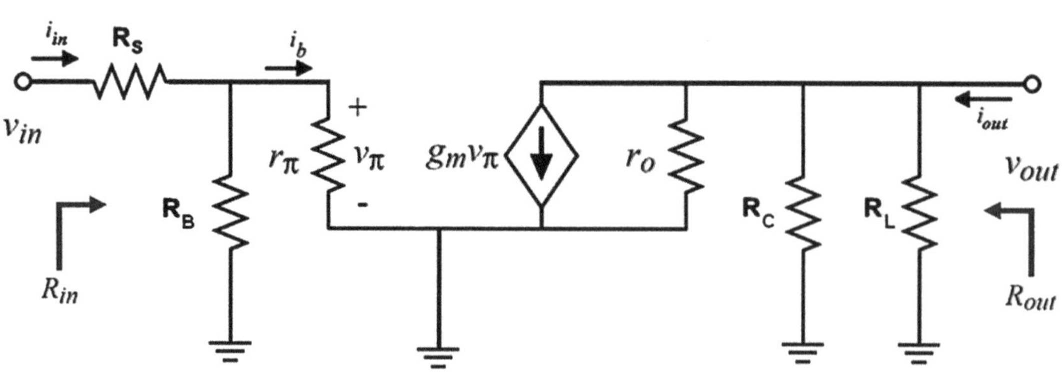

Also, for the input part, it can be written:

$$v_{in} = V_{\pi}\left(\frac{R_S + R_B \parallel r_{\pi}}{R_B \parallel r_{\pi}}\right) \tag{65}$$

in which $R_B=R_1||R_2$. Now, the voltage gain denoted as A_V will be equal to:

$$A_V = \frac{v_{out}}{v_{in}} = -g_m\left(R_C \parallel R_L \parallel r_o\right)\frac{R_B \parallel r_{\pi}}{R_S + R_B \parallel r_{\pi}} \tag{66}$$

In most cases $R_B||r_{\pi} \gg R_s$ and $r_o \gg R_C||R_L$. Hence, Equation (66) can be simplified as:

$$A_V = \frac{v_{out}}{v_{in}} = -g_m\left(R_C \parallel R_L\right) \tag{67}$$

According to practical results, A_V could have a great value. Here, of the intrinsic factors of BJT can be defined as inherent gain. If two ideal current sources are put instead of collector and load resistors, we'll have $R_C||R_L \gg r_o$. Therefore, Equation (66) can be simplified as:

$$A_{V_{inherent}} = \frac{v_{out}}{v_{in}} = -g_m r_o \tag{68}$$

Substituting from (52) and (60), Equation (68) would be simplified as:

$$A_{V_{inherent}} = -\frac{I_C}{V_T} \times \frac{V_A}{I_C} = -\frac{V_A}{V_T} \tag{69}$$

According to typical values of V_A and V_T, the inherent gain would be greater than 4000. For current gain which is defined as the ratio of output signal I_{out} to input signal I_{in}, we have:

$$A_I = \frac{i_{out}}{i_{in}} = -g_m r_{\pi}\left(\frac{R_C \parallel R_L}{r_o + R_C \parallel R_L}\right) \tag{70}$$

For $r_o \gg R_C||R_L$, Equation (70) tends to $-g_m r_{\pi}$ in which $g_m r_{\pi}=\beta$. For the input impedance V_{out} is set to zero. Thus, it can be written:

$$R_{in} = \frac{v_{in}}{i_{in}} = R_S + R_B \parallel r_{\pi} \tag{71}$$

For V_{in}=0, the output impedance is equal to:

$$R_{out} = \frac{v_{out}}{i_{out}} = R_C \parallel R_L \parallel r_o \cong R_C \parallel R_L \tag{72}$$

On the calculations, the capacitors are assumed as big as possible so that at the operating frequency of the circuit could be substituted by low impedances.

To show the effect of emitter resistance, let's suppose the circuit of Figure 36 without bypass capacitor C_b. The small signal equivalent circuit is illustrated in Figure 38.

Since r_o has a large value, the passing current across it will have a negligible value. For simplification, it is considered as zero. Therefore, v_e is equal to:

$$v_e = R_E\left(i_b + g_m V_\pi\right) = R_E\left(i_b + g_m r_\pi i_b\right) \tag{73}$$

$$g_m r_\pi = \beta .$$

Therefore:

$$v_e = i_b R_E\left(1+\beta\right) \tag{74}$$

which indicates that from the base viewpoint, the emitter resistance has been enlarged (1+β) times. But from collector, the emitter resistor is almost unchanged because $I_C \cong I_E$. For accurate calculations, one can say:

$$\alpha = \frac{\beta}{1+\beta} \tag{75}$$

Figure 38. Small signal equivalent circuit for circuit of Figure 36

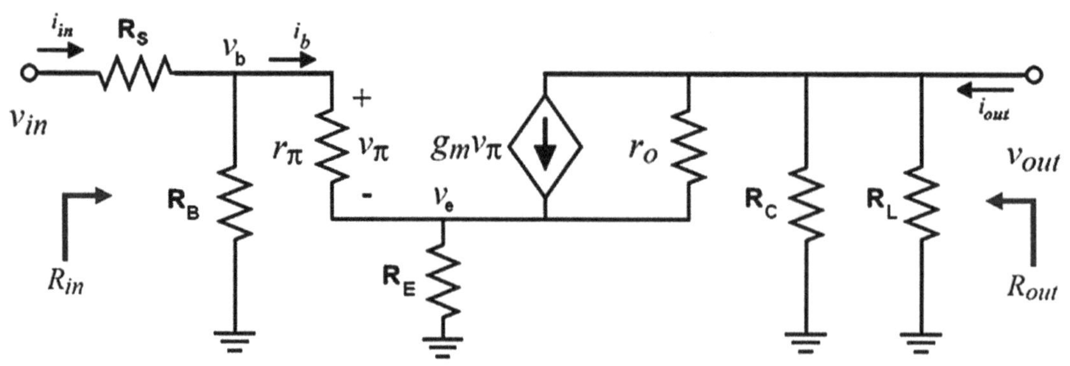

and in order to show the reflection of emitter resistance in collector, R_E will be multiplied in $\frac{1}{\alpha}$. This reflection is known as the dual form of Miller's Theorem. According to this theorem, in a linear circuit if an impedance Z connects two nodes with voltage values of V_1 and V_2 and we can independently calculate the ratio of $\frac{V_2}{V_1}$ which is equal to parameter G, then we can replace the branch consisting of impedance Z with two branches connecting the nodes to the ground with the values $\frac{Z}{1-G}$ and $\frac{GZ}{G-1}$, respectively for V_1 and V_2. The dual form applies for the ratio of the currents in which for $\frac{I_2}{I_1} = K$, the impedance Z will be replaced with two impedances from the corresponding nodes to ground having values of $(1+K)Z$ and $\frac{K+1}{K}Z$, respectively for I_1 and I_2. Here for Figure 38, K is the β of the transistor.

Meanwhile, the reflected resistor in collector is in series with the VCCS. Therefore, it can be omitted and the simplified circuit can be realized as the one shown in Figure 39.

Now, the voltage gain can be calculated as follows:

$$v_{out} = -i_{out}\left(R_C \parallel R_L \parallel r_o\right) = -g_m V_\pi \left(R_C \parallel R_L \parallel r_o\right) \tag{76}$$

$$V_\pi = v_b \left(\frac{r_\pi}{r_\pi + \left(1+\beta\right) R_E}\right) \tag{77}$$

In bias circuit design it is considered that $R_B \ll (1+\beta)R_E$. Therefore, v_b will equal to:

$$v_b = \frac{R_B}{R_B + R_S} v_{in} \tag{78}$$

and as said before, for $R_B \gg R_S$, $v_b \cong v_{in}$. Now, we obtain:

Figure 39. Simplified circuit of Figure 38 by Miller's Theorem

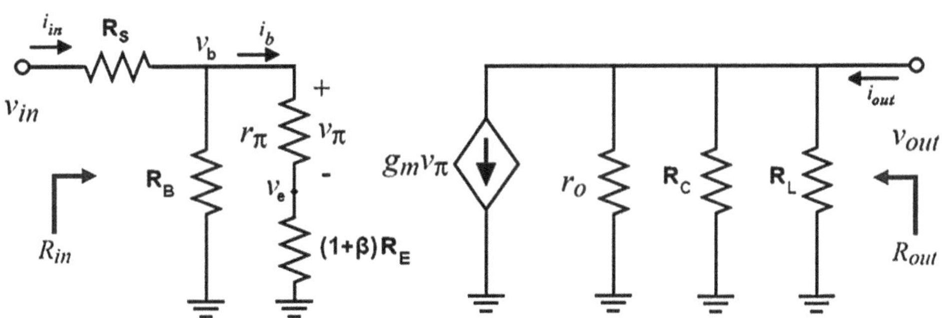

$$A_V = \frac{v_{out}}{v_{in}} = -\frac{g_m r_\pi \left(R_C \parallel R_L \parallel r_o\right)}{r_\pi + \left(1+\beta\right) R_E} \tag{79}$$

in which for $r_o \gg R_C \| R_L$, we have:

$$A_V = \frac{v_{out}}{v_{in}} = -\frac{\beta\left(R_C \parallel R_L\right)}{r_\pi + \left(1+\beta\right) R_E} \tag{80}$$

With the exception of low bias currents, usually $(1+\beta)R_E \gg r_\pi$. Thus, the following simplification for Equation (80) can be done:

$$A_V = -\frac{\beta\left(R_C \parallel R_L\right)}{\left(1+\beta\right) R_E} \cong -\frac{R_C \parallel R_L}{R_E} \tag{80}$$

and in simple words, the gain which was before equal to $-\frac{R_C \parallel R_L}{\frac{1}{g_m}}$, now is reduced to $-\frac{R_C \parallel R_L}{R_E}$. This expresses why the emitter resistor is bypassed whilst it doesn't affect the current gain as well as output impedance. But the input impedance is now equal to:

$$R_{in} = \frac{v_{in}}{i_{in}} = R_S + \left(R_B \parallel \left(r_\pi + \left(1+\beta\right) R_E\right)\right) \tag{81}$$

and because of design considerations R_{in} is almost equal to R_B. To increase the input impedance, sometimes the bias level is reduced but the best solution is the use of Bootstrap Technique (IEEE Standards Terms). It is shown in Figure 40.

In DC bias, the capacitor C_2 doesn't have any effect and is considered as open circuit. But for small signal, it shows low impedance and can be considered as short circuit. Therefore, from the signal viewpoint the resistor R_3 is connected to the input voltage from one side, while its other side is attached to out_2 node (the emitter of transistor) (Uyemura, 1999). In CC configuration we'll see that the voltage gain has a value very close to unity. By means of Miller's Theorem, this resistor can be substituted by two resistors in which for the input part the reflected value will be $\frac{R_3}{1-A_{v2}}$ and for the output section the corresponding value will be $\frac{A_{v2} R_3}{A_{v2} - 1}$. Since $A_{v2} \cong 1$, the input part will see a large resistance and the output section will observe a large negative resistance. Therefore, the input impedance will be:

$$R_{in} = Z_1 \parallel \left(r_\pi + \left(1+\beta\right)\left(R_E \parallel R_1 \parallel R_2\right)\right) \tag{82}$$

Figure 40. Bootstrap technique for input impedance enhancement

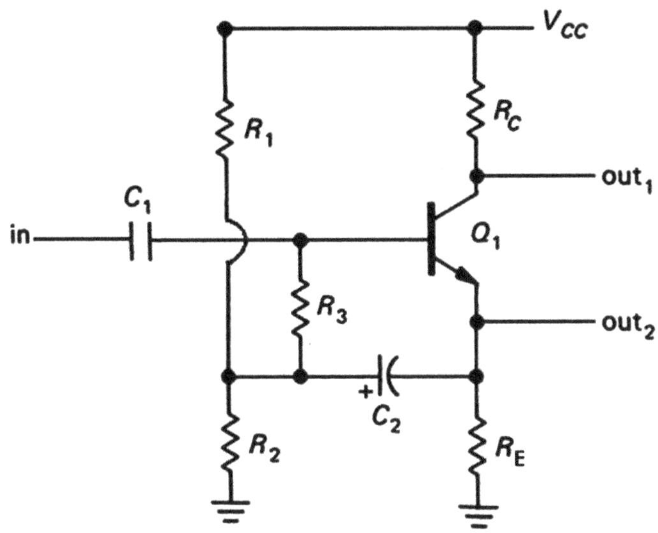

in which, $Z_1 = \dfrac{R_3}{1 - A_{v2}}$. As $Z_1 \gg \beta(R_E || R_1 || R_2) \gg r_\pi$, we will obtain:

$$R_{in} \cong \beta\left(R_E \parallel R_1 \parallel R_2\right) \tag{83}$$

The capacitor C_2 in Figure 40 is known as the Bootstrap Capacitor and its role is to enhance the input impedance of the transistor. The circuit of Figure 40 is also a pretext to introduce the CC configuration. As the name suggests, from the small signal view the collector is held as common terminal between input and output. The relevant circuit is shown in Figure 41.

As it is clear from the structure of Figure 41, in CC configuration it is nonsense to employ the collector resistor. It doesn't have any effect on corresponding voltage ad current gains, but reduces the voltage swing of the output node. For small signal an analysis, the equivalent circuit is illustrated in Figure 42 (Jaeger & Blalock, 2003).

For the output voltage, we have:

$$v_{out} = \left(R_E \parallel R_L \parallel r_o\right)\left(1 + g_m r_\pi\right) i_b \tag{84}$$

Also, the input voltage with the consideration of the venin circuit for the left side of v_b, can be written as:

$$v_{in} = \left(1 + \frac{R_S}{R_B}\right)\left[\left(R_B \parallel R_S\right) + r_\pi + \left(1 + g_m r_\pi\right)\left(R_E \parallel R_L \parallel r_o\right)\right] i_b \tag{85}$$

But as discussed before, $R_B \gg R_S$ which simplifies Equation (85) as:

Figure 41. CC configuration with load resistor

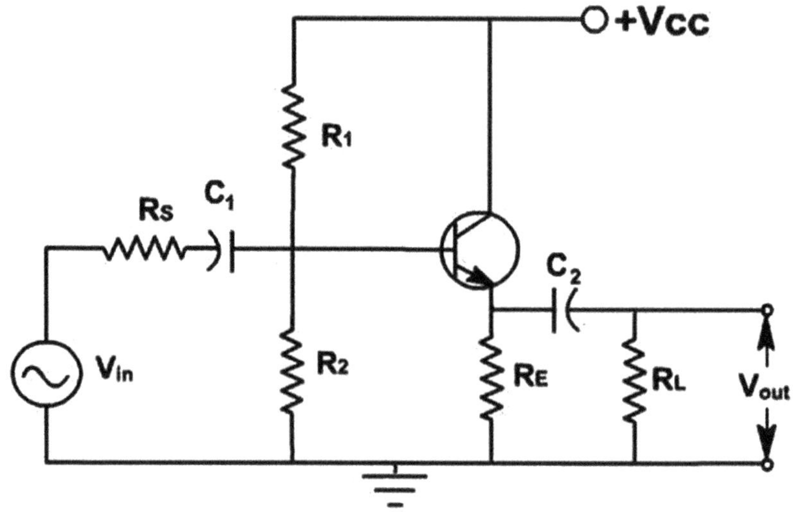

Figure 42. Small signal equivalent circuit for Figure 41

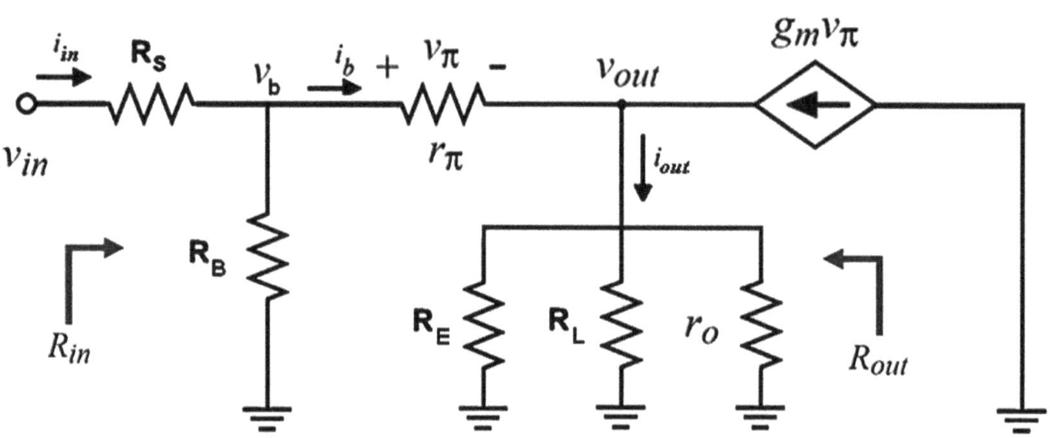

$$v_{in} = \left[R_S + r_\pi + \left(1 + g_m r_\pi\right)\left(R_E \parallel R_L \parallel r_o\right)\right] i_b \tag{86}$$

Now, the voltage gain can be calculated as follows:

$$A_V = \frac{v_{out}}{v_{in}} = \frac{\left(R_E \parallel R_L \parallel r_o\right)\left(1 + g_m r_\pi\right)}{R_S + r_\pi + \left(1 + g_m r_\pi\right)\left(R_E \parallel R_L \parallel r_o\right)} \tag{87}$$

For $r_o \gg R_E \| R_L$ and $r_\pi \gg R_S$, and by approximating $1+g_m r_\pi=\beta$, one can obtain:

$$A_V = \frac{v_{out}}{v_{in}} = \frac{\left(1+g_m r_\pi\right)\left(R_E \parallel R_L\right)}{r_\pi + \left(1+g_m r_\pi\right)\left(R_E \parallel R_L\right)} \cong \frac{\beta\left(R_E \parallel R_L\right)}{r_\pi + \beta\left(R_E \parallel R_L\right)} \tag{88}$$

then, by dividing the numerator and denominator to β, the expression for voltage gain can be simplified as:

$$A_V \cong \frac{R_E \parallel R_L}{\frac{r_\pi}{\beta} + \left(R_E \parallel R_L\right)} \tag{89}$$

But $\frac{r_\pi}{\beta}$ is equal to $\frac{1}{g_m}$. The $\frac{1}{g_m}$ parameter is the dynamic resistance of base-emitter diode and is denoted as r_e. It is also the reflected value of r_π from the emitter viewpoint, which indicates that any impedance in base if is to travel to the emitter, it must be divided to 1+β or as an approximation, it must be divided to β.

As Equation (89) illustrates, the voltage gain is less than unity. In order to make it as close as possible to unity value, the collector current must be large. For example, if we set I_C=2*mA* and R_E=R_L=2*K*Ω, then the voltage gain will be:

$$A_V = \frac{1000}{12.5+1000} = 0.988 \tag{90}$$

As discussed in voltage divider bias circuit, the emitter resistor provides a local feedback to adjust the base-emitter diode voltage variations. Since a same process occurs in CC configuration and the emitter trails the base voltage changes, this configuration is also known as Emitter Follower amplifier.

For the current gain, we have:

$$A_I = \frac{i_{out}}{i_{in}} = 1 + g_m r_\pi = 1 + \beta \tag{91}$$

Also, by means of impedance reflections in base and emitter, it can simply be obtained:

$$R_{out} = \frac{R_s + r_\pi}{1+\beta} \parallel R_E \parallel R_L \parallel r_o \tag{92}$$

$$R_{in} = R_S + \left(R_B \parallel \left(r_\pi + \left(1+\beta\right)\left(R_E \parallel R_L \parallel r_o\right)\right)\right) \tag{93}$$

and with proper approximations used for voltage gain calculations, Equations (92) and (93) will be simplified as:

$$R_{out} \cong \frac{R_s}{\beta} + \frac{1}{g_m} = \frac{R_s}{\beta} + r_e \quad (94)$$

$$R_{in} \cong R_S + R_B \quad (95)$$

With Bootstrap Technique the input impedance can be enlarged significantly which shows that this configuration with good design considerations, can show the characteristics of an ideal voltage buffer (an ideal voltage buffer has high input impedance along with low output impedance and voltage gain equal to unity).

The last single stage amplifier of BJT which has to be analyzed is CB configuration with the schematic shown in Figure 43.

In this circuit the emitter is the input node for signal while the collector is the output. Just like CE circuit, C_b is the bypass capacitor which bypasses the bias resistors so that they won't affect the voltage gain value. The main difference of this circuit with previous configurations is that for DC and AC load curves the KVL must be applied to output loop consisting of collector and base (Boylestad & Nashelsky, 2002; Horowitz & Hill, 2015) We know that:

$$V_{BE} = V_{BC} + V_{CE} \quad (96)$$

By means of (96), the prerequisites can be obtained for bias circuit design. The rest is the same as described for self-bias circuit of Figure 28. But for small signal an analysis, we have the equivalent circuit of Figure 44.

Figure 43. Scheme of CB configuration

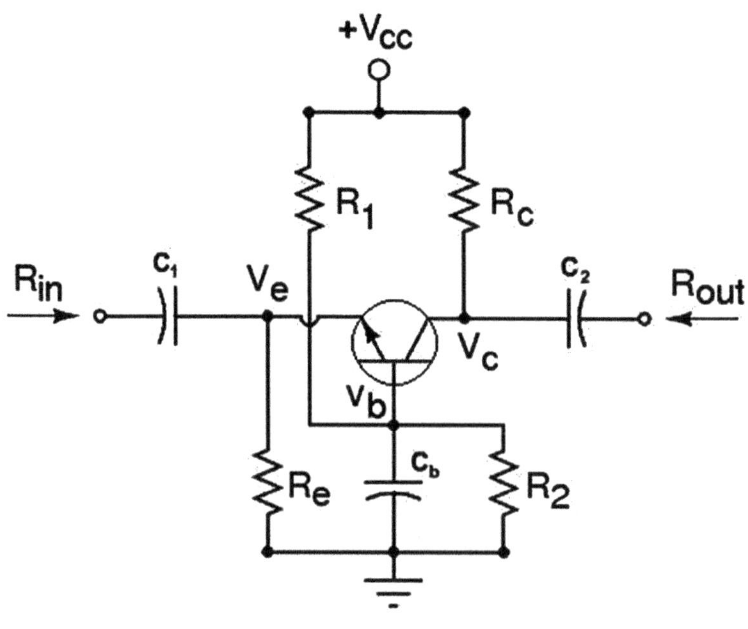

Figure 44. Small signal equivalent circuit for Figure 43

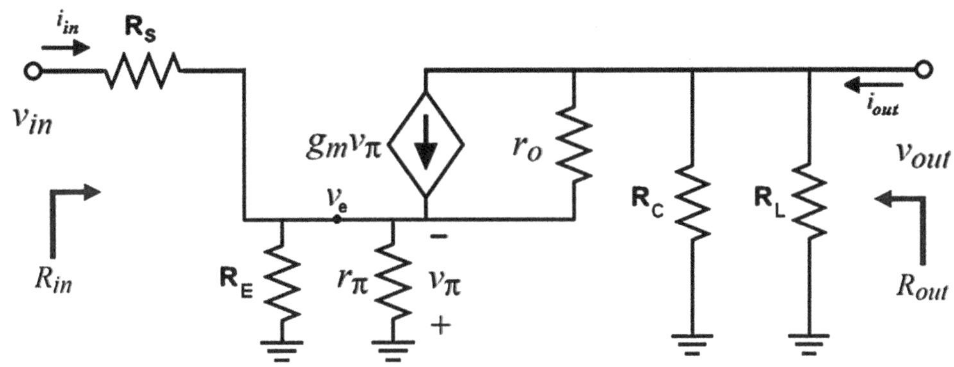

Again, R_S is the source resistor and R_L is the load. With the following calculations we'll see that unlike CE and CC, the source and load impedances directly affect the small signal values except current gain. By applying Kirchhoff's Current Law (KCL) to emitter node and knowing that $V_e=-V_\pi$, we obtain:

$$\frac{v_{in}+v_\pi}{R_S}+\frac{v_\pi}{R_E}+\frac{v_\pi}{r_\pi}+g_m v_\pi+\frac{v_{out}+v_\pi}{r_o}=0 \tag{97}$$

Also, the KCL in output node results in:

$$\frac{v_{out}+v_\pi}{r_o}+g_m v_\pi+\frac{v_{out}}{R_C}+\frac{v_{out}}{R_L}=0 \tag{98}$$

Therefore, from Equation (98), the relation between v_π and v_{out} is:

$$v_\pi=-\frac{r_o}{\left(1+g_m r_o\right)\left(r_o \parallel R_C \parallel R_L\right)}v_{out} \tag{99}$$

In order to simplify (97), instead of $g_m v_\pi+\frac{v_\pi}{r_o}$ term we substitute the equivalent from Equation (98) which results in:

$$v_\pi\left(\frac{1}{R_S}+\frac{1}{R_E}+\frac{1}{r_\pi}\right)+\frac{v_{in}}{R_S}+\frac{v_{out}}{r_o}-v_{out}\left(\frac{1}{R_L}+\frac{1}{R_C}+\frac{1}{r_o}\right)=0 \tag{100}$$

Now, the calculated value for in (99) must be put in (100). The result is:

$$\frac{v_{in}}{R_S} = v_{out}\left(\frac{1}{R_C \parallel R_L} + \frac{1}{r_\pi \parallel R_S \parallel R_E}\left(\frac{r_o}{\left(1+g_m r_o\right)\left(r_o \parallel R_C \parallel R_L\right)}\right)\right) \tag{101}$$

From Equation (68) we know that $g_m r_o$ is the inherent gain of transistor in which $A_{V_{inherent}} \gg 1$. Hence, (101) can be simplified as:

$$\frac{v_{in}}{v_{out}} = \frac{R_S}{R_C \parallel R_L} + \frac{R_S}{r_\pi \parallel R_S \parallel R_E}\left(\frac{1}{g_m\left(r_o \parallel R_C \parallel R_L\right)}\right) \tag{102}$$

As we observe, the right side of (102) contains two fractional terms. To calculate the voltage gain based upon the value of , two different situations may happen. If $R_S \gg \frac{1}{g_m}$, then the first term determines the voltage gain as follows:

$$A_V = \frac{v_{out}}{v_{in}} = \frac{R_C \parallel R_L}{R_S} \tag{103}$$

which shows that the gain is no longer depended to transistor parameters, but to the source and load impedances. For $R_S \ll \frac{1}{g_m}$ the second term is dominant which simplifies the gain expression as follows:

$$A_V = \frac{v_{out}}{v_{in}} = g_m\left(r_o \parallel R_C \parallel R_L\right) \tag{104}$$

and for $r_o \gg R_C \| R_L$, just like the one we obtained for CE configuration without emitter resistor, we'll obtain:

$$A_V = \frac{v_{out}}{v_{in}} = g_m\left(R_C \parallel R_L\right) \tag{105}$$

But here unlike CE circuit, the emitter and collector are at the same phase and the amplification has a positive value. A point which must be considered, is that for $R_S = \frac{1}{g_m}$ both of the terms in right side of Equation (102) will be equal and the gain will be:

$$A_V = \frac{v_{out}}{v_{in}} = \frac{g_m}{2}\left(R_C \parallel R_L\right) \tag{106}$$

For the input impedance we consider the structure of Figure 45.

Figure 45. The circuit for input and output impedance calculation

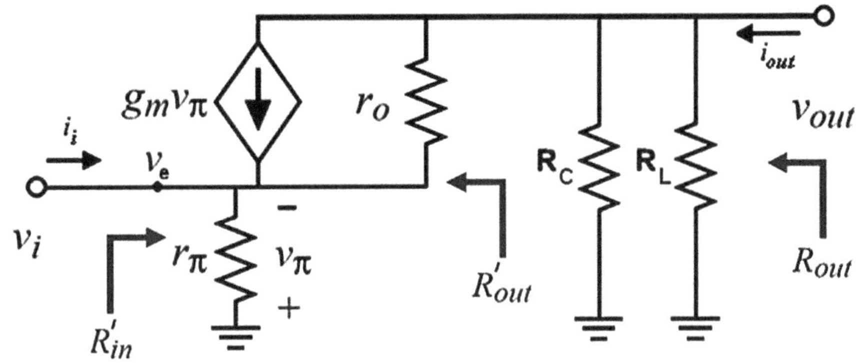

At first R'_{in} must be calculated and it is clear that $v_i = -v_\pi$. Therefore, we can write:

$$v_i = r_o\left(i_i - \frac{v_i}{r_\pi} - g_m v_i\right) + \left(R_C \parallel R_L\right)\left(i_i - \frac{v_i}{r_\pi}\right) \tag{107}$$

$$v_i\left(1 + \frac{\left(1 + g_m r_\pi\right) r_o + \left(R_C \parallel R_L\right)}{r_\pi}\right) = i_i\left(r_o + \left(R_C \parallel R_L\right)\right) \tag{108}$$

in which it will result in:

$$R'_{in} = \frac{v_i}{i_i} = \frac{r_\pi\left(r_o + \left(R_C \parallel R_L\right)\right)}{r_\pi + \left(1 + \beta\right) r_o + \left(R_C \parallel R_L\right)} \tag{109}$$

We know that $\beta \gg 1$ and $\beta r_o \gg r_\pi$. Thus, Equation (109) can be simplified as:

$$R'_{in} \cong \frac{r_\pi\left(r_o + \left(R_C \parallel R_L\right)\right)}{\beta r_o + \left(R_C \parallel R_L\right)} \tag{110}$$

Again, two states may occur. If active loads are employed instead of the load resistors, then $R_C \| R_L$ tends to infinity and this factor determines the value of R'_{in}. The result is:

$$R'_{in} \cong r_\pi \tag{111}$$

But for normal cases we usually have $r_o \gg (R_C \| R_L)$ which means:

$$R'_{in} \cong \frac{r_\pi}{\beta} = \frac{1}{g_m} \tag{112}$$

For the input impedance we have:

$$R_{in} = R_S + \left(R_E \parallel R'_{in}\right) \tag{113}$$

Following the same procedure for output impedance calculation in Figure 45 with putting into consideration that the emitter resistor is in parallel with r_π, we'll obtain:

$$R'_{out} = \left(r_\pi \parallel R_E\right) + r_o\left(1 + g_m\left(r_\pi \parallel R_E\right)\right) \tag{114}$$

in which for $R_E \gg r_\pi$, Equation (114) can be simplified as:

$$R'_{out} = r_\pi + r_o\left(1+\beta\right) \cong \beta r_o \tag{115}$$

and the output impedance is:

$$R_{out} = R'_{out} \parallel R_C \parallel R_L \cong R_C \parallel R_L \tag{116}$$

As the final parameter, the current gain for R_L=0, is:

$$A_I = \frac{i_{out}}{i_{in}} = \frac{r_\pi + \beta r_o}{r_\pi + \left(1+\beta\right) r_o} \tag{117}$$

in which by approximation we'll obtain:

$$A_I \cong \frac{\beta r_o}{\left(1+\beta\right) r_o} = \frac{\beta}{\beta + 1} = \alpha \tag{118}$$

As Equation (118) expresses, CB configuration has a current gain less than unity. Since the input impedance has a small value whilst the output impedance contains a large value, this amplifier is a current buffer (current follower) or in other words, it is a bilateral which connects the output to the input.

A comparison between the three different configurations of BJT is summarized in Table 1. As Table 1 suggests, CE configuration features lies between CC and CB. It is more suitable for gain stage design than those two. In multistage amplifier design section we'll see that with some modifications, CB is sometimes employed as the name of Cascode amplifier. CC structure is good to be used as the final stage (output stage) because of voltage buffering property. The modified structure of CC known as Darlington amplifier is sometimes used in output stages in which high output current is required.

Table 1. The comparison between CE, CC and CB configurations

Amplifier	A_V	A_I	R_{in}	R_{out}
CE (without R_E)	$-g_m(R_C\|\|R_L)$	$-\beta$	$R_S+R_B\|\|r_\pi$	$R_C\|\|R_L$
CE (with R_E)	$-\dfrac{\beta(R_C \parallel R_L)}{r_\pi+(1+\beta)R_E}$	$-\beta$	R_S+R_B	$R_C\|\|R_L$
CC	$\dfrac{g_m(R_E \parallel R_L)}{1+g_m(R_E \parallel R_L)}$	$1+\beta$	R_S+R_B	$\dfrac{R_s}{\beta}+\dfrac{1}{g_m}$
CB	$g_m(R_C\|\|R_L)$	$\dfrac{r_\pi+\beta r_o}{r_\pi+(1+\beta)r_o}$	$R_S+\dfrac{1}{g_m}$	$R_C\|\|R_L$

BJTs and Sensors

Just like diodes, BJTs have their applications allocated with sensors. The simplest example for this purpose is Silicon bandgap temperature sensor which is widely used in bandgap voltage reference circuits (Widlar, 1967). But here we want to introduce some simpler configurations which are widely used in non-expensive circuit design criteria. One example is the transistor based version of temperature sensor in Figure 13 which was previously discussed. The circuitry of this sensor is illustrated in Figure 46.

Figure 46. Transistor based version of temperature sensor shown in Figure 13

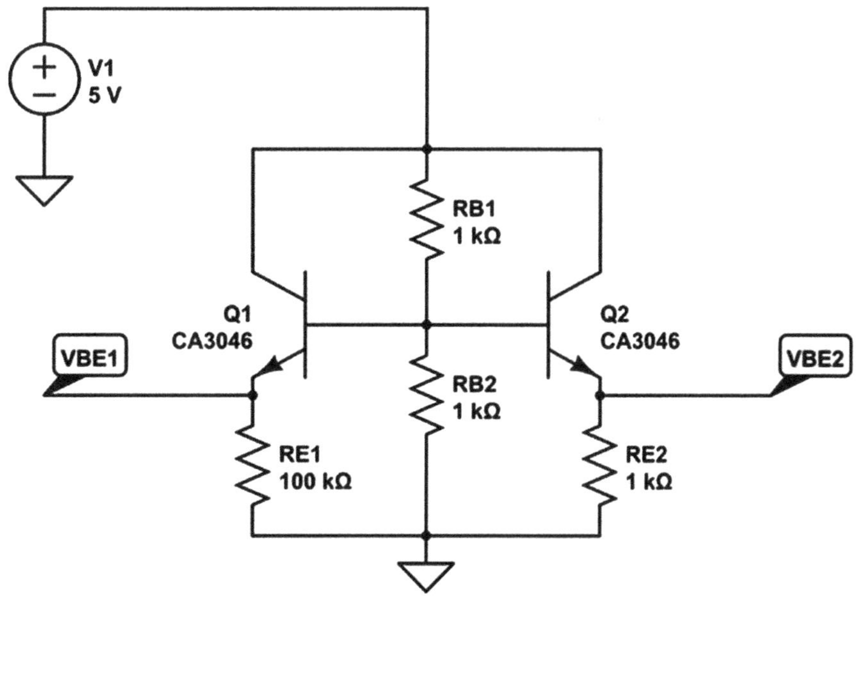

Another example is simple touch alarm circuit of Figure 47. At normal conditions no current flows through switching transistors T1 and T2 and the potential across capacitor C1 is almost 0V. When the touch plates are touched together by a finger, the alarm circuit will be enabled by T1 and T2 (Hareendran, n.p.).

As the last example, Light Dark Sensor of Figure 48 is provided (Electrical Engineer, 2013).

Figure 47. Simple touch alarm circuit
Hareendran, n.p.

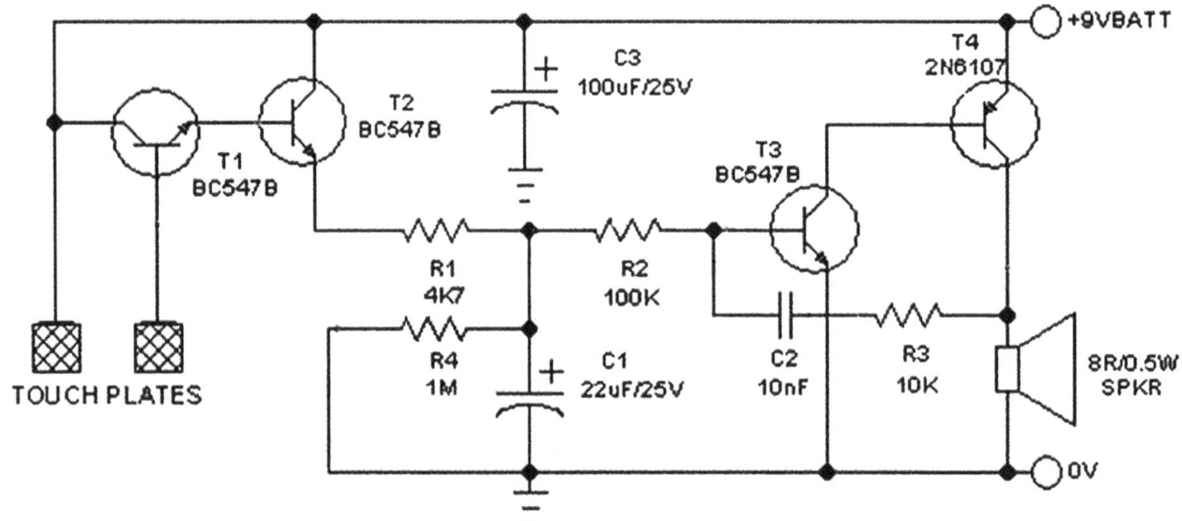

Figure 48. Light dark sensor
Electrical Engineer, 2013.

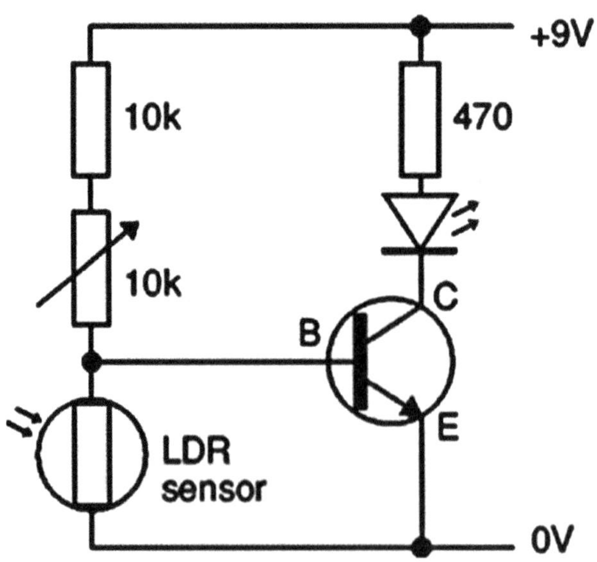

FIELD EFFECT TRANSISTORS

Structure

As the name field-effect transistor (FET) suggests, the group comprises another kind of semiconductor transistors in which by means of an electric field the electrical conductivity of the channel pertaining to the corresponding charge carrier is controlled. As a result, in FETs only one of the charge carriers involve on the flow of electric current and this is why they are known as unipolar transistors (Millman, 1985). The most important different between FETs and BJTs is that as discussed in previous section, in bipolar transistors the output current (I_C) is adjusted by input current but in a FET, the output current is regulated by means of the applied voltage to an isolated terminal from the transistor denoted as the gate (G). The applied voltage induces an electric field inside the device which in turn by attracting or repelling the charge carriers to or from the area between the two other terminals known as the source (S) and the drain (D), controls the flow of current inside FET. As the first property belonging to all types of FETs, since the gate is insulated we expect a high value for input impedance where by the reverse bias of this terminal it can be easily achieved.

For a comparison between FET and BJT, one can say that the input impedance for FET is much more than that of BJT and because the FET is a voltage controlled device, it can provide good isolation between input and output. Because of unipolarity and also because of current noise increment in base of BJT along with shaping time, the produced noise by FET is less than BJT making it better choice for communication equipment design. Also, the thermal stability of FET is better than BJT. But the Gain-Bandwidth Product (GBWP or GBW) of FET is relatively less than BJT. Meanwhile, the FET and especially MOSFET are more vulnerable to electrostatic damage and because of intense differences between ON and OFF resistances the power dissipation over switching is large. But the smaller active area consumption during fabrication process along with lower production costs made the FET the first choice for today's IC design technologies (Jaeger & Blalock, 2003).

Two main classifications of FETs which are mostly employed in amplifier design criteria are JFETs and MOSFETs. Although the operating principle and the electrical properties are almost the same for both classes, their difference comes from the structure due to different fabrication rules. As shown in Figure 49, an N-channel JFET contains two gates (so as P-channel).

For the structure of MOSFET as it can be observed from Figure 50, in addition of gate, drain and source terminals, it has a fourth terminal named as the bulk.

The main differences between JFET and MOSFET are as follows:

1. JFETs can only operate in the depletion mode whilst MOSFETs can work in both depletion and enhancement mode.
2. Since the leakage current is negligible in MOSFETs, they have much higher input impedance than that of JFETs.
3. The characteristic curves for JFETs are more flat than those of MOSFETs along with higher drain resistance for JFETs.
4. When operated with a reverse bias on the junction, the gate current for JFET is larger in comparison with MOSFET.

Figure 49. The structure of N-channel JFET with schematic symbol

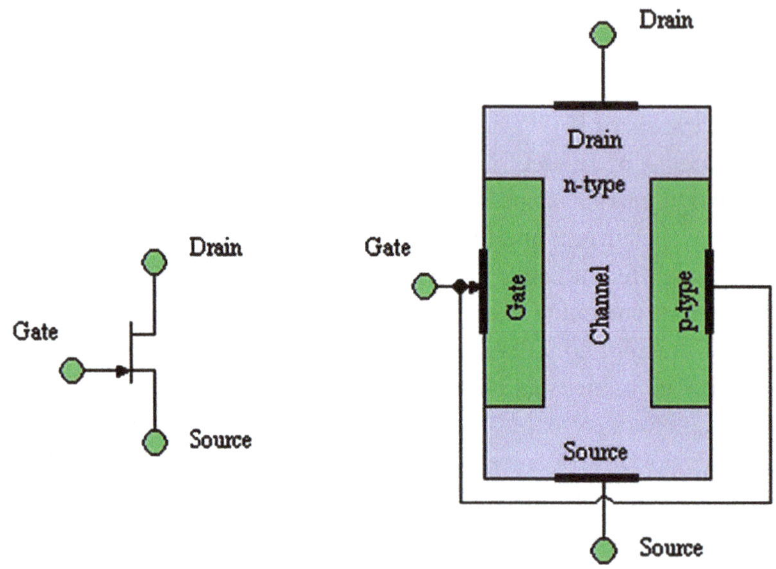

Figure 50. The structure of a MOSFET with schematic symbol

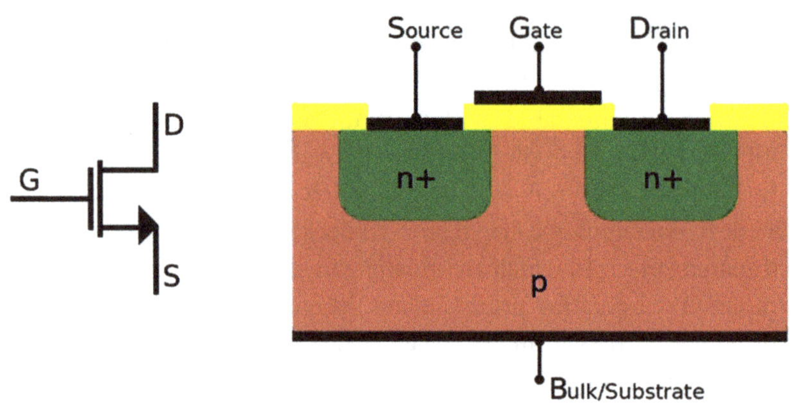

Considering the above differences and based on the fact that the fabrication of MOSFET is somewhat easier than JFET, MOSFETs are widely used in amplifier circuits and IC design.

Operating Principle of FET

In this section we first study the operating principle of JFET. Then it will be expanded for MOSFET. To operate as an amplifier like the one introduced for BJT, certain conditions must be fulfilled for FETs. Thus, as shown in Figure 51, the source is selected as the common terminal and is connected to the ground. Despite of the channel type (N-channel or P-channel) in both cases the applied voltage between the gate and source must be in a way to set the gate in reverse bias region (Galup-Montoro & Schneider, 2007). Because of symmetric structure, the drain and source terminals can interchangeably used instead of each other. This interesting feature suggests that either end can be employed as a source and the other

Figure 51. JFET bias in saturation region

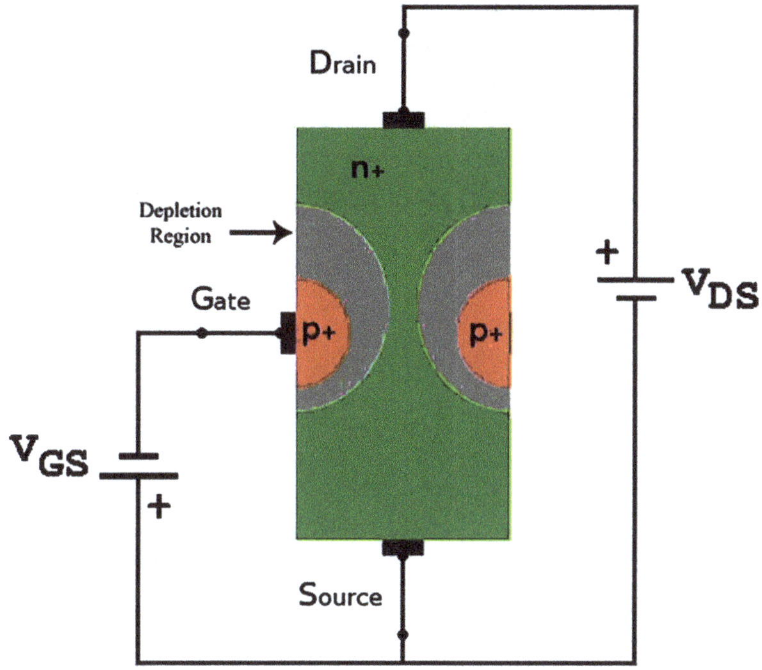

end as a drain. The point must be considered is that source terminal must always be connected to that end of the drain voltage supply to obtain the necessary charge carriers. As a result, in an N-channel JFET the source terminal is attached to the negative end of the drain voltage supply.

Now, according to Figure 51 just like the procedure discussed for diode, along with the increment of reverse bias across the p-n junction, the depletion region grows. Therefore, this phenomenon constricts the cross section of the conducting N-channel material and on the other side increases the channel resistance. The drain current (I_D) in the channel originates from the applied voltage between the drain and source and the applied voltage between the gate and source controls it (Malik, 1995). A same process goes on for the MOSFET with the exception that MOSFET can be biased in enhancement mode as shown in Figure 52.

Figure 52. Enhancement type N-channel MOSFET: (a) gate bias; (b) positive gate bias

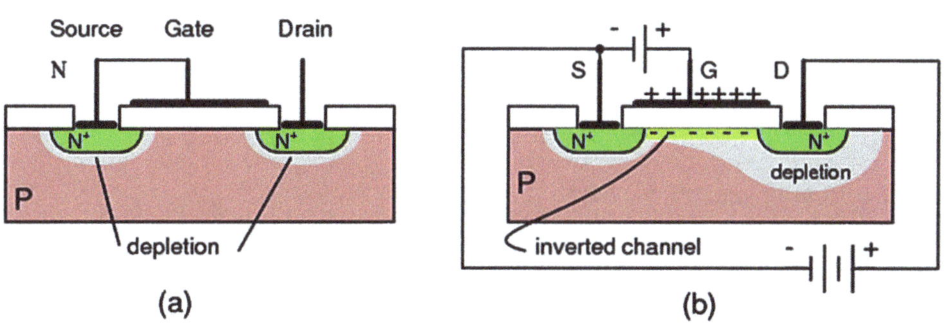

As shown in Figure 52, in enhancement mode the gate-source voltage can be positive with the consideration that we interpret N-channel FETs. Based on the bias voltages applied to gate and drain, three working regions will be defined for the FET. They are as follows:

1. Cut Off mode in which the JFET and MOSFET are OFF. For JFET the drain current is equal to zero but for MOSFET there exists a subthreshold current. In this region the gate-source voltage is less than a margin voltage denoted as threshold voltage. Thus for this mode, we can write:

$$V_{GS} < V_{th} \tag{119}$$

For the subthreshold current of MOSFET in which the source and bulk are tied together, we have:

$$I_D = I_{D0} e^{\frac{V_{GS} - V_{th}}{nV_T}} \tag{120}$$

where I_D is the drain current, I_{D0} is the current for $V_{GS}=V_{th}$, n is the process coefficient and V_T is the thermal voltage defined by equation (2).

2. Linear or Ohmic mode (also known as Triode region) in which we have:

$$V_{GS} > V_{th} \text{ and } V_{GD} > V_{th} \tag{121}$$

Based on Equation (121), it can easily be obtained:

$$-V_{GD} < -V_{th} \tag{122}$$

By adding V_{GS} to both sides of the inequality (122), we can write:

$$V_{DS} < V_{GS} - V_{th} \tag{123}$$

which is the main condition along with $V_{GS}>V_{th}$ for FET operation in triode mode. As a definition, when the gate-source voltage exceeds the threshold voltage the difference is known as the overdrive voltage of MOSFET and is explained by following expression (Sedra & Smith, 2004):

$$V_{OV} = \Delta V = V_{GS} - V_{th} \tag{124}$$

For the drain current in general, we have:

$$I_D = K\left[\left(V_{GS} - V_{th}\right)V_{DS} - \frac{V_{DS}^2}{2}\right] \tag{125}$$

in which K is a constant depending on type of FET and also type of channel. For JFET, the expression will be as follows:

$$I_D = \frac{2I_{DSS}}{V_{th}^2}\left[\left(V_{GS} - V_{th}\right)V_{DS} - \frac{V_{DS}^2}{2}\right] \tag{126}$$

where I_{DSS} is the saturation current at zero gate-source voltage. Also for MOSFET, we have the following expression:

$$I_D = \mu_n C_{ox} \frac{W}{L}\left[\left(V_{GS} - V_{th}\right)V_{DS} - \frac{V_{DS}^2}{2}\right] \tag{127}$$

where μ_n is the charge-carrier effective mobility, C_{ox} is the gate oxide capacitance per unit area and W and L are the gate width and length, respectively.

3. Saturation region or Active mode in which $V_{GS}>V_{th}$ and the drain-source voltage is higher than overdrive voltage ($V_{DS}>V_{GS}-V_{th}$) which means that $V_{GD}<V_{th}$. This is the desired region where the FET shows amplification properties like BJT. For the drain current in general, we have:

$$I_D = \frac{K}{2}\left(V_{GS} - V_{th}\right)^2 \tag{128}$$

For JFET, the expression will be as follows (Kumar & Jain, 2013):

$$I_D = \frac{I_{DSS}}{V_{th}^2}\left(V_{GS} - V_{th}\right)^2 \tag{129}$$

Also with some approximation for MOSFET, we have the following expression (Sedra & Smith, 2004; Gray, Hurst, Lewis & Meyer, 2001):

$$I_D = \frac{1}{2}\mu_n C_{ox} \frac{W}{L}\left(V_{GS} - V_{th}\right)^2 \tag{130}$$

For more accurate calculations the drain current for MOFSET will be as follows:

$$I_D = \frac{1}{2}\mu_n C_{ox} \frac{W}{L}\left(V_{GS} - V_{th}\right)^2\left[1 + \lambda\left(V_{DS} - V_{DS(sat)}\right)\right] \tag{131}$$

in which λ is the channel-length modulation parameter and $V_{DS(sat)}=V_{GS}-V_{th}$ is defined to account for the transition between triode and saturation region.

Just as discussed for BJTs, if we apply KVL to the drain-source loop, we'll obtain the FET's output characteristics function for I_D versus V_{DS} which is a straight line with negative slope. If we draw the characteristics function known as DC load curve at the same plane with the gate-source voltage, we can easily obtain the transistor's operating point. Figure 53 shows this characteristics curve along with three operating regions.

In Figure 53, we can see that based upon the gate-source voltage, different operating points may be obtained. Just like before, choosing the best operating point is one of the important principles which must be emphasized in the bias circuit design of a transistor. For example, the point Q_1 is located near the linear region and Q_3 is located near the cut off region. Any undesired factor may replace the location of these points. Therefore, the best operating point must be selected in middle of the output characteristics curve so that we can obtain a good swing of the drain voltage. It must be mentioned that thermal runaway doesn't happen for FETs.

The Bias of FETs

By means of selecting a proper operating point, we will guaranty that our circuit meets the design requirements such as swing, voltage gain, GBW and etc. From the viewpoint of DC operation except for current equations, FETs are similar to BJTs. In a BJT the base current or the base voltage determines the operating point parameters and now for a FET the gate voltage carries the same responsibility. Despite of the structural and fabrication differences between BJT and FET, in most cases for circuit analysis the gate is treated as the base while the same parity exists between drain-collector and source-emitter. Therefore, for the bias circuit design the same procedure used for BJT can be applied to the FET with one exception which is the zero current for gate. To cut a long story short, for circuit modeling the FET can be assumed as BJT with $\beta \cong \infty$.

This itself simplifies our job for bias circuit design of FET knowing that in order to obtain certain drain current, the DC bias of a FET device only requires setting of the gate-source voltage to a desired value. Thus, unlike BJT the fixed bias of Figure 54 can be employed in which the only demerit is the use of two voltage supplies (Jaeger & Blalock, 2003; Malik, 1995).

Figure 53. I_D–V_{DS} characteristics curvature of a MOSFET

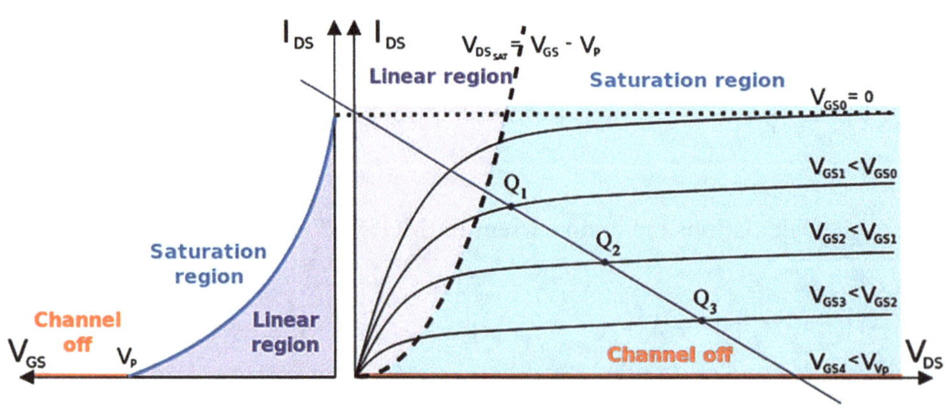

Figure 54. Fixed bias circuit for N-channel JFET

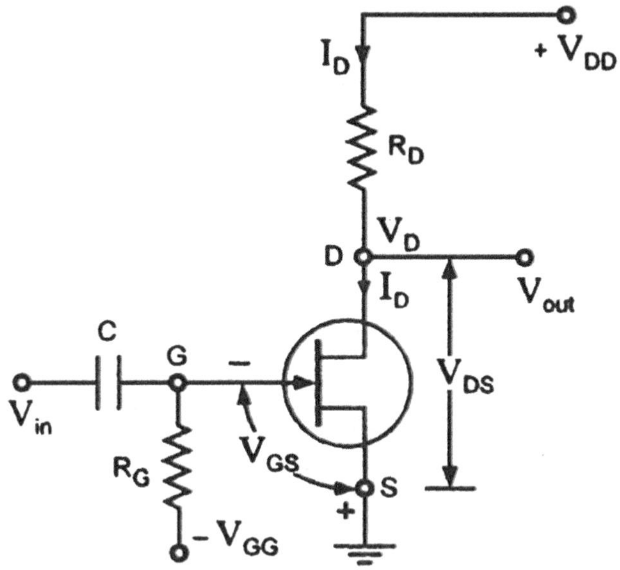

To solve this, as discussed for BJT, the voltage divider circuit can be employed for FET as shown in Figure 55. We can also set the V_{GG} to zero in Figure 54 and put a resistor in the source in which the gate voltage will be equal to zero and the gate-source voltage will be:

$$V_{GS} = -I_D R_S \tag{132}$$

Figure 55. Voltage divider bias circuit for N-channel JFET

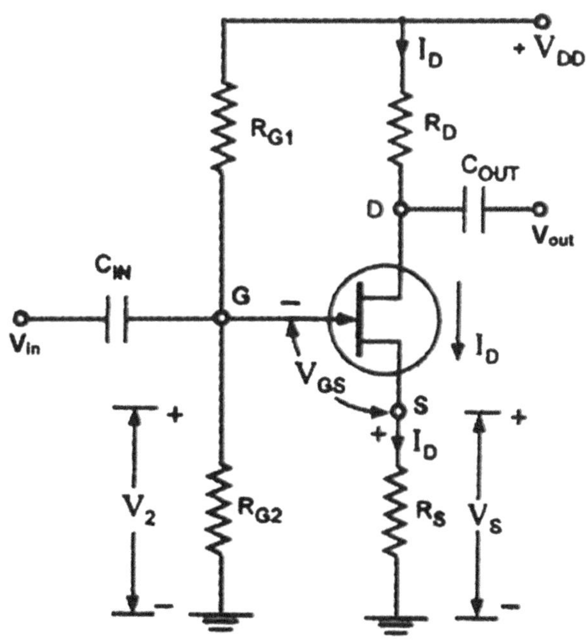

But in Figure 55 we have one more degree of freedom which is the ability to set the gate voltage to the favorable value. Therefore, we will obtain:

$$V_G = \frac{R_{G2}}{R_{G1} + R_{G2}} V_{DD} \tag{133}$$

For gate-source voltage we have:

$$V_{GS} = V_G - I_D R_S \tag{134}$$

It must be mentioned that for this circuit the value of $I_D R_S$ must be greater than V_G so that the correct bias can be obtained (because of JFET). Then the drain current will be obtained by means of:

$$I_D = \frac{V_G - V_{GS}}{R_S} \tag{135}$$

For calculations, at first the parametric form of Equation (134) must be obtained. Then it must be substituted in the current equation of (128) and depending on transistor type, the proper K must be selected. By solving the obtained quadratic equation for I_D, the drain current will be calculated. For drain-source voltage, we have:

$$V_{DS} = V_{DD} - I_D \left(R_D + R_S\right) \tag{136}$$

which is the DC load curve equation. For the AC load curve, the details are the same as discussed for BJT.

Amplification of FET

Just like the defined two port model for BJT, by considering the FET as a three terminal component, a proper two port model must be provided based on Figure 33 in order to analyze the amplification feature of FET.

According to the structure of Figure 33, 6 different models would be available for a FET. But since the drain and source can interchangeably used instead of each other and the gate isn't used as the output, there will exist 3 patterns as follows:

1. Common Source (CS) configuration in which the gate is used as the input node and the drain is employed as the output.
2. Common Drain (CD) configuration in which the gate is used as the input node and the source is employed as the output.
3. Common Gate (CG) configuration in which the source is used as the input node and the drain is employed as the output.

Based on these structures, 3 different models can be defined for a FET. But similar to the discussions for BJT, we'll define the π model for CS configuration which has no limitations for use in CD and CG structures. Based on hybrid parameters, the π model of CS configuration for a MOSFET is shown in Figure 56.

The main difference of this model with the one shown in Figure 35 comes from the fact that since I_G=0, then equivalent r_π of Figure 56 will be equal to infinity and there must be an open circuit. Therefore, two important parameters must be calculated which are g_m and r_o. From Equation (128) for g_m we have:

$$g_m = \frac{\partial I_D}{\partial V_{GS}} = K\left(V_{GS} - V_{th}\right) \tag{137}$$

In Equation (137) if the numerator and the denominator are multiplied to (V_{GS}–V_{th}), then it will be simplified as:

$$g_m = \frac{2I_D}{V_{GS} - V_{th}} = \frac{2I_D}{V_{OV}} \tag{138}$$

For JFET the equation can be rewritten as:

$$g_m = \left|\frac{2I_{DSS}}{V_{th}^2}\left(V_{GS} - V_{th}\right)\right| \tag{139}$$

Just as before, g_m is known as the transistor transconductance. As the second parameter, we have to calculate r_o. Since it is the output impedance in order to obtain it, the input should be set to zero. Therefore, v_{gs} becomes zero and there will be no Voltage Controlled Current Source (VCCS) in output node. Now, one can say:

$$g_{ds} = \frac{1}{r_o} = \frac{\partial I_D}{\partial V_{DS}} \tag{140}$$

Figure 56. π model of CS configuration
Jaeger & Blalock, 2003; Horowitz & Hill, 2015; Galup-Montoro & Schneider, 2007.

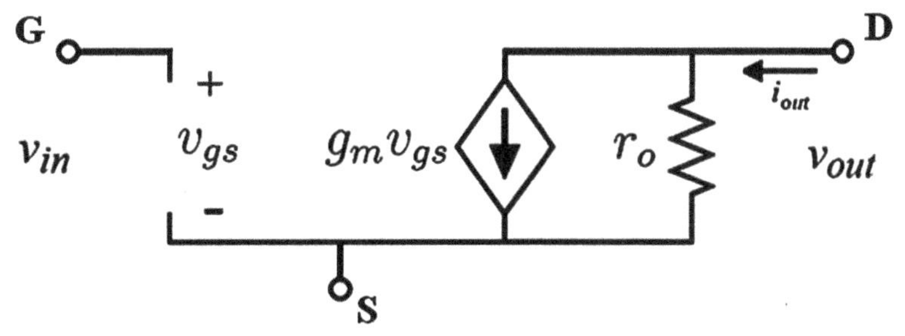

By means of Equation (131), we'll obtain:

$$\frac{\partial I_D}{\partial V_{DS}} = \frac{1}{2}\lambda\mu_n C_{ox}\frac{W}{L}\left(V_{GS} - V_{th}\right)^2 \tag{141}$$

Comparing with (131) which is a result of Shichman–Hodges model (*The Shichman-Hodges Enhancement MOSFET Model*), one can say:

$$\frac{\partial I_D}{\partial V_{DS}} = \frac{\lambda I_D}{1+\lambda\left(V_{DS} - V_{DS(sat)}\right)} \tag{142}$$

Usually $\lambda(V_{DS}-V_{DS(sat)})\ll 1$, which simplifies Equation (142) as follows:

$$r_o = \frac{1}{g_{ds}} = \frac{1}{\lambda I_D} \tag{143}$$

Now with the use of the attained model, we are going to analyze the amplification feature of FET briefly because the results are almost the same as the obtained ones for BJT. In Figure 57 the π model for the CS configuration of Figure 55 which is to drive a load like R_L is shown. R_S here indicates on source terminal attached resistor while R_{Sig} represents signal source resistor.

By defining the gate resistor as $R_G=R_{G1}\|R_{G2}$, for the input part it can be written:

$$v_g = \frac{R_G}{R_{sig} + R_G} v_{in} \tag{144}$$

Figure 57. Small signal equivalent circuit for circuit of Figure 55

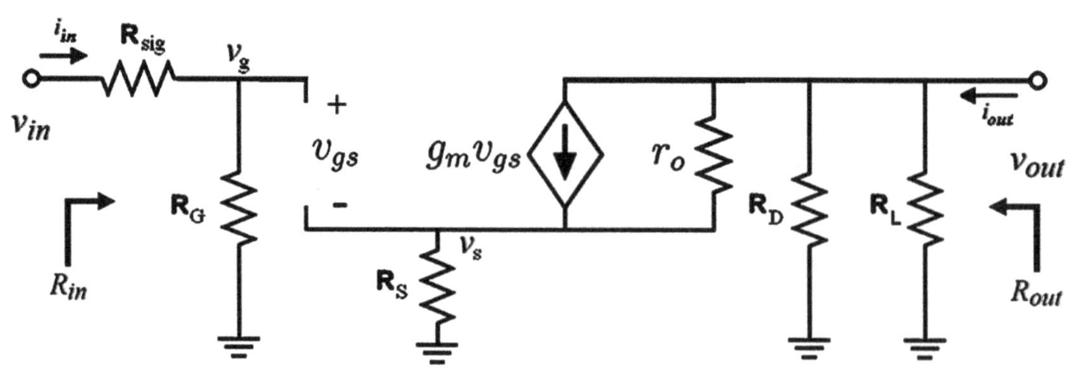

and because $R_G \gg R_{sig}$, we have $v_g \cong v_{in}$. Also, for the source node it can be written:

$$v_s = R_S\left(g_m v_{gs} + \frac{v_{out} - v_s}{r_o}\right) \tag{145}$$

where:

$$v_s = v_g - v_{gs} = v_{in} - v_{gs} \tag{146}$$

Substitution of (146) in (145) will result in:

$$v_{in} - v_{gs} = R_S\left(g_m v_{gs} + \frac{v_{out} - v_{in} + v_{gs}}{r_o}\right) \tag{147}$$

and by simplification, we have:

$$\frac{v_{in}}{R_S \parallel r_o} = v_{gs}\left(\frac{1}{R_S} + g_m + \frac{1}{r_o}\right) + \frac{v_{out}}{r_o} \tag{148}$$

$$\frac{v_{in}}{R_S \parallel r_o} = v_{gs}\left(\frac{1 + g_m\left(R_S \parallel r_o\right)}{R_S \parallel r_o}\right) + \frac{v_{out}}{r_o} \tag{149}$$

Also, for the output node we have:

$$v_{out} = \left(R_D \parallel R_L\right)\left[-g_m v_{gs} + \frac{v_{in} - v_{gs} - v_{out}}{r_o}\right] \tag{1150}$$

in which by the simplification, we'll obtain:

$$\frac{v_{out}}{R_D \parallel R_L \parallel r_o} = -v_{gs}\left(\frac{1 + g_m r_o}{r_o}\right) + \frac{v_{in}}{r_o} \tag{151}$$

The parameter $g_m r_o$ is defined as the intrinsic gain of MOSFET and is equal to:

$$A_{V_{inherent}} = g_m r_o \tag{152}$$

By substituting from (138) and (143), we have:

$$A_{V_{inherent}} = \frac{2I_D}{V_{OV}} \times \frac{1}{\lambda I_D} = \frac{2}{\lambda V_{OV}} \tag{153}$$

According to typical values of V_{OV} and λ, the inherent gain for MOSFET contains a value at the range of 20 which is much less than its counterpart for BJT but it is still much greater than unity. In addition, because r_o has a great value the terms $\frac{v_{out}}{r_o}$ in Equation (149) and $\frac{v_{in}}{r_o}$ in Equation (151) will have very small values and could be neglected. Therefore, Equation (149) will be simplified as follows:

$$\frac{v_{in}}{R_S} = v_{gs}\left(\frac{1+g_m R_S}{R_S}\right) \tag{154}$$

For (151) we can write:

$$\frac{v_{out}}{R_D \parallel R_L} = -v_{gs}\left(\frac{1+g_m r_o}{r_o}\right) \cong -g_m v_{gs} \tag{155}$$

By division of (155) to (154), we will obtain:

$$A_V = \frac{v_{out}}{v_{in}} = \frac{-g_m\left(R_D \parallel R_L\right)}{1+g_m R_S} \tag{156}$$

which is the same result we have previously obtained for CE configuration containing emitter resistor. The only difference is that in (156) the voltage gain is defined in terms of the transistor transconductance. If the numerator and denominator of (156) are divided to g_m, we have:

$$A_V = \frac{v_{out}}{v_{in}} = \frac{-\left(R_D \parallel R_L\right)}{\frac{1}{g_m} + R_S} \tag{157}$$

and here, $\frac{1}{g_m}$ is the reflection of the resistor seen by source of MOSFET . The results for the other configurations are almost the same as obtained for BJT and are beyond the scope of discussion.

There are only two points which must take into account. First is that in CD configuration which is known as Source Follower configuration, unlike the emitter follower, we cannot easily obtain unity gain since the corresponding g_m for MOSFET is much less than that of BJT. The second one is that in CG configuration because the gate-source part is an open circuit, the input impedance for this configuration is $\frac{1}{g_m}$ (Holberg & Allen, 2002).

Table 2 summarizes the characteristics for different MOS amplifiers. As the table suggests, CS configuration features lies between that of CD and CG. Except for CG configuration, there will be no current gain for other structures. The rest is similar as explained for BJT.

MULTISTAGE AMPLIFIERS

General Considerations

As seen in previous sections, a single stage amplifier cannot provide the voltage amplification as large as needed for a signal obtained from environment. As a result, in order to achieve higher amplification, more than a single stage amplifier is required. As illustrated in Figure 58, the general idea is to cascade various stages of amplifiers which enables us to attain the needed amplification. Therefore, it is necessary to peruse the features of multistage amplifiers to design a good circuitry for this purpose.

Table 2. The comparison between CS, CD, and CG configurations

Amplifier	A_V	A_I	R_{in}	R_{out}
CS (without R_S)	$-g_m(R_D\|\|R_L)$	∞	$R_{sig}+R_G$	$R_D\|\|R_L$
CS (with R_S)	$\frac{-g_m(R_D \parallel R_L)}{1+g_m R_S}$	∞	$R_{sig}+R_G$	$R_D\|\|R_L$
CD	$\frac{g_m(R_S \parallel R_L)}{1+g_m(R_S \parallel R_L)}$	∞	$R_{sig}+R_G$	$\frac{1}{g_m}$
CG	$g_m(R_D\|\|R_L)$	1	$R_{sig}+\frac{1}{g_m}$	$R_D\|\|R_L$

Figure 58. General architecture of a multistage amplifier

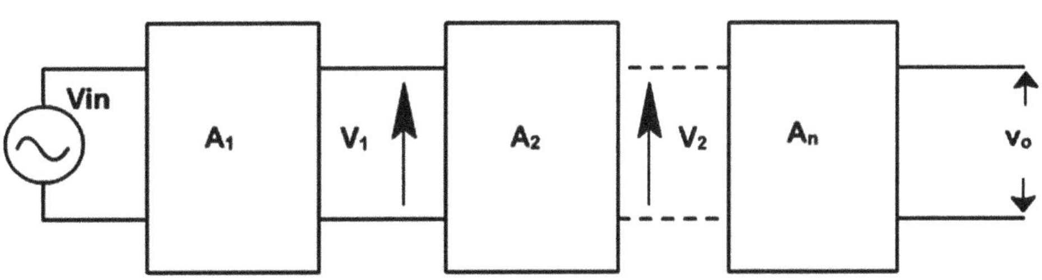

On the discussion of two port networks one can see that in order to analyze the performance of cascaded amplifiers, the transfer matrix with ABCD parameters must be employed (Bakshi & Bakshi, 2009). This matrix is defined as below:

$$\begin{bmatrix} V_1 \\ I_1 \end{bmatrix} = \begin{bmatrix} A & B \\ C & D \end{bmatrix} \begin{bmatrix} V_2 \\ -I_2 \end{bmatrix} \tag{158}$$

Here, the goal is not to investigate this topic. In order to analyze such networks, the basic knowledge of circuit theory will be enough. It is clear that in such a structure, input performance of the first amplifier determines the input performance of the resulted system which means that the input impedance of the whole system belongs to the input stage. Following the same way, the output performance of the whole system is specified by the last amplifier which indicates that it is the output stage which determines the output impedance of the multistage amplifier.

Also, in a multistage amplifier the output of one stage behaves as the input of the next stage. Therefore, to achieve the minimum loss of voltage when the signal passes through one network to the next stage, normally a matching network is utilized between two stages. In some cases in order to avoid bias disturbing, the DC voltage at the output of one stage should not be permitted to go to the input of the next level. There are three general coupling methods known as:

1. RC coupling,
2. Impedance coupling,
3. Transformer coupling.

Usually, we use RC coupling especially for cascaded CE amplifiers and sometimes no coupling will be used. In the next sections we'll introduce some of these amplifiers. But the important point in cascaded amplifiers is the total gain calculation. From the properties of two port networks one can simply obtain:

$$A_V = \frac{V_o}{V_{in}} = \frac{V_1}{V_{in}} \times \frac{V_2}{V_1} \times \ldots \times \frac{V_o}{V_{o-1}} \tag{159}$$

which results in:

$$A_V = A_{V1} \times A_{V2} \times \ldots \times A_{Vn} \tag{160}$$

and illustrates that with perfect matching, the total gain is the product of the gains of the individual stages. There exists a same result for current amplification as:

$$A_I = A_{I1} \times A_{I2} \times \ldots \times A_{In} \tag{161}$$

In the case of non-perfect matching, the loading effect must be considered which imposes some approximations to Equations (160) and (161). For precise calculation, let's consider the two stage amplifier of Figure 59. Because of loading effect, the gain of first stage is:

$$A_{V1} = \frac{V_{in2}}{V_{in1}} = A_1 \left(\frac{R_{in2}}{R_{in2} + R_{out1}} \right) \tag{162}$$

Also, for the gain of second stage, we have:

$$A_{V2} = \frac{V_{out}}{V_{in2}} = A_2 \left(\frac{R_L}{R_L + R_{out2}} \right) \tag{163}$$

and the total gain by means of Equation (159) will be as follows:

$$A_V = A_1 A_2 \left(\frac{R_{in2}}{R_{in2} + R_{out1}} \right) \left(\frac{R_L}{R_L + R_{out2}} \right) \tag{164}$$

which shows the impact of loading. For $R_{in2} \gg R_{out1}$ and $R_L \gg R_{out2}$, the matching is perfect.

In addition of the above discussions, by means of proper feedback techniques in multistage amplifiers the degree of freedom for the design can be increased. Now, before the topic of Operational Amplifiers (OpAmps), we have to get familiar with some important two stage circuits needed in OpAmp design.

Darlington Pair

As the first two stage amplifier, we are about to discuss about the Darlington pair or Darlington transistor. As shown in Figure 60, two common configurations of this amplifier known as Darlington and Sziklai are composed of two BJTs cascaded in such a way that the amplified current of the first transistor is

Figure 59. Two stage cascaded amplifier
Jaeger & Blalock, 2003.

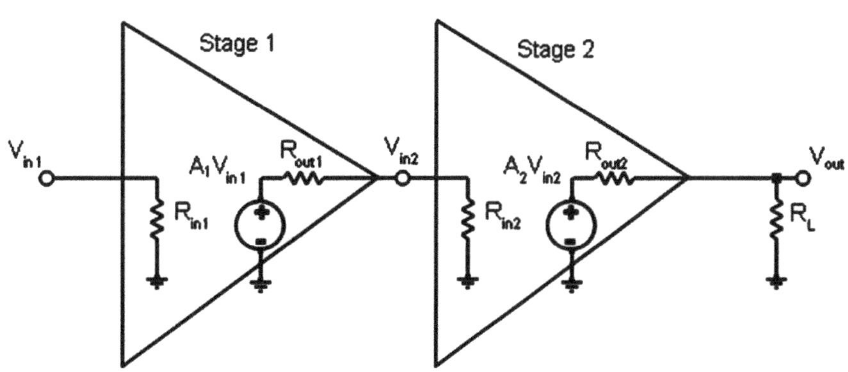

Figure 60. Darlington pair: (a) main form; (b) Sziklai pair

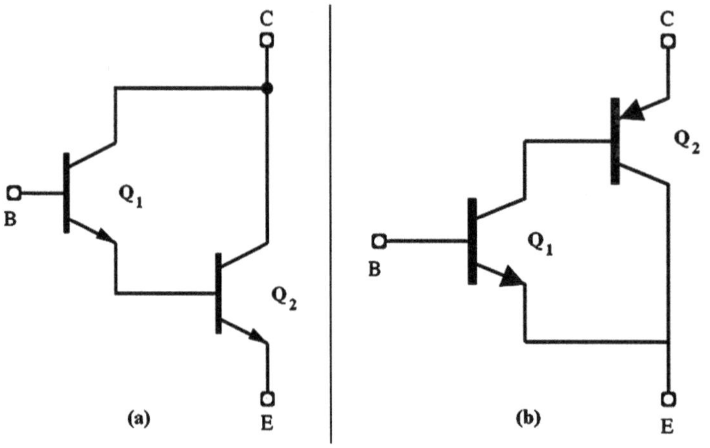

amplified again by the second one (Horowitz & Hill, 2015). Therefore, its main feature is the current amplification larger than any other transistor configuration according to Equation (161).

As Figure 60 indicates, if Darlington transistors will be manufactured in IC scale, they will contain three leads. For the equivalent β we have:

$$\beta_{Darlington} = \beta_1 \times \beta_2 \tag{165}$$

Because of high β, as discussed for CC configuration, if Darlington pair is used in Figure 41 as voltage buffer, the gain will be:

$$A_V \cong \frac{\left(R_E \parallel R_L \parallel r_{o2}\right)\beta_{Darlington}}{R_S + r_{\pi 1} + r_{\pi 2} + \left(R_E \parallel R_L \parallel r_{o2}\right)\beta_{Darlington}} \tag{166}$$

In order to make this configuration a good buffer, the input impedance can be enhanced by Bootstrap Technique of Figure 40. But the input impedance cannot exceed $r_{\pi 1}+\beta_1 r_{o1}$ as r_o of Q_1 becomes parallel with $\beta_1 r_{\pi 2} + \beta_1\beta_2\left(R_E \parallel R_L\right)$ which is the input impedance of second stage transistor. Thus, in discrete implementation of Darlington pair a secondary bootstrap capacitor is needed as shown in Figure 61. This capacitor is installed between collector of first stage and emitter of second stage to enlarge the value of r_{o1} by means of Miller's Theorem.

However, the Darlington pair has some drawbacks which must be taken into account for amplifier design. The ON voltage for this transistor is $V_{BE1}+V_{BE2}$ in which for similar transistors it will be $2V_{BE}$. Also, the saturation voltage which is determined by Q_2, will be higher than $0.2V$ as the collector of second stage constitutes a loop with collector-emitter of first stage and base-emitter of second stage and this will be expressed by:

$$V_{CE2} = V_{CE1} + V_{BE2} \tag{167}$$

Figure 61. The secondary bootstrap capacitor for Darlington pair input impedance enhancement

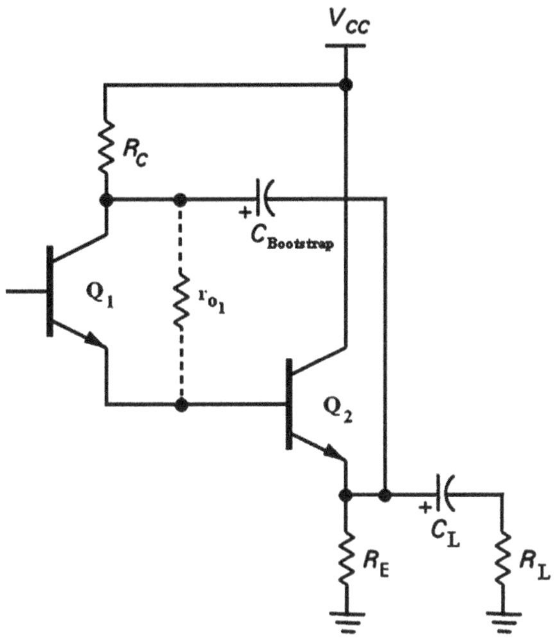

For $V_{BE(ON)}$=0.7, the saturation voltage of second stage will be 0.9*V*.

In addition, the leakage current of first stage (I_{SC1}) is being amplified by second stage with β_2 coefficient which can mislead the operating point parameters via thermal runaway. But because of good buffering features, this configuration is widely used in output stage design of OpAmps and other amplifying circuits. It must be mentioned that CC configuration is also used in output stage design if it is driven with a current source instead of emitter resistor.

Cascode Amplifier

In order to solve the problem of low input impedance and also, to enhance the current gain of CB configuration, a two stage amplifier consisting of a transconductance amplifier as first stage and a current buffer as second stage is employed which is known as Cascode amplifier (Hickman & Hunt, 1939). The main form is obtained by CE-CB configurations in which CE structure has the extra role of current mirror for CB configuration. Also, CS-CD is used on the implementation of this two stage amplifier in which the input resistance is almost ∞, but there will be no current amplification (Gray, Hurst, Lewis & Meyer, 2001). As shown in Figure 62, since there is no direct coupling form the output to input, a good isolation between input and output will be obtained.

Because of CE structure as input stage, just like Equation (71) the input impedance will be:

$$R_{in} = R_S + \left(R_B \parallel r_\pi\right) \tag{168}$$

For the current gain we have:

Figure 62. Cascode amplifier: (a) CE-CB configuration; (b) small signal equivalent

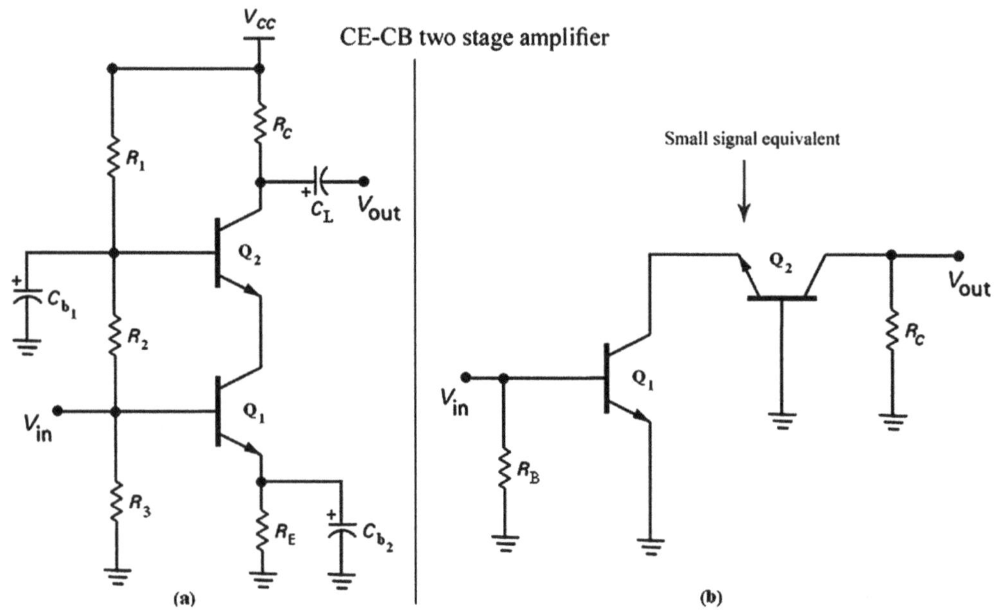

$$A_I = A_{I1} \times A_{I2} = A_{I_{CE}} \times A_{I_{CB}} \tag{169}$$

With proper approximations $A_{I_{CE}} \cong \beta_1$ and $A_{I_{CB}} \cong \dfrac{\beta_2}{\beta_2 + 1}$, which simplifies Equation (169) as:

$$A_I \cong \beta_1 \tag{170}$$

For the output impedance, by means of Equation (114) we will obtain:

$$R'_{out} = \left(r_{\pi 2} \parallel r_{o1}\right) + r_{o2}\left(1 + g_{m2}\left(r_{\pi 2} \parallel r_{o1}\right)\right) \cong \beta_2 r_{o2} \tag{171}$$

and the final result is:

$$R_{out} = R_C \parallel R'_{out} \cong R_C \tag{172}$$

For the voltage gain of first stage, the loading effect of second stage must be considered which results in:

$$A_{V1} = -g_{m1}\left(\frac{1}{g_{m2}} \parallel r_{o1}\right) \cong -\frac{g_{m1}}{g_{m2}} \tag{173}$$

However, $I_{E2} = I_{C1} \cong I_{E1}$ and as a result, $A_{V1} \cong -1$ which illustrates that CE stage behaves as a voltage buffer. Also, for the game of second stage we have:

$$A_{V2} = g_{m2}\left(R_C \parallel r_{o2}\right) \cong g_{m2}R_C \tag{174}$$

Therefore, by means of (160) we obtain:

$$A_V = A_{V1} \times A_{V2} = -g_{m1}R_C \tag{175}$$

which illustrates that unlike single stage CB amplifier, the input impedance and current gain are enhanced considerably. Meanwhile, since $A_{V1} \cong -1$, the Miller effect for C_μ of first stage which is the base-collector capacitor of CE amplifier is reduced significantly which enhances GBW of whole amplifier. But the voltage swing is reduced as two transistors are in series for DC load curve. For higher output impedances, more than two stage Cascode is needed.

Differential Pair

Although CE amplifier shows good characteristics for voltage amplification, but as previously discussed in large signal analysis, the calculated requirement for small signal behavior in Equation (43) implies that if v_{be} won't be small enough, in addition of nonlinearity it will cause an effect known as Harmonic Distortion (HD) in output signal. For each harmonic component in output signal, the corresponding HD is defined as the voltage amplitude of that harmonic to the amplitude of main component and is expressed as:

$$HD_2 = \frac{Voltage\ Amplitude\ of\ second\ Harmoic}{Voltage\ Amplitude\ of\ main\ Harmoic} \tag{176}$$

$$HD_3 = \frac{Voltage\ Amplitude\ of\ third\ Harmoic}{Voltage\ Amplitude\ of\ main\ Harmoic} \tag{177}$$

For CE structure of Figure 36, the voltage amplitude of second harmonic by means of Maclaurin Series is equal to $\frac{1}{2}\left(\frac{v_{be}}{V_T}\right)^2$. Also, for the main component the amplitude is equal to $\frac{v_{be}}{V_T}$. Thus, we'll obtain:

$$HD_2 = \frac{v_{be}}{2V_T} \tag{178}$$

In order to avoid the existence of other harmonic components, the amplitude of each harmonic must be small enough which indicates that to suppress the effect of second harmonic, we must have $HD_2 \ll 1$. This itself guaranties that the higher order harmonics because of their smaller amplitudes will be suppressed, too.

Based on this fact, a factor known as Total Harmonic Distortion (THD) is defined as the ratio of the sum of the powers of all harmonic components to the power of the fundamental frequency and is expressed as (Kester, 2010):

$$THD = \frac{\sqrt{V_2^2 + V_3^2 + V_4^2 + \ldots}}{V_1} \tag{179}$$

Calculations illustrate that single stage amplifiers as the viewpoint of HD, must contain the emitter resistor in order to show good behavior and it should not be bypassed. Therefore, the voltage gain will drop considerably and the swing will be reduced. To solve this problem, differential pair transistors which are also known as Emitter Coupled Logic (ECL) in digital circuit design have been designed and implemented.

General block diagram of a differential amplifier is illustrated in Figure 63. The voltages V_A and V_B are input voltages while V_{out} is the single ended output voltage. In some cases the output voltage is being obtained separately for each input which is then called differential output voltage. $+V_{cc}$ and $-V_{cc}$ are the bias voltages (Laplante, 2005).

Each differential amplifier has two types of inputs as defined below:

$$V_{id} = V_A - V_B, \quad V_{ic} = \frac{1}{2}\left(V_A + V_B\right) \tag{180}$$

in which V_{id} is the differential input mode and V_{ic} is the common mode input mode. Therefore, we can write:

Figure 63. General architecture of a differential amplifier

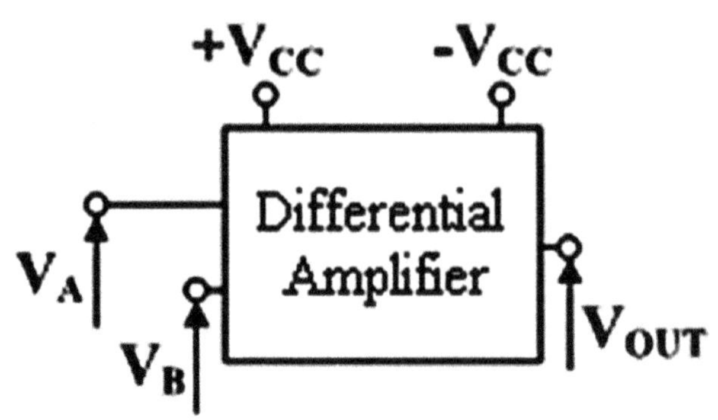

$$V_A = V_{ic} + \frac{1}{2}V_{id}, \quad V_B = V_{ic} - \frac{1}{2}V_{id} \tag{181}$$

For the output voltage, we have:

$$V_{out} = A_d V_{id} \pm A_c V_{ic} \tag{182}$$

in which A_d is known as differential gain and A_c is the common mode gain. Substituting from (180) into (182) results in:

$$V_{out} = A_d\left(V_A - V_B\right) \pm A_c\left(\frac{V_A + V_B}{2}\right) \tag{183}$$

If we want to draw a block diagram to embody Equation (183), then we will easily obtain Figure 64. This configuration must be considered carefully specially in simulation of a differential amplifier because it shows how the inputs must be applied to the amplifier.

Since we have two kinds of amplifications and we only want to amplify the differential input, a new parameter is defined as Common Mode Rejection Ratio (CMRR) which is known as a Figure of Merit (FOM) on the operation of differential pairs (Holberg & Allen, 2002). It is explained as:

$$CMRR = \left|\frac{A_d}{A_c}\right| \quad \text{or} \quad CMRR = 20\log\left|\frac{A_d}{A_c}\right|\left(dB\right) \tag{184}$$

For an ideal differential amplifier CMRR tends to infinity which means we must have $A_c \cong 0$. There are other factors which are beyond the scope of discussion.

Figure 64. Different input types for a differential amplifier

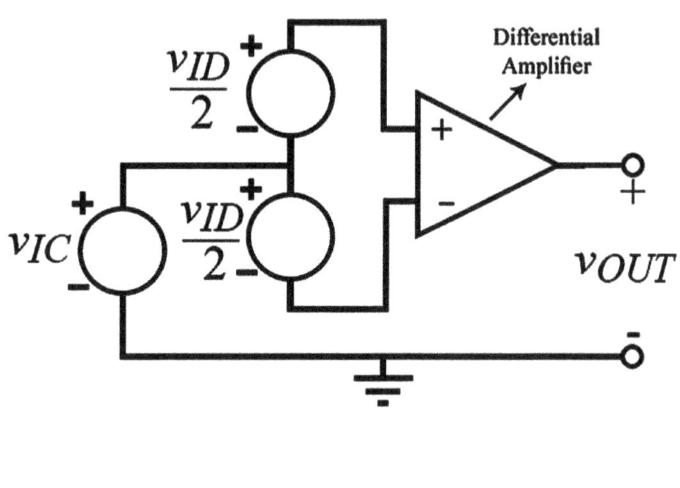

As shown in Figure 65, the simplest form of differential pair is composed of two transistors in which their emitter is coupled and the output is taken differentially. At first view, it seems that if two CE configurations are joined together at their emitters, they will form a differential pair.

To analyze large signal behavior of Figure 65 structure, at first the KVL must be applied to input loop which results in:

$$v_{i1} - v_{BE1} + v_{BE2} - v_{i2} = v_{id} - v_{BE1} + v_{BE2} = 0 \tag{185}$$

We want to keep both of the transistors in active region. Thus, by means of Equation (17) we can write:

$$v_{BE1} = V_T Ln\left(\frac{i_{C1}}{I_{S1}}\right), \quad v_{BE2} = V_T Ln\left(\frac{i_{C2}}{I_{S2}}\right) \tag{186}$$

If both of the transistors have the same structural characteristics, then $I_{S1}=I_{S2}$. If the calculated values for v_{BE1} and v_{BE2} are substituted in (185), it results in:

$$v_{id} = V_T Ln\left(\frac{i_{C1}}{I_{S1}}\right) - V_T Ln\left(\frac{i_{C2}}{I_{S2}}\right) = V_T Ln\left(\frac{i_{C1}}{I_{C2}}\right) \tag{187}$$

$$\rightarrow \frac{i_{C1}}{I_{C2}} = e^{\frac{v_{id}}{V_T}} \tag{188}$$

Figure 65. BJT differential pair

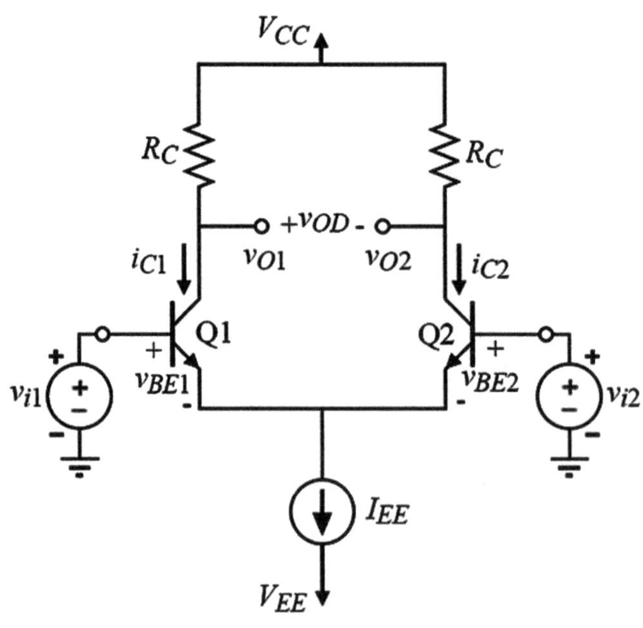

In addition, by applying KCL to the joint emitter nodes the second equation for collector currents will be obtained as follows:

$$i_{E1} + i_{E2} = I_{EE} = \frac{1}{\alpha}\left(i_{C1} + i_{C2}\right) \tag{189}$$

By combining (188) and (189), the result will be:

$$i_{C1} = \frac{\alpha I_{EE}}{1 + e^{\frac{-v_{id}}{V_T}}}, \quad i_{C2} = \frac{\alpha I_{EE}}{1 + e^{\frac{v_{id}}{V_T}}} \tag{190}$$

If we draw the current-voltage characteristics for both collector currents, Figure 66 will be achieved. For the output differential voltage only a KVL from first to second collector is needed. The result is:

$$v_{OD} = v_{O1} - v_{O2} = R_C\left(i_{C2} - i_{C1}\right) \tag{191}$$

By substituting from (190) and with simplification, the following equation will be obtained for output differential voltage:

$$v_{OD} = \alpha R_C I_{EE} \tanh\left(\frac{-v_{id}}{2V_T}\right) \tag{192}$$

Figure 66. Current-Voltage characteristics of differential pair

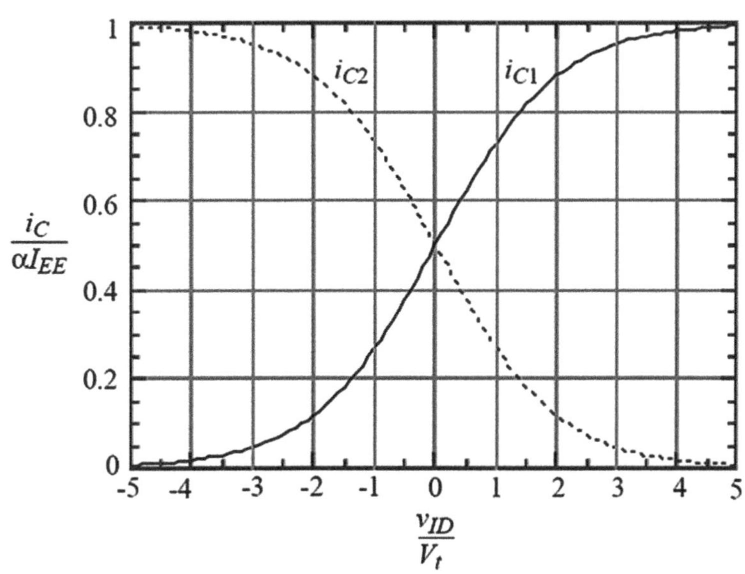

which has a tangent hyperbolic form. According to obtained value and by means of Maclaurin Series, we will obtain:

$$v_{OD} = \alpha R_C I_{EE} \left(-\frac{v_{id}}{2V_T} + \frac{v_{id}^3}{24V_T^3} - \dots \right) \tag{193}$$

According to (193), we'll see that for a symmetric differential pair, the second harmonic is completely eliminated. The nearest harmonic component is the third one. Thus, we'll obtain:

$$HD_3 = \frac{1}{12}\left(\frac{v_{be}}{V_T}\right)^2 \tag{194}$$

which illustrates that for $v_{be}=V_T$, the distortion of third harmonic is less than 10% which is a considerable improvement compared with CE configuration. In addition, if emitter resistances are used, the linear operating range can be enhanced significantly.

For small signal analysis at first the transconductance parameters of both transistors for $v_{id}=0$ must be calculated. For this purpose, we have:

$$g_{m1} = \left.\frac{\partial i_{C1}}{\partial v_{id}}\right|_Q = \frac{I_{C1}}{2V_T}, \quad g_{m2} = \left.\frac{\partial i_{C2}}{\partial v_{id}}\right|_Q = \frac{-I_{C2}}{2V_T} \tag{195}$$

To interpret the small signal behavior of a differential amplifier, one can divide it into the differential mode and common mode analysis. Because of symmetric form, we can use half circuit concept. Therefore, in differential mode analysis because the small signal currents have opposite polarity, the joint node for emitters can be considered as virtual ground. Also, in common mode analysis the output resistance of bottom current source will be doubled in each emitter forming a virtual open circuit in joint node of the emitters. As a result, the configuration of Figure 67 will be obtained.

For the differential mode gain, we have:

$$v_{od} = -g_m R_C \left(\frac{v_{id}}{2} - \left(\frac{-v_{id}}{2} \right) \right) = -g_m R_C v_{id} \tag{196}$$

$$\rightarrow A_d = \frac{v_{od}}{v_{id}} = -g_m R_C \tag{197}$$

For the common mode gain, we will obtain:

Figure 67. Half circuit concept for small signal analysis of differential pair

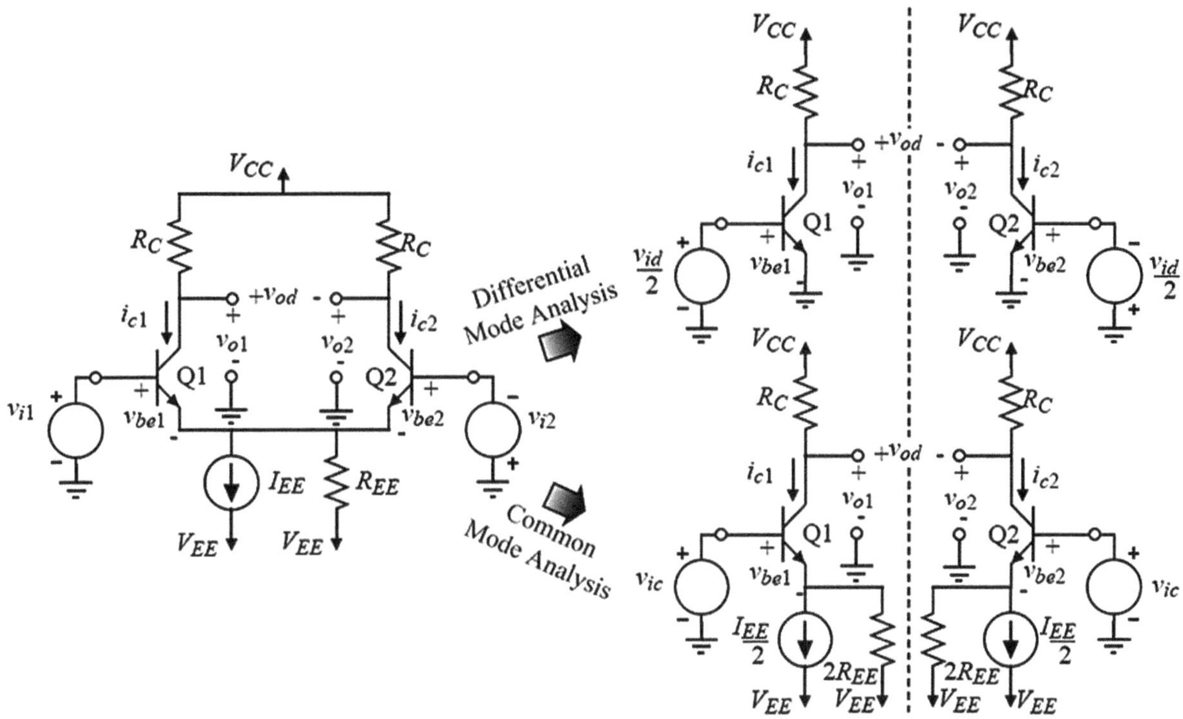

$$A_c = \frac{v_{o1}}{v_{ic}} = -\frac{\beta R_C}{r_\pi + \left(1+\beta\right)2R_{EE}} \tag{198}$$

For the input impedance of differential mode, we have:

$$R_{id} = r_\pi + \frac{\beta}{g_m} = 2r_\pi \tag{199}$$

Also, for the input impedance of common mode, we will obtain:

$$R_{ic} = R_{ic1} \parallel R_{ic2} = \frac{r_\pi}{2} + \left(1+\beta\right)R_{EE} \tag{200}$$

As the final parameter, CMRR will be equal to:

$$CMRR = \left|\frac{A_d}{A_c}\right| = \frac{g_m R_C}{\dfrac{\beta R_C}{r_\pi + \left(1+\beta\right)2R_{EE}}} \cong 1 + 2g_m R_{EE} \tag{201}$$

As the results illustrate, the differential pair is a very nice choice for the input stage of OpAmps. It suppresses all even harmonics and rejects common mode signals. It also has a larger linear range than CE configuration. However, the existence of voltage offset which occurs as a result of transistor mismatches makes the designer regard careful design considerations. In order to achieve higher input impedances, the FET based differential pairs can be used. For BJT pairs this can be achieved by biasing the transistors for low currents which magnifies the value of r_π. Any of the discussed amplifiers can be employed on the structure of differential pair to achieve more advantages.

OPERATIONAL AMPLIFIERS

General Definitions

The OpAmp is a high gain architecture consisting of multistage transistor amplifiers which has a variety of applications in the criteria of electric and electronic circuits. In DC amplifying applications done for analog computers it performs mathematical operations such as summation, subtraction, integration and differentiation. But the main usages of OpAmp are in electronic circuits which are Active Filters, Instrumentation, ADCs, Voltage Regulators and etc. Here, we want to analyze the usage of OpAmps on the amplification of acquired signals by sensors. But before that, some of the important properties of OpAmps must be introduced.

The schematic symbol and structural block diagram of an OpAmp is illustrated in Figure 68 (Laplante, 2005). As this block diagram suggests, an OpAmp contains three or four stage amplifiers, although a single stage differential amplifier can be considered as an OpAmp, too.

In schematic symbol, V_+ is known as non-inverting input while V_- is known as inverting input. V_{S+} and V_{S-} are positive and negative supply voltages, respectively. V_{out} is the single ended output. Differential output OpAmps contain two outputs in which the output can either be taken individually or differentially.

As the structural block diagram suggests, an OpAmp with high DC gain will contain all or most of the four building blocks shown in Figure 68. The input stage is surely a differential amplifier because

Figure 68. Schematic symbol and structural block diagram of an OpAmp

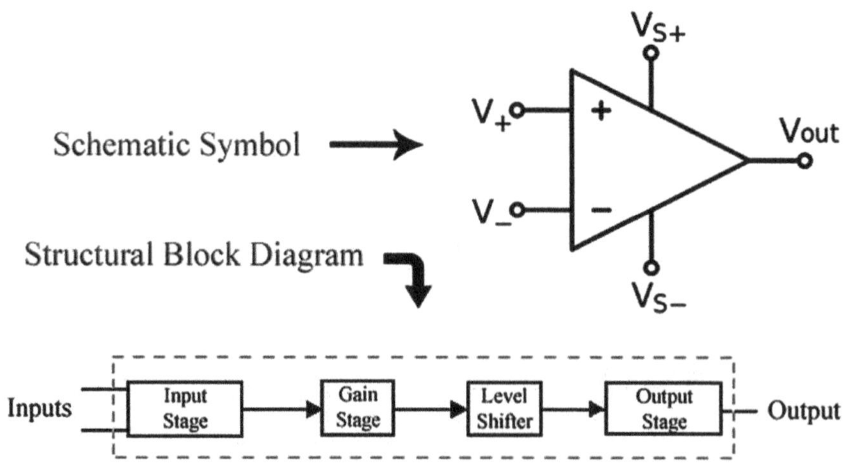

of the merits mentioned for that amplifier. For the gain stage, depending on the operation, amplification and other factors, any of the discussed amplifiers may be used. For high gain, usually Darlington pair in voltage amplification mode or Cacode amplifier or even a differential amplifier can be utilized. Since the OpAmp is being designed in IC form, usually no coupling is used between the different stages. In order to properly fit the DC output of one stage to another one, sometimes voltage shifting is needed which is done by level shifting circuits. These circuits are amplifiers usually with very high input impedance and low output impedances. The final important block is the output stage which is usually a good voltage buffer such as emitter follower or Darlington in buffer mode. This stage has the responsibility of reducing the output impedance to make a good matching network and setting the output DC voltage in the desired value for needed voltage swing.

For circuit modeling and calculations, the non-ideal model of Figure 69 is used for OpAmp (Jaeger & Blalock, 2003). It contains three important parameters which are input resistance R_{in}, output resistance R_{out} and DC gain denoted as G.

An OpAmp contains other parameters but among them input bias current expressed as:

$$I_{bias} = \frac{I_1 + I_2}{2} \tag{202}$$

and input offset current defined as:

$$I_{os} = \left| I_1 - I_2 \right| \tag{203}$$

are important because in non-ideal modeling and feedback calculations sometimes their effect should be considered. But before using the non-ideal model, the ideal OpAmp features must be defined in order

Figure 69. Equivalent circuit of OpAmp which is used for non-ideal modeling

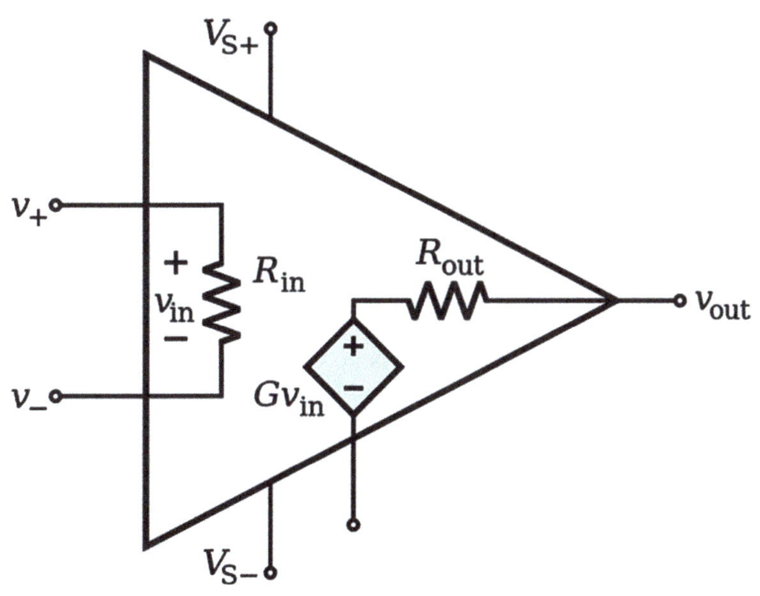

to obtain a good knowledge of its behavior. As a start point, the model of Figure 69 can be considered. To make it look like the ideal model, we must have:

1. Infinite Open Loop Gain in which as the definition,

$$G = \frac{v_{out}}{v_{+} - v_{-}} \tag{204}$$

for infinite G, we must have:

$$v_{+} = v_{-} \tag{205}$$

which says that the input terminals of an ideal OpAmp can be treated as virtual short circuit.

2. Infinite Input Impedance (Horowitz & Hill, 2015) in which as the definition,

$$R_{in} = \frac{v_{+} - v_{-}}{I_{bias}} \tag{206}$$

for infinite R_{in}, we must have:

$$I_{bias} = 0 \tag{207}$$

which says that the input terminals of an ideal OpAmp can be treated as virtual open circuit.

3. Zero Output Impedance in which as the definition,

$$R_{out} = 0 \tag{208}$$

4. Infinite CMRR and PSRR (Power Supply Rejection Ratio).
5. Infinite GBW with zero phase shift and infinite slew rate.

From the above properties, the first two features are mostly used in circuit analysis and are valid for most of the calculations as the DC gain of OpAmp has a great value. In particular cases, the designer itself specifies which of the mentioned parameters is non-ideal.

Feedback

Because of high DC gain, one of the most important applications of OpAmp is well suited in feedback systems. As a definition, a feedback system is a system in which the output or a fraction of it is routed back to the input to constitute a loop (Ford, 2010). As described in Figure 70, if the backward signal is used for comparison which means that it is subtracted from input signal, then we have a negative feedback system. For summation, it is said that the system has a positive feedback (Chen, 2005).

Most of the electronic systems as well as organic systems use negative feedback because it has unique features such as reducing system sensitivity to the disturbances, bandwidth extension, variable input/output impedances and self adjustment. But the main cost is loss of high open loop gain of the amplifier. In Figure 70, *G* is the open loop gain, *H* is the feedback network gain and θ_e is the error signal. For error signal we can write:

$$\theta_e = \theta_i - H\theta_o \tag{209}$$

in which θ_e and θ_o are input and output signals, respectively. Also, we have:

$$\theta_o = G\theta_e \tag{210}$$

The combination of (209) and (210) results in:

$$\theta_o = \frac{G}{1+GH}\theta_i \tag{211}$$

which illustrates that the open loop gain has dropped by $1+GH$ factor in which GH is called the loop gain. For positive feedback, the subtraction turns to summation which results in:

$$\theta_o = \frac{G}{1-GH}\theta_i \tag{212}$$

and illustrates that the positive feedback pushes the system to instability because for GH=1, the denominator becomes zero. The negative feedback involves a large area of circuit analysis and we are not interested in covering it. One main reason is that most of the electronic feedback systems can be analyzed

Figure 70. Negative feedback system

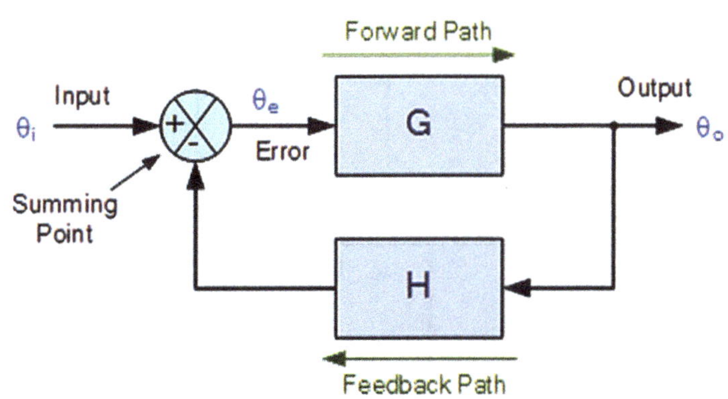

by means Miller's Theorem. For example, we want to analyze the inverting amplifier of Figure 71 which is a negative feedback system (Jaeger & Blalock, 2003).

Considering the properties of ideal OpAmp and setting the only exception of finite gain G, we can use equivalent circuit of Figure 69 with $R_{in}=\infty$ and $R_{out}=0$. By means of Miller's Theorem, the reflected value of R_2 on input side will be $\frac{R_2}{1+G}$. By defining $v_d=v_+-v_-$, we can write:

$$v_d = \frac{\frac{R_2}{1+G}}{R_1 + \frac{R_2}{1+G}} v_{in} \tag{213}$$

also, because of inverting amplifier and by means of Figure 69 model, we have:

$$v_{out} = -Gv_d \tag{214}$$

Therefore, for the gain of feedback system defined as A_f we have:

$$A_f = \frac{v_{out}}{v_{in}} = \frac{-GR_2}{GR_1 + R_1 + R_2} \tag{215}$$

and for infinite open loop gain, there will be:

$$\lim_{G\to\infty} A_f = \frac{-R_2}{R_1} \tag{216}$$

In order to write Equation (215) in the main form of negative feedback system equation, since $GR_1 \gg R_1$, we will obtain:

Figure 71. Inverting amplifier

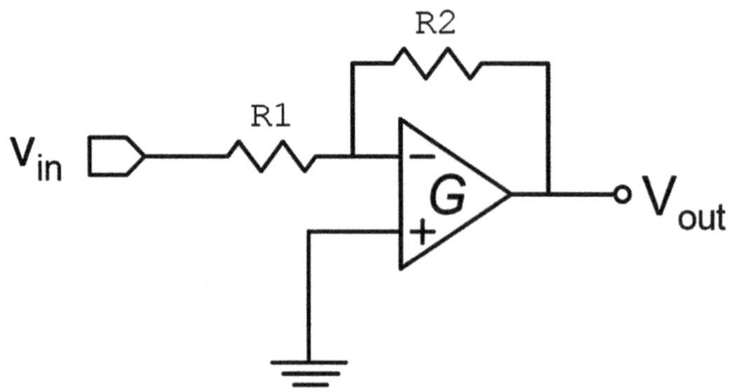

$$\frac{v_{out}}{v_{in}} = \frac{-G}{1+G\frac{R_2}{R_1}} \tag{217}$$

in which the feedback network gain is $\frac{-R_2}{R_1}$. For $GH \gg 1$, Equation (216) will easily be obtained.

Applications in the Field of Sensors

In the field of sensors, OpAmps and other transistor amplifiers are mostly used for amplification. However, other operations such as biasing, feedback systems, signal conversion, filtering and etc are also carried out by transistors and other circuit elements. Here, we will show some examples of the applications of amplifiers in conjunction with the sensors.

As the first application, we want to convert the output voltage of a pressure sensor in order to apply it to the ADC. The circuitry for this purpose is shown in Figure 72.

In this application, the signal conditioner requires two amplifiers configured as a two stage amplifiers in which the first OpAmp is used as a buffer and the second OpAmp is employed as inverting amplifier.

As the second application, LTC6081 Shock Sensor Amplifier (Accelerometer) circuitry is introduced here. A shock sensor is a capacitive piezoelectric element which produced a charge under physical acceleration. This charge can be converted to a voltage which is then filtered by high valued resistors and a high input impedance CMOS amplifier such as the LTC6081. This circuit which is shown in Figure 73, has a gain of 100.

As the third application, Low Noise Non-Inverting Shock Sensor Amplifier circuitry is shown in Figure 74. For this sensor, the charge is converted to a voltage and is then amplified with a precision resistor ratio for gain accuracy. Paralleling of the amplifiers reduces noise.

Figure 72. Pressure sensor application

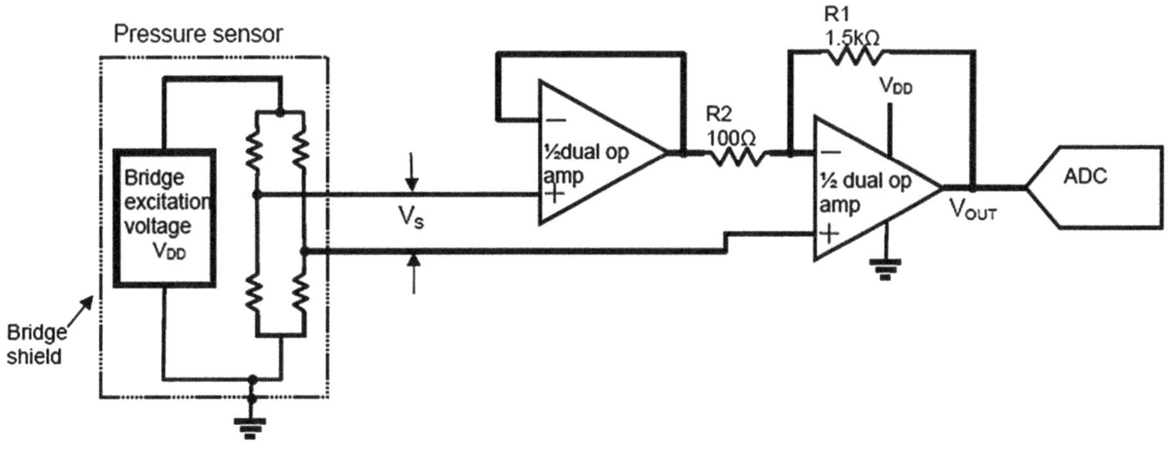

Figure 73. Shock sensor amplifier circuit

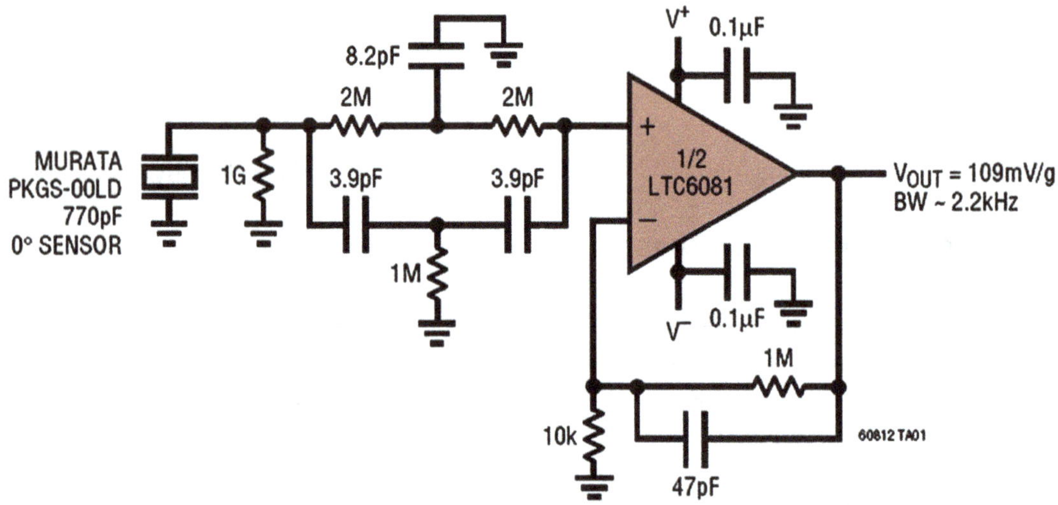

Figure 74. Low noise non-inverting shock sensor amplifier circuit

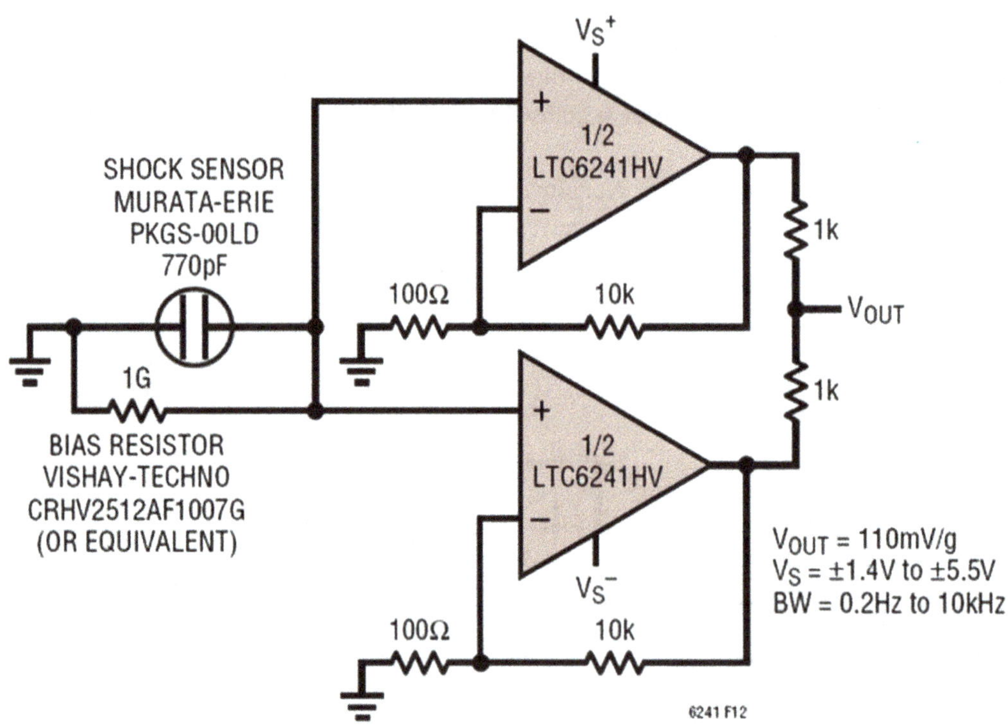

CONCLUSION

In this chapter the most useful information about the semiconducting based amplifiers which are largely used in the design of electronic circuits and devices is provided to simplify the readers work on the analysis of sensor amplifying circuits. Introducing the semiconductors, their derivatives such as diode,

BJT and MOSFET are studied in order to interpret their amplification features. Three important models for BJT and FET were studied as the basis of multistage amplifiers. By expanding the idea, OpAmp with its unique properties was introduced. With the brief definition of negative feedback, some of the applications of amplifiers in conjunction with the sensors were discussed.

REFERENCES

Alioto, M., & Palumbo, G. (2005). *Model and Design of Bipolar and Mos current-mode logic: CML, ECL and SCL digital circuits*. Springer.

Bakshi, A. V., & Bakshi, U. A. (2009). *Network analysis*. Technical Publications.

Boylestad, R. L., & Nashelsky, L. (2002). *Electronic devices and circuit theory*. Prentice Hall.

Chang, C. Y., & Sze, S. M. (1996). *ULSI technology* (International Editions). Singapore: McGraw-Hill.

Chen, W.-K. (2005). *General feedback theory, "Circuit analysis and feedback amplifier theory"*. CRC Press. doi:10.1201/9781420037272

Electrical Engineering. (2013, February 22). *Light Dark Sensor*. Retrieved from http://electrical-engineer10.blogspot.nl/2013/02/light-dark-sensor.html

Ford, A. (2010). *Information feedback and causal loop diagrams, "Modeling the environment*". Island Press.

Galup-Montoro, C., & Schneider, M. C. (2007). *MOSFET modeling for circuit analysis and design*. London: World Scientific. doi:10.1142/6111

Gray, P. R., Hurst, P. J., Lewis, S. H., & Meyer, R. G. (2001). *Analysis and design of analog integrated circuits*. New York: Wiley.

Hareendran, T. K. (n.d.). *Top 10 Articles*. Retrieved from http://www.electroschematics.com/6129/touch-alarm

Hickman, R. W., & Hunt, F. V. (1939). On electronic voltage stabilizers. *The Review of Scientific Instruments*, *10*(1), 6–21. doi:10.1063/1.1751443

Holberg, D. R., & Allen, P. E. (2002). *CMOS analog circuit design*. OxfordUniversity Press.

Hook, J. R., & Hall, H. E. (2001). *Solid state physics*. John Wiley & Sons.

Horowitz, P. (1989). The art of electronics (2nd ed.). Cambridge University Press.

Horowitz, P. (2015). The art of electronics. Cambridge University Press.

Hu, C. C. (2010). Modern semiconductor devices for integrated circuits. Academic Press.

IEEE Standard 100 authoritative dictionary of IEEE standards terms (2000). IEEE Press.

Jaeger, R. C. (1988). *Introduction to microelectronic fabrication*. New York: Addison-Wesley.

Jaeger, R. C., & Blalock, T. N. (2003). *Microelectronic circuit design*. McGraw-Hill.

Kester, W. (2010). *003: Understand SINAD, ENOB, SNR, THD, THD + N, and SFDR so you don't get lost in the noise floor*. Analog Devices. Retrieved.

Kumar, B., & Jain, S. B. (2013). *Electronic devices and circuits*. PHI Learning Pvt. Ltd.

Laplante, P. A. (2005). *Comprehensive dictionary of electrical engineering* (2nd ed.). CRC Press. doi:10.1201/9781420037807

Lowe, D. (2013). *Electronics components: Diodes- Electronics All-In-One Desk Reference For Dummies*. John Wiley & Sons.

Malik, N. R. (1995). *Electronic circuits: Analysis, simulation, and design*. Englewood Cliffs, NJ: Prentice Hall.

McKay, K. (1954). Avalanche breakdown in silicon. *Physical Review*, *94*(4), 877–884. doi:10.1103/PhysRev.94.877

Mehta, V. K. (2008). *Principles of electronics*. Chand.

Mijoe, J. (2015). *Fundamentals of semiconductor physics*. Anchor Academic Publishing.

Millman, J. (1985). *Electronic devices and circuits*. Singapore: McGraw-Hill International.

Mohamed, A. H. E. (1966). Nonlinear distortion in common-emitter transistor amplifiers operated in the normal domain. *Electronics Letters*, *2*(2).

Neamen, D. A. (2003). *Semiconductor physics and devices: Basic principles* (3rd ed.). McGraw-Hill Higher Education.

Ocaya, R. O. (2013, January 18). *Circuit's RMS output is linearly proportional to temperature over wide range*. Retrieved from http://www.edn.com/design/components-and-packaging/4405267/Circuit-s-RMS-output-is-linearly-proportional-to-temperature-over-wide-range

Popadic, M., Lorito, G., & Nanver, L. K. (2009). Analytical model of I–V characteristics of arbitrarily shallow p-n junctions. *IEEE Transactions on Electron Devices*, *56*(1), 116–125. doi:10.1109/TED.2008.2009028

Sedra, A. S., & Smith, K. C. (2004). *Microelectronic circuits*. Oxford University Press.

Sze, S. M. (1985). *Semiconductor devices physics and technology*. New York: Wiley.

The Shichman-Hodges Enhancement MOSFET Model and SwitcherCAD III SPICE. (2007). Report NDT14-08-2007. NanoDotTek.

Uyemura, J. P. (1999). *CMOS logic circuit design*. Springer Science & Business Media.

Widlar, R. J. (1967, January). An exact expression for the thermal variation of the emitter base voltage of bi-polar transistors. *Proceedings of the IEEE*, *55*(1), 96–97. doi:10.1109/PROC.1967.5396

Chapter 17
Wireless Nanosensor Networks:
Prospects and Challenges

Sohail Anwar
Penn State University, USA

ABSTRACT

Nanotechnology is enabling the development of devices in a scale ranging from one to a few hundred nanometers, which can perform tasks such as sensing, data storing, computing, and actuation. These nano devices will be able to cover larger areas and perform more complex tasks through communication. Wireless nanosensor networks (WNSNs) are collections of nanosensor devices with communication capability. The key components of a WNSN include nano-nodes, nano-router, nano-micro interface, and gateway. WNSNs have numerous potential biomedical, environmental, industrial, and military applications. This chapter provides an overview of the architecture, applications, and issues associated with the development of WNSNs.

INTRODUCTION

Wireless nanosensor networks (WNSNs) are collections of nanosensors with communication capability. They can be used for sensing and data gathering with very high resolution and low power consumption. There are numerous potential applications for WNSNs. In the biomedical field, the WNSNs can be used in the health monitoring systems. They can also be used in the drug delivery systems. In industry, the WNSNs can be used to develop ultra high sensitivity touch surfaces and haptic technology interfaces. The network architecture of WNSNs is based on the following components:

- Nano-nodes,
- Nano-routers,
- Nano-micro interface devices,
- Gateway.

DOI: 10.4018/978-1-5225-0736-9.ch017

Communication among nanosensors is of paramount importance in expanding the capabilities and applications of individual nanosensors. However, the development of effective communication among nanosensors currently faces several research challenges. This chapter provides an overview of the architecture, applications, and the research issues associated with wireless nanosensor networks (WNSNs). The chapter begins with a presentation of background information regarding nanosensors and WNSN. Next, the current developments pertaining to the architecture of WNSN and the communication among nanosensors are described. Following this description, an overview of WNSN applications is presented. Finally, the research issues associated with the development of WNSN are highlighted.

BACKGROUND

Nanosensors constitute one of the most promising applications of nanotechnology. A nanosensor is a device that is able to sense a physical or environmental property at the nanoscale level. Nanosensors make use of the unique properties of nanomaterials to detect and measure events in the nanoscale. However, due to its extremely small size, the sensing range of a single nanosensor is limited to its close nano-environment. A nanosensor network is needed to cover larger areas. A Wireless Nanosensor Network (WNSN), which consists of a collection of nanosensors, can be used to cover larger areas and to perform sensing and data collection with extremely high resolution and low power consumption (Kocaoglu & Akan, 2013).

Development of novel communication techniques is essential for WNSN design. Communication among nanosensors will expand the capabilities and applications of individual nano-devices. Two key approaches used for communication at the nanoscale level are nano-electromagnetic communication and molecular communication (Usibe, Menkiti, Onuu, & Ogbulezie, 2013). Nano-electromagnetic communication uses electromagnetic carriers for transmission and reception of information among components based on nanomaterials. The unique properties of nanomaterials will determine the specific bandwidths for electromagnetic radiation emission, the time log of the emission, or the emitter power strength for a given amount of input energy (Akyildiz & Jornet, 2010).

Molecular communication involves the transmission and reception of information encoded in molecules (Akyildiz, Brunetti & Blazquez, 2008). Molecular communication of nanoscale elements is an existing natural phenomenon and provides opportunities for developing solutions through modeling nanonetworks (Malak and Akan, 2012).

The WNSN applications are categorized into environmental, medical, industrial, and military. The WNSNs can be used in plant monitoring and plagues defeating systems. Biological nanosensor networks have potential applications in remote monitoring of physiological data. Distributed networks of nanosensors and nanoactuators can be used to release a specific drug in unreachable locations within a human body (Anwar, 2015). In the context of military applications, WNSNs can be used for armor damage detection and for the battlefield surveillance. In the arena of industrial applications, networks of physical nanosensors and nanoactuators can be used to enhance the remote control of machines of complex machines and devices. Moreover, the physical sensors can be used in a distributed arrangement to develop touch surfaces with high sensitivity and precision.

The development of WNSNs will need to overcome many challenges. The key WNSN research challenges are as follows:

- Communication between nanodevices is a key issue in WNSN research. To overcome this challenge, proper communication scheme, channel model, and routing mechanism are needed.
- Nanosensor has limited energy. Thus, an energy harvesting process is required by nanosensors to cope with the limited size and flexibility of their energy storage (Pierobon, Journet, Akkari, Almasri, and Akyildiz, 2014).
- The nanosensor network efficiency is affected by its size to manage the network size and other important network parameters. Thus, efficient medium access control (MAC) protocol is needed.

NANOSENSORS AND WIRELESS NANOSENSOR NETWORKS

A nanosensor is a device that collects and transmits information about data on the nanoscale. The nanomaterials possess novel electronic, magnetic, mechanical, thermal, optical, and physicochemical properties which provide the nanosensors with superior sensing capabilities. Nanosensors have numerous potential applications, such as, environmental monitoring, food safety, health diagnosis, drug delivery, and damage detection systems. Nanosensors can monitor physical parameters, such as, temperature, displacement, and flow. Nanosensors can be used for pollution monitoring and as accelerometers in microelectromechanical system (MEMS) devices, such as, airbag sensors.

Nanosensors can be broadly categorized into the following types:

- Physical nanosensors,
- Chemical nanosensors,
- Biological nanosensors.

Physical nanosensors are used to measure physical quantities, such as, mass, pressure, displacement, or force. The operation of physical nanosensors is based on the changes in the electronic properties of nanotubes and nanoribbons when these are bent or deformed (Hierold, Jungen, Stampfer, & Helbling, 2007). Chemical nanosensors are used to detect the presence of a specific type of molecules, molecular composition of a substance, or the concentration of a given gas. Their operation is based on the changes in the electronic properties of carbon nanotubes and graphene nanoribbons when different types of molecules are absorbed on top of them (Leenaerts, Partoens, & Peeters, 2009). Biological nanosensors are used to monitor biomolecular processes, such as, DNA interaction and enzymatic interactions. A biological nanosensor consists of a biological recognition system or a bioreceptor and a transduction mechanism. There are two key sub classes of biological nanosensors. These are electrochemical biological nanosensors and photometric biological nanosensors.

Communication among nanosensors will enhance the capabilities and applications of the individual nano devices. There are four key strategies for communication among nanosensors. These are classified as nano-mechanical, acoustic, molecular, and electromagnetic (Piro, Grieco, Boggia, & Camarda, 2013). Wireless Nanosensor Networks (WNSNs) consist of integrated nanosensor devices which communicate via electromagnetic communication. The key elements of a WNSN are nano-nodes, nano-routers, nano-micro interface devices, and gateways (Rupani, Kargathara, & Sureja, 2015). Nano-nodes are nano devices that are capable of performing simple computations, have limited energy, and can only transmit over very short distances. Nano-routers are nano devices having more computational power than nano-nodes. They can also control the behavior of nano-nodes. However, this enhancement of capabilities means an increase in their size. Nano-micro interface devices aggregate the information coming from nano-routers and transmit this information to the micro-scale and vice versa. Gateways are micro scale devices that enable the remote control of entire system over the Internet Gateways are the devices that collect the information from the nano networks and direct it to the remote monitor via Internet.

It is important to mention here the salient features of WNSNs. These networks consist of nano devices having a size of one to few hundreds of nanometers. The WNSNs use graphene based nano antennas that support the electromagnetic communication in Terahertz band. A nano antenna is an electromagnetic collector designed to absorb specific wavelengths that are proportional to the size of the nano antenna. Bit rates in WNSNs are extremely high and the transmission ranges are very small.

WNSN APPLICATIONS

The potential applications of WNSNs range from intra-body wireless nanosensor networks for advanced health monitoring systems to terabit wireless network-on-chip for high performance computer architectures. The applications of WNSNs are broadly classified into medical, environmental, industrial and military.

A large number of WNWN applications are in the medical field. Biological nanosensors provide an interface between biological phenomena and nano devices. Nanosensors can be used to monitor sodium, glucose and other ions in blood. The presence of different infectious agents can also be monitored using nanosensors. A human body nanosensor network can be used to gather important data regarding the level of different substances in human body. A wireless interface between these nanosensors and a micro-device such as a cell phone can be used to collect medical data and transmit them to the health care provider. A distributed network of nanosensors and nanoactuators may be used to decide whether or not to release a given drug to control the intra-cranial pressure. Thus, instead of just releasing a fixed amount of drug, the decisions to act or not can be taken through an effective coordination among nanosensors and nanoactuators.

WNSNs can also be used in numerous environmental applications. Chemical nanosensors can be used to detect the presence of chemical compounds that are being released and exchanged between plants. The large diversity and high sensibility of chemical nanosensors can be exploited in environmental applications such as plant monitoring and plaques defeating systems (Pieterse & Dicke, 2007)

WNSNs have many potential industrial applications. Physical nanosensors can be used in a distributed arrangement to develop touch surfaces having high sensitivity and precision. Touch nanosensors can be randomly placed and fixed over a relatively large surface area. A nano-controller based calibration process may be used to define the boundaries and the positions of the different nanosensors.

A wireless network of nanosensors and nanoactuators can be used to enhance the capabilities of a peripheral, such as, a cell-phone keypad. Another potential application of the interconnection of nanosensors and nanoactuators is in the form of an interconnected office where every single element uses a nanosensor device and it can be permanently connected to the Internet. As a result, a user will be able to keep track of all professional belongings.

Many applications of WNSNs are envisioned in the military field. Chemical and biological nanosensors can be used to detect harmful chemicals and biological weapons in a distributed manner (Akyildiz & Jornet, 2010). Using chemical nanosensors, it is possible to recover the molecular composition of the air in a specific location. Also, physical nanosensors can be used to detect minute cracks in bridges, soldiers' armor, and vehicles.

RESEARCH CHALLENGES FOR ELECTROMAGNETI WIRELESS NANOSENSOR NETWORKS

There are several research challenges that need to be overcome for wireless nanosensor network (WNSN) development. The major challenge is related to the communication among nanosensor devices. The nanosensors will potentially communicate among themselves in the terahertz band (01 – 10.0THz). This is one of the least explored frequency zones in the electromagnetic spectrum. Thus, a key issue for nanosensor communication is the development of new channel models for the terahertz band. The key characteristics of the terahertz channel at nanoscale are as follows:

1. The terahertz channel is very frequency selective.
2. The absorption from molecules present in the terahertz channel medium not only attenuates the transmitted signal, it also introduces noise. This molecular noise is neither white nor Gaussian.
3. The available bandwidth of the terahertz channel depends on the molecular composition of the channel and the transmission distance (Jornet & Akyildiz, 2010) The predicted channel capacity of wireless nanosensor networks in the terahertz band is expected to be very large (several terabits per second)
4. The signal propagation in terahertz channel is affected by multi-path fading and nanoparticle scattering.

Nanosensor devices need unique information coding and modulation techniques that are capable of exploiting the huge bandwidth mode possible by terahertz channel. A possible solution will be the use of sub-pico second long pulses for communication among nanosensor devices. The power of a sub-pico second long pulse is contained within the terahertz frequency band.

There are several open issues pertaining to communication among nanosensor devices. These issues can be handled using proper communication scheme, channel model, and routing mechanism. Also, there are reliability issues in electromagnetic wireless nanosensor networks. Nanosensor device failure and transient molecular interference can affect the WNSN reliability.

Nanosensor device has limited amount of energy. Thus energy harvesting is an issue in WNSN research. One of the WNSN research topics deals with energy harvesting in WNSNs. A routing framework for energy harvesting WNSNs in the terahertz frequency band is proposed by Pierobon, Jornet, Akkari, Almasri, and Akyildiz (2014).

Size of WNSN affects its efficiency. To manage the size of WNSN, there is a need for the development of efficient Media Access Control (MAC) protocols and routing algorithms are required. An energy-aware MAC protocol for perpetual WNSNs in the terahertz band is proposed by Wang, Jornet, Malik, Akkari, and Akyildiz (2013).

CONCLUSION

Wireless nanosensor networks (WNSNs) are collections of nanosensors with the ability to communicate. These networks can be used for sensing and data collection with very high resolution and low power consumption. WNSNs have numerous potential applications which range from advanced health monitoring systems to environmental protection. Molecular communication and electromagnetic communication constitute two key communication techniques for WNSNs. Graphene based nano antennas are used for propagation of electromagnetic waves. WNSNs have extremely high bit rates. The key elements of a WNSN consist of nano nodes, nano router, nano-micro interface, and gateway. Development of WNSNs faces several research challenges. The most prominent among them is communication among nanosensor devices. Other issues include reliability, energy harvesting, and the need for novel information coding and modulation techniques. Active research is in progress to overcome the challenges faced by WNSN development.

REFERENCES

Akyildiz, I. F., Brunetti, F., & Blazquez, C. (2008). Nanonetworks: A new communication paradigm. *Computer Networks*, *52*(12), 2260–2279. doi:10.1016/j.comnet.2008.04.001

Akyildiz, I. F., & Jornet, J. M. (2010). Electromagnetic wireless nanosensor networks. *Nano Communication Networks*, *1*(1), 3–19. doi:10.1016/j.nancom.2010.04.001

Anwar, S. (2015). Wireless nanosensaor networks: A basic review. *International Journal of Emergeing Technology and Advanced Engineering*, *5*(12), 151–154.

Hierold, C., Jungen, A., Stampfer, C., & Helbling, T. (2007). Nano electromechanical sensors based on carbon nanotubes. *Sensors and Actuators. A, Physical*, *136*(1), 51–61. doi:10.1016/j.sna.2007.02.007

Jornet, J. M., & Akyildiz, I. F. (2010). Channel capacity of electromagnetic nanonetworks in the terhertz band.*Proceedings of IEEE International Conference on Communications.*

Kacaoglu, M., & Akan, O. B. (2013). Minimum energy channel codes for nanoscale wireless communications. *IEEE Transactions on Wireless Communications*, *12*(4), 1492–1500. doi:10.1109/TWC.2013.022113.2190

Leenaerts, O., Partoens, B., & Peeters, F. (2009). Adsorption of small molecules on graphene. *Microelectronics Journal*, *40*(4-5), 860–862. doi:10.1016/j.mejo.2008.11.022

Malak, D., & Akan, O. B. (2012). Molecular communication nanonetworks inside human body. *Nano Communication Networks*, *3*(1), 19–35. doi:10.1016/j.nancom.2011.10.002

Pierobon, M., Jornet, J. M., Akkari, N., Almasri, S., & Akyildiz, I. F. (2014). A routing framework for energy harvesting wireless nanosensor networks in the Terahertz Band. *Wireless News*, *20*(5), 1169–1183. doi:10.1007/s11276-013-0665-y

Pieterse, C. M., & Dicke, M. (2007). Plant interactions with microbes and insects: From molecular mechanisms to ecology. *Trends in Plant Science*, *12*(12), 564–569. doi:10.1016/j.tplants.2007.09.004 PMID:17997347

Piro, G., Grieco, L. A., Boggia, G., & Camarda, P. (2013). Simulating wireless nano sensor networks in the NS – 3 platform. *IEEE Advanced Information Networking and Applications Workshops (WAINA).*

Rupani, V., Kargathara, S., & Sureja, J. (2015). A review on wireless nanosensor networks based on electromagnetic communication. *International Journal of Computer Science and Information Technologies*, *6*(2), 1019–1022.

Usibe, B. E., Menkiti, A. I., Onuu, M. U., & Ogbulezie, J. C. (2013). Development and analysis of a potential nanosensor communication network using carbon nanotubes. *International Journal of Materials Engineering*, *3*(1), 4–10.

Wang, P., Jornet, J. M., Malik, M. G., Akkari, N., & Akyildiz, I. F. (2013). Energy and spectrum-aware MAC protocol for perpetual wireless nanosensor networks in the Terahertz Band. *Ad Hoc Networks*, *11*(8), 2541–2555. doi:10.1016/j.adhoc.2013.07.002

Compilation of References

(2013). Pejovi, #x107, M., #x107, Ciraj-Bjelac, O., Kova, . . ., G. (2013). Sensitivity of P-Channel MOSFET to X- and Gamma-Ray Irradiation. *International Journal of Photoenergy*, *6*. doi:10.1155/2013/158403

Abadi, H. K. F., Yusof, R., Eshrati, S. M., Naghib, S., Rahmani, M., Ghadiri, M., & Ahmadi, M. et al. (2013). Current–voltage modeling of graphene-based DNA sensor. *Neural Computing & Applications*, 1–5.

Abdel-Baky, A., El-Ghazaly, M., & Mansour, M. (2000). Spectroscopic and electric probe measurements of plasma parameters in the negative glow region of helium discharge plasma. *Journal of Quantitative Spectroscopy & Radiative Transfer*, *67*(3), 217–225. doi:10.1016/S0022-4073(99)00209-5

Abdel-Hamid, I., Ivnitski, D., Atanasov, P., & Wilkins, E. (1999). Flow-through immunofiltration assay system for rapid detection of E. coli O157: H7. *Biosensors & Bioelectronics*, *14*(3), 309–316. doi:10.1016/S0956-5663(99)00004-4 PMID:10230031

Abdelwahab, A. A., Koh, W. C. A., Noh, H.-B., & Shim, Y.-B. (2010). A selective nitric oxide nanocomposite biosensor based on direct electron transfer of microperoxidase: Removal of interferences by co-immobilized enzymes. *Biosensors & Bioelectronics*, *26*(3), 1080–1086. doi:10.1016/j.bios.2010.08.070 PMID:20864332

Abergel, D. S. L., Apalkov, V., Berashevich, J., Ziegler, K., & Chakraborty, T. (2010). Properties of graphene: A theoretical perspective. *Advances in Physics*, *59*(4), 261–482. doi:10.1080/00018732.2010.487978

Abolgasim, A. A. (2008). *Classification of capital expenditures and revenue expenditures using neural network model*. Universiti Utara Malaysia.

Acik, M., & Chabal, Y. J. (2011). Nature of graphene edges: A review. *Japanese Journal of Applied Physics, 50*(7). doi: 07010110.1143/jjap.50.070101

Acik, M., & Chabal, Y. J. (2011). Nature of graphene edges: A review. *Japanese Journal of Applied Physics*, 50.

Adamiec, G., Iñiguez, M. P., Lorente, A., Voytchev, M., & Gallego, E. (2004). Response of a silicon PIN photodiode to an Am-Be neutron source. *Nuclear Instruments & Methods in Physics Research. Section A, Accelerators, Spectrometers, Detectors and Associated Equipment*, *534*(3), 544–550. doi:10.1016/j.nima.2004.06.145

Adam, S., Hwang, E., Galitski, V., & Sarma, S. D. (2007a). A self-consistent theory for graphene transport. *Proceedings of the National Academy of Sciences of the United States of America*, *104*(47), 18392–18397. doi:10.1073/pnas.0704772104 PMID:18003926

Adenilson, J. C., Cleber, A. A., Olivia, M. B., Luana, S. A., Eric, P. B., & Edson, R. L. (2012). Back-to-back Schottky diodes: The generalization of the diode theory in analysis and extraction of electrical parameters of nanodevices. *Journal of Physics Condensed Matter*, *24*(22), 225303. doi:10.1088/0953-8984/24/22/225303 PMID:22556197

Adonkin, V. T., Gorelov, B. M., Dyakin, V. V., Karpov, I. I., & Prikhodko, G. P. (1994). Effect of intercalation and exfoliation on positron-annihilation lifetimes in graphites. *Journal of Physics and Chemistry of Solids*, *55*(5), 443–446. doi:10.1016/0022-3697(94)90171-6

Afrin, R., Khaliq, J., Islam, M., Gul, I. H., Bhatti, A. S., & Manzoor, U. (2012). Synthesis of multiwalled carbon nanotube-based infrared radiation detector. *Sensors and Actuators. A, Physical*, *187*(0), 73–78. doi:10.1016/j.sna.2012.08.028

Agosteo, S., D'Angelo, G., Fazzi, A., Para, A. F., Pola, A., Ventura, L., & Zotto, P. (2005). Performance of a neutron spectrometer based on a PIN diode. *Radiation Protection Dosimetry*, *116*(1-4), 180–184. doi:10.1093/rpd/nci102 PMID:16604623

Ahammad, A. J. S., Lee, J. J., & Rahman, M. A. (2009). Article. *Sensors (Basel, Switzerland)*, *9*(4), 2289–2319.

Ahin, Y. S., Aydın, A., Arslan Udum, Y., Pekmez, K., & Yıldız, A. (2004). Electrochemical synthesis of sulfonated polypyrrole in FSO3H/acetonitrile solution. *Journal of Applied Polymer Science*, *93*, 526–533. doi:10.1002/app.20473

Ahmadi, M. T., & Ismail, R. 2010. Graphene Nanoribbon Fermi energy model in parabolic band structure. In *Intelligent Systems, Modelling and Simulation (ISMS), 2010 International Conference on*, (pp. 401-405). IEEE. doi:10.1109/ISMS.2010.78

Ahmadi, M. T., Johari, Z., Amin, N. A., Fallahpour, A. H., & Ismail, R. (2010). Graphene nanoribbon conductance model in parabolic band structure. *Journal of Nanomaterials.*

Ahmadi, M. T., Johari, Z., Amin, N. A., Mousavi, S. M., & Ismail, R. (2010). *Carbon nanotube conductance model in parabolic band structure*. Academic Press.

Ahmadi, M. T., Johari, Z., Amin, N. A., Mousavi, S. M., & Ismail, R. (2010c). Carbon nanotube conductance model in parabolic band structure. In *Semiconductor Electronics (ICSE), 2010 IEEE International Conference on*, (pp. 256-259). IEEE. doi:10.1109/SMELEC.2010.5549582

Ahmadi, M. T., Johari, Z., Amin, N. A., Mousavi, S. M., & Ismail, R. (2010d). Carbon nanotube conductance model in parabolic band structure. IEEE.

Ahmadi, M. T., Karamdel, J., Ismail, R., Dee, C., & Majlis, B. Y. (2008). *Modelling of the current-voltage characteristics of a carbon nano tube field effect transistor.* Paper presented at the Semiconductor Electronics, 2008. ICSE 2008. IEEE International Conference on. doi:10.1109/SMELEC.2008.4770391

Ahmadi, S., Hassani, F., Korman, C., Rahaman, M., & Zaghloul, M. (2004). *Characterization of multi-and single-layer structure SAW sensor [gas sensor].* Paper presented at the Sensors, 2004.

Ahmadi, M. T., Ismail, R., Michael Tan, L. P., & Aroral Vijay, K. (2008). The Ultimate Ballistic Drift Velocity in Carbon Nanotubes. *Journal of Nanomaterials*, *769250*. doi:10.1155/2008/769250

Ahmadi, M. T., Ismail, R., Tan, M. L., & Arora, V. K. (2008). The ultimate ballistic drift velocity in carbon nanotubes. *Journal of Nanomaterials.*

Ahmadi, M. T., Johari, Z., Amin, N. A., Fallahpour, A. H., & Ismail, R. (2010a). Graphene nanoribbon conductance model in parabolic band structure. *Journal of Nanomaterials*, *2010*, 12. doi:10.1155/2010/753738

Ahmadi, M. T., Johari, Z., Amin, N. A., & Ismail, R. (2012). Band energy effect on carrier velocity limit in graphene nanoribbon. *Journal of Experimental Nanoscience*, *7*(1), 62–73. doi:10.1080/17458080.2010.501826

Ahmadi, M. T., Rahmani, M., Ghadiry, M. H., & Ismail, R. (2012). Graphene Nanoribbon Homojunction Charactristics. *Sci. Adv. Mater.*, *4*(7), 753–756. doi:10.1166/sam.2012.1367

Ahmadi, M. T., Tan, M. L. P., Ismail, R., & Arora, V. K. (2009). Article. *International. Journal of Nanotechnology*, *6*(7-8), 601–617. doi:10.1504/IJNT.2009.025299

Ahmadi, M. T., Tan, M. L. P., Ismail, R., & Arora, V. K. (2009). The high-field drift velocity in degenerately-doped silicon nanowires. *International Journal of Nanotechnology*, *6*(7-8), 601–617. doi:10.1504/IJNT.2009.025299

Ahn, Y. H., Tsen, A. W., Kim, B., Park, Y. W., & Park, J. (2007). Photocurrent imaging of p-n junctions in ambipolar carbon nanotube transistors. *Nano Letters*, *7*(11), 3320–3323. doi:10.1021/nl071536m PMID:17939725

Akbari, E., Ahmadi, M., Kiani, M., Feizabadi, H. K., Rahmani, M., & Khalid, M. (2013a). Monolayer graphene based co2 gas sensor analytical model. *Journal of Computational and Theoretical Nanoscience*, *10*(6), 1301–1304. doi:10.1166/jctn.2013.2846

Akbari, E., Ahmadi, M., Yusof, R., Ghadiry, M., & Saeidmanesh, M. (2013b). Gas concentration effect on channel capacitance in graphene based sensors. *Journal of Computational and Theoretical Nanoscience*, *10*(10), 2449–2452. doi:10.1166/jctn.2013.3229

Akbari, E., Arora, V. K., Enzevaee, A., Ahmadi, M. T., Saeidmanesh, M., Khaledian, M., & Yosof, R. et al. (2014a). An analytical approach to evaluate the performance of graphene and carbon nanotubes for NH_3 gas sensor applications. *Beilstein Journal of Nanotechnology*, *5*, 726–734. doi:10.3762/bjnano.5.85 PMID:24991510

Akbari, E., Buntat, Z., Ahmad, M. H., Enzevaee, A., Yousof, R., Iqbal, S. M. Z., & Karimi, H. et al. (2014b). Analytical calculation of sensing parameters on carbon nanotube based gas sensors. *Sensors (Basel, Switzerland)*, *14*(3), 5502–5515. doi:10.3390/s140305502 PMID:24658617

Akbari, E., Buntat, Z., Ensevaee, A., Ebrahimi, M., Yazdavar, A. H., & Yusof, R. (2014c). Analytical modeling and simulation of I-V characteristics in carbon nanotube gas sensors using ANN and SVR methods. *Chemometrics and Intelligent Laboratory Systems*, *137*, 173–180. doi:10.1016/j.chemolab.2014.07.001

Akbari, E., Yousof, R., Ahmadi, M. T., Kiani, M. J., Rahmani, M., Feiz Abadi, H. K., & Saeidmanesh, M. (2014e). The effect of concentration on gas sensor model based on graphene nanoribbon. *Neural Computing & Applications*, *24*(1), 143–146. doi:10.1007/s00521-013-1463-2

Akbari, E., Yusof, R., Ahmadi, M. T., Enzevaee, A., Kiani, M. J., Karimi, H., & Rahmani, M. (2014d). Bilayer graphene application on NO_2 sensor modelling. *Journal of Nanomaterials*, *534105*. doi:10.1155/2014/534105

Akdim, B., & Pachtert, R. (2011). Switching behavior of carbon chains bridging graphene nanoribbons: Effects of uniaxial strain. *ACS Nano*, *5*(3), 1769–1774. doi:10.1021/nn102403j PMID:21344875

Akhavan, O., & Ghaderi, E. (2012). Escherichia coli bacteria reduce graphene oxide to bactericidal graphene in a self-limiting manner. *Carbon*, *50*(5), 1853–1860. doi:10.1016/j.carbon.2011.12.035

Akundy, G. S., & Iroh, J. (2001). Polypyrrole coatings on aluminum synthesis and characterization. *Polymer*, *42*(24), 9665–9669. doi:10.1016/S0032-3861(01)00529-8

Akundy, G. S., Rajagopalan, R., & Iroh, J. O. (2002). Electrochemical deposition of polyaniline-polypyrrole composite coatings on aluminum. *Journal of Applied Polymer Science*, *83*(9), 1970–1977. doi:10.1002/app.10123

Akyildiz, I. F., Brunetti, F., & Blazquez, C. (2008). Nanonetworks: A new communication paradigm. *Computer Networks*, *52*(12), 2260–2279. doi:10.1016/j.comnet.2008.04.001

Akyildiz, I. F., & Jornet, J. M. (2010). Electromagnetic wireless nanosensor networks. *Nano Communication Networks*, *1*(1), 3–19. doi:10.1016/j.nancom.2010.04.001

Aldrich, C., & Auret, L. (2013). Artificial neural networks. In Unsupervised Process Monitoring and Fault Diagnosis with Machine Learning Methods, (pp. 71-115). Springer. doi:10.1007/978-1-4471-5185-2_3

Alioto, M., & Palumbo, G. (2005). *Model and Design of Bipolar and Mos current-mode logic: CML, ECL and SCL digital circuits*. Springer.

Alizadeh, N., Ataei, A., & Pirsa, S. (2015). Nanostructured conducting polypyrrole film prepared by chemical vapor deposition on the interdigital electrodes at room temperature under atmospheric condition and its application as gas sensor. *Journal of the Indian Chemical Society*, *12*, 1585–1594.

Alizadeh, N., Pirsa, S., Mani-Varnosfaderani, A., & Alizadeh, M. S. (2015). Design and fabrication of open-tubular array gas sensors based on conducting polypyrrole modified with crown ethers for simultaneous determination of alkylamines. *IEEE Sensors Journal*, *15*(7), 4130–4136. doi:10.1109/JSEN.2015.2411515

Alizadeh, N., Shahdousti, P., Nabavi, S., & Tabrizchi, M. (2011). Atmospheric pressure gas-phase ammonium/alkyl ammonium exchange studies of some crown ethers complexes using ion mobility spectrometry: A thermodynamic investigation and collision cross section measurements. *International Journal of Mass Spectrometry*, *308*(1), 18–25. doi:10.1016/j.ijms.2011.07.001

Allen, B. L., Kichambare, P. D., & Star, A. (2007). Carbon nanotube field-effect-transistor-based biosensors. *Advanced Materials*, *19*(11), 1439–1451. doi:10.1002/adma.200602043

Alwarappan, S., Boyapalle, S., Kumar, A., Li, C.-Z., & Mohapatra, S. (2012). Comparative study of single-, few-, and multilayered graphene toward enzyme conjugation and electrochemical response. *The Journal of Physical Chemistry C*, *116*(11), 6556–6559. doi:10.1021/jp211201b

Amani, A., & Mohammadyani, D. (2011). *Artificial neural networks: applications in nanotechnology. In Artificial neural networks—application*. Rijeka: INTECH.

Amaya, T., Saio, D., & Hirao, T. (2007). Template synthesis of polyaniline/Pd nanoparticles and its catalytic application. *Tetrahedron Letters*, *48*(15), 2729–2732. doi:10.1016/j.tetlet.2007.02.059

Amin, N. A., Ahmadi, M. T., Johari, Z., Mousavi, S. M., & Ismail, R. (2011). Effective mobility model of graphene nanoribbon in parabolic band energy. *Modern Physics Letters B*, *25*(10), 739–745. doi:10.1142/S0217984911026073

Amin, N. A., Ahmadi, M. T., Johari, Z., Webb, J. F., Mousavi, M., & Razali, R. (2010). Drift Velocity and Mobility of a Graphene Nanoribbon in a High Magnitude Electric Field.*Fourth Global Conference on Power Control and Optimization*, Sarawak, Malaysia.

Amin, N. A., Ahmadi, M. T., Johari, Z., Webb, J. F., Mousavi, S. M., & Ismail, R. (2011). Drift velocity and mobility of a graphene nanoribbon in a high magnitude electric field. In N. W. J. F. V. P. Barsoum (Ed.), *Proceedings of the Fourth Global Conference on Power Control and Optimization* (Vol. 1337, pp. 177-179). doi:10.1063/1.3592462

Amiri, I. S., Afroozeh, A., & Bahadoran, M. (2011). Simulation and analysis of multisoliton generation using a panda ring resonator system. *Chinese Physics Letters*, *28*, 104205. doi:10.1088/0256-307X/28/10/104205

Amiri, I. S., Ahsan, R., Shahidinejad, A., Ali, J., & Yupapin, P. P. (2012). Characterisation of bifurcation and chaos in silicon microring resonator. *IET Communications*, *6*(16), 2671–2675. doi:10.1049/iet-com.2011.0773

Amiri, I. S., & Ali, J. (2013). Optical buffer application used for tissue surgery using direct interaction of nano optical tweezers with nano cells. *Quantum Matter*, *2*(6), 484–488. doi:10.1166/qm.2013.1086

Amiri, I. S., Nikoukar, A., Shahidinejad, A., Anwar, T., & Ali, J. (2014). Quantum Transmission of optical tweezers via fiber optic using half-panda system. *Life Science Journal*, *10*, 391–400.

Anand, K., Singh, O., Singh, M. P., Kaur, J., & Singh, R. C. (2014). Hydrogen sensor based on graphene/ZnO nanocomposite. *Sensors and Actuators. B, Chemical*, *195*, 409–415. doi:10.1016/j.snb.2014.01.029

Anantram, M., & Leonard, F. (2006). Physics of carbon nanotube electronic devices. *Reports on Progress in Physics*, *69*(3), 507–561. doi:10.1088/0034-4885/69/3/R01

Anderson, H., Atkinson, Bremner, Carrington, & Peacock. (2012). *Quantitative systematic review of short term associations between ambient air pollution (particulate matter, ozone, nitrogen dioxide, sulphur dioxide and carbon monoxide), and mortality and morbidity*. Report to Department of Health revised following first review.

Anderson, B. A., & Nowak, E. J. (2009). *Graphene-based field effect transistor has pair of graphene layers located on pair of sidewalls of silicon carbide fin. Anderson B a.* Nowak E J.

Anderson, M. R., Mattes, B. R., Reiss, H., & Kaner, R. B. (1991). Conjugated polymer films for gas separations. *Science*, *252*(5011), 1412–1415. doi:10.1126/science.252.5011.1412 PMID:17772913

Anderson, P., Lee, P., & Saitoh, M. (1973). Remarks on giant conductivity in TTF-TCNQ. *Solid State Communications*, *13*(5), 595–598. doi:10.1016/S0038-1098(73)80020-1

Andle, J., & Vetelino, J. (1994). Acoustic wave biosensors. *Sensors and Actuators. A, Physical*, *44*(3), 167–176. doi:10.1016/0924-4247(94)00801-9

Ando, Y., & Zhao, X. (2006). Synthesis of carbon nanotubes by arc-discharge method. *New Diamond and Frontier Carbon Technology*, *16*, 123–138.

Ang, P. K., Jaiswal, M., Lim, C. H. Y. X., Wang, Y., Sankaran, J., Li, A., & Loh, K. P. et al. (2010). A bioelectronic platform using a graphene– lipid bilayer interface. *ACS Nano*, *4*(12), 7387–7394. doi:10.1021/nn1022582 PMID:21067155

An, K. H., Jeong, S. Y., Hwang, H. R., & Lee, Y. H. (2004). Enhanced sensitivity of a gas sensor incorporating single-walled carbon nanotube-polypyrrole nanocomposites. *Advanced Materials*, *16*(12), 1005–1009. doi:10.1002/adma.200306176

Anselmetti, D. (2012). Nanopores: Tiny holes with great promise. *Nature Nanotechnology*, *7*(2), 81–82. doi:10.1038/nnano.2012.11 PMID:22306893

Anupama, G., Late Dattatray, J., Panchakarla, L. S., Govindaraj, A. & Rao, C. N. R. (2009). NO_2 and humidity sensing characteristics of few-layer graphene. *Physics, Chemical Physics,* 560064.

Anwar, S. (2015). Wireless nanosensaor networks: A basic review. *International Journal of Emergeing Technology and Advanced Engineering*, *5*(12), 151–154.

Ao, Z., Yang, J., Li, S., & Jiang, Q. (2008). Enhancement of CO detection in Al doped graphene. *Chemical Physics Letters*, *461*(4-6), 276–279. doi:10.1016/j.cplett.2008.07.039

Aphesteguy, J. C., & Jacobo, S. E. (2004). Composite of polyaniline containing iron oxides. *Physica B, Condensed Matter*, *354*(1-4), 224–227. doi:10.1016/j.physb.2004.09.053

Arakelyan, V. M., Aleksanyan, M. S., Hovhannisyan, R. V., Shahnazaryan, G. E., Aroutiounian, V. M., Hernadi, K., & Forro, L. et al. (2013). Gas sensors made of multiwall carbon nanotubes modified by tin dioxide. *Journal of Contemporary Physics-Armenian Academy of Sciences*, *48*(4), 176–183. doi:10.3103/S1068337213040063

Arash, B., & Wang, Q. (2012). Detection of gas atoms with graphene sheets. *Computational Materials Science*, *60*, 245–249. doi:10.1016/j.commatsci.2012.03.053

Arash, B., & Wang, Q. (2013). Detection of gas atoms with carbon nanotubes. *Scientific Reports*, *3*, 1782. doi:10.1038/srep01782

Arduini, F., Cassisi, A., Amine, A., Ricci, F., Moscone, D., & Palleschi, G. (2009). Electrocatalytic oxidation of thiocholine at chemically modified cobalt hexacyanoferrate screen-printed electrodes. *Journal of Electroanalytical Chemistry*, *626*(1), 66–74. doi:10.1016/j.jelechem.2008.11.003

Arora, N., & Sharma, N. N. (2014). Arc discharge synthesis of carbon nanotubes: Comprehensive review. *Diamond and Related Materials*, *50*(0), 135–150. doi:10.1016/j.diamond.2014.10.001

Arora, V. K., Tan, M. L. P., Saad, I., & Ismail, R. (2007). Ballistic quantum transport in a nanoscale metal-oxide-semiconductor field effect transistor. *Applied Physics Letters*, 91.

Arshak, K., Moore, G. M., Lyons, J., & Clifford, S. (2004). A review of gas sensors employed in electronic nose applications. *Sensor Review*, *24*(2), 181–198. doi:10.1108/02602280410525977

Assadi, A., & Farshi, B. (2011). Stability analysis of graphene based laminated composite sheets under non-uniform inplane loading by nonlocal elasticity. *Applied Mathematical Modelling*, *35*(9), 4541–4549. doi:10.1016/j.apm.2011.03.020

Atashbar, M. Z., Bazuin, B. J., Simpeh, M., & Krishnamurthy, S. (2005). 3D FE simulation of H(2)SAW gas sensor. *Sensors and Actuators. B, Chemical*, *111*, 213–218. doi:10.1016/j.snb.2005.06.054

Aufray, B., Kara, A., Vizzini, S., Oughaddou, H., Leandri, C., Ealet, B., & Le Lay, G. (2010). Graphene-like silicon nanoribbons on Ag(110): A possible formation of silicene. *Applied Physics Letters*, *96*(18), 183102. doi:10.1063/1.3419932

Auld, B. A. (1973). *Acoustic fields and waves in solids*. Рипол Классик.

Avdoshenko, S. M., Nozaki, D., Gomes da Rocha, C., González, J. W., Lee, M. H., Gutierrez, R., & Cuniberti, G. (2013). Dynamic and Electronic Transport Properties of DNA Translocation through Graphene Nanopores. *Nano Letters*, *13*(5), 1969–1976. doi:10.1021/nl304735k PMID:23586585

Avetisyan, A. A., Partoens, B., & Peeters, F. M. (2010). Stacking order dependent electric field tuning of the band gap in graphene multilayers. *Physical Review B: Condensed Matter and Materials Physics*, *81*(11), 115432. doi:10.1103/PhysRevB.81.115432

Avila, J., Padova, P. D., Cho, S., Colambo, I., Lorcy, S., & Quaresima, C. et al. (2013). Presence of gapped silicene-derived band in the prototypical (3 × 3) silicene phase on silver (111) surfaces. *Journal of Physics Condensed Matter*, *25*(26), 262001. doi:10.1088/0953-8984/25/26/262001 PMID:23759650

Avouris, P. (2002). Molecular electronics with carbon nanotubes. *Accounts of Chemical Research*, *35*(12), 1026–1034. doi:10.1021/ar010152e PMID:12484790

Avouris, P., Chen, Z. H., & Perebeinos, V. (2007). Carbon-based electronics. *Nature Nanotechnology*, *2*(10), 605–615. doi:10.1038/nnano.2007.300 PMID:18654384

Avouris, P., & Xia, F. (2012). Graphene applications in electronics and photonics. *MRS Bulletin*, *37*(12), 1225–1234. doi:10.1557/mrs.2012.206

Avramescu, V., Bostan, C., Serban, B., Georgescu, I., Costea, S., & Varachiu, N. (2009). Surface acoustic wave devices and their sensing capabilities Cas. *2009 International Semiconductor Conference*. doi:10.1109/SMICND.2009.5336615

Babcock, R. (1965). Radiation Damage in SiC. *IEEE Transactions on Nuclear Science*, *12*(6), 43–47. doi:10.1109/TNS.1965.4323922

Baby, T. T., Aravind, S. S. J., Arockiadoss, T., Rakhi, R. B., & Ramaprabhu, S. (2010). Metal decorated graphene nanosheets as immobilization matrix for amperometric glucose biosensor. *Sensors and Actuators. B, Chemical*, *145*(1), 71–77. doi:10.1016/j.snb.2009.11.022

Badhulika, S., Myung, N. V., & Mulchandani, A. (2014). Conducting polymer coated single-walled carbon nanotube gas sensors for the detection of volatile organic compounds. *Talanta*, *123*, 109–114. doi:10.1016/j.talanta.2014.02.005 PMID:24725871

Bakshi, A. V., & Bakshi, U. A. (2009). *Network analysis*. Technical Publications.

Balandin, A. A. (2013). *Review of the low-frequency 1/f noise in graphene devices.* arXiv preprint arXiv:1307.4797

Balandin, A. A. (2011). Thermal properties of graphene and nanostructured carbon materials. *Nature Materials*, *10*(8), 569–581. doi:10.1038/nmat3064 PMID:21778997

Balandin, A. A., Kotchetkov, D., & Ghosh, S. (2009). *Lateral graphene heat spreaders for electronic and optoelectronic devices and circuits*. Google Patents.

Balapanuru, J., Yang, J.-X., Xiao, S., Bao, Q., Jahan, M., Polavarapu, L., & Loh, K. P. et al. (2010). A Graphene oxide–organic dye ionic complex with dna-sensing and optical-limiting properties. *Angewandte Chemie International Edition*, *49*(37), 6549–6553. doi:10.1002/anie.201001004 PMID:20669201

Ballantine, D. S., Jr., White, R. M., Martin, S. J., Ricco, A. J., Zellers, E. T., Frye, G. C., & Wohltjen, H. (1996). Acoustic wave sensors: Theory, design, & physico-chemical applications. Academic Press.

Ballato, A. (1996). *Piezoelectricity: History and new thrusts*. Paper presented at the Ultrasonics Symposium. doi:10.1109/ULTSYM.1996.584046

Bao, Q., & Loh, K. P. (2012). Graphene photonics, plasmonics, and broadband optoelectronic devices. *ACS Nano*, *6*(5), 3677–3694. doi:10.1021/nn300989g PMID:22512399

Barbier, M., Vasilopoulos, P., Peeters, F. M., & Pereira, J. M. (2009). Bilayer graphene with single and multiple electrostatic barriers: Band structure and transmission. *Physical Review B: Condensed Matter and Materials Physics*, *79*(15), 155402. doi:10.1103/PhysRevB.79.155402

Barholm-Hansen, C., Bentzon, M., & Hansen, J. B. (1994). Optical emission spectroscopy during growth of diamond-like carbon from a methane plasma. *Diamond and Related Materials*, *3*(4-6), 564–568. doi:10.1016/0925-9635(94)90225-9

Bartlett, P. N., & Ling-Chung, S. K. (1989). Conducting polymer gas sensors part III: Results for four different polymers and different vapors. *Sensors and Actuators*, *20*(3), 287–292. doi:10.1016/0250-6874(89)80127-1

Basu, S., & Bhattachayya, P. (2012). Recent developments on graphene and graphene oxide based solid state gas sensors. *Sensors and Actuators. B, Chemical*, *173*, 1–21. doi:10.1016/j.snb.2012.07.092

Baughman, R. H. 2000. Carbon nanotube actuators. In *APS Meeting Abstracts*, 10001. doi:10.1126/science.284.5418.1340

Baughman, R. H., Zakhidov, A. A., & de Heer, W. A. (2002). Carbon nanotubes--the route toward applications. *Science*, *297*(5582), 787–792. doi:10.1126/science.1060928 PMID:12161643

Bell, B. P., Franks, P., Hildreth, N., & Melius, J. (1991). Methylene chloride exposure and birthweight in Monroe County, New York. *Environmental Research*, *55*(1), 31–39. doi:10.1016/S0013-9351(05)80138-0 PMID:1855488

Bellucci, S. (2005). Carbon nanotubes: Physics and applications. *Physica Status Solidi (c), 2*, 34-47.

Beloqui Redondo, A., Troussard, E., & van Bokhoven, J. (2012). Non-oxidative methane conversion assisted by corona discharge. *Fuel Processing Technology*, *104*, 265–270. doi:10.1016/j.fuproc.2012.05.021

Benjamin, P. W., Kemshall, C. D., & Redfearn, J. (1968). A high resolution spherical proportional counter. *Nuclear Instruments and Methods*, *59*(1), 77–85. doi:10.1016/0029-554X(68)90347-9

Berger, C., Song, Z., Li, T., Li, X., Ogbazghi, A. Y., Feng, R., & First, P. N. (2004). Ultrathin epitaxial graphite: 2D electron gas properties and a route toward graphene-based nanoelectronics. *The Journal of Physical Chemistry B*, *108*(52), 19912–19916. doi:10.1021/jp040650f

Berger, C., Song, Z., Li, X., Wu, X., Brown, N., Naud, C., & Marchenkov, A. N. et al. (2006). Electronic confinement and coherence in patterned epitaxial graphene. *Science*, *312*(5777), 1191–1196. doi:10.1126/science.1125925 PMID:16614173

Berland, K., & Hyldgaard, P. (2014). Exchange functional that tests the robustness of the plasmon description of the van der Waals density functional. *Physical Review B: Condensed Matter and Materials Physics*, *89*(3), 035412. doi:10.1103/PhysRevB.89.035412

Bertini, I., Lee, Y. M., Luchinat, C., Piccioli, M., & Poggi, L. (2001). Locating the metal ion in calcium-binding proteins by using cerium(III) as a probe. *ChemBioChem*, *2*(7-8), 550–558. doi:10.1002/1439-7633(20010803)2:7/8<550::AID-CBIC550>3.0.CO;2-T PMID:11828488

Bertuccio, G., Casiraghi, R., Cetronio, A., Lanzieri, C., & Nava, F. (2004). Silicon carbide for high resolution X-ray detectors operating up to 100°C. *Nuclear Instruments & Methods in Physics Research. Section A, Accelerators, Spectrometers, Detectors and Associated Equipment*, *522*(3), 413–419. doi:10.1016/j.nima.2003.11.413

Besteman, K., Lee, J.-O., Wiertz, F. G., Heering, H. A., & Dekker, C. (2003). Enzyme-coated carbon nanotubes as single-molecule biosensors. *Nano Letters*, *3*(6), 727–730. doi:10.1021/nl034139u

Betti, A., Fiori, G., & Iannaccone, G. (2011). Drift velocity peak and negative differential mobility in high field transport in graphene nanoribbons explained by numerical simulations. *Applied Physics Letters*, *99*(24), 242108. doi:10.1063/1.3664091

Beulens, J., Gastineau, C., Guerrassimov, N., Koulidiati, J., & Schram, D. (1994). Atomic and molecular emission spectroscopy on an expanding argon/methane plasma. *Plasma Chemistry and Plasma Processing*, *14*(1), 15–42. doi:10.1007/BF01448735

Bhadra, S., & Khastgir, D. (2008). Extrinsic and intrinsic structural change during heat treatment of polyaniline. *Polymer Degradation Stability*, *93*, 1094–9.

Bhat, N. V., Gadre, A. P., & Bambole, V. A. (2003). Investigation of electropolymerized polypyrrole composite film: Characterization and application to gas sensors. *Journal of Applied Polymer Science*, *88*(1), 22–29. doi:10.1002/app.11641

Bi, H., Yin, K., Xie, X., Ji, J., Wan, S., Sun, L., & Dresselhaus, M. S. et al. (2013). Ultrahigh humidity sensitivity of graphene oxide. *Scientific Reports*, *3*, 2714. doi:10.1038/srep02714 PMID:24048093

Bin, X., Sargent, E. H., & Kelley, S. O. (2010). Nanostructuring of sensors determines the efficiency of biomolecular capture. *Analytical Chemistry*, *82*(14), 5928–5931. doi:10.1021/ac101164n PMID:20568723

Bishop, C. (2001). *The relationship between loss, conductivity, and dielectric constant*. Academic Press.

Bittle, E. G., Brill, J. W., & Anthony, J. E. (2010). Electro-optic measurement of carrier mobility in an organic thin-film transistor. *Applied Physics Letters*, *97*(1), 013302. doi:10.1063/1.3460526

Boeringer, D. W., & Werner, D. H. (2004). Particle swarm optimization versus genetic algorithms for phased array synthesis. *Antennas and Propagation. IEEE Transactions on*, *52*(3), 771–779. doi:10.1109/TAP.2004.825102

Bois, FY, & Smith, AE. (1995). *Evaluation of variability in predictions of cancer risk from occupational exposure to methylene chloride as a consequence of variability in model input parameters*. OSHA.

Bo, L., Reddy, C. D., Jinwu, J., Hongwei, Z., Julia, A. B., Sergey, V. D., & Kun, Z. (2014). Thermal conductivity of silicene nanosheets and the effect of isotopic doping. *Journal of Physics. D, Applied Physics*, *47*(16), 165301. doi:10.1088/0022-3727/47/16/165301

Bolz, A., Lang, V., Merkely, B., & Schaldach, M. (1997). Paper. *Conference Proc. Ann. Int. Conf. IEEE Eng. Med. and Biol*, *5*, 2341.

Bonanni, A. L., Loo, A. H., & Pumera, M. (2012). Graphene for impedimetric biosensing. *Trac-Trends in Analytical Chemistry*, *37*, 12–21. doi:10.1016/j.trac.2012.02.011

Bondavalli, P., Legagneux, P., & Pribat, D. (2009). Carbon nanotubes based transistors as gas sensors: State of the art and critical review. *Sensors and Actuators. B, Chemical*, *140*(1), 304–318. doi:10.1016/j.snb.2009.04.025

Bousrih, S., Ershov-Pavlov, E., Megy, S., & Baronnet, J.-M. (1995). Hydrogen/argon plasma jet with methane addition. *Plasma Chemistry and Plasma Processing*, *15*(2), 333–351. doi:10.1007/BF01459702

Bousse, L. B., & Bergveld, P. (1984). The role of buried OH sites in the response mechanism of inorganic-gate pH-sensitive ISFETs. *Sensors and Actuators*, *6*(1), 65–78. doi:10.1016/0250-6874(84)80028-1

Boyd, A., Dube, I., Fedorov, G., Paranjape, M., & Barbara, P. (2014). Gas sensing mechanism of carbon nanotubes: From single tubes to high-density networks. *Carbon*, *69*, 417–423. doi:10.1016/j.carbon.2013.12.044

Boylestad, R. L., & Nashelsky, L. (2002). *Electronic devices and circuit theory*. Prentice Hall.

Bradley, D. (2012). Graphene gas sensor CARBON. *Materials Today*, *15*(6), 233–233. doi:10.1016/S1369-7021(12)70105-0

Bratov, A. M. o., Dominguez, & , & Bartroli, . (1995). Photocurable polymers applied as encapsulating materials for ISFET production. *Sensors and Actuators. B, Chemical*, *25*(1–3), 823–825. doi:10.1016/0925-4005(95)85183-6

Brey, L., & Fertig, H. (2006). Electronic states of graphene nanoribbons studied with the Dirac equation. *Physical Review B: Condensed Matter and Materials Physics*, *73*(23), 235411. doi:10.1103/PhysRevB.73.235411

Brooks, F. D., & Klein, H. (2002). Neutron spectrometry—historical review and present status. *Nuclear Instruments & Methods in Physics Research. Section A, Accelerators, Spectrometers, Detectors and Associated Equipment*, *476*(1–2), 1–11. doi:10.1016/S0168-9002(01)01378-X

Byfield, M., & Abuknesha, R. (1994). Biochemical aspects of biosensors. *Biosensors & Bioelectronics*, *9*(4), 373–399. doi:10.1016/0956-5663(94)80038-3 PMID:8068232

Cadogan, A. G., Lewenstam, Ivaska, Diamond, & Dermot. (1992). All-solid-state sodium-selective electrode based on a calixarene ionophore in a poly(vinyl chloride) membrane with a polypyrrole solid contact. *Analytical Chemistry, 64*(21), 2496-2501.

Cahangirov, S., Topsakal, M., Aktürk, E., Şahin, H., & Ciraci, S. (2009). Two- and One-Dimensional Honeycomb Structures of Silicon and Germanium. *Physical Review Letters*, *102*(23), 236804. doi:10.1103/PhysRevLett.102.236804 PMID:19658958

Campbell, C. (1998). *Surface acoustic wave devices for mobile and wireless communications*. Academic press.

Cantalini, C., Valentini, L., Lozzi, L., Armentano, I., Kenny, J., & Santucci, S. (2003). NO< sub> 2</sub> gas sensitivity of carbon nanotubes obtained by plasma enhanced chemical vapor deposition. *Sensors and Actuators. B, Chemical*, *93*(1-3), 333–337. doi:10.1016/S0925-4005(03)00224-7

Cao, Q., Han, S.-, Tulevski, G. S., Zhu, Y., Lu, D. D., & Haensch, W. (2013). Arrays of single-walled carbon nanotubes with full surface coverage for high-performance electronics. *Nature Nanotechnology*, *8*(3), 180–186. doi:10.1038/nnano.2012.257 PMID:23353673

Carrillo, I., Rangel, E., & Magaña, L. (2009). Adsorption of carbon dioxide and methane on graphene with a high titanium coverage. *Carbon*, *47*(11), 2758–2760. doi:10.1016/j.carbon.2009.06.022

Castro Neto, A. H., Guinea, F., Peres, N. M. R., Novoselov, K. S., & Geim, A. K. (2009). The electronic properties of graphene. *Reviews of Modern Physics*, *81*(1), 109–162. doi:10.1103/RevModPhys.81.109

Cecilia, J.-J. J., Antoni, & Baldi. (2009). ISFET Based Microsensors for Environmental Monitoring. *Sensors (Basel, Switzerland)*, *10*(1). PMID:22315527

Chadwick, J. (1932). The existence of a neutron. *Proceedings of the Royal Society of London A: Mathematical, Physical and Engineering Sciences*, *136*(830), 692-708. doi:10.1098/rspa.1932.0112

Chang, C. Y., & Sze, S. M. (1996). *ULSI technology* (International Editions). Singapore: McGraw-Hill.

Chang, P.-H., & Nikolic, B. K. (2012). Edge currents and nanopore arrays in zigzag and chiral graphene nanoribbons as a route toward high-ZT thermoelectrics. *Physical Review B: Condensed Matter and Materials Physics*, *86*(4), 041406. doi:10.1103/PhysRevB.86.041406

Chang, P.-Y., Huang, W.-M., & Lin, H.-H. (2009). Impurity-induced conductance anomaly in zigzag carbon nanotubes. In P. Kes, & R. Jochemsen (Eds.), *25th International Conference on Low Temperature Physics* (Vol. 150). doi:10.1088/1742-6596/150/2/022007

Chaniotakis, N. A. (2004). Enzyme stabilization strategies based on electrolytes and polyelectrolytes for biosensor applications. *Analytical and Bioanalytical Chemistry*, *378*(1), 89–95. doi:10.1007/s00216-003-2188-3 PMID:13680062

Chaplin, M. F., & Bucke, C. (1990). *Enzyme technology*. CUP Archive.

Chavadej, S., Supat, K., Lobban, L. L., & Mallinson, R. G. (2005). Partial oxidation of methane and carbon dioxide reforming with methane in corona discharge with/without Pt/KL catalyst. *Journal of Chemical Engineering of Japan*, *38*(3), 163–170. doi:10.1252/jcej.38.163

Chek, D. C., Tan, M. L., Ahmadi, M. T., Ismail, R., & Arora, V. K. (2010). Analytical modeling of high performance single-walled carbon nanotube field-effect-transistor. *Microelectronics Journal*, *41*(9), 579–584. doi:10.1016/j.mejo.2010.05.008

Chen Yan-Kuin, S. C. (1951). The fabrication and characterisation of ion-sensitive field-effect transistors with a silicon dioxide gate. *Journal of Physics. D, Applied Physics*, *19*(10).

Chen, Z., & Appenzeller, J. (2008). Mobility extraction and quantum capacitance impact in high performance graphene field-effect transistor devices. *Electron Devices Meeting, IEEE International, IEDM*, *509*, 1-4.

Chen, C., Liu, L., Lu, Y., Kong, E. S.-W., Zhang, Y., Sheng, X., & Ding, H. (2007). A method for creating reliable and low-resistance contacts between carbon nanotubes and microelectrodes. *Carbon*, *45*(2), 436–442. doi:10.1016/j.carbon.2006.08.021

Chen, F., Qing, Q., Xia, J., & Tao, N. (2010). Graphene field-effect transistors: Electrochemical gating, interfacial capacitance, and biosensing applications. *Chemistry, an Asian Journal*, *5*(10), 2144–2153. doi:10.1002/asia.201000252 PMID:20715049

Chen, G., Paronyan, T. M., & Harutyunyan, A. R. (2012). Sub-ppt gas detection with pristine graphene. *Applied Physics Letters*, *101*(5), 053119. doi:10.1063/1.4742327

Cheng, Y. C., Zhu, Z. Y., & Schwingenschlögl, U. (2011). Doped silicene: Evidence of a wide stability range.[Europhysics Letters]. *EPL*, *95*(1), 17005. doi:10.1209/0295-5075/95/17005

Chen, H., Josowicz, M., & Janatax, J. (2004). Chemical effects in organic electronics. *Chemistry of Materials*, *16*(23), 4728–4735. doi:10.1021/cm040101m

Chen, J. H., Jang, C., Adam, S., Fuhrer, M. S., Williams, E. D., & Ishigami, M. (2008). Charged-impurity scattering in graphene. *Nature Physics*, *4*(5), 377–381. doi:10.1038/nphys935

Chen, J., Hamon, M. A., Hu, H., Chen, Y., Rao, A. M., Eklund, P. C., & Haddon, R. C. (1998). Solution Properties of Single-Walled Carbon Nanotubes. *Science*, *282*(5386), 95–98. doi:10.1126/science.282.5386.95 PMID:9756485

Chen, J.-L., & Yan, X.-P. (2010). A dehydration and stabilizer-free approach to production of stable water dispersions of graphene nanosheets. *Journal of Materials Chemistry*, *20*(21), 4328–4332. doi:10.1039/c000177e

Chen, L., Feng, B., & Wu, K. (2013a). Observation of a possible superconducting gap in silicene on Ag(111) surface. *Applied Physics Letters*, *102*(8), 081602. doi:10.1063/1.4793998

Chen, L., Liu, C.-C., Feng, B., He, X., Cheng, P., & Ding, Z. et al. (2012). Evidence for Dirac Fermions in a Honeycomb Lattice Based on Silicon. *Physical Review Letters*, *109*(5), 056804. doi:10.1103/PhysRevLett.109.056804 PMID:23006197

Chen, N., Li, X., Wang, X., Yu, J., Wang, J., Tang, Z., & Akbar, S. A. (2013). Enhanced Room Temperature Sensing of Co_3O_4-intercalated reduced graphene oxide based gas sensors. *Sensors and Actuators. B, Chemical*, *188*, 902–908. doi:10.1016/j.snb.2013.08.004

Chen, R. J., Bangsaruntip, S., Drouvalakis, K. A., Wong Shi Kam, N., Shim, M., Li, Y., & Dai, H. et al. (2003). Noncovalent functionalization of carbon nanotubes for highly specific electronic biosensors. *Proceedings of the National Academy of Sciences of the United States of America*, *100*(9), 4984–4989. doi:10.1073/pnas.0837064100 PMID:12697899

Chen, S. Z., Zhang, Z.-B., Ma, L., Ahlberg, P., Gao, X., Qiu, Z., & Zhang, S.-L. et al. (2012). A graphene field-effect capacitor sensor in electrolyte. *Applied Physics Letters*, *101*(15), 154106–154105. doi:10.1063/1.4759147

Chen, W.-K. (2005). *General feedback theory, "Circuit analysis and feedback amplifier theory"*. CRC Press. doi:10.1201/9781420037272

Chen, Y., Gao, B., Zhao, J.-X., Cai, Q.-H., & Fu, H.-G. (2011). Si-doped graphene: An ideal sensor for NO- or NO2-detection and metal-free catalyst for N2O-reduction. *Journal of Molecular Modeling*, *18*(5), 2043–2054. doi:10.1007/s00894-011-1226-x PMID:21881853

Chen, Z., Lin, Y.-M., Rooks, M. J., & Avouris, P. (2007). Graphene nano-ribbon electronics. *Physica E, Low-Dimensional Systems and Nanostructures*, *40*(2), 228–232. doi:10.1016/j.physe.2007.06.020

Cheon, S., Kihm, K. D., Park, J. S., Lee, J. S., Lee, B. J., Kim, H., & Hong, B. H. (2012). How to optically count graphene layers. *Optics Letters*, *37*(18), 3765–3767. doi:10.1364/OL.37.003765 PMID:23041852

Cherf, G. M., Lieberman, K. R., Rashid, H., Lam, C. E., Karplus, K., & Akeson, M. (2012). Automated forward and reverse ratcheting of DNA in a nanopore at 5-A precision. *Nature Biotechnology*, *30*(4), 344–348. doi:10.1038/nbt.2147 PMID:22334048

Chiang, J., & MacDiarmid, A. (1986). Polyaniline: Protonic acid doping of the emeraldine form to the metallic regime. *Synthetic Metals*, *13*(1-3), 193–205. doi:10.1016/0379-6779(86)90070-6

Chiu, N.-F., Huang, T.-Y., Lai, H.-C., & Liu, K.-C. (2014). Graphene oxide-based SPR biosensor chip for immunoassay applications. *Nanoscale Research Letters*, *9*(1), 1–7. doi:10.1186/1556-276X-9-445 PMID:25232298

Cho, T. S., Lee, K. J., Kong, J., & Chandrakasan, A. R. (2008) The design of a low power carbon nanotube chemical sensor system. *2008 45th ACM/IEEE Design Automation Conference*.

Cho, B., Yoon, J., Hahm, M. G., Kim, D.-H., Kim, A. R., Kahng, Y. H., & Ko, H. C. et al. (2014). Graphene-based gas sensor: Metal decoration effect and application to a flexible device. *Journal of Materials Chemistry C*, *2*(27), 5280–5285. doi:10.1039/c4tc00510d

Choi, S. H., Kim, Y. L., & Byun, K. M. (2011). Graphene-on-silver substrates for sensitive surface plasmon resonance imaging biosensors. *Optics Express*, *19*(2), 458–466. doi:10.1364/OE.19.000458 PMID:21263585

Choi, W., Lahiri, I., Seelaboyina, R., & Kang, Y. S. (2010). Synthesis of graphene and its applications: A review. *Critical Reviews in Solid State and Material Sciences*, *35*(1), 52–71. doi:10.1080/10408430903505036

Chopra, S., Pham, A., Gaillard, J., Parker, A., & Rao, A. M. (2002). Carbon-nanotube-based resonant-circuit sensor for ammonia. *Applied Physics Letters*, *80*(24), 4632–4634. doi:10.1063/1.1486481

Choudhury, B., Shinar, R., & Shinar, J. (2004). Glucose biosensors based on organic light-emitting devices structurally integrated with a luminescent sensing element. *Journal of Applied Physics*, *96*(5), 2949–2954. doi:10.1063/1.1778477

Chu, B. H., Lo, C. F., Nicolosi, J., Chang, C. Y., Chen, V., Strupinski, W., & Ren, F. et al. (2011). Hydrogen detection using platinum coated graphene grown on SiC. *Sensors and Actuators. B, Chemical*, *157*(2), 500–503. doi:10.1016/j.snb.2011.05.007

Chung, W.-Y. C., Szu, Pijanowska, Dawgul, Torbicz, Grabiec, … Wei-Po. (2009). *ISFET electronic tongue system for environmental multi-ion sensing with independent component analysis signal processing*. doi: 10.1117/12.821635

Chung, M. G., Kim, D. H., Lee, H. M., Kim, T., Choi, J. H., Seo, D. K., & Kim, Y. H. et al. (2012). Highly sensitive NO2 gas sensor based on ozone treated graphene. *Sensors and Actuators. B, Chemical*, *166*, 172–176. doi:10.1016/j.snb.2012.02.036

Chung, M. G., Kim, D.-H., Seo, D. K., Kim, T., Im, H. U., Lee, H. M., & Kim, Y. H. et al. (2012). Flexible hydrogen sensors using graphene with palladium nanoparticle decoration. *Sensors and Actuators. B, Chemical*, *169*, 387–392. doi:10.1016/j.snb.2012.05.031

Chung, W.-Y. L., Lin, Y.-T., Pijanowska, D. G., Yang, C.-H., Wang, M.-C., Krzyskow, A., & Torbicz, W. (2006). New ISFET interface circuit design with temperature compensation. *Microelectronics Journal*, *37*(10), 1105–1114. doi:10.1016/j.mejo.2006.05.001

Ci, L., Xu, Z., Wang, L., Gao, W., Ding, F., Kelly, K. F., & Ajayan, P. M. (2008). Controlled nanocutting of graphene. *Nano Research*, *1*(2), 116–122. doi:10.1007/s12274-008-8020-9

Clarke, J., Wu, H.-C., Jayasinghe, L., Patel, A., Reid, S., & Bayley, H. (2009). Continuous base identification for single-molecule nanopore DNA sequencing. *Nature Nanotechnology*, *4*(4), 265-270. Retrieved from http://www.nature.com/nnano/journal/v4/n4/suppinfo/nnano.2009.12_S1.html

Clerc, M. (2006). *Particle swarm optimization*. Wiley-ISTE. doi:10.1002/9780470612163

Clermont, O., Bonacorsi, S., & Bingen, E. (2000). Rapid and simple determination of the Escherichia coli phylogenetic group. *Applied and Environmental Microbiology*, *66*(10), 4555–4558. doi:10.1128/AEM.66.10.4555-4558.2000 PMID:11010916

Cobben, P. L. H. M. E., Bomer, Bergveld, Verboom, & Reinhoudt. (1992). Transduction of selective recognition of heavy metal ions by chemically modified field effect transistors (CHEMFETs). *Journal of the American Chemical Society*, *114*(26), 10573-10582.

Cohen-Karni, T., Qing, Q., Li, Q., Fang, Y., & Lieber, C. M. (2010). Graphene and nanowire transistors for cellular interfaces and electrical recording. *Nano Letters*, *10*(3), 1098–1102. doi:10.1021/nl1002608 PMID:20136098

Cole, B. E., & Zook, D. J. (2005). *Carbon nanotube sensor*. Google Patents.

Collins, P. G., Arnold, M. S., & Avouris, P. (2001). Engineering Carbon Nanotubes and Nanotube Circuits Using Electrical Breakdown. *Science*, *292*(5517), 706–709. doi:10.1126/science.1058782 PMID:11326094

Collins, P. G., Bradley, K., Ishigami, M., & Zettl, A. (2000). Extreme Oxygen Sensitivity of Electronic Properties of Carbon Nanotubes. *Science*, *287*(5459), 1801–1804. doi:10.1126/science.287.5459.1801 PMID:10710305

Comini, E. (2006). Metal oxide nano-crystals for gas sensing. *Analytica Chimica Acta*, *568*(1), 28–40. doi:10.1016/j.aca.2005.10.069 PMID:17761243

Cooper, J. S., Myers, M., Chow, E., Hubble, L. J., Cairney, J. M., Pejcic, B., & Raguse, B. et al. (2014). Performance of graphene, carbon nanotube, and gold nanoparticle chemiresistor sensors for the detection of petroleum hydrocarbons in water. *Journal of Nanoparticle Research*, *16*(1), 2173. doi:10.1007/s11051-013-2173-5

Costamagna, S., & Dobry, A. (2011). From graphene sheets to graphene nanoribbons: Dimensional crossover signals in structural thermal fluctuations. *Physical Review B*, *83*(23). doi: 23340110.1103/PhysRevB.83.233401

Couto, S. R. M., & Sanromán, D. (2006). Optimum stability conditions of pH and temperature for ligninase and manganese-dependent peroxidase from Phanerochaete chrysosporium. Application to in vitro decolorization of Poly R-478 by MnP. *World Journal of Microbiology & Biotechnology*, *22*(6), 607–612. doi:10.1007/s11274-005-9078-0

Craciun, M. F., Russo, S., Yamamoto, M., Oostinga, J. B., Morpurgo, A. F., & Tarucha, S. (2009). Trilayer graphene is a semimetal with a gate-tunable band overlap. *Nature Nanotechnology*, *4*(6), 383–388. doi:10.1038/nnano.2009.89 PMID:19498401

Craciun, M. F., Russo, S., Yamamoto, M., & Tarucha, S. (2011). Tuneable electronic properties in graphene. *NanoToday Press*, *6*(1), 42–60. doi:10.1016/j.nantod.2010.12.001

Cunha, C., Panseri, S., Iannazzo, D., Piperno, A., Pistone, A., Fazio, M., & Galvagno, S. et al. (2012). Hybrid composites made of multiwalled carbon nanotubes functionalized with Fe 3 O 4 nanoparticles for tissue engineering applications. *Nanotechnology*, *23*(46), 465102. doi:10.1088/0957-4484/23/46/465102 PMID:23093179

Cuong, T. V., Pham, V. H., Chung, J. S., Shin, E. W., Yoo, D. H., Hahn, S. H., & Kohl, P. A. et al. (2010). Solution-Processed ZnO-chemically converted graphene gas sensor. *Materials Letters*, *64*(22), 2479–2482. doi:10.1016/j.matlet.2010.08.027

Curreli, M., Zhang, R., Ishikawa, F. N., Chang, H.-K., Cote, R. J., Zhou, C., & Thompson, M. E. (2008). Real-Time, Label-Free Detection of Biological Entities Using Nanowire-Based FETs. *IEEE Transactions on Nanotechnology*, *7*(6), 651–667. doi:10.1109/TNANO.2008.2006165

d'AMico, A., Palma, A., & Verona, E. (1982). *Hydrogen sensor using a palladium coated surface acoustic wave delay-line*. Paper presented at the 1982 Ultrasonics Symposium.

Daghestani, H. N., & Day, B. W. (2010). Theory and applications of surface plasmon resonance, resonant mirror, resonant waveguide grating, and dual polarization interferometry biosensors. *Sensors (Basel, Switzerland)*, *10*(11), 9630–9646. doi:10.3390/s101109630 PMID:22163431

Dai, H. J. (2001). *Carbon Nanotubes*. Berlin: Springer.

Dalosto, S. D., & Tinte, S. (2011). Fluctuation effects of the electric field induced by water on a graphene dot band gap. *The Journal of Physical Chemistry C*, *115*(11), 4381–4386. doi:10.1021/jp109297p

Dankerl, M., Eick, S., Hofmann, B., Hauf, M., Ingebrandt, S., Offenhäusser, A., & Garrido, J. A. (2009). Diamond transistor array for extracellular recording from electrogenic cells. *Advanced Functional Materials*, *19*(18), 2915–2923. doi:10.1002/adfm.200900590

Dankerl, M., Hauf, M. V., Lippert, A., Hess, L. H., Birner, S., Sharp, I. D., & Garrido, J. A. et al. (2010). Graphene solution-gated field-effect transistor array for sensing applications. *Advanced Functional Materials*, *20*(18), 3117–3124. doi:10.1002/adfm.201000724

Das, A., Chakraborty, B., & Sood, A. K. (2011). Probing single and bilayer graphene field effect transistors by raman spectroscopy. *Modern Physics Letters B*, *25*(08), 511–535. doi:10.1142/S0217984911025973

Das, S., Abraham, A., & Konar, A. (2008). Particle swarm optimization and differential evolution algorithms: Technical analysis, applications and hybridization perspectives. *Advances of Computational Intelligence in Industrial Systems*, *116*, 1–38. doi:10.1007/978-3-540-78297-1_1

Datta, S. (1997). *Electronic transport in mesoscopic systems*. Cambridge university press.

Datta, S. (2002). *Electronic Transport in Mesoscopic Systems*. Cambridge, UK: Cambridge University Press.

Datta, S. (2005). *Quantum Transport: Atom to Transistor*. New York: Cambridge University Press. doi:10.1017/CBO9781139164313

Davies, J. H. (2005). *Low-dimensional semiconductors: An introduction*. New York, NY, USA: Cambridge University Press.

Dekker, C. (2007). Solid-state nanopores. *Nature Nanotechnology*, *2*(4), 209–215. doi:10.1038/nnano.2007.27 PMID:18654264

del Valle, Y., Venayagamoorthy, G. K., Mohagheghi, S., Hernandez, J. C., & Harley, R. G. (2008). Particle swarm optimization: Basic concepts, variants and applications in power systems. *Evolutionary Computation. IEEE Transactions on*, *12*(2), 171–195.

Dellabetta, B., & Gilbert, M. J. (2010). *The effect of disorder in superfluid graphene bilayers*. Academic Press.

Derycke, V., Martel, R., Appenzeller, J., & Avouris, P. (2001). Carbon nanotube inter- and intramolecular logic gates. *Nano Letters*, *1*(9), 453–456. doi:10.1021/nl015606f

Dhall, S., & Jaggi, N. (2015). Highly dispersed platinum sputtered multiwall carbon nanotubes based hydrogen gas sensor at room temperature. *Sensors and Actuators. A, Physical*, *224*, 50–56. doi:10.1016/j.sna.2015.01.010

DHS. (1989). *Proposed Identification of Methylene Chloride as a Toxic Air Contaminant. Part B- Health Effects of Methylene Chloride*. Berkeley, CA: DHS.

Di, C.-A., Liu, Y., Yu, G., & Zhu, D. (2009). Interface engineering: An effective approach toward high-performance organic field-effect transistors. *Accounts of Chemical Research*, *42*(10), 1573–1583. doi:10.1021/ar9000873 PMID:19645474

Diederich, F., & Rubin, Y. (1992). Synthetic approaches toward molecular and polymeric carbon allotropes. *Angewandte Chemie International Edition in English*, *31*(9), 1101–1123. doi:10.1002/anie.199211013

Ding, Y., & Ni, J. (2009). Electronic structures of silicon nanoribbons. *Applied Physics Letters*, *95*(8), 083115. doi:10.1063/1.3211968

Ding, L., Wang, S., Zhang, Z. Y., Zeng, Q. S., Wang, Z. X., Pei, T., & Peng, L. M. et al. (2009). Y-Contacted high-performance n-type single-walled carbon nanotube field-effect transistors: Scaling and comparison with sc-contacted devices. *Nano Letters*, *9*(12), 4209–4214. doi:10.1021/nl9024243 PMID:19995085

Dingle, R. B., & Dingle, R. (1973). *Asymptotic expansions: Their derivation and interpretation*. Academic Press London.

Ding, W. D., Hayashi, R., Suehiro, J., Zhou, G. B., Imasaka, K., & Hara, M. (2006). Calibration methods of carbon nanotube gas sensor for partial discharge detection in SF6. *IEEE Transactions on Dielectrics and Electrical Insulation*, *13*(2), 353–361. doi:10.1109/TDEI.2006.1624280

Ding, Y., & Wang, Y. (2012a). Electronic structures of silicene fluoride and hydride. *Applied Physics Letters*, *100*(8), 083102–083102. doi:10.1063/1.3688035

Ding, Y., & Wang, Y. (2012b). Electronic structures of zigzag SiC nanoribbons with asymmetric hydrogen-terminations. *Applied Physics Letters*, *101*(1), 013102–013104. doi:10.1063/1.4732786

Do, J. S., & Chang, W. B. (2001). Amperometric nitrogen dioxide gas sensor: Preparation of PAn/Au/SPE and sensing behavior. *Sensors and Actuators. B, Chemical*, *72*(2), 101–107. doi:10.1016/S0925-4005(00)00607-9

Dong, X., Huang, W., & Chen, P. (2011). In Situ Synthesis of Reduced Graphene Oxide and Gold Nanoelectronics and Biosensing. Nanoscale Research Letters, 6. doi:10.1007/s11671-010-9806-8

Dongjin, L. T. (2012). An electric detection of immunoglobulin G in the enzyme-linked immunosorbent assay using an indium oxide nanoparticle ion-sensitive field-effect transistor. *Journal of Micromechanics and Microengineering*, *22*(1), 015009. doi:10.1088/0960-1317/22/1/015009

Dong, L. X., & Chen, Q. (2010). Properties, synthesis, and characterization of graphene. *Frontiers of Materials Science in China*, *4*(1), 45–51. doi:10.1007/s11706-010-0014-3

Dong, X., Fu, D., Fang, W., Shi, Y., Chen, P., & Li, L.-J. (2009). Doping single-layer graphene with aromatic molecules. *Small*, *5*(12), 1422–1426. doi:10.1002/smll.200801711 PMID:19296561

Dong, X., Shi, Y., Huang, W., Chen, P., & Li, L. J. (2010). Electrical detection of DNA hybridization with single-base specificity using transistors based on cvd-grown graphene sheets. *Advanced Materials*, *22*(14), 1649–1653. doi:10.1002/adma.200903645 PMID:20496398

Dorgan, V. E., Bae, M. H., & Pop, E. (2010). Mobility and saturation velocity in graphene on SiO_2. *Applied Physics Letters*, *97*(8), 082112–082113. doi:10.1063/1.3483130

Drafts, B. (2001). Acoustic wave technology sensors. *IEEE Transactions on Microwave Theory and Techniques*, *49*(4), 795–802. doi:10.1109/22.915466

Dragoman, M., & Dragoman, D. (2009). Graphene-based quantum electronics. *Quantum Electronics*, *33*(6), 165–214. doi:10.1016/j.pquantelec.2009.08.001

Dresselhaus, M. S., Dresselhaus, G., & Eklund, P. C. (1996). *Science of Fullerenes and Carbon Nanotubes*. San Diego, CA: Academic Press.

Dresselhaus, M. S., Dresselhaus, G., Eklund, P. C., & Saito, R. (1998). Carbon nanotubes. *Physics World*, *11*(1), 33–38. doi:10.1088/2058-7058/11/1/32

Dresselhaus, M. S., Dresselhaus, G., Eklund, P., & Rao, A. (2000). *Carbon nanotubes*. Springer. doi:10.1007/978-94-011-4038-6_9

Dresselhaus, M., Dresselhaus, G., & Saito, R. (1995). Physics of carbon nanotubes. *Carbon*, *33*(7), 883–891. doi:10.1016/0008-6223(95)00017-8

Drummond, N. D., Zólyomi, V., & Fal'ko, V. I. (2012). Electrically tunable band gap in silicene. *Physical Review B: Condensed Matter and Materials Physics*, *85*(7), 075423. doi:10.1103/PhysRevB.85.075423

Drummond, T. G., Hill, M. G., & Barton, J. K. (2003). Electrochemical DNA sensors. *Nature Biotechnology*, *21*(10), 1192–1199. doi:10.1038/nbt873 PMID:14520405

Dube, I., Jiménez, D., Fedorov, G., Boyd, A., Gayduchenko, I., Paranjape, M., & Barbara, P. (2015). Understanding the electrical response and sensing mechanism of carbon-nanotube-based gas sensors. *Carbon*, *87*, 330–337. doi:10.1016/j.carbon.2015.01.060

Du, D., Zou, Z., Shin, Y., Wang, J., Wu, H., Engelhard, M. H., & Lin, Y. et al. (2010). Sensitive immunosensor for cancer biomarker based on dual signal amplification strategy of graphene sheets and multienzyme functionalized carbon nanospheres. *Analytical Chemistry*, *82*(7), 2989–2995. doi:10.1021/ac100036p PMID:20201502

Dulloo, A. R., Ruddy, F. H., Seidel, J. G., Davison, C., Flinchbaugh, T., & Daubenspeck, T. (1999). Simultaneous measurement of neutron and gamma-ray radiation levels from a TRIGA reactor core using silicon carbide semiconductor detectors. *IEEE Transactions on Nuclear Science*, *46*(3), 275–279. doi:10.1109/23.775527

Eberhart, R., & Shi, Y. (2001). *Particle swarm optimization: Developments, applications and resources.* Paper presented at the Evolutionary Computation. doi:10.1109/CEC.2001.934374

Eberhart, R. C., Shi, Y., & Kennedy, J. (2001). *Swarm intelligence.* Indianapolis, IN: Elsevier.

Eda, G., & Chhowalla, M. (2010). Chemically derived graphene oxide: Towards large-area thin-film electronics and optoelectronics. *Advanced Materials*, *22*(22), 2392–2415. doi:10.1002/adma.200903689 PMID:20432408

El Gowini, M. M., & Moussa, W. A. (2010). A finite element model of a MEMS-based surface acoustic wave hydrogen sensor. *Sensors (Basel, Switzerland)*, *10*(2), 1232–1250. doi:10.3390/s100201232 PMID:22205865

Electrical Engineering. (2013, February 22). *Light Dark Sensor.* Retrieved from http://electrical-engineer10.blogspot.nl/2013/02/light-dark-sensor.html

Elhadidy, H., Sikula, J., & Franc, J. (2012). Symmetrical current–voltage characteristic of a metal–semiconductor–metal structure of Schottky contacts and parameter retrieval of a CdTe structure. *Semiconductor Science and Technology*, *27*(1), 015006. doi:10.1088/0268-1242/27/1/015006

Elhaes, H., & Ibrahim, M. (2013). Fullerene as sensor for Halides: Modeling approach. *Journal of Computational and Theoretical Nanoscience*, *10*(9), 2026–2028. doi:10.1166/jctn.2013.3164

Elias, D. C., Nair, R. R., Mohiuddin, T. M. G., Morozov, S. V., Blake, P., Halsall, M. P., & Novoselov, K. S. et al. (2009). Control of graphene's properties by reversible hydrogenation: Evidence for graphane. *Science*, *323*(5914), 610–613. doi:10.1126/science.1167130 PMID:19179524

Elmazria, O. (2008). *Surface acoustic wave devices for wireless sensing applications: Principle, technology and market.* Academic Press.

Enoki, T., Kobayashi, Y., & Fukui, K.-I. (2007). Electronic structures of graphene edges and nanographene. *International Reviews in Physical Chemistry*, *26*(4), 609–645. doi:10.1080/01442350701611991

Enriquez, H., Vizzini, S., Kara, A., Lalmi, B., & Oughaddou, H. (2012). Silicene structures on silver surfaces. *Journal of Physics Condensed Matter*, *24*(31), 314211. doi:10.1088/0953-8984/24/31/314211 PMID:22820837

Farmer, D. B., Chiu, H.-Y., Lin, Y.-M., Jenkins, K. A., Xia, F., & Avouris, P. (2009). Utilization of a buffered dielectric to achieve high field-effect carrier mobility in graphene transistors. *Nano Letters*, *9*(12), 4474–4478. doi:10.1021/nl902788u PMID:19883119

Fazzi, A., Agosteo, S., Pola, A., Varoli, V., & Zotto, P. (2003). *Pulse discrimination between recoil protons and secondary electrons for a silicon diode based neutron spectrometer.* Paper presented at the Nuclear Science Symposium Conference Record. doi:10.1109/NSSMIC.2003.1351811

Feather, N. (1932). The collisions of neutrons with nitrogen nuclei. *Proceedings of the Royal Society of London A: Mathematical, Physical and Engineering Sciences, 136*(830), 709-727. doi:10.1098/rspa.1932.0113

Feng, B., Ding, Z., Meng, S., Yao, Y., He, X., & Cheng, P. et al. (2012). Evidence of Silicene in Honeycomb Structures of Silicon on Ag(111). *Nano Letters, 12*(7), 3507–3511. doi:10.1021/nl301047g PMID:22658061

Feng, B., Li, H., Liu, C.-C., Shao, T., Cheng, P., & Yao, Y. et al. (2013). Observation of Dirac cone warping and chirality effects in silicene. *ACS Nano, 7*(10), 9049–9054. doi:10.1021/nn403661h PMID:24003914

Ferrer, J., Lambert, C., Garcia-Suarez, V., Zsolt Manrique, D., Visontai, D., & Oroszlani, L. et al. (2014). GOLLUM: A next-generation simulation tool for electron, thermal and spin transport. *New Journal of Physics, 16*(9), 093029. doi:10.1088/1367-2630/16/9/093029

Fiori, G., & Iannaccone, G. (2009). Performance analysis of graphene bilayer transistors through tight-binding simulations. *Iwce-13: 2009 13th International Workshop on Computational Electronics*, (pp. 301-304).

Fiori, G., Yoon, Y., Hong, S., Jannacconet, G., & Guo, J. (2007). Performance Comparison of Graphene Nanoribbon Schottky Barrier and MOSFETs. *IEEE International Electron Devices Meeting.*

Fiori, G., & Iannaccone, G. (2007). Simulation of graphene nanoribbon field-effect transistors. *Electron Device Letters, IEEE, 28*(8), 760–762. doi:10.1109/LED.2007.901680

Fleurence, A., Friedlein, R., Ozaki, T., Kawai, H., Wang, Y., & Yamada-Takamura, Y. (2012). Experimental Evidence for Epitaxial Silicene on Diboride Thin Films. *Physical Review Letters, 108*(24), 245501. doi:10.1103/PhysRevLett.108.245501 PMID:23004288

Ford, A. (2010). *Information feedback and causal loop diagrams, "Modeling the environment"*. Island Press.

Fraden, J. (2010). *Handbook of modern sensors.* Springer. doi:10.1007/978-1-4419-6466-3

Franklin, A. D., Koswatta, S. O., Farmer, D. B., Smith, J. T., Gignac, L., Breslin, C. M., & Haensch, W. et al. (2013). Carbon nanotube complementary wrap-gate transistors. *Nano Letters, 13*(6), 2490–2495. doi:10.1021/nl400544q PMID:23638708

Frank, R. (2013). *Understanding smart sensors.* Artech House.

Freund, M. S., & Lewis, N. S. (1995). A chemically diverse conducting polymer based electronic nose. *Proceedings of the National Academy of Sciences of the United States of America, 92*(7), 2652–2656. doi:10.1073/pnas.92.7.2652 PMID:11607521

Friedlein, R., Fleurence, A., Sadowski, J. T., & Yamada-Takamura, Y. (2013). Tuning of silicene-substrate interactions with potassium adsorption. *Applied Physics Letters, 102*(22), 221603. doi:10.1063/1.4808214

Fritz, J., Cooper, E. B., Gaudet, S., Sorger, P. K., & Manalis, S. R. (2002). Electronic detection of DNA by its intrinsic molecular charge. *Proceedings of the National Academy of Sciences of the United States of America, 99*(22), 14142–14146. doi:10.1073/pnas.232276699 PMID:12386345

Fromherz, P., Offenhausser, A., Vetter, T., & Weis, J. (1991). A neuron-silicon junction: A Retzius cell of the leech on an insulated-gate field-effect transistor. *Science, 252*(5010), 1290–1293. doi:10.1126/science.1925540 PMID:1925540

Fu, D., & Li, L. J. (2010). Label-free electrical detection of DNA hybridization using carbon nanotubes and graphene. *Nano Reviews, 1.*

Fu, D., & Li, L.J. (2010). Label-free electrical detection of DNA hybridization using carbon nanotubes and graphene. *Nano Reviews, 1*.

Fu, W. N., Knopfmacher, Tarasov, Weiss, Calame, & Schonenberger, C. (2011). Graphene transistors are insensitive to pH changes in solution. *Nano Letters, 11*(9), 3597-3600.

Fu, W. N., Knopfmacher, Tarasov, Weiss, Calame, & Schonenberger. (2011). Graphene Transistors Are Insensitive to pH changes in Solution. *Nano Letters, 11*(9), 3597-3600.

Fuhrer, M. S., Lau, C. N., & MacDonald, A. H. (2010). Graphene: Materially better carbon. *MRS Bulletin, 35*(04), 289–295. doi:10.1557/mrs2010.551

Fyta, M., Melchionna, S., & Succi, S. (2011). Translocation of Biomolecules Through Solid-State Nanopores: Theory Meets Experiments. *Journal of Polymer Science. Part B, Polymer Physics, 49*(14), 985–1011. doi:10.1002/polb.22284

Gabor, N. M., Zhong, Z., Bosnick, K., Park, J., & McEuen, P. L. (2009). Extremely efficient multiple electron-hole pair generation in carbon nanotube photodiodes. *Science, 325*(5946), 1367–1371. doi:10.1126/science.1176112 PMID:19745146

Galib, O. (2012). *Graphene grown by chemical vapor deposition on evaporated copper thin films*. Academic Press.

Galipeau, D. W., Story, P. R., Vetelino, K. A., & Mileham, R. D. (1997). Surface acoustic wave microsensors and applications. *Smart Materials and Structures, 6*(6), 658–667. doi:10.1088/0964-1726/6/6/002

Galup-Montoro, C., & Schneider, M. C. (2007). *MOSFET modeling for circuit analysis and design*. London: World Scientific. doi:10.1142/6111

Ganesh, E. (n.d.). *Single walled and multi walled carbon nanotube structure, synthesis and applications*. Academic Press.

Gangopadhyay, R., & De, A. (2001). Conducting polymer composites: Novel materials for gas sensing. *Sensor and Actuator B, 77*(1-2), 326–329. doi:10.1016/S0925-4005(01)00719-5

Gao, N., Zheng, W. T., & Jiang, Q. (2012). Density functional theory calculations for two-dimensional silicene with halogen functionalization. *Physical Chemistry Chemical Physics, 14*(1), 257–261. doi:10.1039/C1CP22719J PMID:22083171

Garaj, S., Liu, S., Golovchenko, J. A., & Branton, D. (2013). Molecule-hugging graphene nanopores. *Proceedings of the National Academy of Sciences of the United States of America, 110*(30), 12192–12196. doi:10.1073/pnas.1220012110 PMID:23836648

Gardner, J. W., & Bartlett, P. N. (1995). Application of conducting polymer technology in microsystems. *Sensors and Actuators. A, Physical, 51*(1), 57–66. doi:10.1016/0924-4247(95)85004-X

Gardner, J. W., Bartlett, P. N., & Pratt, K. F. E. (1995). Modeling of gas-sensitive conducting polymer devices. *IEE Proceedings. Circuits, Devices and Systems, 142*(5), 321–333. doi:10.1049/ip-cds:19952170

Gautam, M., & Jayatissa, A. H. (2012). Detection of organic vapors by graphene films functionalized with metallic nanoparticles. *Journal of Applied Physics, 112*(11), 114326. doi:10.1063/1.4768724

Geetha, P., & WahidaBanu, R. S. D. (2014). A compact modelling of double-walled gate wrap around carbon nanotube array field effect transistors. *Journal of Computational Electronics, 13*(4), 900–916. doi:10.1007/s10825-014-0607-7

Geim, A. K. (2009). Graphene: Status and prospects. *Science, 324*(5934), 1530–1534. doi:10.1126/science.1158877 PMID:19541989

Geim, A. K., & Novoselov, K. S. (2007). The rise of graphene. *Nature Materials, 6*(3), 183–191. doi:10.1038/nmat1849 PMID:17330084

Ghosh, S., Sood, A., & Kumar, N. (2003). Carbon Nanotube Flow Sensors. *Science*, *299*(5609), 1042–1044. doi:10.1126/science.1079080 PMID:12532025

Giannopoulos, G. I. (2014). Fullerenes as mass sensors: A numerical investigation. *Physica E, Low-Dimensional Systems and Nanostructures*, *56*, 36–42. doi:10.1016/j.physe.2013.07.024

Gnani, E., Gnudi, A., Reggiani, S., & Baccarani, G. (2010). Effective Mobility in Nanowire FETs under quasi-ballistic conditions. *IEEE Transactions on Electron Devices*, *57*(1), 336–344. doi:10.1109/TED.2009.2035545

Goldoni, A., Larciprete, R., Petaccia, L., & Lizzit, S. (2003). Single-wall carbon nanotube interaction with gases: Sample contaminants and environmental monitoring. *Journal of the American Chemical Society*, *125*(37), 11329–11333. doi:10.1021/ja034898e PMID:16220955

Gordillo-Vázquez, F. J., Camero, M., & Gomez-Aleixandre, C. (2006). Spectroscopic measurements of the electron temperature in low pressure radiofrequency Ar/H2/C2H2 and Ar/H2/CH4 plasmas used for the synthesis of nanocarbon structures. *Plasma Sources Science & Technology*, *15*(1), 42–51. doi:10.1088/0963-0252/15/1/007

Gotoh, M. S., Suzuki, M., Kubo, I., Tamiya, E., & Karube, I. (1989). Immuno-FET sensor. *Journal of Molecular Catalysis*, *53*(3), 285–292. doi:10.1016/0304-5102(89)80063-X

Goyanes, S., Rubiolo, G. R., Salazar, A., Jimeno, A., Corcuera, M. A., & Mondragon, I. (2007). Carboxylation treatment of multiwalled carbon nanotubes monitored by infrared and ultraviolet spectroscopies and scanning probe microscopy. *Diamond and Related Materials, 16*(2), 412-417. doi: 10.1016/j.diamond.2006.08.021

Grassi, R., Gnudi, A., Gnani, E., Reggiani, S., & Baccarani, G. (2008). Graphene nanoribbons FETs for high-performance logic applications: Perspectives and challenges. *2008 9th International Conference on Solid-State and Integrated-Circuit Technology*.

Grate, J. W., Patrash, S. J., Abraham, M. H., & Du, C. M. (1996). Selective vapor sorption by polymers and cavitands on acoustic wave sensors: Is this molecular recognition? *Analytical Chemistry*, *68*(5), 913–917. doi:10.1021/ac950518z PMID:21619189

Gray, P. R., Hurst, P. J., Lewis, S. H., & Meyer, R. G. (2001). *Analysis and design of analog integrated circuits*. New York: Wiley.

Grieshaber, D., MacKenzie, R., Vörös, J., & Reimhult, E. (2008). Electrochemical biosensors-Sensor principles and architectures. *Sensors (Basel, Switzerland)*, *8*(3), 1400–1458. doi:10.3390/s8031400

Grossi, E., & Buscema, M. (2007). Introduction to artificial neural networks. *European Journal of Gastroenterology & Hepatology*, *19*(12), 1046–1054. doi:10.1097/MEG.0b013e3282f198a0 PMID:17998827

Gu, F., Zhang, J.-H., Yang, L.-J., & Gu, B. (2011). Molecular dynamics simulation of resonance properties of strain graphene nanoribbons. *Acta Physica Sinica*, 60.

Guinea, F., Castro Neto, A. H., & Peres, N. M. R. (2006). Electronic states and Landau levels in graphene stacks. *Physical Review B: Condensed Matter and Materials Physics*, *73*(24), 245426. doi:10.1103/PhysRevB.73.245426

Guisan, J. M. (2006). *Immobilization of enzymes and cells* (Vol. 22). Springer. doi:10.1007/978-1-59745-053-9_1

Gulotty, R., Castellino, M., Jagdale, P., Tagliaferro, A., & Balandin, A. A. (2013). Effects of functionalization on thermal properties of single-wall and multi-wall carbon nanotube–polymer nanocomposites. *ACS Nano*, *7*(6), 5114–5121. doi:10.1021/nn400726g PMID:23672711

Gunlycke, D., Areshkin, D., & White, C. (2007). Semiconducting graphene nanostrips with edge disorder. *Applied Physics Letters*, *90*(14), 142104. doi:10.1063/1.2718515

Guo, S., & Dong, S. (2011). Graphene and its derivative-based sensing materials for analytical devices. *Journal of Materials Chemistry*, *21*(46), 18503–18516. doi:10.1039/c1jm13228h

Gutés, A., Hsia, B., Sussman, A., Mickelson, W., Zettl, A., Carraro, C., & Moboudian, R. (2012). Graphene decoration with metal nanoparticles: Towards easy integration for sensing applications. *Nanoscale*, *4*(2), 438–440. doi:10.1039/C1NR11537E PMID:22147241

Guth, U. G. F., Decker, M., Oelßner, W., & Vonau, W. (2009). Solid-state reference electrodes for potentiometric sensors. *Journal of Solid State Electrochemistry*, *13*(1), 27–39. doi:10.1007/s10008-008-0574-7

Gu, X., & Yang, R. (2015). First-Principles Prediction of Phononic Thermal Conductivity of Silicene: A Comparison with Graphene. *Journal of Applied Physics*, *117*(2), 025102. doi:10.1063/1.4905540

Guzmán-Verri, G. G., & Lew Yan Voon, L. C. (2007). Electronic structure of silicon-based nanostructures. *Physical Review B: Condensed Matter and Materials Physics*, *76*(7), 075131. doi:10.1103/PhysRevB.76.075131

Haddon, R. C. (2002). Carbon nanotubes. *Accounts of Chemical Research*, *35*(12), 997–997. doi:10.1021/ar020259h PMID:12484786

Haes, A. J., Chang, L., Klein, W. L., & Van Duyne, R. P. (2005). Detection of a biomarker for Alzheimer's disease from synthetic and clinical samples using a nanoscale optical biosensor. *Journal of the American Chemical Society*, *127*(7), 2264–2271. doi:10.1021/ja044087q PMID:15713105

Hafaiedh, I., Elleuch, W., Clement, P., Llobet, E., & Abdelghani, A. (2013). Multi-walled carbon nanotubes for volatile organic compound detection. *Sensors and Actuators. B, Chemical*, *18*, 344–350. doi:10.1016/j.snb.2013.03.020

Haigang Yang, H. S., Han, Wei, Lin, Xia, & Zhong. (2005). *A pH-ISFET based micro sensor system on chip using standard CMOS technology*. Paper presented at the Fifth International Workshop on System-on-Chip for Real-Time Applications. doi:10.1109/IWSOC.2005.21

Hamid, T., Jameil, Taghi, Ahmadi, & Centeno. (2015). Investigation on optical and electrical properties of bilayer graphene. *Diyala Journal of Engineering Sciences*, 583-545.

Hamker, F. (2009). *Artificial neural networks*. Academic Press.

Hamon, M. A., Hu, H., Bhowmik, P., Niyogi, S., Zhao, B., Itkis, M. E., & Haddon, R. C. (2001). End-group and defect analysis of soluble single-walled carbon nanotubes. *Chemical Physics Letters*, *347*(1-3), 8–12. doi:10.1016/S0009-2614(01)01035-1

Han, J.-W., Kim, B., Li, J., & Meyyappan, M. (2013). A carbon nanotube based ammonia sensor on cotton textile. *Applied Physics Letters*, *102*(19), 193104, 193104–4. doi:10.1063/1.4805025

Han, M. Y., Özyilmaz, B., Zhang, Y., & Kim, P. (2007). Energy band-gap engineering of graphene nanoribbons. *Physical Review Letters*, *98*(20), 206805. doi:10.1103/PhysRevLett.98.206805 PMID:17677729

Harada, N., Sato, S., & Yokoyama, N. (2013). Theoretical Investigation of graphene nanoribbon field-effect transistors designed for digital applications. *Japanese Journal of Applied Physics*, *52*(9R), 094301. doi:10.7567/JJAP.52.094301

Hareendran, T. K. (n.d.). *Top 10 Articles*. Retrieved from http://www.electroschematics.com/6129/touch-alarm

Härtl, A., Schmich, E., Garrido, J. A., Hernando, J., Catharino, S. C., Walter, S., & Stutzmann, M. (2004). Protein-modified nanocrystalline diamond thin films for biosensor applications. *Nature Materials*, *3*(10), 736–742. doi:10.1038/nmat1204 PMID:15359341

Hatami, H., Abedpour, N., Qaiumzadeh, A., & Asgari, R. (2011). Conductance of bilayer graphene in the presence of a magnetic field: Effect of disorder. *Physical Review B: Condensed Matter and Materials Physics*, 83.

Hatef, S., SeyedMahdi, M., Meisam, R., MohammadTaghi, A., & Razali, I. (2012). Bilayer graphene nanoribbon transport model. In Advanced Nanoelectronics (pp. 203-222). CRC Press.

Hayati, M., Rezaei, A., & Seifi, M. (2010). CNT-MOSFET modeling based on artificial neural network: Application to simulation of nanoscale circuits. *Solid-State Electronics*, *54*(1), 52–57. doi:10.1016/j.sse.2009.09.027

Hays, S. P., & Hays, B. D. (1989). *Beauty, health, and permanence: Environmental politics in the United States, 1955-1985*. Cambridge University Press.

Held, M., Schuhmann, W., Jahreis, K., & Schmidt, H.-L. (2002). Microbial biosensor array with transport mutants of Escherichia coli K12 for the simultaneous determination of mono-and disaccharides. *Biosensors & Bioelectronics*, *17*(11), 1089–1094. doi:10.1016/S0956-5663(02)00103-3 PMID:12392960

Heller, A., & Feldman, B. (2008). Electrochemical Glucose Sensors and Their Applications in Diabetes Management. *Chemical Reviews*, *108*(7), 2482–2505. doi:10.1021/cr068069y PMID:18465900

Heller, I., Janssens, A. M., Mannik, J., Minot, E. D., Lemay, S. G., & Dekker, C. (2008). Identifying the mechanism of biosensing with carbon nanotube transistors. *Nano Letters*, *8*(2), 591–595. doi:10.1021/nl072996i PMID:18162002

Heller, I., Kong, J., Williams, K. A., Dekker, C., & Lemay, S. G. (2006). Electrochemistry at single-walled carbon nanotubes: The role of band structure and quantum capacitance. *Journal of the American Chemical Society*, *128*(22), 7353–7359. doi:10.1021/ja061212k PMID:16734491

He, Q., Wu, S., Yin, Z., & Zhang, H. (2012). Graphene-based electronic sensors. *Chemical Science*, *3*(6), 1764–1772. doi:10.1039/c2sc20205k

Hernandez-Perez, T., Morales, M., Batina, N., & Salmon, M. (2001). Effect of the electrosynthesis method on the surface morphology of the polypyrrole film an atomic force microscopy study. *Journal of the Electrochemical Society*, *148*(5), 369–375. doi:10.1149/1.1365143

He, X., Arsat, R., Sadek, A. Z., Wlodarski, W., Kalantar-zadeh, K., & Li, J. (2010). Electrospun PVP fibers and gas sensing properties of PVP/36 YX LiTaO 3 SAW device. *Sensors and Actuators. B, Chemical*, *145*(2), 674–679. doi:10.1016/j.snb.2010.01.013

Hickman, R. W., & Hunt, F. V. (1939). On electronic voltage stabilizers. *The Review of Scientific Instruments*, *10*(1), 6–21. doi:10.1063/1.1751443

Hidayet, C., & Enise, A. (2005). Temperature dependence of electrical parameters of the Au/n-InP Schottky barrier diodes. *Semiconductor Science and Technology*, *20*(6), 625–631. doi:10.1088/0268-1242/20/6/025

Hierold, C., Jungen, A., Stampfer, C., & Helbling, T. (2007). Nano electromechanical sensors based on carbon nanotubes. *Sensors and Actuators. A, Physical*, *136*(1), 51–61. doi:10.1016/j.sna.2007.02.007

Hieu, N. V., Duc, N. A. P., Trung, T., Tuan, M. A., & Chien, N. D. (2010). Gas-sensing properties of tin oxide doped with metal oxides and carbon nanotubes: A competitive sensor for ethanol and liquid petroleum gas. *Sensors and Actuators. B, Chemical*, *144*(2), 450–456. doi:10.1016/j.snb.2009.03.043

Hijikata, K., K. Ogawa & N. Miyakawa (1999). Methanol conversion from methane and water vapor by electric discharge (effect of electric discharge process on methane conversion). *Heat Transfer—Asian Research, 28*, 404-417.

Hill, C. M., Zhu, Y., & Pan, S. (2011). Fluorescence and electroluminescence quenching evidence of interfacial charge transfer in poly (3-hexylthiophene): Graphene oxide bulk heterojunction photovoltaic devices. *ACS Nano*, *5*(2), 942–951. doi:10.1021/nn1022457 PMID:21294531

Hill, E. W., Vijayaragahvan, A., & Novoselov, K. (2011). Graphene sensors. *IEEE Sensors Journal*, *11*(12), 3161–3170. doi:10.1109/JSEN.2011.2167608

Hirsch, A., Eigler, R., & Heinemann, F. W. (2013). Selective oxidative conversion of Triaryldihydro [C59N] fullerenes: A model case for carbon allotrope oxygenation. *Chemical Communications*.

Ho, C. K., Lindgren, E. R., Rawlinson, K. S., McGrath, L. K., & Wright, J. L. (2003). Development of a surface acoustic wave sensor for in-situ monitoring of volatile organic compounds. *Sensors (Basel, Switzerland)*, *3*(7), 236–247. doi:10.3390/s30700236

Ho, K.-I., Liao, J.-H., Huang, C.-H., Su, C.-Y., & Lai, C.-S. (2013). Electrical probing of multi-ions solution by using graphene-based sensor. In *Nanoelectronics Conference (INEC), 2013 IEEE 5th International*, (pp. 37-39). IEEE. doi:10.1109/INEC.2013.6465946

Holberg, D. R., & Allen, P. E. (2002). *CMOS analog circuit design*. OxfordUniversity Press.

Homola, J. I., & Dostálek, J. (2006). *Surface plasmon resonance based sensors*. Academic Press.

Hone, J., Llaguno, M. C., Nemes, N. M., Johnson, A. T., Fischer, J. E., & Walters, D. A. et al. (2000). Electrical and thermal transport properties of magnetically aligned single wall carbon nanotube films. *Applied Physics Letters*, *77*(5), 666–668. doi:10.1063/1.127079

Hook, J. R., & Hall, H. E. (2001). *Solid state physics*. John Wiley & Sons.

Horowitz, P. (1989). The art of electronics (2nd ed.). Cambridge University Press.

Horowitz, P. (2015). The art of electronics. Cambridge University Press.

Hosono, Y., Sjafruddin, , Iguchi, T., & Nakazawa, M. (1995). Fast neutron detector using PIN-type silicon photodiode. *Nuclear Instruments & Methods in Physics Research. Section A, Accelerators, Spectrometers, Detectors and Associated Equipment*, *361*(3), 554–557. doi:10.1016/0168-9002(95)00206-5

Houssa, M., Pourtois, G., Afanas'Ev, V. V., & Stesmans, A. (2010). Can silicon behave like graphene? A first-principles study. *Applied Physics Letters, 97*(11), 112106 - 112106-112103

Houssa, M., Scalise, E., Sankaran, K., Pourtois, G., Afanas'Ev, V. V., & Stesmans, A. (2011). Electronic properties of hydrogenated silicene and germanene. *Applied Physics Letters*, *98*(22), 223107–223107. doi:10.1063/1.3595682

Hribšek, M., & Tošić, D. (n.d.). *Analysis and modelling of surface acoustic wave chemical vapour sensors*. Academic Press.

Hu, C. C. (2010). Modern semiconductor devices for integrated circuits. Academic Press.

Hua, B., & Gaoquan, S. (2007). Gas sensors based on conducting polymers. *Sensors (Basel, Switzerland)*, *7*(3), 267–307. doi:10.3390/s7030267

Huang, B., Li, Z., Liu, Z., Zhou, G., Hao, S., Wu, J., & Duan, W. et al. (2008a). Adsorption of gas molecules on graphene nanoribbons and its implication for nanoscale molecule sensor. *The Journal of Physical Chemistry C*, *112*(35), 13442–13446. doi:10.1021/jp8021024

Huang, S., Kang, W., & Yang, L. (2013). Electronic Structure and Quasiparticle Band Gap of Silicene Structures. *Applied Physics Letters*, *102*(13), 133106. doi:10.1063/1.4801309

Huang, S., Pang, P., Xiao, X., He, L., Cai, Q., & Grimes, C. A. (2008). A wireless, remote-query sensor for real-time detection of Escherichia coli O157: H7 concentrations. *Sensors and Actuators. B, Chemical*, *131*(2), 489–495. doi:10.1016/j.snb.2007.12.036

Huang, X., Qi, X., Boey, F., & Zhang, H. (2012). Graphene-based composites. *Chemical Society Reviews*, *41*(2), 666–686. doi:10.1039/C1CS15078B PMID:21796314

Huang, X., Yin, Z., Wu, S., Qi, X., He, Q., Zhang, Q., & Zhang, H. et al. (2011). Graphene-based materials: Synthesis, characterization, properties, and applications. *Small*, *7*(14), 1876–1902. doi:10.1002/smll.201002009 PMID:21630440

Huang, X., Zeng, Z., Fan, Z., Liu, J., & Zhang, H. (2012). Graphene-based electrodes. *Advanced Materials*, *24*(45), 5979–6004. doi:10.1002/adma.201201587 PMID:22927209

Huang, Y., Dong, X., Liu, Y., Li, J., & Chen, P. (2011). Graphene-based biosensors for detection of bacteria and their metabolic activities. *Journal of Materials Chemistry*, *21*(33), 12358–12362. doi:10.1039/c1jm11436k

Huang, Y., Dong, X., Shi, Y., Li, C. M., Li, L.-J., & Chen, P. (2010). Nanoelectronic biosensors based on CVD grown graphene. *Nanoscale*, *2*(8), 1485–1488. doi:10.1039/c0nr00142b PMID:20820739

Hughes, M., & Spinks, G. M. (2005). Multiwalled carbon nanotube actuators. *Advanced Materials*, *17*(4), 443–446. doi:10.1002/adma.200401076

Hu, J., Ruan, X., & Chen, Y. P. (2009). Thermal Conductivity and Thermal Rectification in Graphene Nanoribbons: A Molecular Dynamics Study. *Nano Letters*, *9*(7), 2730–2735. doi:10.1021/nl901231s PMID:19499898

Hu, N., Wang, Y., Chai, J., Gao, R., Yang, Z., Kong, E. S.-W., & Zhang, Y. (2012). Gas sensor based on p-phenylenediamine reduced graphene oxide. *Sensors and Actuators. B, Chemical*, *163*(1), 107–114. doi:10.1016/j.snb.2012.01.016

Husain, M., & Husain, Q. (2007). Applications of redox mediators in the treatment of organic pollutants by using oxidoreductive enzymes: A review. *Critical Reviews in Environmental Science and Technology*, *38*(1), 1–42. doi:10.1080/10643380701501213

Hutchison, J., Kiselev, N., Krinichnaya, E., Krestinin, A., Loutfy, R., Morawsky, A., & Terekhov, S. et al. (2001). Double-walled carbon nanotubes fabricated by a hydrogen arc discharge method. *Carbon*, *39*(5), 761–770. doi:10.1016/S0008-6223(00)00187-1

Hu, Y., Li, F., Bai, X., Li, D., Hua, S., Wang, K., & Niu, L. (2011). Label-free electrochemical impedance sensing of DNA hybridization based on functionalized graphene sheets. *Chemical Communications*, *47*(6), 1743–1745. doi:10.1039/C0CC04514D PMID:21125081

Hwang, W. S., Zhao, P., Tahy, K., Nyakiti, L. O., Wheeler, V. D., Myers-Ward, R., . . . Haensch, W. (2013). *Graphene nanoribbon field-effect transistors on wafer-scale epitaxial graphene on sic substrates.* arXiv preprint arXiv:1310.6823

Hwang, W. S., Zhao, P., Tahy, K., Nyakiti, L. O., Wheeler, V. D., Myers-Ward, R., . . . Haensch, W. (2013). *Graphene nanoribbon field-effect transistors on wafer-scale epitaxial graphene on SiC substrates.* arXiv preprint arXiv:1310.6823

Hwang, E., Adam, S., & Sarma, S. D. (2007). Carrier transport in two-dimensional graphene layers. *Physical Review Letters*, *98*(18), 186806. doi:10.1103/PhysRevLett.98.186806 PMID:17501596

IEEE Standard 100 authoritative dictionary of IEEE standards terms (2000). IEEE Press.

Iijima, S. (1991). Helical microtubules of graphite carbon. *Nature*, *354*(6348), 56–58. doi:10.1038/354056a0

Ilani, S., Donev, L. A., Kindermann, M., & McEuen, P. L. (2006). Measurement of the quantum capacitance of interacting electrons in carbon nanotubes. *Nature Physics*, *2*(10), 687–691. doi:10.1038/nphys412

Indarto, A., Choi, J., Lee, H., & Song, H. K. (2007). A brief catalyst study on direct methane conversion using a dielectric barrier discharge. *Journal-Chinese Chemical Society Taipei*, *54*(4), 823–828. doi:10.1002/jccs.200700120

Islam, M. S. (2008). The all pervading nanosensors. *International Journal of Nanotechnology*, *5*(4/5), 393–401. doi:10.1504/IJNT.2008.017445

ISO 8529-1. (2001). *Reference Neutron Radiations. Part1: Characteristics and methods of production*. International Organization for Standardization ISO8529-1.

Jablan, M., Soljacic, M., & Buljan, H. (2013). Plasmons in graphene: Fundamental properties and potential applications. *Proceedings of the IEEE*, *101*(7), 1689–1704. doi:10.1109/JPROC.2013.2260115

Jaeger, R. C. (1988). *Introduction to microelectronic fabrication*. New York: Addison-Wesley.

Jaeger, R. C., & Blalock, T. N. (2003). *Microelectronic circuit design*. McGraw-Hill.

Jain, S., Chakane, S., Samui, A. B., Krishnamurthy, V. N., & Bhoraskar, S. V. (2003). Humidity sensing with weak acid-doped polyaniline and its composites. *Sensors and Actuators. B, Chemical*, *96*(1-2), 124–129. doi:10.1016/S0925-4005(03)00511-2

Jamasb, S. C., & Smith, S. D., R. L. (1997). *Correction of instability in ion-selective field effect transistors (ISFETs) for accurate continuous monitoring of pH*. Paper presented at the Engineering in Medicine and Biology Society.

Jamgotchian, H., Colignon, Y., Hamzaoui, N., Ealet, B., Hoarau, J., Aufray, B., & Bibérian, J. (2012). Growth of silicene layers on Ag (111): Unexpected effect of the substrate temperature. *Journal of Physics Condensed Matter*, *24*(17), 172001. doi:10.1088/0953-8984/24/17/172001 PMID:22487603

Janata, J. (2002). Electrochemical sensors and their impedances. *Analytical Chemistry*, *32*, 109–120.

Janata, J. (2003). Electrochemical microsensors. *Proceedings of the IEEE*, *91*(6), 864–869. doi:10.1109/JPROC.2003.813576

Janata, J., & Josowicz, M. (2003). Conducting polymers in electronic chemical sensors. *Nature Materials*, *2*(1), 19–24. doi:10.1038/nmat768 PMID:12652667

Javey, A., Guo, J., Wang, Q., Lundstrom, M., & Dai, H. (2003). Ballistic carbon nanotube field-effect transistors. *Nature*, *424*(6949), 654–657. doi:10.1038/nature01797 PMID:12904787

Jena, D. (2009). A theory for the high-field current-carrying capacity of one-dimensional semiconductors. *Journal of Applied Physics*, *105*(12), 123701–123705. doi:10.1063/1.3147877

Jeong, H. Y., Lee, D.-S., Choi, H. K., Lee, D. H., Kim, J.-E., Lee, J. Y., & Choi, S.-Y. et al. (2010). Flexible room-temperature NO gas sensors based on carbon nanotubes/reduced graphene hybrid films. *Applied Physics Letters*, *96*(21), 213105. doi:10.1063/1.3432446

Jesse, D. F., Matthew, J. A., Vincent, C. T., Yang, Y., Richard, B. K., & Bruce, H. W. (2009). Practical chemical sensors from chemically derived graphene. *ACS Nano*, *3*(2), 301–306. doi:10.1021/nn800593m PMID:19236064

Jiao, L., Zhang, L., Wang, X., Diankov, G., & Dai, H. (2009). Narrow graphene nanoribbons from carbon nanotubes. *Nature*, *458*(7240), 877–880. doi:10.1038/nature07919 PMID:19370031

Jilin Xia, F. C. (2009). Measurement of the quantum capacitance of graphene. *Nature Nanotechnology*. doi:10.1038/NNANO.2009.177

Jiménez, C. B., & Bartroli, J. (1997). Development of an ion-sensitive field effect transistor based on PVC membrane technology with improved long-term stability. *Electroanalysis*, *9*(4), 316–319. doi:10.1002/elan.1140090411

Jimenez-Cadena, G., Riu, J., & Rius, F. X. (2007). Gas sensors based on nanostructured materials. *Analyst (London)*, *132*(11), 1083–1099. doi:10.1039/b704562j PMID:17955141

Jin, S. H., Dunham, S. N., Song, J., Xie, X., Kim, J.-, Lu, C., & Felts, J. et al. (2013). Using nanoscale thermocapillary flows to create arrays of purely semiconducting single-walled carbon nanotubes. *Nature Nanotechnology*, *8*(5), 347–355. doi:10.1038/nnano.2013.56 PMID:23624697

Johari, P., & Shenoy, V. B. (2011). Modulating optical properties of graphene oxide: Role of prominent functional groups. *ACS Nano*, *5*(9), 7640–7647. doi:10.1021/nn202732t PMID:21875075

Johnson, A.T., Jr., Gelperin, A., & Staii, C. (2011). *Single walled carbon nanotubes functionally adsorbed to biopolymers for use as chemical sensors: Google Patents.*

Jon, W. (2005). *Sensor Technology Handbook*. Oxford, UK: Elsevier Inc.

Jornet, J. M., & Akyildiz, I. F. (2010). Channel capacity of electromagnetic nanonetworks in the terhertz band.*Proceedings of IEEE International Conference on Communications.*

Jose, D., & Datta, A. (2011). Structures and electronic properties of silicene clusters: A promising material for FET and hydrogen storage. *Physical Chemistry Chemical Physics*, *13*(16), 7304–7311. doi:10.1039/c0cp02580a PMID:21359313

Joshi, R. K., Gomez, H., Alvi, F., & Kumar, A. (2010). Graphene films and ribbons for sensing of O_2, and 100 ppm of CO and NO_2 in practical conditions. *The Journal of Physical Chemistry C*, *114*(14), 6610–6613. doi:10.1021/jp100343d

Journet, C., & Bernier, P. (1998). Production of carbon nanotubes. *Applied Physics. A, Materials Science & Processing*, *67*(1), 1–9. doi:10.1007/s003390050731

Jun-Feng, Z., Xin-Chao, B., Qiang, C., Fu-Ping, L., & Zhong-Wei, L. (2009). Diagnosis of methane plasma generated in an atmospheric pressure dbd micro-jet by optical emission spectroscopy. *Chinese Physics Letters*, *26*(3), 035203. doi:10.1088/0256-307X/26/3/035203

Jung, D., Ham, M., & Lee, G. S. (2014). Room-temperature gas sensor using carbon nanotube with cobalt oxides. *Sensors and Actuators. B, Chemical*, *204*, 596–601. doi:10.1016/j.snb.2014.08.020

Jung, J. (2011). Nonlocal exchange effects in zigzag-edge magnetism of neutral graphene nanoribbons. *Physical Review B: Condensed Matter and Materials Physics*, 83.

Kacaoglu, M., & Akan, O. B. (2013). Minimum energy channel codes for nanoscale wireless communications. *IEEE Transactions on Wireless Communications*, *12*(4), 1492–1500. doi:10.1109/TWC.2013.022113.2190

Kal, S. B. P. V. (2007). *Design and modeling of ISFET for pH sensing.* Paper presented at the TENCON 2007 - 2007 IEEE Region 10 Conference.

Kamal, C., Chakrabarti, A., Banerjee, A., & Deb, S. (2013). Silicene Beyond Mono-layers-Different Stacking Configurations And Their Properties. *Journal of Physics Condensed Matter*, *25*(8), 085508. doi:10.1088/0953-8984/25/8/085508 PMID:23370369

Kaniyoor, A., Jafri, R. I., Arockiadoss, T., & Ramaprabhu, S. (2009). Nanostructured Pt decorated graphene and multi walled carbon nanotube based room temperature hydrogen gas sensor. *Nanoscale*, *1*(3), 382–386. doi:10.1039/b9nr00015a PMID:20648277

Kara, A., Enriquez, H., Seitsonen, A. P., Lew Yan Voon, L. C., Vizzini, S., Aufray, B., & Oughaddou, H. (2012). A review on silicene — New candidate for electronics. *Surface Science Reports*, *67*(1), 1–18. doi:10.1016/j.surfrep.2011.10.001

Karamdel, J., Ahmadi, M. T., Majlis, B. Y., Dee, C. F., & Ismail, R. (2008). Formulation and Simulation for Electrical Properties of a (5,3) Single Wall Carbon Nanotube. *ICSE: 2008 IEEE International Conference on Semiconductor Electronics Proceedings*.

Karamitaheri, H., Neophytou, N., Pourfath, M., Faez, R., & Kosina, H. (2012). Engineering enhanced thermoelectric properties in zigzag graphene nanoribbons. *Journal of Applied Physics*, *111*(5), 054501. doi:10.1063/1.3688034

Karamitaheri, H., Pourfath, M., Faez, R., & Kosina, H. (2011). Geometrical effects on the thermoelectric properties of ballistic graphene antidot lattices. *Journal of Applied Physics*, *110*(5), 054506. doi:10.1063/1.3629990

Karimi, H., Yusof, R., Rahmani, R., Hosseinpour, H., & Ahmadi, M. T. (2014). Development of solution-gated graphene transistor model for biosensors. *Nanoscale Research Letters, 9*.

Kasani, H., Taghi Ahmadi, M., Khoda-bakhsh, R., RezaeiOchbelagh, D., & Ismail, R. (2016). Influences of Sr-90 beta-ray irradiation on electrical characteristics of carbon nanoparticles. *Journal of Applied Physics, 119*(12), 124510. doi:10.1063/1.4944901

Katsnelson, M. I. (2007). Graphene: Carbon in two dimensions. *Materials Today*, *10*(1), 20–27. doi:10.1016/S1369-7021(06)71788-6

Kažukauskas, V., V. Kalendra, C. Bumby, B. Ludbrook & A. Kaiser (2008) Electrical conductivity of carbon nanotubes and polystyrene composites. *Physica Status Solidi (c), 5*, 3172-3174.

Kester, W. (2010). *003: Understand SINAD, ENOB, SNR, THD, THD + N, and SFDR so you don't get lost in the noise floor*. Analog Devices. Retrieved.

Kiani, M. J. A., Rahmani, Harun, & Che. (2013). Degeneracy effect on carrier transport in bilayer graphene nanoribbon. *International Journal of Nano Devices, Sensors and Systems (IJ-Nano), 2*(1).

Kiani, M. J., Ahmadi, M. T., Abadi, H. K. F., & Rahmani, M. (2013). Analytical modelling of monolayer graphene-based ion-sensitive FET to pH changes. *Nanoscale Research Letters*, *8*(1), 1–9. doi:10.1186/1556-276X-8-173 PMID:23590751

Kiani, M. J., Ahmadi, M., Akbari, E., Karimi, H., & Che Harun, F. (2013). Graphene Nanoribbon Based Gas Sensor. *Key Engineering Materials*, *553*, 7–11. doi:10.4028/www.scientific.net/KEM.553.7

Kiani, M. J., Ahmadi, M., & Harun, F. (2013). Quantum capacitance effect on bilayer graphene nanoribbon based nanoscale transistors. *Journal of Nanoengineering and Nanomanufacturing*, *3*(2), 138–141. doi:10.1166/jnan.2013.1119

Kim, J., Fischetti, M. V., & Aboud, S. (2012a). *Electronic and transport properties of armchair and zigzag sp3-hybridized silicane nanoribbons.* Paper presented at the Computational Electronics (IWCE), 2012 15th International Workshop on.

Kim, R., & Lundstrom, M. (2008). Notes on Fermi-Dirac integrals (2nd ed.). Academic Press.

Kim, B., Lee, J., Namgung, S., Kim, J., Park, J. Y., Lee, M.-S., & Hong, S. (2012). DNA sensors based on CNT-FET with floating electrodes. *Sensors and Actuators. B, Chemical*, *169*, 182–187. doi:10.1016/j.snb.2012.04.063

Kim, J., Fischetti, M. V., & Aboud, S. (2012b). Structural, electronic, and transport properties of silicane nanoribbons. *Physical Review B: Condensed Matter and Materials Physics*, *86*(20), 205323. doi:10.1103/PhysRevB.86.205323

Kim, S. J. (2009). Flexible alcohol vapor sensors using multiple spray-coated SWNTs on PES substrates. *Journal of the Korean Physical Society*, *54*(5), 1779–1783. doi:10.3938/jkps.54.1779

Kim, S.-S., Lee, H., Na, B.-K., & Song, H. K. (2003). Reaction pathways of methane conversion in dielectric-barrier discharge. *Korean Journal of Chemical Engineering*, *20*(5), 869–872. doi:10.1007/BF02697290

Kinoshita, K. (1988). *Carbon: Electrochemical and physicochemical properties*. Academic Press.

Kirsch, J., Siltanen, C., Zhou, Q., Revzin, A., & Simonian, A. (2013). Biosensor technology: Recent advances in threat agent detection and medicine. *Chemical Society Reviews*, *42*(22), 8733–8768. doi:10.1039/c3cs60141b PMID:23852443

Kishore, K., & Asmus, K. D. (1989). Radical cations from one-electron oxidation of aliphatic sulfoxides in aqueous solution – A radiation chemical study. *Journal of the Chemical Society*, *2*, 2079–2084.

Knoll, G. F. (2000). *Radiation Detection and Measurement* (3rd ed.). Wiley.

Koehler, C., Thielen, S., Ginzkey, C., Hackenberg, S., Scherzed, A., Burghartz, M., & Kleinsasser, N. H. et al. (2013). Nitrogen dioxide is genotoxic in urban concentrations. *Inhalation Toxicology*, 1–7. PMID:23701639

Ko, G., Kim, H. Y., Ahn, J., Park, Y. M., Lee, K. Y., & Kim, J. (2010). Graphene-based nitrogen dioxide gas sensors. *Current Applied Physics*, *10*(4), 1002–1004. doi:10.1016/j.cap.2009.12.024

Kokott, S., Pflugradt, P., Matthes, L., & Bechstedt, F. (2014). Nonmetallic substrates for growth of silicene: An ab initio prediction. *Journal of Physics Condensed Matter*, *26*(18), 185002. doi:10.1088/0953-8984/26/18/185002 PMID:24728001

Kong, J., Franklin, N. R., Zhou, C., Chapline, M. G., Peng, S., Cho, K., & Dai, H. (2000). Nanotube Molecular Wires as Chemical Sensors. *Science*, *287*(5453), 622–625. doi:10.1126/science.287.5453.622 PMID:10649989

Kordas, K., Mustonen, T., Toth, G., Jantunen, H., Lajunen, M., & Soldano, C. et al. (2006). Inkjet Printing of Electrically Conductive Patterns of Carbon Nanotubes. *Small*, *2*(8-9), 1021–1025. doi:10.1002/smll.200600061 PMID:17193162

Korotcenkov, G. (2014). Handbook of gas sensor materials: Properties, advantages and shortcomings for applications volume 2 new trends and techniques. Springer. doi:10.1007/978-1-4614-7388-6

Korotcenkov, G. (2013). *Handbook of gas sensor materials*. Springer. doi:10.1007/978-1-4614-7165-3

Kozlov, V., Starik, A., Titova, N., & Vedishchev, I. Y. (2013). On mechanisms of formation of environmentally harmful compounds in homogeneous combustors. *Combustion, Explosion, and Shock Waves*, *49*(5), 520–535. doi:10.1134/S0010508213050031

Kraus, M., Egli, W., Haffner, K., Eliasson, B., Kogelschatz, U., & Wokaun, A. (2002). Investigation of mechanistic aspects of the catalytic CO2 reforming of methane in a dielectric-barrier discharge using optical emission spectroscopy and kinetic modeling. *Physical Chemistry Chemical Physics*, *4*(4), 668–675. doi:10.1039/b108040g

Krcma, F., Klohnova, K., Polachova, L., & Horvath, G. (2010). Optical emission spectroscopy of abnormal glow discharge in nitrogen-methane mixtures at atmospheric pressure. *Publications de l'Observatoire Astronomique de Beograd*, *89*, 371–374.

Kreibig, U., Gartz, M., & Hilger, A. (1997). Mie resonances: Sensors for physical and chemical cluster interface properties. *Berichte der Bunsengesellschaft fur Physikalische Chemie*, *101*(11), 1593–1604. doi:10.1002/bbpc.19971011107

Kriz, D., Ramström, O., & Mosbach, K. (1997). Peer Reviewed: Molecular Imprinting: New Possibilities for Sensor Technology. *Analytical Chemistry*, *69*(11), 345A–349A. doi:10.1021/ac971657e

Kuan, A. T., & Golovchenko, J. A. (2012). Nanometer-thin solid-state nanopores by cold ion beam sculpting. *Applied Physics Letters*, *100*(21), 213104. doi:10.1063/1.4719679 PMID:22711913

Kuang, B. M., & Hafiz, S. (2012). Sensing soil properties in the laboratory, in situ, and on-line: A Review. In L. S. Donald (Ed.), Advances in Agronomy (Vol. 114, pp. 155-223). Academic Press.

Kumar, B., & Jain, S. B. (2013). *Electronic devices and circuits*. PHI Learning Pvt. Ltd.

Kumar, B., Min, K., Bashirzadeh, M., Farimani, A. B., Bae, M.-H., Estrada, D., & Salehi-Khojin, A. et al. (2013). the role of external defects in chemical sensing of graphene field-effect transistors. *Nano Letters*, *13*(5), 1962–1968. doi:10.1021/nl304734g PMID:23586702

Kupny, S. (2011). *Artificial neural networks*. Academic Press.

Kuzmenko, A., Van Heumen, E., Carbone, F., & Van Der Marel, D. (2008). Universal optical conductance of graphite. *Physical Review Letters*, *100*(11), 117401. doi:10.1103/PhysRevLett.100.117401 PMID:18517825

Lakshmi, G. B. V. S., & Anju Dhillon, D. K. (1672–1673). Avasthi, c., Azher, M., & Siddiqui, M. (2010). Synthesis and characterization of thin films of poly(3-methyl thiophene) by rf-plasma polymerization. *Materials Letters*, 64.

Lalmi, B., Oughaddou, H., Enriquez, H., Kara, A., Vizzini, S., Ealet, B., & Aufray, B. (2010). Epitaxial growth of a silicene sheet. *Applied Physics Letters*, *97*(22), 223109. doi:10.1063/1.3524215

Lam, K.-T., Lee, C., & Liang, G. (2009). Bilayer graphene nanoribbon nanoelectromechanical system device: A computational study. *Applied Physics Letters*, 95.

Lange, U., Hirsch, T., Mirsky, V. M., & Wolfbeis, O. S. (2011). Hydrogen sensor based on a graphene-palladium nanocomposite. *Electrochimica Acta*, *56*(10), 3707–3712. doi:10.1016/j.electacta.2010.10.078

Laplante, P. A. (2005). *Comprehensive dictionary of electrical engineering* (2nd ed.). CRC Press. doi:10.1201/9781420037807

Laurent, C., Flahaut, E., & Peigney, A. (2010). The weight and density of carbon nanotubes versus the number of walls and diameter. *Carbon*, *48*(10), 2994–2996. doi:10.1016/j.carbon.2010.04.010

Lee, J.-H., & Grossman, J. C. (2009). Thermoelectric properties of nanoporous Ge. *Applied Physics Letters, 95*(1), 013106. doi:10.1063/1.3159813

Lee, W.-C., Kuo, C.-C., & Chiu, N.-F. (2012). Simple fabrication of glucose biosensor based on graphene-nafion composite by amperometric detections. *2012 IEEE Sensors Proceedings* (pp. 1625-1628).

Lee, C. Y., Baik, S., Zhang, J. Q., Masel, R. I., & Strano, M. S. (2006). Charge Transfer from Metallic Single-Walled Carbon Nanotube Sensor Arrays. *The Journal of Physical Chemistry B*, *110*(23), 11055–11061. doi:10.1021/jp056425v PMID:16771365

Lee, C., Wei, X., Kysar, J. W., & Hone, J. (2008). Measurement of the elastic properties and intrinsic strength of monolayer graphene. *Science*, *321*(5887), 385–388. doi:10.1126/science.1157996 PMID:18635798

Lee, D., & Cui, T. (2009a). Layer-by-layer self-assembled single-walled carbon nanotubes based ion-sensitive conductometric glucose biosensors. *Sensors Journal, IEEE*, *9*(4), 449–456. doi:10.1109/JSEN.2009.2014414

Lee, D., & Cui, T. (2009b). pH-dependent conductance behaviors of layer-by-layer self-assembled carboxylated carbon nanotube multilayer thin-film sensors. *Journal of Vacuum Science & Technology B Microelectronics and Nanometer Structures*, *27*(2), 842. doi:10.1116/1.3002386

Lee, D., & Cui, T. (2010). Low-cost, transparent, and flexible single-walled carbon nanotube nanocomposite based ion-sensitive field-effect transistors for pH/glucose sensing. *Biosensors & Bioelectronics*, *25*(10), 2259–2264. doi:10.1016/j.bios.2010.03.003 PMID:20417088

Lee, H. K., Lee, J., Choi, N. J., Moon, S. E., Lee, H., & Yang, W. S. (2011). Efficient reducing method of graphene oxide for gas sensor applications. In G. Kaltsas & C. Tsamis (Eds.), *Eurosensors Xxv*. doi:10.1016/j.proeng.2011.12.219

Lee, H., Vashaee, D., Wang, D., Dresselhaus, M. S., Ren, Z., & Chen, G. (2010). Effects of nanoscale porosity on thermoelectric properties of SiGe. *Journal of Applied Physics*, *107*(9), 094308. doi:10.1063/1.3388076 PMID:20473369

Lee, J. H., Joo, B. S., Huh, J. S., & Lee, D. D. (2006). Fabrication of SAW sensor for detecting chemical agent. *Rare Metal Materials and Engineering*, *35*, 433–435.

Lee, J.-H., Galli, G. A., & Grossman, J. C. (2008). Nanoporous Si as an Efficient Thermoelectric Material. *Nano Letters*, *8*(11), 3750–3754. doi:10.1021/nl802045f PMID:18947211

Lee, J.-H., Kang, W.-S., Najeeb, C. K., Choi, B.-S., Choi, S.-W., Lee, S. S., & Kim, J.-H. (2013a). A hydrogen gas sensor using single-walled carbon nanotube Langmuir-Blodgett films decorated with palladium nanoparticles. *Sensors and Actuators. B, Chemical*, *188*, 169–175. doi:10.1016/j.snb.2013.06.066

Lee, K., Scardaci, V., Kim, H.-Y., Hallam, T., Nolan, H., Bolf, B. E., & Duesberg, G. S. et al. (2013b). Highly sensitive, transparent, and flexible gas sensors based on gold nanoparticle decorated carbon nanotubes. *Sensors and Actuators. B, Chemical*, *188*, 571–575. doi:10.1016/j.snb.2013.07.048

Leenaerts, O., Partoens, B., & Peeters, F. (2008). Adsorption of H_{2} O, NH_{3}, CO, NO_{2}, and NO on graphene: A first-principles study. *Physical Review B: Condensed Matter and Materials Physics*, *77*(12), 125416. doi:10.1103/PhysRevB.77.125416

Leenaerts, O., Partoens, B., & Peeters, F. (2009). Adsorption of small molecules on graphene. *Microelectronics Journal*, *40*(4-5), 860–862. doi:10.1016/j.mejo.2008.11.022

Leghrib, R., Felten, A., Demoisson, F., Reniers, F., Pireaux, J.-J., & Llobet, E. (2010). Room-temperature, selective detection of benzene at trace levels using plasma-treated metal-decorated multiwalled carbon nanotubes. *Carbon*, *48*(12), 3477–3484. doi:10.1016/j.carbon.2010.05.045

Leghrib, R., Felten, A., Pireaux, J. J., & Llobet, E. (2011). Gas sensors based on doped-CNT/SnO2 composites for NO2 detection at room temperature. *Thin Solid Films*, *520*(3), 966–970. doi:10.1016/j.tsf.2011.04.186

Lehman, J. H., Terrones, M., Mansfield, E., Hurst, K. E., & Meunier, V. (2011). Evaluating the characteristics of multiwall carbon nanotubes. *Carbon*, *49*(8), 2581–2602. doi:10.1016/j.carbon.2011.03.028

Leonenko, Z., Cramb, D. T., Amrein, M., & Finot, E. (2008). Atomic force microscopy: Interaction forces measured in phospholipid monolayers, bilayers and cell membranes. *Applied Scanning Probe Methods*, *IX*, 207–234.

Lew Yan Voon, L. C., Sandberg, E., Aga, R. S., & Farajian, A. A. (2010). Hydrogen compounds of group-IV nanosheets. *Applied Physics Letters*, *97*(16), 163114–163114. doi:10.1063/1.3495786

Li, X.-S. L., Shi, C., Xu, Y., Wang, Y.-N., & Zhu, A.-M. (2008). Stable kilohertz spark discharges for high-efficiency conversion of methane to hydrogen and acetylene. *Journal Of Physics -London- D Applied Physics*, *41*(17).

Liang, G., Neophytou, N., Nikonov, D. E., & Lundstrom, M. S. (2007). Performance projections for ballistic graphene nanoribbon field-effect transistors. *Electron Devices. IEEE Transactions on*, *54*(4), 677–682. doi:10.1109/TED.2007.891872

Liang, S.-Z., Chen, G., Harutyunyan, A. R., Cole, M. W., & Sofo, J. O. (2013). Analysis and optimization of carbon nanotubes and graphene sensors based on adsorption-desorption kinetics. *Applied Physics Letters*, *103*(23), 233108. doi:10.1063/1.4841535

Liao, K.-H., Lin, Y.-S., Macosko, C. W., & Haynes, C. L. (2011). Cytotoxicity of graphene oxide and graphene in human erythrocytes and skin fibroblasts. *ACS Applied Materials & Interfaces*, *3*(7), 2607–2615. doi:10.1021/am200428v PMID:21650218

Li, C., Thostenson, E. T., & Chou, T.-W. (2008). Sensors and actuators based on carbon nanotubes and their composites: A review. *Composites Science and Technology*, *68*(6), 1227–1249. doi:10.1016/j.compscitech.2008.01.006

Lich Quang, N., Pho Quoc, P., Huyen Ngoc, D., Chien Duc, N., & Lam Huu, N. (2013). Enhancement of NH3 gas sensitivity at room temperature by carbon nanotube-based sensor coated with co nanoparticles. *Sensors (Basel, Switzerland)*, *13*(2), 1754–1762. doi:10.3390/s130201754 PMID:23364198

Li, D., Jiang, Y., Wu, Z., Chen, X., & Li, Y. (2000). Self-assembly of polyaniline ultrathin films based on doping-induced deposition effect and applications for chemical sensors. *Sensors and Actuators. B, Chemical*, *66*(1-3), 125–127. doi:10.1016/S0925-4005(00)00315-4

Li, L., Lu, Y., Ye, Q., Cinke, M., Han, J., & Meyyappan, M. (2003). Carbon Nanotube Sensors for Gas and Organic Vapor Detection. *Nano Letters*, *3*(7), 929–933. doi:10.1021/nl034220x

Lillie, G., Payne, P., & Vadgama, P. (2001). Electrochemical impedance spectroscopy as a platform for reagentless bioaffinity sensing. *Sensors and Actuators. B, Chemical*, *78*(1), 249–256. doi:10.1016/S0925-4005(01)00821-8

Li, M., & Liu, Y. (2009). Underground coal mine monitoring with wireless sensor networks. *ACM Transactions on Sensor Networks*, *5*(2), 10. doi:10.1145/1498915.1498916

Lin, Y.-M., Appenzeller, A., Chen, Z., Chen, Z.-G., Cheng, H.-M., & Avouris, P. (2005). *Demonstration of a high performance 40-nm-gate carbon nanotube field-effect transistor.* Paper presented at the Device Research Conference Digest.

Lin, C.-L., Arafune, R., Kawahara, K., Tsukahara, N., Minamitani, E., & Kim, Y. et al. (2012). Structure of silicene grown on Ag (111). *Applied Physics Express*, *5*(4), 045802. doi:10.1143/APEX.5.045802

Lin, H., Lu, Q., Ge, S., Cai, Q., & Grimes, C. A. (2010). Detection of pathogen Escherichia coli O157: H7 with a wireless magnetoelastic-sensing device amplified by using chitosan-modified magnetic nanoparticles. *Sensors and Actuators. B, Chemical*, *147*(1), 343–349. doi:10.1016/j.snb.2010.03.011

Lin, L., Liu, Y., Tang, L., & Li, J. (2011). Electrochemical DNA sensor by the assembly of graphene and DNA-conjugated gold nanoparticles with silver enhancement strategy. *Analyst (London)*, *136*(22), 4732–4737. doi:10.1039/c1an15610a PMID:21952074

Lin, Q., Li, Y., & Yang, M. (2012). Tin oxide/graphene composite fabricated via a hydrothermal method for gas sensors working at room temperature. *Sensors and Actuators. B, Chemical*, *173*, 139–147. doi:10.1016/j.snb.2012.06.055

Lin, Y., Kan, K., Song, W., Zhang, G., Dang, L., Xie, Y., & Shi, K. et al. (2015). Controllable synthesis of Co_3O_4/polyethyleneimine-carbon nanotubes nanocomposites for CO and NH_3 gas sensing at room temperature. *Journal of Alloys and Compounds*, *639*, 187–196. doi:10.1016/j.jallcom.2015.03.139

Lin, Y.-M., Valdes-Garcia, A., Han, S.-J., Farmer, D. B., Meric, I., Sun, Y., & Avouris, P. (2011). Wafer-scale graphene integrated circuit. *Science*, *332*(6035), 1294–1297. doi:10.1126/science.1204428 PMID:21659599

Li, T., Yang, M., & Li, H. (2011). Label-free electrochemical detection of cancer marker based on graphene–cobalt hexacyanoferrate nanocomposite. *Journal of Electroanalytical Chemistry*, *655*(1), 50–55. doi:10.1016/j.jelechem.2011.02.009

Liu, C. C., Jiang, H., & Yao, Y. (2011). Low-energy effective Hamiltonian involving spin-orbit coupling in silicene and two-dimensional germanium and tin. *Physical Review B: Condensed Matter and Materials Physics*, *84*(19), 195430. doi:10.1103/PhysRevB.84.195430

Liu, D. M., Aguilar-Hernandez, J., Potje-Kamloth, K., & Liess, H. D. (1997). A new carbon monoxide sensor using a polypyrrole film grown on an interdigital-capacitor substrate. *Sensors and Actuators. B, Chemical*, *41*(1-3), 203–206. doi:10.1016/S0925-4005(97)80296-1

Liu, F., Chu, X., Dong, Y., Zhang, W., Sun, W., & Shen, L. (2013). Acetone gas sensors based on graphene-$ZnFe_2O_4$ composite prepared by solvothermal method. *Sensors and Actuators. B, Chemical*, *188*, 469–474. doi:10.1016/j.snb.2013.06.065

Liu, H., Liu, Y., & Zhu, D. (2011). Chemical doping of graphene. *Journal of Materials Chemistry*, *21*(10), 3335–3345. doi:10.1039/C0JM02922J

Liu, H., Wang, G., Chen, D., Zhang, W., Li, C., & Fang, B. (2008). Fabrication of polythionine/NPAu/MWNTs modified electrode for simultaneous determination of adenine and guanine in DNA. *Sensors and Actuators. B, Chemical*, *128*(2), 414–421. doi:10.1016/j.snb.2007.06.028

Liu, J., Fu, S., Yuan, B., Li, Y., & Deng, Z. (2010). Toward a universal "adhesive nanosheet" for the assembly of multiple nanoparticles based on a protein-induced reduction/decoration of graphene oxide. *Journal of the American Chemical Society*, *132*(21), 7279–7281. doi:10.1021/ja100938r PMID:20462190

Liu, S., & Guo, X. (2012). Carbon nanomaterials field-effect-transistor-based biosensors. *NPG Asia Materials*, *4*(8), e23. doi:10.1038/am.2012.42

Liu, S., Tian, J., Wang, L., & Sun, X. (2011). A method for the production of reduced graphene oxide using benzylamine as a reducing and stabilizing agent and its subsequent decoration with Ag nanoparticles for enzymeless hydrogen peroxide detection. *Carbon*, *49*(10), 3158–3164. doi:10.1016/j.carbon.2011.03.036

Liu, S., Wang, Z., Zhang, Y., Zhang, C., & Zhang, T. (2015). High performance room temperature NO_2 sensors based on reduced graphene oxide-multiwalled carbon nanotubes-tin oxide nanoparticles hybrids. *Sensors and Actuators. B, Chemical*, *211*, 318–324. doi:10.1016/j.snb.2015.01.127

Liu, S., Zhao, Q., Xu, J., Yan, K., Peng, H., & Yang, F. et al. (2012). Fast and controllable fabrication of suspended graphene nanopore devices. *Nanotechnology*, *23*(8), 085301–085301. doi:10.1088/0957-4484/23/8/085301 PMID:22293107

Liu, Z., Robinson, J. T., Sun, X., & Dai, H. (2008). PEGylated Nanographene Oxide for Delivery of Water-Insoluble Cancer Drugs. *Journal of the American Chemical Society*, *130*(33), 10876–10877. doi:10.1021/ja803688x PMID:18661992

Li, X. S., Cai, W. W., Jung, I. H., An, J. H., Yang, D. X., Velamakanni, A., & Ruoff, R. S. (2009). Synthesis, characterization, and properties of large-area graphene films. *Graphene and Emerging Materials for Post-Cmos Applications*, *19*(5), 41–52. doi:10.1149/1.3119526

Li, X., Wang, X., Zhang, L., Lee, S., & Dai, H. (2008). Chemically derived, ultrasmooth graphene nanoribbon semiconductors. *Science*, *319*(5867), 1229–1232. doi:10.1126/science.1150878 PMID:18218865

Li, Y., Wang, H. C., Cao, X. H., Yuan, M. Y., & Yang, M. J. (2008). A composite of polyelectrolyte grafted multi-walled carbon nanotubes and in situ polymerized polyaniline for the detection of low concentration triethylamine vapor. *Nanotechnology*, *19*(1), 015503. doi:10.1088/0957-4484/19/01/015503 PMID:21730534

Llobet, E. (2013). Gas sensors using carbon nanomaterials: A review. *Sensors and Actuators. B, Chemical*, *179*, 32–45. doi:10.1016/j.snb.2012.11.014

Llobet, E., Vilanova, X., Brezmes, J., López, D., & Correig, X. (2001). Electrical equivalent models of semiconductor gas sensors using PSpice. *Sensors and Actuators. B, Chemical*, *77*(1-2), 275–280. doi:10.1016/S0925-4005(01)00710-9

Lo Giudice, A., Fasolo, F., Durisi, E., Manfredotti, C., Vittone, E., Fizzotti, F., & Rosi, G. et al. (2007). Performances of 4H-SiC Schottky diodes as neutron detectors. *Nuclear Instruments & Methods in Physics Research. Section A, Accelerators, Spectrometers, Detectors and Associated Equipment*, *583*(1), 177–180. doi:10.1016/j.nima.2007.08.241

Lowe, D. (2013). *Electronics components: Diodes- Electronics All-In-One Desk Reference For Dummies*. John Wiley & Sons.

Luan, B., Stolovitzky, G., & Martyna, G. (2012). Slowing and controlling the translocation of DNA in a solid-state nanopore. *Nanoscale*, *4*(4), 1068–1077. doi:10.1039/C1NR11201E PMID:22081018

Luican, A., Li, G. H., Reina, A., Kong, J., Nair, R. R., Novoselov, K. S., & Andrei, E. Y. et al. (2011). Single-layer behavior and its breakdown in twisted graphene layers. *Physical Review Letters, 106*(12). doi. *Artn, 126802*. doi:10.1103/Physrevlett.106.126802

Lundstrom, M. S., & Guo, J. (2006). *Nanoscale Transistors: Device Physics, Modeling and Simulation.* Springer Science+Business Media Inc.

Lundstrom, M., & Guo, J. (2006). *Nanoscale transistors: Device physics, modeling and simulation.* Springer Science & Business Media.

Lu, X., & Chen, Z. (2005). Curved Pi-Conjugation, Aromaticity, and the Related Chemistry of Small Fullerenes (<C_{60}) and Single-Walled Carbon Nanotubes. *Chemical Reviews*, *105*(10), 3643–3696. doi:10.1021/cr030093d PMID:16218563

Lu, Y. H., Wu, R. Q., Shen, L., Yang, M., Sha, Z. D., Cai, Y. Q., & Feng, Y. P. et al. (2009). Effects of edge passivation by hydrogen on electronic structure of armchair graphene nanoribbon and band gap engineering. *Applied Physics Letters*, 94.

MacDiarmid, A. G., & Epstein, A. J. (1995). Secondary doping in polyaniline. *Synthetic Metals*, *69*(1-3), 85–92. doi:10.1016/0379-6779(94)02374-8

MacDonald, R. A., Laurenzi, B. F., Viswanathan, G., Ajayan, P. M., & Stegemann, J. P. (2005). Collagen–carbon nanotube composite materials as scaffolds in tissue engineering. *Journal of Biomedical Materials Research. Part A*, *74*(3), 489–496. doi:10.1002/jbm.a.30386 PMID:15973695

Machala, Z., Marode, E., Laux, C. O., & Kruger, C. H. (2004). DC glow discharges in atmospheric pressure air. *Journal of Advanced Oxidation Technologies*, *7*, 133–137.

Maharana, P. K., Jha, R., & Palei, S. (2014). Sensitivity enhancement by air mediated graphene multilayer based surface plasmon resonance biosensor for near infrared. *Sensors and Actuators. B, Chemical*, *190*, 494–501. doi:10.1016/j.snb.2013.08.089

Mahmood, S., Burhanudin, Z. A., & Hamid, N. H. (2013). Field emission model of CNT based ionization gas sensor. In M. R. Mahmood, T. Soga, M. H. Mamat, Z. Khusaimi, & A. M. Nor (Eds.), *Nanosynthesis and Nanodevice* (pp. 135–143). doi:10.4028/www.scientific.net/AMR.667.135

Mahmoudian, M. R., Alias, Y., Basirum, W. J., & Ebadi, M. (2011). Poly (N-methyl pyrrole) and its copolymer with o-toluidine electrodeposited on steel in mixture of DBSA and oxalic acid electrolytes. *Current Applied Physics*, *11*(3), 368–375. doi:10.1016/j.cap.2010.08.006

Ma, J., Alfè, D., Michaelides, A., & Wang, E. (2009). Stone-Wales defects in graphene and other planar sp(2)-bonded materials. *Physical Review B: Condensed Matter and Materials Physics*, *80*(3), 033407. doi:10.1103/PhysRevB.80.033407

Majzik, Z., Tchalala, M. R., Švec, M., Hapala, P., Enriquez, H., & Kara, A. et al. (2013). Combined AFM and STM measurements of a silicene sheet grown on the Ag(111) surface. *Journal of Physics Condensed Matter*, *25*(22), 225301. doi:10.1088/0953-8984/25/22/225301 PMID:23674193

Mak, K. F., Lui, C. H., Shan, J., & Heinz, T. F. (2009). Observation of an electric-field-induced band gap in bilayer graphene by infrared spectroscopy. *Physical Review Letters*, *102*(25), 256405. doi:10.1103/PhysRevLett.102.256405 PMID:19659105

Mak, K. F., Shan, J., & Heinz, T. F. (2010). Electronic Structure of Few-Layer Graphene: Experimental demonstration of strong dependence on stacking sequence. *Physical Review Letters*, *104*(17), 176404. doi:10.1103/PhysRevLett.104.176404 PMID:20482122

Malak, D., & Akan, O. B. (2012). Molecular communication nanonetworks inside human body. *Nano Communication Networks*, *3*(1), 19–35. doi:10.1016/j.nancom.2011.10.002

Malhotra, B. D., Chaubey, A., & Singh, S. (2006). Prospects of conducting polymers in biosensors. *Analytica Chimica Acta*, *578*(1), 59–74. doi:10.1016/j.aca.2006.04.055 PMID:17723695

Malik, N. R. (1995). *Electronic circuits: Analysis, simulation, and design*. Englewood Cliffs, NJ: Prentice Hall.

Mangeney, C., Lacroix, J., Chane, K., & Aeiyach, S. (2000). Macroscopic modulation of the π-electron density of pendant groups grafted on conductive polymers. *Synthetic Metals*, *108*(3), 237–243. doi:10.1016/S0379-6779(99)00260-X

Manohar, S., Mantz, A. R., Bancroft, K. E., Hui, C. Y., Jagota, A., & Vezenov, D. V. (2008). Peeling single-stranded DNA from graphite surface to determine oligonucleotide binding energy by force spectroscopy. *Nano Letters*, *8*(12), 4365–4372. doi:10.1021/nl8022143 PMID:19368004

Manrao, E. A., Derrington, I. M., Laszlo, A. H., Langford, K. W., Hopper, M. K., & Gillgren, N. et al. (2012). Reading DNA at single-nucleotide resolution with a mutant MspA nanopore and phi29 DNA polymerase. *Nature Biotechnology*, *30*(4), 349–353. doi:10.1038/nbt.2171 PMID:22446694

Mao, S., Lu, G., & Chen, J. (2014). Nanocarbon-based gas sensors: Progress and challenges. *Journal of Materials Chemistry A*, *2*(16), 5573–5579. doi:10.1039/c3ta13823b

Martel, R., Schmidt, T., Shea, H. R., Hertel, T., & Avouris, P. (1998). Single- and multi-wall carbon nanotube field-effect transistors. *Applied Physics Letters*, *73*(17), 2447–2449. doi:10.1063/1.122477

Martina Cheli, G. F. a. G. I. (2009). *A semi-analytical model of bilayer-graphene field effect transistor*. Academic Press.

Martinoia, S. M., & Massobrio, P. (2004). ISFET–neuron junction: Circuit models and extracellular signal simulations. *Biosensors & Bioelectronics*, *19*(11), 1487–1496. doi:10.1016/j.bios.2003.12.003 PMID:15093221

Martoccia, D., Björck, M., Schlepütz, C., Brugger, T., Pauli, S. A., Patterson, B. D., & Willmott, P. R. et al. (2010). Graphene on Ru (0001): A corrugated and chiral structure. *New Journal of Physics*, *12*(4), 043028. doi:10.1088/1367-2630/12/4/043028

Massera, E., La Ferrara, V., Miglietta, M., Polichetti, T., Nasti, I., & Di Francia, G. (2012). Gas sensors based on graphene:Comparison of two different fabrication approaches. *Chimica Oggi-Chemistry Today*, *30*, 29–31.

Massereau-Guilbaud, V., Géraud-Grenier, I., & Plain, A. (2009). Determination of the electron temperature by optical emission spectroscopy in a 13.56 MHz dusty methane plasma: Influence of the power. *Journal of Applied Physics*, *106*(11), 113305, 113305–113307. doi:10.1063/1.3267292

Mastrangelo, C. H. (1999). DNA analysis systems on a chip. *Adv. Sci. Technol*, *26*, 465–476.

Matsuguchi, M., Io, J., Sugiyama, G., & Sakai, Y. (2002). Effect of NH3 gas on the electrical conductivity of polyaniline blend films. *Synthetic Metals*, *128*(1), 15–19. doi:10.1016/S0379-6779(01)00504-5

McEuen, P. L., Fuhrer, M. S., & Park, H. (2002). Single-walled carbon nanotube electronics. *Nanotechnology. IEEE Transactions on*, *1*, 78–85.

McGinn, S., & Gut, I. G. (2012). DNA sequencing – spanning the generations. *New Biotechnology*, (0). doi:10.1016/j.nbt.2012.11.012 PMID:23165096

McKay, K. (1954). Avalanche breakdown in silicon. *Physical Review*, *94*(4), 877–884. doi:10.1103/PhysRev.94.877

Mehta, V. K. (2008). *Principles of electronics*. Chand.

Meixner, H. (2008). *Sensors, micro-and nanosensor technology: Trends in sensor markets* (Vol. 8). John Wiley & Sons.

Mendoza, F., Hernández, D. M., Makarov, V., Febus, E., Weine, B. R., & Morell, G. (2014). Room temperature gas sensors based on tin dioxide-carbon nanotubes composite films. *Sensors and Actuators. B, Chemical*, *190*, 227–233. doi:10.1016/j.snb.2013.08.050

Meng, F.-L., Guo, Z., & Huang, X.-J. (2015). Graphene-based hybrids for chemiresistive gas sensors. *Trends in Analytical Chemistry*, *68*, 37–47. doi:10.1016/j.trac.2015.02.008

Meng, L., Wang, Y., Zhang, L., Du, S., Wu, R., & Li, L. et al. (2013). Buckled Silicene Formation on Ir(111). *Nano Letters*, *13*(2), 685–690. doi:10.1021/nl304347w PMID:23330602

Meshginqalam, B., Toloue, H. A. T., Ahmadi, M. T., & Sabatyan, A. (2016). Engineer-able optical properties of trilayer graphene nanoribbon. *Physica Scripta*, *91*(3), 035802. doi:10.1088/0031-8949/91/3/035802

Meshginqalam, B., Toloue, H., Ahmadi, M. T., Sabatyan, A., Centeno, A., & Ismail, R. (2016). Graphene embedded surface plasmon resonance based sensor prediction model. *Optical and Quantum Electronics*, *48*(6), 1–14. doi:10.1007/s11082-016-0597-8

Mhamed, C. (2005). Synthesis and characterization of poly(thiophene sulfides) prepared via chemically initiated oxidative polymerization methods. *Synthetic Metals*, *155*(3), 474–479. doi:10.1016/j.synthmet.2005.06.023

Miasik, J., Hopper, A., & Tofield, B. (1986). Conducting polymer gas sensors. *Journal of the Chemical Society*, *82*, 1117.

Michalska, A., & Lewenstam, A. (2000). Potentiometric selectivity of p-doped polymer films. *Analytica Chimica Acta*, *406*(2), 159–169. doi:10.1016/S0003-2670(99)00774-6

Mijoe, J. (2015). *Fundamentals of semiconductor physics*. Anchor Academic Publishing.

Mingo, N., & Broido, D. A. (2007). Thermoelectric power factor of nanoporous semiconductors. *Journal of Applied Physics, 101*(1), 014322. doi:10.1063/1.2405232

Min, S. K., Cho, Y., Mason, D. R., Lee, J. Y., & Kim, K. S. (2011). Theoretical design of nanomaterials and nanodevices: Nanolensing, supermagnetoresistance, and ultrafast DNA sequencing. *The Journal of Physical Chemistry C*, *115*(33), 16247–16257. doi:10.1021/jp203748r

Mishra, L. N., Shibata, K., Ito, H., Yugami, N., & Nishida, Y. (2004). Conversion of methane to hydrogen via pulsed corona discharge. *Journal of Natural Gas Chemistry*, *13*, 82–86.

Modi, A., Koratkar, N., Lass, E., Wei, B., & Ajayan, P. (2003). Miniaturized gas ionization sensors using carbon nanotubes. *Nature*, *424*(6945), 171–174. doi:10.1038/nature01777 PMID:12853951

Mohamed, A. H. E. (1966). Nonlinear distortion in common-emitter transistor amplifiers operated in the normal domain. *Electronics Letters*, *2*(2).

Mohammad Javad Kiani, M. T. A. (2012). Analytical modeling of bilayer graphene based biosensor. *Journal of Computational and Theoretical Nanoscience (CTN).*

Mohanty, N., & Berry, V. (2008). Graphene-Based Single-Bacterium Resolution Biodevice and DNA Transistor: Interfacing Graphene Derivatives with Nanoscale and Microscale Biocomponents. *Nano Letters*, *8*(12), 4469–4476. doi:10.1021/nl802412n PMID:19367973

Molle, A., Grazianetti, C., Chiappe, D., Cinquanta, E., Cianci, E., Tallarida, G., & Fanciulli, M. (2013). Hindering the Oxidation of Silicene with Non-Reactive Encapsulation. *Advanced Functional Materials*, *23*(35), 4340–4344. doi:10.1002/adfm.201300354

Monthioux, M., & Kuznetsov, V. (2006). Who should be given the credit for the discovery of carbon nanotubes? *Carbon*, *44*(9), 1621–1623. doi:10.1016/j.carbon.2006.03.019

Moradian, R., Mohammadi, Y., & Ghobadi, N. (2008). Investigation of gas sensing properties of armchair graphene nanoribbons. *Journal of Physics Condensed Matter*, 20.

Morgan, D. (1991). Surface-Wave Devices for Signal Processing. Amsterdam: Academic Press.

Morgan, D. P. (1985). *Surface-wave devices for signal processing* (Vol. 19). Elsevier Publishing Company.

Morgenshtein, A. D. U., Jakobson, C. G., & Nemirovsky, Y. (2004). Wheatstone-Bridge readout interface for ISFET/REFET applications. *Sensors and Actuators. B, Chemical*, *98*(1), 10–10. doi:10.1016/j.snb.2003.07.017

Moriconi, L. N. D., & Niemeyer, D. (2011). Graphene conductivity near the charge neutral point. *Physical Review B: Condensed Matter and Materials Physics*, *84*(19), 193401. doi:10.1103/PhysRevB.84.193401

Morozov, S. V., Novoselov, K. S., Katsnelson, M. I., Schedin, F., Elias, D. C., Jaszczak, J. A., & Geim, A. K. (2008). Giant intrinsic carrier mobilities in graphene and its bilayer. *Physical Review Letters*, *100*(1), 016602. doi:10.1103/PhysRevLett.100.016602 PMID:18232798

Mortazavi, B., Potschke, M., & Cuniberti, G. (2014). Multiscale modeling of thermal conductivity of polycrystalline graphene sheets. *Nanoscale*, *6*(6), 3344–3352. doi:10.1039/c3nr06388g PMID:24518878

Mousavi, H. (2011). Graphene as gas sensors. *Communications in Theoretical Physics*, *56*(2), 373–376. doi:10.1088/0253-6102/56/2/30

Mousavi, S. M., Ahmadi, M. T., Sadeghi, H., Nilghaz, A., Amin, A., Johari, Z., & Ismail, R. (2011). Bilayer graphene nanoribbon carrier statistic in degenerate and non degenerate limit. *Journal of Computational and Theoretical Nanoscience*, *8*(10), 2029–2032. doi:10.1166/jctn.2011.1921

Mousavi, S. M., Ahmadi, M. T., Webb, J. F., Sadeghi, H., Nilghaz, A., Amin, A., & Ismail, R. et al. (2011). Bilayer graphene nanoribbon carrier statistics in the degenerate regime. In *Proceedings of the Fourth Global Conference on Power Control and Optimization*. doi:10.1063/1.3592463

Moyo, M., Okonkwo, J. O., & Agyei, N. M. (2012). Recent advances in polymeric materials used as electron mediators and immobilizing matrices in developing enzyme electrodes. *Sensors (Basel, Switzerland)*, *12*(1), 923–953. doi:10.3390/s120100923 PMID:22368503

Murali, R. (2012). *Graphene nanoelectronics: From materials to circuits*. Springer. doi:10.1007/978-1-4614-0548-1

Mustafa, M. G., & Tierney, D. F. (1978). Biochemical and metabolic changes in the lung with oxygen, ozone, and nitrogen dioxide toxicity. *The American Review of Respiratory Disease*, *118*, 1061–1090. PMID:369411

Myung, S., Solanki, A., Kim, C., Park, J., Kim, K. S., & Lee, K.-B. (2011). Graphene-encapsulated nanoparticle-based biosensor for the selective detection of cancer biomarkers. *Advanced Materials*, *23*(19), 2221–2225. doi:10.1002/adma.201100014 PMID:21469221

Naeemi, A., & Meindl, J. D. (2007). Conductance modeling for graphene nanoribbon (GNR) interconnects. *IEEE Electron Device Letters*, *28*(5), 428–431. doi:10.1109/LED.2007.895452

Nakamura, K., Kazumi, M., & Shimizu, H. (1977). *SH-type and Rayleigh-type surface waves on rotated Y-cut LiTaO 3*. Paper presented at the Ultrasonics Symposium. doi:10.1109/ULTSYM.1977.196951

Nakano, H., Mitsuoka, T., Harada, M., Horibuchi, K., Nozaki, H., & Takahashi, N. et al. (2006). Soft synthesis of single-crystal silicon monolayer sheets. *Angewandte Chemie International Edition*, *45*(38), 6303–6306. doi:10.1002/anie.200600321 PMID:16941712

Ndlovu, T., Arotiba, O. A., Krause, R. W., & Mamba, B. B. (2010). Electrochemical detection of o-nitrophenol on a poly (propyleneimine)-gold nanocomposite modified glassy carbon electrode. *International Journal of Electrochemical Science*, *5*, 1179–1186.

Neamen, D. A. (2003). *Semiconductor physics and devices: Basic principles* (3rd ed.). McGraw-Hill Higher Education.

Neamen, D. A. (2011). *Semiconductor physics and devices*. University of New Mexico.

Neamen, D. A. (2011). *Semiconductor Physics and Devices*. University of New Mexico.

Nelson, T., Zhang, B., & Prezhdo, O. V. (2010). Detection of Nucleic Acids with Graphene Nanopores: Ab Initio Characterization of a Novel Sequencing Device. *Nano Letters*, *10*(9), 3237–3242. doi:10.1021/nl9035934 PMID:20722409

Nguyen, V. C., & Potje-Kamloth, K. (1999). Electrical and chemical sensing properties of doped polypyrrole/gold Schottky barrier diodes. *Thin Solid Films*, *338*(1-2), 142–148. doi:10.1016/S0040-6090(98)01060-8

Nicolay, P., Elmazria, O., Sarry, F., Bouvot, L., Kambara, H., Singh, K. J., & Alnot, P. (2009). *SAW pressure sensor for vacuum control applications*. Paper presented at the Frequency Control Symposium, 2009 Joint with the 22nd European Frequency and Time forum. doi:10.1109/FREQ.2009.5168151

Nika, D. L., & Balandin, A. A. (2012). Two-dimensional phonon transport in graphene. *Journal of Physics Condensed Matter*, *24*(23), 233203. doi:10.1088/0953-8984/24/23/233203 PMID:22562955

Nilsson, J. (2006) Electronic properties of graphene multilayers. *Physical Review Letters*. doi:10.1103/PhysRevLett.97.266801

Nilsson, J., Castro Neto, A. H., Guinea, F., & Peres, N. M. R. (2007). Transmission through a biased graphene bilayer barrier. *Physical Review B: Condensed Matter and Materials Physics*, *76*(16), 165416–165410. doi:10.1103/PhysRevB.76.165416

Nilsson, J., Neto, A. C., Guinea, F., & Peres, N. (2008). Electronic properties of bilayer and multilayer graphene. *Physical Review B: Condensed Matter and Materials Physics*, *78*(4), 045405. doi:10.1103/PhysRevB.78.045405

Ni, Z., Liu, Q., Tang, K., Zheng, J., Zhou, J., & Qin, R. et al. (2011). Tunable Bandgap in Silicene and Germanene. *Nano Letters*, *12*(1), 113–118. doi:10.1021/nl203065e PMID:22050667

Novoselov, K. S., Geim, A. K., Morozov, S. V., Jiang, D., Katsnelson, M. I., Grigorieva, I. V., & Firsov, A. A. et al. (2005). Two-dimensional gas of massless Dirac fermions in graphene. *Nature*, *438*(7065), 197–200. doi:10.1038/nature04233 PMID:16281030

Novoselov, K. S., Geim, A. K., Morozov, S. V., Jiang, D., Zhang, Y., Dubonos, S. V., & Firsov, A. A. et al. (2004). Electric Field Effect in Atomically Thin Carbon Films. *Science*, *306*(5696), 666–669. doi:10.1126/science.1102896 PMID:15499015

Novoselov, K. S., Jiang, D., Schedin, F., Booth, T. J., Khotkevich, V. V., Morozov, S. V., & Geim, A. K. (2005). Two-dimensional atomic crystals. *Proceedings of the National Academy of Sciences of the United States of America*, *102*(30), 10451–10453. doi:10.1073/pnas.0502848102 PMID:16027370

Novoselov, K. S., Jiang, Z., Zhang, Y., Morozov, S., Stormer, H., Zeitler, U., & Geim, A. (2007). Room-temperature quantum Hall effect in graphene. *Science*, *315*(5817), 1379–1379. doi:10.1126/science.1137201 PMID:17303717

Novoselov, K., Fal, V., Colombo, L., Gellert, P., Schwab, M., & Kim, K. (2012). A roadmap for graphene. *Nature*, *490*, 192–200. doi:10.1038/nature11458 PMID:23060189

Nozaki, T. (2011). *Electron Driven Microchemistry for Green Technology: Direct Conversion of Methane to Methanol*. Academic Press.

Nozaki, T., Unno, Y., Miyazaki, Y., & Okazaki, K. (2001). Optical diagnostics for determining gas temperature of reactive microdischarges in a methane-fed dielectric barrier discharge. *Journal of Physics. D, Applied Physics*, *34*(16), 2504–2511. doi:10.1088/0022-3727/34/16/316

O'Hare, A., Kusmartsev, F. V., & Kugel, K. I. (2012). A Stable "Flat" Form of Two-Dimensional Crystals: Could Graphene, Silicene, Germanene Be Minigap Semiconductors? *Nano Letters*, *12*(2), 1045–1052. doi:10.1021/nl204283q PMID:22236130

Oberlin, A., Endo, M., & Koyama, T. (1976). Filamentous growth of carbon through benzene decomposition. *Journal of Crystal Growth*, *32*(3), 335–349. doi:10.1016/0022-0248(76)90115-9

Ocaya, R. O. (2013, January 18). *Circuit's RMS output is linearly proportional to temperature over wide range*. Retrieved from http://www.edn.com/design/components-and-packaging/4405267/Circuit-s-RMS-output-is-linearly-proportional-to-temperature-over-wide-range

Ohno, Y., Maehashi, K., Yamashiro, Y., & Matsumoto, K. (2009). Electrolyte-gated graphene field-effect transistors for detecting pH protein adsorption. *Nano Letters*, *9*(9), 3318–3322. doi:10.1021/nl901596m PMID:19637913

Olthuis, W., Faber, Krommenhoek, & van den Berg. (2007). *Sensing with FETs-once, now and future*. Academic Press.

Ong, K. G., Zeng, K., & Grimes, C. A. (2002). A wireless, passive carbon nanotube-based gas sensor. *IEEE Sensors Journal*, *2*(2), 82–88. doi:10.1109/JSEN.2002.1000247

Osborn, T. H., Farajian, A. A., Pupysheva, O. V., Aga, R. S., & Voon, L. C. L. Y. (2011). Ab initio simulations of silicene hydrogenation. *Chemical Physics Letters*, *511*(1-3), 101–105. doi:10.1016/j.cplett.2011.06.009

Ouyang, Y., Dai, H., & Guo, J. (2009). Multilayer Graphene Nanoribbon for 3D Stacking of the Transistor Channel. *IEEE International Electron Devices Meeting*.

Ouyang, Y., Yoon, Y., & Guo, J. (2007). Scaling behaviors of graphene nanoribbon FETs: A three-dimensional quantum simulation study. *Electron Devices. IEEE Transactions on*, *54*(9), 2223–2231. doi:10.1109/TED.2007.902692

Ovchinnikov, V., & Shevchenko, A. (2013). Self-organization-based fabrication of stable noble-metal nanostructures on large-area dielectric substrates. *Journal of Chemistry*, *10*. doi:10.1155/2013/158431

Owens, F. J. (2008). Electronic and magnetic properties of armchair and zigzag graphene nanoribbons. *The Journal of Chemical Physics*, *128*(19), 194701. doi:10.1063/1.2905215 PMID:18500880

Padova, P. D., Leandri, C., Vizzini, S., Quaresima, C., Perfetti, P., & Olivieri, B. et al. (2008). Burning Match Oxidation Process of Silicon Nanowires Screened at the Atomic Scale. *Nano Letters*, *8*(8), 2299–2304. doi:10.1021/nl800994s PMID:18624391

Padova, P. D., Vogt, P., Resta, A., Avila, J., Razado-Colambo, I., & Quaresima, C. et al. (2013). Evidence of Dirac fermions in multilayer silicene. *Applied Physics Letters*, *102*(16), 163106. doi:10.1063/1.4802782

Padova, P., Quaresima, C., Olivieri, B., Perfetti, P., & Le Lay, G. (2011). Strong resistance of silicene nanoribbons towards oxidation. *Journal of Physics. D, Applied Physics*, *44*(31), 312001. doi:10.1088/0022-3727/44/31/312001

Pallas-Areny, R., J. G. Webster & R. Areny. (2001). *Sensors and signal conditioning*. Wiley.

Pan, L., Liu, H. J., Tan, X. J., Lv, H. Y., Shi, J., Tang, X. F., & Zheng, G. (2012a). Thermoelectric properties of armchair and zigzag silicene nanoribbons. *Physical Chemistry Chemical Physics*, *14*(39), 13588–13593. doi:10.1039/c2cp42645e PMID:22965156

Pan, L., Liu, H. J., Wen, Y. W., Tan, X. J., Lv, H. Y., Shi, J., & Tang, X. F. (2012b). Magnetic and electronic properties of silicane with hydrogen vacancies on the surface. *Applied Surface Science*, *258*(24), 10135–10139. doi:10.1016/j.apsusc.2012.06.092

Pantano, A., Parks, D. M., & Boyce, M. C. (2004). Mechanics of deformation of single-and multi-wall carbon nanotubes. *Journal of the Mechanics and Physics of Solids*, *52*(4), 789–821. doi:10.1016/j.jmps.2003.08.004

Panzer, M. A., Zhang, Mann, Hu, Pop, Dai, & Goodson. (2008). Thermal properties of metal-coated vertically aligned single-wall nanotube arrays. *Journal of Heat Transfer-Transactions of the Asme, 130.*

Park, H., Afzali, A., Han, S.-J., Tulevski, G. S., Franklin, A. D., Tersoff, J., & Haensch, W. (2012). High-density integration of carbon nanotubes via chemical self-assembly. *Nature Nanotechnology*, *7*(12), 787–791. doi:10.1038/nnano.2012.189 PMID:23103933

Passlack, M. (2008). III-V metal-oxide-semiconductor technology. *2008 IEEE 20th International Conference on Indium Phosphide and Related Materials (IPRM).*

Passlack, M. (2008). III-V Metal-oxide-semiconductor technology. *2008 IEEE 20th International Conference on Indium Phosphide and Related Materials (IPRM).*

Patacsil, C., Malapit, & Ramos. (2006). Optical emission spectroscopy of low temperature CVD diamond. *J Plasma Fusion Res Ser*, *7*, 145–149.

Patil, S., Mahajan, J. R., More, M. A., & Patil, P. P. (1999). Electrochemical synthesis of poly(o-methoxyaniline) thin films: Effect of post treatment. *Materials Chemistry and Physics*, *58*(1), 31–36. doi:10.1016/S0254-0584(98)00245-4

Patolsky, F., Timko, B. P., Yu, G., Fang, Y., Greytak, A. B., Zheng, G., & Lieber, C. M. (2006). Detection, stimulation, and inhibition of neuronal signals with high-density nanowire transistor arrays. *Science*, *313*(5790), 1100–1104. doi:10.1126/science.1128640 PMID:16931757

Patolsky, F., Zheng, G., & Lieber, C. M. (2006). Fabrication of silicon nanowire devices for ultrasensitive, label-free, real-time detection of biological and chemical species. *Nature Protocols*, *1*(4), 1711–1724. doi:10.1038/nprot.2006.227 PMID:17487154

Patra, M. K., Manzoor, K., Manoth, M., Negi, S. C., Vadera, S. R., & Kumar, N. (2008). Nanotechnology applications for chemical and biological sensors. *Defence Science Journal*, *58*(5), 636–649. doi:10.14429/dsj.58.1686

Patterson, D. W. (1998). *Artificial neural networks: theory and applications*. Prentice Hall PTR.

Patterson, D. W. (1998). *Artificial neural networks: Theory and applications*. Prentice Hall PTR.

Pearce, R., Iakimov, T., Andersson, M., Hultman, L., Spetz, A. L., & Yakimova, R. (2011). Epitaxially grown graphene based gas sensors for ultra sensitive NO2 detection. *Sensors and Actuators. B, Chemical*, *155*(2), 451–455. doi:10.1016/j.snb.2010.12.046

Peng, Y., & Li. (2013). Ammonia adsorption on graphene and graphene oxide: A first-principles study. *Frontiers of Environmental Science & Engineering*, 1-9.

Peng, H., Zhang, L. J., Soeller, C., & Travas-Sejdic, J. (2009). Conducting polymers for electrochemical DNA sensing. *Biomaterials*, *30*(11), 2132–2148. doi:10.1016/j.biomaterials.2008.12.065 PMID:19147223

Peng, N., Zhang, Q., Chow, C. L., Tan, O. K., & Marzari, N. (2009). Sensing mechanisms for carbon nanotube based NH3 gas detection. *Nano Letters*, *9*(4), 1626–1630. doi:10.1021/nl803930w PMID:19281216

Peng, N., Zhang, Q., Lee, Y. C., Tan, O. K., & Marzani, N. (2008). Gate modulation in carbon nanotube field effect transistors-based NH_3 gas sensors. *Sensors and Actuators. B, Chemical*, *132*(1), 191–195. doi:10.1016/j.snb.2008.01.025

Penza, M., Antolini, F., & Antisari, M. (2004). Carbon nanotubes as SAW chemical sensors materials. *Sensors and Actuators. B, Chemical*, *100*(1), 47–59. doi:10.1016/j.snb.2003.12.019

Penza, M., Cassano, G., Rossi, R., Alvisi, M., Rizzo, A., Signore, M. A., & Giorgi, R. et al. (2007b). Enhancement of sensitivity in gas chemiresistors based on carbon nanotube surface functionalized with noble metal (Au, Pt) nanoclusters. *Applied Physics Letters*, *90*(17), 173123. doi:10.1063/1.2722207

Penza, M., Cassano, G., Rossi, R., Rizzo, A., Signore, M. A., Alvisi, M., & Giorgi, R. et al. (2007a). Effect of growth catalysts on gas sensitivity in carbon nanotube film based chemiresistive sensors. *Applied Physics Letters*, *90*(10), 103101. doi:10.1063/1.2456258

Penza, M., Rossi, R., Alvisi, M., Cassano, G., Signore, M. A., Serra, E., & Giorgi, R. (2008). Pt- and Pd-nanoclusters functionalized carbon nanotubes networked films for sub-ppm gas sensors. *Sensors and Actuators. B, Chemical*, *135*(1), 289–297. doi:10.1016/j.snb.2008.08.024

Penza, M., Rossi, R., Alvisi, M., Signore, M. A., Serra, E., Paolesse, R., & Di Natale, C. et al. (2010). Metalloporphyrins-modified carbon nanotubes networked films-based chemical sensors for enhanced gas sensitivity. *Sensors and Actuators. B, Chemical*, *144*(2), 387–394. doi:10.1016/j.snb.2008.12.060

Perdew, J. P., Burke, K., & Ernzerhof, M. (1996). Generalized Gradient Approximation Made Simple. *Physical Review Letters*, *77*(18), 3865–3868. doi:10.1103/PhysRevLett.77.3865 PMID:10062328

Peres, N., Neto, A. H. C., & Guinea, F. (2006). Conductance quantization in mesoscopic graphene. *Physical Review B: Condensed Matter and Materials Physics*, *73*(19), 195411. doi:10.1103/PhysRevB.73.195411

Pesetski, A. A., Baumgardner, J. E., Folk, E., Przybysz, J. X., Adam, J. D., & Zhang, H. (2006). Carbon nanotube field-effect transistor operation at microwave frequencies. *Applied Physics Letters*, 88.

Pham, M. T. H., Kunath, S., Kurth, C., & Köhler, E. (1994). Backside membrane structures for ISFETs applied in miniature analysis systems. *Sensors and Actuators. B, Chemical*, *19*(1–3), 333–335. doi:10.1016/0925-4005(93)00985-8

Phan, A. D., & Viet, N. A. (2012). A new type of optical biosensor from DNA wrapped semiconductor graphene ribbons. *Journal of Applied Physics*, *111*(11). doi: 11470310.1063/1.4728196

Picaud, F., Girardet, C., & Rao, A. M. (2009). A comparative study of single- and multiwalled carbon nanotube sensitivity to ammonia. *Journal of Applied Physics*, *105*(1), 014315. doi:10.1063/1.3060242

Pierobon, M., Jornet, J. M., Akkari, N., Almasri, S., & Akyildiz, I. F. (2014). A routing framework for energy harvesting wireless nanosensor networks in the Terahertz Band. *Wireless News*, *20*(5), 1169–1183. doi:10.1007/s11276-013-0665-y

Pieterse, C. M., & Dicke, M. (2007). Plant interactions with microbes and insects: From molecular mechanisms to ecology. *Trends in Plant Science*, *12*(12), 564–569. doi:10.1016/j.tplants.2007.09.004 PMID:17997347

Pijanowska, D., & Torbicz, W. (1997). *Simple method of enzyme immobilization for pH-ISFET-based urea biosensors.* doi: 10.1117/12.266713

Pike, A., & Gardner, J. W. (1997). Thermal modelling and characterisation of micropower chemoresistive silicon sensors. *Sensors and Actuators. B, Chemical*, *45*(1), 19–26. doi:10.1016/S0925-4005(97)00261-X

Piro, G., Grieco, L. A., Boggia, G., & Camarda, P. (2013). Simulating wireless nano sensor networks in the NS – 3 platform. *IEEE Advanced Information Networking and Applications Workshops (WAINA).*

Pirsa, S. (in press). Fast Determination of water content of some organic solvents by smart sensor based on PPy-Ag nanocomposite. *Nanoscience & Nanotechnology-Asia.*

Pirsa, S., Afshar Asl, A., Khani, A., & Allahverdi Pur, A. (2013). Fabrication of 1, 1-dimethylhydrazine gas sensor based on nano structure conducting polyaniline. *Journal of Sciences. Islamic Republic of Iran*, *24*, 209–215.

Pirsa, S., & Alizadeh, N. (2010). Design and fabrication of gas sensor based on nanostructure conductive polypyrrole for determination of volatile organic solvents. *Sensors and Actuators. B, Chemical*, *147*(2), 461–466. doi:10.1016/j.snb.2010.03.026

Pirsa, S., & Alizadeh, N. (2011). Nanoporous conducting polypyrrole gas sensor coupled to a gas chromatograph for determination of aromatic hydrocarbons using dispersive liquid–liquid microextraction method. *IEEE Sensors Journal*, *11*(12), 3400–3405. doi:10.1109/JSEN.2011.2159970

Pirsa, S., & Alizadeh, N. (2012). A selective DMSO gas sensor based on nanostructured conducting polypyrrole doped with sulfonate anion. *Sensors and Actuators. B, Chemical*, *168*, 303–309. doi:10.1016/j.snb.2012.04.027

Pirsa, S., Alizadeh, N., Zandi, M., Almasi, H., & Heidari, H. (2014). Chemically synthesize nanostructure polypyrrole derivatives and fabrication of gas sensor based on synthesized polymers. *Sensor Letters*, *12*(12), 1–8. doi:10.1166/sl.2014.3386

Pirsa, S., Heidari, H., & Lotfi, J. (2016). Design selective gas sensors based on nano sized polypyrrole/ polytetrafluoroethylene and polypropylene membranes. *IEEE Sensors Journal*, *16*(9), 2922–2928. doi:10.1109/JSEN.2016.2527712

Pirsa, S., Zandi, M., Almasi, H., & Hasanlu, S. (2015). Selective hydrogen peroxide gas sensor based on nanosized polypyrrole modified by CuO nanoparticles. *Sensor Letters*, *13*(7), 1–6. doi:10.1166/sl.2015.3506

Pitcher, G. (2011). Scientists create silicene. *New Electronics*, *44*(7), 8–8.

Pokhrel, S., Joo, J. C., Kim, Y. H., & Yoo, Y. J. (2012). Rational design of a Bacillus circulans xylanase by introducing charged residue to shift the pH optimum. *Process Biochemistry*, *47*(12), 2487–2493. doi:10.1016/j.procbio.2012.10.011

Poli, R., Kennedy, J., & Blackwell, T. (2007). Particle swarm optimization. *Swarm Intelligence*, *1*(1), 33–57. doi:10.1007/s11721-007-0002-0

Popadic, M., Lorito, G., & Nanver, L. K. (2009). Analytical model of I–V characteristics of arbitrarily shallow p-n junctions. *IEEE Transactions on Electron Devices*, *56*(1), 116–125. doi:10.1109/TED.2008.2009028

Postma, H. W. C., Teepen, T., Yao, Z., Grifoni, M., & Dekker, C. (2001). Carbon nanotube single-electron transistors at room temperature. *Science*, *293*(5527), 76–79. doi:10.1126/science.1061797 PMID:11441175

Potyrailo, R. A., & Mirsky, V. M. (2008). Combinatorial and high-throughput development of sensing materials. *Chemical Reviews*, *108*(2), 770–813. doi:10.1021/cr068127f PMID:18211102

Pradeep Kumar, M., Triranjita, S., & Rajan, J. (2014). Low index dielectric mediated surface plasmon resonance sensor based on graphene for near infrared measurements. *Journal of Physics. D, Applied Physics*, *47*(38), 385102. doi:10.1088/0022-3727/47/38/385102

Pramanik, A., & Seok Kang, H. (2011). Density functional theory study of O_2 and NO adsorption on heteroatom-doped graphenes including the van der waals interaction. *PhysChemComm*, *115*, 10971–10978.

Prause, C. A. (2006). *External detection and measurement of inhaled radionuclides using thermoluminescent dosimeters.* (Master's thesis). Texas A&M University.

Pu, H. H., Rhim, S. H., Gajdardziksa-Josifovska, M., Hirschmugl, C. J., Weinert, M., & Chen, J. H. (2014). A statistical thermodynamics model for monolayer gas adsorption on graphene-based materials: Implications for gas sensing applications. *RSC Advances*, *4*(88), 47481–47487. doi:10.1039/C4RA08124B

Putzbach, W., & Ronkainen, N. J. (2013). Immobilization techniques in the fabrication of nanomaterial-based electrochemical biosensors: A review. *Sensors (Basel, Switzerland)*, *13*(4), 4811–4840. doi:10.3390/s130404811 PMID:23580051

Qi, P., Vermesh, O., Grecu, M., Javey, A., Wang, Q., Dai, H., & Cho, K. (2003). Toward large arrays of multiplex functionalized carbon nanotube sensors for highly sensitive and selective molecular detection. *Nano Letters*, *3*(3), 347–351. doi:10.1021/nl034010k

Qiu, Y., Qu, X., Dong, J., Ai, S., & Han, R. (2011). Electrochemical detection of DNA damage induced by acrylamide and its metabolite at the graphene-ionic liquid-Nafion modified pyrolytic graphite electrode. *Journal of Hazardous Materials*, *190*(1-3), 480–485. doi:10.1016/j.jhazmat.2011.03.071 PMID:21497017

Rahmani, M., Ahmadi, M. T., Karimi, F. A., Saeidmanesh, M., Akbari, E., & Ismail, R. (2013). *Analytical Modeling of Trilayer Graphene Nanoribbon Schottky-Barrier FET for High Speed*. Academic Press.

Rahmani, M. (2012). *Advanced Nanoelectronics: Trilayer Graphene Nanoribbon Field Effect Transistor Modeling*. Boca Raton, FL: Taylor and Francis.

Rahmani, M., Ahmadi, M. T., Abadi, H. K. F., Saeidmanesh, M., Akbari, E., & Ismail, R. (2013). Analytical modeling of trilayer graphene nanoribbon Schottky-barrier FET for high-speed switching applications. *Nanoscale Research Letters*, *8*(1), 1–13. doi:10.1186/1556-276X-8-55 PMID:23363692

Rahmani, M., Ahmadi, M. T., Ghadiry, M. H., Anwar, S., & Ismail, R. (2012). Performance of Bilayer Graphene Nanoribbon Schottky Diode in Comparison with Conventional Diodes. *Journal of Computational and Theoretical Nanoscience*, *9*, 1618–1621. doi:10.1166/jctn.2012.2254

Rahmani, M., Ahmadi, M. T., Ghadity, M. H., Anwar, S., & Ismail, R. (2011). the effect of applied voltage on the carrier effective mass in ABA trilayer graphene nanoribbon. *Journal of Computational and Theoretical Nanoscience*, *9*, 1–4.

Rahmani, M., Ahmadi, M. T., Ismail, R., & Ghadiry, M. H. (2013). Performance of Bilayer Graphene Nanoribbon Schottky Diode in Comparison with Conventional Diodes. *Journal of Computational and Theoretical Nanoscience*, *10*(2), 323–327. doi:10.1166/jctn.2013.2699

Rahmani, M., Ahmadi, M. T., Karimi, F. A., Kiani, M. J., Akbari, E., & Ismail, R. (2013). Analytical Modeling of Monolayer Graphene-based NO_2 Sensor. *Sensor Letters*, *11*(2), 270–275. doi:10.1166/sl.2013.2742

Rahmani, M., Ismail, R., Ahmadi, M. T., & Ghadiry, M. H. (2013). Quantum Confinement Effect on Trilayer Graphene Nanoribbon Carrier Concentration. *Journal of Experimental Nanoscience.*, *794309*. doi:10.1080/17458080

Rahmani, M., Ismail, R., Ahmadi, M. T., Kiani, M. J., Saeidmanesh, M., Karimi, F., & Rahmani, K. et al. (2013). The effect of bilayer graphene nanoribbon geometry on Schottky-Barrier diode performance. *Journal of Nanomaterials.*

Ramanavicius, A., Ramanaviciene, A., & Malinauskas, A. (2006). Electrochemical sensors based on conducting polymer-polypyrrole. *Electrochimica Acta*, *51*(27), 6025–6037. doi:10.1016/j.electacta.2005.11.052

Ram, M. K., Yavuz, O., & Aldissi, M. (2005). NO_2 gas sensing based on ordered ultrathin films of conducting polymer and its nanocomposite. *Synthetic Metals*, *151*(1), 77–84. doi:10.1016/j.synthmet.2005.03.021

Rana, F., George, P. A., Strait, J. H., Dawlaty, J., Shivaraman, S., Chandrashekhar, M., & Spencer, M. G. (2009). Carrier recombination and generation rates for intravalley and intervalley phonon scattering in graphene. *Physical Review B: Condensed Matter and Materials Physics*, *79*(11), 115447. doi:10.1103/PhysRevB.79.115447

Rao, C. N. R., & Sood. (2013). *Graphene: Synthesis, properties, and phenomena*. Wiley.

Rao, C. N. R., & Voggu, R. (2010). Charge-transfer with graphene and nanotubes. *Materials Today*, *13*(9), 34–40. doi:10.1016/S1369-7021(10)70163-2

Rao, Y. L., & Zhang, G. (2006). Enhancing the sensitivity of SAW sensors with nanostructures. *Current Nanoscience*, *2*(4), 311–318. doi:10.2174/157341306778699374

Rapp, M., & Rupp, S. (2007). New sensing materials for SAW sensors using polymer-silicate composites. *IEEE Sensors,* (pp. 155-157).

Ratcliffe, N. M. (1990). Polypyrrole-based sensor for hydrazine and ammonia. *Analytica Chimica Acta*, *239*, 257–262. doi:10.1016/S0003-2670(00)83859-3

Ratinac, K. R., Yang, W., Ringer, S. P., & Braet, F. (2010). Toward ubiquitous environmental gas sensors capitalizing on the promise of graphene. *Environmental Science & Technology*, *44*(4), 1167–1176. doi:10.1021/es902659d PMID:20099803

Recher, P., & Trauzettel, B. (2011). A defect controls transport in graphene. *Physics*, *4*, 25. doi:10.1103/Physics.4.25

Redondo, M. I., Sanchez de la Blanca, E., Garcı'a, M. V., Raso, M. A., Tortajada, J., & Gonzalez-Tejera, M. J. (2001). FTIR study of chemically synthesized poly (N-methylpyrrole). *Synthetic Metals*, *122*(2), 431–435. doi:10.1016/S0379-6779(00)00563-4

Reeder, T. M., Cullen, D. E., & Gilden, M. (1975). *SAW oscillator pressure sensors*. Paper presented at the 1975 Ultrasonics Symposium. doi:10.1109/ULTSYM.1975.196515

Reich, S., Maultzsch, J., Thomsen, C., & Ordejon, P. (2002). Tight-binding description of graphene. *Physical Review B: Condensed Matter and Materials Physics*, *66*(3), 035412. doi:10.1103/PhysRevB.66.035412

Reina, A., Jia, X., Ho, J., Nezich, D., Son, H., Bulovic, V., & Kong, J. et al. (2008). Large area, few-layer graphene films on arbitrary substrates by chemical vapor deposition. *Nano Letters*, *9*(1), 30–35. doi:10.1021/nl801827v PMID:19046078

Reina, A., Thiele, S., Jia, X., Bhaviripudi, S., Dresselhaus, M. S., Schaefer, J. A., & Kong, J. (2009). Growth of large-area single-and bi-layer graphene by controlled carbon precipitation on polycrystalline Ni surfaces. *Nano Research*, *2*(6), 509–516. doi:10.1007/s12274-009-9059-y

Reinhoudt, D. N. S., & Ernst, J. R. (1990). The transduction of host-guest interactions into electronic signals by molecular systems. *Advanced Materials*, *2*(1), 23–32. doi:10.1002/adma.19900020105

Ricco, A. J., & Martin, S. J. (1991). Thin metal film characterization and chemical sensors: Monitoring electronic conductivity, mass loading and mechanical properties with surface acoustic wave devices. *Thin Solid Films*, *206*(1), 94–101. doi:10.1016/0040-6090(91)90399-I

Ricco, A. J., Martin, S. J., & Zipperian, T. E. (1985). Surface acoustic wave gas sensor based on film conductivity changes. *Sensors and Actuators*, *8*(4), 319–333. doi:10.1016/0250-6874(85)80031-7

Richard, S., Rajasekhar, M., & Subramania, A. (2009). Synthesis of polythiophene nanoparticles by surfactant assisted dilute polymerization method for high performance redox supercapacitors. *International Journal of Electrochemical Science*, *4*, 1289–1301.

Rigoni, F., Drera, G., Pagliara, S., Goldoni, A., & Sangaletti, L. (2014). High sensitivity, moisture selective, ammonia gas sensors based on single-walled carbon nanotubes functionalized with indium tin oxide nanoparticles. *Carbon*, *80*, 356–363. doi:10.1016/j.carbon.2014.08.074

Rijal, K., & Mutharasan, R. (2013). A method for DNA-based detection of E. coli O157:H7 in a proteinous background using piezoelectric-excited cantilever sensors. *Analyst (London)*, *138*(10), 2943–2950. doi:10.1039/c3an36814a PMID:23535914

Riu, J., Maroto, A., & Rius, F. X. (2006). Nanosensors in environmental analysis. *Talanta*, *69*(2), 288–301. doi:10.1016/j.talanta.2005.09.045 PMID:18970568

Rivera, I. F., Joshi, & Wang. (2010). Graphene-based ultra-sensitive gas sensors. In *2010 IEEE Sensors*, (pp. 1534-1537).

Rodrigues, J. N. B., Gonçalves, P. A. D., Rodrigues, N. F. G., Lopes dos Santos, J. M. B., & Peres, N. M. R. (2011). Zigzag graphene nanoribbon edge reconstruction with Stone-Wales defects. *Physical Review B: Condensed Matter and Materials Physics*, *84*(15), 155435. doi:10.1103/PhysRevB.84.155435

Romero, H. E., Joshi, P., Gupta, A. K., Gutierrez, H. R., Cole, M. W., Tadigadapa, S. A., & Eklund, P. C. (2009). Adsorption of ammonia on graphene. *Nanotechnology*, 20. PMID:19468162

Ruddy, F. H., Dulloo, A. R., Seidel, J. G., Seshadri, S., & Rowland, L. B. (1998). Development of a silicon carbide radiation detector. *Nuclear Science. IEEE Transactions on*, *45*(3), 536–541. doi:10.1109/23.682444

Rumyantsev, S., Liu, G., Shur, M. S., Potyrailo, R. A., & Balandi, A. A. (2012). Selective gas sensing with a single pristine graphene transistor. *Nano Letters*, *12*(5), 2294–2298. doi:10.1021/nl3001293 PMID:22506589

Rupani, V., Kargathara, S., & Sureja, J. (2015). A review on wireless nanosensor networks based on electromagnetic communication. *International Journal of Computer Science and Information Technologies*, *6*(2), 1019–1022.

Rushi, A. D., Datta, K. P., Ghosh, P. S., Mulchandani, A., & Shirsat, M. D. (2014). Selective discrimination among benzene, toluene, and xylene: Probing metalloporphyrin-functionalized single-walled carbon nanotube-based field effect transistors. *The Journal of Physical Chemistry C*, *118*(41), 24034–24041. doi:10.1021/jp504657c

Russo, C. J., & Golovchenko, J. A. (2012). Atom-by-atom nucleation and growth of graphene nanopores. *Proceedings of the National Academy of Sciences of the United States of America*, *109*(16), 5953–5957. doi:10.1073/pnas.1119827109 PMID:22492975

Rutter, G. M., Crain, J., Guisinger, N., First, P. N., & Stroscio, J. A. (2008). Optoelectronic properties of ABC-stacked trilayer graphene. *Journal of Vacuum Science & Technology. A, Vacuum, Surfaces, and Films*, *26*, 938–943. doi:10.1116/1.2944257

S. M. Sze & K. K. Ng, (2007). *Physics of semiconductor devices*. Wiley-Blackwell.

Sadeghi, H., Ahmadi, M. T., Mousavi, S. M., & Ismail, R. (2012). Channel conductance of ABA stacking trilayer graphene nanoribbon field-effect transistor. *Modern Physics Letters B*, *26*(8). doi: 125004710.1142/s0217984912500479

Sadeghi, H., Lambert, C. J., Algharagholy, L., Bailey, S. W. D., Ferrer, J., & Garcia-Suarez, V. M. (2013b). *Patent: Nanopore arrangement for DNA sequencing*. WO/2015/092411/A1.

Sadeghi, H. (2014a). Electrical Transport Model of Silicene as a Channel of Field Effect Transistor. *Journal of Nanoscience and Nanotechnology*, *14*(6), 4178–4184. doi:10.1166/jnn.2014.8914 PMID:24738367

Sadeghi, H., Ahmadi, M. T., Ishak, B., Mousavi, S., & Ismail, R. (2011). Ballistic Conductance Model of Bilayer Graphene Nanoribbon (BGN). *Journal of Computational and Theoretical Nanoscience*, *8*(10), 1993–1998. doi:10.1166/jctn.2011.1915

Sadeghi, H., Ahmadi, M. T., Mousavi, S. M., Ghadiry, M. H., & Ismail, R. (2012). Channel Conductance of ABA Stacking Trilayer Graphene Nanoribbon Field Effect Transistor. *Modern Physics Letters B*, *26*(8), 1250047. doi:10.1142/S0217984912500479

Sadeghi, H., Ahmadi, M. T., Mousavi, S. M., & Ismail, R. (2012). Channel conductance of aba stacking trilayer graphene nanoribbon field-effect transistor. *Modern Physics Letters B*, 26.

Sadeghi, H., Ahmadi, M. T., Webb, J. F., Mousavi, S. M., Ismail, R., & Ismail, I. (2011). CARRIER statistics model for a bilayer graphene nanoribbon in the nondegenerate regime. In N. W. J. F. V. P. Barsoum (Ed.), *Proceedings of the Fourth Global Conference on Power Control and Optimization* (Vol. 1337, pp. 184-187). doi:10.1063/1.3592464

Sadeghi, H., Algaragholy, L., Pope, T., Bailey, S., Visontai, D., & Manrique, D. et al. (2014b). Graphene Sculpturene Nanopores for DNA Nucleobase Sensing. *The Journal of Physical Chemistry B*, *118*(24), 6908–6914. doi:10.1021/jp5034917 PMID:24849015

Sadeghi, H., Bailey, S., & Lambert, C. J. (2014c). Silicene-based DNA nucleobase sensing. *Applied Physics Letters*, *104*(10), 103104. doi:10.1063/1.4868123

Sadeghi, H., Lai, D. T., Redoute, J.-M., & Zayegh, A. (2013a). Classic and quantum capacitances in Bernal bilayer and trilayer graphene field effect transistor. *Journal of Nanomaterials*, *2013*, 1–7. doi:10.1155/2013/127690

Sadeghi, H., Mol, J. A., Lau, C. S., Briggs, G. A. D., Warner, J., & Lambert, C. J. (2015a). Conductance enlargement in picoscale electroburnt graphene nanojunctions. *Proceedings of the National Academy of Sciences of the United States of America*, *112*(9), 2658–2663. doi:10.1073/pnas.1418632112 PMID:25730863

Sadeghi, H., Sangtarash, S., & Lambert, C. J. (2015b). Enhanced Thermoelectric Efficiency of Porous Silicene Nanoribbons. *Scientific Reports*, *5*, 9514. doi:10.1038/srep09514 PMID:25820162

Sadrolhosseini, A. R. (n.d.). *Surface Plasmon resonance characterization of biodiesel*. (PhD thesis). Universiti Putra Malaysia.

Sadrzadeh, A., Farajian, A. A., & Yakobson, B. I. (2008). Electron transport of nanotube-based gas sensors: An *ab initio* study. *Applied Physics Letters*, *92*(2), 022103. doi:10.1063/1.2829609

Saeidmanesh, M. (2013). Perpendicular electric field effect on bilayer graphene carrier statistic. *JCTN*.

Saeidmanesh, M., Ahmadi, M., Ghadiry, M., Akbari, E., Kiani, M., & Ismail, R. (2013). Perpendicular electric field effect on bilayer graphene carrier statistic. *Journal of Computational and Theoretical Nanoscience*, *10*(9), 1975–1978. doi:10.1166/jctn.2013.3158

Sagas, J. C., Neto, A. H., Maciel, H., & Lacava, P. (2011). Basic characteristics of gliding-arc discharges in air and natural gas. *Plasma Science. IEEE Transactions on*, *39*, 775–780.

Saha, K. K., Drndic, M., & Nikolic, B. K. (2012). DNA Base-Specific Modulation of Microampere Transverse Edge Currents through a Metallic Graphene Nanoribbon with a Nanopore. *Nano Letters*, *12*(1), 50–55. doi:10.1021/nl202870y PMID:22141739

Saito, R. (1998). *Physical properties of carbon nanotubes*. Academic Press.

Salavagione, H. J., Díez-Pascual, A. M., Lázaro, E., Vera, S., & Gómez-Fatou, M. A. (2014). Chemical sensors based on polymer composites with carbon nanotubes and graphene: The role of the polymer. *Journal of Materials Chemistry A*, *2*(35), 14289–14328. doi:10.1039/C4TA02159B

Salehi-Khojin, A., Field, C. R., Yeom, J., & Masel, R. I. (2010). Sensitivity of nanotube chemical sensors at the onset of Poole-Frenkel conduction. *Applied Physics Letters*, *96*(16), 163110. doi:10.1063/1.3379291

Salihoglu, O., Balci, S., & Kocabas, C. (2012). Plasmon-polaritons on graphene-metal surface and their use in biosensors. *Applied Physics Letters, 100*(21), 213110. doi:10.1063/1.4721453

Sangtarash, S., Sadeghi, H., Ahmadi, M.T., Ghadiry, M.H., Anwar, S., & Ismail, R. (2012). Article. *Journal of Computational and Theoretical Nanoscience*, *9*(10), 1554-1557.

Sangtarash, S., Sadeghi, H., Ahmadi, M., Ghadiry, M., Anwar, S., & Ismail, R. (2012). Schottky current in carbon nanotube-metal contact. *Journal of Computational and Theoretical Nanoscience*, *9*(10), 1554–1557. doi:10.1166/jctn.2012.2243

Santangelo, S., Faggio, G., Messina, G., Fazio, E., Neri, F., & Neri, G. (2013). On the hydrogen sensing mechanism of Pt/TiO2/CNTs based devices. *Sensors and Actuators. B, Chemical*, *178*, 473–484. doi:10.1016/j.snb.2013.01.005

Sarma, A. K., Vatsyayan, P., Goswami, P., & Minteer, S. D. (2009). Recent advances in material science for developing enzyme electrodes. *Biosensors & Bioelectronics*, *24*(8), 2313–2322. doi:10.1016/j.bios.2008.09.026 PMID:19022645

Sauviat, M., & Cohen-Addad, J. P. (1981). Polyisobutylene: Fractional precipitation from binary solvent induced by non-solvent addition. *Polymer*, *22*(4), 461–464. doi:10.1016/0032-3861(81)90160-9

Sayago, I., Fernandez, M., Fontecha, J., Horrillo, M., Terrado, E., Seral-Ascaso, A., & Munoz, E. (2013). Carbon nanotube-based SAW sensors. In *Electron Devices (CDE), 2013 Spanish Conference on*, (pp. 127-130). IEEE. doi:10.1109/CDE.2013.6481359

Sayago, I., Fernandez, M. J., Fontecha, J. L., Horrillo, M. C., Vera, C., Obieta, I., & Bustero, I. (2011). Surface acoustic wave gas sensors based on polyisobutylene and carbon nanotube composites. *Sensors and Actuators. B, Chemical*, *156*(1), 1–5. doi:10.1016/j.snb.2011.04.047

Schaller, R. R. (1997). Moore's law: Past, present and future. Spectrum, IEEE, 34, 52-59.

Schedin, F., Geim, A., Morozov, S., Hill, E., Blake, P., Katsnelson, M., & Novoselov, K. (2007). Detection of individual gas molecules adsorbed on graphene. *Nature Materials*, *6*(9), 652–655. doi:10.1038/nmat1967 PMID:17660825

Schlesinger, R. B. M., Becht, R., Dosenbach, S., Hoffmann, W., & Ache, H. J. (1996). ISFETs with sputtered sodium alumino-silicate glass membranes. *Fresenius' Journal of Analytical Chemistry*, *354*(7-8), 852–856. PMID:15048401

Schmiedt, L., Kaňka, & Hrachová. (n.d.). *Optical emission spectroscopy–Passive method for plasma diagnostics in dc glow oxygen discharge*. Academic Press.

Schneider, G. F., & Dekker, C. (2012). DNA sequencing with nanopores. *Nature Biotechnology*, *30*(4), 326–328. doi:10.1038/nbt.2181 PMID:22491281

Schuhmann, W. (1995). Conducting polymer based amperometric enzyme electrodes. *Mikrochimica Acta*, *121*(1-4), 1–29. doi:10.1007/BF01248237

Schweyer, M. G., Andle, J. C., McAllister, D. J., French, L. A., & Vetelino, J. F. (1996). *An acoustic plate mode sensor for aqueous mercury*. Paper presented at the Ultrasonics Symposium. doi:10.1109/ULTSYM.1996.583991

Schwierz, F. (2010a). Graphene transistors. *Nature Nanotechnology*, *5*(7), 487–496. doi:10.1038/nnano.2010.89 PMID:20512128

Schwierz, F. (2010b). graphene transistors. *nature. Nanotechnology*. doi:10.1038/NNANO.2010.89

Scott, B. W., & Leburton, J. (2010). *High-field carrier velocity and current saturation in graphene field-effect transistors*. 10th IEEE Conference on Nanotechnology (IEEE-NANO), Seoul, South Korea. doi:10.1109/NANO.2010.5697808

Sedeghi, H., Lai, D. T. H., Redoute, J. M., & Zayegh, A. (2013). Classic and Quantum Capacitances in Bernal Bilayer and Trilayer Graphene Field Effect Transistor. *Journal of Nanomaterials*.

Sedra, A. S., & Smith, K. C. (2004). *Microelectronic circuits*. Oxford University Press.

Selapinar, F., Toppare, L., Akbulut, U., Yalcin, T., & Suzer, S. (1995). A conducting composite of polypyrrole as a gas sensor. *Synthetic Metals*, *68*(2), 109–116. doi:10.1016/0379-6779(94)02299-E

Selvaraj, K., Kumar, S., & Lakshmanan, R. (2014). Analytical expression for concentration and sensitivity of a thin film semiconductor gas sensor. *Ain Shams Engineering Journal*, *5*(3), 885–893. doi:10.1016/j.asej.2014.01.004

Sena, D., Thapab, R., & Chattopadhyaya, K. K. (2013). Small Pd cluster adsorbed double vacancy defect graphene sheet for hydrogen storage: A first-principles study. *International Journal of Hydrogen Energy*, *38*(7), 3041–3049. doi:10.1016/j.ijhydene.2012.12.113

Sevik, C., Sevinçli, H., Cuniberti, G., & Cagın, T. (2011). Phonon engineering in carbon nanotubes by controlling defect concentration. *Nano Letters*, *11*(11), 4971–4977. doi:10.1021/nl2029333 PMID:21967464

Seymour, R. B. (1966). Plastics. *Industrial & Engineering Chemistry, 58*(8), 61-73.

Shandas, R., & Lanning, C. (2003). Article. *Medical & Biological Engineering & Computing*, *41*, 416–424. doi:10.1007/BF02348084 PMID:12892364

Shang, J., Ma, L., Li, J., Ai, W., Yu, T., & Gurzadyan, G. G. (2012). The Origin of fluorescence from graphene oxide. *Scientific Reports*, *2*, 792. doi:10.1038/srep00792 PMID:23145316

Shao, L., Chen, G., Ye, H., Niu, H., Wu, Y., Zhu, Y., & Ding, B. (2014). Sulfur dioxide molecule sensors based on zigzag graphene nanoribbons with and without Cr dopant. *Physics Letters. [Part A]*, *378*(7-8), 667–671. doi:10.1016/j.physleta.2013.12.042

Sharifi, F., Moaiyeri, M. H., Navi, K., & Bagherzadeh, N. (2015). Quaternary full adder cells based on carbon nanotube FETs. *Journal of Computational Electronics*, *14*(3), 762–772. doi:10.1007/s10825-015-0714-0

Shendure, J., & Aiden, E. L. (2012). The expanding scope of DNA sequencing. *Nature Biotechnology*, *30*(11), 1084–1094. doi:10.1038/nbt.2421 PMID:23138308

Shendure, J., & Ji, H. (2008). Next-generation DNA sequencing. *Nature Biotechnology*, *26*(10), 1135–1145. doi:10.1038/nbt1486 PMID:18846087

Shengjun, Y., Raedt, H. D., & Katsnelson, M. I. (2010). *Electronic Transport in Disordered Bilayer*. Academic Press.

Shen, Y., & Wan, M. (1998). In situ doping polymerization of pyrrole with sulfonic acid as a dopant. *Synthetic Metals*, *96*(2), 127–132. doi:10.1016/S0379-6779(98)00076-9

Shepherd, L. T., & Toumazou, C. (2005). Weak Inversion ISFETs for ultra-low power biochemical sensing and real-time analysis. *Sensors and Actuators. B, Chemical*, *107*(1), 468–473. doi:10.1016/j.snb.2004.11.006

Shim, J., Humphreys, G. I., Venkatesan, B. M., Munz, J. M., Zou, X., Sathe, C., et al. (2013). Detection and Quantification of Methylation in DNA using Solid-State Nanopores. *Sci. Rep., 3*. Retrieved from http://www.nature.com/srep/2013/130306/srep01389/abs/srep01389.html#supplementary-information

Shim, J., Solovyeva, V., Estrada, D., Banerjee, S., Rivera, J., & Pop, E. (2012). *Graphene Nanopores for Nucleic Acid Analysis*. IEEE.

Shishir, R. S., Ferry, D. K., & Goodnick, S. M. (2009). Room-temperature velocity saturation in intrinsic graphene. *Journal of Physics: Conference Series, 193*, 012118. doi:10.1088/1742-6596/193/1/012118

Shi, Y. B., Xiang, J. J., Feng, Q. H., Hu, Z. P., Zhang, H. Q., & Guo, J. Y. (2006). Binary channel SAW mustard gas sensor based on PdPc(0.3)PANI(0.7)hybrid sensitive film. In J. Tan (Ed.), *4th International Symposium on Instrumentation Science and Technology* (Vol. 48, pp. 292-297).

Shi, Y., Dong, X., Chen, P., Wang, J., & Li, L. J. (2009). Effective doping of single-layer graphene from underlying SiO_ {2} substrates. *Physical Review B: Condensed Matter and Materials Physics, 79*(11), 115402. doi:10.1103/PhysRevB.79.115402

Shrivastava, I. H., & Sansom, M. S. (2000). Simulations of ion permeation through a potassium channel: Molecular dynamics of KcsA in a phospholipid bilayer. *Biophysical Journal, 78*(2), 557–570. doi:10.1016/S0006-3495(00)76616-1 PMID:10653771

Shukla, S., & Saxena, S. (2011). Spectroscopic investigation of confinement effects on optical properties of graphene oxide. *Applied Physics Letters, 98*(7), 073104. doi:10.1063/1.3555438

Shults, M. C., Rhodes, R. K., Updike, S. J., Gilligan, B. J., & Reining, W. N. (1994). A telemetry-instrumentation system for monitoring multiple subcutaneously implanted glucose sensors. *IEEE Transactions on Bio-Medical Engineering, 41*(10), 937–942. doi:10.1109/10.324525 PMID:7959800

Shylau, A., Kłos, J., & Zozoulenko, I. (2009). Capacitance of graphene nanoribbons. *Physical Review B: Condensed Matter and Materials Physics, 80*(20), 205402. doi:10.1103/PhysRevB.80.205402

Simsek, E. (2013). Improving tuning range and sensitivity of localized spr sensors with graphene. *Photonics Technology Letters, IEEE, 25*(9), 867–870. doi:10.1109/LPT.2013.2253316

Singh, R. J. (2012). *Solid State Physics*. Dorling Kindersley.

Singh, V., Joung, D., Zhai, L., Das, S., Khondaker, S. I., & Seal, S. (2011). Graphene based materials: Past, present and future. *Progress in Materials Science, 56*(8), 1178–1271. doi:10.1016/j.pmatsci.2011.03.003

Sinha, N., Ma, J., & Yeow, J. T. W. (2006). Carbon Nanotube-Based Sensors. *Journal of Nanoscience and Nanotechnology, 6*(3), 573–590. doi:10.1166/jnn.2006.121 PMID:16573108

Sivasathya, S., & Thiruvadigal, D. J. (2013). Ab Initio study of carbon nanotube transistor-based gas sensor for NO2 detection. *Asian Journal of Chemistry, 25*, S411–S413.

Skoog, D. A., West, D. M., & Holler, F. J. (1988). *Fundamentals of analytical chemistry*. Academic Press.

Skotheim, T. A. (1986). *Handbook of Conducting Polymers*. New York: Marcel Dekker.

Smith, A. D., Elgammal, K., Niklaus, F., Delin, A., Fischer, A. C., Vaziri, S., & Lemme, M. C. et al. (2015). Resistive graphene humidity sensors with rapid and direct electrical readout. *Nanoscale, 7*(45), 19099–19109. doi:10.1039/C5NR06038A PMID:26523705

Snow, E. S., & Perkins, F. K. (2005). Capacitance and Conductance of Single-Walled Carbon Nanotubes in the Presence of Chemical Vapors. *Nanoletter*, *5*(12), 2414–2417. doi:10.1021/nl051669c

Snow, E. S., Perkins, F. K., Houser, E. J., Badescu, S. C., & Reinecke, T. L. (2005). Chemical Detection with a Single-Walled Carbon Nanotube Capacitor. *Science*, *307*(5717), 1942–1945. doi:10.1126/science.1109128 PMID:15790850

Soler, J. M., Artacho, E., Gale, J. D., García, A., Junquera, J., Ordejón, P., & Sánchez-Portal, D. (2002). The SIESTA method for ab initio order- N materials simulation. *Journal of Physics Condensed Matter*, *14*(11), 2745–2779. doi:10.1088/0953-8984/14/11/302

Sols, F., Guinea, F., & Neto, A. C. (2007). Coulomb blockade in graphene nanoribbons. *Physical Review Letters*, *99*(16), 166803. doi:10.1103/PhysRevLett.99.166803 PMID:17995278

Some, S., Xu, Y., Kim, Y., Yoon, Y., Qin, H., Kulkarni, A., & Lee, H. et al. (2013). Highly sensitive and selective gas sensor using hydrophilic and hydrophobic graphenes. *Scientific Reports*, *3*, 1868. doi:10.1038/srep01868 PMID:23736838

Someya, T., Small, J., Kim, P., Nuckolls, C., & Yardley, J. T. (2003). Alcohol Vapor Sensors Based on Single-Walled Carbon Nanotube Field Effect Transistors. *Nano Letters*, *3*(7), 877–881. doi:10.1021/nl034061h

Sonawane, M. R., Habale, Nagare, & Gharde. (n.d.). *Interaction of O2, CO2, NO2 and SO2 on Si-doped Carbon Nanotube*. Academic Press.

Song, D. W., Shen, W.-N., Dunn, B., Moore, C. D., Goorsky, M. S., Radetic, T., . . . Chen, G. (2004). Thermal conductivity of nanoporous bismuth thin films. *Applied Physics Letters*, *84*(11), 1883-1885. doi:10.1063/1.1682679

Song, P., Wang, Q., & Yang, Z. (2011). Ammonia gas sensor based on PPy/ZnSnO3 nanocomposites. *Materials Letters*, *65*(3), 430–432. doi:10.1016/j.matlet.2010.10.087

Son, Y.-W., Cohen, M. L., & Louie, S. G. (2006). Energy gaps in graphene nanoribbons. *Physical Review Letters*, *97*(21), 216803. doi:10.1103/PhysRevLett.97.216803 PMID:17155765

Sotiropoulou, S., Gavalas, V., Vamvakaki, V., & Chaniotakis, N. (2003). Novel carbon materials in biosensor systems. *Biosensors & Bioelectronics*, *18*(2), 211–215. doi:10.1016/S0956-5663(02)00183-5 PMID:12485767

Spinney, P. S., Collins, S. D., Howitt, D. G., & Smith, R. L. (2012). Fabrication and characterization of a solid-state nanopore with self-aligned carbon nanoelectrodes for molecular detection. *Nanotechnology*, *23*(13), 135501–135501. doi:10.1088/0957-4484/23/13/135501 PMID:22421078

Star, A., Han, T. R., Gabriel, J. C. P., Bradley, K., & Grüner, G. (2003). Interaction of aromatic compounds with carbon nanotubes: Correlation to the Hammett parameter of the substituent and measured carbon nanotube FET response. *Nano Letters*, *3*(10), 1421–1423. doi:10.1021/nl0346833

Star, A., Stoddart, J. F., Steuerman, D., Diehl, M., Boukai, A., & Wong, E. W. et al. (2001). Preparation and Properties of Polymer-Wrapped Single-Walled Carbon Nanotubes. *Angewandte Chemie*, *113*(9), 1771–1775. doi:10.1002/1521-3757(20010504)113:9<1771::AID-ANGE17710>3.0.CO;2-Y

Star, A., Tu, E., Niemann, J., Gabriel, J. C. P., Joiner, C. S., & Valcke, C. (2006). Label-free detection of DNA hybridization using carbon nanotube network field-effect transistors. *Proceedings of the National Academy of Sciences of the United States of America*, *103*(4), 921–926. doi:10.1073/pnas.0504146103 PMID:16418278

Stebunov, Y. V., Aftenieva, O. A., Arsenin, A. V., & Volkov, V. S. (2015). Highly sensitive and selective sensor chips with graphene-oxide linking layer. *ACS Applied Materials & Interfaces*, *7*(39), 21727–21734. doi:10.1021/acsami.5b04427 PMID:26358000

Steinhoff, G. H. M., Schaff, W. J., Eastman, L. F., Stutzmann, M., & Eickhoff, M. (2003). pH response of GaN surfaces and its application for pH-sensitive field-effect transistors. *Applied Physics Letters*, *83*(1), 177–179. doi:10.1063/1.1589188

Stern, E., Klemic, J. F., Routenberg, D. A., Wyrembak, P. N., Turner-Evans, D. B., Hamilton, A. D., & Reed, M. A. (2007). Label-free immunodetection with CMOS-compatible semiconducting nanowires. *Nature*, *445*(7127), 519–522. doi:10.1038/nature05498 PMID:17268465

Stern, E., Vacic, A., & Reed, M. A. (2008). Semiconducting nanowire field-effect transistor biomolecular sensors. *IEEE Transactions on Electron Devices*, *55*(11), 3119–3130. doi:10.1109/TED.2008.2005168

Sternschulte, H., Schreck, M., & Stritzker, B. (2002). In situ characterisation of CVD diamond growth under H< sub> 2</sub> S addition by optical emission spectroscopy, mass spectroscopy and laser reflection interferometry. *Diamond and Related Materials*, *11*(3-6), 296–300. doi:10.1016/S0925-9635(01)00675-6

Stobinski, L., Lesiak, B., Kövér, L., Tóth, J., Biniak, S., Trykowski, G., & Judek, J. (2010). Multiwall carbon nanotubes purification and oxidation by nitric acid studied by the FTIR and electron spectroscopy methods. *Journal of Alloys and Compounds*, *501*(1), 77–84. doi:10.1016/j.jallcom.2010.04.032

Strutt, J. W. (1885). On waves propagated along the plane surface of an elastic solid. *Proceedings of the London Mathematical Society*, *17*(1), 4–11.

Subrahmanyam, K., Panchakarla, L., Govindaraj, A., & Rao, C. (2009). Simple method of preparing graphene flakes by an arc-discharge method. *The Journal of Physical Chemistry C*, *113*(11), 4257–4259. doi:10.1021/jp900791y

Subramanian, P., Barka-Bouaifel, F., Bouckaert, J., Yamakawa, N., Boukherroub, R., & Szunerits, S. (2014). Graphene-coated surface plasmon resonance interfaces for studying the interactions between bacteria and surfaces. *ACS Applied Materials & Interfaces*, *6*(8), 5422–5431. doi:10.1021/am405541z PMID:24433135

Suehiro, J., Zhou, G. B., Imakiire, H., Ding, W. D., & Hara, M. (2005). Controlled fabrication of carbon nanotube NO2 gas sensor using dielectrophoretic impedance measurement. *Sensors and Actuators. B, Chemical*, *108*(1-2), 398–403. doi:10.1016/j.snb.2004.09.048

Sun, Y. P., Fu, K., Lin, Y., & Hang, W. (2002). Functionalized Carbon Nanotubes: Properties and Applications. *Accounts of Chemical Research*, *35*(12), 1096–1104. doi:10.1021/ar010160v PMID:12484798

Sutter, P., Hybertsen, M. S., Sadowski, J. T., & Sutter, E. (2009). Electronic structure of few-layer epitaxial graphene on Ru(0001). *Nano Letters*, *9*(7), 2654–2660. doi:10.1021/nl901040v PMID:19505134

Suzuki, T., & Yokomizo, Y. (2010). Energy bands of atomic monolayers of various materials: Possibility of energy gap engineering. *Physica E, Low-Dimensional Systems and Nanostructures*, *42*(10), 2820–2825. doi:10.1016/j.physe.2010.04.031

Sven Kochmann, T. H., & Wolfbeis. (2012). Graphenes in chemical sensors and biosensors. *TrAC Trends in Analytical Chemistry, 39.*

Svozil, D., Kvasnicka, V., & Pospichal, J. (1997). Introduction to multi-layer feed-forward neural networks. *Chemometrics and Intelligent Laboratory Systems*, *39*(1), 43–62. doi:10.1016/S0169-7439(97)00061-0

Sydlik, S. A., Lee, J.-H., Walish, J. J., Thomas, E. L., & Swager, T. M. (2013). Epoxy functionalized multi-walled carbon nanotubes for improved adhesives. *Carbon*, *59*(0), 109–120. doi:10.1016/j.carbon.2013.02.061

Sze, S. M. (1981). *Physics of Semiconductor Devices*. New York: Wiley.

Sze, S. M. (1985). *Semiconductor devices physics and technology*. New York: Wiley.

Takayanagi, A., Yamanouchi, K., & Shibayama, K. (1970). Piezoelectric leaky surface wave in LiNbO3. *Applied Physics Letters*, *17*(5), 225–227. doi:10.1063/1.1653375

Tameev, A. R., Jiménez, L. L., Pereshivko, L. Y., Rychwalski, R. W., & Vannikov, A. V. (2007). Charge carrier mobility in films of carbon-nanotube- polymer composites. *Journal of Physics: Conference Series*, *61*(1), 1152–1156. doi:10.1088/1742-6596/61/1/228

Taneja, L. (2012). A fractal analysis of the detection of glucose molecules on biosensor surfaces. (1515330 M.S.). The University of Mississippi. Retrieved from ProQuest Dissertations & Theses Full Text database.

Tang, S., & Cao, Z. (2011). Adsorption of nitrogen oxides on graphene and graphene oxides: Insights from density functional calculations. *Phys*, *134*, 044710. PMID:21280788

Tang, Y., Zeng, X., & Liang, J. (2010). Surface plasmon resonance: An introduction to a surface spectroscopy technique. *Journal of Chemical Education*, *87*(7), 742–746. doi:10.1021/ed100186y PMID:21359107

Taniguchi, M., Tsutsui, M., Yokota, K., & Kawai, T. (2009). Fabrication of the gating nanopore device. *Applied Physics Letters*, *95*(12), 123701, 123701–123703. doi:10.1063/1.3236769

Tans, S. J., Verschueren, A. R. M., & Dekker, C. (1998). Room-temperature transistor based on a single carbon nanotube. *Nature*, *393*(6680), 49–52. doi:10.1038/29954

Tan, Y. W., Zhang, Y., Stormer, H. L., & Kim, P. (2007). Temperature dependent electron transport in graphene. *The European Physical Journal. Special Topics*, *148*(1), 15–18. doi:10.1140/epjst/e2007-00221-9

Tao, L., Cinquanta, E., Chiappe, D., Grazianetti, C., Fanciulli, M., Dubey, M., et al. (2015). Silicene field-effect transistors operating at room temperature. *Nat Nano, 10*(3), 227-231. Retrieved from http://www.nature.com/nnano/journal/v10/n3/abs/nnano.2014.325.html#supplementary-information

Tao, X., Bai, M., Li, X., Long, H., Shang, S., Yin, Y., & Dai, X. (2011). CH< sub> 4</sub>–CO< sub> 2</sub> reforming by plasma–challenges and opportunities. *Progress in Energy and Combustion Science*, *37*(2), 113–124. doi:10.1016/j.pecs.2010.05.001

Tchalal, M. R., Enriquez, H., Mayne, A. J., Kara, A., Roth, S., & Silly, M. G. et al. (2013). Formation of one-dimensional self-assembled silicon nanoribbons on Au (110)-(2x1). *Applied Physics Letters*, *102*(8), 083107. doi:10.1063/1.4793536

Teerapanich, P., Myint, M. T. Z., Joseph, C. M., Hornyak, G. L., & Dutta, J. (2013). Development and improvement of carbon nanotube-based ammonia gas sensors using ink-jet printed interdigitated electrodes. *IEEE Transactions on Nanotechnology*, *12*(2), 255–262. doi:10.1109/TNANO.2013.2242203

Tel-Vered, R., Willner, B., & Willner, I. (2010). Biohybrid Electrochemical Devices. *Electrochemistry of Functional Supramolecular Systems, 6*, 333.

Terranova, M. L., Orlanducci, S., & Rossi, M. (2012). *Carbon nanomaterials for gas adsorption*. Boca Raton, FL: Pan Stanford Publishing.

The Shichman-Hodges Enhancement MOSFET Model and SwitcherCAD III SPICE. (2007). Report NDT14-08-2007. NanoDotTek.

Tigli, O., & Zaghloul, M. E. (2007). A novel SAW device in CMOS: Design, modeling, and fabrication. *Sensors Journal, IEEE*, *7*(2), 219–227. doi:10.1109/JSEN.2006.883771

Toloue, H. M. B., Ahmadi, M. T., & Centeno, A. (2015). *Graphene based surface plasmon resonance bio-sensor response modeling*. Nanomeghyas.

Toloue, H., Ahmadi, M., Meshginqalam, B., Centeno, A., & Sabatyan, A. (2015). Graphene based surface plasmon resonance bio-sensor modeling. *Annual International RIAPA Meeting on Low Dimensional Systems.*

Treacy, M. M. J., Ebbesen, T. W., & Gibson, J. M. (1996). Exceptionally high Young's modulus observed for individual carbon nanotubes. *Nature*, *381*(6584), 678–680. doi:10.1038/381678a0

Tüken, T., Tansuğ, G., Yazıcı, B., & Erbil, M. (2007). Poly(N-methyl pyrrole) and its copolymer with pyrrole for mild steel protection. *Surface and Coatings Technology*, *202*(1), 146–154. doi:10.1016/j.surfcoat.2007.05.022

Tu, X., & Whitehead, J. (2012). Plasma-catalytic dry reforming of methane in an atmospheric dielectric barrier discharge: Understanding the synergistic effect at low temperature. *Applied Catalysis B: Environmental*, *125*, 439–448. doi:10.1016/j.apcatb.2012.06.006

Tzamalis, G., Zaidi, N. A., Homes, C. C., & Monkman, A. P. (2002). Doping-dependent studies of the Anderson–Mott localization in polyaniline at the metal–insulator boundary. *Physical Review B: Condensed Matter and Materials Physics*, *66*(8), 1–7. doi:10.1103/PhysRevB.66.085202

Uchida, K., Saitoh, M., & Kobayashi, S. (2008). Carrier transport and stress engineering in advanced nanoscale transistors from (100) and (110) transistors to carbon nanotube fets and beyond. *IEEE International Electron Devices Meeting 2008, Technical Digest.*

Uhrig, R. E. (1995). Introduction to artificial neural networks. In *Industrial Electronics, Control, and Instrumentation, 1995.,Proceedings of the 1995 IEEE IECON 21st International Conference on*, (pp. 33-37). IEEE.

Uhrig, R. E. 1995. Introduction to artificial neural networks. In *Industrial Electronics, Control, and Instrumentation, 1995.,Proceedings of the 1995 IEEE IECON 21st International Conference on*. IEEE.

Ulrich, L., Nataliya, V., & Vladimir, M. (2008). Conducting polymers in chemical sensors and arrays. *Analytical Chimica Acta, 6,* 1–26.

Umadevi, D., & Sastry, G. N. (2014). Feasibility of carbon nanomaterials as gas sensors: A computational study. *Current Science*, *106*(9), 1224–1234.

Ünal, M., Ak, A., Topuz, V., & Erdal, H. (2013). Artificial neural networks. In Optimization of PID Controllers Using Ant Colony and Genetic Algorithms, (pp. 5-17). Springer.

Usibe, B. E., Menkiti, A. I., Onuu, M. U., & Ogbulezie, J. C. (2013). Development and analysis of a potential nanosensor communication network using carbon nanotubes. *International Journal of Materials Engineering*, *3*(1), 4–10.

Uyemura, J. P. (1999). *CMOS logic circuit design*. Springer Science & Business Media.

Valentini, F., Carbone, M., & Palleschi, G. (2013). Graphene oxide nanoribbons (GNO), reduced graphene nanoribbons (GNR), and multi-layers of oxidized graphene functionalized with ionic liquids (GO-IL) for assembly of miniaturized electrochemical devices. *Analytical and Bioanalytical Chemistry*, *405*(11), 3449–3474. doi:10.1007/s00216-012-6615-1 PMID:23274557

Valentini, L., Cantalini, C., Armentano, I., Kenny, J. M., Lozzi, L., & Santucci, S. (2003). Investigation of the NO[sub 2] sensitivity properties of multiwalled carbon nanotubes prepared by plasma enhanced chemical vapor deposition. *Journal of Vacuum Science & Technology B Microelectronics and Nanometer Structures*, *21*(5), 1996–2000. doi:10.1116/1.1599858

Vargas-Bernal, R., & Herrera-Pérez, G. (2014). Importance of the nanostructured ceramic materials on gas sensing. In M. Aliofkhazraei (Ed.), Nanotechnology for Optics and Sensors (pp. 1-34). Manchester, UK: One Central Press.

Vargas-Bernal, R. (2015). Advances in computational modeling of electronic devices based on graphene. *IEEE Journal on Emerging and Selected Topics in Circuits and Systems*, *5*(1), 109–116. doi:10.1109/JETCAS.2015.2398219

Vargas-Bernal, R., & Herrera-Pérez, G. (2012). Carbon nanotube- and graphene based devices, circuits and sensors for VLSI design. In E. Tlelo-Cuautle & S. Tan (Eds.), *VLSI Design* (pp. 41–66). Rijeka, Croatia: Intech; doi:10.5772/38743

Varghese, S. S., Lonkar, S., Singh, K. K., Swaminathan, S., & Abdala, A. (2015b). Recent advances in graphene based gas sensors. *Sensors and Actuators. B, Chemical*, *218*, 160–183. doi:10.1016/j.snb.2015.04.062

Varguese, S. S., Varghese, S. H., Swaminathan, S., Singh, K. K., & Mittal, V. (2015a). Two-dimensional materials for sensing: Graphene and beyond. *Electronics*, *4*(3), 651–687. doi:10.3390/electronics4030651

Velasco, J. Jr, Lee, Y., Jing, L., Liu, G., Bao, W., & Lau, C. N. (2012). Quantum transport in double-gated graphene devices. *Solid State Communications*, *152*(15), 1301–1305. doi:10.1016/j.ssc.2012.04.024

Venkatesan, B. M., & Bashir, R. (2011). Nanopore sensors for nucleic acid analysis. *Nature Nanotechnology*, *6*(10), 615–624. doi:10.1038/nnano.2011.129 PMID:21926981

Venkatesan, B. M., Estrada, D., Banerjee, S., Jin, X., Dorgan, V. E., & Bae, M.-H. et al. (2012). Stacked Graphene-Al_2O_3 Nanopore Sensors for Sensitive Detection of DNA and DNA-Protein Complexes. *ACS Nano*, *6*(1), 441–450. doi:10.1021/nn203769e PMID:22165962

Vianello, F., Ragusa, S., Cambria, M. T., & Rigo, A. (2006). A high sensitivity amperometric biosensor using laccase as biorecognition element. *Biosensors & Bioelectronics*, *21*(11), 2155–2160. doi:10.1016/j.bios.2005.10.006 PMID:16293408

Vincent-Randonnier, A., Larigaldie, S., Magre, P., & Sabel'nikov, V. (2007). Plasma assisted combustion: Effect of a coaxial DBD on a methane diffusion flame. *Plasma Sources Science & Technology*, *16*(1), 149–160. doi:10.1088/0963-0252/16/1/020

Voelker, M., & Fromherz, P. (2005). Signal transmission from individual mammalian nerve cell to field-effect transistor. *Small*, *1*(2), 206–210. doi:10.1002/smll.200400077 PMID:17193431

Vogt, P., De Padova, P., Quaresima, C., Avila, J., Frantzeskakis, E., & Asensio, M. C. et al. (2012). Silicene: Compelling Experimental Evidence for Graphenelike Two-Dimensional Silicon. *Physical Review Letters*, *108*(15), 155501. doi:10.1103/PhysRevLett.108.155501 PMID:22587265

Voigt, H. S. F., Lange, T., Kullick, T., & Ferretti, R. (1997). Diamond-like carbon-gate pH-ISFET. *Sensors and Actuators. B, Chemical*, *44*(1–3), 441–445. doi:10.1016/S0925-4005(97)00236-0

von Wenckstern, H., Biehne, G., Rahman, R. A., Hochmuth, H., Lorenz, M., & Grundmann, M. (2006). Mean barrier height of Pd Schottky contacts on ZnO thin films. *Applied Physics Letters*, *88*(9), 092102. doi:10.1063/1.2180445

Voytchev, M., Iñiguez, M. P., Méndez, R., Mañanes, A., Rodríguez, L. R., & Barquero, R. (2003). Neutron detection with a silicon PIN photodiode and 6LiF converter. *Nuclear Instruments & Methods in Physics Research. Section A, Accelerators, Spectrometers, Detectors and Associated Equipment*, *512*(3), 546–552. doi:10.1016/S0168-9002(03)02013-8

Wakabayashi, K., Fujita, M., Ajiki, H., & Sigrist, M. (1999). Electronic and magnetic properties of nanographite ribbons. *Physical Review B: Condensed Matter and Materials Physics*, *59*(12), 8271–8282. doi:10.1103/PhysRevB.59.8271

Wang, W., Lim, C., Lee, K., & Yang, S. (2009). Wireless surface acoustic wave chemical sensor for simultaneous measurement of CO(2) and humidity. *Journal of Micro-Nanolithography Mems and Moems*, *8*(3). doi: 03130610.1117/1.3158610

Wang, X. R., Ouyang, Y. J., Li, X. L., Wang, H. L., Guo, J., & Dai, H. J. (2008). Room-temperature all-semiconducting sub-10-nm graphene nanoribbon field-effect transistors. *Physical Review Letters*, *100*(20).

Wang, D., Yu, Z., McKernan, S., & Burke, P. J. (2007). Ultrahigh frequency carbon nanotube transistor based on a single nanotube. *IEEE Transactions on Nanotechnology*, *6*(4), 400–403. doi:10.1109/TNANO.2007.901179

Wang, J. (2005). Carbon-nanotube based electrochemical biosensors: A review. *Electroanalysis*, *17*(1), 7–14. doi:10.1002/elan.200403113

Wang, J. (2008). Electrochemical Glucose Biosensors. *Chemical Reviews*, *108*(2), 814–825. doi:10.1021/cr068123a PMID:18154363

Wang, J., Liu, G., & Jan, M. R. (2004). Ultrasensitive Electrical Biosensing of Proteins and DNA: Carbon-Nanotube Derived Amplification of the Recognition and Transduction Events. *Journal of the American Chemical Society*, *126*(10), 3010–3011. doi:10.1021/ja031723w PMID:15012105

Wang, J., & Lundstrom, M. (2003). Ballistic transport in high electron mobility transistors. *Electron Devices. IEEE Transactions on*, *50*(7), 1604–1609. doi:10.1109/TED.2003.814980

Wang, J., & Musameh, M. (2003). Carbon nanotube/teflon composite electrochemical sensors and biosensors. *Analytical Chemistry*, *75*(9), 2075–2079. doi:10.1021/ac030007+ PMID:12720343

Wang, L. J., Cao, G., Tu, T., Li, H. O., Zhou, C., Hao, X.-J., & Guo, G.-P. et al. (2010). A graphene quantum dot with a single electron transistor as an integrated charge sensor. *Applied Physics Letters*, *97*(26), 262113–262115. doi:10.1063/1.3533021

Wang, P., Jornet, J. M., Malik, M. G., Akkari, N., & Akyildiz, I. F. (2013). Energy and spectrum-aware MAC protocol for perpetual wireless nanosensor networks in the Terahertz Band. *Ad Hoc Networks*, *11*(8), 2541–2555. doi:10.1016/j.adhoc.2013.07.002

Wang, Q., & Arash, B. (2014). A review on applications of carbon nanotubes and graphenes as nano-resonator sensors. *Computational Materials Science*, *82*, 350–360. doi:10.1016/j.commatsci.2013.10.010

Wang, W., & He, S. (2009). Theoretical analysis on response mechanism of polymer-coated chemical sensor based Love wave in viscoelastic media. *Sensors and Actuators. B, Chemical*, *138*(2), 432–440. doi:10.1016/j.snb.2009.03.006

Wang, X., & Dai, H. (2010). Etching and narrowing of graphene from the edges. *Nature Chemistry*, *2*(8), 661–665. doi:10.1038/nchem.719 PMID:20651729

Wang, X.-N., & Hu, P.-A. (2012). Carbon nanomaterials: Controlled growth and field-effect transistor biosensors. *Frontiers of Materials Science*, *6*(1), 26–46. doi:10.1007/s11706-012-0160-x

Wang, X., Ouyang, Y., Li, X., Wang, H., Guo, J., & Dai, H. (2008). Room-temperature all-semiconducting sub-10-nm graphene nanoribbon field-effect transistors. *Physical Review Letters*, *100*(20), 206803. doi:10.1103/PhysRevLett.100.206803 PMID:18518566

Wang, X.-Q., Li, H.-D., & Wang, J.-T. (2012). Induced ferromagnetism in one-side semihydrogenated silicene and germanene. *Physical Chemistry Chemical Physics*, *14*(9), 3031–3036. doi:10.1039/c2cp23385a PMID:22286024

Wang, Y. X., Ping, J. F., Ye, Z. Z., Wu, J., & Ying, Y. B. (2013). Impedimetric immunosensor based on gold nanoparticles modified graphene paper for label-free detection of Escherichia coli O157:H7. *Biosensors & Bioelectronics*, *49*, 492–498. doi:10.1016/j.bios.2013.05.061 PMID:23811484

Wang, Y., Huang, B.-C., Zhang, M., & Woo, J. C. S. (2012). Optimizing the fabrication process for high performance graphene field effect transistors. *Microelectronics and Reliability*, *52*(8), 1602–1605. doi:10.1016/j.microrel.2011.09.036

Wang, Y., Lv, J., Zhu, L., & Ma, Y. (2010). Crystal structure prediction via particle-swarm optimization. *Physical Review B: Condensed Matter and Materials Physics*, *82*(9), 094116. doi:10.1103/PhysRevB.82.094116

Wang, Y., & Yeow, J. T. W. (2009). A review of carbon nanotubes-based gas sensors. *Journal of Sensors, 493904*. doi:10.1155/2009/493904

Wang, Y., Ye, Z., Si, C., & Ying, Y. (2011). Subtractive inhibition assay for the detection of E. coli O157: H7 using surface plasmon resonance. *Sensors (Basel, Switzerland), 11*(3), 2728–2739. doi:10.3390/s110302728 PMID:22163763

Wang, Z. F., Zheng, H., Shi, Q. W., & Chen, J. (2009). Emerging nanodevice paradigm: Graphene-Based electronics for nanoscale computing. *Acm Journal on Emerging Technologies in Computing Systems*, *5*(1), 1–19. doi:10.1145/1482613.1482616

Wang, Z., Zhou, X., Zhang, J., Boey, F., & Zhang, H. (2009). Direct electrochemical reduction of single-layer graphene oxide and subsequent functionalization with glucose oxidase. *The Journal of Physical Chemistry C*, *113*(32), 14071–14075. doi:10.1021/jp906348x

Wanunu, M. (2012). Nanopores: A journey towards DNA sequencing. *Physics of Life Reviews*, *9*(2), 125–158. doi:10.1016/j.plrev.2012.05.010 PMID:22658507

Wan, Y., Wang, Y., Wu, J., & Zhag, D. (2011). Graphene oxide sheet-mediated silver enhancement for application to electrochemical biosensors. *Analytical Chemistry*, *83*(3), 648–653. doi:10.1021/ac103047c PMID:21175166

Wehling, T. O., Novoselov, K. S., Morozov, S. V., Vdovin, E. E., Katsnelson, M. I., Geim, A. K., & Lichtenstein, A. I. (2008). Molecular doping of graphene. *Nano Letters*, *8*(1), 173–177. doi:10.1021/nl072364w PMID:18085811

Wei, L., Frank, D. J., Chang, L., & Wong, H. S. P. (2011). A fully anlytical model for carbon nanotube FETs including quantum capacitances and electrostatics. Boston, MA: Academic Press.

Wei, B. Q., Vajtai, R., & Ajayan, P. M. (2001). Reliability and current carrying capacity of carbon nanotubes. *Applied Physics Letters*, *79*(8), 1172–1174. doi:10.1063/1.1396632

Weiss, L. A., Sakai, N., Ghebremariam, B., Ni, C., & Matile, S. (1997). Rigid rod-shaped polyols: Functional nonpeptide models for transmembrane proton channels. *Journal of the American Chemical Society*, *119*(50), 12142–12149. doi:10.1021/ja973126d

Wei, X.-L., Chen, Y.-P., Liu, W.-L., & Zhong, J.-X. (2012). Enhanced gas sensor based on nitrogen-vacancy graphene nanoribbons. *Physics Letters. [Part A]*, *376*(4), 559–562. doi:10.1016/j.physleta.2011.10.055

Wells, D. B., Belkin, M., Comer, J., & Aksimentiev, A. (2012). Assessing Graphene Nanopores for Sequencing DNA. *Nano Letters*, *12*(8), 4117–4123. doi:10.1021/nl301655d PMID:22780094

Wen, W., He, S., Li, S., Liu, M., & Yong, P. (2007). Enhanced sensitivity of SAW gas sensor coated molecularly imprinted polymer incorporating high frequency stability oscillator. *Sensors and Actuators. B, Chemical*, *125*(2), 422–427. doi:10.1016/j.snb.2007.02.037

Wen, Y., Xing, F., He, S., Song, S., Wang, L., Long, Y., & Fan, C. et al. (2010). A graphene-based fluorescent nanoprobe for silver (I) ions detection by using graphene oxide and a silver-specific oligonucleotide. *Chemical Communications*, *46*(15), 2596–2598. doi:10.1039/b924832c PMID:20449319

Werner, J. H., & Güttler, H. H. (1991). Barrier inhomogeneities at Schottky contacts. *Journal of Applied Physics, 69*(3), 1522-1533. doi:10.1063/1.347243

White, P. C., Lawrence, N. S., Davis, J., & Compton, R. G. (2002). Electrochemical determination of thiols: A perspective. *Electroanalysis*, *14*(2), 89–98. doi:10.1002/1521-4109(200201)14:2<89::AID-ELAN89>3.0.CO;2-Y

White, R. M., & Voltmer, F. W. (1965). Direct piezoelectric coupling to surface elastic waves. *Applied Physics Letters*, *7*(12), 314–316. doi:10.1063/1.1754276

Wibawa, G., Khoiroh, I., Afrizal, D., & Suki, G. (2010). Solubilities of Dichloromethane, Diethyl Ether, Ethyl Acetate, and Nitrobenzene in three polymers using the Piezoelectric Quartz Sorption method. *Journal of Chemical & Engineering Data*, *55*(12), 5581–5586. doi:10.1021/je100389d

Widlar, R. J. (1967, January). An exact expression for the thermal variation of the emitter base voltage of bi-polar transistors. *Proceedings of the IEEE*, *55*(1), 96–97. doi:10.1109/PROC.1967.5396

Williams, K. A., Veenhuizen, P. T. M., Torre, B. G., Eritja, R., & Dekker, C. (2002). Nanotechnology: Carbon nanotubes with DNA recognition. *Nature*, *420*(6917), 761. doi:10.1038/420761a PMID:12490938

Winters, M., Hassan, J., Zirath, H., Janze, E., & Rorsman, N. (2013). A temperature dependent measurement of the carrier velocity vs. electric field characteristic for as-grown and H-intercalated epitaxial graphene on SiC. *Journal of Applied Physics*, *113*(19), 193708. doi:10.1063/1.4807162

Woan, G. (n.d.). *The Cambridge Handbook of Physics Formulas*. Cambridge University Press.

Wohltjen, H. (1984). Mechanism of operation and design considerations for surface acoustic wave device vapour sensors. *Sensors and Actuators*, *5*(4), 307–325. doi:10.1016/0250-6874(84)85014-3

Wongchoosuk, C., Wisitsoraat, A., Tuantranont, A., & Kerdcharoen, T. (2010). Portable electronic nose based on carbon nanotube-SnO_2 gas sensors and its application for detection of methanol contamination in whiskeys. *Sensors and Actuators. B, Chemical*, *147*(2), 392–399. doi:10.1016/j.snb.2010.03.072

Wong, H. S. P. (2002). Beyond the conventional transistor. *IBM Journal of Research and Development*, *46*(2.3), 133–168. doi:10.1147/rd.462.0133

World Health Organization. (2012). *Estimated deaths and DA-LYs attributable to selected environmental risk factors*. WHO.

Wouter Olthuis, E. F. (2007). *Sensing with FETs - once, now and future.* UT Publications.

Wu, J., Li, M., Jiang, Y., Li, J., Zhang, Y., Gao, H., & Lei, J. et al. (2015). Performance of a 4H-SiC Schottky diode as a compact sized detector for neutron pulse form measurements. *Nuclear Instruments & Methods in Physics Research. Section A, Accelerators, Spectrometers, Detectors and Associated Equipment*, *771*, 17–20. doi:10.1016/j.nima.2014.10.032

Wu, L., Chu, H. S., Koh, W. S., & Li, E. P. (2010). Highly sensitive graphene biosensors based on surface plasmon resonance. *Optics Express*, *18*(14), 14395–14400. doi:10.1364/OE.18.014395 PMID:20639924

Wu, S., He, Q., Tan, C., Wang, Y., & Zhang, H. (2013a). Graphene-based electrochemical sensors. *Small*, *9*(8), 1160–1172. doi:10.1002/smll.201202896 PMID:23494883

Wu, T. C., Chen, M. K., Lee, Y. W., Kuo, M. Y., & Wu, Y. T. (2013b). Bowl-shaped fragments of C70 or higher fullerenes: Synthesis, structural analysis, and inversion dynamics. *Angewandte Chemie International Edition*, *52*(4), 1289–1293. doi:10.1002/anie.201208200 PMID:23225309

Wu, W., Palaniapan, M., & Wong, W. K. (2008). Multiwall carbon nanotube resonator for ultra-sensitive mass detection. *Electronics Letters*, *44*(18), 1060–1061. doi:10.1049/el:20081932

Wu, Z., Chen, X., Zhu, S., Zhou, Z., Yao, Y., Quan, W., & Liu, B. (2013). Enhanced sensitivity of ammonia sensor using graphene/polyaniline nanocomposite. *Sensors and Actuators. B, Chemical*, *178*, 485–493. doi:10.1016/j.snb.2013.01.014

Wu, Z.-S., Zhou, G., Yin, L. C., Ren, W., Li, F., & Cheng, H.-M. (2012). Graphene/metal oxide composite electrode materials for energy storage. *Nano Energy*, *1*(1), 107–131. doi:10.1016/j.nanoen.2011.11.001

Xia, H., Wang, Q., & Qiu, G. (2003). Polymer-Encapsulated Carbon Nanotubes Prepared through Ultrasonically Initiated In Situ Emulsion Polymerization. *Chemistry of Materials*, *15*(20), 3879–3886. doi:10.1021/cm0341890

Xia, J. L., Chen, F., Li, J. H., & Tao, N. J. (2009). Measurement of the quantum capacitance of graphene. *Nature Nanotechnology*, *4*(8), 505–509. doi:10.1038/nnano.2009.177 PMID:19662012

Xie, H., Hu, M., & Bao, H. (2014). Thermal conductivity of silicene from first-principles. *Applied Physics Letters, 104*(13), 131906. doi:10.1063/1.4870586

Xie, H., Sheng, C., Chen, X., Wang, X., Li, Z., & Zhou, J. (2012). Multi-wall carbon nanotube gas sensors modified with amino-group to detect low concentration of formaldehyde. *Sensors and Actuators. B, Chemical*, *168*, 34–38. doi:10.1016/j.snb.2011.12.112

Xie, P., Xiong, Q., Fang, Y., Qing, Q., & Lieber, C. M. (2012). Local electrical potential detection of DNA by nanowire-nanopore sensors. *Nature Nanotechnology*, *7*(2), 119–125. doi:10.1038/nnano.2011.217 PMID:22157724

Xie, Y. E., Chen, Y. P., & Zhong, J. X. (2009). Electron transport of folded graphene nanoribbons. *Journal of Applied Physics*, *106*(10), 103714. doi:10.1063/1.3261757

Xie, Y., Chen, A., Du, D., & Lin, Y. (2011). Graphene-based immunosensor for electrochemical quantification of phosphorylated p53 (S15). *Analytica Chimica Acta*, *699*(1), 44–48. doi:10.1016/j.aca.2011.05.010 PMID:21704756

Xu, C., Wei, B., Ma, R., Liang, J., Ma, X., & Wu, D. (1999). Fabrication of aluminum–carbon nanotube composites and their electrical properties. *Carbon*, *37*(5), 855–858. doi:10.1016/S0008-6223(98)00285-1

Xu, G., Bai, J., Torres, C., Song, E., Tang, J., Zhou, Y., & Wang, K. et al. (2010). Nanowire-mask based fabrication of high mobility and low noise graphene nanoribbon short-channel field-effect transistors. In *Device Research Conference (DRC)*. IEEE. doi:10.1109/DRC.2010.5551935

Xu, M., Gao, Y., Yang, X., & Chen, H. (2012). Unique synthesis of graphene-based materials for clean energy and biological sensing applications. *Chinese Science Bulletin*, *57*(23), 3000–3009. doi:10.1007/s11434-012-5128-9

Yamacli, S., & Avci, M. (2010). Neural network modeling of voltage-dependent resistance of metallic carbon nanotube interconnects: An< i> ab initio</i> study. *Expert Systems with Applications*, *37*(12), 8014–8018. doi:10.1016/j.eswa.2010.05.089

Yamamoto, H., Hatakeyama, S., Norimura, T., & Tsuchiya, T. (1984). A radiation detector fabricated from silicon photodiode. *Radioisotopes*, *33*(12), 864–866. doi:10.3769/radioisotopes.33.12_864 PMID:6528065

Yang, A., Tao, X., Wang, R., Lee, S., & Surya, C. (2007). Room temperature gas sensing properties of SnO_2/multiwall-carbon-nanotube composite nanofibers. *Applied Physics Letters*, *91*(13), 133110. doi:10.1063/1.2783479

Yang, K., Cahangirov, S., Cantarero, A., Rubio, A., & D'Agosta, R. (2014). Thermoelectric properties of atomically thin silicene and germanene nanostructures. *Physical Review B: Condensed Matter and Materials Physics*, *89*(12), 125403. doi:10.1103/PhysRevB.89.125403

Yang, L., Park, C.-H., Son, Y.-W., Cohen, M. L., & Louie, S. G. (2007). Quasiparticle energies and band gaps in graphene nanoribbons. *Physical Review Letters*, *99*(18), 186801. doi:10.1103/PhysRevLett.99.186801 PMID:17995426

Yang, M.-Q., Zhang, N., & Xu, Y.-J. (2013). Synthesis of Fullerene–, Carbon Nanotube–, and Graphene–TiO2 Nanocomposite Photocatalysts for Selective Oxidation: A Comparative Study. *ACS Applied Materials & Interfaces*, *5*(3), 1156–1164. doi:10.1021/am3029798 PMID:23324025

Yang, M., Teo, K., Milne, W., & Hasko, D. (2005). Carbon nanotube Schottky diode and directionally dependent field-effect transistor using asymmetrical contacts. *Applied Physics Letters*, *87*(25), 253116. doi:10.1063/1.2149991

Yang, W., Auciello, O., Butler, J. E., Cai, W., Carlisle, J. A., Gerbi, J. E., & Russell, J. N. (2002). DNA-modified nano-crystalline diamond thin-films as stable, biologically active substrates. *Nature Materials*, *1*(4), 253–257. doi:10.1038/nmat779 PMID:12618788

Yang, Y. X., & Murali, R. (2010). Impact of size effect on graphene nanoribbon transport. *IEEE Electron Device Letters*, *31*(3), 237–239. doi:10.1109/LED.2009.2039915

Yang, Y., Brenner, K., & Murali, R. (2012). The influence of atmosphere on electrical transport in graphene. *Carbon*, *50*(5), 1727–1733. doi:10.1016/j.carbon.2011.12.008

Yao, X. (1999). Evolving artificial neural networks. *Proceedings of the IEEE*, *87*(9), 1423–1447. doi:10.1109/5.784219

Yao, Y., Chen, X., Guo, H., Wu, Z., & Li, X. (2012). Humidity sensing behaviors of graphene oxide-silicon bi-layer flexible structure. *Sensors and Actuators. B, Chemical*, *161*(1), 1053–1058. doi:10.1016/j.snb.2011.12.007

Yazyev, O. V. (2013). A guide to the design of electronic properties of graphene nanoribbons. *Accounts of Chemical Research*, *46*(10), 2319–2328. doi:10.1021/ar3001487 PMID:23282074

Yegnanarayana, B. (2004). *Artificial neural networks*. PHI Learning Pvt. Ltd.

Ye, J.-S., Liu, X., Cui, H. F., Zhang, W.-D., Sheu, F.-S., & Lim, T. M. (2005). Electrochemical oxidation of multi-walled carbon nanotubes and its application to electrochemical double layer capacitors. *Electrochemistry Communications*, *7*(3), 249–255. doi:10.1016/j.elecom.2005.01.008

Yi, J., Lee, J. M., & Park, W. I. (2011). Vertically aligned ZnO nanorods and graphene hybrid architectures for high-sensitive flexible gas sensors. *Sensors and Actuators. B, Chemical*, *155*(1), 264–269. doi:10.1016/j.snb.2010.12.033

Yin, G., Liang, Y., Jiang, F., Chen, H., Wang, P., Mizuseki, H., & Kawazoe, Y. (2009). Polarization-induced switching effect in graphene nanoribbon edge-defect junction. *The Journal of Chemical Physics*, *131*(23), 234706. doi:10.1063/1.3273312 PMID:20025341

Yin, P., Kim, T.-H., Choi, J.-W., & Lee, K. (2013). Prospects for graphene-nanoparticle-based hybrid sensors. *Physical Chemistry Chemical Physics*, *15*(31), 12785. doi:10.1039/c3cp51901e PMID:23828095

Yoneda, T. (n.d.). *Carbon nano tube (CNT) group temperature-change detection element for hydrogen gas sensor, has CNT group arranged perpendicular with respect to collector electrode surfaces and electrically connected between collector electrodes*. Nippon Valqua Ind Ltd.

Yoon, H. J., Jun, D. H., Yang, J. H., Zhou, Z., Yang, S. S., & Cheng, M. M.-C. (2011). Carbon dioxide gas sensor using a graphene sheet. *Sensors and Actuators. B, Chemical*, *157*(1), 310–313. doi:10.1016/j.snb.2011.03.035

Yu, S.-S., Wang, Zheng, & Jiang. (2011). Mechanical and electron-transport properties of graphene nanoribbons under tensile strain: A first-principles study. *Physica Status Solidi A-Applications and Materials Science, 208*, 2328-2331.

Yu, T., Lee, Briggs, Nagabhirava, & Yu. (2010). *Reliability study of bilayer graphene - material for future transistor and interconnect*. IEEE.

Yu, M. F., Lourie, O., Dyer, M. J., Moloni, K., Kelly, T. F., & Ruoff, R. S. (2000). Strength and Breaking Mechanism of Multiwalled Carbon Nanotubes Under Tensile Load. *Science*, *287*(5453), 637–640. doi:10.1126/science.287.5453.637 PMID:10649994

Yun, Y. H., Bange, A., Heineman, W. R., Halsall, H. B., Shanov, V. N., & Dong, Z. et al. (2007). A nanotube array immunosensor for direct electrochemical detection of antigen–antibody binding. *Sensors and Actuators. B, Chemical*, *123*(1), 177–182. doi:10.1016/j.snb.2006.08.014

Yu, S. S., Zheng, W. T., & Jiang, Q. (2008). Oxidation of graphene nanoribbon by molecular oxygen. *IEEE Transactions on Nanotechnology*, *7*(5), 628–635. doi:10.1109/TNANO.2008.2002267

Zaharah, J., Mohammad Taghi, A., Desmond Chang Yih, C., Aziziah, A. N., & Razali, I. (2010). Modelling of graphene nanoribbon fermi energy. *Journal of Nanomaterials*.

Zaharah, J., Mohammad Taghi, A., Desmond Chang Yih, C., & Aziziah, N. (2010). A., & Razali, I. (2010). Modelling of Graphene Nanoribbon Fermi Energy. *Journal of Nanomaterials*.

Zaki Dizaji, H., Kakavand, T., & Abbasi Davani, F. (2014). Spectrometry and dosimetry of fast neutrons using pin diode detectors. *Nuclear Instruments & Methods in Physics Research. Section A, Accelerators, Spectrometers, Detectors and Associated Equipment*, *741*, 84–87. doi:10.1016/j.nima.2013.12.018

Zanolli, Z., Leghrib, R., Felten, A., Pireaux, J.-J., Llobet, E., & Charlier, J.-C. (2011). Gas sensing with Au-decorated carbon nanotubes. *ACS Nano*, *5*(6), 4592–4599. doi:10.1021/nn200294h PMID:21553864

Zaragoza-Contreras, E. A., Lozano-Rodríguez, E. D., Román-Aguirre, M., Antunez-Flores, W., Hernández-Escobar, C. A., Flores-Gallardo, S. G., & Aguilar-Elguezabal, A. (2009). Evidence of multi-walled carbon nanotube fragmentation induced by sonication during nanotube encapsulation via bulk-suspension polymerization. *Micron (Oxford, England)*, *40*(5–6), 621–627. doi:10.1016/j.micron.2009.02.007 PMID:19299150

Zberecki, K., Wierzbicki, M., Barnaś, J., & Swirkowicz, R. (2013). Thermoelectric effects in silicene nanoribbons. *Physical Review B: Condensed Matter and Materials Physics*, *88*(11), 115404. doi:10.1103/PhysRevB.88.115404

Zeng, H., Zhao, J., Wei, J. W., & Hu, H. F. (2011). Effect of N doping and Stone-Wales defects on the electronic properties of graphene nanoribbons. *The European Physical Journal B*, *79*(3), 335–340. doi:10.1140/epjb/e2010-10725-4

Zhang, X.-S. (2000). Introduction to artificial neural network. In Neural Networks in Optimization, (pp. 83-93). Springer. doi:10.1007/978-1-4757-3167-5_5

Zhang, B., & Cui, T. (2011). An ultrasensitive and low-cost graphene sensor based on layer-by-layer nano self-assembly. *Applied Physics Letters*, *98*(7), 073116. doi:10.1063/1.3557504

Zhang, C.-, & Yan, S.-. (2012). First-Principles Study of Ferromagnetism in Two-Dimensional Silicene with Hydrogenation. *The Journal of Physical Chemistry C*, *116*(6), 4163–4166. doi:10.1021/jp2104177

Zhang, J., Hu, P., Zhang, R., Wang, X., Yang, B., Cao, W., & O'Neill, W. et al. (2012). Soft-lithographic processed soluble micropatterns of reduced graphene oxide for wafer-scale thin film transistors and gas sensors. *Journal of Materials Chemistry*, *22*(2), 714–718. doi:10.1039/C1JM14071J

Zhang, J., Sun, Y., Wu, Q., Gao, Y., Zhang, H., Bai, Y., & Song, D. (2014). Preparation of graphene oxide-based surface plasmon resonance biosensor with Au bipyramid nanoparticles as sensitivity enhancer. *Colloids and Surfaces. B, Biointerfaces*, *116*, 211–218. doi:10.1016/j.colsurfb.2014.01.003 PMID:24480068

Zhang, J., Wang, S. R., Wang, Y. M., & Wu, S. H. (2009). NO_2 sensing performance of SnO_2 hollow-sphere sensor. *Sensors and Actuators. B, Chemical*, *135*(2), 610–617. doi:10.1016/j.snb.2008.09.026

Zhang, J., Zhang, R., Wang, X., Feng, W., Hu, P., O'Neill, W., & Wang, Z. (2013). Fabrication of highly oriented reduced graphene oxide microbelts array for massive production of sensitive ammonia gas sensors. *Journal of Micromechanics and Microengineering*, *23*(9), 095031. doi:10.1088/0960-1317/23/9/095031

Zhang, M., Smith, A., & Gorski, W. (2004). Carbon nanotube-chitosan system for electrochemical sensing based on dehydrogenase enzymes. *Analytical Chemistry*, *76*(17), 5045–5050. doi:10.1021/ac049519u PMID:15373440

Zhang, P., Li, X. D., Hu, C. H., Wu, S. Q., & Zhu, Z. Z. (2011). First-principles studies of the hydrogenation effects in silicene sheets. *Physics Letters. [Part A]*, *376*(14), 1230–1233. doi:10.1016/j.physleta.2012.02.030

Zhang, T., Mubeen, S., Myung, N. V., & Deshusses, M. A. (2008). Recent progress in carbon nanotube-based gas sensors. *Nanotechnology*, *19*(33), 332001. doi:10.1088/0957-4484/19/33/332001 PMID:21730614

Zhang, W., Zhu, Z. Y., Wang, F., Wang, T. T., Sun, L. T., & Wang, Z. X. (2004). Chirality dependence of the thermal conductivity of carbon nanotubes. *Nanotechnology*, *15*(8), 936–939. doi:10.1088/0957-4484/15/8/010

Zhang, X., Xie, H., Hu, M., Bao, H., Yue, S., Qin, G., & Su, G. (2014). Thermal conductivity of silicene calculated using an optimized Stillinger-Weber potential. *Physical Review B: Condensed Matter and Materials Physics*, *89*(5), 054310. doi:10.1103/PhysRevB.89.054310

Zhang, Y. B., Brar, V. W., Girit, C., Zettl, A., & Crommie, M. F. (2009). Origin of spatial charge inhomogeneity in graphene. *Nature Physics*, *5*(10), 722–726. doi:10.1038/nphys1365

Zhang, Y.-H., Chen, Y.-B., Zhou, K.-G., Liu, C.-H., Zeng, J., Zhang, H.-L., & Peng, Y. (2009). Improving gas sensing properties of graphene by introducing dopants and defects: A first-principles study. *Nanotechnology*, *20*(18), 185504. doi:10.1088/0957-4484/20/18/185504 PMID:19420616

Zhang, Y.-H., Han, L.-F., Xiao, Y.-H., Jia, D.-Z., Guo, Z.-H., & Li, F. (2013). Understanding dopant and defect effect on H< sub> 2</sub> S sensing performances of graphene: A first-principles study. *Computational Materials Science*, *69*, 222–228. doi:10.1016/j.commatsci.2012.11.048

Zhang, Y.-Q., Tao, M.-L., Shen, W.-D., Zhou, Y.-Z., Ding, Y., Ma, Y., & Zhou, W.-L. (2004). Immobilization of L-asparaginase on the microparticles of the natural silk sericin protein and its characters. *Biomaterials*, *25*(17), 3751–3759. doi:10.1016/j.biomaterials.2003.10.019 PMID:15020151

Zhang, Y., Xu, G., Han, F., Wang, Z., & Ge, C. (2010). Preparation and Thermoelectric Properties of Nanoporous Bi2Te3-Based Alloys. *Journal of Electronic Materials*, *39*(9), 1741–1745. doi:10.1007/s11664-010-1199-z

Zhao, G.-B., John, S., Zhang, J.-J., Wang, L., Muknahallipatna, S., Hamann, J. C., & Plumb, O. A. et al. (2006). Methane conversion in pulsed corona discharge reactors. *Chemical Engineering Journal*, *125*(2), 67–79. doi:10.1016/j.cej.2006.08.008

Zhao, J., Buldum, A., Han, J., & Lu, J. P. (2002). Gas molecule adsorption in carbon nanotubes and nanotube bundles. *Nanotechnology*, *13*(2), 195–200. doi:10.1088/0957-4484/13/2/312

Zhao, Y. S., Song, X., Song, Q., & Yin, Z. (2012). A facile route to the synthesis copper oxide/reduced graphene oxide nanocomposites and electrochemical detection of catechol organic pollutant. *CrystEngComm*, *14*(20), 6710–6719. doi:10.1039/c2ce25509j

Zhao, Z., Tian, F., Dong, X., Li, Q., Wang, Q., & Wang, H. et al. (2012). Tetragonal Allotrope of Group 14 Elements. *Journal of the American Chemical Society*, *134*(30), 12362–12365. doi:10.1021/ja304380p PMID:22803841

Zheng, M., Jagota, A., Semke, E. D., Diner, B. A., McLean, R. S., Lustig, S. R., & Tassi, N. G. et al. (2003). DNA-assisted dispersion and separation of carbon nanotubes. *Nature Materials*, *2*(5), 338–342. doi:10.1038/nmat877 PMID:12692536

Zhou, C., Kong, J., Yenilmez, E., & Dai, H. (2000). Modulated chemical doping of individual carbon nanotubes. *Science*, *290*(5496), 1552–1555. doi:10.1126/science.290.5496.1552 PMID:11090348

Zhou, M., Lu, Y.-H., Cai, Y.-Q., Zhang, C., & Feng, Y.-P. (2011). Adsorption of gas molecules on transition metal embedded graphene: A search for high-performance graphene-based catalysts and gas sensors. *Nanotechnology*, 22. PMID:21869463

Zhou, Y., Bao, Q., Tang, L. A. L., Zhong, Y., & Loh, K. P. (2009). Hydrothermal dehydration for the "green" reduction of exfoliated graphene oxide to graphene and demonstration of tunable optical limiting properties. *Chemistry of Materials*, *21*(13), 2950–2956. doi:10.1021/cm9006603

Zhou, Y., Jiang, Y., Xie, G., Du, X., & Tai, H. (2014). Gas sensors based on multiple-walled carbon nanotubes-polyethylene oxide films for toluene vapor detection. *Sensors and Actuators. B, Chemical*, *191*, 24–30. doi:10.1016/j.snb.2013.09.079

Zhu, A., Zhang, X., Li, X., & Gong, W. (2002). "Beyond-thermal-equilibrium" conversion of methane to acetylene and hydrogen under pulsed corona discharge. *Science in China Series B: Chemistry*, *45*(4), 426–434. doi:10.1360/02yb9055

Zhu, L., Nicovich, J. M., & Wine, P. H. (2003). Temperature-dependent kinetics studies of aqueous phase reactions of SO_4^- radicals with dimethylsulfoxide, dimethyl-sulfone, and methanesulfonate. *Journal of Photochemistry and Photobiology A Chemistry*, *157*(2-3), 311–319. doi:10.1016/S1010-6030(03)00064-9

Zhu, Y., Murali, S., Cai, W., Li, X., Suk, J. W., Potts, J. R., & Ruoff, R. S. (2010). Graphene and graphene oxide: Synthesis, properties, and applications. *Advanced Materials*, *22*(35), 3906–3924. doi:10.1002/adma.201001068 PMID:20706983

Zólyomi, V., & Kürti, J. (2004). First-principles calculations for the electronic band structures of small diameter single-wall carbon nanotubes. *Physical Review B: Condensed Matter and Materials Physics*, *70*(8), 085403. doi:10.1103/PhysRevB.70.085403

Zuppella, P., Tosatto, S., Corso, A. J., Zuccon, S., & Pelizzo, M. G. (2013). Graphene–noble metal bilayers for inverted surface plasmon resonance biosensors. *Journal of Optics*, *15*(5), 055010. doi:10.1088/2040-8978/15/5/055010

Zurada, J. M. (1992). *Introduction to artificial neural systems*. Academic Press.

About the Contributors

Mohammad Taghi Ahmadi, tenured Associate Professor of Electrical Engineering at Urmia univercity, received his Ph.D. degree in Electrical Engineering and completed his Post Doctoral program, at Makmal Nano Universiti Teknologi Malaysia. Dr. Ahmadi was awarded Universiti Teknologi Malaysia (UTM) chancellor award, UTM Academic excellence award, UTM Post Doctoral Fellowship Award, Best PhD student Award (international student center Iranian student society). His research interests include nanoscale non-classical device simulation, modeling and characterization. He has published more than 170 papers in internationally refereed indexed journals and conference proceedings.

Razali Ismail received the B.Sc. and M.Sc. degrees in Electrical and Electronic Engineering from the University of Nottingham, Nottingham, U.K. in 1980 and 1983 respectively and the Ph.D. degree from Cambridge University, Cambridge, U.K., in 1989. In 1984 he joined the Faculty of Electrical Engineering, Universiti Teknologi Malaysia as a lecturer in Electrical and Electronic Engineering. He has held various faculty positions including as Head of Department and Chief Editor of the university journal. In 1985, he began his PhD work at the Department of Electronics and Computer Science, University of Southampton, Southampton, U.K. From 1987 to 1989, he continued his research work at the Department of Engineering, Cambridge University, Cambridge, U.K. where he completed his Ph.D. degree in microelectronics. His main research interest is in the field of microelectronics which includes the Modeling and Simulation of IC Fabrication Process and Modeling of Semiconductor Devices. His current research interest is in the emerging area of nanoelectronics devices focusing on the use of carbon based materials and novel device structures. He has published more than 180 papers in internationally refereed indexed journals and conference proceedings.

Sohail Anwar is currently serving as a Professor of engineering at the Pennsylvania State University, Altoona College. He has served as an invited professor of electrical engineering in several universities located in Poland, Malaysia, France, and China. Such professorships have focused on research and training in nanotechnology, energy, and electrical power systems. Recently, he has been granted the status of Associate Researcher at the LGCgE Laboratory in France. This laboratory is a prestigious energy and environment research facility. Dr. Anwar edited "Nanotechnology for Telecommunications" and "Advanced Nanoelectronics" published by Taylor and Francis Group/CRC Press. He also edited "Handbook of Research on Solar Energy Systems and Technologies"published by IGI Global Pressin August 2012. Dr. Anwar is serving as the Editor-in-Chief of the Encyclopedia of Energy Engineering & Technology published by Taylor and Francis Group. Dr. Anwar is a Senior Member of IEEE, and a

Member of ASEE and PAS. He has also served as a Commissioner of the Engineering Technology Accreditation Commission (ETAC) of ABET.

* * *

Sabra Anbari is a master holder of Industrial Engineering from the Unviersity of Technology in Malaysia.

Sarkis Azizian was born in Urmia, Iran, in 1982. He received his B.Sc. and M.Sc. degrees in electronics engineering from Urmia University, Urmia, Iran, in 2006 and 2011, respectively. He is currently Ph.D. student of electronics engineering at K. N. Toosi University of Technology, Tehran, Iran. His research interests are high speed analog and digital integrated circuit design, data converters, digital signal processing and hardware implementation of neural networks.

Zolkafle Buntat was born in Pontian, Johor, Malaysia on July 26, 1964. He is now an Associate Professor at Faculty of Electrical Engineering, Universiti Teknologi Malaysia. Currently he holds the post for the Deputy Director (R&D) of Institute of High Voltage and High Current (IVAT), Faculty of Electrical Engineering, Universiti Teknologi Malaysia. He received the Diploma in Electrical (Communication) in 1985, B.Sc in Electronics and Electrical. Eng. degree from University of Strathclyde, United Kingdom in 1988 and M.E.E in High Voltage Insulator study from Universiti Teknologi Malaysia, in 1996. In February 2005, he successfully completed his PhD in High Voltage and Pulsed Power Engineering from Loughborough University, United Kingdom. His teaching and research responsibilities are in the areas of Electrical Power Engineering, High Voltage Technology, Electrical Discharges and Ozone Generation and Technology. He had also given lectures at many high voltage and Ozone technology short courses for industries. His research interests include the generation of electrical discharges operating at atmospheric pressure, ozone generation, ozone application in medical, water and wastewater treatment, rain water harvesting, removal of NOx and SOx from diesel engine vehicles and process of agriculture produce. He has published many research papers in indexed journals and conferences. He is now a member of the International Ozone Association (IOA) and a collective member of The International Council on Large Electric Systems (CIGRE).

Amir Fathi was born in Urmia, Iran, in 1985. He received his B.Sc. and M.Sc. degrees in electronics engineering from Urmia University, Urmia, Iran, in 2008 and 2011, respectively. He is currently Ph.D. student of electronics engineering at Urmia University. His research interests are high speed analog and digital integrated circuit design, data converters, RF circuits and hardware implementation of neural networks.

Fauzan Khairi Che Harun received his B.Sc degree in Electrical-Electronics Engineering from Universiti Teknologi Malaysia in 2003. He then further his study in the area of biomedical engineering in University of Warwick where he was awarded with MSc in Advanced Electronic (2005) and PhD in Engineering (2010). His research interests are mainly but not restricted to biomedical electronics, Biological Inspired Micro-system and MEMS/NEMS. Currently he is a lecturer at the Faculty of Electrical Engineering, UTM.

Mina Hassanzadazar was born in Urmia, Iran, in 1986. She received her B.S. and M.S. degrees in Electrical Engineering from Urmia University, Urmia, Iran, in 2008, 2012 and is currently Ph.D. student of electrical engineering at Urmia University. Her research interests are high speed analog and digital integrated circuit design, data converters and Fuzzy Systems.

Rasoul Khodabakhsh was born in 1325 in Ardabil. He spent his primary and secondary education in Ardabil. Professor Khodabakhsh received his MA in the field of Nuclear power Technology and his Ph.D in the field of Atomics in Birmengam University of England.

Mohammad Javad Kiani is currently running an associated professor in Electrical Engineering at Azad University of Yasooj. He is a member of Computational Nanoelectronics (CoNE) Research Group in UTM. His main research interest is in the field of nanoelectronics, which includes the modeling of nanoscale devices. He has good knowledge in designing and developing the graphene nanoribbon (GNR) based nanostructures and nanosensor.

Sajad Pirsa received the Ph.D. degree in analytical chemistry from Tarbiat Modares University, Tehran, Iran. He is currently an Assistant Professor of Food Science and Technology with Urmia University, Urmia, Iran. His research interests cover chemical sensors, separation methods, and food analysis.

Ali Hosseingholi Pourasl received his Master. degree in in Eng. (Electronic and Telecommunication) from the Universiti Teknologi Malaysia (UTM), Malaysia in 2014. At the moment, he is pursuing his PhD's degree in Electronic engineering from Universiti Teknologi, Malaysia. He is currently a member of the Computational Nanoelectronics (CoNE) Research Group in UTM. His current research interests are in circuit simulation and nanotransistors and device modeling.

Hatef Sadeghi is a Research Associate at Lancaster University. His main research area is theoretical nano and molecular electronics. He is the GOLLUM (a next generation quantum transport simulation tool) administrator and one of the Gollum authors. In Sep. 2012, he appointed as Marie Curie Early Stage Researcher at Lancaster University within Nanoelectronics: Concepts, Theory and Modeling EU Innovative Training Network (EU ITN: NanoCTM). He won the Physics Department Best International PhD Student Award in July 2015.

Meisam Rahmani graduated his doctoral degree program in Electrical Engineering at Universiti Teknologi Malaysia (UTM). He is a member of Computational Nanoelectronics (CoNE) Research Group in UTM. His main research interest is in the field of nanoelectronics, which includes the modeling of nanoscale devices. He has good knowledge in designing and developing the graphene nanoribbon (GNR) based nanostructures and nanodevices.

Shahrokh received his PhD in Electronics Engineering from University of Putra Malaysia in 2010. He joined to the Hakim Sabzevari University, Sabzevar, Iran as Assistant Professor in 2012. Most of his works are related to the "Semiconductor Material Characterization" and "Chemical Sensor Design and Applications". He is also interested in Semiconductor Devices: Simulation, Analysis, Design, and Fabrication. Currently he is working on "Wide Band Gap Semiconductor Materials", "Biosensor for

Detection of Dityrosine", "Graphene Assisted Polyaniline for Semiconductor Applications", and "Detection of Explosive Material Based-on 2D Graphene Assisted MOS".

Nastaran Sharifan was born in Tehran, Iran, in 1990. She received her B.S. in law from Azad university in 2014 and M.S. degrees in International Economic and Commercial law in University Of Tehran, Iran, in 2016. Her research interests are divided to Law and Electronic both.

Michael Loong Peng Tan is a Senior Lecturer and researcher in the field of low dimensional nanoelectronics devices in the Computational Nanoelectronics (CoNE) Research Group, Faculty of Electrical Engineering, Universiti Teknologi Malaysia (UTM). He obtained his Bachelor of Engineering and was appointed as a tutor in the same university in September 2003. He started his Master by research on June 2004 and carried out his research as a graduate intern in Penang Design Centre (PDC) Intel Technology Sdn. Bhd, Penang. He was awarded Masters of Engineering from UTM in 2006. Subsequently, Michael pursued his PhD in Engineering at Queens' College, Cambridge from October 2007 until June 2011. He was awarded his PhD degree in July 2011 at the University of Cambridge, United Kingdom. He is a Chartered Engineer registered under the Engineering Council, United Kingdom. He is a lifetime corporate member of the Institution of Engineers Malaysia (IEM), member of the Institution of Engineering and Technology (IET) and Institute of Physics (IOP), life member of the American Physical Society (APS) and senior member of IEEE (SMIEEE). His research interests are in device modeling and circuit simulation of silicon and carbon-based nanotransistor.

Hamid Toloue is a PhD candidate in Malaysia-Japan International Institute of Technology (MJIIT) at the University Teknologi Malaysia, Department of Electronic System Engineering. Mr. Toloue graduated from faculty of electrical engineering with a Master's of Science in Electrical, Electronic and Telecommunication Engineering. His research interests are Plasmonics, 2D materials and their applications in biosensor. In addition to pursuing his PhD, he has been working in Department of Applied Physics, Graduate School of Engineering, Osaka University as a special research student since November 2015.

Rafael Vargas-Bernal was born in Irapuato, Guanajuato, México in September 30th, 1972. He received the B. E. degree in Electronics in October 6th, 1995 from the University of Guanajuato. After, the M. S. degree in Electronics from INAOE in September 4th, 1997. Lastly, the Sc. D. degree in Electronics from the same institute in July 19th, 2000. During October 2000 - September 2001, he worked as professor of the University of Guanajuato in the Department of Communications and Electronics Engineering. Since January 2002 to December 2009, he has been with the Department of Electronics Engineering of ITESI as Professor-Researcher. From August 2005 until today, he is also working in the Department of Materials Science in the area of Advanced Materials. He is technical reviewer of standards on microtechnology and MEMS in SEMI from November 2006 to the date. Moreover, he is reviewer of the magazine IEEE Latin America Transactions of IEEE from August 2003, Sensors and Actuators B: Chemical of Elsevier from February 2010, IEEE Sensors Journal from October 2012, Nanoscale of RSC November 2012, and RSC Advances published by RSC from April 2013 until today.

Index

F

G

H

I

M

N

www.ingramcontent.com/pod-product-compliance
Lightning Source LLC
LaVergne TN
LVHW080530300926
849162LV00007B/1574
* 9 7 8 1 5 2 2 5 0 7 3 6 9 *